AIRFRAME STRUCTURAL DESIGN

SECOND EDITION

*Practical Design Information and Data
on Aircraft Structures*

By

MICHAEL CHUN-YUNG NIU

Lockheed Aeronautical Systems Company
Burbank, California

HONG KONG CONMILIT PRESS LTD.

MICHAEL C. Y. NIU'S AIRFRAME BOOK SERIES:

AIRFRAME STRUCTURAL DESIGN (1988)

COMPOSITE AIRFRAME STRUCTURES (1992)

AIRFRAME STRESS ANALYSIS AND SIZING (1997)

About the author

Prof. Michael C. Y. Niu is the president of AD Airframe Consulting Company and is a metallic and composite airframe consultant. He was a Senior Research and Development Engineer (management position), Lockheed Aeronautical Systems Co. Prof. Niu has acquired more than over 30 years experience in aerospace and airframe structural analysis and design. At Lockheed, he was the acting Department Manager and program manager in charge of various structural programs, including development of innovative structural design concepts for both metallic and composite materials that are applicable to current and/or future advanced tactical fighters and advanced transports. During his service at Lockheed, he was deeply involved all aspects of preliminary design including aircraft layout, integration, configuration selection and airworthiness. He was lead engineer responsible for the L-1011 wide body derivative aircraft wing and empennage stress analysis. During 1966 and 1968, he served as stress engineer for the B727 and B747 at The Boeing Company.

Prof. Niu is the author of the texts, AIRFRAME STRUCTURAL DESIGN (1988), COMPOSITE AIRFRAME STRUCTURES (1992) and AIRFRAME STRESS ANALYSIS AND SIZING (1997). He has also written Lockheed's Composites Design Guide and Composites Drafting Handbook. He received the Lockheed Award of Achievement and Award of Product Excellence in 1973 and 1986 respectively. He is listed in Who's Who is Aviation, 1973.

He is a Consulting professor in Beijing University of Aeronautics and Astronautics and a Chair professor in Nanjing University of Aeronautics and Astronautics, The People's Republic of China.

He teaches the UCLA (University of California in Los Angeles) Engineering Short Courses "Airframe Design and Repairs", "Composite Airframe Structures" and "Airframe Stress Analysis and Sizing".

Prof. Niu received a B.S. in Hydraulic Engineering from Chungyuan University, Taiwan, in 1962 and a M.S. in Civil Engineering from the University of Wyoming in 1966.

©**HONG KONG CONMILIT PRESS LTD.**
All right reserved. No part of this book may be
reproduced in any form or by any electronic or mechanical means,
including information storage and retrieval devices or systems,
without prior written permission from the publisher.
Second edition, January, 1999 with part of corrections

All inquiries should be directed to: HONG KONG CONMILIT PRESS LTD.
Flat A, 10TH Floor,   or   P. O. Box 23250
Shing Loong Court,   Wanchai Post Office
No.13, Dragon Terrace,   HONG KONG
North Point,   Tel: (852) 2578 9529
HONG KONG   Fax:(852) 2578 1183

U.S. order/inquiry: TECHNICAL BOOK COMPANY
2056 Westwood Blvd.
Los Angeles, CA 90025 U.S.A.
Tel: (310) 475 5711
FAX: (310) 470 9810

Please forward any suggestions or comments to:
Prof. Michael C. Y. Niu
P. O. Box 3552
Granada Hills, CA 91394 U.S.A.
Fax: (818) 701 0298
E-mail: mniu@ worldnet.att.net

ISBN 962-7128-09-0   *Printed in Hong Kong*

Preface

This book is intended to advance the technical understanding and practical knowledge of both practicing engineers and students. The book represents several decades of data collection, research, conversations with different airframe specialists, plus the author's more than twenty years' experience in airframe structural design. In addition, the text is partly based on the author's lecture (Structures Symposium of Airframe Design) for the Lockheed Extension Education Program (LEEP). It is, therefore, equally useful to those with primary degrees in engineering fields as a reference for designing advanced structures.

From a structural standpoint, the book is intended to be used as a tool to help achieve structural integrity according to government regulations, specifications, criteria, etc., for designing commercial or military transports, military fighters, as well as general aviation aircraft. It can also be considered as a troubleshooting guide for airline structural maintenance and repair engineers or as a supplementary handbook in teaching aircraft structural design in college. Aircraft design encompasses almost all the engineering disciplines and it is not practical to cover all the information and data within one book. Instead, relevant references are presented at the end of each chapter so that the reader can explore his own personal interests in greater detail. This book does not cover basic strength of materials and structural (or stress) analysis. It is assumed that the reader already has this background knowledge.

This book is divided into a total of sixteen chapters and emphasizes itemized write-ups, tables, graphs and illustrations to lead directly to points of interest. The data can be used for designing and sizing airframe structures and, wherever needed, example calculations are presented for clarification. As technology continues to progress, basic technical data hold true, however, to suit today's design such as advanced composite structure some modification to the analysis may be required.

In preparing this book, it was necessary to obtain and collect vast amounts of information and data from many sources. (Information and data used in this book does not constitute official endorsement, either expressed or implied, by the manufacturers or the Lockheed Aeronautical Systems Company.) Sincere appreciation is given to the Technical Information Center of Lockheed Aeronautical Systems Company (LASC) for their gracious help. Thanks also to those who contributed to this book, my colleagues at LASC and other specialists from various companies. Special thanks to Mr. Richard W Baker (Research and Development Engineer of LASC) for his valuable comments in reviewing most of the drafts. Also, I would like to express my appreciation to Mr. Anthony C Jackson (Composite Design Department Manager of LASC) for his comments on Chapter 14.0, Advanced Composite Structures; and to my daughter Nina Niu for her help with this book.

Lastly, it is my hope that this book, with its wide scope and information on the application of technology on aircraft structural design, will prove not only to be a valuable reference tool for designing sound airframes with structural integrity but also as a "bridge" to carry over the valuable experience and knowledge from those who have retired from the aircraft industry to the next generation of engineers. However, any suggestions and comments for revision would be greatly appreciated by the author.

Michael Chun-yung Niu
牛春匀
California
U.S.A.
March, 1988

CONTENTS

Preface

CHAPTER	1.0	**GENERAL INFORMATION**	1
	1.1	Introduction	1
	1.2	Development Progress	3
	1.3	Planning and Structural Weight	5
	1.4	Computer Aid	6
CHAPTER	2.0	**DESIGN FOR MANUFACTURING**	11
	2.1	Introduction	11
	2.2	Engineer's Responsibility	12
	2.3	Producibility	12
	2.4	Maintainability	14
	2.5	Tooling	15
	2.6	Other Considerations	19
CHAPTER	3.0	**AIRCRAFT LOADS**	21
	3.1	Introduction	21
	3.2	Aeroelasticity	30
	3.3	Flight Maneuvers	33
	3.4	Basic Data	36
	3.5	Wing Design Loads	53
	3.6	Empennage Loads	59
	3.7	Fuselage Loads	66
	3.8	Propulsion Loads	67
	3.9	Landing Gear Loads	69
	3.10	Miscellaneous Loads	73
	3.11	Example of An Airplane Loads Calculation	74
CHAPTER	4.0	**MATERIALS**	90
	4.1	Introduction	90
	4.2	Material Selection Criteria	95
	4.3	Aluminum Alloys	101
	4.4	Titanium Alloys	109
	4.5	Steel Alloys	110
	4.6	Composite Materials	112
	4.7	Corrosion Prevention and Control	113
CHAPTER	5.0	**BUCKLING AND STABILITY**	118
	5.1	Introduction	118
	5.2	Columns and Beam-Columns	120
	5.3	Crippling Stress	134
	5.4	Buckling of Thin Sheets	137
	5.5	Thin Skin-Stringer Panels	141
	5.6	Skin-Stringer Panels	146
	5.7	Integrally Stiffened Panels	155

CHAPTER	6.0	CUTOUTS	162
	6.1	Introduction	162
	6.2	Lightly Loaded Beams	165
	6.3	Heavily Loaded Beams	173
	6.4	Cutouts in Skin-stringer Panels	177
	6.5	Cutouts in Curved Skin-stringer Panels	186
	6.6	Fuselage Cutouts for Big Cargo Doors	204

CHAPTER	7.0	FASTENERS AND STRUCTURAL JOINTS	207
	7.1	Introduction	207
	7.2	Rivets	210
	7.3	Bolts and Screws	214
	7.4	Fastener Selection	218
	7.5	Lug Design and Analysis	219
	7.6	Welded and Adhesive Bond	227
	7.7	Fatigue Design Considerations	230
	7.8	Shim Control and Requirements	243

CHAPTER	8.0	WING BOX STRUCTURE	247
	8.1	Introduction	247
	8.2	Wing Box Design	251
	8.3	Wing Covers	256
	8.4	Spars	269
	8.5	Ribs and Bulkheads	277
	8.6	Wing Root Joints	282
	8.7	Variable Swept Wings	288
	8.8	Wing Fuel Tank Design	296

CHAPTER	9.0	WING LEADING AND TRAILING EDGES	303
	9.1	Introduction	303
	9.2	Leading Edges	326
	9.3	Trailing Edges	335
	9.4	Wing Control Surfaces	347
	9.5	Fixed Leading and Trailing Edges	352
	9.6	Design Considerations	355

CHAPTER	10.0	EMPENNAGE STRUCTURE	358
	10.1	Introduction	358
	10.2	Horizontal Stabilizer	363
	10.3	Vertical Stabilizer	369
	10.4	Elevator and Rudder	371

CHAPTER	11.0	FUSELAGE	376
	11.1	Introduction	376
	11.2	Fuselage Configurations	379
	11.3	Fuselage Detail Design	380
	11.4	Forward Fuselage	398
	11.5	Wing and Fuselage Intersection	406
	11.6	Stabilizer and Aft Fuselage Intersection	412
	11.7	Fuselage Opening	417

CHAPTER 12.0	LANDING GEAR	430
12.1	Introduction	430
12.2	Development and Arrangements	442
12.3	Stowage and Retraction	449
12.4	Selection of Shock Absorbers	454
12.5	Wheels and Brakes	461
12.6	Detail Design	464
12.7	Testing	466

CHAPTER 13.0	ENGINE MOUNTS	471
13.1	Introduction	471
13.2	Propeller-driven Engine Mounts	475
13.3	Inlet of Jet Engines	478
13.4	Wing-pod (Pylon) Mounts	479
13.5	Rear Fuselage Mounts and Tail Mounts	487
13.6	Fuselage Mounts (Fighters)	489

CHAPTER 14.0	ADVANCED COMPOSITE STRUCTURES	492
14.1	Introduction	492
14.2	Composite Materials	500
14.3	Design	509
14.4	Structural Joint Design	520
14.5	Manufacturing	526

CHAPTER 15.0	FATIGUE, DAMAGE TOLERANCE AND FAIL-SAFE DESIGN	538
15.1	Introduction	538
15.2	Performance and Functions	543
15.3	Design Criteria and Ground Rules	547
15.4	Structural Life Estimation	548
15.5	Fail-safe Design	554
15.6	Detail Design	561
15.7	Sonic Fatigue Design and Prevention	567
15.8	Verification Tests	570

CHAPTER 16.0	WEIGHT CONTROL AND BALANCE	581
16.1	Introduction	581
16.2	Weight Prediction	585
16.3	Performance and Configuration Influences	590
16.4	Balance and Loadability	591

Appendix A:	CONVERSION FACTORS (U.S. unit vs. SI unit)	599
Appendix B:	LIST OF CUTAWAY DRAWINGS	602
Index		609

CHAPTER 1.0

GENERAL

1.1 Introduction

There are many aspects of design of aircraft structures as shown in Fig. 1.1.1. For modern jet aircraft, the design must incorporate clear aerodynamic shapes for long range flight near or at supersonic speeds, and/or wings to open up like parachutes at very low speeds. The wings must serve as fuel tanks and engine support structures. All structures must withstand hail and lightning strikes, and must operate in, and be protected against, corrosive environments indigenous to all climates. The structure must be serviceable from 15 to 20 years with minimum maintenance and still be light enough to be economically competitive. The design must incorporate new materials and processes that advance the state of the art. Using new techniques often require developing still newer processes.

A good overall structural concept incorporating all these factors is initiated during preliminary design. At the very beginning of a preliminary design effort, the designer writes a set of specifications consistent with the needs. It should be clearly understood that during preliminary design it is not always possible for the designer to meet all the requirements of a given set of specifications. In fact, it is not at all uncommon to find certain minimum requirements unattainable. It is then necessary to compromise. The extent to which compromises can be made must be left to the judgment of the designer. However, it must be kept in mind that to achieve a design most adaptable to the specified purpose of the airplane, sound judgment must be exercised in considering the value of the necessary modifications and/or compromises.

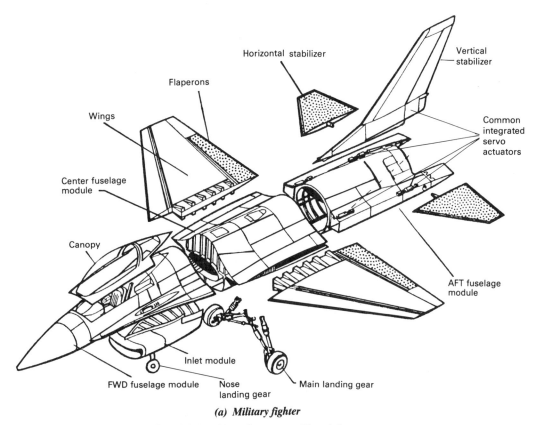

(a) Military fighter

Fig. 1.1.1 Aircraft structural breakdown.

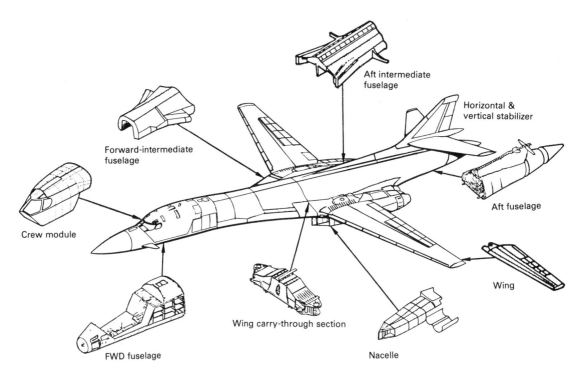

(b) Military bomber

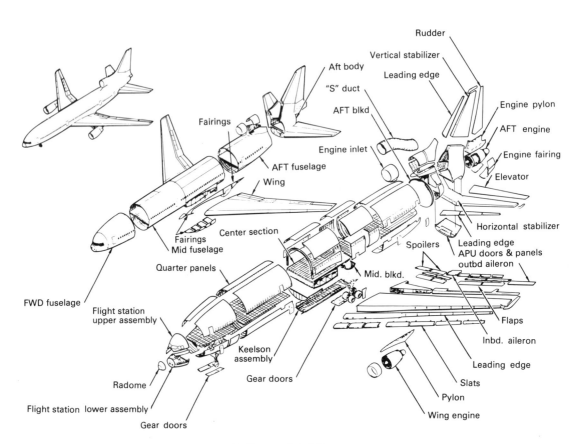

(c) Commercial airliner

Fig. 1.1.1 *(continued).*

The first task of the designer is to familiarize himself thoroughly with the specifications of the airplane upon which the design is to be based. Also, if the airplane is to be sold to more than one customer, all available information should be obtained to minimize the design that might be required in the future. There should be no thought of making a general purpose airplane, suitable for any purchaser or any use, because that is an impossibility. However, it is frequently possible to arrange a design which would simplify future changes without sacrificing either structural or aerodynamic efficiency or taking a weight penalty.

Next, the designer should familiarize himself with all existing airplanes of the same general type as that proposed. If possible, it is advisable to collect all comments, both positive and negative, of pilots, passengers, maintenance groups, and operators using the existing equipment. The designer should not blindly copy any existing design just because it happens to be available. On the other hand, not to take advantage both of the successes and mistakes of others is inefficient. Today's jet airplanes have much greater payloads at longer ranges and at higher speeds than past models. A great many shapes and sizes of wings and empennages were considered. Various wing shapes should be examined in depth for aerodynamic high and low speed performance, fuel capacity, range, torsional and weight characteristics, and system compatibility.

High lift and lateral control devices, pitch and yaw devices on the empennage, and sizes were established. As the final configuration was determined, statements of work describing the structure were supplied to manufacturing for scheduling purposes. Joints in the structure were established based on manufacturing's facilities, subcontract programs, raw material availabilty, and schedules.

The aircraft industry has for the past two decades spent considerable research and development effort to exploit the very attractive structural efficiencies achievable through the use of advanced composite structures. Advanced composites offer promise of substantial weight savings relative to current metallic structures. Further, the number of parts required to build a composite component may be dramatically less than the number of parts needed to construct the same component of metal alloy. This can lead to significant labor savings, sometimes offsetting the somewhat higher price of the present composite materials. These features, together with the inherent resistance to corrosion, make composites very attractive candidate materials for future aircraft structures.

1.2 Development Progress

The modern aeronautical engineering of aircraft design has been an evolutionary process accelerated tremendously in recent times from the demanding requirements for safety and the pressures of competitive economics in structural design. For example:

1900—1915 In this period, the Wright Brothers' demonstration of practical mechanical flight, power requirements, stability and control were overriding considerations. A successful flight was one which permitted repair and turn around in a few weeks or days. Strength considerations were subordinate and ultimate strength of a few critical parts was the extent of structural analysis.

1915—1930 World War I accelerated the solution of power plants and stability and control problems. Engine reliability was improved by ground qualification (fatigue) testing.

1930—1940 Commercial development of metal aircraft for public transport took place in this era. Design and analysis emphasized static ultimate strength and, except for the engine, had little or no consideration for airframe fatigue.

1940—1955 During this period, there grew an increasing awareness of the fatigue potential in airframe safety. A large increase in performance capability resulted from WW II technology. Higher material static strengths were developed without a corresponding increase in fatigue strength. Static ultimate design alone was not sufficient; it was joined by fatigue design.

1955—present Safety from fatigue alone was recognized to be inadequate; fail-safe and damage tolerance, i.e., static strength of damaged requires adequate inspection intervals to discover and repair fatigue and other damage before cracks reach catastrophic proportions.

So today we design for:
- Static ultimate (and yield) strength.
- Fatigue life of the airframe (crack initiation).
- Static residual strength of damaged structure.
- Fatigue life of damaged structure (inspection intervals).
- Thermal stress analysis and design of supersonic aircraft.

The primary objective of the structural designer is to achieve the maximum possible safety margin and achieve a "reasonable" lifetime of the aircraft structure. Economic obsolescence may not come as soon as anticipated. For example, some of the old DC-3's still flying today are approaching or exceeding 100,000 hours of service. This record is achieved only by fail-safe structure, knowledge of when and where to look for cracks, and replacement of a few vital parts.

It is the purpose of this book to discuss, in some detail, the design procedures, analysis methods, material properties and experimental data necessary to equal or better the past structural safety record in the face of ever increasing performance, adverse environments, and complexities of future aircraft.

All airframe structural design goes through these phases:
- Specification of function and design criteria.
- Determination of basic external applied loads.
- Calculation of internal element loads.

- Determination of allowable element strengths and margins of safety.
- Experimental demonstration or substantiation test program.(Fig. 1.2.1).

Engineering is experimental, empirical, and theoretical in that order. The physical facts must be known first; they may be empirically manipulated before the "perfect theory" is available. Failure to recognize this order of priority can lead to disaster if theoretical analyses are relied upon without thorough and careful experimental substantiation.

loads are relatively predictable from model data and substantiation for a certification program should be more or less routine. The laboratory development test program is an important feature of any new vehicle program; both to develop design data on materials and shapes, and to substantiate any new theory or structural configuration. Assuming clear-cut objectives, design criteria, and adherence to design rules and development test evaluations, the certification test program will demonstrate success without degenerating into more and expensive development work.

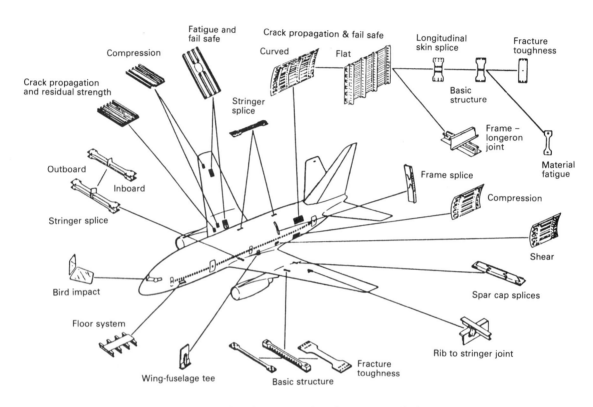

Fig. 1.2.1   Development testing of a transport airplane.

This does not mean we do not need theory. It does mean that we may and do need to progress beyond the capability of theory. But when we do, we must recognize and account for this fact wherever it may be critical. The engineering system evolved to handle any situation, be it a method of design, a component, or a vehicle design is illustrated in Fig. 1.2.2. Of primary significance are the three boxes in heavy outlines.

- Laboratory development testing
- Flight test data
- Certification or substantiation test program.

The dotted arrows indicate feed-back where experimental data is utilized to modify the design as necessary. The feed back loop for flight data on basic loads for conventional airplane vehicles is not as important as it once was simply because a vast store of research, experimental data, empirical know-how, and substantiated theory on the subject exists. Basic

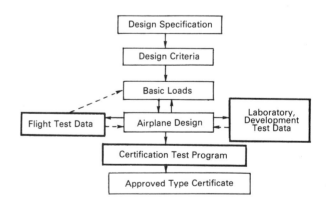

Fig. 1.2.2   Airplane design, development and certification.

4   Airframe Structural Design

1.3 Planning and Structural Weight

A good design is the result of proper planning and scheduling. This means scheduling not only the direct responsibility but also scheduling the data from other groups or staffs. It must have all the data required, such as loads from stress, aerodynamic requirements, systems data such as controls, electrical, fuel, and hydraulic interfaces, interchangeability, maintainability, serviceability, special airline requirements, and much more. All must be available in a timely manner to ensure a good integrated design. The designer is the only one who knows when he needs this data based on schedules he is committed to. He is the one who must make his requirements known and when he needs them. He is the one who must follow up as often as necessary to ensure receiving the data. He must recognize that someone else's performance is unsatisfactory in time to take steps through his superiors to remedy the situation. There is no other satisfactory way to schedule interchange of design data since it is a "give and take" situation, particularly between design groups. Loads and aerodynamic data requirements can be scheduled somewhat more precisely, but it is still the designer's responsibility to maintain a constant monitoring of stress and aerodynamic progress. Remember, it is the designer's ultimate responsibility to release designs on schedule to the shop. Fig. 1.3.1 shows an example of bar charting design purposes. It is particularly useful in getting management stirred into action.

The fact that structural designers or stress engineers who determine structural sizes should be concerned with weight should not strike one as strange. Nevertheless, in today's specialization, there is a tendency to narrow one's viewpoint to the mechanics of his job and to forget the fundamental reasons for that job. It has been said, sometimes in jest and again in earnest, that the weight engineer is paid to worry about weight. However, unless every aircraft engineer in a company is concerned about weight, that company may find it difficult to meet competition or, in other words, to design a good performance airplane. Weight engineers can estimate or calculate the weight of an airplane and its component parts. Actual weight savings, however, are always made by designers or stress engineers. A very small margin of weight can determine the difference between excellent and poor performance of an airplane. If the structure and equipment of a successful model were increased only 5% of its gross weight, the consequent reduction in fuel or pay load may well mean cancellation of a contract. In transport aircraft the gross-weight limit is definitely stipulated; thus, any increase in empty weight will be offset by a reduction in fuel or pay load (Fig. 1.3.2).

The weight break-down of aircraft structure over the years shows a remarkable consistency in the values of the structure weight expressed as 20–40% of the take-off gross weight (or all-up weight) realized in service, irrespective of whether they were driven by propellers or by jet engines.

At the project stage, if performance and strength are kept constant, a saving of structure weight is also accompanied by savings in fuel, the use of smaller engines, smaller wings to keep the same wing loading and so on, such that the savings in take-off weight of the aircraft to do the same job is much greater than the weight saved on the structure alone.

The object of structural design is to provide the structure that will permit the aircraft, whether military or civilian, to do the job most effectively; that is with

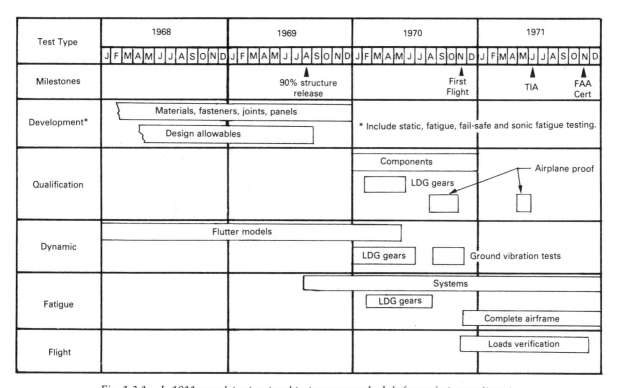

Fig. 1.3.1 L-1011 complete structural test program schedule for project commitment.

the least total effort, spread over the whole life of the aircraft from initial design until the aircraft is thrown on the scrap heap. There is thus, an all-embracing simple criterion by which the success of the structural design can be judged. It is not sufficient to believe that percentage structural weight is of itself an adequate measure of effective design, either of the complete airplane or of the structure itself. A well-known example is the provision of increased aspect ratio at the expense of structural weight, which may give increased fuel economy at cruise and reduction in the total aircraft weight. Nevertheless, percentage structural weight is a useful measure, provided its limitations are recognized.

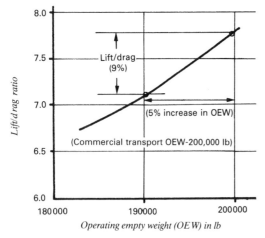

Fig. 1.3.2 Example of 5% increase in OEW and 25% reduction of payload, or to keep the same payload and redesign to achieve a 9% lift drag ratio.

1.4 Computer Aid

The requirements of structural analysis are undergoing changes due to different environments (i.e., altitude, speed, etc.), different construction, refined detail, expanded analysis coverage, and broadened analytical concepts as shown in Fig. 1.4.1. Motivated by these changes in analytical requirement, the digital computer and its effective use have assumed paramount importance. Environmental changes have precipitated the search for improved structures. The changes in altitude and speed have not only motivated a search for lighter structures but have necessitated the considerations of heated structure.

- Different environment
- Different construction
- Refined detail
- Expanded analysis coverage
- Broadened analytical concepts
- Stress analysis for composite structures

Fig. 1.4.1 Changed requirements for modern airplane design.

Finite Element Modeling (FEM)

Probably the most versatile tool in structural analysis is the use of finite element modeling (FEM). Before FEM, industrial stress analysis was largely an approximate science. Equations were available for determining stress and strain exactly in simple beams shown in Fig. 1.4.2. A major structural discontinuity occurs at the juncture of components such as the wing and body. At such structural junctions, a major redistribution of stresses must occur and the flexural dissimilarities of the wing and body must be designed for. Regardless of the construction details at this juncture, the major components affect each other. In those cases where the proportions of a component are such that beam analysis (or theory) can be employed, it is common practice to assume the behavior of one of the components and correspondingly analyze the other component. In the case of the wing/body juncture as mentioned previously, such a procedure could assume the body to provide cantilevered support boundary for the wing and then analyze the wing by beam bending theory.

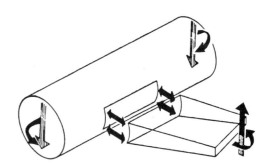

Fig. 1.4.2 Equilibrium and compatibility analysis.

Such conventional procedures will essentially ensure that the analytical forces that occur between the wing/body will be in equilibrium. Unfortunately, the actual elastic structural compatibility that is present usually enforces a different distribution of forces between these two major structures. Thus a change in the analytical requirements is present and the analytical technology must be powerful enough to ensure both force equilibrium and deflection compatibility at the structural discontinuity.

Most of the aircraft construction is such that not only must the force equilibruim be satisfied, but the elastic deflection compatibility must also be represented. This represents a broadening of the analytical concepts, wherein it is required to change the analytical technology to include both equilibrium and compatibility concepts which give the actual structural load or stress distribution.

However, most practical structures, especially aircraft structures, are very redundant or indeterminate which means the analysis of redundant structures leads to the need to solve sets of simultaneous linear algebraic equations. If the actual redundant structure is large, the set of simultaneous equations will also be large.

The finite element modeling represents a part with a mesh-like network of simple geometric shapes combined in building-block fashion as shown in

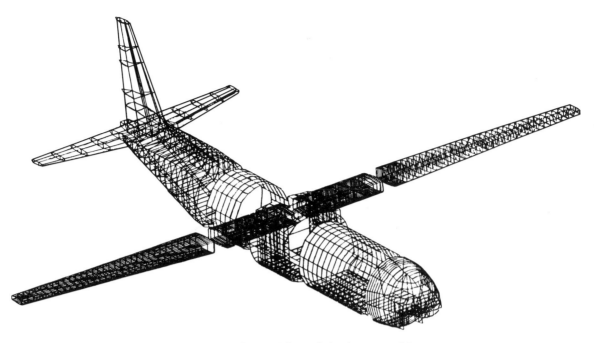

Fig. 1.4.3 Entire airframe finite element model.

Fig. 1.4.3. Each element has characteristics easily found from simple equations. So the behavior of the entire structure is determined by solving the resulting set of simultaneous equations (or by matrix techniques) for all the elements.

In the early days, finite element models were built manually (the element mesh was drawn by hand). Then node coordinates, element connectivity and other pertinent data were written on lengthy tabulation sheets and transferred to computer cards via a keypunch machine. Because of the huge amount of data to be handled, manual model-building is tedious, time-consuming, costly, and error-prone. In addition, the resulting model may be less than optimal because the lengthy construction time prohibits the analyst from refining the model with alternative mesh densities and configurations. To overcome these deficiencies, preprocessors or 3-D mesh graphics program from CAD/CAM (Computer-Aided Design and Computer-Aided Manufacturing) system were developed to aid in model building. These programs reportedly reduce model-building time and costs by as much as more than 80%.

Many FEM programs have been written. However, only a few are satisfactory for general use. These programs handle linear static and dynamic structural analysis problems. Some of them perform a type of linear heat-transfer analysis and damage tolerance analysis.

One of the earliest FEM programs and probably the most well-known is NASTRAN (NASA STRuctural ANalysis), developed by NASA in the mid-1960s to handle the analysis of missiles and aircraft structures. NASTRAN is one of few major programs with public domain versions available. Several major aerospace companies have modified NASTRAN for their particular applications.

Various types of graphic scopes are available showing model deformed-shape and displacements. For dynamic analysis, these deformed shapes are sometimes animated in slow motion to show how the structure bends, twists, and rocks during operation.

CAD/CAM System

The use of interactive computer graphics for data handling in design, manufacturing and product support programs quickly spread from aerospace to automotive applications and gradually permeated the broad base of general industry during the past decade. CAD/CAM (Computer-Aided-Design and Computer-Aided-Manufacturing) provides a common data base of the elected design for all disciplines such as preliminary design studies, lofting production design, tool design, numerical control (NC) design, quality control, product support, manufacturing, etc. Data can be accessed to other design groups like airplane payloads, controls, hydraulics, powerplant, electrical, maintenance, etc. to narrow the gap between design and manufacturing and also to open up a new opportunity for excellence in all design. It is no exaggeration that CAD/CAM system has been considered as the second industrial revolution of this century.

The large-scale computer (mainframe) is the heart of the CAD/CAM system that can provide a rapid and uncomplicated interface between engineer and machine to produce the geometric and mathematical definition for different programs. The large-scale computer is the center of this system and the graphics terminal as shown in Fig. 1.4.4 provides the engineer and machine interface. The other components of the hardware system provide fast response and recovery of work in the event of a hardware failure or power failure, the ability for the computer to operate on multiple tasks, magnetic tapes for control of machines

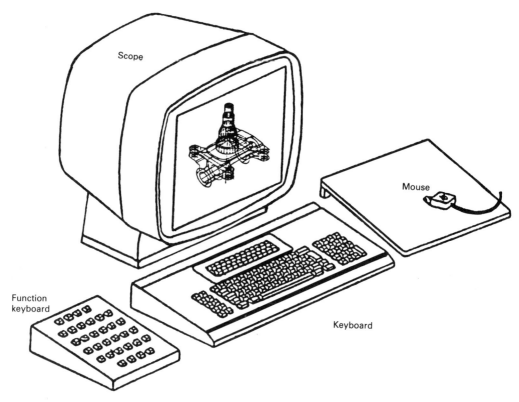

Fig. 1.4.4   Typical graphics terminal.

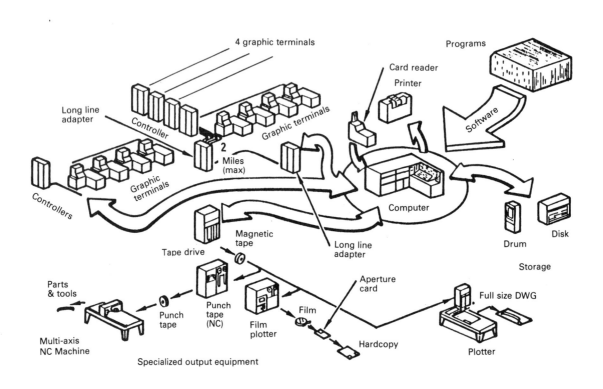

Fig. 1.4.5   CADAM hardware system.

to make parts or plot drawings, and disk drives for accounting data and storage of the work accomplished. Fig. 1.4.5 shows the typical hardware system components and the data flow.

CAD/CAM system is a broad data base system and most of the companies simply do not need this complete system. These companies find automated drafting system a more reasonable way to get started in CAD/CAM. The automated drafting system increases in productivity from two to six-fold and sometimes more. The drawing is stored in computer memory and can easily be changed and replotted in a few minutes to accommodate engineering modifications.

Automated drafting increases productivity by coupling the creativity of the engineer with the computer's high speed and huge memory. This frees the engineer from performing the time-consuming, repetitive tasks such as drawing the same shape many times, making the same change to several drawings, or painstakingly measuring and dimensioning a part. And the drawing is produced in minutes with the push of a button. As a result, the engineer-machine team can produce a drawing more quickly and more accurately than would otherwise be possible.

REFERENCES

1.1. Anon.: *Airframe & Powerplant Mechanics.* AC65-15, Department of Transportation: Federal Aviation Administration.
1.2. McKinley J.L. and Bent R.D.: *Basic Science for Aerospace Vehicles.* McGraw-Hill Book Company, New York, N.Y. 1963.
1.3. Torenbeek E.: *Synthesis of Subsonic Airplane Design.* Delft University Press, Netherlands. 1976.
1.4. Nicolai, L.M.: *Fundamentals of Aircraft Design.* University of Dayton, Dayton, Ohio 45469. 1975.
1.5. Williams, D.: *An Introduction to The Theory of Aircraft Structures.* Edward Arnold (Publishers) Ltd., London. 1960.
1.6. Bruhn, E.F.: *Analysis and Design of Flight Vehicle Structures.* Tri-State Offset Company, Cincinnati, Ohio 45202. 1965.
1.7. Corning, G.: *Supersonic and Subsonic CTOL and VTOL, Airplane Design.* P.O. Box No. 14, College, Maryland. 1976.
1.8. Wood, K. D.: *Aerospace Vehicle Design, Vol.I — Aircraft Design.* Johnson Publishing Company, Boulder, Colorado. 1968.
1.9. Newell, A.F.: 'Impressions of the Structural Design of American Civil Aircraft.' *The Aeroplane,* (Aug. 15, 1958), 229–232.
1.10. Megson, T.H.G.: *Aircraft Structures for Engineering Students.* Edward Arnold (Publishers) Ltd, London WIX8LL. 1972.
1.11. Sandifer, R.H.: 'Aircraft Design Philosophy.' *Journal of The Royal Aeronautical Society,* (May, 1956), 301–330.
1.12. Newell, A.F. and Howe, D.: 'Aircraft Design Trends.' *Aircraft Engineering,* (May, 1962), 132–139.
1.13. Way, W.W.: 'Advantages of Aircraft System Maurity.' *SAE paper No. 730907,* (1973).
1.14. North, D.M.: 'Soviets Test Two-seat MIG-25 Version.' *Aviation Week & Space Technology,* (Mar. 28, 1977), 17.
1.15. Sweetman, B.: 'Foxbat — Something Completely Different.' *Flight International,* (April 23, 1977), 1121–1124.
1.16. Smyth, S.J.: 'CADAM Data Handling from Conceptual Design Through Product Support.' *AIAA paper No. 79-1846,* (1979).
1.17. Wehrman, M.D.: 'CAD Produced Aircraft Drawings.' *AIAA paper No. 80-0732R,* (1981).
1.18. Stinton, D.: *The Anatomy of The Aeroplane.* American Elsevier Publishing Company Inc. New York, N.Y., 1966.
1.19. Wells, E.C.: 'Operational Aspects of Jet Transport Design.' *Aero Digest,* (May 1953).
1.20. Epstein, A.: 'Some Effects of Structural Deformation in Airplane Design.' *Aero Digest,* (Feb. 1949), 17.
1.21. Anon.: 'Tomorrow's Fighter: Updated or outdated.' *Aviation Week & Space Technology,* (Aug. 10, 1987), 61-108.
1.22. Wood, C.: 'Cargo Plane Design Considerations.' *Aero Digest,* (Dec. 1943), 203.
1.23. Rieger, N.F.: 'Basic Course in Finite-element Analysis? *Machine Design,* (Jun.–Jul. 1981).
1.24. Anon.: 'Russia's New Long-Hauler.' *Flight International,* (August 20, 1977), 524.
1.25. Laser: 'Design Probe — Another Look at the Square Cube Law.' *Flight International,* (Oct. 17, 1968), 615–616.
1.26. Cleveland, F.A.: 'Size Effects in Conventional Aircraft Design.' *Journal of Aircraft,* (Nov.–Dec., 1970), 483–512.
1.27. Black, R.E. and Stern, J.A.: 'Advanced Subsonic Transports — A Challenge for the 1990's'. *Journal of Aircraft,* (May 1976), 321–326.
1.28. Anon.: 'Two or Three Engines?' *Flight International,* (Sept. 18, 1969), 446–447.
1.29. Higgins, R.W.: 'The Choice Between One Engine or Two — An Appreciation of the Factors involved in Choosing A Single or Twin-Engine Layout for Military Tactical Aircraft.' *Aircraft Engineering,* (Nov. 1968.)
1.30. Higgins, R.W.: 'The Choice Between One Engine or Two for Tactical Strike/Close Support Aircraft.' *The Aeronautical Journal of The Royal Aeronautical Society,* (July, 1964), 620–622.
1.31. Lachmann, G.V.: *Boundary Layer and Flow Control — Its Principles and Application.* Vol. 1 and Vol. 2. Pregamon Press. New York, N.Y. 1961.
1.32. Teichmann, F.K.: *Airplane Design Manual.* Pitman Publishing Corporation, New York, N.Y., 4th Edition, 1958.
1.33. Anon.: *NATO's Fifteen Nations.* Frank O'Shanohun Ass. Ltd, London EC4A3JD.
1.34. Foody, J.J.: 'The Air Force/Boeing Advanced Medium STOL Transport Prototype.' *SAE paper No. 730365,* (April 1973).
1.35. Bates, R.E.: 'Structural Development of The DC-10.' *Douglas paper No. 6046,* (May 1972).
1.36. Spulding, E.H.: 'Trends in Modern Aircraft Structural Design.' *SAE paper No. 92,* (May, 1957).
1.37. Mackey, D.J. and Simons, H.: 'Structural Development of the L-1011 Tri-star.' *AIAA paper No. 72-776,* (April 1972).
1.38. Marks, M.D.: 'Technical Basis for The STOL Characteristics of The McDonnell Douglas/USAF YC-15 Prototype Airplane.' *SAE paper No. 730366,* (April 1973).

1.39. Magruder, W.M.: 'Development of Requirement, Configuration and Design for the Lockheed 1011 Jet Transport.' *SAE paper No. 680688*, (Oct. 1968).
1.40. Morisset, J.: 'Tupolev 144 and Concorde — The official performances are compared for the first time.' *NASA TT F 15446*, (April 1974).
1.41. Krogull, B. and Herbst, W.B.: 'Design for Air Combat.' *AIAA paper No. 72-749*, (Aug. 1972).
1.42. Woolsey J.P.: 'U.S. Airplane Builders Finding Many Barriers to New Programs.' *Air Transport World*, (Feb. 1976), 12—17.
1.43. Sandoz, P.L.: 'Structural Design of Future Commercial Transports'. *AIAA paper No. 73-20*, (Jan. 1973).
1.44. Foottit, H.R.: 'Design of Fighter Aircraft.' *Aero Digest*, (Feb. 1948).
1.45. Anon.: 'Aircraft Design at the AIAA.' *Flight International*, (Sept. 8, 1972).
1.46. Barton, C.: 'Spruce Gosse — Pterodactyl of World war II.' *Popular Mechanics*, (Nov. 1977).
1.47. Aronson, R.B.: 'Blown-wing STOLs on Trial.' *Machine Design*, (Oct. 1977), 26—31.
1.48. Satre, P.: 'Supersonic Air Transport — True Problems and Misconceptions.' *Journal of Aircraft*, (Jan.—Feb. 1970), 3—12.
1.49. Raymer, D.P.: Aircraft Design: A Conceptual Approach. AIAA Education Series, 1989.
1.50. Fisher, F.A. and Plumer, J.A.: Lightning Protection of Aircraft. Lightning Technologies Inc, 10 Downing Parkway, Pittsfield, MAOIZOI, 1990.

CHAPTER 2.0

DESIGN FOR MANUFACTURING

2.1 Introduction

Design and manufacturing are successive phases of a single operation; the ultimate objective of which is the emergence of an acceptable final product. In aerospace context, such acceptability has several components: market viability, operational efficiency, capacity for further development and structural integrity. Less obvious but just as important, a structure must not be so complex or difficult in concept that its realization will create great difficulties, or increase the cost of the manufacturing process.

Design has always carried with it — indeed — a degree of prestige; because its effectiveness can be seen in the final product and a successful design can confer something approaching glamour upon those responsible. Production, on the other hand, emerged later as a specialized branch of engineering and is sandwiched between the designer's drawings and the final product. Consequently, its achievement is less apparent and frequently, in the past, it has not been accorded a like degree of consideration or credit. Yet, it is the production phase of the operation that translates the design into hardware (see Fig. 2.1.1).

An aircraft is conceived as a complete structure, but for manufacturing purposes must be divided into sections, or main components, which are in turn split into sub-assemblies of decreasing size that are finally resolved into individual detail parts. Each main component is planned, tooled and built as a separate unit and joined with the others in the intermediate and final assembly stages (see Fig. 2.1.2).

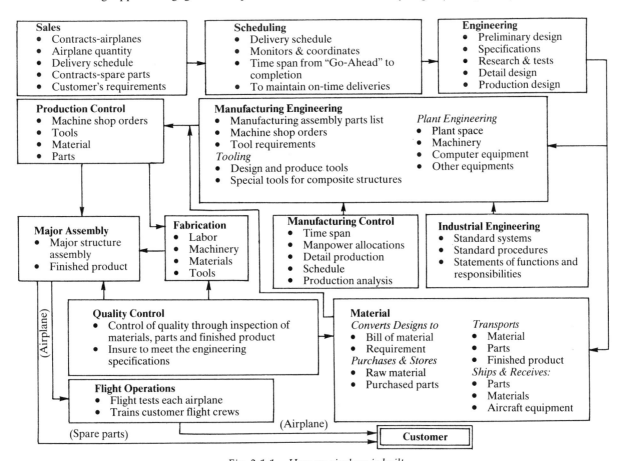

Fig. 2.1.1   How an airplane is built.

Airframe Structural Design 11

Tooling is required for each stage of the building of each component — detail tooling of individual parts, of which there may be many thousands, followed by assembly tooling of increasing size and complexity for the stages of progressive assembly.

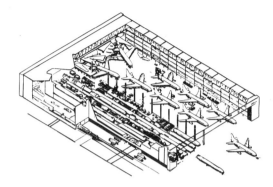

Fig. 2.1.2   Final assembly of airplane.

There is nothing new in attempting to design an aircraft to give trouble-free operation. This has, of course, always been one of the major parts of a designer's job. In recent years there has been incorporated the maintenance requirements in every engineering drawing before going into production. If aircraft have become too difficult to maintain, it is not entirely due to lack of appreciation of the problem in the drawing, but mainly to the very great increase in complexity of modern aircraft, particularly in the past few decades. A great deal of this complication is due to equipment and automatic gadgetry.

Between 25 to 40 per cent of the total direct operating cost (DOC) of an airplane is due to maintenance, quite apart from the losses due to airplanes not being serviceable when required.

2.2   Engineer's Responsibility

The design engineer is the "general practitioner" to the engineering profession as compared with staff or research types or engineers. Design engineers must diagnose the symptoms of potential structural failures and service problems early in the design stage. On the shoulders of the design engineer rests the final total design integration responsibility. To do a thorough job, the design engineer must:

(1) Coordinate thoroughly and integrate the design package into the overall structure:
   - Use design data type diagram to make co-ordination definite
   - Do not depend on oral cooordination
(2) Establish basics as early as possible:
   - Based on functional requirements
   - Loads and materials — make sure to use the right loads and materials to avoid needless drawing changes
   - Aerodynamic requirements
   - Geometry and jig information, interchangeability, producibility, repairability, replaceability, maintainability, etc.
   - Identification of problem details early to avoid backtracking in design
(3) Spend adequate time to plan the job:
   - Plan layouts and drawings to represent the work
   - Make schedules realistic
   - Study the requirements and select an optimum solution
(4) If you encounter interface problems, make adequate sections to show:
   - Where clearances are required and the requirement for such clearances
   - The interface to which the detail attaches
(5) Review processes, finishes, assembly procedures. etc:
   - Heat treatment requirements
   - Prevent stress risers which cause structural fatigue
   - Rough machine requirements to eliminate mocking stresses
   - Cold work treatment on machined surfaces to increase structural fatigue life, such as shot-peening
   - Surface plating
   - Forming and machining techniques
(6) Subcontractor-built production joints for assemblies that must conform to shipping limitations.
(7) Production joints resulting from raw material size restrictions or size of fabrication tools, ie. skin mills, stringer mills, stretch presses, protective finish tank size.
(8) The subassembly plan and how these subassemblies are loaded into the final assembly fixtures. The design engineer is essentially involved in putting together a structural jigsaw puzzle.
(9) After all, the most important one, is that the engineer should dedicate himself (or herself) to the job.

2.3   Producibility

In aircraft design, time is always the essence of the contract, and the pressure is always on. There is never sufficient time (and, frequently, not the information) to consider objectively all the possible ways of doing a job and thereby to arrive, on the first occasion, at the optimum method.

Because of the difficulties of assessing (under the combined pressures of technical requirements; economic realities and time) the best method of achieving a given objective in terms of function and cost, the initial process of collaboration and analysis has, over the last several decades, tended to develop almost into a new form of specialized engineering activities under the name of producibility engineering. The name is no more than a convenient label for an analytical process — now, perhaps, more concentrated and consciously applied than formerly — that has always been a part of good design and engineering practice. The first stage is to make available in the earliest days of the design, before anything is committed to materials or methods, all available information that will contribute to the conception of an efficient design at minimum costs. It is not possible, however, to have all the necessary information available at the moment when it is needed, regardless of the amount of intelligent forecasting that is applied, and problems of materials, processing, or continuing design development usually compel reconsideration of the methods used for reasons of time and cost.

Principles of Producibility Design

(1) General configuration:
- Rectangular vs. tapered wing sections, flaps and control surfaces
- Minimum number of major structures
- Cylindrical, straight, or conical surfaces vs. compound curvature (see Fig. 2.3.1)
- Extend of fairing and filleting required, see Fig.2.3.2

(2) Major breakdowns:
- Adequate access for assembly
- Ease of handling and transportation
- Completeness of master breakdown units
- Assembly joints
- Effectiveness (see Fig. 2.3.3)

with the aid of controlled media, and require only the application of attaching means for their installation. Interchangeable items shall be capable of being readily installed, removed, or replaced without alteration, misalignment, or damage to items being installed or to adjoining items or structure
— Adjustment and take-up (see Fig. 2.3.6)
— Tolerance
— Adequate clearances (see Fig. 2.5.6)
— Rework margins
— Adequate fastener edge distance
— Machining economy

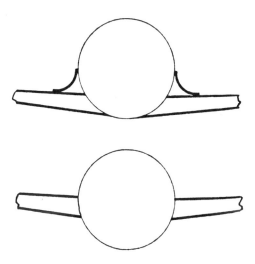

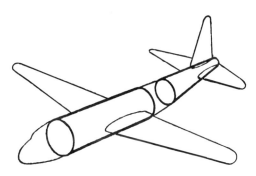

Fig. 2.3.1   Requires no forming operations since the flat skin may be wrapped onto its supporting frames.

Fig. 2.3.2   Fairings and fillets: interchangeability is difficult to achieve because of the tolerance accumulation at these points and should be used only when essential.

(3) Structure and equipment: Structure includes all primary and secondary structure. Equipment includes everything within the structural frame, i.e. controls, furnishings, instruments, powerplant, accessories, and all functional installations.
- Simplicity
  — Adequate access for fabrication and sub-assembly
  — Avoid compound curvature
  — Free body principles (see Fig. 2.3.4)
  — Alignment relations (see Fig. 2.3.5)
  — Minimum fabricating and processing operations
  — Straight line systems
  — Mechanical simplicity (avoid "Gadgetry")
- Parts
  — Multiple use and minimum number of different parts
  — Minimum total number of parts
  — Minimum amount and types of attachments
  — Effective use of standard parts, materials, and material sizes
- Detail design
  — Interchangeability, which applies to interchangeable items that are manufactured

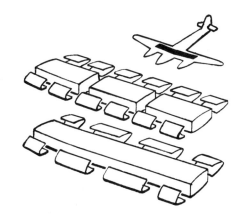

Fig. 2.3.3   The degree of breakdown should be dictated solely on the basis of its overall effect upon producibility. Improved producibility does not necessarily follow an increase in degree of breakdown. The effective breakdown of the center section in the lower view is preferable to that shown in the upper view.

Airframe Structural Design 13

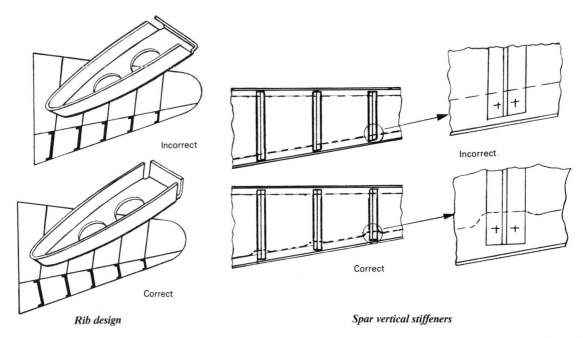

Fig. 2.3.4 Ribs parallel with the airplane of symmetry often result in acute angle of attachment to the spar or leading edge as illustrated in the left view. By ignoring unrelated airplane datum lines, improved producibility is achieved. Another example, as illustrated in right view, is that the lower view design results not only in saving fabricating operations, but the parts can be used on the right and left hand side.

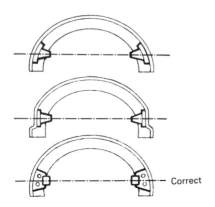

Fig. 2.3.5 When it is impossible to avoid the location of such related parts in different assemblies, these fittings should not be mounted on surfaces parallel with or normal to the centerline, but means of adjustment should usually be provided. The lower view illustrates the greater simplicity by using surfaces normal or parallel to the bearing axis.

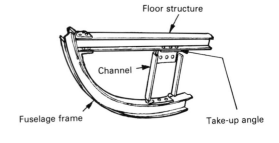

Fig. 2.3.6 If one flange of a channel is to be installed against the under surface of a floor structure and the other flange is to be against the upper surface of a lower fuselage frame, difficulties would undoubtedly be encountered. Therefore, one flange of the channel should be replaced with a separate so-called take-up angle drilled for the channel at assembly.

2.4 Maintainability

The proper outlook on maintenance must be instilled into everyone from the start. It is believed that in the old-fashioned system of two or three years maintainability training must be in the works before going into the drawing office. This period cannot be skipped if a person is to become a designer. It is very important that designers and draftsmen should be in constant contact with the shops and service departments so that they can see the difficulties which occur in the parts that they have designed.

It is of the greatest importance that designers should be in close touch with the engineering and maintenance department of the users for whom the airplane is being designed. It is not sufficient to rely on some published book of requirements. A large proportion of these are sure to be out of date or do not refer to the problem in hand.

Between military and civil transport aircraft, the first and most important difference is in the amount of flying done by the two types. A modern civil transport aircraft will probably have to last for 15–20 years, and fly between 2000 and 3000 hours a year, that is a total life of 30,000 to 45,000 hours. The average flying life of a fighter during peacetime will probably have a total of 4000 to 8000 hours. These times do not, of course, apply to transport command aircraft or to trainers, and probably not to most types used by coastal command, which do long periods of patrol. It is evident that the whole problem of design for maintenance presents an entirely different aspect in the two classes of aircraft.

In the case of transport aircraft, life of components and wear of the aircraft generally become very important. The aim will be to keep the airplane flying every day thus putting a premium on the ability to change components quickly. Military aircraft, however, will spend long periods on the ground. Even in wartime a military aircraft would spend a great deal of its time on the ground between relatively short periods of intense activity.

Facilities for maintenance in the case of military aircraft, especially under wartime conditions, are very much inferior to those available to the civil operator. It would seem logical, therefore, to cut down to the very minimum the amount of maintenance to be carried out by the squadrons of operational machines. With today's types of fast military aircraft having very thin wings and slim fuselages, it may be quite impossible to provide the same degree of accessibility (see Fig. 2.4.1) for easy removal of components that it has been used to in the past. It is suggested that much of this equipment can be built-in and, provided it is properly developed and tested, should be capable of functioning satisfactorily between major overhauls.

Fig. 2.4.1 Accessable openings for F-16 maintenance.
(Courtesy of General Dynamics Corp.)

In general, a low-wing transport layout is better from a servicing standpoint than high wing since engines and refuelling points are more readily accessible without the use of steps. The low-wing layout has considerable advantages from the point of view of installation of control cables, hydraulic pipes, electric cables and equipment, etc., all of which can be run under the cabin floor and reached for inspection and maintenance through doors in the underside of the fuselage without having to disturb cabin upholstery, and minimizes the need for maintenance personnel to work in the passenger cabin.

Position of service joints in the main structure can be of considerable importance to the operator; if possible, a fighter aircraft wing joint should be located at the side of the fuselage. This allows wings and fuselages to be transported more easily than the conventional stub wing which is more or less permanently attached to the fuselage.

A modern airline technique is to change the component or items of equipment which have given trouble or have reached their service life. This obviously means providing good accessibility to items which may have to be changed. It is very desirable to separate routine maintenance and changing points, so that a number of men can work on the airplane at the same time without all being crowded into one small space.

In civil transport, Air Transport Association of America (ATA) specification number 100 titled *"Specification for Manufacturer's Technical Data"* was devised by the airline industry in order to standardize the treatment of all subject matter and to simplify the user's problem in locating technical information for designing transport aircraft to meet maintainability requirements.

2.5 Tooling

30 years ago, manufacturing and interchangeability specifications were not taken too seriously in the presentation of a tool estimate, but such cannot be the case today. These increased customer requirements are supplemented by closer engineering tolerances dictated by higher performance requirements. All of this adds up to higher costs of tooling which can be reduced only by a tempering of specifications or by improved production engineering and tooling. Thus it is recognized that the closest coordination among sales, engineering and tooling is essential in order that all are traveling the same path.

It is of prime importance that the type of engineering be determined at the beginning in order that the tooling and manufacturing plan be properly established. Of course, the type of engineering is based on the requirements specified in the bid proposal. If the airplane is to be experimental with no production engineering, the airplane will be fabricated in the experimental department with tooling of a type suitable for experimental manufacturing. If the requirements specify by quantity or by quality that the contract shall be on a production basis, the engineering and tooling must be adapted to suit. This is the all-important time when the pattern is made for all to follow.

The company makes or loses money on the ability of each department to live up to its estimate. Any

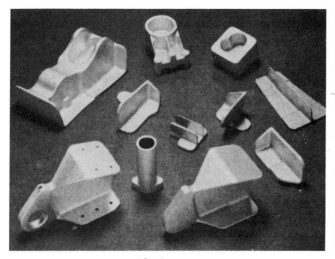
Aluminum

Titanium

Fig. 2.5.1 Precision forging products.

changes in the original specifications or bid proposal must eventually require renegotiation with all departments concerned. If such decisions are made without this consideration, some department may stray from the agreed upon path of direction and an unbalanced operation results with an end result of possible financial losses.

In order to reduce the cost of tools and manufacturing time, the design engineer must consider the method of manufacturing of each part individually and the quantity per airplane, plus a careful look at the specification. The evidence of the work that can be done in this field is the number of tools that must be made in order to meet drawing requirements. There are as many or more tools on nearly every contract than there are parts. This means that all of these tools must be catalogued, numbered, stored and continually kept up-to-date to all changes. The savings resulting from the elimination of each unnecessary tool is very apparent.

Low cost production is only possible with an engineering drawing that has incorporated in it every possible manufacturing advantage. It is understood that the practicability of the extent to which this can be approached is dependent upon many factors which inevitably restrict the design engineer.

It is noted that an objective attitude towards the engineer's problem by the tool engineer, serving as manufacturing division's representative, should result in a more standard, consistent working plan. This resultant standardization should enable the design engineer to become better acquainted with the "best" production design. While it is true that an airplane is not built until it is designed, it is also a fact that an airplane is not designed until it is built.

An aircraft is conceived as a complete structure but obviously cannot be built as a single unit. The structure is divided into a number of main components, which are further broken down into sub-assemblies that are finally resolved into individual parts. Individual parts may be forged, cast, extruded, press-formed or machined from solid or pre-forged shape. Today, two new processes have been developed, ie. precision forging as shown in Fig. 2.5.1 which requires little or no further machining, and super-plastic forming as shown in Fig. 2.5.2. These two processes have attracted great attention to reduce both cost and weight. The complexity of tooling and its durability and the degree of performance are determined by the number of parts to be made, the rate of production called for, and the quality required in the part. If the result can be obtained only with expensive tooling, that method must be adopted even if the number required is small.

Fig. 2.5.2 Superplastic forming of aluminum, 0.030 inch thick.

Numerical Control

Numerically controlled (NC) machining of large integral components constitutes a considerable proportion of the NC techniques employed in the production of the modern airframe. Programming the machining of a component is an extended process of transcribing the information that defines the shape of the finished part into the particular programming language (or tapes) that has been used extensively today.

The need to speed up the programming of machines and to develop procedures less prone to human error has led to the increased use of computers in the early stages of NC program preparation. The computer-graphic programming of components, or parts, is a technique for communicating with computers by means of words and diagrams. In this way a picture of the shape or part required can be created and its machining simulated on the screen of a display console linked to a computer in order to check the operations involved.

The principal objectives of these program developments are to simplify part programming procedures, to diminish programming errors, to reduce the cost of producing accurate control tapes, to shorten the time taken to produce these tapes and, in so doing, to shorten the preproduction period.

NC is a method of simulating functions of a machine operator through the use of a control tape. The geometry of a part is mathematically described and related via the computer to the coordinate system of an NC machine. When machining instructions are added, this new complete information, on tape, is used to control the machining operation.

Robots

Robots are machines which are programmable and can be reassigned tasks by changing its instructions. Robots can improve the quality, productivity, safety and reliability of a manufacturing process.

Robots have an arm that functions as a human arm; ie, the arm can pick up objects, move them around and put them down again, all with great precision and repeatability. A robot arm is able to move in at least three directions: in and out, up and down, and around. When a robot hand is added, another three axes of motion are yaw, pitch and roll, as shown in Fig. 2.5.3.

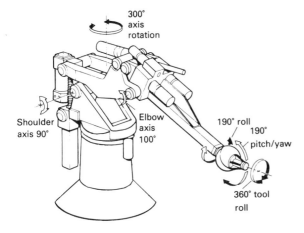

Fig. 2.5.3 Robot movement capability.

Robots, by functions, fall into four basic categories:
Pick and place (PNP): A PNP robot is the simplest one, its function is to pick up an object and move it to another location. Typical applications include machine loading and unloading and general materials handling tasks.
- Point to point (PTP): Similar to PNP robots, a PTP unit moves from point to point, but it can move to literally hundreds of points in sequence. At each point, it stops and performs an action, such as spot welding, glueing, drilling, deburring, or similar tasks.
- Continuous path (CP): A CP robot also moves from point to point, but the path it takes is critical. This is because it performs its task while it is moving. Paint spraying, seam welding, cutting and inspection are typical applications.
- Robotic assembly (RA): The most sophisticated robot of all, a RA combine the path control of CP robots with the precision of machine tools. RA often work faster than PNP and perform smaller, smoother, and more intricate motions than CP Robots.

Shot-peen Forming

An example illustrates the design of the long, narrow, and slightly contoured wing skins such as for the Boeing 747, Lockheed L-1011, DC-10 etc, which takes advantage of shot-peen forming technique, and is much less expensive to produce than the earlier "creep-formed" airplane wing skins.

Fig. 2.5.4 shows an integrally stiffened wing skin being peened in a peenmatic gantry machine which has the necessary controls to compensate for varying curvature requirements along the skin, varying thickness, cut-outs and reinforcements, as well as distortion caused by machining or heat treatment. No dies are required for peen-forming. However, for severe forming applications, stress-peen fixtures are sometimes used. Peen-forming is effective on all metals.

In the same manner in which shot-peening has been used to straighten parts, it can be used to form certain parts in production. Integrally stiffened wing skins are an excellent application of shot-peen-forming, as frequently no other forming process can be used to produce the required chordwise curvature

Although peen-formed parts usually require shot-peening on one side only, both sides will have compressive stresses in the surface. Besides forming and increasing fatigue strength, these compressive stresses serve to prevent stress corrosion.

Some parts should be shot-peened all over prior to peen-forming to further improve these characteristics. Parts of this kind which have been cold formed are often shot-peened to overcome the harmful tensile stresses set up by bending.

Assembly Fixture

Basically, all assembly tooling locates and clamps individual workpieces or units while they are being joined together. The resultant assembly is a larger section of the final airframe structure and, in the next stage, may itself be one of the units to be joined with others. Any assembly fixture must be sufficiently accurate and stable to guarantee that the assembly built within it conforms to design tolerances and will retain them after removal; it must be sufficiently rigid to withstand distortion by the assembly itself. All mating holes and pick-up points with other assemblies or units must be accurately located and maintained; wing-root, fin and tailplane fittings are obvious for fighter airplane examples of such parts whose interchangeability must be assured to meet military requirements.

A problem that arises in the design of large assembly fixtures is that of thermal expansion and contraction. With large airframes, movement can be considerable and it is essential to match the thermal distortion of the fixture to that of the airframe unit assembled in it. The method usually adopted is to match the coefficient of expansion of the airframe assembly with that of the fixture or, at least, to use material for the locations on the fixture that have the same coefficient of expansion as the assembly. This may not solve the problem entirely, because thermal inertia can cause the expansion of the fixture to lag behind that of the assembly. In these circumstances it may be necessary to incorporate some provision for free movement in the fixture location to accommodate this differential.

Fixture design is governed by the accurate location of parts, rigidity of clamping, accessibility to all parts, and provision for rapid removal of the completed assembly. It is probably quite important to make every part of a component in its assembly fixture instantly and effortlessly accessible.

Early in the design, surfaces on the structure are agreed upon by both manufacturing and engineering as a tooling surface. The agreement is that manufacturing will tool to the designated surfaces where engineering agrees never to change these surfaces. Examples are shown in Fig. 2.5.5 for reference. The advantages of establishing tooling surfaces are as follows:

- Tooling can get started on design early.
- Documentation of tooling surfaces early in the design to reduce subsequent coordination effort.
- Structural component can be easily and cheaply changed for structural revisions or beef-up for future growth.

In addition, designers should understand the manufacturing and assembly sequence to have excess edge margin in order to allow float as shown in Fig. 2.5.6.

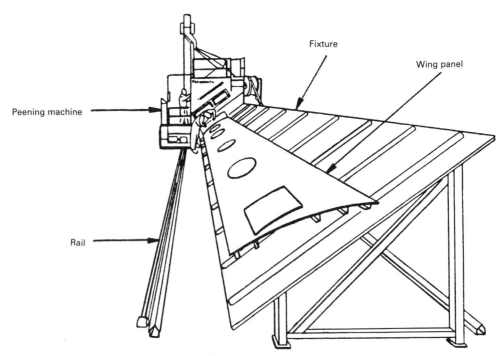

Fig. 2.5.4   Shot-peening operation.

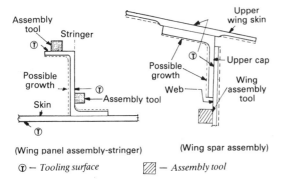

Fig. 2.5.5   Tooling surface examples.

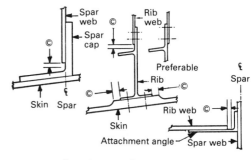

Fig. 2.5.6   Design for assembly tolerances.

2.6 Other Considerations

(1) If liaison finds trouble with your design, don't go on the defensive — find out if it is a true problem and fix it early.
(2) Mixing of fastener materials and types in any one fastener pattern or area should be avoided.
(3) Ends of rib and spar web stiffeners should be square-cut to allow saw cutting to length and to allow turning the stiffener end for end (see Fig. 2.3.4), thus using a common part on both the left and right hand. The outstanding leg should be chamfered to reduce weight.
(4) All designs should consider supplier capability, particularly in sizes and kinds of raw materials or standards, so that at least two sources are available. Competition for orders is thus maintained, and not dependent on a single source in case of emergency.
(5) Tolerances less than ±0.03 for length, depth and width, and ±0.01 for machine thickness should be coordinated with the manufacturing.
(6) Make ribs normal to the front or rear spars where practical to minimize tooling and master tooling template problems.
(7) Crawl holes through ribs and spars should be a minimum of 12 inches by 18 inches. Larger holes should be used where allowed by shear stress in minimum gage areas. Consideration should be given to hole locations in adjacent ribs for maintenance. Sharp corners and protrusions around crawl holes should be eliminated.
(8) Structural stiffeners or plates should not be incorporated across crawl holes unless a failsafe condition exists, in case they are inadvertently left off during maintenance. Producibility will also be enhanced with an unencumbered hole.
(9) Aluminum alloy upset head rivets or pull-type lockbolts should be used for web and stiffener riveting wherever possible. Design should consider automatic riveting.
(10) Edge margins of rib cap to panel stringer attach bolts shall be standardized for each diameter of fastener. A standard tool can then be used for drilling these holes in the wing assembly fixture.
(11) Web stiffeners on ribs should be located to allow use of either bolts and nuts or pull-type lockbolts for rib cap to panel stringer attachment as far as is practical. This should be kept in mind for all areas. Make room for lockbolt and Hi-Lok fastener equipment if possible.
(12) Angle rib cap should be flanged inboard whenever practical for ease of fabricating open angles.
(13) Skin and stringer detail drawings should show stations or some datum lines (reference) as well as direction arrows such as FWD, INBD, UP for orientation of the part in relation to the airplane.
(14) All length dimensions should originate from a machine reference plane. This should include step cuts and end trims.
(15) On machine parts consider commonality of tools. The most important of these is cutter radius. Use as few different radii as possible consistent with the function of the part and weight consideration.
(16) All edge trims should be dimensioned from a machine reference plane on the straight tooling surface (see Fig. 2.5.5) side of the web. The reference plane surface should be maintained as flat plane whenever possible.
(17) Manufacturing recommends stringer tolerances as follows: Thickness ±0.01, width and length ±0.03, height ±0.03. Special deviations may be made on basic gage taper dimensions and cutter radii. All tolerances should be reviewed for weight savings within the established economic limits.
(18) All skin tolerances should be as follows: Thickness ±0.005, edge trim and critical location coordinates ±0.03.
(19) The minimum machine gage on stringers should be 0.08. The maximum flange thickness-to-width ratio may be 1 to 15 when the flange can be supported during machining. Otherwise, use a maximum 1 to 10 ratio unsupported. The maximum flange runout angle should be 15°. The maximum thickness taper angle should be 6°.
(20) See Reference 2.17.

Reference

2.1 Petronic, G.: 'Producibility in Design.' *SAE paper No. 710748*, (1971).
2.2 Brooks, Jr., D.T.: 'Designing for Producibility: Design-Influenced Production Cost Program.' *SAE paper No. 710747*, (1971).
2.3 Torget, S.J.: 'Producibility Considerations in Production Planning for New Aircraft.' *SAE Paper No. 710746*, (1971).
2.4 Jones, W. R. and Harrison, H.M.: 'Design Producibility Serves the Industry, (Part I).' *Aero Digest*, (Aug., 1948) 40.
'Design Producibility Serves the Industry, (part II).' *Aero Digest*. (Sept., 1948).
2.5 Green, E.A.: 'Simplicity — key to producibility.' *Machine Design*, (Nov., 1953). 112–123.
2.6 Epstein, A.: 'Some Effects of Structural Deformation in Airplane Design.' *Aero Digest*, (Feb., 1949).
2.7 Goff, W.E.: 'Design for Production.' *Flight International*, (Oct. 25, 1973), 707.
2.8 Dougherty, J.E.: 'FAA Appraisal of Aircraft Design for Maintainability.' *SAE paper No. 710431*, (1971).
2.9 Gilmaer, G.W. and Olson, B.D.: 'Maintenance — The designer's Stepchild.' *AIAA Paper No. 68-210*, (1968).
2.10 Adams, C.W. and Hinders, U.A.: 'Design-to-cost for the A-10 Close Air Support Aircraft.' *AIAA paper No. 74-963*, (1974).
2.11 Knowles, Jr, G.E.: 'Material Thickness Control Through Manufacturing Process Refinement.' *J. of Aircraft*, (July 1976), 467–469.
2.12 Anon.: 'Elements of Aircraft Production Design.'

Engineering Training Materials issued by Lockheed Aircraft Corp.
2.13 Anon.: 'Lockheed Engineering Developments for Better Production Design.' *Issued by Lockheed Aircraft Corp.*, (1954).
2.14 Foster, R.B.: 'An Approach to the Problem of Airplane Production.' *Aero Digest*, (Mar. 1, 1944).
2.15 Anon.: *Vol. 2 — Heat Treating, Cleaning and Finishing; Vol. 3 — Machining; Vol. 4 — Forming; Vol. 5 — Forging and Casting.* American Society for Metals, Metals Park, Ohio 44073.
2.16 Anon.: *Assembly Engineering.* Hitchcock Publishing Company, Wheaton, Illinois 60188.
2.17 Anon.: *ATA Specification.* Air Transport Association of America, Washington, D.C. 20006.
2.18 Anon.: 'Interchangeability and Replaceability of Component Parts for Aerospace Vehicles.' *Mil-I-8500 D*, (Mar., 1980).
2.19 Lermer, E.J.: 'Unplugging CAD/CAM from the Mainframes.' *Aeropace America*, (Oct., 1986).
2.20 Grand, R.L.: *Manufacturing Engineer's Manual.* McGraw-Hill Book Co., New York, N.Y. 1971.
2.21 Witt, R.H. and Ferreri, A.L.: 'Titanium Near Net Shape Components for Demanding Airframe Applications.' *SAMPE Quarterly,* (Apr., 1986), 55—61.

CHAPTER 3.0

AIRCRAFT LOADS

3.1 Introduction

Aircraft loads are those forces and loadings applied to the airplane structural components to establish the strength level of the complete airplane. These loadings may be caused by air pressure, inertia forces, or ground reactions during landing. In more specialized cases, design loadings may be imposed during other operations such as catapulted take-offs, arrested landings, or landings in water.

The determination of design loads involves a study of the air pressures and inertia forces during certain prescribed maneuvers, either in the air or on the ground. Since the primary objective is an airplane with a satisfactory strength level, the means by which this result is obtained is sometimes unimportant. Some of the prescribed maneuvers are therefore arbitrary and empirical which is indicated by a careful examination of some of the criteria. Rational maneuvers and loadings are considered when the "state-of-the-art" permits their use. The increasing availability of electronic computing equipment will permit investigations of airplane and loading behavior as realistically as can be described mathematically. With the advent of faster aircraft, the need for rational analyses is becoming more important.

The amount of analysis used in derivation of aircraft loads is dependent on the size, complexity, and knowledge and data available. The time element is also important during a design. The structural design is dependent on loads. Therefore, the loads must be determined early in the design to preclude the possibility of delaying design work. The time available often governs the amount of load analysis that can be made. Another consideration in determining the extent of the load analysis is the amount of structural weight involved. Since weight is always of great importance, a refinement of the methods used to compute loads may be justified. A fairly detailed analysis may be necessary when computing operating loads on such items as movable surfaces, doors, landing gears, etc. Proper operation of the system requires an accurate prediction of the loads. In cases where loads are difficult to predict, recourse must be made to wind tunnel measurements. The final basic loads must be an acceptable compromise of all the considerations.

Aircraft loads is the science of determining the loads that an aircraft structure must be designed to withstand. A large part of the forces that make up design loads are the forces resulting from the flow of air about the airplane's surfaces — the same forces that enable flight and control of the aircraft. The study of the behavior of these forces is the subject of aerodynamics that can be found from any book of the fundamentals of aerodynamics and will not be discussed in this Chapter.

Load Factors

In normal straight and level flight the wing lift supports the weight of the airplane. During maneuvers or flight through turbulent (gusty) air, however, additional loads are imposed which will increase or decrease the net loads on the airplane structure. The amount of additional load depends on the severity of the maneuvers or the turbulence, and its magnitude is measured in terms of *load factor*.

Load factor is a multiplying factor which defines a load in terms of weight. This definition is illustrated in the accompanying diagram. In Fig. 3.1.1(a) the airplane is in straight and level flight. In Fig. 3.1.1(b) the airplane is in accelerated flight; the airplane is being subjected to an acceleration normal to its flight path. The accelerating force, $\triangle L$, is an increment of lift which is caused either by encountering a gust or as a result of an intentional maneuver. The additional force causes an acceleration, A_z, and equilibrium is provided by the inertia force, F_i.

$$\triangle L = F_i = \left(\frac{W}{g}\right) A_z$$

$$\triangle L + L = W \left(1 + \frac{A_z}{g}\right)$$

$$\text{Load factor, } n = 1 + \frac{A_z}{g} = 1 + \frac{\triangle L}{W}$$

(1) The maximum *maneuvering load factor* to which an airplane is designed depends on its intended usage. Fighters, which are expected to execute violent maneuvers, are designed to withstand loads commensurate with the accelerations a pilot can physically withstand. Long range, heavily loaded bombers, on the other hand, are designed to low load factors and must be handled accordingly. Fig. 3.1.2 shows the relative strength (in terms of design load factor) of various airplane types. Actual values of design maneuvering load factors are based on statistical data. Transports have low values of the order of 2 to 3, while fighters are designed to higher values, such as 6 to 8. The magnitudes to be used for design are specified by the licensing or procuring agencies in their specifications.

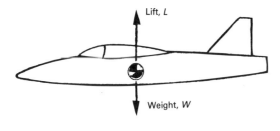

(a) L = W case

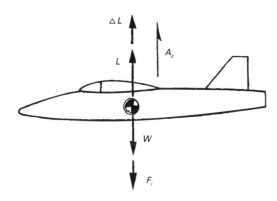

(b) $L + \triangle L = W + F_i$ case
where $L = W$ and $\triangle L = F_i$

Fig. 3.1.1 *Airplane load factor.*

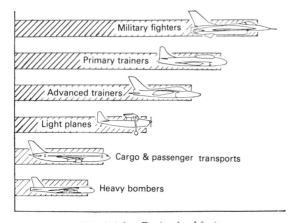

Fig. 3.1.2 *Design load factor.*

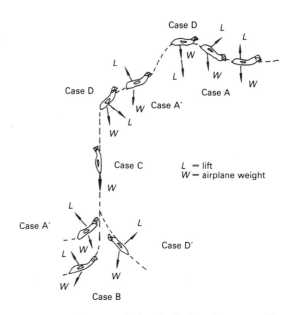

(a) Position of the aircraft along its flight path corresponding to the basic wing load cases

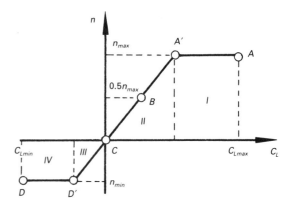

(b) Basic wing load cases

Fig. 3.1.3 *Typical fighter wing maneuvering flight path vs. wing load cases.*

Fig. 3.1.3 shows the characteristic regions of C_L and n combinations that are possible in flight (where C_L = wing lift coefficient, n = load factor) and the corresponding positions of the aircraft along its flight path in all these cases;

- Case A is reaching large angles of attack corresponding to $C_{L\,max}$ while reaching a load factor of n_{max} (steep climb).
- Case A' is reaching a load factor of n_{max} at maximum permissible airspeed, V_{max} (recovery from a dive or glide, ascending current of rough air).
- Case B is reaching the mean load factor ($0.5\,n_{max}$) for maximum permissible airspeed with ailerons deflected (advanced aerobatics and roll maneuvers).
- Case C is that of ailerons deflected at maximum airspeed when $C_L = 0$ (dive).
- Cases D and D' correspond to entering a dive to the elements of aerobatic maneuvers with a negative load factor, and to the action of a descending current of rough air.

Typical wing load cases are within the limits of these regions:

Region I $n = n_{max}$ = constant
  (n_{max} = maximum load factor)

Region II $n_{max} \geq n \geq 0$;

22 Airframe Structural Design

Region III $\quad 0 \geqslant n \geqslant n_{min}$;
(where n_{min} = minimum load factor)

Region IV $\quad n = n_{min}$ = constant

(2) The *gust load factors* experienced by an airplane are caused by atmospheric disturbances and are beyond the control of the pilot. The loads engineer is faced with the problem of how to establish criteria that will lead to the best compromise between an adequate strength level and an unconservative design.

The nature of atmospheric turbulence and its effects on the aircraft structure have been subjects of much research effort for a number of years. The turbulence characteristic of most interest relative to structural design loads is the intensity of turbulence likely to be encountered; in other words, the magnitude of gusts that the structure must be designed to withstand. Design gust intensities used to determine design gust load factors are determined from statistical data and are generally specified by the licensing, or procuring agencies.

The gust load factors on an airplane are greater when the airplane is flying at the minimum flying weight than they are at the gross weight condition. While this is seldom critical for the wings, since they have less weight to carry, it is critical for a structure such as the engine mounts which carries the same weight at a higher load factor. It is therefore necessary to calculate gust load factors at the minimum weight at which the airplane with be flown (see Section 3.4 for further discussion).

Limit and Ultimate Loads

Limit loads are the maximum loads anticipated on the aircraft during its service life. The aircraft structure shall be capable of supporting the limit loads without suffering detrimental permanent deformation. For all loads up to "limit" the deformation of the structure shall be such as not to interfere with the safe operation of the aircraft.

Ultimate loads (or design loads) are equal to the limit loads multiplied by a factor of safety,

Ultimate load = Limit load × Factor of safety

In general the ultimate factor of safety is 1.5. The requirements also specify that these ultimate loads be carried by the structure without failure.

Although aircraft are not supposed to undergo greater loads than the specified limit loads, a certain amount of reserve strength against complete structural failure of a unit is necessary in the design of practically any part of the structure. This is due to many factors such as: (1) The approximations involved in aerodynamic theory and also structural stress analysis theory; (2) Variation in physical properties of materials; (3) Variation in fabrication and inspection standards. Possibly the most important reason for the factors of safety for aircraft is due to the fact that practically every aircraft is limited to the maximum velocity it can be flown and the maximum acceleration it can be subjected to in flight or landing. Since these are under the control of the pilot it is possible in emergency conditions that the limit loads may be slightly exceeded but with a reserve factor of safety against failure this exceeding of the limit load should not prove serious from an aircraft safety standpoint, although it might cause permanent structural deformation that might require repair or replacement of small units or portions of the structure.

Loads due to aircraft gusts are arbitrary in that the gust velocity is assumed. Although this gust velocity is based on years of experience in measuring and recording gust force in flight all over the world, it is quite possible that during the lifetime of an aircraft, turbulent conditions from storm areas or over mountains or water areas might produce air gust velocities slightly greater than that specified in the load requirements; thus the factor of safety insures safety against failure if this situation should arise.

Structural Design Criteria

The structural criteria define the types of maneuvers, speeds, useful loads, and gross weights which are to be considered for structural design analysis. These are items which are under the control of the airplane operator. In addition, the structural criteria must consider such items as inadvertent maneuvers, effects of turbulent air, and severity of ground contact during landing. The basic structural design criteria, from which the loadings are determined, are based largely on the type of airplane and its intended use.

The criteria which are under control of the airplane operator are based on conditions for which the pilot will expect the airplane to be satisfactory. Loadings for these conditions are calculated from statistical data on what pilots do with and expect from an airplane. The strength provided in the airplane structure to meet these conditions must be adequate for the airplane to perform its intended mission. The commercial airplane must be capable of performing its mission in a profitable as well as safe manner and is assumed to be operated by qualified personnel under well regulated conditions. Military types do not usually lend themselves as readily to a definition of strength level based on a mission, because the mission may not be well defined. Military airplanes are not always operated under well regulated conditions and thus necessitate a wider range of design limits.

The latter group of criteria include those loadings over which the pilot may have little or no control. The strength necessary for these conditions is based almost entirely on statistical data. Statistics are constantly being gathered and structural design conditions being modified or formulated from interpretations of these data. These statistics can easily be used to design an airplane similar to the one on which the statistics were collected. Unfortunately, the data cannot always be successfully extrapolated to a new and different design.

Regardless of airplane type, the criteria are based primarily on previous experience and statistics. Occasionally, a condition is experienced which is outside the statistics and a failure occurs. Therefore, a 100 percent safety level is difficult, if not impossible, to obtain and still have an economical airplane.

When the type and use of the airplane is defined, reference is made to the requirements of the customer involved and the requirements of the licensing agency. The minimum structural requirements for airplanes for the various agencies are presented in the following documents:

Civil Aircraft
1. Federal Aviation Regulations (FAR), Volume III, Part 23 — Airworthiness Standards: Normal, Utility, and Aerobatic Category Airplanes
2. FAR, Volume III, Part 25 — Airworthiness Standards: Transport Category
3. British Civil Airworthiness Requirements, Section D — Aeroplanes

Military Aircraft
The Military Specification MIL-A-8860 (ASG) Series consisting of:
(a) MIL-A-8860 (ASG) — General Specification for Airplane Strength and Rigidity
(b) MIL-A-8861 (ASG) — Flight Loads
(c) MIL-A-8862 (ASG) — Landplane Landing and Ground Handling Loads
(d) MIL-A-8863 (ASG) — Additional Loads for Carrier-Based Landplanes
(e) MIL-A-8864 (ASG) — Water and Handling Loads for Seaplanes
(f) MIL-A-8865 (ASG) — Miscellaneous Loads
(g) MIL-A-8866 (ASG) — Reliability Requirements, Repeated Loads, and Fatigue
(h) MIL-A-8867 (ASG) — Ground Tests
(i) MIL-A-8868 (ASG) — Data and Reports
(j) MIL-A-8869 (ASG) — Special Weapons Effects
(k) MIL-A-8870 (ASG) — Vibration, Flutter, and Divergence

The requirements are general in nature and do not always apply to a new type of airplane. Consequently, interpretations of and deviations from the requirements are necessary. These deviations and interpretations are then negotiated with the licensing or procuring agency. In some cases, special requirements may be necessary to cover unusual airplane configurations. The manufacturers requirements are usually the result of an unhappy experience or an advancement in the "state of the art" by that manufacturer. The trend on commercial airplanes in recent years has been toward the establishment of specially designated *Special Conditions* for each individual airplane design. The FAA specifies these conditions with some negotiation with the airframe and engine manufacturer.

Atmospheric Properties
As altitude increases the density of the air decreases as does the temperature. Other properties of the air change with these items, also. In order that all aerodynamic calculations should be made on the same basis a standard variation with altitude of the various parameters has been established.

(1) Dynamic Pressure
Dynamic pressure is a measure of the kinetic energy in a moving gas or fluid. It is indicated by the symbol q and is defined as,

$$q = \frac{1}{2}\rho V_e^2$$

V_e = true airspeed in ft/sec.

ρ = mass density = $\frac{w}{g}$ in slugs/ft^3

w = density in lb/ft^3

$$q = \frac{V_e^2}{296} \quad (V_e \text{ in knots}) \quad (3.1.1)$$

$$q = \frac{V_e^2}{391} \quad (V_e \text{ in mph}) \quad (3.1.2)$$

(2) Equivalent Airspeed
At altitude the difference in air density must be considered. The equation for q is ($\frac{1}{2}\rho_{alt} V_t^2$) where ρ_{alt} is air density at the altitude being considered, V_t is true airspeed. Rather than considering different values of ρ, the concept of equivalent airspeed is introduced. *The equivalent airspeed, V_e is the true airspeed corrected for the difference between sea level air density and the air density at the altitude being considered.* At sea level $V_t = V_e$.

At altitude, $V_e = V_t \left(\frac{\rho_{alt}}{\rho_{S.L.}}\right)^{\frac{1}{2}}$. This density ratio, $\frac{\rho_{alt}}{\rho_{S.L.}}$, is defined as σ. Then

$$V_e = V_t \sigma^{\frac{1}{2}} \quad \text{or} \quad V_t = \frac{V_e}{\left(\frac{\rho_{alt}}{\rho_{S.L.}}\right)^{\frac{1}{2}}} \quad (3.1.3)$$

Substituting in the q equation

$$q = \frac{1}{2}\rho_{alt} \frac{V_e^2}{\left(\frac{\rho_{alt}}{\rho_{S.L.}}\right)} = \frac{1}{2}\rho_{S.L.} V_e^2 \quad (3.1.4)$$

(3) Compressibility Correction to Calibrated Airspeeds
Starting with the *isentropic* relationship

$$\frac{p_0}{p} = \left(1 + \frac{k-1}{2}M^2\right)^{\frac{k}{k-1}} \quad (3.1.5)$$

where p_0 = total stagnation pressure
p = static pressure
k = 1.4, ratio of specific heats
M = Mach number

Substituting $k = 1.4$ in the above equation and subtracting 1 from both sides of the equation.

$$\frac{p_0 - p}{p} = (1 + .2M^2)^{3.5} - 1 \quad (3.1.6)$$

At sea level,

$$p_0 - p = P_{S.L.}\left\{\left[1 + .2\left(\frac{V_c}{661.5}\right)^2\right]^{3.5} - 1\right\} \quad (3.1.7)$$

The above equation is used to calibrate the airspeed V_c at sea level, measured in knots.
At other altitudes, a correction is needed as shown below:

$$\Delta V_C = V\sigma^{\frac{1}{2}} - V_C \quad (3.1.8)$$

which is plotted on Fig. 3.1.4

(4) Tables and Figures
Fig. 3.1.5 is a typical summary of atmospheric properties useful in evaluating equivalent and true airspeeds, etc.

Fig. 3.1.6 is a plot of altitude versus equivalent airspeed with families of constant Mach lines and constant calibrated airspeeds.

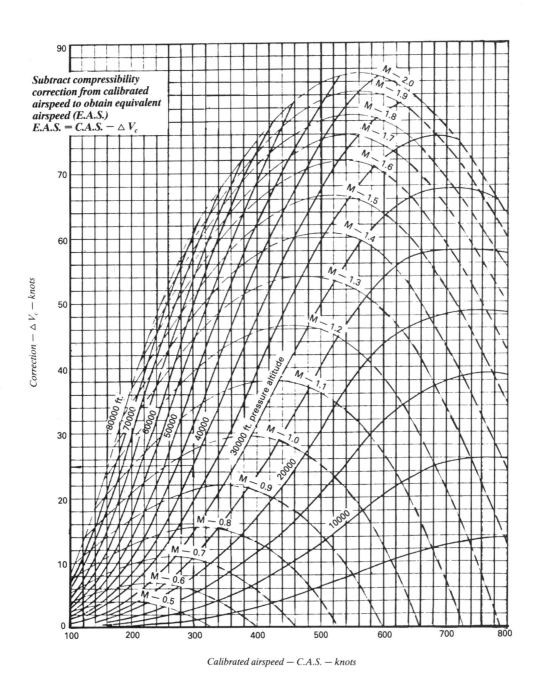

Fig. 3.1.4 Compressibility correction to calibrated airspeed.

Altitude		Pressure			Density			Temperature			
Geom ft	Pressure ft	P $\frac{lb}{ft^2}$	δ $\frac{P}{P_o}$	$\frac{q}{M^2}$ $0.7\,P$ $\frac{lb}{ft^2}$	ρ $\frac{Slug}{ft^3}$	σ $\frac{\rho}{\rho_0}$ $(\rho_0 = \rho_{S.L.})$	$\sqrt{\sigma}$	T (°R)	t (°F)	θ $\frac{T}{T_0}$	t_c (°C)
0	0	2116.22 + 0	1.00000 + 0	1481.35 + 0	0.237689 − 2	1.0000 + 0	1.0000 + 0	518.670	59.000	1.00000	15.000
1000	1000	2040.86	9.64389 − 1	1428.60	0.230812	9.7107 − 1	9.8543 − 1	515.104	55.434	0.99312	13.019
2000	2000	1967.69	9.29816	1377.38	0.224088	9.4278	9.7097	511.538	51.868	0.98625	11.038
3000	3000	1896.67	8.96256	1327.67	0.217516	9.1513	9.5662	507.973	48.303	0.97938	9.057
4000	4000	1827.75	8.63687	1279.42	0.211093	8.8811	9.4239	504.408	44.738	0.97256	7.077
5000	5000	1760.87	8.32085	1232.61	0.204817	8.6170	9.2828	500.843	41.173	0.96563	5.096
6000	6000	1696.00	8.01430	1187.20	0.198685	8.3590	9.1428	497.279	37.609	0.95876	3.116
7000	7000	1633.08	7.71698	1143.16	0.192695	8.1070	9.0039	493.715	34.045	0.95189	1.136
8000	8000	1572.07	7.42869	1100.45	0.186845	7.8609	8.8662	490.152	30.482	0.94502	− 0.844
9000	9000	1512.93	7.14921	1059.05	0.181133	7.6206	8.7296	486.588	26.918	0.93815	− 2.823
10000	10000	1455.60 + 0	6.87832 − 1	1018.92 + 0	0.175555 − 2	7.3859 − 1	8.5941 − 1	483.025	23.355	0.93128	− 4.803
11000	11000	1400.05	6.61584	9800.38 − 1	0.170110	7.1568	8.4598	479.463	19.793	0.92441	− 6.782
12000	12000	1346.24	6.36155	9423.69	0.164796	6.9333	8.3266	475.901	16.231	0.91754	− 8.761
13000	13000	1294.12	6.11525	9058.84	0.159610	6.7151	8.1946	472.339	12.669	0.91067	− 10.740
14000	14000	1243.65	5.87676	8705.55	0.154551	6.5022	8.0636	468.777	9.107	0.90381	− 12.718
15000	15000	1194.79	5.64587	8363.52	0.149616	6.2946	7.9338	465.216	5.546	0.89694	− 14.697
16000	16000	1147.50	5.42241	8032.50	0.144802	6.0921	7.8052	461.655	1.985	0.89007	− 16.675
17000	17000	1101.74	5.20619	7712.19	0.140109	5.8946	7.6776	458.095	− 1.575	0.88321	− 18.653
18000	18000	1057.48	4.99702	7402.34	0.135533	5.7021	7.5512	454.535	− 5.135	0.87635	− 20.631
19000	19000	1014.67	4.79472	7102.67	0.131072	5.5145	7.4259	450.975	− 8.695	0.86948	− 22.609
20000	20000	9732.75 − 1	4.59913 − 1	6812.93 − 1	0.126726 − 2	5.3316 − 1	7.3018 − 1	447.415	− 12.255	0.86262	− 24.586
21000	21000	9332.66	4.41007	6532.86	0.122491	5.1534	7.1787	443.858	− 15.814	0.85576	− 26.563
22000	22000	8946.02	4.22736	6262.21	0.118365	4.9798	7.0568	440.297	− 19.373	0.84890	− 28.540
23000	23000	8572.49	4.05086	6000.74	0.114347	4.8108	6.9360	436.739	− 22.931	0.84204	− 30.517
24000	24000	8211.72	3.88038	5748.21	0.110435	4.6462	6.8163	433.181	− 26.489	0.83518	− 32.494
25000	25000	7863.38	3.71577	5504.37	0.106626	4.4859	6.6977	429.623	− 30.047	0.82832	− 34.471
26000	26000	7527.14	3.55688	5269.00	0.102919	4.3300	6.5802	426.065	− 33.605	0.82146	− 36.447
27000	27000	7202.66	3.40356	5041.86	0.993112 − 3	4.1782	6.4639	422.508	− 37.162	0.81460	− 38.423
28000	28000	6889.64	3.25564	4822.75	0.958016	4.0305	6.3487	418.951	− 40.719	0.80774	− 40.399
29000	29000	6587.75	3.11298	4611.42	0.923880	3.8869	6.2345	415.395	− 44.275	0.80089	− 42.375
30000	30000	6296.69 − 1	2.97545 − 1	4407.68 − 1	0.890686 − 3	3.7473 − 1	6.1215 − 1	411.839	− 47.831	0.79403	− 44.351
31000	31000	6016.15	2.84288	4211.31	0.858416	3.6115	6.0096	408.283	− 51.387	0.78717	− 46.326
32000	32000	5745.85	2.71515	4022.10	0.827050	3.4795	5.8988	404.728	− 54.742	0.78032	− 48.301
33000	33000	5485.50	2.59213	3839.85	0.796572	3.3513	5.7891	401.173	− 58.497	0.77346	− 50.275
34000	34000	5234.80	2.47366	3664.36	0.766963	3.2267	5.6804	397.618	− 62.052	0.76661	− 52.251

Fig. 3.1.5 Standard atmospheric properties.

Viscosity		RN "K" Factory			Speed of sound					
μ $\dfrac{\text{Slug}}{\text{ft-sec}}$	ν $\dfrac{\text{ft}^2}{\text{sec}}$	K_{PPS} $RN/\ell V_{FPS}$ $\dfrac{1}{\text{ft-FPS}}$	K_{KT} $RN/\ell V_{KT}$ $\dfrac{1}{\text{ft-kt}}$	K_M $RN/\ell M$ $\dfrac{1}{\text{ft-Mach}}$	C_S $\dfrac{\text{ft}}{\text{sec}}$	C_{se} $C_S\sqrt{\sigma}$ $\dfrac{\text{ft}}{\text{sec}}$	C_S mph	C_{se} $C_S\sqrt{\sigma}$ mph	C_S kt	C_{se} $C_S\sqrt{\sigma}$ kt
0.37372 − 6	0.15723 − 3	6.360 + 3	1.073 + 4	7.101 + 6	1116.45	1116.45 + 0	761.22	761.22 + 0	661.48	661.48 + 0
0.37172	0.16105	6.209	1.048	6.908	1112.61	1096.39	758.59	747.54	659.20	649.60
0.36971	0.16499	6.061	1.023	6.720	1108.75	1076.56	755.96	734.02	656.92	637.84
0.36770	0.16905	5.916	9.984 + 3	6.536	1104.88	1056.95	753.33	720.65	654.62	626.23
0.36568	0.17323	5.773	9.743	6.356	1100.99	1037.57	750.68	707.43	652.32	614.74
0.36366	0.17755	5.632	9.506	6.179	1097.10	1018.41	748.02	694.37	650.01	603.39
0.36162	0.18201	5.494	9.273	6.006	1093.19	9994.76 − 1	745.35	681.46	647.70	592.17
0.35958	0.18661	5.359	9.045	5.837	1089.26	9807.61	742.68	668.70	645.37	581.09
0.35754	0.19136	5.226	8.820	5.672	1085.32	9622.67	739.99	656.09	643.04	570.13
0.35549	0.19626	5.095	8.600	5.510	1081.37	9439.92	737.30	643.63	640.70	559.30
0.35343 − 6	0.20132 − 3	4.967 + 3	8.384 + 3	5.352 + 6	1077.40	9259.36 − 1	734.59	631.32 + 0	638.35	548.60 + 0
0.35136	0.20655	4.841	8.172	5.197	1073.42	9080.96	731.88	619.16	635.99	538.03
0.34928	0.21195	4.718	7.963	5.046	1069.43	8904.73	729.16	607.14	633.62	527.59
0.34720	0.21753	4.597	7.759	4.898	1065.42	8730.65	726.42	595.27	631.25	517.28
0.34512	0.22330	4.478	7.558	4.753	1061.40	8558.71	723.68	583.55	628.86	507.09
0.34302	0.22927	4.362	7.362	4.612	1057.36	8388.90	720.92	571.97	626.47	497.03
0.34092	0.23544	4.247	7.169	4.474	1053.30	8221.21	718.16	560.54	624.07	487.09
0.33881	0.24182	4.135	6.980	4.339	1049.23	8055.63	715.39	549.25	621.65	477.28
0.33669	0.24842	4.025	6.794	4.207	1045.15	7892.14	712.60	538.10	619.23	467.60
0.33457	0.25525	3.918	6.612	4.078	1041.05	7730.74	709.80	527.10	616.80	458.03
0.33244 − 6	0.26233 − 3	3.812 + 3	6.434 + 3	3.953 + 6	1036.93	7571.42 − 1	707.00	516.23 + 0	614.37	448.60 + 0
0.33030	0.26965	3.709	6.259	3.830	1032.80	7414.16	704.18	505.51	611.92	439.28
0.32815	0.27723	3.607	6.088	3.710	1028.65	7258.96	701.35	494.93	609.46	430.08
0.32599	0.28509	3.508	5.920	3.594	1024.48	7105.80	698.51	484.49	606.99	421.01
0.32383	0.29323	3.410	5.756	3.479	1020.30	6954.67	695.66	474.18	604.51	412.05
0.32166	0.30167	3.315	5.595	3.368	1016.10	6805.56	692.80	464.02	602.03	403.22
0.31948	0.31042	3.221	5.437	3.260	1011.89	6658.47	689.92	453.99	599.53	394.50
0.31730	0.31950	3.130	5.283	3.154	1007.65	6513.37	687.04	444.09	597.02	385.91
0.31510	0.32891	3.040	5.131	3.051	1003.40	6370.27	684.14	434.34	594.50	377.43
0.31290	0.33868	2.953	4.983	2.950	999.14	6229.14	681.23	424.71	591.97	369.07
0.31069 − 6	0.34882 − 3	2.867 + 3	4.839 + 3	2.852 + 6	994.85	6089.97 − 1	678.31	415.23 + 0	589.43	360.82 + 0
0.30847	0.35935	2.783	4.697	2.756	990.55	5952.77	675.37	405.87	586.88	352.69
0.30625	0.37029	2.701	4.558	2.663	986.22	5817.50	672.43	396.65	584.32	344.68
0.30401	0.38165	2.620	4.422	2.573	981.88	5684.17	669.47	387.56	581.75	336.78
0.30177	0.39346	2.542	4.290	2.484	977.52	5552.77	666.49	378.60	579.17	328.99

Fig. 3.1.5 (continued)

Altitude		Pressure			Density			Temperature			
Geom ft	Pressure ft	P $\frac{lb}{ft^2}$	δ $\frac{P}{P_o}$	$\frac{q}{M^2}$ $0.7P$ $\frac{lb}{ft^2}$	ρ $\frac{Slug}{ft^3}$	σ $\frac{\rho}{\rho_0}$ ($\rho_0 = \rho_{S.L.}$)	$\sqrt{\sigma}$	T (°R)	t (°F)	θ $\frac{T}{T_0}$	t_c (°C)
35000	35000	4993.49	2.35963	3495.44	0.738206	3.1058	5.5729	394.064	−65.606	0.75976	−54.226
36000	36000	4761.28	2.24990	3332.90	0.710284	2.9883	5.4665	390.510	−69.160	0.75291	−56.200
36152	36152	4726.80 − 1	2.23361 − 1	3308.76 − 1	0.706117 − 3	2.9708 − 1	5.4505 − 1	389.970	−69.700	0.75187	−56.500
37000	37000	4536.63	2.14469	3177.04	0.678007	2.8525	5.3409	389.970	−69.700	0.75187	−56.500
38000	38000	4326.40	2.04440	3028.48	0.646302	2.7191	5.2145	389.970	−69.700	0.75187	−56.500
39000	39000	4124.10	1.94881	2886.87	0.616082	2.5920	5.0911	389.970	−69.700	0.75187	−56.500
40000	40000	3931.29 − 1	1.85770 − 1	2751.90 − 1	0.587278 − 3	2.4708 − 1	4.9707 − 1	389.970	−69.700	0.75187	−56.500
41000	41000	3747.50	1.77085	2623.25	0.559823	2.3553	4.8531	389.970	−69.700	0.75187	−56.500
42000	42000	3572.33	1.68807	2500.63	0.533655	2.2452	4.7383	389.970	−69.700	0.75187	−56.500
43000	43000	3405.36	1.60917	2383.75	0.508711	2.1402	4.6263	389.970	−69.700	0.75187	−56.500
44000	44000	3246.20	1.53397	2272.34	0.484936	2.0402	4.5169	389.970	−69.700	0.75187	−56.500
45000	45000	3094.50	1.46228	2166.15	0.462274	1.9449	4.4101	389.970	−69.700	0.75187	−56.500
46000	46000	2949.90	1.39395	2064.93	0.440673	1.8540	4.3058	389.970	−69.700	0.75187	−56.500
47000	47000	2812.08	1.32882	1968.45	0.420084	1.7674	4.2040	389.970	−69.700	0.75187	−56.500
48000	48000	2680.70	1.26674	1876.49	0.400458	1.6848	4.1046	389.970	−69.700	0.75187	−56.500
49000	49000	2555.47	1.20757	1783.83	0.381751	1.6061	4.0076	389.970	−69.700	0.75187	−56.500
50000	50000	2436.11 − 1	1.15116 − 1	1705.27 − 1	0.363919 − 3	1.5311 − 1	3.9129 − 1	389.970	−69.700	0.75187	−56.500
51000	51000	2322.33	1.09740	1625.63	0.346922	1.4956	3.8204	389.970	−69.700	0.75187	−56.500
52000	52000	2213.87	1.04615	1549.71	0.330721	1.3914	3.7301	389.970	−69.700	0.75187	−56.500
53000	53000	2110.49	9.97295 − 2	1477.34	0.315277	1.3264	3.6420	389.970	−69.700	0.75187	−56.500
54000	54000	2011.95	9.50729	1408.36	0.300556	1.2645	3.5560	389.970	−69.700	0.75187	−56.500
55000	55000	1918.01	9.06341	1342.61	0.286524	1.2055	3.4720	389.970	−69.700	0.75187	−56.500
56000	56000	1828.47	8.64030	1279.93	0.273148	1.1492	3.3900	389.970	−69.700	0.75187	−56.500
57000	57000	1743.12	8.23691	1220.19	0.260397	1.0955	3.3099	389.970	−69.700	0.75187	−56.500
58000	58000	1661.76	7.85251	1163.23	0.248243	1.0444	3.2317	389.970	−69.700	0.75187	−56.500
59000	59000	1584.21	7.48603	1108.94	0.236658	9.9566 − 2	3.1554	389.970	−69.700	0.75187	−56.500
60000	60000	1510.28 − 1	7.13669 − 2	1057.19 − 1	0.225614 − 3	9.4920 − 2	3.0809 − 1	389.970	−69.700	0.75187	−56.500
61000	61000	1439.81	6.80368	1007.86	0.215086	9.0491	3.0082	389.970	−69.700	0.75187	−56.500
62000	62000	1372.63	6.48623	9608.39 − 2	0.205051	8.6269	2.9372	389.970	−69.700	0.75187	−56.500
63000	63000	1308.59	6.18363	9160.13	0.195485	8.2244	2.8678	389.970	−69.700	0.75187	−56.500
64000	64000	1247.55	5.89517	8732.62	0.186365	7.8407	2.8001	389.970	−69.700	0.75187	−56.500
65000	65000	1189.35	5.62019	8325.48	0.177673	7.4750	2.7340	389.970	−69.700	0.75187	−56.500
65824	65824	1143.45 − 1	5.40330 − 2	8004.18 − 2	0.170816 − 3	7.1865 − 2	2.6808 − 1	389.970	−69.700	0.75187	−56.500
66000	66000	1133.88	5.35807	7937.19	0.169344	7.1246	2.6692	390.066	−69.604	0.75205	−56.447
67000	67000	1081.05	5.10842	7567.37	0.161229	6.7832	2.6045	390.611	−69.059	0.75310	−56.144
68000	68000	1030.76	4.87076	7215.30	0.153513	6.4586	2.5414	391.156	−68.514	0.75415	−55.841
69000	69000	9828.71 − 2	4.64447	6880.10	0.146178	6.1500	2.4799	391.701	−67.969	0.75520	−55.536

Fig. 3.1.5 (continued)

Viscosity		RN "K" Factory			Speed of sound					
μ $\frac{\text{Slug}}{\text{ft-sec}}$	ν $\frac{\text{ft}^2}{\text{sec}}$	K_{PPS} $\frac{RN/}{\ell V_{FPS}}$ $\frac{1}{\text{ft-FPS}}$	K_{KT} $\frac{RN/}{\ell V_{KT}}$ $\frac{1}{\text{ft-kt}}$	K_M $RN/\ell M$ $\frac{1}{\text{ft-Mach}}$	C_S $\frac{\text{ft}}{\text{sec}}$	C_{se} $C_S\sqrt{\sigma}$ $\frac{\text{ft}}{\text{sec}}$	C_S mph	C_{se} $C_S\sqrt{\sigma}$ mph	C_S kt	C_{se} $C_S\sqrt{\sigma}$ kt
0.29951	0.40573	2.465	4.160	2.398	973.14	5423.27	663.51	369.77	576.57	321.32
0.29725	0.41850	2.389	4.033	2.315	968.75	5295.67	660.51	361.07	573.97	313.76
0.29691 − 6	0.42048 − 3	2.378 + 3	4.014 + 3	2.302 + 6	968.08	5276.47 − 1	660.05	359.76 + 0	573.57	312.62 + 0
0.29691	0.43792	2.284	3.854	2.211	968.08	5170.37	660.05	352.53	573.57	306.34
0.29691	0.45940	2.177	3.674	2.107	968.08	5048.04	660.05	344.18	573.57	299.09
0.29691	0.48193	2.075	3.502	2.009	968.08	4928.61	660.05	336.04	573.57	292.01
0.29691 − 6	0.50557 − 3	1.978 + 3	3.338 + 3	1.915 + 6	968.08	4812.01 − 1	660.05	328.09 + 0	573.57	285.10 + 0
0.29691	0.53036	1.885	3.182	1.825	968.08	4698.19	660.05	320.33	573.57	278.36
0.29691	0.55637	1.797	3.034	1.740	968.08	4587.07	660.05	312.75	573.57	271.78
0.29691	0.58365	1.713	2.892	1.659	968.08	4478.58	660.05	305.36	573.57	265.35
0.29691	0.61227	1.633	2.757	1.581	968.08	4372.67	660.05	298.14	573.57	259.07
0.29691	0.64228	1.557	2.628	1.507	968.08	4269.28	660.05	291.09	573.57	252.93
0.29691	0.67376	1.484	2.505	1.437	968.08	4168.34	660.05	284.21	573.57	246.97
0.29691	0.70679	1.415	2.388	1.370	968.08	4069.80	660.05	277.49	573.57	241.13
0.29691	0.74143	1.349	2.276	1.306	968.08	3973.59	660.05	270.93	573.57	235.43
0.29691	0.77776	1.286	2.170	1.245	968.08	3879.67	660.05	264.52	573.57	229.86
0.29691 − 6	0.81587 − 3	1.226 + 3	2.069 + 3	1.187 + 6	968.08	3767.98 − 1	660.05	258.27 + 0	573.57	224.43 + 0
0.29691	0.85584	1.168	1.972	1.131	968.08	3698.46	660.05	252.17	573.57	219.13
0.29691	0.89777	1.114	1.880	1.078	968.08	3611.07	660.05	246.21	573.57	213.95
0.29691	0.94174	1.062	1.792	1.028	968.08	3525.75	660.05	240.39	573.57	208.90
0.29691	0.98787	1.012	1.709	9.800 + 5	968.08	3442.45	660.05	234.71	573.57	203.96
0.29691	0.10362 − 2	9.650 + 2	1.629	9.342	968.08	3361.13	660.05	229.17	573.57	199.14
0.29691	0.10870	9.200	1.553	8.906	968.08	3281.74	660.05	223.75	573.57	194.44
0.29691	0.11402	8.770	1.480	8.490	968.08	3204.23	660.05	218.47	573.57	189.85
0.29691	0.11960	8.361	1.411	8.094	968.08	3128.55	660.05	213.31	573.57	185.36
0.29691	0.12546	7.791	1.345	7.716	968.08	3054.68	660.05	208.27	573.57	180.98
0.29691 − 6	0.13160 − 2	7.599 + 2	1.283 + 3	7.356 + 5	968.08	2982.55 − 1	660.05	203.36 + 0	573.57	176.71 + 0
0.29691	0.13804	7.244	1.223	7.013	968.08	2912.13	660.05	198.55	573.57	172.54
0.29691	0.14480	6.906	1.166	6.686	968.08	2843.39	660.05	193.87	573.57	168.47
0.29691	0.15188	6.584	1.111	6.374	968.08	2776.27	660.05	189.29	573.57	164.49
0.29691	0.15932	6.277	1.059	6.076	968.08	2710.74	660.05	184.82	573.57	160.61
0.29691	0.16711	5.984	1.010	5.793	968.08	2646.76	660.05	180.46	573.57	156.82
0.29691 − 6	0.17382 − 2	5.753 + 2	9.710 + 2	5.569 + 5	968.08	2595.19 − 1	660.05	176.94 + 0	573.57	153.76 + 0
0.29697	0.17537	5.702	9.625	5.521	968.20	2584.30	660.13	176.20	573.64	153.12
0.29732	0.18441	5.423	9.153	5.254	968.87	2523.38	660.59	172.05	574.04	149.51
0.29767	0.19390	5.157	8.704	5.000	969.55	2463.98	661.05	168.00	574.44	145.99
0.29801	0.20387	4.905	8.279	4.759	970.22	2406.07	661.52	164.05	574.84	142.56

Fig. 3.1.5 (continued)

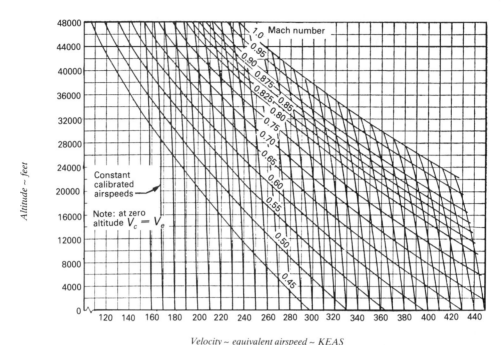

Fig. 3.1.6 Constant calibrated airspeeds.

Summary of Primary Design Conditions for Transport
(1) Flight: Pilot induced maneuvers and airplane system malfunctions
  • Combination stabilizer-elevator maneuvers
  • Aileron and/or asymmetric spoiler maneuvers
  • Rudder maneuvers
  • Combinations of above three items
  • Engine failure
(2) Flight: Atmospheric turbulence
  • Power spectral approach
    (a) *Mission Analysis* based on operational spectra
    (b) *Design Envelope Analysis* based on extremes of operational envelopes.
  • Discrete Gust Approach
    (a) FAR gust load factor equation
    (b) Tuned discrete gusts (i.e., one-minus-cosine shape gusts with varying wavelengths)
(3) Landing
  • FAR landing condition (i.e., LL1, LL2, LL3, DL)
  • Rational landing time histories (based on FAR-designated sink speeds, wing lifts, pitch altitudes, and speeds)
(4) Ground handling
  • FAR ground handling (i.e., BR2, BR3, turning, pivoting, unsymmetrical braking, towing)
  • Rational taxiing (i.e., bump encountered during take-off runs, rejected take-offs, landing runout, taxiway taxiing)
  • Rational ground maneuvering
  • Jacking
(5) Fail-safe and breakaway design

3.2 Aeroelasticity

Aeroelastic effects in loading calculations are those associated with structural deflection. When certain parts of the structure deflect, significant changes may occur in the airplane stability and distribution of airloads. (Ref. 3.20 and 3.33).

The effects of aeroelasticity are usually not found without a large amount of calculations. These calculations must be based on estimated stiffness of the structure. Also, the effect of structural deflection on the aerodynamics is difficult to determine accurately. Consequently, wherever possible, the aeroelastic effects are neglected or simple assumptions made when the effects are significant. The simplest method is to estimate the angle of attack change from structural deflection and calculate an increment of loading from this change.

Unswept Wing
An accurate method of determining wing-twist can be used where the loads are readily determined. Each increment of load caused by structural deflection produces another deflection as shown in Fig. 3.2.1 and requires an iteration to determine the final wing loads.

A reasonable assumption is that the iteration follows a geometric series. The final equilibrium increment of angle of attack α of the wing takes the form

$$\Delta \alpha = \Delta \alpha_1 [1 + R + R^2 + \ldots]$$

where $R = \dfrac{\Delta \alpha_2}{\Delta \alpha_1}$

The limit of the sum of this series is

$$\Delta \alpha = \Delta \alpha_1 \left[\frac{1}{1-R} \right] \quad \text{where } R < 1$$

The lift distribution on the wing is a basic type of distribution. The change in angle of attack of the complete wing to maintain constant lift is equal to the $\bar{\epsilon}$ (see Fig. 3.2.2) corrected by the iteration series.

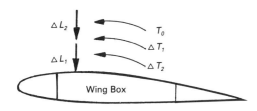

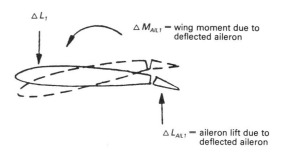

Fig. 3.2.3 Illustration of aileron reversal.

where: T_0 is wing torsion and causes $\Delta \alpha_1$
$\Delta \alpha_1$ is initial deflection and causes ΔL_1
ΔL_1 causes ΔT_1, ΔT_1 causes $\Delta \alpha_2$
$\Delta \alpha_2$ causes ΔL_2, etc.

Fig. 3.2.1 Wing loads vs. structural deflections.

A considerable loss in roll performance is experienced as shown in Fig. 3.2.4.

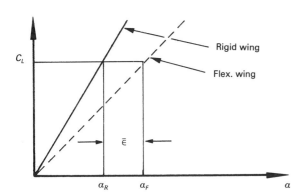

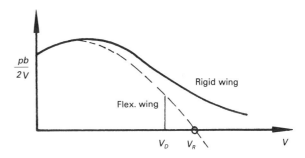

Fig. 3.2.4 Roll performance of a wing.

Fig. 3.2.2 Wing lift comparision between a rigid and flexible wing.

From the value of $\bar{\epsilon}$, the change in lift curve can be found for a flexible wing as follows:

$$(C_{L\alpha})_R = \frac{1}{\alpha_R}$$

$$(C_{L\alpha})_F = \frac{1}{\alpha_F} = \frac{1}{\alpha_R + \bar{\epsilon}}$$

$$= (C_{L\alpha})_R \left[\frac{\alpha_R}{\alpha_R + \bar{\epsilon}} \right]$$

Divergence of a wing or control surface occurs when the value of $\Delta \alpha_2$ approaches $\Delta \alpha_1$ in the iteration equation. When this situation exists, $R \to 1.0$ and $\Delta \alpha \to \infty$. The result is rapid destruction of the wing or surface.

The largest source of torsion on a wing and, consequently, the most serious aeroelastic problem is aileron deflection. Assuming a down aileron deflection, the wing is twisted nose down. The aileron effectiveness is decreased and a down load occurs on the forward section of the wing. This down load further aggravates the situation as shown in Fig. 3.2.3.

At the speed, V_R, the rolling moment caused by the wing deflection is equal to that caused by the aileron. This speed is the aileron reversal and V_D is the airplane design speed which is smaller than V_R.

Swept Wing

A swept wing of a control surface is especially important with respect to aeroelastic effects. In this case, the bending and torsion deflection contribute considerably to the twist angle of the airfoil sections. Twist angle caused by these is shown in Fig. 3.2.5.

The change in slope of the elastic curve between any two points on the beam is equal to the area under the $\frac{M}{EI}$ diagram between points as shown in Fig. 3.2.6,

$$\phi_A - \phi_B = \int_B^A \frac{M}{EI} dy$$

E = modulus of elasticity (ratio of stress/strain) (e.g., 30×10^6 lb/in^2 for steel; 10.5×10^6 lb/in^2 for some aluminum alloys)
I = moment of inertia of wing box
G = modulus of rigidity (e.g., 12×10^6 lb/in^2 for steel; 4×10^6 lb/in^2 for aluminum alloys)
J = torsion constant

Airframe Structural Design 31

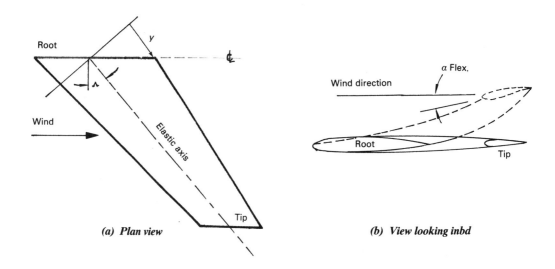

Fig. 3.2.5 Wing twist angle caused by swept.

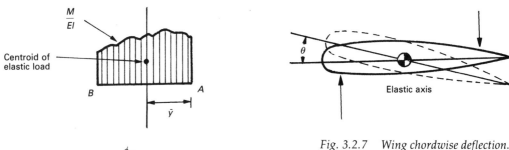

Fig. 3.2.7 Wing chordwise deflection.

The angular deflection due to torsion about the elastic axis between point A and B (see Fig. 3.2.8) is

$$d\theta = \frac{T}{GJ} dy$$

$$\theta_A - \theta_B = \int_B^A \frac{T}{GJ} dy$$

The change in angle of attack on the strip due to structural deflection is

$$\triangle \alpha_{FLEX} = \theta \cos \Lambda - \phi \sin \Lambda$$

where:

$\theta =$ twist angle due to torsion $= \int_0^B \frac{T}{GJ} dy$

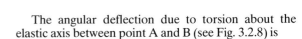

Fig. 3.2.6 Wing spanwise pure structural bending (no torsion).

The deflection of point A on the elastic curve of the beam from a *tangent* to the elastic curve at point B is equal to the *moment* of the area under the $\frac{M}{EI}$ diagram between points taken *about point A*.

$$\delta_{AB} = \int_B^A \frac{M\bar{y}}{EI} dy$$

where $\bar{y} =$ distance between centroid of elastic load $\left(\dfrac{M}{EI}\right)$ and point A

To consider the chordwise deflection (torsion) the concept of the structural elastic axis (also called shear center) is introduced and shown in Fig. 3.2.7.

$\phi =$ slope of bending deflection $= \int_0^A \frac{M}{EI} dy$

It is obvious from the above equation that changes in airload due to structural deflection on a straight wing are mainly due to torsional loads and that bending loads contribute more and more as the sweep

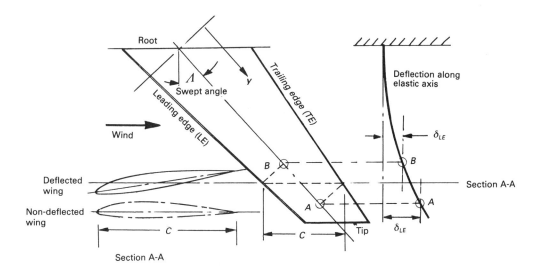

Fig. 3.2.8 *Relation between wing deflection and local streamwise angle of attack.*

increases. Less obvious is the fact that the bending induced deflection on an aft swept wing tends to decrease the bending moment on the wing (shifting the center of pressure inboard) and conversely the bending induced deflection on a swept forward wing tends to shift the loads outboard and increases the bending moment on the wing. Interpreted another way, the swept forward wings are more susceptible to divergence at high speeds.

3.3 Flight Maneuvers

Flight maneuvers are those displacements, velocities, and accelerations resulting from a control surface movement. Maneuvers which create critical structural loadings are the results of a single or combined predetermined control surface movements. The control surface movements and the forces applied to the control systems are determined from statistical data. In most airplanes, however, restrictions on piloting techniques may be necessary in order to prevent structural damage. Few airplanes are designed to a strength level high enough to sustain loads resulting from the maximum output of the pilot.

Pitching Maneuvers
Pitching maneuvers are those maneuvers involving motion and equilibrium along the vertical axis and about the lateral axis and are a result of elevator motion shown in Fig. 3.3.1.

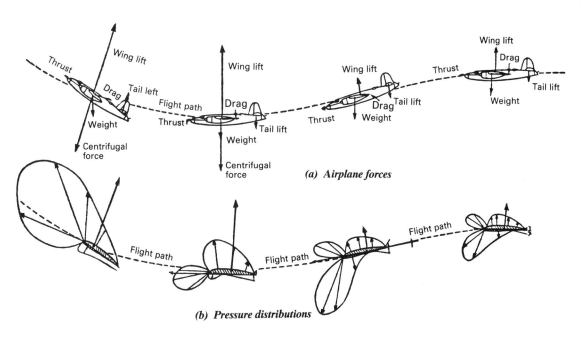

Fig. 3.3.1 *Typical airplane pitching maneuver.*

Steady level flight is a form of maneuver where the forces on the airplane are put into equilibrium with the airplane which is subject to $1.0g$ acceleration. This type of maneuver is known as balanced flight. The aerodynamic moments are balanced to zero about the center-of-gravity. This balancing is accomplished by deflecting the elevators until a balancing tail load is developed. In level flight the airplane is then considered trimmed.

This type of equilibrium is also associated with accelerated flight when the load factor is a value other than one. The balanced flight conditions exist in a slow pull-out or steady coordinated turn. In these cases the angular motion about the lateral axis is so small that it can be considered zero except where gyroscopic moments are involved.

A force system on an airplane in balanced flight is shown in Fig. 3.3.2.

Another type maneuver exists where the aerodynamic moments are not balanced to zero about the airplane center-of-gravity. This maneuver is associated with a rapid movement of the elevator and the maneuver is for very short durations. Pitch maneuver is the term commonly used to describe this condition.

The time involved in this elevator motion is very short and the unbalanced aerodynamic moments are balanced by the airplane inertia.

$$\Sigma \text{ (aerodynamic moments)} = I\ddot{\theta}$$

where I = airplane moment of inertia
$\ddot{\theta}$ = airplane pitching acceleration

If the elevator could be deflected instantaneously, the force is as shown in Fig. 3.3.3.

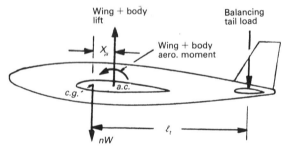

(tail load) ℓ_t = *(wing + body lift)* X_a + *(wing + body)* moment

Fig. 3.3.2 *Airplane in balanced flight.*

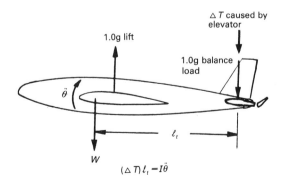

Fig. 3.3.3 *Force diagram due to deflected elevator.*

Usually a variation of the elevator motion with time is established and the airplane response can be readily computed. In lieu of this calculation, some designers have established arbitrary pitching accelerations which give satisfactory strength levels.

Rolling

Rolling maneuvers are usually associated with aileron deflection. Rolling may be induced by rudder maneuvers and asymmetric engine thrust. Roll attitudes and velocities are usually accompanied by yawing motion and often the two motions must be considered simultaneously. Structurally, rolling is of importance since the following design conditions are encountered in roll maneuvers.
(1) Critical wing torsion caused by aileron loading.
(2) Vertical tail loads caused by induced yaw.
(3) Centrifugal force on wing-mounted stores and internal fuel tanks.

Steady roll is that part of the roll maneuver when the aerodynamic forces about the rolling axis are zero and the airplane is no longer accelerating. The rolling velocity is reduced to a dimensionless quantity by measuring the roll as the wing tip helix angle. This angle in terms of roll velocity is $\frac{pb}{2V}$ where

p = roll velocity in radians/second
b = wing span in feet
V = airplane velocity in feet/second

Accelerated roll occurs immediately after the ailerons are deflected. While the velocity is building up the aerodynamic moments about the roll axis are not zero and the forces are shown in Fig. 3.3.4.

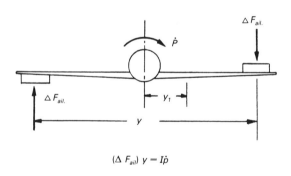

$(\Delta F_{ail}) y = I\dot{p}$

Fig. 3.3.4 *Forces due to deflected ailerons.*

The increment of load factor at any point, y_1, on the wing is

$$n_{y1} = \frac{\dot{p}y_1}{g}$$

Yawing

Several maneuvers involving rudder motion are prescribed for structural analysis.
(1) Rapid movement of the rudder with a given pedal force.
(2) Oscillating rudder motion.

These conditions may result in critical vertical tail loads. The magnitude of the tail loads produced is proportional to the rudder mechanical advantage built into the airplane. In each case the subsequent airplane motion must be investigated to determine the tail loads.

Inadvertent yawing maneuvers are experienced when a wing-mounted engine fails or a gust strikes the vertical tail. These conditions are not under the pilot's control, but may cause considerable yawing and produce critical vertical tail loads.

Equations of Motion

The previous sections define the various maneuvers required to establish a strength level. In order to determine loads on various structural components, the motion of the airplane must be studied. This section presents an approach to the subject of airplane dynamics as related to structural loadings. The foundation of airplane dynamics lies in the studies of dynamic stability and control.

The six degrees of freedom and the positive senses of forces, moments and airplane velocities along and about the various axes are shown in Fig. 3.3.5. The body axes are used as reference and all subsequent motion equations are referred to these axes. Note that the positive sense of the axes are opposite those normally used and moments and rotations are determined from the right hand rule.

The yawing maneuver considered here is associated with rudder deflection. Two degrees of freedom (sideslip and yaw) are used. A third degree of freedom, roll, may be necessary. The ailerons are considered to be neutral or deflected sufficiently to maintain a level roll attitude. If the ailerons create asymmetric drag these terms must be included in the equations.

Because a moving set of axes is used, the curvature of the flight path introduces another inertia term in the sideslip equation.

Sideslip acceleration $= \dot{v} + V\dot{\psi}$
Rolling acceleration, $\ddot{\phi} = 0$

Aerodynamic stability derivatives used in the equations are defined as follows:

$C_{y\beta} = \dfrac{dC_y}{d\beta}$ = change in side force with sideslip angle (negative)

$C_{n\beta} = \dfrac{dC_n}{d\beta}$ = change in yawing moment with sidelipe angle (positive)

$C_{n\gamma} = \dfrac{dC_n}{d\left(\dfrac{\gamma b}{2V}\right)}$ = damping moment in yaw caused by yawing velocity (negative)

$C_{n\delta_\gamma} = \dfrac{dC_n}{d\delta_\gamma}$ = change in yawing moment with rudder angle (negative)

For $\Sigma F_v = 0$

$$-m(\dot{v} + V\dot{\psi}) + \beta \dfrac{\partial F_v}{\partial \beta} = 0$$

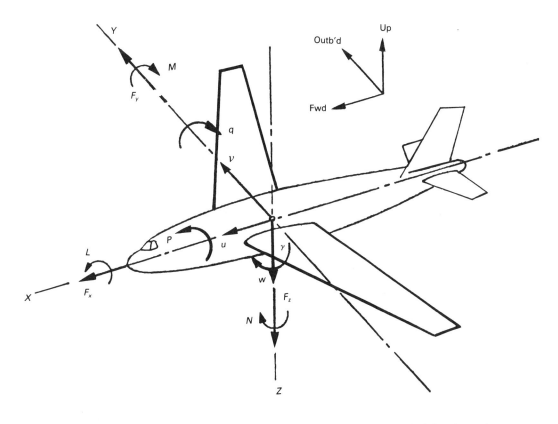

Fig. 3.3.5   Notations used in airplane motions (see appendix at the end of this chapter).

For $\Sigma N = 0$

$$-I_z \ddot{\psi} + \dot{\psi}\frac{\partial N}{\partial \dot{\psi}} + \beta \frac{\partial N}{\partial \beta} + \delta_\gamma \frac{\partial N}{\partial \delta_\gamma} = 0$$

where $\dot{v} = \dot{\beta}V$

$\dfrac{\partial F_y}{\partial \beta} = C_{y\beta} q_p S_w$ (side force caused by sideslip)

$\dfrac{\partial N}{\partial \dot{\psi}} = \dfrac{C_{n_y} b}{2V} q_p S_w b$ (damping moment caused by yaw velocity)

$\dfrac{\partial N}{\partial \beta} = C_{n\beta} q_p S_w b$ (yawing moment caused by sideslip)

$\dfrac{\partial N}{\partial \delta_\gamma} = C_{n\delta_\gamma} q_p S_w b$ (yawing moment caused by rudder)

Equations of motion then become
For $\Sigma F_y = 0$

$$\dot{\beta} + \dot{\psi} - \beta \left[C_{y\beta} \frac{q_p S_w}{mV} \right] = 0$$

For $\Sigma N = 0$

$$\ddot{\psi} - \dot{\psi}\left[\frac{C_{n_y} b}{2V} \frac{q_p S_w b}{I_z} \right] - \beta \left[C_{n\beta} \frac{q_\gamma S_w b}{I_z} \right] = \delta_\gamma(t) \left[C_{n\delta r} \frac{q_\gamma S_w b}{I_z} \right]$$

The vertical tail load is determined from the following equation:

$$P_{y vt} = \beta \left[\Delta C_{n\beta vt} \frac{q_p S_w b}{\ell_{vt}} \right]$$
$$+ \dot{\psi} \left[\Delta C_{n\gamma vt} \frac{b}{2V} \frac{q_p S_w b}{\ell_{vt}} \right]$$
$$+ \delta_\gamma(t) \left[C_{n\delta r} \frac{q_\gamma S_w b}{\ell_{vt}} \right]$$

where $\Delta C_{n\gamma vt}$ may be put in terms of vertical tail lift curve slope.

The previous equations assume linear stability derivatives through the range of sideslip and rudder angles which will be experienced. Rudder effectiveness and yaw stability curves should be examined to determine the linear ranges. Where nonlinearities are encountered, assumptions can usually be made to linearize the derivatives. No general procedure can be adopted since the method of handling each set of data will be different. The method chosen will depend on the accuracy of the data, the accuracy desired of the results, and the characteristics of the particular data.

3.4 Basic Data

The primary sources of basic data to the loads analysis are:

- Airplane geometry
- Aerodynamic data
- Weight data and design speeds
- Stiffness data
- Miscellaneous systems data
- Operational data
- $V - n$ diagrams

Much of the basic data is expressed in matrix format in which unit distribution data (unit airloads, weights, stiffnesses, etc.) are represented over the entire airplane on defined grid systems. The term "unit distribution" is very common in loads work and this manner of expressing data has been, and still is, fundamental to almost any loads analysis. These distributions can be various load quantities such as panel loads, shears, bending moments, torsions, hinge moments, total empennage loads, etc. Basically, these load quantities are defined for unit changes in key variables such as airplane angle of attack (i.e., $\Delta \alpha_{FRL} = 1$), unit control surface deflections, unit changes in velocities and accelerations, etc. It follows that, in general, total loads can be obtained by multiplying each unit distribution by the variable appropriate to the maneuver being analyzed and adding these (i.e., superposition appropriate to linear analysis).

Airplane Geometry

The airplane geometry must be established in order to start the analysis. This includes such items as the planform, the inboard profile, the camber and twist distributions on the wing (see Fig. 3.4.1) and horizontal and vertical tail, the wing and tail incidence with the fuselage, control surfaces locations and geometry, etc.

In order to clarify the concepts involved, a simplified grid system is shown in Fig. 3.4.2 with far fewer points than would be used on an actual airplane model. A further simplification is that the loads, stress, SIC (Structural Influence Coefficients), and control points will be considered coincident.

Typical grid system established include as follows;
(1) The basic loads grid on which basic weight and airload unit distributions are established.
(2) The previous mentioned SIC grid system at which the structural deflections are defined and at which the external panel loads are defined on computer tapes as input to the redundant structural "finite element."
(3) The control point grid defining the collocation points used in the lifting surface aerodynamic theories. The theoretical airloads over the airplane are expressed as functions of unit changes in angle of attack at each of these collocation points.
(4) The wind tunnel pressure model orifice locations.
(5) Other miscellaneous grids which are generally established by reducing (lumping) the above systems. These reduced order systems are usually selected to minimize the size of the computation on the computers or because of computer program limitations.

Aerodynamic Data

Aerodynamic data is of primary importance in the loads analysis. The typical types of data classified under aerodynamic data include:

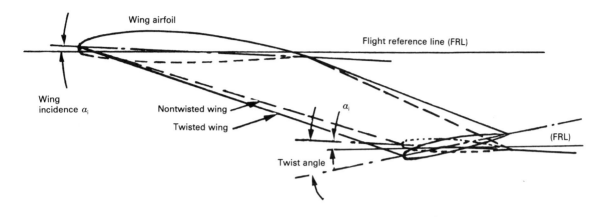

Fig. 3.4.1 Twisted wing vs non-twisted wing.

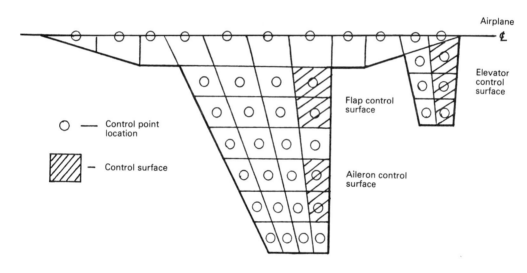

Fig. 3.4.2 Simple grid system of an airplane.

- Wind tunnel or flight measured aerodynamic force data.
- Wind tunnel or flight measured surface pressure data.
- Predicted flight profiles for optimum performance.
- Control surface requirements for stability and control (Ref. 3.33).
- Limitations due to stall, flutter (Ref. 3.41), buffeting, and stability and control (Ref. 3.33).
- Predicted aerodynamic force and pressure data based on theoretical methods (e.g., lifting surface theory per Ref. 3.9, 3.10 and 3.11).

Wind Tunnel Data

Basic wind tunnel force data is usually measured on scale models along and about the wind axis. The forces and moments are reduced to coefficient form by dividing by reference areas and references distances. The coefficients measured usually include:

C_L = lift; C_ℓ = roll
C_D = drag; C_m = pitch
C_Y = side force; C_n = yaw

Complete ranges of angle of attack (α_{FRL}) and sideslip (β) are measured for most operational configurations such as clean, flaps extended, high drag, gear down, in and out of ground effect, etc. In addition, selected variations in control surface deflections such as aileron (δ_a), rudder (δ_r), elevator (δ_e), etc. are measured in combination with variation in α and β. The above combinations can constitute a substantial amount of wind tunnel data especially when Mach number effects must be considered.

Certain basic data are measured on a model build-up basis. These might include such combinations as body, body-wing, body-wing-nacelles, airplane-less-tail, and airplane. These component built-ups are an aid in estimating component load distributions as well as performance effects. It should be kept in mind, however, that the presence of each component usually induces interference effects on other components and vice versa. Therefore, incremental measured forces cannot always be added exclusively to the component that caused the increment. The application of airplane-less-tail and airplane data to aid in estimating empennage loads is discussed further in later sections. Strain gages are often used to measure hinge moments and pylon loads.

The airplane derivatives due to airplane translational and rotational velocities (i.e., $\frac{Pb}{2V}$, $\frac{\gamma b}{2V}$, $\frac{\dot{\theta}c}{2V}$ and $\frac{\dot{\alpha}c}{2V}$) are usually not obtained from wind tunnel measurements but are estimated from theory or from extrapolations from past airplanes.

The data acquisition and reduction problems with wind tunnel force data are less complex than the problems associated with wind tunnel pressure model data. A typical pressure model contains hundreds of orifices which measure surface pressures over the entire airplane. This type of data provides much useful loads information. In addition to local surface design pressures, the integrated pressures provide accurate distributions and over-all force data on each major component of the airplane. This type of data is essential for complex configurations, and/or for designs that are weight critical, and/or for configurations designed to operate in the transonic mixed flow environment where theoretical potential flow lifting surface theories are inadequate.

Chordwise pressure distributions have been discussed previously in connection with various structural components. Methods are available for calculation of these pressure distributions. However, in reports such as Ref. 3.34, 3.36 & 3.37 the theoretical pressure distributions are already calculated for a large number of airfoil sections. The references listed in these reports may be used for further calculations.

In the transonic speed range, estimations of chordwise pressure distribution should be based on experimental data.

Chordwise pressure distributions are of considerable importance in the study of the loads on leading and trailing edge flaps. With flaps in the neutral position, the leading edge flap experiences maximum loading at high angles of attack in subsonic speed ranges. Trailing edge flaps are most highly loaded in transonic and supersonic ranges. Leading edges will have pressures of 10 to 12 psi, while trailing edges of high speed airplanes will have pressures of 3 to 5 psi.

Leading edge flaps are used on thin wings to delay the flow breakdown as the angle of attack is increased. Consequently, the high pressures mentioned will be on the flap even when it is depressed 4 or 5 degrees.

When studying trailing edge flaps, two increments of pressure are considered. First is the basic pressure on the wing with undeflected flap. Second is the increment caused by the flap which is superposed on the basic pressure distribution. (See Fig. 3.4.3).

Airplane Weight Data

Design loads are determined in the early stages of airplane design; design loads affect the weight of the airframe structure; the weight of the airplane influences the magnitude of design loads. This interdependence suggests that a judicious selection of preliminary design weights is mandatory to the economical design of an airframe.

Airplane weight and detailed distribution of weight both have a large influence on structural design loads. Obviously, the lifting air loads on an airplane wing must be as great as the weight in order to support the airplane in flight as shown in Fig. 3.4.4. Another effect of weight on design loads is the vertical load imposed on the landing gear when it contacts the ground at a given sink speed.

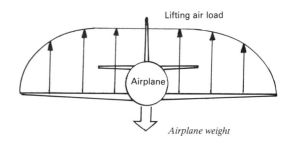

Fig. 3.4.4   Airplane weight and lifting air loads.

Airplane weight to be used for structural design are determined from the mission requirements of the airplane as mentioned previously in Section 3.1. Mission requirements and/or design criteria spell out maximum and minimum amounts of fuel and payload to be considered at various stages of airplane operation. Typical design weights are:
- **Maximum take-off weight (MTOW):** weight to perform the specified mission. The airplane is generally considered as containing capacity fuel and maximum payload. This is varied in some instances to suit the particular needs of the design.

Take-off weight is considered for taxiing conditions and flight conditions, but the airplane need

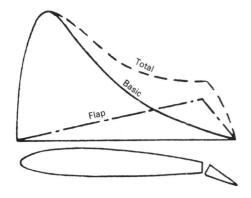

Fig. 3.4.3   Subsonic chordwise pressure distribution with deflected flap.

not be designed for landing at this weight.
- **Maximum landing weight (MLW):** weight for the landing operation. It is reasonable to expect that a predetermined amount of fuel is used up before the airplane is expected to land. It would be unnecessarily penalizing the airplane structure to design it to loads imposed during a landing at gross weights above the Design Landing Weight. In many instances fuel dump provisions are installed in an airplane so that gross weight may be reduced to the Design Landing Weight in emergency landings.
- **Zero fuel weight (ZFW):** weight with zero fuel aboard. This somewhat confusing title for a weight is very popular in structures because it is descriptive. It defines the summation of weight of empty airplane, items necessary for operation (crew, engine oil, etc.), and cargo and/or passengers.
- **Operation empty weight (OEW):** weight with no payload on board. Fuel considered is a predetermined or specified minimum, generally something between 0 and 10% of total fuel. Flight maneuvers at this weight can produce some critical loadings.

All design gross weights are negotiated during the formulation of design criteria. Using past experience, an attempt is made to establish the design weights at a level that will include the natural growth of gross weights between preliminary and final design, and yet not be so excessive as to penalize the design. See Chapter 16.0 for more detailed discussion.

(1) Center of gravity envelope

The combination of gross weight and airplane center-of-gravity location are of considerable importance. A plot is made of the variation of c.g. with different airplane weight. An envelope enclosing all extremes of the variation is determined for design. An attempt is made to set the c.g. limits such that it will include the extremes that can result from changes that occur as the design progresses. A typical example for a commercial transport is shown in Fig. 3.4.5.

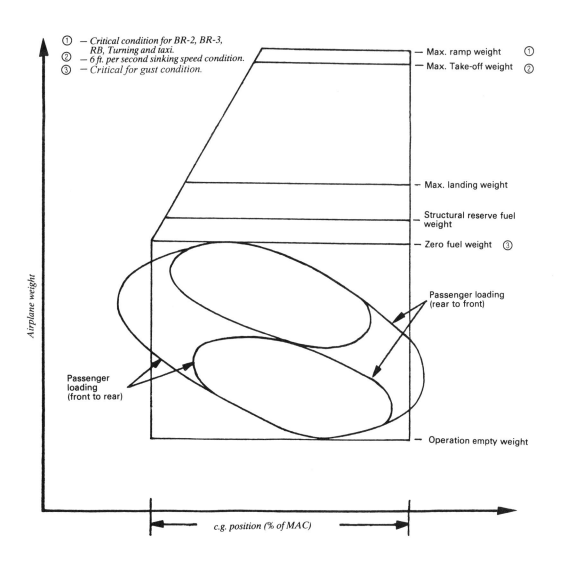

Fig. 3.4.5   Typical c.g. envelope of a commercial transport.

Airframe Structural Design

(2) Weight distribution

Once the critical loadings are determined, a distribution of the dead weight is necessary. An accurate distribution of weight is important because the dead weight of fuselage, wing, cargo, etc., contribute a large part of the loading. The effect of distribution of weight can be realised by considering the distribution of mass items of cargo within a fuselage. Whether these large masses are condensed about the center body or placed at extreme forward or aft location will influence greatly the magnitude of downbending experienced by the fuselage forebody or aftbody during a hard landing.

The amount and disposition of fuel weight in the wing is particularly important in that it can be utilized to provide bending relief during flight. This is easily explained in terms of wing bending, considering an airplane containing fuel in wing cells. The Fig. 3.4.6 illustrates how the weight of fuel in the wing acts to relieve the bending caused by airload. The illustration points out that placing fuel as far outboard as possible and using fuel from the most outboard tanks last provides the optimum arrangement for wing bending during flight.

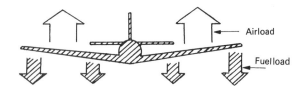

(a) Fuel weight provides relief to wing bending.

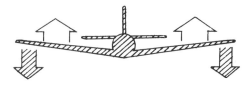

(b) Inboard fuel expended. Airload reduced because of reduced gross weight. Outboard fuel providing relief.

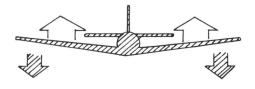

(c) Outboard fuel nearly expended. Bending relief decaying faster than airload bending; therefore, net bending increasing slightly.

Fig. 3.4.6 *Illustration of the effect of fuel weight in wing.*

However, thinking in terms of the large inertia effects in a downbending direction during a hard landing points out that this extreme arrangement of wing fuel is not necessarily the best solution. The illustration does make it apparent that fuel management (sequence of usage) should be given considerable attention in order that the final design be the best compromise.

Design Speeds

Structural design speeds must be chosen such that they are compatible with the airplane performance and operational requirements. These airspeeds are generally chosen by the contractor, but must meet the design requirements of the certifying or procuring agency. The speeds are generally selected as low as possible to avoid penalizing airplane weight.

Basic design speeds generally considered are:

Design Cruising Speed, V_C A minimum value of the maximum speed to be considered for straight and level flight. The speed is determined empirically (based on statistics), but need not exceed the maximum speed available in straight and level flight.

Design Dive Speed, V_D The maximum speed for structural design. A statistically determined speed sufficiently greater than V_c to provide for safe recovery from inadvertant upsets.

Design Flap Speed, V_F The maximum speed for design during flaps- extended flight; generally arrived at empirically.

The above speeds are named and defined per FAA Regulations. Similar design speeds with similar definitions are presented in each of the military specifications.

Fig. 3.4.7 shows altitude vs airplane cruise speed, dive speed and maximum power.

Speeds for extension and retraction of flaps, landing gears, etc., are dependent on the operational requirements of the airplane. Minimum values are found in the specifications of the licensing and procuring agencies.

Stiffness Data

The stiffness data is provided in the form of Structural Influence Coefficients (SIC), in terms of deflections per unit load. Two matrices are usually provided: (1) for unit loads applied symmetrically on both the left and right side of the airplane; and (2) for unit loads applied anti-symmetrically (in opposite directions) on right and left sides. In addition, it is sometimes necessary to obtain sets of stiffness data appropriate to low load levels, such as minor excursions from 1.0 g flight and also stiffness data appropriate to limit load levels (e.g., buckled skin). For supersonic airplanes it is also necessary to establish stiffness changes due to temperature.

The SIC format is supplemented by plots of EI and GJ on components of the airplane such as the wing, fuselage, horizontal tail, vertical tail, control surfaces, etc. (See Ref. 3.20).

Miscellaneous Systems Data

Systems which can strongly interact with the primary structure include the following:
- propulsion
- primary flight controls
- automatic flight controls
- environmental control

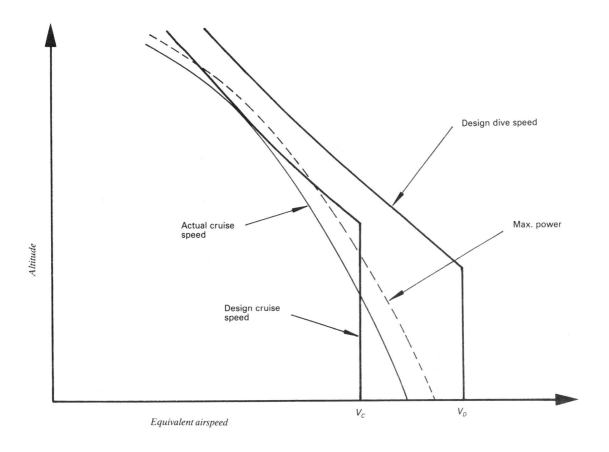

Fig. 3.4.7 Design speed vs. altitude.

- fuel
- hydraulic, and
- brake anti-skid systems

Knowledge of both the normal operating characteristics and failure modes and effects of system malfunctions are required in order to assure the desired level of over-all system/airframe integrity. Some of the above items will be discussed in more detail in further sections.

Operational Data

It is necessary to gather detailed data on anticipated operational usage of the airplane being designed. This data is needed to establish realistic *load spectra for fatigue analysis*, and also to provide necessary inputs to the dynamic gust *mission analysis* unit loads determination. This includes such items as statistical data on landing sinking speeds, gusts, and maneuvering. They also include anticipated operational runway roughnesses and typical pilot ground maneuvers. An important part of the data are the *mission profiles* which predict typical mixes (including the percentage of time) of payloads and ranges.

***V-n* (or *V-g*) Diagrams**

The operating flight strength limitation of an airplane are presented in the form of a *V-n* or *V-g* diagram. A typical *V-n* diagram is shown in Fig. 3.4.8. Each airplane type has its own particular *V-n* diagram with specific V_s and n_s. The flight operating strength of an airplane is presented on a graph whose horizontal scale is airspeed (V) and vertical scale is load factor (n). The presentation of the airplane strength is contingent on four factors being known: (1) the aircraft gross weight; (2) the configuration of the aircraft (clean, external stores, flaps and landing gear position, etc,); (3) symmetry of loading (since a rolling pullout at high speed can reduce the structural limits to approximately two-thirds of the symmetrical load limits); and (4) the applicable altitude. A change in any one of these four factors can cause important changes in operating limits. The limit airspeed is a design reference point for the airplane and an airplane in flight above the limit airspeed may encounter: (a) critical gust, (b) destructive flutter, (c) aileron reversal, (d) wing or surface divergence, (e) critical compressibility effects such as stability and control problems, damaging buffets, etc.

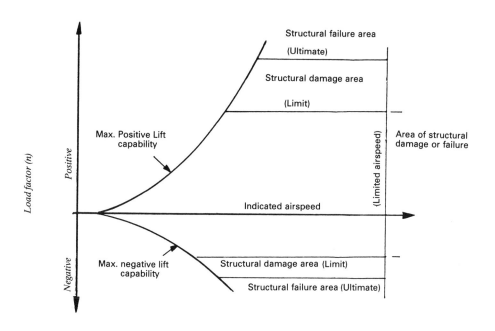

Fig. 3.4.8   Significance of the V-n diagram.

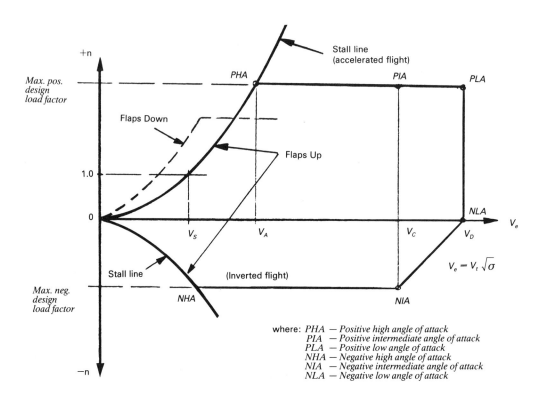

Fig. 3.4.9   Typical V-n Diagram for Maneuver.

42 Airframe Structural Design

(1) Construction of Diagrams: Maneuver envelope
A typical diagram is shown in Fig. 3.4.9.
The stall line indicates the maximum load factor that can be obtained because of the inability of the airplane to produce any more lift at a particular speed.

$$L = W = C_{L\max} \frac{V_{s1}^2 S}{296}, \text{ (all speeds will be considered in knots)}$$

$$V_{s1} = \sqrt{\frac{296 \, W/S}{C_{L\max}}} \qquad (3.4.1)$$

In accelerated flight and assuming that any velocity along stall line higher than V_{s1} is V_{sn};

$$nW = C_{L\max} \frac{V_{sn}^2}{296} S$$

Substituting for W

$$n C_{L\max} \frac{(V_{s1})^2}{296} S = C_{L\max} \frac{(V_{sn})^2}{296} S$$

$$nV_{s1}^2 = V_{sn}^2$$

$$V_{sn} = V_{s1} \sqrt{n} \qquad (3.4.2)$$

If $C_{L\max}$ is $f(M)$ the following procedure can be used in calculating ordinates of the stall line:
1. Assume M
2. Determine $C_{L\max}$ as $f(M)$
3. Calculate V_s from M at altitude being considered
4. $q = \dfrac{V_s^2}{296}$
5. From equation for accelerated flight

$$n = \frac{C_{L\max} q}{\frac{W}{S}}$$

(2) Construction of Diagrams — Gust envelope
The preceding section shows the construction of the maneuvering V-n diagram, which presents the limits to which the airframe must be designed for adequate strength to withstand all maneuvers to which the airplane may by subjected. Before proceeding with loads analyses, a similar diagram must be constructed to provide an envelope of velocities and load factors that might be encountered in flight through turbulent air. This envelope is referred to as the gust envelope or the gust V-n diagram. In the final analysis the airframe strength must be based on both the maneuver and gust envelopes or a composite diagram, representing both.
To better understand the meaning of the gust envelope it will be of value to discuss the gust loads problem. The problem resolves itself into investigating the source of loads, the airplane's behavior during application of the loads, and conditions likely to lead to critical loads (loads affecting structural design).

The problem of a loads change when encountering a gust stems from the very rapid change in direction of relative wind. This angle of attack change with virtually no change in forward velocity results in a change in wing lift. A simple idealization of the problem lends itself readily to representing the load change with a simple equation. The concept simply pictures the airplane flying along at a constant velocity, V, and suddenly encountering a sharp-edged gust. The gust velocity, U_{de} shown below is assumed to act vertically, normal to the flight path and is of uniform intensity across the wing span. The steady flow relation between lift and angle of attack is assumed to hold (see Fig. 3.4.10 for design gust velocities vs. altitude).

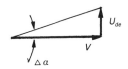

Angle of attack change,

$$\triangle \alpha = \frac{U_{de}}{V} \quad \text{(for } U_{de} \text{ and } V \text{ in feet per second)}$$

$$= \frac{U_{de}}{1.68 V_e}$$

where V_e = equivalent airspeed in knots

$$\triangle L \text{ (caused by gust)} = \triangle \alpha \left(C_{L\alpha A} \frac{V_e^2}{296} S \right)$$

Substituting for $\triangle \alpha$,

$$\triangle L = \frac{C_{L\alpha A} V_e}{498} S$$

Note that $C_{L\alpha A}$ must be per radian when the equation is in this form.

$$\triangle n = \frac{(C_{L\alpha A}) U_{de} V_e}{498 \frac{W}{S}} \qquad (3.4.3)$$

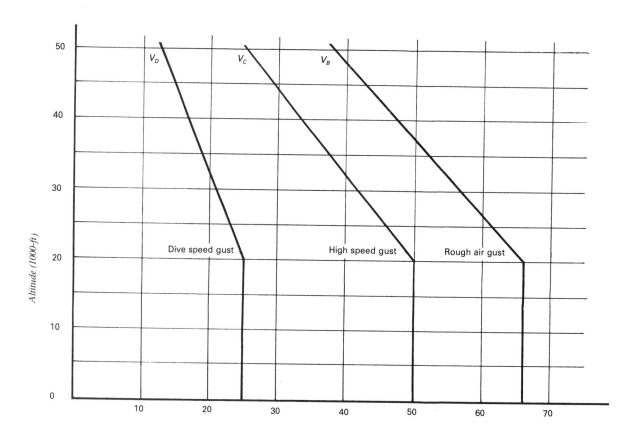

Equivalent Gust Velcoity U_{de} — $\frac{ft}{sec}$

(From FAR-25, Airworthiness Standards — Transport category airplanes per Ref. 3.2)

Fig. 3.4.10 Design gust velocities vs. altitude.

By use of this simple formula, measured accelerations and velocities can be converted to gust velocities for use in design. Because of the extreme simplicity of the equation, a gust intensity determined in this manner would not be a true gust velocity, but rather, an effective gust velocity.

Note that from the simple equation for Δn (see Fig. 3.4.11) that the only parameter used for transferring these measured sharp-edged-gust intensities from one airplane design to another is the wing loading. This assumes that all airplanes of a given wing loading will have the same dynamic response and the same lag in lift increment due to $\Delta \alpha$, and that all gusts encountered are sharp-edged rather than possessing a velocity gradient. No serious effects of this simplification were noticed in earlier designs when most airplanes were of similar size and design; however, with the high speed of airplane types and sizes of today, a connection for the unaccounted for properties is in order. This is accomplished by the introduction of the gust alleviation factor, K_g thus:

$$n = 1 \pm \frac{K_g U_{de} V_e C_{L\alpha A}}{498 \left(\frac{W}{S}\right)} \quad (3.4.4)$$

The factor K_g has taken several forms during the development of the gust equation. Recent investigations show it to be related to wing loading, lift-curve slope, air density, and a representative airplane dimension such as wing chord. The alleviation factor K_g is given as an empirical function of the so-called mass parameter,

$$K_g = \frac{0.88\mu}{5.3 + \mu} \quad (3.4.5)$$

where $\mu = \dfrac{2\left(\dfrac{W}{S}\right)}{g\bar{c}\rho C_{L\alpha A}}$

44 Airframe Structural Design

The above K_g factor is an attempt to account for: (1) gust gradient; (2) airplane response; (3) lag in lift increment caused by $\triangle \alpha$.

The inertia factor or total airload factor on the airplane is determined by using C_{LaA} in the equation, whereas the airload factor on the wing is determined by using C_{LaW} that can be estimated from Ref. 3.5 by using $C_L = C_{LaW}$. For convenience, curves from Ref. 3.5 are shown in Fig. 3.4.12–Fig. 3.4.14.

Determination of C_{LaW} from these curves is quite accurate for thin airfoils. The curves correct a theoretical two-dimensional flow value of 2π for the effects of three-dimensional flow and for compressibility effects. The effects of three-dimensional flow are accounted for by considering wing geometry; aspect ratio, taper ratio, and sweep. Definitions for the various symbols appearing on the curves are as follows:

β = compressibility parameter = $\sqrt{1 - M^2}$

C_{La} = lift-curve slope

The previous discussion of gust load formula was developed from Ref. 3.45 that the derived gust velocities are based on a rigid airplane. However, due to its simplicity, this formula is good enough for the preliminary loads. The recent dynamic gust analysis of a flexible airframe, discrete gust analysis and power spectral density (PSD) analysis, will be discussed in more details in Section 3.5.

Gust Envelope (see Fig. 3.4.15):
1. The stall line is constructed in the same manner as for the maneuver envelope.
2. Gust lines, radiating from $n = 1.0$ at $V = 0$ are constructed using the equation 3.4.4 and the gust intensities given by equation 3.4.5 (Note that design gust intensities reduce as velocity increases. It is intended that the airplane will be flown accordingly, i.e., that as intensity of turbulence increases the pilot will reduce speed).

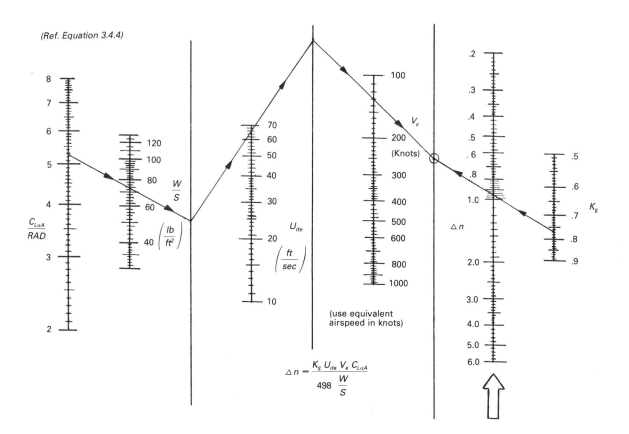

Fig. 3.4.11 Gust load factor nomogram.

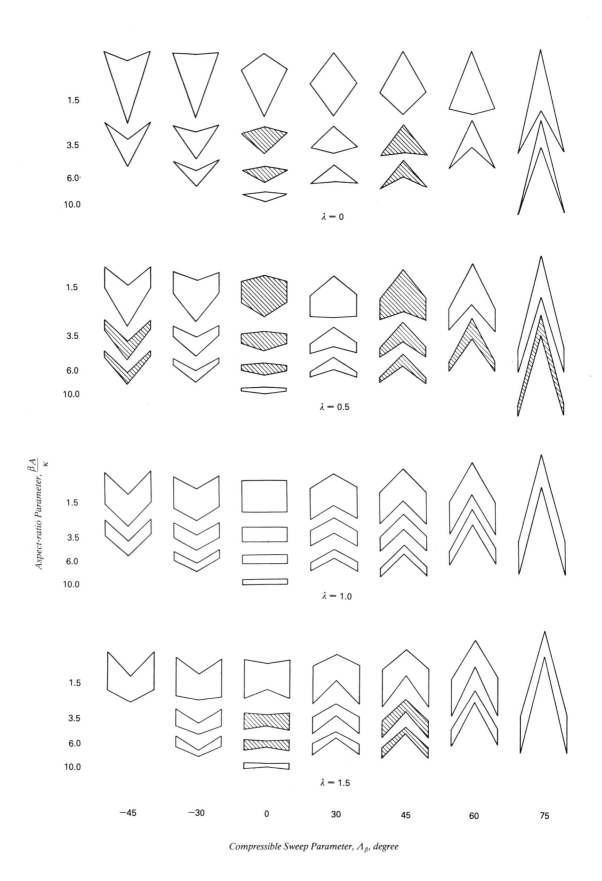

Fig. 3.4.12 Range of plan forms included in the study of the characteristics of straight-tapered wings (Ref. 3.5).

46 Airframe Structural Design

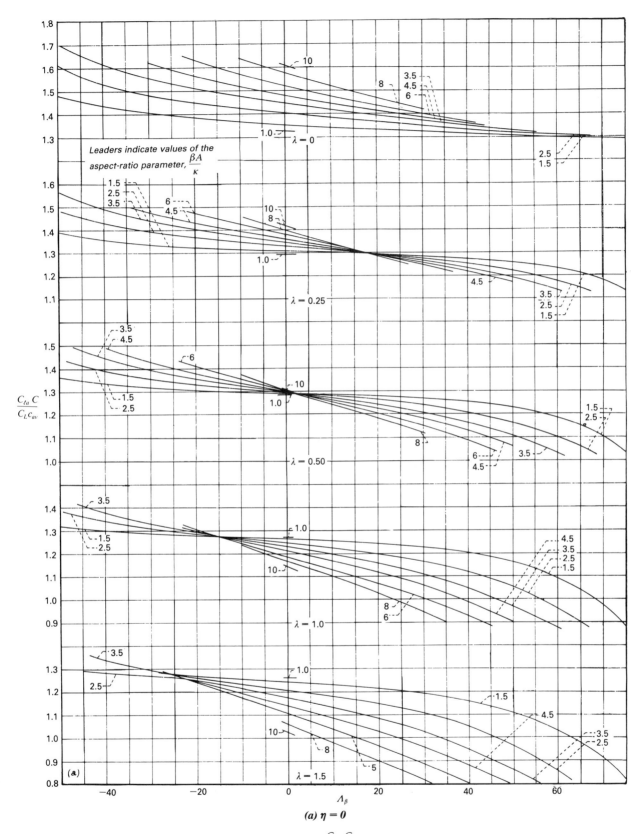

Fig. 3.4.13 Variation of the spanwise loading coefficient $\dfrac{C_{\ell a} C}{C_L c_{av}}$ with the compressible sweep parameter Λ_β (degrees) for various values of the aspect-ratio parameter $\dfrac{\beta A}{\kappa}$ and taper-ratio λ (Ref. 3.5).

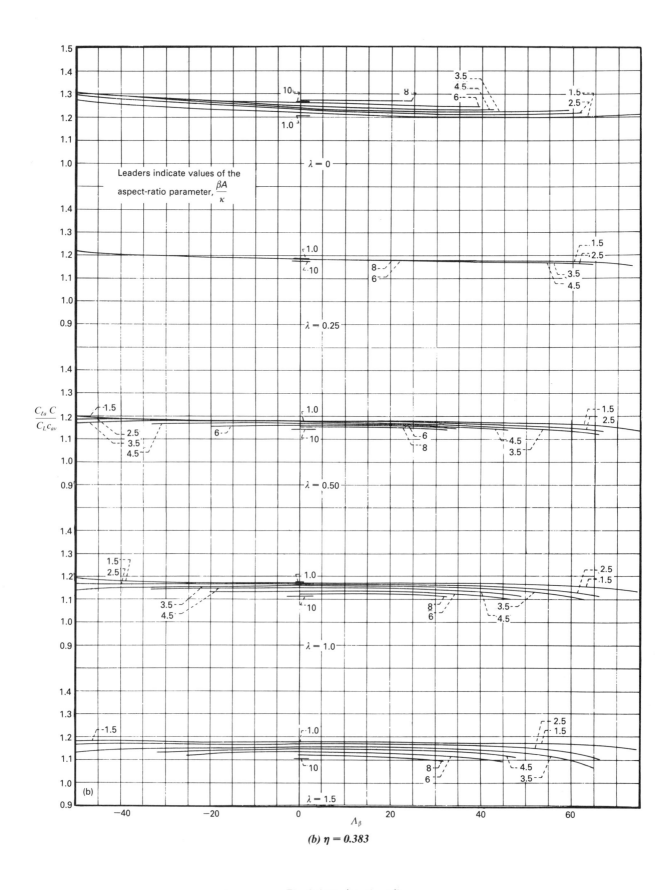

Fig. 3.4.13 (continued)

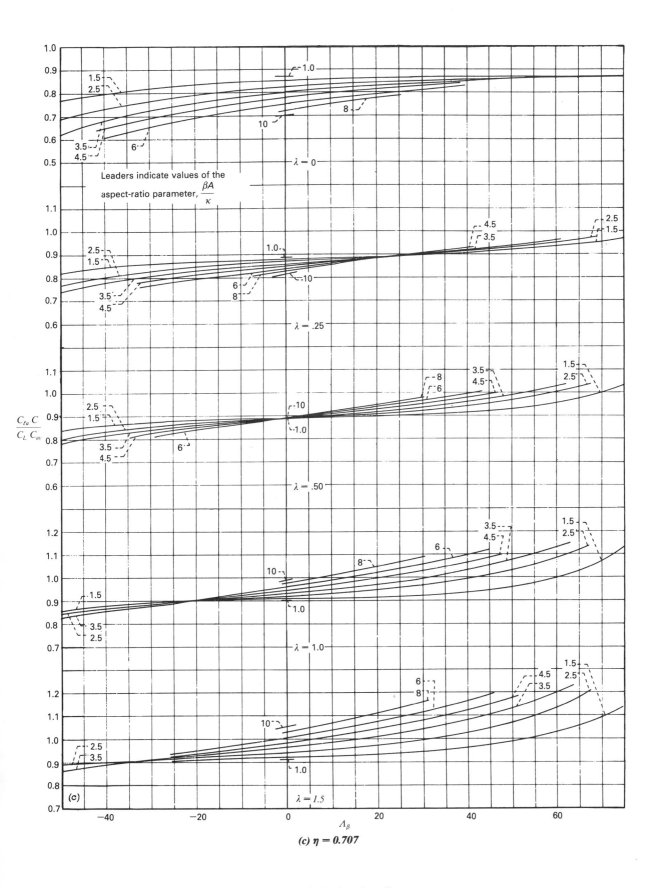

Fig. 3.4.13 (continued)

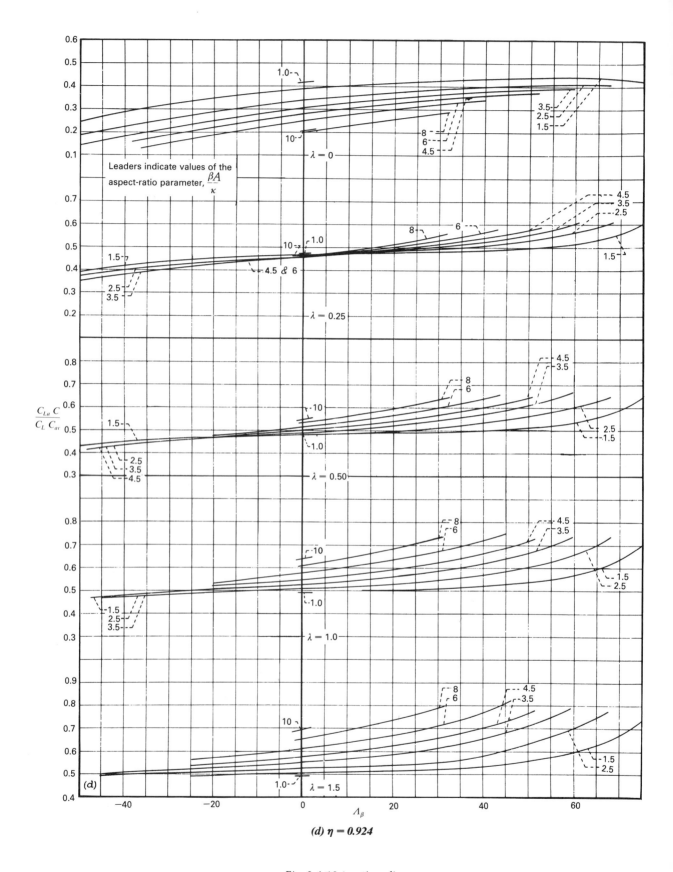

Fig. 3.4.13 (continued)

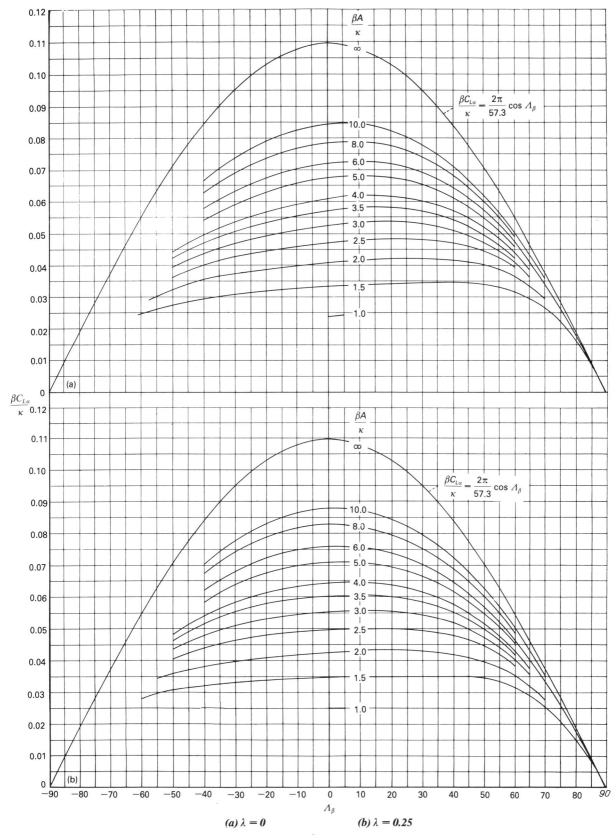

(a) $\lambda = 0$ **(b)** $\lambda = 0.25$

Fig. 3.4.14 Variation of the lift-curve-slope parameter $\frac{\beta C_{L\alpha}}{\kappa}$ per degree, with the compressible sweep parameter Λ_β (degrees), for various values of the aspect-ratio parameter $\frac{\beta A}{\kappa}$ (Ref. 3.5).

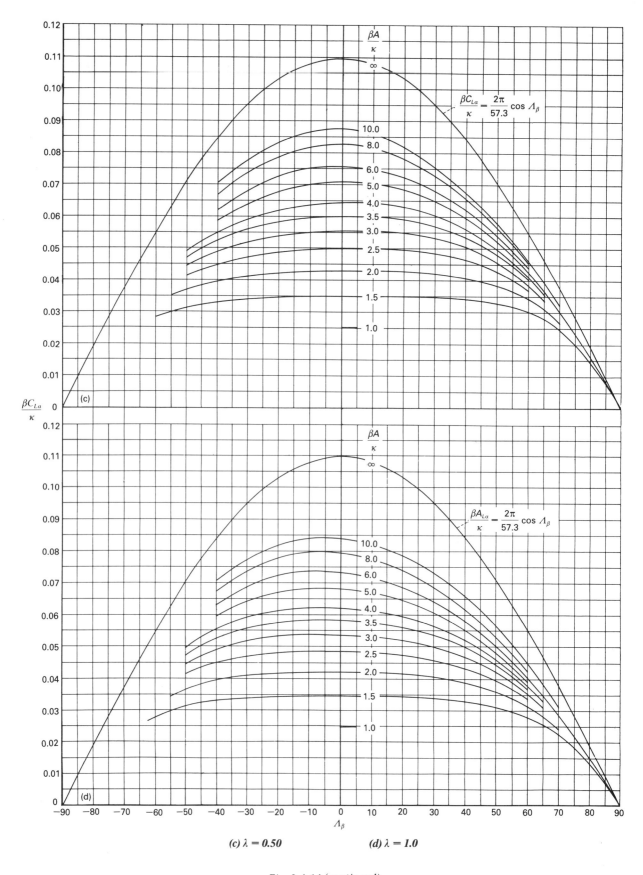

(c) $\lambda = 0.50$ (d) $\lambda = 1.0$

Fig. 3.4.14 (continued)

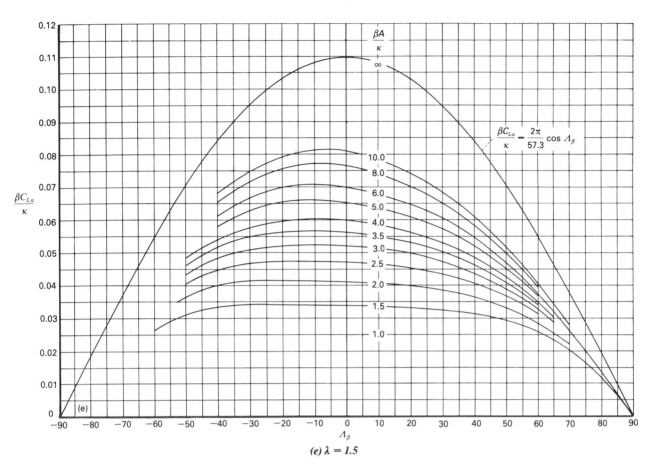

(e) $\lambda = 1.5$

Fig. 3.4.14 (continued)

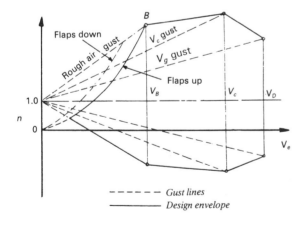

Fig. 3.4.15 V-n Diagram (Gust Envelope).

(3) Combined Flight Envelope (V-n diagram)

It is often the practice to present a combined flight envelope which is merely a composite of the gust and maneuver envelope, as shown in Fig. 3.4.16. A calculation example is given in Section 3.11.

3.5 Wing Design Loads

Design wing loads consist of the shears, bending moments, and torsions which result from air pressures and inertia loadings. Flight loads are those experienced when maneuvering to the limits of the V-n diagram or those caused by gusts. Other flight conditions are those associated with control surface deflections. In addition, wing design loads must be determined for the landing and taxi conditions.

Airload Spanwise Distribution

Clean configuration: The air loading on a wing consists of two parts, *additional loading* and *basic loading*. The additional air loading is caused by angle of attack. On normal aspect ratio wings (> 3) this lift and its distribution varies directly with angle of attack. The basic loading is that distribution of air load on the wing when the total lift is zero. This type of loading is caused by wing twist.

The distribution of the *additional lift* will take the following form:

$C_{\ell a}$ is the section or two dimensional lift coefficient. The distribution of the additional section lift coefficient may be readily estimated from Ref. 3.5 (see curves in Fig. 3.4.12–Fig. 3.4.14.) The distribution is usually carried to the airplane center line. The

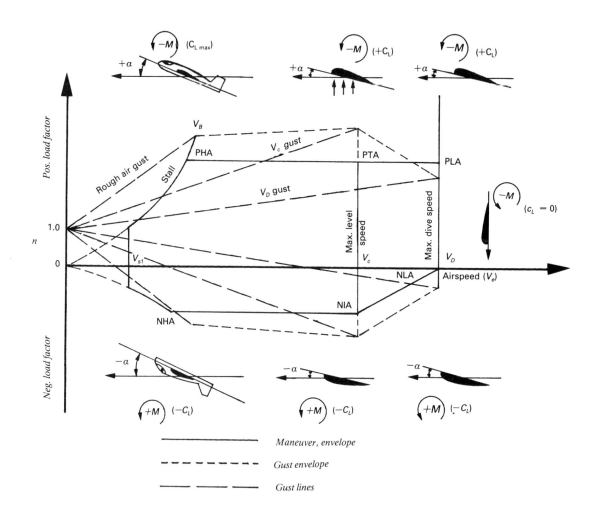

Fig. 3.4.16 Combined flight envelope (V-n diagram) — fighter.

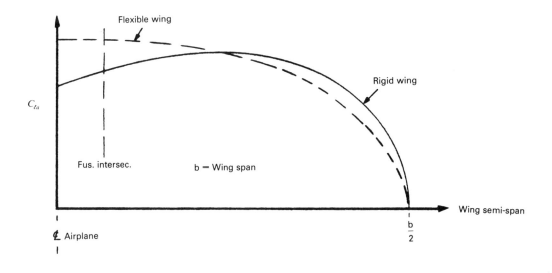

Fig. 3.5.1 Wing additional lift distribution.

54 Airframe Structural Design

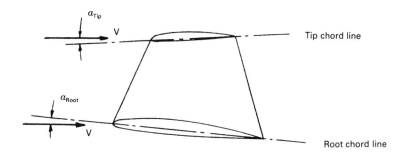

Fig. 3.5.2 Basic lift distribution.

fuselage is assumed to carry the same amount of lift that would be on the blanketed wing area. Subscript 1 (i.e. $C_{\ell a1}$) denotes the distribution for a wing lift coefficient (C_L) of 1.0.

The concept of *basic lift* distribution is illustrated by Fig. 3.5.2 which is an end view of the wing with the tip section twisted down relative to the root (washout). Assuming the angles of attack shown result in zero lift on the wing, the lift curve slopes are shown in Fig. 3.5.3.

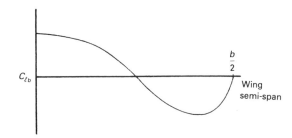

Fig. 3.5.4 Section lift coefficient distribution of basic lift case.

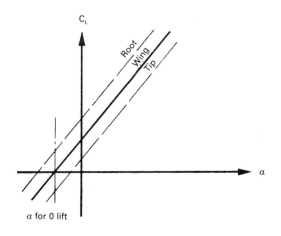

Fig. 3.5.3 Lift curve slopes on zero lift wing.

The root section is producing positive lift and the tip section is producing negative lift. The section lift coefficient, $C_{\ell b}$, distribution across the span for this case will appear as Fig. 3.5.4.

The distribution of the basic lift for both symmetrical or antisymmetrical twists can be estimated from Ref. 3.6. The symmetrical distribution of the lift takes the form of

$$\frac{C_{\ell b} c}{b} \approx \frac{\epsilon - \bar{\epsilon}}{1 + \frac{360\, a}{\pi^2 A}} \left(\frac{C_{\ell a1}\, c}{b} a \right) \qquad (3.5.1)$$

where
c = local chord of wing
ϵ = angle of twist, degree

$$\bar{\epsilon} = A \int_{-1}^{1} \epsilon\, \frac{C_{\ell a1}\, C}{b}\, d\left(\frac{2y}{b}\right) \text{ average angle}$$

of twist in, degree

a = three dimensional lift slope per degree

$$\approx \frac{A}{\frac{A \varrho}{\cos \Lambda} + 2} \left(\frac{\pi^2}{90} \right)$$

or $C_{L\alpha}$ similar to that of Fig. 3.5.3
A = aspect ratio
$C_{\ell a1}$ = section lift coeff. for additional lift (Ref. 3.5) distribution at a wing lift coef. of unity.
b = wing span
y = spanwise coordinate
Λ = sweep angle of wing quarter-chord line, degrees
ϱ = edge-velocity factor for symmetrical lift distributions $\sqrt{1 + \frac{4 \cos^2 \Lambda}{A^2}}$

For detailed discussion on the basic lift, see Ref. 3.6.

Fuselage, Nacelle and Wing Stores Effect on Wing Loads

Effects of these items are not readily determined as quantitative values. Qualitatively, the effects appear as shown in Fig. 3.5.5.

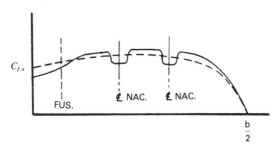

Fig. 3.5.5 Additional lift distribution due to nacelle or stores.

Stores located near the wing tip or large protuberances on the fuselage, such as radomes, have the greatest effect on wing loads. Both tend to move the center-of-pressure outboard. The tip tank will effectively increase the aspect ratio. The radome will interfere with the lift near the fuselage causing the airplane to fly at a higher angle of attack and result in higher loads on the outboard sections of the wing.

In the stress analysis of a conventional wing, it will be necessary to investigate each cross section for each of the four conditions shown in Fig. 3.5.6. Each stringer or spar flange will then be designed for the maximum tension and the maximum compression obtained in any of the conditions.

Dynamic Gust Loads

The advent of high speed, structurally efficient, flexible airplanes has made necessary the determination of the dynamic response of these airplanes to atmospheric turbulence. Early techniques for evaluating the dynamic effects of turbulence involved representing the gust as a discrete phenomenon (Ref. 3.45) and calculating the time response of the airplane to this discrete function.

Characteristically, the gust shape used in the discrete analysis was chosen to be $1 - \cos(at)$, where a is a function of chord-length. The amplitude of the gust (i.e. 50 fps) is assumed invariant with the wavelength, and the wavelength is varied to *tune* (*tune* means let the airplane natural frequency including damping to match the maximum gust frequency) the airplane's response to achieve a maximum dynamic overtravel. Fig. 3.5.7 represents a typical response curve, showing the envelope of the peak load as a function of gust wavelength.

A program was developed in the 1950s to evaluate the airplane's response in two degrees of freedom, vertical translation and wing bending (discrete plunge-bending type). Various publications such as one shown in Ref. 3.26 indicated these to be the major response freedoms in airplanes, an indiction that was probably correct for relatively stiff, straight-winged airplanes. This, again, was a time history solution considering the response to a discrete gust.

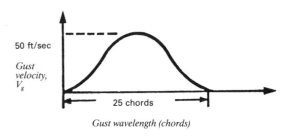

Fig. 3.5.7 Typical discrete gust.

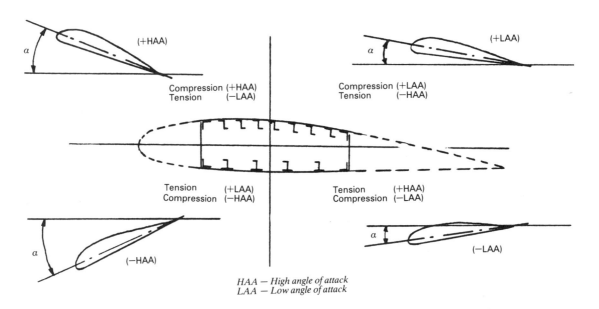

HAA — High angle of attack
LAA — Low angle of attack

Fig. 3.5.6 Critical conditions for wing box structure.

56 Airframe Structural Design

The technology in loads has advanced considerably since the 1950s such that the discreted plunge-bending type of analysis has lost its applicability. Lifting-surface aerodynamics combined with panel weight and stiffness presentations are much more amenable to power-spectral analysis (Ref. 3.43) such that to perform the discrete analysis today requires use of a more sophisticated and less direct approach.

Previous discussion showed that the discrete gust approach loads are less critical than the design loads which are due to other conditions including gust loads obtained on a power-spectral basis as shown in Ref. 3.43.

The industry has recognized for many years that turbulence is a continuous phenomenon, not discrete as assumed in the discrete gust analysis. Accordingly, the dynamic gust analysis methods accommodate the continuous nature of turbulence in the power-spectral approach. In this approach the gust is represented as a continuous function in the frequency domain rather than the time domain.

Briefly, the criteria are defined in two forms, namely mission analysis and design envelope (refer to Section 3.4). The mission analysis considers normal or typical expected utilization of the airplane during its life. The design envelope ignores anticipated utilization and considers instead the actual design envelope of weight, speed, altitude, and c.g. location. Ref. 3.43 describes in depth the development of both forms of the power-spectral gust loads criteria.

Airplanes that have been analyzed using power-spectral methods have gernerally been critical for loads based on mission analysis criteria, so major emphasis is placed on application of this mission analysis.

As noted before, the critical design envelope requires consideration of the airplane's actual design envelope of weight, c.g., speed, and altitude. The complete design envelope analysis can require consideration of as many analysis conditions as the mission analysis, particularly when various combinations of fuel and payload are included to insure adequate coverage of the envelope. Engineering judgement and experience must be relied upon to guide the analyst and to limit to a reasonable number the conditions to be investigated. For example, the vertical gust loads have been found to be most critical for high gross weights at forward c.g. and low weights at aft c.g. Wing loads appear to be critical for high payloads, low fuel cases, while the fuselage loads seem to be critical for high fuel, low payload. Therefore, the loads were generally less than as calculated by mission analysis procedures, or, in the few locations where they exceeded the mission analysis values, some other conditions being designed.

Detailed discussion and development of the power-spectral criteria are found in Ref. 3.20 and 3.43, and are not covered in depth here.

Landing and Taxi

Landing wing loads are of importance because of the down loads experienced and because of the large concentrated loads applied to the wing if the gear is located on the wing.

Taxiing must be accomplished at maximum take-off weights when the load in the wing is a maximum and wing stores are most likely to be installed. Rough runways will cause fairly large load factors on the airplane (in the vicinity of 1.7 to 2.0). Because of the flexibility of the wing, a whipping of the wing will occur and a considerable magnification of this load factor will be experienced at the wing tip. This magnification is a function of the wing weight and stiffness. Variation in load factor along the wing as shown in Fig. 3.5.8.

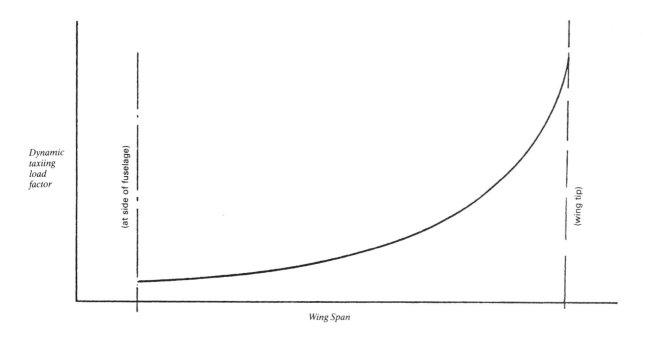

Fig. 3.5.8 Dynamic taxiing load factor vs. wing span.

Additional landing and ground handling conditions that are of particular importance to the strength of the wing and wing-mounted stores are the dynamic conditions; particularly, consideration of the effect of large masses located in an elastic wing structure during landings or taxiing over rough runways.

Wing Control Surfaces

For airplanes equipped with wing flaps, slats, ailerons, spoilers, dive brakes or other high-lift devices, additional flight loading conditions must be investigated for these control surfaces extended. These conditions are usually not critical for wing bending stresses, since the specified load factors are not large, but may be critical for wing torsion, shear in the rear spar, or down tail loads, since the negative pitching moments may be quite high. The aft portion of the wing, which forms the flap supporting structure, will be critical for the condition with flaps extended.

(1) Flaps

This span load distributions as shown in Fig. 3.5.9 may be estimated from data contained in Ref. 3.8 and 3.44.

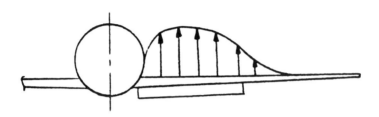

(a) Load distribution from flap

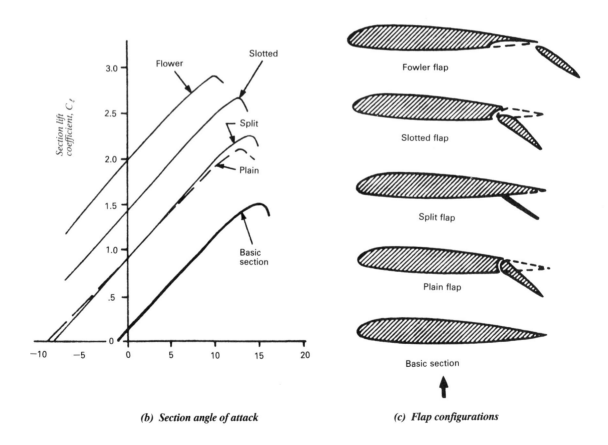

(b) Section angle of attack   *(c) Flap configurations*

Fig. 3.5.9   Effect on section lift characteristics of a 25% chord flap deflected 30°.

58 Airframe Structural Design

Since the flapped airfoil has different values for the magnitude and location of the airfoil characteristics, the wing structure must be checked for all possible flap conditions within the specified requirement relative to maximum speed at which the flaps may be operated. Generally speaking, the flap conditions will effect only the wing portion inboard of the flap and it is usually only critical for the rear beam web and for the upper and lower skin panels of the wing box near to the rear beam. This is due to the fact that the deflection flap moves the center of pressure considerably aft thus producing more shear load on the rear beam as well as torsional moment on the conventional cantilever wing box.

(2) Ailerons

This spanwise load distribution as shown in Fig. 3.5.10 may be estimated from data contained in Ref. 3.7 and 3.22.

Load distribution from aileron

Fig. 3.5.10 *Wing load distribution due to deflected ailerons.*

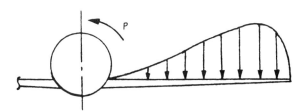

Fig. 3.5.11 *Damping load distribution on wing span due to rolling.*

Rolling involves a damping load distribution as shown in Fig. 3.5.11 on the wing which is a function of the rolling velocity $\left(\frac{Pb}{2V}\right)$. This distribution can be found in Ref. 3.7 and 3.22.

Wing Weight Distribution

To determine design loadings the weight of the wing must be taken into account. The weight distribution consists of (1) basic wing structure and (2) power plants and landing gear, if mounted on the wing, and removable stores such as fuel, tip tanks, etc.

Since in flight conditions the weight is a relieving load, considerable attention is given to the placement of fuel tanks and the sequence of using fuel.

3.6 Empennage Loads

Empennage loads include both horizontal tail and vertical tail loads and are described as follow:

- The horizontal tail is put on an airplane for two primary reasons:
  - Balance the moments caused by aerodynamic and inertia forces of other parts of the airplane.
  - Provide control about the pitching axis.
- Loads on the vertical tail are caused by the following:
  - Rudder deflection
  - Aileron deflection
  - Lateral gust
  - Asymmetric engine thrust

Horizontal Tail Loads

To determine loads on the horizontal tail in this condition requires a knowledge of the aerodynamic forces and moments on the other components of the airplane. This information is obtained by measuring these forces and moments in the wind tunnel with the horizontal tail removed. The force and moment coefficients are as shown in Fig. 3.6.1.

The value of C_L measured in these tests is that of the airplane less the tail. This fact is important when calculating the horizontal tail loads for balance.

The system of force and moment coefficients and the method of balancing an airplane are shown in Fig. 3.6.2.

For equilibrium: $\Sigma F_z = 0 = C_{L_{A-t}} + C_{L_t} + \dfrac{n\dfrac{W}{S}}{q}$

where lift coefficient on entire airpline,

$$C_{L_A} = C_{L_{A-t}} + C_{L_t} = \dfrac{n\dfrac{W}{S}}{q} \quad (3.6.1)$$

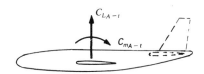

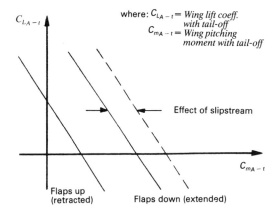

Fig. 3.6.1 *Wing lift coeff. with tail-off case (subsonic speeds)*

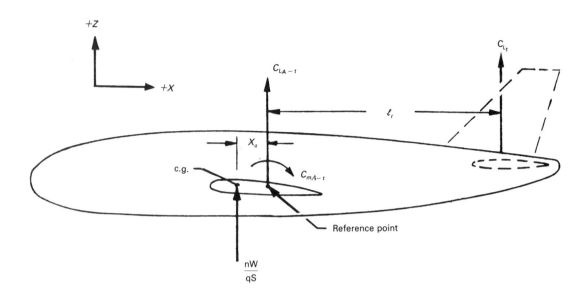

Fig. 3.6.2 *Forces of balancing an airplane.*

and $\Sigma M_{cg} = 0$

$$= C_{mA-t} - C_{L_A-t}\left(\frac{X_a}{\bar{c}}\right) - C_{L_t}\left(\frac{\ell_t}{\bar{c}}\right)$$

or
$$C_{L_t} = \frac{C_{mA-t} C_{L_A-t}\left(\frac{X_a}{\bar{c}}\right)}{\frac{\ell_t}{\bar{c}}} \quad (3.6.2)$$

The $C_{L_A-t}\left(\frac{X_a}{\bar{c}}\right)$ term can be eliminated by manipulating the axis of the C_{mA-t} vs. C_{L_A-t} curve. Assume a C_{L_A-t} of 1.0. The term is then $\frac{X_a}{\bar{c}}$ which is the moment increment of C_{L_A-t} at $C_{L_A-t} = 1.0$. On the curve of tail-off moment coefficient this shift of axis is as shown in Fig. 3.6.3.

Using the new axis, the equations for all tail lift coefficients become

$$C_{L_t} = \frac{(C_{mA-t})_{c.g.}}{\frac{\ell_t}{\bar{c}}} \quad (3.6.3)$$

$$C_{L_A} = \frac{n\frac{W}{S}}{q} = C_{L_A-t} + C_{L_t}$$

The value of C_{L_t} can be found by trial and error, or by obtaining the variation of C_{L_t} with C_{L_A} as follows:
1. Assume value of C_{L_A-t}
2. Obtain $(C_{mA-t})_{c.g.}$ from curves shown in Fig. 3.6.3.

3. $C_{L_t} = \frac{(C_{mA-t})_{c.g.}}{\left(\frac{\ell_t}{\bar{c}}\right)}$

4. $C_{L_A} = C_{L_A-t} + C_{L_t}$
5. Plot C_{L_A} vs. C_{L_t}

The tail load is obtained from given value of $\frac{n\frac{W}{S}}{q}$ and then,

$$C_{L_A} = \frac{n\frac{W}{S}}{q}$$

therefore, the vertical load on horizontal tail is

$$P_{Z_t} = C_{L_t} q S \quad (3.6.4)$$

(see Section 3.11 for sample calculation)

(1) Steady Maneuvers

A steady pitch maneuver such as a steady turn, steady pull-up or steady push-over is defined to have zero values of $\ddot{\theta}$ and $\dot{\alpha}$ but does have a steady value of $\dot{\theta}$. The value of $\dot{\theta}$ is a function of maneuver load factor and type of maneuver and is stated below

$$\dot{\theta} = \frac{g(n-1)}{V} \quad \text{for pull-ups and push-overs from level flight}$$

$$\dot{\theta} = \frac{g}{V}\left(n - \frac{1}{n}\right) \quad \text{for steady level turns}$$

60 Airframe Structural Design

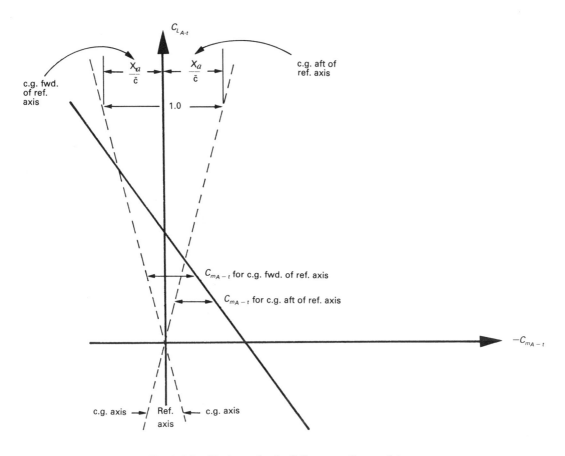

Fig. 3.6.3   Horizontal tail-off $C_{L_{A-t}}$ vs. $C_{m_{A-t}}$ plots

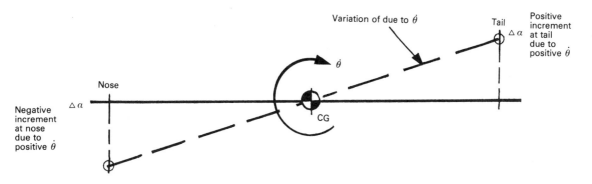

Fig. 3.6.4   Angular velocities vs angle of attack.

where $\dot{\theta}$ = angular velocity in pitch (radian/second)
V = airplane forward velocity (ft/sec)
n = airplane load factor
g = 32.2 ft/sec²

It should be remembered that the above load factor is opposite to the inertia load factor [i.e., a positive 2.5g pull-up maneuver would have a positive value of $\dot{\theta} = \frac{g}{V}(1.5)$].

The above angular velocities impart a change in angle of attack over the entire airplane which varies linearly from the tail to the nose with a sign change at the center of gravity as shown in Fig. 3.6.4.

The effect of $\dot{\theta}$ on airload redistribution is small on total lift since the lifts are opposite on either side of the c.g. However, the effect on moment can be large since the moments add in the same direction on either side of the c.g. A negative pitch moment (i.e., $C_{m_{\dot{\theta}}}$ is negative) results

Airframe Structural Design 61

from a positive $\dot{\theta}$. This can be thought of as a damping term tending to counteract $\dot{\theta}$. The negative $C_{m\dot\theta}$ can also be thought of as moving the aerodynamic center aft thus providing a further aft neutral point for maneuvering static stability. The incremental balancing tail load due to $\dot\theta$ can be estimated by

$$\Delta C_{L_t} \approx \frac{\left(C_{m\frac{\dot\theta C}{2V}}\right)_{A-t}\left(\frac{\dot\theta C}{2V}\right)}{\left(\frac{X_{cPht}-X_{c.g.}}{C}\right)} \quad (3.6.5)$$

To balance $\dot\theta$ effect

For positive values of $\dot\theta$ the incremental balancing tail load increment will be negative.

(2) Transient Maneuvers

The previous sections discuss the use of airplane-less-tail data to estimate horizontal tail loads for steady flight conditions. The increment of horizontal tail load due to transient motion of the elevator and airplane (to be added to the level flight balancing tail load) can be estimated from equations such as

$$\Delta C_{L_t} = C_{L_{a_{ht}}}(\Delta a_{ht}) \quad (3.6.6)$$

Transient

where

$$\Delta a_{ht} = \Delta a_{FRL}\left(1-\frac{d\epsilon}{d\alpha}\right) + \dot\alpha\left(\frac{\ell_{ht}}{V}\frac{d\epsilon}{d\alpha}\right)$$
$$+ \dot\theta\left(\frac{\ell_{ht}}{V\sqrt{\eta_{ht}}}\right) + \delta_e(t)\frac{d a_{ht}}{d\delta_e}$$

(η_{ht} = efficiency of the horizontal tail and use 0.8 as a good approximation.)

$C_{L_{a_{ht}}}$ can be estimated theoretically or may be estimated from the difference between airplane and airplane-less-tail pitching moment data in conjunction with a reliable estimate of $d\epsilon/d\alpha$, or using $C_{m_{\delta c}}$ wind tunnel data in conjunction with a reliable estimate of $\frac{da_{ht}}{d\delta_e}$.

Theoretical lifting surface methods are often used to aid the correlation of the above parameters.

(3) Pitch Maneuvers

The pitch maneuver involves a rapid motion of the elevator (see Fig. 3.3.3). A dynamic analysis of the airplane is necessary to determine the tail load accurately. The equations of motion are similar to those derived for the yawing maneuver in Section 3.3. Typical curves of pitching maneuver time history are shown in Fig. 3.6.5

The increment of tail load shown here must be added to the level flight balancing tail load. This tail load increment may be estimated from the eq-

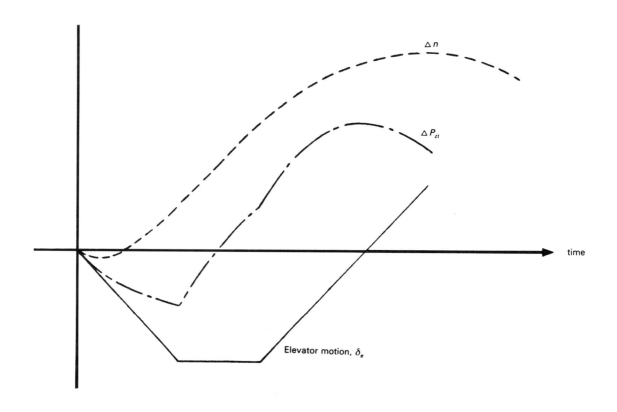

Fig. 3.6.5 Typical pitching maneuver time history. (Ref: 3-38.)

uilibrium equation of the pitching moment.

$$I\ddot{\theta} = \Delta P_Z (\ell_t)$$

$$P_{Z_t} = \frac{I\ddot{\theta}}{\ell_t} \qquad (3.6.7)$$

where $\ddot{\theta}$ = pitching acceleration in rad/sec^2

Approximate values of $\ddot{\theta}$ are ± 3.0 for fighters and ± 0.8 to ± 1.0 for transports and bombers.

(4) Unsymmetrical Loads

Unsymmetrical loads on the horizontal tail may occur from
- Buffet
- Misalignment
- Roll and yaw

Buffet and misalignment cannot be determined analytically and are taken into account arbitrarily.

Roll induces a damping load on the tail which can be determined from the same method used on a wing.

Yawing will induce a rolling moment on the horizontal tail which is difficult to estimate because of fuselage interference. The rolling moment caused by yaw is particularly sensitive to dihedral on the horizontal tail.

(5) Gust

The gust load on the horizontal tail is similar to the gust load on the wing or vertical tail. However, the horizontal tail is operating in the wing downwash. The correction factor for this effect takes the form of $(1 - \frac{d\epsilon}{d\alpha})$, where $\frac{d\epsilon}{d\alpha}$ is the rate of change of downwash angle with angle of attack. This value may vary from 0.5 to 0.8. The gust load equation becomes (approximately)

$$\Delta P_{Z_t} = \frac{K_g U_{de} V [\eta_{ht} C_{L_{\alpha ht}}] S_{ht} \left(1 - \frac{d\epsilon}{d\alpha}\right)}{498} \qquad (3.6.8)$$

where $C_{L_{\alpha ht}}$ can be estimated the same as for the wing and

$$\eta_{ht} \approx 0.8 \text{ to } 0.9$$

Lateral gust velocities and speeds are similar to those for vertical load factors. The lateral gust load on the vertical tail is shown below.

$$P_{y_{vt}} = \frac{K_g U_{de} V (\eta_{vt} C_{N_{\beta vt}}) S_{vt}}{498} \qquad (3.6.9)$$

for V in knots and
K_g = alleviation factor
$\eta_{vt}(C_{N_\beta})_{vt}$ is lift curve slope of tail

Vertical Tail Loads

The loads imposed by rudder deflection are a direct function of the rudder power. The rudder power consists of the mechanical advantage built in the control system between the pilot and the surface. The rudder effectiveness must also be included in considering rudder power.

The rudder hinge moment for boosted rudder system and a simple unboosted system are as shown in Fig. 3.6.6.

The 300 lb pedal force is the maximum pedal force considered for structural design. The boost cut-off is usually in the vicinity of 160–180 lb. The hinge moment coefficient of the rudder versus rudder angle will appear as shown in Fig. 3.6.7.

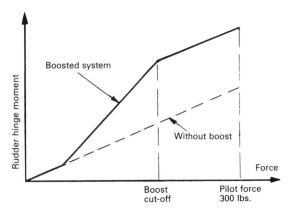

Fig. 3.6.6 Rudder hinge moment.

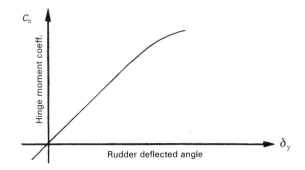

Fig. 3.6.7 Rudder hinge moment coeff. vs. rudder angle.

The non-linearity at high rudder angles usually cannot be neglected. Any non-linearity at low rudder angles is usually small and can, in general, be neglected. From these curves the rudder angle at any speed may be determined from

$$\delta_\gamma = \frac{H.M.}{C_{h_{\delta_\gamma}} q S_\gamma \overline{C}_\gamma} \qquad (3.6.10)$$

where $C_{h_{\delta_\gamma}}$ is slope of C_h vs δ_γ curve
S_γ = rudder area
$\overline{C}_\gamma$ = rudder m.a.c.

The other important parameter that must be considered is the variation of rudder hinge moment due to sideslip (C_{h_β}). The hinge moment will be

$$C_h = C_{h_\beta} \beta + C_{h_{\delta_\gamma}} \delta_\gamma \qquad (3.6.11)$$

The rudder deflection induces loads on the fin as well as the rudder. However, the primary loading on the vertical tail is caused by the resulting side slip angle when the rudder is deflected. The side slip angle

Airframe Structural Design 63

can be determined from the directional stability. A cross plot of C_n vs δ_y at constant β angles yields the rudder effectiveness as shown in Fig. 3.6.8.

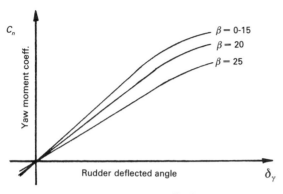

Fig. 3.6.8 Rudder effectiveness.

These curves indicate the efficiency of the rudder in producing side slip angles.

The side slip angle is found by equating the yaw moment caused by the rudder to the yaw moment of the airplane.

$$C_{n\delta_y} \delta_r q S b = C_{n\beta_A} \beta_0 q S b$$

or

$$\beta_0 = \frac{C_{n\delta_y} \delta_y}{C_{n\beta_A}} \quad (3.6.12)$$

where
β_0 = the equilibrium yaw angle
$C_{n\delta_y}$ = slope of C_n vs δ_y curve
$C_{n\beta_A}$ = slope of C_n vs β curve for complete airplane

The approximate vertical tail load at β_0 with rudder returned instantaneously to neutral is found as follows:

$$P_{y_{vt}} = \eta_{vt} (C_{n\beta})_{vt} \beta_0 q S_{vt} \quad (3.6.13)$$

where $(C_{n\beta})_{vt}$ is the slope of the tail lift curve and η_{vt} is the efficiency of the vertical tail. The product of these values is found by equating the yaw moment from the tail load to the moment increment between tail-on and tail-off as shown in Fig. 3.6.9.

Then

$$\eta_{vt}(C_{n\beta})_{vt} \beta q S_{vt} \ell_{vt} = (C_{n\beta_A} - C_{n\beta_{A-t}}) \beta q S b$$

where $C_{n\beta_A} - C_{n\beta_{A-t}} = (\Delta C_{n\beta})_{vt}$

$$\eta_{vt}(C_{n\beta})_{vt} = \frac{(\Delta C_{n\beta})_{vt}}{\dfrac{S_{vt} \ell_{vt}}{S b}}$$

The quality $\dfrac{S_{vt} \ell_{vt}}{S b}$ is known as tail volume, $\overline{V}_{vt}$

The equation for tail load then becomes

$$P_{y_{vt}} = (\Delta C_{n\beta})_{vt} \beta_0 q \frac{S_{vt}}{\overline{V}_{vt}} \quad (3.6.14)$$

If the rudder is deflected, the load from the rudder will subtract from the side slip load. The load caused by the rudder is derived similarly as

$$P_{y_{vt}} = C_{n\delta_y} \delta_y q \frac{S_{vt}}{\overline{V}_{vt}} \quad (3.6.15)$$

(due to deflected rudder)

The dynamic consideration of a sudden rudder

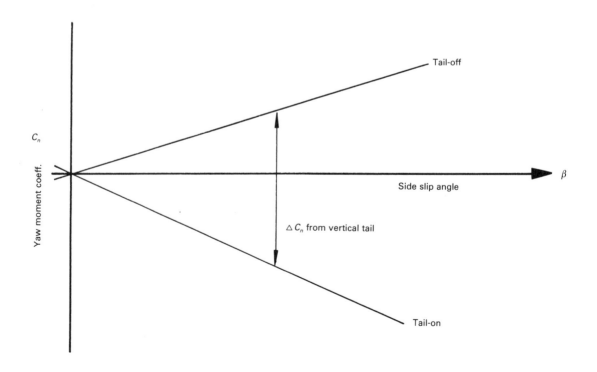

Fig. 3.6.9 Yawing moment coeff. between tail-on and tail-off.

movement will cause an overswing of the airplane past the equilibrium side slip angle. The magnification of the side slip, K_β, will be approximately 1.4 to 1.6. In this condition the rudder is considered to be deflected. The tail load equation becomes

$$P_{y_{vt}} = [(\Delta C_{n\beta})_{vt} K_\beta \beta_0 - C_{n\delta\gamma} \delta_\gamma] q \frac{S_{vt}}{\overline{V}_{vt}} \quad (3.6.16)$$

(overswing force)

(1) Aileron roll case

In this condition the ailerons are considered to be suddenly deflected while the rudder is held neutral. The equations for tail load are similar to the ones shown previously for rudder deflection. In this case, the side slip angle is being caused by the ailerons and the airplane rolling velocity.

The equilibrium side slip angle is

$$\beta_0 = \frac{C_{n_p}\frac{pb}{2V} + C_{n\delta a} \delta_a}{C_{n\beta_A}} \quad (3.6.17)$$

The numerator terms can be approximated very roughly by $\frac{C_L}{8}$. The magnification, K_β, from dynamic overswing is again in the vicinity of 1.4−1.6. The exact value can be determined only from a dynamic analysis of the airplane. The tail load equation becomes

$$P_{y_{vt}} = K_\beta \left[\frac{C_{n_p}\frac{pb}{2V} + C_{n\delta a} \delta_a}{C_{n\beta_A}} \right] (q)\frac{S_{vt}}{\overline{V}_{vt}}(\Delta C_{n\beta})_{vt}$$

$$(3.6.18)$$

(2) Asymmetric engine thrust case

When engines are located away from the airplane center line, a loss of thrust from engine failure must be balanced by a vertical tail load. For static conditions this is easily determined. However, with the failure of a turbine engine, a large amount of drag on the failed engine will be experienced. Time history of the engine thrust and drag is shown in fig. 3.6.10.

This asymmetric thrust will induce large side slip angles. Because of the rapid application of this force, a considerable magnification of the side slip angle will occur.

(3) Distribution of load on vertical tail

For spanwise variation of airload, a plan form distribution is usually satisfactory. Chordwise pressure distributions for subsonic speeds will assume the shapes as shown in Fig. 3.6.11.

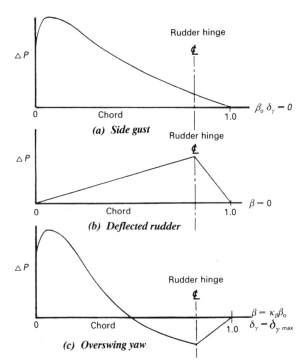

Fig. 3.6.11 Vertical tail chordwise pressure distributions (subsonic).

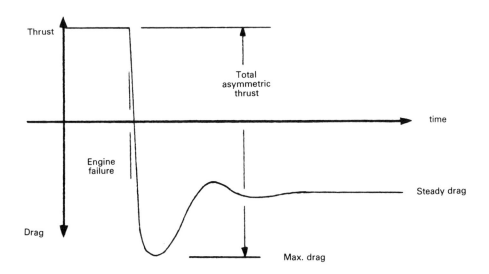

Fig. 3.6.10 Time history of the engine thrust and failure engine drag.

3.7 Fuselage Loads

Loads affecting fuselage design can result from flight maneuvers, landings or ground handling conditions. Fuselage loads are primarily a problem of determining the distribution of weight, tail loads, and nose landing gear loads. Weight distribution is important because a large part of fuselage loads stems from the inertia of mass items acted upon by accelerations, both translational and rotational. Tail loads, which are generally quite large, contribute heavily to bending the aft portion of the fuselage. Disymmetry of tail loads cause significant aft-body torsions. In the same sense that tail loads affect the aft-body, loads acting on the nose landing gear will contribute significantly to net loads on the forward portion of the fuselage, the forebody.

As implied above, the various portions of the fuselage can be most critically loaded by completely different flight or handling conditions. For expedience of analysis the airplane is divided into three sections and each of these is analyzed separately. Of course, eventually the structure must be considered for effect of loads carrying through from one section to the other. For discussion here, the fuselage will be divided into sections in the same manner as is generally applied in analysis. For a typical airframe, the fuselage is divided into three sections (see Fig. 3.7.1.):

(cargo and passengers) is considered to include such items as seats, galleys, lavatories, etc. An arbitrary floor loading is then used for transport airplane design. This arbitrary loading is selected to envelope all possible variations of cabin loading. The floor loading for this equipment plus passengers runs approximately 45 lb per sq. ft. Baggage weight is satisfactorily approximated by a maximum of 20 lb per cu. ft. of cargo space.

The arbitrary loading used is then distributed in the fuselage to obtain the gross weights and center-of-gravity locations on the c.g. envelope.

Forebody Loads

The forebody loads incurred during symmetrical flight are obviously in a vertical direction only; there are no side loads. This vertical direction is normal to the airplane reference axis, or in the "z" direction. These vertical loads are determined simply by multiplying the weight loads by the load factor.

Vertical airloads are generally neglected in forebody loads calculations except for wide body fuselage or their affect on local structure. Neglecting them is generally conservative because they are in the direction to relieve inertia loads. Also, they are usually small relative to net loads. Another reason contributing to their omission is the fact that accurate

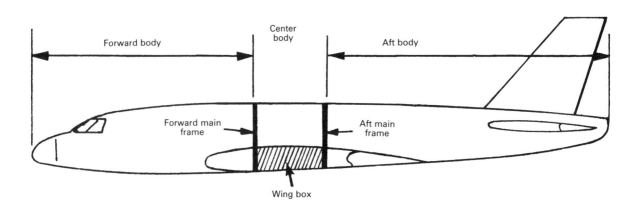

Fig. 3.7.1 Divided fuselage into three sections.

(1) Forebody: that portion of the fuselage forward of the forward main frame.
(2) Attbody: that portion of the fuselage aft of the aft main frame; including the empennage.
(3) Centerbody: that portion of the fuselage between main frames.

Distribution of Weight

The fuselage weight distribution consists of the fixed weight of the structure and equipment, and the removable load. The removable load in military types is relatively small and concentrated. Passenger and cargo airplanes, however, are required to carry loads of varying quantity and location in the fuselage.

Because of the various possible arrangements of the cabin for different customers, the removable load

distributions are difficult to determine because of irregularities in the fuselage profile. Good distributions would have to be determined by wind tunnel pressure measurements. The cost involved is usually not warranted. Assumed, linearized distributions are sometimes used, but it should be borne in mind that the probable inaccuracy, coupled with the relative magnitude of the airload, results in a load component of questionable reliability.

Side loads (in the y direction) are caused by side and yawing accelerations and airloads incurred during unsymmetrical maneuvers. Here the airloads make up a large part of the net loads and therefore cannot be neglected. As reasonable a distribution as possible is estimated based on best available data.

Critical forebody loadings may also be experienced

from application of nose landing gear loads. Design loadings might arise from landing or application of main wheel brakes during taxiing, particularly unsymmetrical brake application.

Aftbody Loads
Aftbody vertical flight loads are a critical combination of inertia loads and horizontal tail balancing loads. The horizontal tail loads are determined for the various conditions on the V-n diagram and center-of-gravity locations. Since the distribution of weight in the fuselage as well as the tail loads are a function of c.g. location, the problem is one of determining the critical combinations.

Lateral loadings result from application of airloads acting on the vertical tail in combination with side inertia loads.

Airloads on the fuselage aftbody are generally neglected, both in the vertical and side directions. In this case the airloads are not necessarily relieving, therefore it is not conservative to neglect them. However, they are generally quite small, and their distribution in the unpredictable flow behind the wing is impossible to determine.

External Pressures
External pressures on the fuselage, other than in wing vicinity, are usually significant only around protuberances. In the area of the wing, the pressures on the wing are carried onto the fuselage as shown in Fig. 3.7.2.

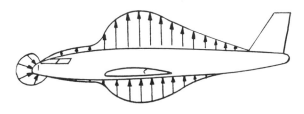

Fig. 3.7.2 Pressure distribution on fuselage.

The pressure on the fuselage will be of the order of magnitude of the pressure on the wing.

Internal Pressures (Cabin Pressure)
The fuselage internal pressure depends on the cruise altitude and the comfort desired for the occupants. Pressure differential may be readily determined from the altitude charts as shown in Fig. 3.7.3, if the actual altitude and the desired cabin pressure are known.

Fuselage pressurization is an important structural loading. It induces hoop and longitudinal stresses in the fuselage which must be combined with flight and ground loading conditions. The important consideration for establishing the fuselage design pressures is the cabin pressure differential or present in altitude and the fuselage is designed to maintain.

Example: Assume an airplane
(a) provide an 8000 ft. altitude cabin pressure
(b) the max. flight altitude at 43,000 ft.
The pressures used with flight and ground conditions on the airplane are:

Condition	Max. Positive (Burst) Pressure	Max. Negative (Collapsing) Pressure
Flight (All)	8.85 psi[1]	−0.50 psi[2]
Ground	1.0 psi	−0.5 psi
In addition, a pressure of $1.33 \times 8.85 = 11.77$ psi is considered to act alone.		

Note:
(1) Upper limit of positive relief valve setting = 8.6 psi (cabin pressure differential, see Fig. 3.7.3) + 0.25 psi (relief valve tolerance) = 8.85 psi
(2) Upper limit of negative relief valve setting = 0.50 psi.

3.8 Propulsion Loads

The propulsion system is one of the major systems that requires extensive integration with the design of the airplane. The following illustrates the magnitude of the task;

"The magnitude of this task can perhaps be best appreciated by partial listing of the necessary data exchanges and coordinated analyses and tests required for the propulsion system alone on the airplane. These included: net thrust, gross thrust, thrust distribution, engine mass data, engine deflection data and tests, engine vibration data, sonic environment (near field), bleed air pressures and temperatures, thrust reverser system safety characteristics, design load and loads spectra for the engine, fan blade containment analyses and tests, fan blade unbalance analyses and tests, fan whirl mode analyses and tests, bird ingestion criteria and tests, and asymmetric thrust due to inadvertent thrust reverser deployment in flight."

The items mentioned above are obviously too numerous and involved to cover here, so the discussion will be limited to the major propulsion system loads that affect design of the pylon (engine support structure) and the airframe structure. These loads are usually referred to as engine mount loads. The major considerations are thrust loads, aerodynamic loads, inertia loads, and gyroscopic loads.

Thrust
The thrust values used must account for installation on the aircraft. Various thrust levels are used in design, maximum take-off, maximum climb, maximum cruise, maximum continuous flight idle and reverse. These thrust levels depend on such parameters as temperature, bleed air, altitude, and Mach number. Most of the thrust from a turbo-fan engine is produced by the fan and the primary jet thrust; however, some thrust is produced by the inlet and fan nozzle.

Airloads
This term refers to the various aerodynamic forces developed on the installation.

The inlet portion of the pod is a highly loaded portion of structure since its function is to straighten the airflow into the engine. Thus, the inlet is changing the momentum of the airflow into the engine and this develops high loads and pressures. A typical inlet pressure distribution for a symmetric pitch condition

Airframe Structural Design 67

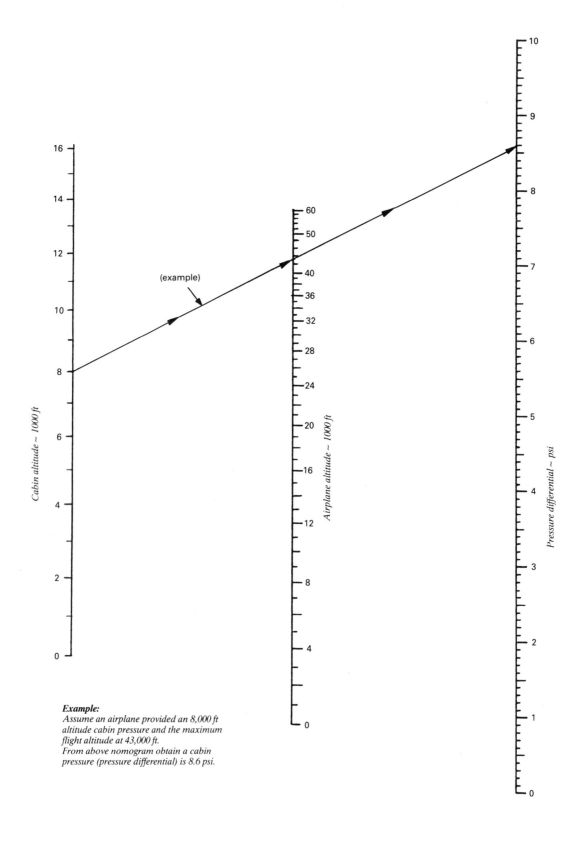

Example:
Assume an airplane provided an 8,000 ft altitude cabin pressure and the maximum flight altitude at 43,000 ft.
From above nomogram obtain a cabin pressure (pressure differential) is 8.6 psi.

Fig. 3.7.3   *Pressure differential nomogram.*

is shown in Fig.3.8.1.

Sizeable moments and forces (besides thrust) can be produced by the fan. The inlet will not straighten the flow into the fan entirely for conditions with large angles of attack or yaw. Therefore, the fan will experience forces and moments similar to those of a propeller.

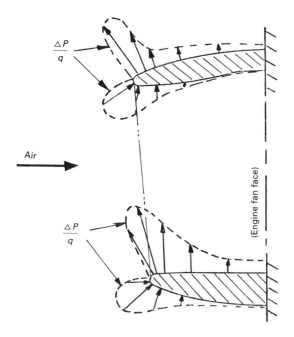

Fig. 3.8.1 Jet engine inlet pressure distribution.

Inertia Loads

Load factors at the engine c.g. for airplane maneuvering conditions can be determined using the following equations and its notation as shown in Fig. 3.8.2.

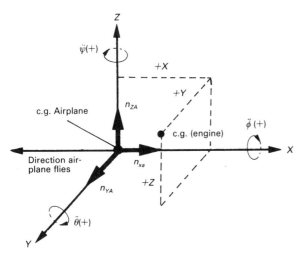

Fig. 3.8.2 Notations and sign conveniences for propulsion system.

$$n_{x_N} = +n_{x_A} - \frac{z\ddot{\theta}}{g} + \frac{y\ddot{\psi}}{g} + \frac{x}{g}(\dot{\theta}^2 + \dot{\psi}^2)$$

$$n_{y_N} = +n_{y_A} - \frac{x\ddot{\psi}}{g} + \frac{z\ddot{\phi}}{g} + \frac{y}{g}(\dot{\psi}^2 + \dot{\phi}^2)$$

$$n_{z_N} = +n_{z_A} - \frac{y\ddot{\phi}}{g} + \frac{x\ddot{\theta}}{g} + \frac{z}{g}(\dot{\phi}^2 + \dot{\theta}^2)$$

The side loads are arbitrarily applied and the load factor is $1.33g$ or $\frac{n_{z\,max}}{4}$.

It should be noted that dynamic landing, taxi, and gust are conditions that often produce the largest engine load factors, in the order of 5.0 to 6.0.

Gyroscopic Loads

A gyroscopic moment is developed when the rotating parts of the propulsion system are rotated about either the airplane pitch or yaw axis. The equations for these moments are:

$$M_Z = -IN\dot{\theta}$$
$$M_Y = IN\dot{\psi}$$

where I = polar moment of inertia in-lb-sec^2
N = engine angular velocity in rad/sec^2
$\dot{\theta}, \dot{\psi}$ = airplane pitch and yaw velocity, respectively in rad/sec

Note: These equations are used for engines that rotate clockwise looking aft. For fighter airplanes, the gyroscopic loads from spin velocity should be considered.

Combining the thrust, aerodynamic, inertia and gyroscopic loads together into the maximum design loads is an involved task since some loads can be relieving other loads.

3.9 Landing Gear Loads

Loads imposed on the various airframe components during landing and ground handling operation are of necessity dependent on characteristics of the airplane's landing gear. A conventional landing gear performs two basic functions. It dissipates the energy associated with vertical descent as the airplane contacts the ground; and, it provides a means of maneuvering the airplane on the ground (taxiing). Analysis of airplane behavior during the landing impact and during the taxiing operation are imperative in order that:
- the landing gear and its attachment be designed to a proper strength level
- other components are investigated for every possible design condition.

It is apparent that landing impact loads are dependent on a number variables, among which are airplane landing contact velocity (airplane design sink speed: 10 ft/sec for transports and land based fighter except trainer; 20 ft/sec for carrier-based airplanes), gear and tire energy absorption characteristics (particularly length of strut stroke), airplane attitude, airplane mass, distribution of airplane mass to the various gears, and magnitude of lift acting on the airplane at time of impact. In addition to the variables contributing to vertical gear loads are those affecting gear drag loads, such as the friction coefficient between the tire and ground surface, and the spring function behavior of the entire gear in the fore and aft direction (see Fig. 3.9.1).

Mathematical prediction of landing gear vertical forces requires that the landing gear behavior at impact be described in equation form with a fair degree of accuracy. The oleo-pneumatic shock strut now in general use displays a rather complicated relation between force, stroke, and rate of closure. Attempting to solve a complex set of equations for loads purposes has been found impractical both because loads predictions are generally made early in the design stages, before enough details have been established to permit accurate mathematical representation, and because the relative accuracy does not warrant the computation time. Instead, equations representing a highly idealized landing gear that can be represented simply are generally used.

The energy absorption during landing is of primary importance. The ability of the landing gear to absorb this energy governs the value of the landing load factor.

The simplest device would be a spring, where variation of the vertical load on the gear with spring deflection is linear, see Fig. 3.9.2.

The energy absorbed would then be

$$\frac{1}{2} P_V X_S = \frac{1}{2} K_S X_S^2$$

The efficiency of the strut is defined as the ratio of a constant vertical load (dash line) over the strut stroke that will absorb the same energy, to $P_{V\,max}$. The efficiency of the spring is, then, 50 percent.

Considerable increase in efficiency can be realized by using a device with non-linear characteristics. The most common type of strut is the air-oil oleo. This oleo is a piston arrangment where oil is forced through an orifice in the piston. The size of the orifice governs the efficiency of the oleo and is usually determined during drop tests. A schematic of an oleo is shown in Fig. 3.9.3.

The efficiency of a well designed strut of this type will be 80–85 percent. The load factor may be approximated by a simple energy relation. The energy of

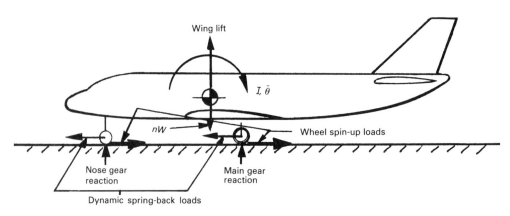

Note: Spin-up and Spring-back loads do not occur simultaneously.

Fig. 3.9.1 Applied loads and accelerations considered during landing.

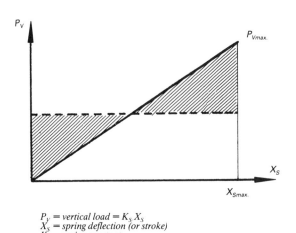

P_V = vertical load = $K_S X_S$
X_S = spring deflection (or stroke)
K_S = spring constant

Fig. 3.9.2 Linear spring deflection with applied load.

the compressed tire is subtracted from the kinetic energy of descent plus the potential energy and equated to the area under the load-stroke curve.

$$\text{Kinetic energy} = \frac{1}{2} \frac{W}{g} V_S^2 \quad (3.9.1)$$

$$\text{Tire energy} = \eta_t \frac{1}{2} K_t X_t^2 \quad (3.9.2)$$

$$\text{Potential energy} = (W - L)(X_t + X_S) \quad (3.9.3)$$

Then $\frac{1}{2} \frac{W}{g} V_S^2 - \eta_t \frac{1}{2} K_t X_t^2 + (W - L)(X_t + X_S)$
$$= \eta_s P_{V\,max} X_S$$

where W = gross weight
V_S = sinking speed (ft/sec)
K_t = tire spring constant (lb/ft)

70 Airframe Structural Design

η_t = factor to account for fact tire deflection non-linear
X_t = tire deflection (ft)
L = wing lift (lb), usually equal to $\frac{2}{3}W$
X_S = strut stroke (ft)
η_s = strut efficiency, 0.8 to 0.85
P_V = vertical load

Landing gear load factor, $n_{LG} = \dfrac{P_V}{W}$

the $n_{LG} = \dfrac{\dfrac{1}{2}\dfrac{V_S^2}{g} - \dfrac{\eta_t K_t X_t^2}{2W} + \left(1 - \dfrac{L}{W}\right)(X_t + X_S)}{\eta_s X_S}$

(3.9.4)

and the inertia factor on the airplane is $n_{LG} + \left(\dfrac{L}{W}\right)$.

The shape of the load-stroke curve can be further modified by the use of a metering pin. This pin changes the size of the orifice during the stroke. If insufficient stroke is available and the strut "bottoms" or high air pressures are experienced, poor efficiency and high load factors result. An example of this situation and a possible cure by use of a metering pin are shown in Fig 3.9.4, and a typical main landing gear strut airspring and load curves are shown in Fig. 3.9.5.

A liquid spring is a similar device where the energy absorption is accomplished by compressing as well as metering the oil (refer to Chapter 12.0).

Landing Conditions

Design loads for the landing operations (Ref. 3.27) are obtained by investigating various airplane altitudes at ground contact at the specified landing and sink speeds. *The procuring service or certificating agency specifies the minimum sink speed and maximum amount of wing lift that may be used.* The various conditions considered are referred to as the landing conditions. These conditions are well defined in Ref. 3.2, 3.3 and 3.4 for both nose wheel type and tail wheel type airplanes. A typical set of landing conditions for an airplane with tricycle gear is summarized below.
- Level landing, three point
- Level landing, two point
- Tail down landing
- One-wheel landing
- Drift landing

Gear reactions and airplane accelerations are determined for each condition. The load on the

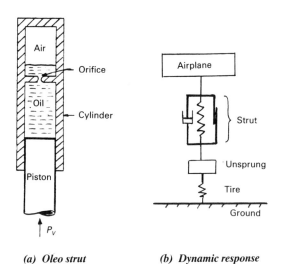

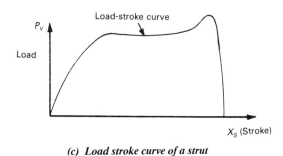

Fig. 3.9.3 A schematic of an oleo gear.

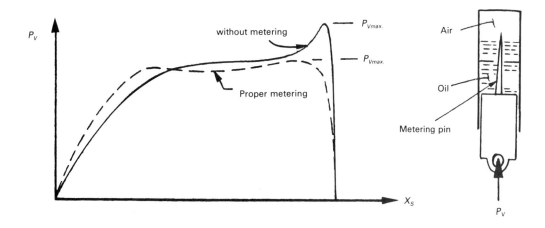

Fig. 3.9.4 Load stroke curve of a strut with a metering pin.

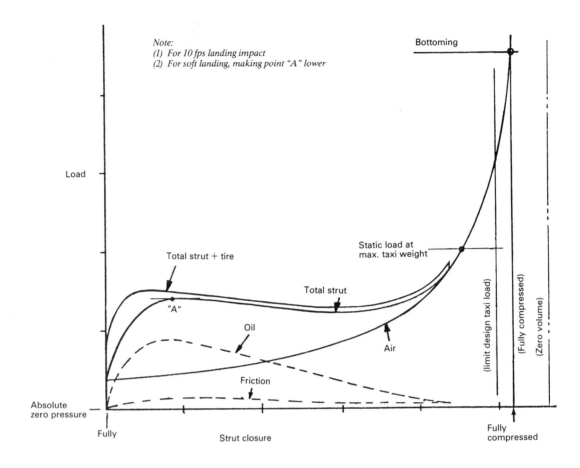

Fig. 3.9.5 Typical main landing gear strut airspring and load curves.

landing gears are externally applied forces and are placed in equilibrium by translational and rotational inertia forces.

In addition to the static loads on the gear, loads associated with accelerating the wheel assembly up to the landing speed must be considered. These "spin-up" loads are difficult to determine rationally and equally difficult to measure in tower drop tests. Ref. 3.2, 3.3 and 3.4 provide semi-empirical equations for calculating the spin-up loads (drag loads) and vertical loads at time of peak spin-up loads. This method of analysis is accepted broadly by load engineers.

The elasticity of the landing gear assembly is considered in determining a forward acting design load. It is assumed that following the wheel spin-up, when sliding friction has reduced to zero, the energy stored in the gear as a result of rearward deformation causes the wheel mass to spring forward resulting in a sizeable forward inertia load. This forward acting dynamic "spring-back" load is considered to occur about the time the vertical load reaches its maximum. Ref. 3.2, 3.3 and 3.4 provide a method of analysis for spring-back along with the spin-up load analysis mentioned in the preceding paragraph.

Generally the spin-up (Ref. 3.29) and spring-back loads can be assumed as very high frequency loading, to which the total airplane mass does not respond.

Consequently these drag loads are used only to design the gear and its attaching structure and do not affect airplane balance. However, care must be taken in that these loads may excite other large masses such as nacelles or tip tanks.

Gear Strut Loads

The ground handling loads are based on the assumption of a rigid airplane with the external loads placed in equilibrium with appropriate linear and angular inertia loads.

(1) Ground handling drag and side loads are applied at ground and vertical loads are applied at the axle.

(2) Landing drag loads are applied at the axle.

Note that the landing and taxi loads are determined by dynamic analysis.

The landing side load condition usually referred to as drift landing is defined from the level landing loads with only the main gear contacting the ground.

Unsymmetrical Loads on Multiple-Wheel Units

Unsymmetrical loads on multiple-wheel units contains the requirements for unsymmetrical distribution of the gear strut loads to individual wheel loads. These requirements cover both flat tire and no flat tire conditions.

(1) Gear/wheel loads with flat tires

The requirements for gear strut loads are specified as percentages of the no tires flat conditions previously discussed. These gear strut loads are then distributed to the wheels with inflated tires.

(2) Gear/wheel loads with no flat tires

The limit gear strut loads for no flats must be distributed to the individual wheels by accounting for the following effects:
- Number of wheels and their physical arrangement.
- Differentials in tire diameters due to manufacturing tolerances, tire growth, and tire wear, (use $\frac{2}{3}$ of the most unfavorable combination).
- Unequal tire inflation pressure (use ±5% of the nominal inflation pressure).
- Runway crown (use zero and $1\frac{1}{2}$% slope).
- Airplane attitude.
- Any structural defection (includes wing, strut, bogie, and axle deflections).

Landing Gear Loads

For a tricycle gear, the landing gear loads associated with ground maneuvers are defined
- Braked roll — 3 wheels
- Braked roll — 2 wheels
- Unsymmetrical braking
- Reverse braking
- Turning
- Pivoting

The maximum coefficient of friction in these conditions is 0.8. The load factor for braked roll conditions is 1.2 at landing weight and 1.0 at take-off weight.

Other ground handling conditions are hoisting, jacking, towing, etc.

Retraction Loads

The strength of the retraction system must be adequate with the gear in any position up to the design speed for retraction. A load factor approximately 2.0 is applied in combination with airloads. In addition, the gyroscopic moment of the wheels and the torque from brake application during retraction must be considered.

3.10 Miscellaneous Loads

The term *miscellaneous loads* is usually applied to loads required to design non-primary airframe structure. As such, it encompasses an extremely wide range of airplane components and systems. There are so many items covered by miscellaneous loads that it is not feasible to discuss many here. The following is a partial list of types of miscellaneous loads:
- Ground handling loads
- Control surface loads
- Door loads (passenger, cargo, landing gear, and access)
- Pressure loads (cabin, fuel tank, and local surface)
- Nose radome loads
- Fluid system requirements
- Seat and floor loads
- Auxiliary power unit loads (APU)
- Environmental control system loads (ECS)
- Jacking and mooring loads
- Fixed leading edge loads
- Engine and gear breakaway
- Antenna Loads
- Ram air turbine loads

Calculation of miscellaneous loads often requires detailed information about a particular system in the airplane in addition to details of the airplane's operating environment. Miscellaneous loads may include structural design loads, operational loads, and fatigue loads. Structural design loads are the maximum loads the component should ever experience. Operating loads are used to size the component or system. These loads are the maximum the system is required to experience and still function. Fatigue loads are determined from normal operational usage and are used to assure that a proper lifetime is achieved for the component.

A limited number of miscellaneous load items will be discussed to provide insight into the types of analysis usually required for miscellaneous components and systems.

Ground Handling Loads

Licensing agencies require that the landing gear and airframe be designed for certain ground handling conditions. These conditions are usually well defined by the agency and in general are based on many years of operational usage of aircraft.

Control Surfaces

Control surfaces are designed by the mechanical advantage of the system. The following pilot forces are required by:
- Aileron: Stick 100 lbs
  Wheel $80D$ (where D = drive wheel diameter)
- Elevator: Stick 250 lb
  Wheel 300 lb
- Rudder: 300 lb

The variation of mechanical advantage with surface deflection should be taken into account. The required output of a system is usually specified at maximum deflection of the surfaces. At lower angles the mechanical advantage may be considerably higher. A typical example of aileron hinge moment is shown in Fig. 3.10.1. In some designs, trim tabs are considered to be deflected in the direction to assist the pilot forces.

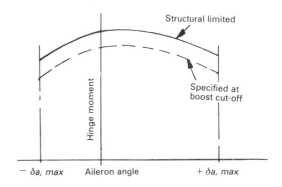

Fig. 3.10.1 Typical aileron hinge moment vs. aileron angles.

The control system is designed for the pilot loads as mentioned above. In the case of a boosted system, the part of the system between the booster and the surfaces is designed for the relief valve setting on the hydraulic cylinder.

Doors

Doors such as weapons bay doors, landing gear doors, etc., are usually required to operate at high design speeds. Bomb bay doors are usually opened at dive speeds. Landing gear doors must be able to extend retract at speeds up to at least 67% V_c.

In addition to the loads associated with opening, doors must be designed stiff enough when closed to prevent deflection and leakage. "Door closed" conditions require an estimation of the pressure distribution over the body to which the doors are attached. Loads on doors in the open position should be based on experimental data, if at all possible.

Structural design loads on the doors should include the static load plus yaw effects plus buffet. The shape of the door will have considerable effect on the hinge moment. For straight clam shell doors with the hinge line parallel to the airstream, the approximate hinge moment coefficient is

$$C_h = 0.05 \pm 0.0064\beta \pm 0.04$$

where $\pm 0.0064\beta$ is from yaw and
± 0.04 is from buffet

(Note: hinge moment = $C_h qSC$)

If the doors are curved fore and aft, the 0.05 value may increase from 0.25 to 0.30. When another body such as a landing gear is near to the doors, the hinge moment will increase again. The chordwise pressure distribution on the doors is usually triangular with peak pressure at the hinge line.

Tip Tanks

Tip tanks are subjected to pressure which are generated by the wing. This pressure carry-over from the wing will cause significant loads on the tanks. The lift on a tank is primarily a function of the diameter of the tank. Consequently, lift coefficients are based on tank cross-section area. The lift curve slope will be approximately 0.17 to 0.20 per degree. The use of a horizontal vane on the tank will almost double the lift of the tank. The C_{Lmax} of the vane, based on vane area, is about 1.2 to 1.5. The center of pressure of the tank load will be around 0.35 of the tank length. With a vane the lift is increased but the c.g. is moved aft considerably.

Radomes, Canopies, ...

Local pressures on protuberances are usually of considerable magnitude. When integrated over the area of the body (i.e. fuselage) a significant load results. These loadings are primarily a function of the shape of the body. Radomes and canopies are among the items affected most. The pressure distribution on a radome is shown in Fig. 3.10.2. The maximum suction $\Delta \frac{p}{q}$ for low speed will be from -0.8 to -1.4. As speed increases this value will increase with the Prandtl-Glauert factor as $\frac{1}{\sqrt{1-M^2}}$

3.11 Example of An Airplane Load Calculation

Airplane data
Geometry (See Fig. 3.11.1).
Wing:
  Area, $S = 1200$ sq ft
  Span, $b = 98$ ft
  Aspect ratio, AR (or A) $= 8.0$
  Taper ratio, $\lambda = \dfrac{C_T}{C_R} = 0.4$
  Root chord, $C_R = 17.5$ ft
  Tip chord, $C_T = 7.0$ ft

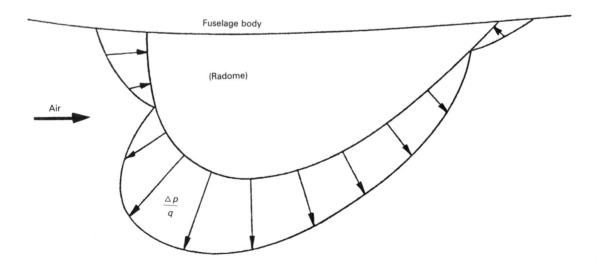

Fig. 3.10.2   Pressure distribution on a radome.

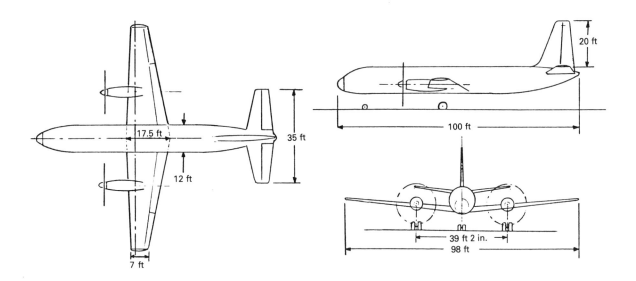

Fig. 3.11.1 Three view diagram of the airplane.

Mean aerodynamic chord, $\bar{c} = 13.0$ ft
Vertical Tail:
  Area, $S_{vt} = 220$ sq ft
  Tail length, $\ell_{vt} = 45$ ft
  Rudder area, $S_y = 60$ sq ft
  Rudder mean chord, $\bar{c}_y = 3.5$ ft
Horizontal Tail:
  Area, $S_{ht} = 200$ sq ft
  Tail length, $\ell_{ht} = 48.0$ ft
Weight Data
  Maximum take-off gross weight = 108 000 lb
  Landing design gross weight = 88 000 lb

Maximum zero fuel weight = 84 000 lb
Aerodynamic Data
  $C_{L\,max}$ (complete airplane) = 1.3
  $C_{L\alpha_A} = 1.10\, C_{L\alpha_W}$
  $C_{h_{\delta_y}} = 0.01$ per degree
  Rudder hinge moment output = 7520 ft-lb
Landing Gear Data
  Strut stroke, $X_S = 18$ in
  Tire defection, $X_t = 4.5$ in
  Tire spring constant = 160 000 lb/ft
  Tire efficiency, $\eta_t = 0.8$

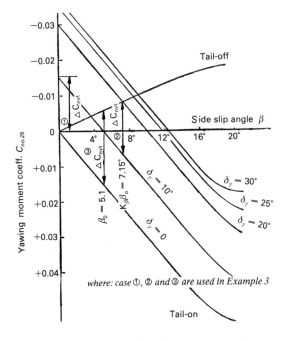

Fig. 3.11.2 Vertical tail yawing moment coeff. vs. side slip curves.

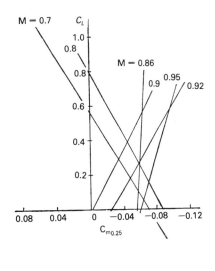

Fig. 3.11.3 Tail-off pitch stability.

Airframe Structural Design 75

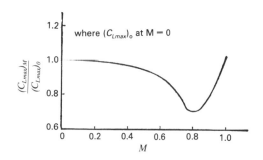

Fig. 3.11.4   Effect of Mach number on $C_{L_{max}}$.

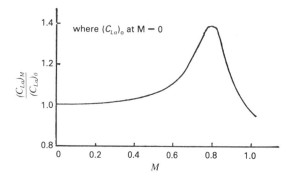

Fig. 3.11.5   Effect of Mach number on $C_{L\alpha}$.

Example 1

Calculate the true airspeed, Mach number, and dynamic pressure for an equivalent airspeed of 325 knots at 10000 feet altitude.

(1) Equivalent airspeed: (Refer to Eq. 3.1.3)

$$V_t = \frac{V_e}{\left(\frac{\rho_{alt}}{\rho_{S.L.}}\right)^{\frac{1}{2}}}$$

From Fig. 3.1.5 at altitude $h = 10000$ ft, find

$$\sqrt{\sigma} = \left(\frac{\rho_{alt}}{\rho_0}\right)^{\frac{1}{2}} = 0.8594$$

therefore,

$$V_t = \frac{325}{0.8594} = 378 \text{ knot}$$

(2) Mach number:
Speed of sound at altitude $h = 10000$ ft is
$$C_{se} = 548.6 \text{ knot} \text{ (from Fig. 3.1.5)}$$
therefore,

$$\text{Mach number } M = \frac{V_e}{C_{se}} = \frac{325}{548.6} = 0.592$$

(3) Dynamic pressure: (Refer to Eq. 3.1.1)

$$q = \frac{V_e^2}{296} = \frac{(325)^2}{296} = 357 \text{ lb/ft}^2$$

Example 2

Calculate and sketch the V-n diagram for maneuver and gust for a gross weight of 88 000 lbs at 10 000 ft altitude assuming that $V_c = 325$ KEAS and $V_D = 400$ KEAS. Use commercial airplane requirements. Obtain wing lift slope ($C_{L\alpha}$) from Ref. 3.5. Assume $C_{L_{max}}$ negative is equal to $C_{L_{max}}$ positive.

Refer to tabular form in Fig. 3.11.6:

Step ②: $M = \dfrac{V_e}{C_{se}}$; where $C_{se} = 548.6$ KEAS at $h = 10\,000$ ft from Fig. 3.1.5

①	②	③	④	⑤	⑥	⑦	⑧	⑨	⑩	⑪	⑫	⑬	⑭
V_e	M	q	$\dfrac{(C_{L\max})_M}{(C_{L\max})_0}$	$C_{L\max}$	$\eta_s + \text{all}$ $= \dfrac{q\, C_{L\max}}{W/S}$	$\dfrac{(C_{L\alpha})_M}{(C_{L\alpha})_0}$	$C_{L\alpha A}$ Per Radian	μ	K_g	$\dfrac{K_g V_e C_{L\alpha A}}{498\, W/S}$	Δn $U_{de} = 66$	Δn $U_{de} = 50$	Δn $U_{de} = 25$
	$\dfrac{①}{548.6}$	$\dfrac{①^2}{296}$	Ref. Fig. 3.11.4	1.3④	$\dfrac{③⑤}{73.2}$	Ref. Fig.3.11.5	5.22⑦	$\dfrac{198.5}{⑧}$	$\dfrac{0.88⑨}{5.3+⑨}$	$\dfrac{①⑩⑧}{36500}$	66⑪	50⑪	25⑪
50	0.091	8.4	0.995	1.290	0.148	1.005	5.250	37.8	0.771	0.006	0.396	0.300	0.150
100	0.182	33.8	0.990	1.288	0.593	1.010	5.260	37.7	0.771	0.011	0.725	0.550	0.275
150	0.274	76.2	0.980	1.275	1.327	1.020	5.320	37.3	0.770	0.017	1.120	0.850	0.425
200	0.364	135.2	0.970	1.260	2.330	1.030	5.370	37.0	0.770	0.0227	1.50	1.138	0.567
250	0.456	211.0	0.955	1.241	3.585	1.042	5.440	36.5	0.768	0.0286	1.89	1.430	0.715
300	0.547	304.0	0.925	1.202	5.0	1.075	5.600	35.5	0.765	0.0352		1.76	0.880
325	0.592	357.0				1.100	5.740	34.6	0.763	0.039		1.95	0.975
350	0.638	413.0				1.140	5.950	33.4	0.759	0.0433			1.082
400	0.730	541.0				1.300	6.780	29.3	0.745	0.0554			1.382
	←———— Maneuver ————→	←———————— Gust ————————→											

Fig. 3.11.6   V-n diagram calculation procedure.

Step ⑤: $C_{L\,max} = 1.3$ at $M = 0$, see given airplane data

Step ⑥: Wing loading $\dfrac{W}{S} = \dfrac{88\,000}{1200} = 73.2$ lb/ft^2

Step ⑧: See given airplane data

AR (or A) $= 8.0$, $\lambda = 0.4$, $\Lambda = 0$,

The compressible sweep parameter

$$\Lambda_\beta = \tan^{-1}\dfrac{\tan \Lambda}{\beta} = 0$$

Compressibility parameter at $M = 0$

$$\beta = \sqrt{1 - M^2} = 1.0$$

Assume $\kappa = 1.0$ for this case and then

$$\dfrac{\beta A}{\kappa} = 8.0,$$

from curves in Fig. 3.4.14 obtain the following values: (let $C_{La} = C_{Law}$).

$$\dfrac{\beta C_{La}}{\kappa} = 0.083 \text{ or } C_{La} = 0.083, \text{ for } \lambda = 0.25$$

$$\dfrac{\beta C_{La}}{\kappa} = 0.0875 \text{ or } L_{La} = 0.0875,$$
$$\text{for } \lambda = 0.50$$

Interpolated above values to $\lambda = 0.40$ obtain

or $\quad C_{Law} = 0.0826$/degree at $M = 0$
$\quad\quad C_{La\omega} = 4.75$/rad at $M = 0$

Step ⑨: Refer to Eq. 3.4.5

$$\mu = \dfrac{2\dfrac{W}{S}}{g\,\overline{C}_p \rho\, C_{LaA}}$$

$$\mu = \dfrac{2\,(73.2)}{(32.2)(13)(0.00176)(C_{LaA})}$$

$$\mu = \dfrac{198.5}{C_{LaA}}$$

Use commercial airplane requirements:

(1) Refer to Ref. 3.2 [FAR 25.337(b)], the positive limit maneuvering factor is as follows:

$$n = 2.1 + \left(\dfrac{24000}{W + 10000}\right)$$

$$= 2.1 + \left(\dfrac{24000}{88000 + 10000}\right)$$

$$= 2.1 + 0.203 = 2.303 < 2.5$$

therefore use $n = 2.5$

(2) Also refer to Ref. 3.2 [FAR 25.337(c)], the negative limit maneuvering factor is as follows:

$$n = -1.0$$

The V-n diagram plot is shown in Fig. 3.11.7.

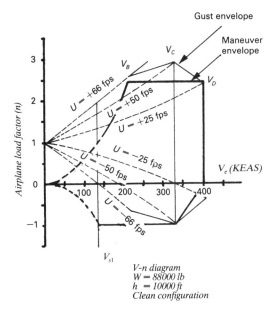

Fig. 3.11.7 V-n diagram.

Example 3

Estimate the loads on the vertical tail for a speed of 325 KEAS for a rudder kick maneuver. Solve for three points in time:
(1) Initiation of maneuver assuming instantaneous rudder deflection;
(2) Dynamic overswing holding constant rudder from initiation of maneuver (use a dynamic overswing factor of $K_\beta = 1.4$);
(3) Instantaneous return of rudder to neutral while at equilibrium yaw angle. Neglect the effect of yaw velocity, and make use of the curves (shown in Fig. 3.11.2) of C_{nA} (tail-on) and C_{nA-t} (tail-off) versus β, assuming that the difference between the airplane and airplane less tail is entirely due to load on the vertical tail and that the airload due to rudder deflection is entirely on the vertical tail.

From the given airplane data:

$C_{h_{\delta_\gamma}} = 0.01$/degree; $S_\gamma = 60$ ft^2; $\overline{C}_\gamma = 3.5$ ft

Rudder hinge moment system output = 7520 ft-lb

$$q = \dfrac{V_e^2}{296} = \dfrac{(325)^2}{296} = 357 \text{ lb/ft}^2$$

From Eq. 3.6.10, the hinge moment H.M. =

$$\delta_\gamma C_{h\delta_\gamma} S_\gamma \overline{C}_\gamma$$

and the obtainable maximum rudder rotation in degree is

$$\delta_\gamma = \dfrac{\text{H.M. (from system output)}}{C_{h\delta_\gamma} q S_\gamma \overline{C}_\gamma}$$

$$= \dfrac{7520}{(0.01)(357)(60)(3.5)} = 10°$$

Airframe Structural Design 77

From curve shown in Fig. 3.11.2, obtain the side slip angle value with $(C_{n\beta})_{vt} = 0$ (the equilibrium yaw angle from Fig. 3.11.2) as

$$\beta_0 = 5.1°$$

(at the intersection point where the tail-on curve with $\delta_\gamma = 10°$ goes through horizontal co-ordinate from Fig. 3.11.2)

and the maximum overswing angle as

$$K_\beta(\beta_0) = (1.4)(5.1°) = 7.15°$$

Case ①: *Initiation of maneuver*

From Eq. 3.6.15, for vertical tail load due to initiation deflected rudder,

$$P_{y_{vt}} = C_{n\delta\gamma} \, \delta_\gamma \, q \frac{S_{vt}}{\overline{V}_{vt}}$$

or $\quad P_{y_{vt}} = C_{n\delta\gamma} \, \delta_\gamma \, q \frac{(S)(b)}{\ell_{vt}}$

where

$$q\frac{(S)(b)}{\ell_{vt}} = \frac{(357)(1200)(98)}{45} = 931000$$

$C_{n\delta\gamma} \, \delta_\gamma = \Delta C_{n_{vt}}$ when $\beta = 0$ and $\delta_\gamma = 10°$

From curves in Fig. 3.11.2 obtain

$$\Delta C_{n_{vt}} = -0.0155$$

Therefore, the vertical tail load

$$P_{y_{vt}} = -0.0155 \,(931000) = -14400 \text{ lb}$$

(due to rudder rotate a $\delta_\gamma = 10°$ only)

Case ②: *Dynamic overswing*

From Eq. 3.6.16 for vertical tail dynamic overswing load as

$$P_{y_{vt}} = [(\Delta C_{n\beta})_{vt} K_\beta \beta_0 - C_{n\delta\gamma}\delta_\gamma] \, q \frac{S_{vt}}{\overline{V}_{vt}}$$

or $\quad P_{y_{vt}} = (\Delta C_{n_{vt}}) \, q \frac{(S)(b)}{\ell_{vt}}$

where $q \dfrac{(S)(b)}{\ell_{vt}} = 931000$

$$\Delta C_{n_{vt}} = 0.014$$

(when $K_\beta \beta_0 = 7.15°$ and $\delta_\gamma = 10°$ from curves shown in Fig. 3.11.2)

Therefore, the vertical tail load

$$P_{y_{vt}} = 0.014\,(931000) = 13000 \text{ lb}$$

[this is approximately equal to 1.4 (19500)* due to side slip (β) less 14400 lb (as calculated in Case ① above) due to rudder $\delta_\gamma = 10°$, a high torsion case]

*Vertical tail load from Case ③ below.

Case ③: *Rudder returned instantaneously to neutral*

From Eq. 3.6.14, for vertical tail load due to rudder returned instantaneously to neutral.

$$P_{y_{vt}} = (\Delta C_{n\beta})_{vt} \, \beta_0 \, q \frac{S_{vt}}{\overline{V}_{vt}}$$

or $\quad P_{y_{vt}} = (\Delta C_{n\beta})_{vt} \, \beta_0 \, q \frac{(S)(b)}{\ell_{vt}}$

where $q\dfrac{(S)(b)}{\ell_{vt}} = 931000$

and $(\Delta C_{n\beta})_{vt} \, \beta_0 = \Delta C_{n_{vt}}$ when $\beta_0 = 5.1°$ and $\delta_\gamma = 10°$ from curves shown in Fig. 3.11.2 obtain $\Delta C_{n_{vt}} = 0.021$

Therefore, the vertical tail load

$$P_{y_{vt}} = 0.021\,(931000) = 19500 \text{ lb}$$

[this load due to side slip (β) only]

Example 4

Solve for the main gear load for a 10 ft/sec sinking speed with assuming a strut efficiency of 85 percent and the airplane lift equal to airplane weight (i.e. $L = W$). What is the airplane inertia load factor at the instant that this load is applied for an airplane weight of 88000 lb?

From airplane data:
 sinking speed, $V_S = 10$ ft/sec
 $W = L = 88000$ lb
 $X_s = 18$ in or 1.5 ft
 $X_t = 4.5$ in or 0.375 ft
 $K_t = 160000$ lb/ft
 $\eta_s = 0.85$
 $\eta_t = 0.8$

From Eq. 3.9.4

$$n_{LG} = \frac{\dfrac{V_s^2}{2g} - \dfrac{\eta_t K_t X_t^2}{2W} + \left(1 - \dfrac{L}{W}\right)(X_t + X_s)}{\eta_s X_s}$$

$$= \frac{\dfrac{10^2}{2(32.2)} - \dfrac{(0.8)(160000)(0.375)^2}{2(88000)} + 0}{(0.85)(1.5)}$$

$$= 1.138$$

The inertia load factor on the airplane is

$$n_z = n_{LG} + \frac{L}{W} = 1.138 + \frac{88000}{88000} = 2.138$$

Example 5

Estimate the balancing horizontal tail load for a steady maneuver at $n = 2.5$, gross weight $= 88000$ lb, c.g. @ 12% MAC, $V = 400$ KEAS, $M = 0.95$. Make use of the pitch curve of airplane-less-tail. (Assume fuselage lift and moment are equal to zero for this case, see Fig. 3.11.8).

Aerodynamic pressure

$$q = \frac{V_e^2}{296} = \frac{(400)^2}{296} = 540 \text{ lb/ft}^2$$

The distance between the 25% MAC and c.g. at 12% MAC

Wing, $\overline{C} = 13$ ft (from airplane data)

Therefore, $\Delta X = (25\% - 12\%)\,\overline{C}$
$\qquad\qquad = (13\%) \times 13 = 1.69$ ft

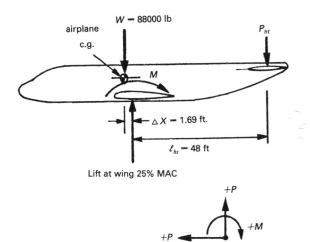

Fig. 3.11.8 Airplane pitch loads.

The airplane lift coefficient calculation is
$$C_{LA} = \frac{nW}{qS} = \frac{(2.5)(88000)}{(540)(1200)} = 0.34$$

with $C_{LA} = 0.34$ and from curve of $M = 0.95$ (Fig. 3.11.3) obtain
$$C_{m_{0.25}} = -0.082$$

The wing pitch moment at 25% MAC is
$$M = (C_{m_{0.25}})(q)(S)(\bar{C})$$
$$= (-0.082)(540)(1200)(13)$$
$$= -690768 \text{ ft-lb}$$

Take moment center at wing 25% MAC, then
$$-(nW)(\Delta X) + M + P_{ht}(\ell_{ht}) = 0$$
$$-(2.5 \times 88000)(1.69) - 690768 + P_{ht}(48) = 0$$
$$P_{ht} = 22137 \text{ lb (balancing horizontal tail load)}$$

Example 6

Use Ref. 3.5 to estimate the "additional lift" on the wing, i.e. $\frac{C_{\ell a} C}{C_L C_{av}}$ versus span, neglecting fuselage effects and considering the wing leading and trailing edge loads carried into centerline.

Considering a strip (Δy) on the wing

The average chord length $C_{av} = \frac{S}{b}$

(see Fig. 3.11.9)

and $\int_{\ell}^{tip} C \, dy = \frac{S}{2}$

let $\Delta \eta = \frac{y}{\frac{b}{2}}$

Lift on half wing $= C_L q \left(\frac{S}{2}\right) = q \int_{\ell}^{tip} C_{\ell a} C \, dy$ \quad (3.11.1)

$$= q \left(\frac{b}{2}\right) \int_{\ell}^{tip} C_{\ell a} C \, d\eta \quad (3.11.2)$$

Dividing both Eq. 3.11.1 and Eq. 3.11.2 by $C_L C_{av}$

$$\frac{C_L q \left(\frac{S}{2}\right)}{C_L C_{av}} = \frac{q \left(\frac{b}{2}\right) \int_{\ell}^{tip} C_{\ell a} C \, d\eta}{C_L C_{av}}$$

Therefore $\int_{\ell}^{tip} \frac{C_{\ell a} C}{C_L C_{av}} d\eta = 1.0$

From curves in Fig. 3.4.13 obtain $\frac{C_{\ell a} C}{C_L C_{av}}$ values as tabulated in Fig. 3.11.10 and the additional calculations are tabulated in Fig. 3.11.11.

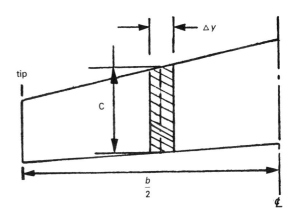

Fig. 3.11.9 Average wing chord length.

A	B	C	D
η $\frac{y}{(b/2)}$	$\frac{C_{\ell a} C}{C_L C_{av}}$ ($\lambda = 0.25$)	$\frac{C_{\ell a} C}{C_L C_{av}}$ ($\lambda = 0.50$)	$\frac{C_{\ell a} C}{C_L C_{av}}$ ($\lambda = 0.4$)*
0	1.400	1.305	1.343
0.383	1.185	1.150	1.164
0.707	0.830	0.890	0.866
0.921	0.460	0.540	0.508
1.000	0.0	0.0	0.0

*Interpolate from B, C.

Note: The above values in Colume D have been plotted in Fig. 3.11.12.

Fig. 3.11.10 Tabulation of $\frac{C_{\ell a} C}{C_L C_{av}}$ values.

n	η	$\dfrac{\Delta\eta}{2}$	$\dfrac{C_{\ell a} C}{C_L C_{av}}$	$\dfrac{S_z}{C_L q\left(\dfrac{S}{2}\right)}$	$\dfrac{M_x}{C_L q\left(\dfrac{S}{2}\right)\left(\dfrac{b}{2}\right)}$
	①	②	③	④	⑤
				↑ ∫ ③ $d\eta$	↑ ∫ ④ $d\eta$
1	0		1.343	0.993*	0.4128
		0.05			
2	0.1		1.325	0.86	0.3202
		0.05			
3	0.2		1.28	0.73	0.2407
		0.05			
4	0.3		1.223	0.604	0.174
		0.05			
5	0.4		1.155	0.485	0.1195
		0.05			
6	0.5		1.072	0.374	0.0766
		0.05			
7	0.6		0.98	0.272	0.0443
		0.05			
8	0.7		0.877	0.179	0.0217
		0.05			
9	0.8		0.738	0.0979	0.0079
		0.05			
10	0.9		0.552	0.0334	0.0013
		0.025			
11	0.95		0.392	0.0098	0.0002
		0.025			
12	1.0		0	0	

(See plots in Fig. 3.11.12)

Step ④:

$$\frac{S_z}{C_L q\dfrac{S}{2}} = \int_{n=0}^{n=1.0} \frac{C_{\ell a} C}{C_L C_{av}} d\eta \approx \sum \left[③_n + ③_{n-1}\right]\left[\frac{\Delta\eta}{2}\right]$$

(approximated by trapezoidal integration)

* $\dfrac{S_z}{C_L q\dfrac{S}{2}}$ should equal to 1.0. Good practice would be to refair $\dfrac{C_{\ell a} C}{C_L C_{av}}$ slightly to converge on 1.0 at $\eta = 0$

Step ⑤:

$$\frac{M_x}{C_L q\left(\dfrac{S}{2}\right)\left(\dfrac{b}{2}\right)} = \int_{n=0}^{n=1.0} \frac{S_z}{C_L q\dfrac{S}{2}} d\eta \approx \sum \left[④_n + ④_{n-1}\right]\left[\frac{\Delta\eta}{2}\right]$$

Fig. 3.11.11 Tabulation of $\dfrac{C_{\ell a} C}{C_L C_{av}}$, $\dfrac{S_z}{C_L q\left(\dfrac{S}{2}\right)}$, $\dfrac{M_x}{C_L q\left(\dfrac{S}{2}\right)\left(\dfrac{b}{2}\right)}$

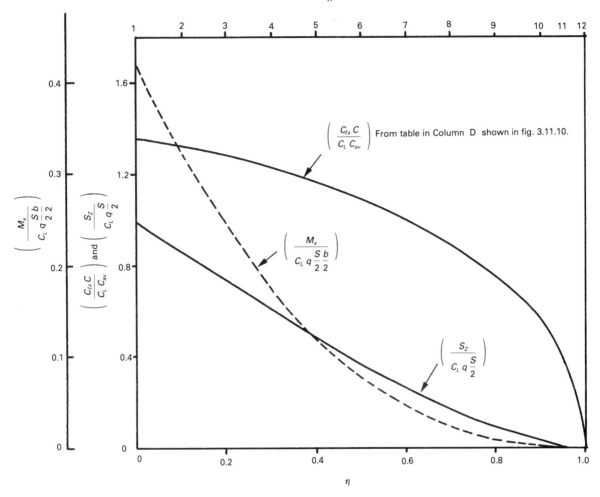

Fig. 3.11.12   *Spanwise additional lift distribution curves.*

Example 7

From the given airplane data plus additional given wing data as shown in Fig. 3.11.13–Fig. 3.11.16, complete the wing loads and plot the net shears, bending moments and torsions versus wing span (η).

This problem is meant to provide a numerical example of the concepts involved in calculating wing loads. The technique illustrated is one of many. The problem is considerably simplified in order to keep the steps to a minimum. The following items should be noted:

(1) The wing information in the airplane data sheets should be used.

(2) The element of wing perpendicular to the airplane centerline is assumed to be the 40 percent wing chord. The local wing *centers of gravity* are assumed at the 40 percent wing chord with the exception of the engine. (the plot in Fig. 3.11.16 reflects this assumption)

(3) The lift distribution on the airplane is assumed to be (wind tunnel data),
+ 0.05 on Forebody
+ 0.05 on Wing propellers
+ 0.00 on Aftbody
− 0.15 on Horizontal tail
+ 1.05 on Wing including the wing area blanketed by the fuselage
$\Sigma = 1.00$ on Total Airplane

(4) The gross weight equals 88000 lb and the maneuver is a steady 2.5g pull-up. (θ effects are ignored). Thus,
$L_A = 2.5 \, (88000) = 220000$ lb
$L_W = 1.05 \, (220000) = 231000$ lb

(5) Use the values of $\dfrac{C_{t_a} C}{C_L C_{av}}$ derived from Example 6.

(6) The value of $\dfrac{C_{t_a} C}{C_{av}}$ is typical of the section properties of a straight wing. It reflects the effects of aerodynamic and geometric section zero lift angles (as discussed in Section 3.5). The method

Airframe Structural Design 81

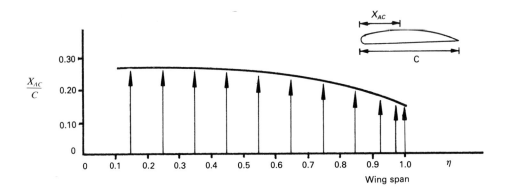

Fig. 3.11.13 Given aerodynamic center.

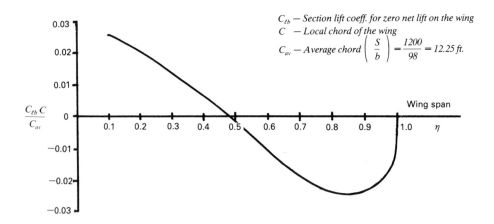

Fig. 3.11.14 Given basic lift.

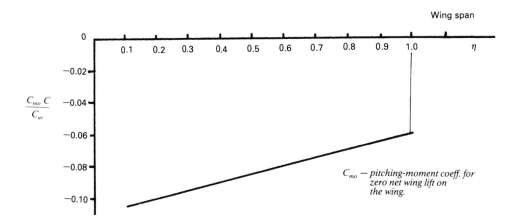

Fig. 3.11.15 Given basic pitching-moment.

of Ref. 3.6 could be used to obtain this type of distribution.

(7) Chordwise (fore & aft) loads effects on torsion are neglected.

(8) Flexible effects are also neglected to simplify the problem.

(9) The speed is 405 knots, then

$$q = \frac{(405)^2}{295} = 556 \text{ lb/ft}^2$$

$$q\frac{S}{2} = 556 \left(\frac{1200}{2}\right) = 33400 \text{ lb}$$

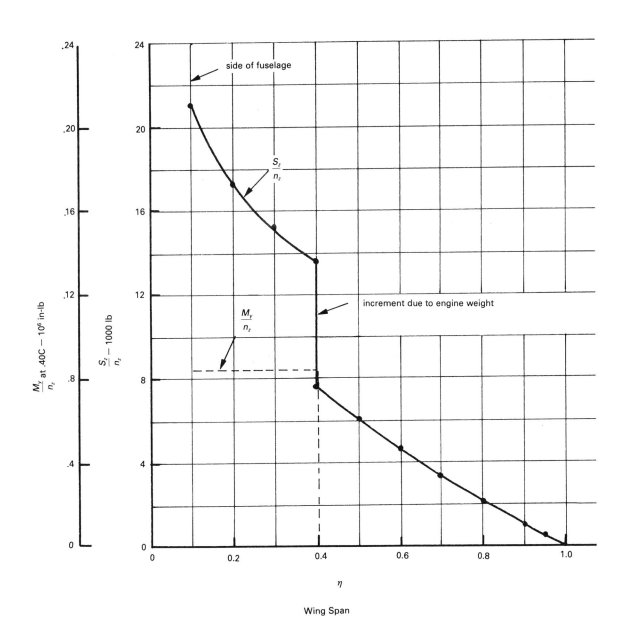

Fig. 3.11.16 Given wing inertia data.

①	②	③	④	⑤	⑥	⑦	Additional Airload		
							⑧	⑨	⑩
n	η	$d\eta = \dfrac{\Delta \eta}{2}$	y (inches)	$\dfrac{\Delta y}{2}$	c (inches)	c Mid. strip (inches)	$\dfrac{C_{\ell a} C}{C_L C_{av}}$	$\dfrac{S_z}{C_L q \dfrac{S}{2}}$	S_z Additional airload
		$\dfrac{②_{n+1} - ②_n}{2}$	$\left[\dfrac{b}{2}\right][\eta]$ $588 \times ②$	$\dfrac{④_{n+1} - ④_n}{2}$	$C_R - (C_R - C_T)\eta$ $210 - 126 \times ②$	$\dfrac{C_n + C_{n-1}}{2}$ $\dfrac{⑥_n + ⑥_{n+1}}{2}$	From Fig. 3.11.11	$\int \dfrac{C_{\ell a} C}{C_L C_{av}} d\eta$ ↑∫ ⑧ $d\eta$	$\dfrac{L_w}{2}$ ⑨ 115500 ⑨
2	0.1	0.05	58.8	29.4	197.4	191.1	1.325	0.86	99330
3	0.2	0.05	117.6	29.4	184.8	178.5	1.28	0.73	84320
4	0.3	0.05	176.4	29.4	172.2	165.9	1.223	0.604	69760
5	0.4⁻	0	235.2	0	159.6	159.6	1.155	0.485	56020
5	0.4⁺	0.05	235.2	29.4	159.6	153.3	1.155	0.485	56020
6	0.5	0.05	294.0	29.4	147.0	140.7	1.072	0.374	43200
7	0.6	0.05	352.8	29.4	134.4	128.1	0.98	0.272	31420
8	0.7	0.05	411.6	29.4	121.8	115.5	0.877	0.179	20670
9	0.8	0.05	470.4	29.4	109.2	102.9	0.738	0.0979	11310
10	0.9	0.025	529.2	14.7	96.6	93.45	0.552	0.0334	3860
11	0.95	0.025	558.6	14.7	90.3	87.15	0.392	0.0098	1130
12	1.0		588.0		84.0		0	0	0

①	Basic Airload			⑭	Airload-Torsion due to Additional and Basic Loads				Inertia Loads			
	⑪	⑫	⑬		⑮	⑯	⑰	⑱	⑲	⑳	㉑	㉒
n	$\dfrac{C_{\ell b} C}{C_{av}}$	$\dfrac{S_z}{q \dfrac{S}{2}}$	S_z Basic airload	S_z additional + basic load ⑩ + ⑬	S_z ⑭$_n$ − ⑭$_{n+1}$	$\dfrac{X_{AC}}{C}$	ΔX $(0.40 - ⑯)$ $\times ⑦$	M_y additional + basic load ↑Σ ⑮ × ⑰	$\dfrac{S_z}{n_z}$	$\dfrac{M_y}{n_z}$	S_z Inertia −2.5 ⑲	M_y Inertia −2.5 ⑳
	Fig. 3.11.14	$\int \dfrac{C_{\ell a} C}{C_L C_{av}} d\eta$ ↑∫ ⑪ $d\eta$	$q \dfrac{S}{2}$ ⑫ 334000 ⑫			From Fig. 3.11.13			From Fig. 3.11.16			
2	0.025	−0.00325	−1086	98244	15771	.261	26.5	2337639	21000	.84 × 10⁶	−52400	−2.1 × 10⁶
3	0.0205	−0.00553	−1847	82473	14804	.260	25	1919707	17100	.84 × 10⁶	−42700	−2.1 × 10⁶
4	0.014	−0.00626	−2091	67669	14074	.257	23.7	1549607	15100	.84 × 10⁶	−37700	−2.1 × 10⁶
5	0.006	−0.00726	−2425	53595	200	.255	23.2	1216053	13500	.84 × 10⁶	−33700	−2.1 × 10⁶
5	0.006	−0.00786	−2625	53395	12887	.248	23.3	1211413	7500	0	−18750	0
6	−0.002	−0.00806	−2692	40508	11580	.240	22.5	911146	6000	0	−15000	0
7	−0.1	−0.00746	−2492	28928	10289	.225	22.5	650596	4700	0	−11750	0
8	−0.0175	−0.00608	−2031	18639	8675	.205	22.5	419093	3400	0	−8480	0
9	−0.0235	−0.00403	−1346	9964	6648	.182	22.5	223905	2100	0	−5240	0
10	−0.024	−0.00163	−544	3316	2363	.160	22.5	74325	1000	0	−2500	0
11	−0.021	−0.00053	−177	953	953	.145	22.2	21157	500	0	−1250	0
12	0	0	0	0					0	0	0	0

Note: $\dfrac{b}{2} = 49$ ft or 588 in

$C_R = 17.5$ ft or 210 in
$C_T = 7.0$ ft or 84 in
$C_R - C_T = 210 - 84 = 126$ in
$C_{av} = \left(\dfrac{C_R + C_T}{2}\right) = 147$ in

Δ Propeller load (including power plant weight),
$\dfrac{0.05 (220000)}{2} = 5500$ lb/side

and assume the above load is located 200 in forward of 0.40C
Where $M_y = 5500 \times 200 = 110000$ in-lb

Fig. 3.11.17 *Tabulation of wing loads.*

①	Propellor air loads △		Torsion load due to C_{mo}			Net wing loads plotted vs. wing span (η)		
	㉓	㉔	㉕	㉖	㉗	㉘	㉙	㉚
n	S_z Propeller air load	M_y Propeller air load	$\dfrac{C_{mo} C}{C_{av}}$ From Fig. 3.11.15	$\dfrac{dM_y}{q\dfrac{S}{2}}$ △ ↑∫㉕⑥$d\eta$	$M_y (10^6)$ due to C_{mo} $334000 \times$ ㉖ (in-lb)	S_z (net) ⑭+㉑+㉓ (lb)	$M_x (10^6)$ (net) ↑∫㉘ dy (in-lb)	$M_y (10^6)$ (net) ⑱+㉒ +㉔+㉗ (in-lb)
2	5500	1100000	−.105			51324	11.434	
3	5500	1100000	−.10	−10.847	−3.623	45273	8.594	−2.286
4	5500	1100000	−.095	− 8.886	−2.986	35469	6.219	−2.05
5	5500	1100000	−.089	− 7.144	−2.386	25395	4.43	−1.836
5	0	0	−.089	0	0	34645	4.43	
6	0	0	−.084	− 5.616	−1.876	25508	2.662	−0.665
7	0	0	−.08	− 4.288	−1.432	17178	1.407	−0.521
8	0	0	−.076	− 3.133	−1.046	10159	0.604	−0.386
9	0	0	−.072	− 2.132	−0.712	4724	0.166	−0.293
10	0	0	−.066	− 1.282	−0.428	816	0.0033	−0.204
11	0	0	−.063	− 0.57	−0.19	−297	−0.004	−0.116
12	0	0	−.06	− 0.268	−0.09	0	0	−0.069

△ $dM_y = C_{mo} C \, dy \, C \, q$

$\quad = \left(\dfrac{C_{mo} C}{C_{av}}\right) \left(\dfrac{dy}{\dfrac{b}{2}}\right) (C) \left(\dfrac{S}{2}\right) (q)$

$\quad = \left(\dfrac{C_{mo} C}{C_{av}}\right) (C) \left(d\eta\right) \left(\dfrac{S}{2}\right) (q)$

therefore $\dfrac{dM_y}{q\dfrac{S}{2}} = \left(\dfrac{C_{mo} C}{C_{av}}\right) (C) (d\eta)$

(Fig. 3.11.17 continued)

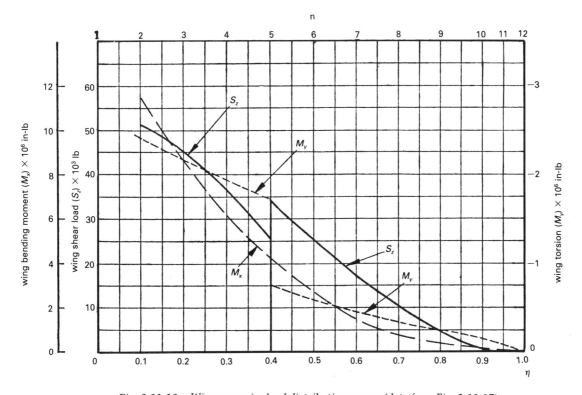

Fig. 3.11.18   Wing spanwise load distribution curves (data from Fig. 3.11.17).

Appendix: Notations

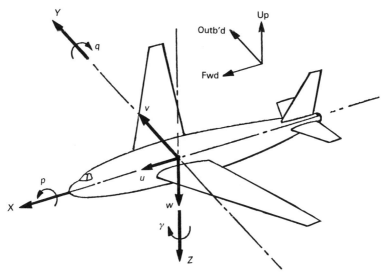

u, v, w — airplane velocity along the X, Y, Z axes, respectively.
p, q, γ — airplane rotational velocities about the X, Y, Z axes, respectively.

(a) Airplane coordinates and motion

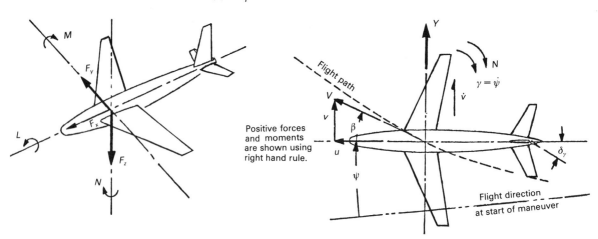

(b) Airplane forces and moments

(c) Airplane yawing motion

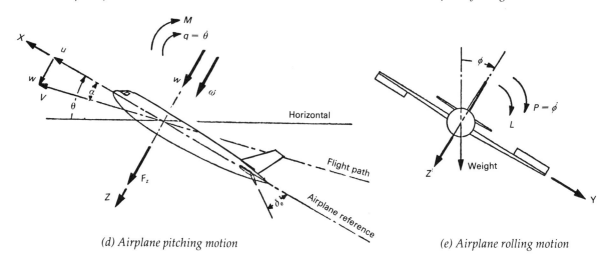

(d) Airplane pitching motion

(e) Airplane rolling motion

86 Airframe Structural Design

A	cross-section area
A, AR	aspect ratio
b	span
C_D	drag coeff.
C_h	hinge moment coeff.
C_L	lift coeff.
C_{Lt}	tail lift coeff.
C_{La}	lift curve slope
C_l	roll moment coeff.
C_ℓ	airfoil section lift coeff.
C_m	pitching moment coeff.
C_{mom}	moment coeff.
C_N	normal force coeff.
C_n	yaw moment coeff.
C_p	pressure coeff.
$\bar{C}$	mean aerodynamic chord
E	modulus of elasticity
G	modulus of rigidity
g	acceleration of gravity
I	moment of inertia
k	ratio of specific heats
L	lift
L	roll moment
M	pitching moment
M	Mach number
N	yaw moment
n	load factor
p	airplane rotational velocity about X axis
p	local pressure
p_v	vertical load
q	rotational velocity about the Y axis
$q = \frac{1}{2}\rho V^2$	dynamic pressure
γ	rotational velocity about the Z axis
S	wing surface area
V	airspeed
V_s	sinking speed
V_s	stall speed
V_c	design cruising speed
V_D	design dive speed
V_e	equivalent airspeed
V_F	design flap speed
u	velocity along the X axis
v	velocity along the Y axis
w	velocity along the Z axis
W	weight
α	the angle of attack
β	the angle of side slip
β	compressibility parameter
θ	angle of pitch
Λ	sweep angle
ρ	mass density
ρ_{alt}	air density at the altitude being considered
$\rho_{S.L.} = \rho_o$	air density at sea level
ϕ	angle of roll
ψ	angle of yaw
$\frac{d\epsilon}{d\alpha}$	change in wing downwash angle with angle of attack
$\frac{d\alpha_{ht}}{d\delta_e}$	change in horizontal tail angle of attack with change elevator angle
ℓ_{ht}	horizontal tail length
δ_e	elevator deflection (degree)
δ_γ	rudder deflection (degree)
η_{ht}	efficiency of the horizontal tail
η_{vt}	efficiency of the vertical tail
S_{ht}	horizontal tail area
S_{vt}	vertical tail area
P_{yvt}	vertical tail load
$C_{N\beta}$	yawing moment coeff. per unit side slip
$H.M.$	hinge moment
C_n	yawing moment
β_0	equilibrium yaw angle
$C_{n\delta\gamma}$	slope of C_n vs. δ_γ curve
$C_{n\beta}$	slope of C_n vs. β curve for complete airplane
$(C_{N\beta})_{vt}$	slope of the tail lift curve
$\bar{V}_{vt}$	vertical tail volume
$\bar{V}_{ht}$	horizontal tail volume
K_β	magnification of slide slip due to rudder overswing
$C.G., c.g.$	center of gravity
$C.P., c.p.$	center of pressure
MAC or mac	mean aerodynamic chord
m	mass
$\triangle p$	pressure increment
V_g	gust velocity
V_A	maneuvering speed
K_g	gust alleviated factor
βA	aspect ratio parameter
κ	
λ	taper ratio
C_R	chord length at wing root
C_T	chord length at the tip
$C_{\ell a}$	section lift coeff. (additonal lift)
$C_{\ell b}$	section lift coeff. (zero lift on wing)
$C_{L_{A-t}}$	wing lift coeff. at tail-off
C_{L_A}	lift coeff. on entire airplane
P_{z_t}	horizontal tail load
$\ddot{\theta}$	pitching angular acceleration
$\dot{\theta}$	angular velocity in pitch (radian per second)
X_t	tire deflection
X_s	strut stroke length
K_t	tire spring constant
n_{LG}	landing gear load factor
LL_1	level landing, one wheel
LL_2	level landing, two wheels
LL_3	level landing, three wheels
DL	draft landing
V_t	true airspeed
V_{sn}	velocity along stall line of a V-n diagram
$C_{L\,max}$	max. lift coeff.
U_{de}	gust velocity
C	chord length
M_c	Mach No. at cruise speed
M_D	Mach No. at dive speed
M_D	divergence Mach number
M_{CR}	critical Mach number
σ	density ratio

Subscript:

ht	horizontal tail
vt	vertical tail
γ	rudder
h	hinge moment
A	airplane
$A-t$	at tail-off
a	aileron
t	tail
e	elevator

Airframe Structural Design

References

3.1 *Federal Aviation Regulations (FAR), Vol. III, Part 23 — Airworthiness Standards: Normal, Utility, and Aerobatic Category Airplanes.*

3.2 *Federal Aviation Regulations (FAR), Vol. III, Part 25 — Airworthiness Standards: Transport Category.*

3.3 *British Civil Airworthiness Requirements, Section D — aeroplanes.*

3.4 *Mil-A-8860 (ASG) — General specification for airplane strength and rigidity.*

3.5 DeYoung, J. and Harper, C.W.: 'Theoretical Symmetric Span Loading at Subsonic Speeds for Wings Having Arbitrary plan form.' *NACA Rep. No. 921,* (1948).

3.6 Sivells, J. C.: 'An Improved Approximate Method for Calculating Lift Distributions due to Twist.' *NACA TN 2282,* (1951).

3.7 DeYoung, J.: 'Theoretical Antisymmetric Span Loading for Wings of Arbitrary Plan Form at Subsonic Speeds.' *NACA Rep. No. 1056,* (1951).

3.8 DeYoung, J.: 'Theoretical Symmetric Span Loading due to Flap Deflection for Wings of Arbitrary Plan Form at Subsonic Speeds.' *NACA Rep. No. 1071,* (1952).

3.9 Weissinger, J.: 'The lift Distribution of Swept-back Wings.' *NACA TM No. 1120,* (1947).

3.10 Falkner, V.M.: 'The Calculation of Aerodynamic Loading on Surfaces of any Shape.' *ARC Rep. R&M No. 1910,* (1943).

3.11 Multhopp, H.: 'Methods for Calculating the lift Distribution of Wings (Subsonic lifting-surface theory).' *ARC Rep. R&M No. 2884,* (1955).

3.12 Watkins, C.E., Woolston, D. S. and Cummingham, H.J.: 'A Systematic Kernel Function Procedure for Determining Aerodynamic Forces on Oscillating or Steady Finite Wings at Subsonic Speeds.' *NASA TR R-48,* (1959).

3.13 Bonney, E.A.: *Engineering Supersonic Aerodynamics.* McGraw-Hill Book Company, Inc., New York, 1950.

3.14 Alford Jr., W.J.: 'Theortical and Experimental Investigation of the Subsonic Flow Fields Beneath Swept and Unswept Wings with Tables of Vortex-induced Velocities.' *NACA Rep. No. 1327.*

3.15 Zartarian, G. and Hsu, P.T.: 'Theoretical Studies on the Prediction of Unsteady Supersonic Airloads on Elastic Wings.' *WADC TR 56—97 Part I,* (1955) and *Part II,* (1956).

3.16 Liepmann, H.W. and Roshko, A.: *Elements of Gas Dynamics.* John Wiley & Sons, Inc., New York, 1957.

3.17 Nielsen, J.N.: *Missile Aerodyamics.* McGraw-Hill Book Company, Inc., New York, 1960.

3.18 Zartarian, G., Heller, A. and Ashley, H.: 'Application of Piston theory to Certain Elementary Aeroelastic Problems.' *Proceedings of the Midwestern Conference on Fluid Mechanics,* Purdue University, West Lafayette, Indiana, U. S. A. 1955.

3.19 Evvard, J.: 'Use of Source Distributions for Evaluating Theoretical Aerodynamics of Thin Finite Wings at Supersonic Speeds.' *NACA TR 951,* (1950).

3.20 Bisplinghoff, R.L., Ashley, H. and Halfman, R.L.: *Aeroelasticity.* Addison-Wesley Publishing Company, Inc., Reading, Massachusetts, 1957.

3.21 Williams, D.: *An Introduction to the Theory of Aircraft Structures.* Edward Arnold (publishers), Ltd., London. 1960.

3.22 Polhamus, E.C.: 'A Simple Method of Estimating the Subsonic lift and Damping in Roll of Sweptback Wings.' *NACA TN No. 1862.*

3.23 Gray, W.L. and Schenk, K.M.: 'A Method for Calculating The Subsonic Steady-State Loading on an Airplane with a Wing of Arbitrary Plan form and Stiffness.' *NACA TN No. 3030,* (1953).

3.24 DeYoung, J.: 'Theoretical Additional Span Loading Characteristics of Wings with Arbitrary Sweep, Aspect Ratio, and Taper Ratio.' *NACA TN No. 1491,* (1947).

3.25 Stevens, V.I.: 'Theoretical Basic Span Loading Characteristics of Wings with Arbitrary Sweep, Aspect Ratio, and Taper Ratio.' *NACA TN No. 1772,* (1948).

3.26 Houbolt, J.C. and Kordes, E.E.: 'Gust-Response Analysis of an Airplane Including Wing Bending Flexibility.' *NACA TN No. 2763,* (1952).

3.27 Biot, M.A. and Bisplinghoff, R.L.: 'Dynamic Loads on Airplane Structures during Landing.' *NACA ARR No. 4H 10,* (1944).

3.28 Flugge, W.: Landing-Gear Impact. *NACA TN No. 2743,* (1952).

3.29 Flugge, W. and Coale, C.W.: 'The Influence of Wheel Spin-up on Landing-Gear Impact.' *NACA TN No. 3217,* (1954).

3.30 Milwitzky, B and Cook, F.E.: 'Analysis of Landing-gear Behavior.' *NACA Rep. No. 1154,* (1953).

3.31 Gudkov, A.I. and Leshakow, P.S.: 'External Loads and Aircraft Strength.' *NASA TT F-753,* (July, 1973).

3.32 Gudkov, A.I. and Leshakov, P.S.: 'External Loads and the Strength of Flight Vehicles.' *FTD-MT-24-64-70,* (Sept., 1970).

3.33 Perkins, C.D. and Hage, R.E.: *Airplane Performance Stability and Control.* John Wiley & Sons, Inc., New York, 1949.

3.34 Abbott, I.H. and Doenhoff, A.E.: *Theory of Wing Sections.* Dover Publications, Inc., New York, 1959.

3.35 Abramson, H.N.: *An Introduction to the Dynamics of Airplanes.* Dover Publications, Inc., New York, 1958.

3.36 Abbott, I.H., Doenhoff, A.E. and Stivers, L.S.: 'Summary of Airfoil Data.' *NACA Rep. No. 824,* (1945).

3.37 Loftin Jr., L.K.: 'Theoretical and Experimental Data for a Number of NACA 6A-Series Airfoil Sections.' *NACA Rep. No. 903,* (1948).

3.38 Kelley, Jr., J. and Missall, J.W.: 'Maneuvering Horizontal Tail Loads.' *A.A.F.T.R. 5185,* (1945).

3.39 Roskam, J.: *Flight Dynamics of Rigid and Elastic Airplanes.* The University of Kansas, Lawrence, Kansas, U.S.A., 1973.

3.40 Roskam, J.: *Methods for Estimating Stability and Control Derivatives of Conventional Subsonic Airplanes.* The University of Kansas, Lawrence, Kansas, U.S.A., 1973.

3.41 Budiansky, B., Kotanchik, J.N. and Chiarito, P.T.: 'A Torsional Stiffness Criterion for Preventing Elutter of Wings of Supersonic Missiles.' *NACA RM L7G02,* (1947).

3.42 Donely, P.: 'Summary of Information Relating to Gust Loads on Airplanes.' *NACA Rep. No. 997,* (1950).

3.43 Hoblit, F.M., Paul, N., Shelton, J.D. and Ashford, F.E.: 'Development of a Power-Spectral Gust Design Procedure for Civil Aircraft.' *FAA-ADS-53,* (1966).

3.44 Paulson Jr., J.W.: 'Wind-Tunnel Investigation of a Fowler Flap and Spoiler for an Advanced Gereral Aviation Wing.' *NASA-TN-D-8236,* (1976).

3.45 Pratt, K.G.: 'A Revised Gust-load Formula and a Re-evaluation of V-G Data Taken on Civil Transport Airplanes From 1033 to 1050.' *NACA Report No. 1206,* (1954).

3.46 Clousing, L.A.: 'Research Studies Directed Toward the Development of Rational Vertical-Tail-Load Criteria.' *J. of the Aeronautical Sciences,* (Mar. 1947), 175—182.

3.47 Sibert, H.W.: 'Aerodynamic Center and Center of Pressure of an Airfoil at Supersonic Speeds.' *J. of the Aeronautical Sciences,* (Mar. 1947), 183—184.

3.48 Anon.: *NACA Conference on Aircraft Loads.* (A compilation of the papers presented), (Feb. 18—19, 1948).

3.49 Pope, A. and Harper, J.J.: *Low-Speed Wind Tunnel*

Testing. John Wiley & Sons, Inc., New York. 1966.
3.50 Pope, A. and Goin, L.K.: *High-Speed Wind Tunnel Testing.* John Wiley & Sons, Inc., New York. 1965.
3.51 Anon.: 'Wind Tunnel Design and Testing Techniques.' *AGARD-CP-174,* (March 1976).
3.52 Anon.: *Characteristics of Nine Research Wind Tunnels of the Langley Aeronautical Laboratory. NACA publication,* 1957.
3.53 Anon.: *Engineering Sciences Data Unit (Aerodynamic Section).* 251–259 Regent Street, London W1R 7AD, England.
3.54 Anon.: *USAF Stability and Control DATCOM.* Flight Control Division, Air Force Flight Dynamics Laboratory, Wright-Patterson Air Force Base, Ohio, U.S.A.
3.55. Taylor, J.: *Manual on Aircraft Loads.* Pergamon Press Inc., New York, 1965.
3.56 Rossman, E.: *Basic Loads Notes.* Unpublished.
3.57 Etkin, B.: *Dynamics of Flight — Stability and Control.* John Wiley & Sons, Inc., New York, N.Y., 4th Edition. 1965.
3.58 Storey R.E.: 'Dynamic Loads and Airplane Structural Weight.' *SAWE Paper No. 1153,* (1977).
3.59 Anon.: 'Combat Loads on Modern Fighters.' *AGARD-R-730,* (1986).
3.60 Anon.: *MIL-A-8861 (ASG) — Airplane Strength and Rigidity, Flight Loads.*
3.61. Anon.: *MIL-A-008861 (USAF) — Airplane Strength and Rigidity, Flight Loads.*
3.62. Hoerner, S.F. and Borst, H.V.: *Fluid-Dynamic Lift.* P.O. Box 342, Brick Town, N.J. 08723.

CHAPTER 4.0

MATERIALS

4.1 Introduction

As airframe design concepts and technology have become more sophisticated, materials' requirements have accordingly become more demanding. The steps from wood to aluminum, and then to titanium and other efficient high strength materials has involved some very extensive development activities and the application of a wide range of disciplines.

Structure weight and therefore the use of light materials has always been important. When a modern full-loaded subsonic transport takes off, only about 20% of its total weight is payload. Of the remaining 80%, roughly half is aircraft empty weight and the other half is fuel. Hence, any saving of structural weight can lead to a corresponding increase in payload. Alternatively, for a given payload, saving in aircraft weight means reduced power requirements. Therefore, it is not surprising that the aircraft manufacturer is prepared to invest heavily in weight reduction.

Airframe designers still demand strong, stiff materials at an acceptable weight and cost. Although significant parts of the airframe are designed on stability considerations, i.e., buckling and therefore a high elastic modulus rather than strength, but high static strength is always sought.

The fact that all materials deteriorate in service has to be taken into account in design and in scheduled maintenance over the expected life, typically 6000–8000 hours for military fighters and over 60,000 hours for transport aircraft. The designer, therefore, requires data on time-dependent properties of materials prior to selection — not only strength and endurance but fatigue crack propagation and fracture toughness, creep data for high speed airframe aircraft materials, and information of resistance to corrosion and stress corrosion cracking.

Aluminum, steel and titanium will probably continue to be the principal materials in airframe design for a while. All these materials, however, will have to be improved. Work is being done to improve the ability of aluminum alloys to resist corrosion and fracture. One of the alloys of major interest is current aluminum-lithium. Its improved properties are based on its lower density. Other aluminum alloys, i.e., aluminum-iron-molybdenum-zirconium, function well enough at high temperature to be competitive with titanium up to near 600°F.

Another material that is having a major effect on aircraft design is the rapidly growing percentage of parts made of composites such as graphite/epoxy materials (thermoset resins). A new trend is toward thermoplastic composites that can be rendered malleable by heat (temperature up to 700°F) and becomes strong and tough after cooling. One of its strongest advantages over thermoset resins is its shorter fabrication cycle, because it does not need a chemical cure and turn out to be cost saving. Composite materials will also be necessary to evolve new structural design concepts to achieve the weight savings of 20 to 30 percent.

New metallic alloys and composite materials give great promise for lighter airframes with longer life and better fatigue resistance. Such weight savings will result in greater payload for a given take-off weight, and hence greater fuel economy, and for combat aircraft, better maneuverability. However, the acceptance of new materials, as always, will depend on both the cost balance throughout the life of the aircraft and the development of new techniques of construction to make the best use of the material's properties.

Stress-strain Data

The significant properties of material for structural design purposes are determined experimentally. There is no successful theory for prediction of these properties, nor is there any likely in the forseeable future. Too many factors in metallurgy, alloying, crystal formation, heat-treatments, and environments, affect the deformation and fracture response of a material to externally applied forces. Tension and compression coupon tests provide the simplest fundamental information on material mechanical properties. Experimental data is recorded on stress-strain curves which record the measured unit strains as a function of the applied force per unit area. Data of importance for structural analysis is derived from the stress-strain curve as shown in Fig. 4.1.1.

(1) Definition of Terms
   Young's Modulus: Slope of the initial, usually straight line portion of the stress-strain curve, E (psi).
   Secant Modulus: Slope of secant line from the origin to any point (stress) on the stress-strain curve — a function of the stress level applied, E_s (psi).
   Tangent Modulus: Slope of the line tangent to stress-strain curve at any point (stress) on the stress-strain curve, E_t. The tangent modulus is the same as the modulus of elasticity in the elastic range.
   Proportional Limit: Stress at which the stress-strain curve deviates from linearity, initiation of

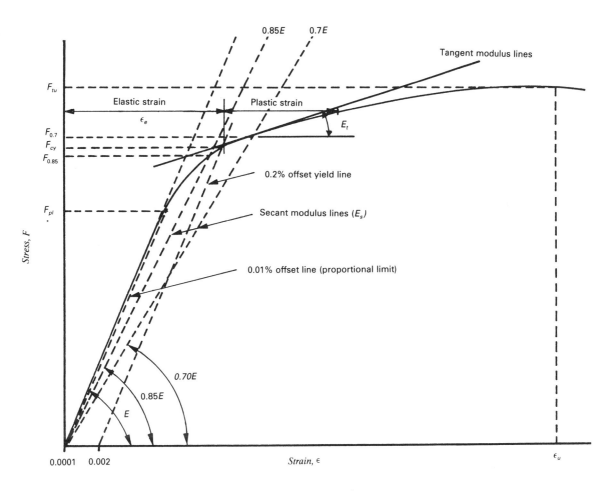

Fig. 4.1.1 *Typical stress-strain curve.*

plasticity or permanent set. Sometimes defined arbitrarily as the stress at 0.01% offset, F_{pl} (psi).

Yield Strength or Stress: At 0.2% offset, F_{cy} (psi).

Ultimate Strength or Stress: Maximum force per unit area sustained during the experiment, F_{tu} (psi).

Secant Stress: Stress at intercept of any given secant modulus line with the stress-strain curve, $F_{0.7}$, $F_{0.85}$ (psi) where subscript 0.7, 0.85 are secant lines of 70 and 85% of Young's Modulus.

Poisson's Ratio: Ratio of the lateral strain measured normal to the loading direction to the normal strain measured in the direction of the load, μ_x, μ_y (non-dimensional ratios) subscripts x, y refer to coordinate directions of principal axes of orthotropy for non-isotropic materials. Poisson's ratio from the elastic value of around $\frac{1}{3}$ to $\frac{1}{2}$ in the plastic range.

Elongation: Permanent strain at fracture (tension) in the direction of loading, e (%). Sometimes this property is considered a measure of ductility or brittleness.

Fracture Strain: Maximum strain at fracture of the material, ϵ_u (in/in).

Elastic Strain: Strains equal to or less than the strain at the proportional limit stress, ϵ_e or the strain to the projected Young's Modulus line of any stress level.

Plastic Strain: Increment of strain beyond the elastic strain at stresses above the proportional limit, ϵ_p (in/in).

(2) Ramberg-Osgood Formula

An algebraic expression for the stress-strain curve shape is of great usefulness for structural analysis. One of the more popular formulas was derived in Ref. 4.7 and is presented here in non-dimensional form:

$$\frac{E_\epsilon}{F_{0.7}} = \frac{F}{F_{0.7}} + \frac{3}{7}\left(\frac{F}{F_{0.7}}\right)^n \quad (4.1.1)$$

where $$n = 1 + \frac{\log\left(\frac{17}{7}\right)}{\log\left(\frac{F_{0.7}}{F_{0.85}}\right)} \quad (4.1.2)$$

(n is shape parameter)

The relation n is plotted in Fig. 4.1.2 as a function of the ratio $\frac{F_{0.7}}{F_{0.85}}$. Three measurements on an experimental stress-strain curve provide E, $F_{0.7}$, $F_{0.85}$ and define all the numbers needed to make the Ramberg-Osgood equation (Eq. 4.1.1) a very good fit to the original stress-strain curve. Non-dimensional stress-strain curves are shown in Fig. 4.1.3.

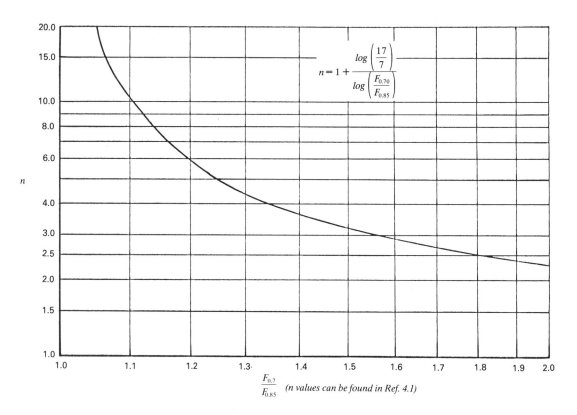

Fig. 4.1.2 Exponent n for Ramberg-Osgood Eq. 4.1.2.

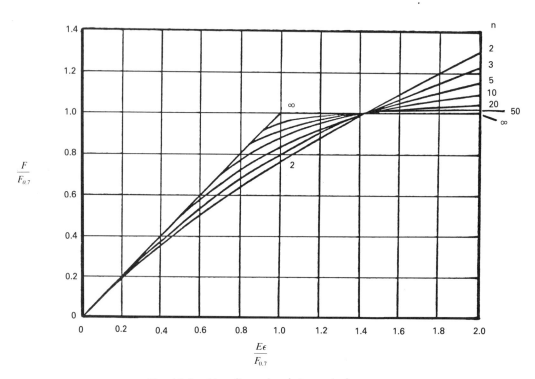

Fig. 4.1.3 Non-dimensional stress-strain curves.

(3) Secant and Tangent Modulus
The secant modulus (ratio to Young's Modulus) and the tangent modulus (ratio to Young's Modulus) can be defined algebraically:

$$\eta_s = \frac{E_s}{E} = \frac{1}{1 + \frac{3}{7}\left(\frac{F}{F_{0.7}}\right)^{n-1}} \quad (4.1.3)$$

$$\eta_t = \frac{E_t}{E} = \frac{1}{1 + \frac{3}{7}n\left(\frac{F}{F_{0.7}}\right)^{n-1}} \quad (4.1.4)$$

The non-dimensional form of the tangent and secant modulus ratios are shown in Fig. 4.1.4 and Fig. 4.1.5 respectively. These curves have been normalized by use of $F_{0.7}$. The characteristics of the materials can be derived without having to reproduce a stress-strain curve and measure secants and tangents graphically which is an inaccurate procedure.

Standard Shapes
- Sheet and Plate
  A rolled rectangular section of thickness 0.006 through 0.249 inch, with sheared, slit or sawed edges (sheet).
  A rolled rectangular section of thickness of 0.250 inch or more, with either sheared or sawed edges (plate).
  Note: Alclad sheet or plate — composite sheet or plate having on both surfaces (if on one side only — alclad one side sheet or plate) a metallurgically bonded aluminum or aluminum alloy coating that is anodic to the core alloy to which it is bonded, thus electrolytically protecting the core alloy against corrosion.
- Foil
  A rolled section of thickness less than 0.006 inch.
- Wire, Rod and Bar
  Wire: A solid section long in relation to its cross-section dimensions, having a symmetrical cross section that is square or rectangular (excluding flattened wire) with sharp or rounded corners of edges, or is round, hexagonal or octagonal, and whose diameter, width, or greatest distance between parallel faces is less than $\frac{3}{8}$ inch.

  Rod: A solid round section $\frac{3}{8}$ inch or greater in diameter, whose length is great in relation to its diameter.

  Bar: A solid section long in relation to its cross-section dimensions, having a symmetrical cross section that is square or rectangular (excluding flattened wire) with sharp or rounded corners or edges, or is a regular hexagon or octagon, and whose width or greatest distance between parallel faces is $\frac{3}{8}$ inch or greater.

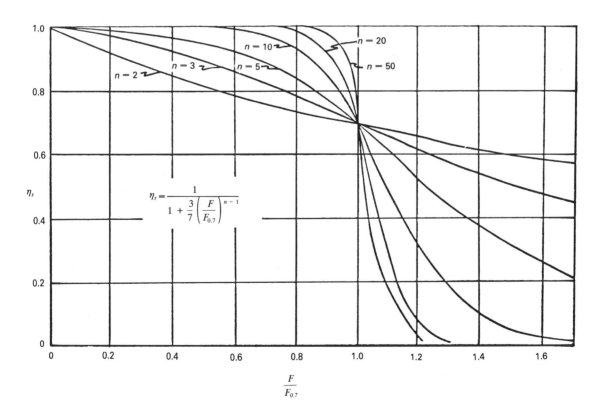

Fig. 4.1.4 Non-dimensional secant modulus.

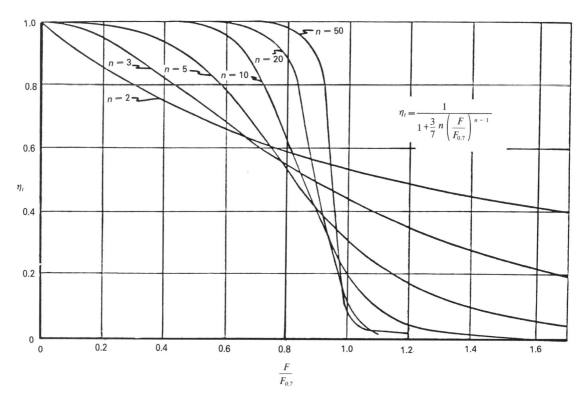

Fig. 4.1.5 *Non-dimensional tangent modulus.*

- Tube
  A hollow section long in relation to its cross-section which is symmetrical and is round, square, rectangular, hexagonal, octagonal or elliptical, with sharp or rounded corners, and having uniform wall thickness except as affected by corner radii.
  Extruded tube: A tube formed by hot extruding.
  Drawn tube: A tube brought to final dimensions by drawing through a die.
- Extruded Shapes
  A section produced by such methods as extruding, rolling, drawing, etc., that is long in relation to its cross-section dimensions and has a cross section other than that of sheet, plate, rod, bar, tube or wire.
- Forging
  A metal part worked to a predetermined shape by one or more such processes as hammering, upsetting, pressing, rolling, etc.
  Die forging: A forging formed to the required shape and size by working in impression dies.
  Hand forging: A forging worked between flat or simply shaped dies by repeated strokes or blows and manipulation of the piece.
  Note: "*Precision forgings*" *is a new manufacturing technique and improved die designs have made it possible in recent years to develop an entirely new type of forging of highest strength with very low draft angles, or no draft at all and these forgings are called Precision forgings as shown in Fig. 2.5.1 of Chapter 2. Such forgings offer greatest advantage where it is desirable (1) to obtain thin web sections and thin ribs while maintaining highest strength: (2) to hold the draft angles to 0° or 1°; and (3) to reduce corner and fillet radii to a minimum. Precision forgings are made to very close dimensional tolerances and have an excellent surface finish. Variations in weight from piece to piece are negligible.*

Material Grain Oriention

Many fabricating operations cause changes in shape of the grains in metals and alloys. This change in shape in the material grain also causes changes in the properties of the material. This behavior results because the properties (mechanical and physical) of the grain are not always equal in all three dimensions. This change in shape of the grains is called distortion or deformation. Severe working of the metal causes large deformation of the grains. Such operations as forging, extrusion, cold drawing, rough machining. etc. can cause such large distortion of the grains that they may be stretched out (elongated) in the direction of application of the working load and shortened in directions perpendicular to the working direction. This effect is called *Grain Flow*.

The three directions of *Grain Orientation* are: *Longitudinal*, parallel to the direction of working or the main direction of grain flow in a worked metal; *Long Transverse and Transverse*, means "across" and is perpendicular to the direction of working; *Short Transverse*, the shortest dimension of the part perpendicular to the direction of working. These grain orientations in a part are shown schematically in Fig. 4.1.6. Because the properties of the material are reduced in the short transverse direction, the tendency toward *Stress Corrosion Cracking* is usually greater in this direction.

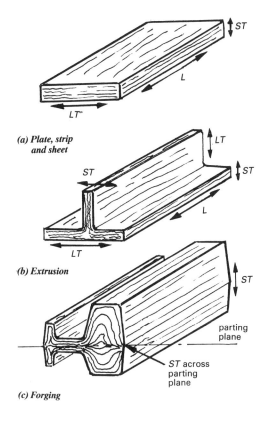

Fig. 4.1.6 Material grain direction.

Basic Sources and Use of Material Allowables

The basic source for the mechanical and physical properties of aluminum alloys shall be MIL-HDBK-5D (Ref. 4.1). For some of the materials more than one set of allowables are given under basic headings of A, B, or S. A and B basis allowables are probability allowables based upon a statistical evaluation of tensile test results. Some typical allowables are shown in Section 4.3, Section 4.4 and Section 4.5.

A basis: The mechanical property value indicated is the value above which at least 99 percent of the population of values is expected to fall with a confidence of 95 percent.

B basis: The mechanical property value indicated is the value above which at least 90 percent of the population of values is expected to fall with a confidence of 95 percent.

S basis: The mechanical property value indicated is the specified minimum value of the governing Military Specification or SAE Aerospace Material Specification for this material. The statistical assurance associated with this value is not known.

MIL-HDBK-5D properties are used to design a single member whose loading is such that its failure would result in loss of structural integrity, the guaranteed minimum design properties, or A or S values, must be used.

On redundant or fail-safe structure analysis, where the loads may be safely distributed to other members, the 90 percent probability, or B values may be used.

4.2 Material Selection Criteria

Materials selection is quite frequently a compromise involving various considerations and the more important considerations have historically been those associated with mechanical properties. A list of selection criteria for materials are as follows:
- Static strength efficiency
- Fatigue
- Fracture toughness and crack growth
- Corrosion and embrittlement
- environmental stability

Other criteria equally important are the criteria associated with producing the basic material in the forms requred and fabricating the end product at a reasonable cost. These criteria are as follows:
- Availability and Producibility
- Material costs
- Fabrication characteristics

All of the criteria listed above are important to the selection of structural materials. In addition to these, the following are a few considerations that are more frequently related to specialized requirements:
- Erosion and abrasion
- Wear characteristics
- Compatibility with other materials
- Thermal and electrical characteristics
- Hard coating to improve wear resistance
- Metallic plating to provide galvanic compatibility

Static Strength Efficiency

For structural applications, the initial evaluation of various materials is a comparison of static strength efficiency that is a satisfactory means of measuring the material relative strength. For certain applications the effect of temparature should be considered, i.e. the lower strength of aluminum alloy 2024-T81 has better strength retention at elevated temperature than 7075-T6 alloy, and for this reason has been considered in Supersonic high speed military aircraft applications.

The material properties and material strength efficiencies for several relatively familiar alloys are shown in Fig. 4.2.1.

There is one of selecting materials for strength and modulus limited applications that is not readily apparent from Fig. 4.2.1. Assume, for example, that in a strength-limited aluminum design application, a titanium alloy with improved strength efficiency is substituted, with a resultant weight savings. Unfortunately the improved strength efficiency of titanium results in a disproportionate loss of cross-sectional area (less moment of Inertia) when considering the almost equal modulus efficiencies of these two materials. The net result is a lighter structure, but also one that has less rigidity and could actually be modulus limited.

Fatigue

The behavior of a materials under conditions of cyclic load can be evaluated as follows:
- The conditions required to initiate a crack
- The conditions required to propagate a crack

Fatigue crack initiation is normally associated with the endurance limit of a material, and is undoubtedly familiar to most in the form of S-N curves (see Chapter 15 and Ref. 4.1). Another major factor

Material	Conditions	Properties (room temp.)				Density ρ (lb/in³)	Structural Efficiency	
		F_{tu} (ksi)	F_{ty} (ksi)	F_{cy} (ksi)	E (10^6 psi)		F_{tu}/ρ (10^3 in)	E/ρ 10^6 in
Aluminum	2014-T6	68	56	48	10.7	.101	673	106
	2024-T4 Extrusion	57	42	38	10.7	.100	570	107
	2024-T81	64	56	57	10.8	.101	634	107
	7075-T6	78	71	70	10.3	.101	772	102
	7075-T6 Extrusion	78	70	70	10.4	.101	772	103
Titanium	6A1-4V Annealed	134	126	132	16.0	.160	838	100
	6A1-4V Heat-treated	157	143	152	16.0	.160	981	100
Steel	4340 180 Ksi H.T.	180	163	173	29.0	.283	636	102
	17-7PH TH1050	177	150	160	29.0	.276	641	105
	AMS6520 Maraging steel	252	242	255	26.5	.283	890	94
	H-11	280	240	240	30.0	.281	996	107
	300M	280	230	247	29.0	.283	989	102
Nickel	Inconel X-750	155	100	100	31.0	.300	517	103
	A-286	130	85	85	29.1	.287	453	101
Beryllium	Be Cross-rolled, SR200D	65	43	43	42.5	.067	970	634
Magnesium	AZ31B-H24	40	30	25	6.5	.064	625	102
Fiber-glass	Glass/Epoxy*	80		60	5	.065	1230	77
Kevlar	Kevlar/Epoxy*	160		40	12	.05	3200	240
Graphite	Graphite/Epoxy*	170		140	22	.056	3040	393

* Unidirectional with 60% of fiber contents.

Fig. 4.2.1 Comparison of current material properties and efficiencies.

effecting the fatigue behavior is the stress concentration and detail design of the structure.

Fracture Toughness and Crack Growth
The fracture toughness and crack growth characteristics of a material have become increasingly important in the evaluation of high strength materials. Analytical techniques involving the application of fracture mechanics and fracture toughness data have become indispensible tools for the design of fail-safe structure.

(1) When a skin in structure containing a crack is subjected to arbitrary loading, the stress field near the crack tip can be divided into three types, each associated with a local mode of deformation as shown in Fig. 4.2.2.

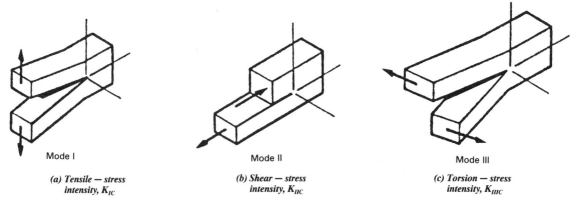

(a) *Tensile — stress intensity, K_{IC}*
(b) *Shear — stress intensity, K_{IIC}*
(c) *Torsion — stress intensity, K_{IIIC}*

Fig. 4.2.2 Three basic types of crack and stress intensities.

Fracture toughness may be defined as the ability of a part with a crack or defect to sustain a load without catastrophic failure. The Griffith-Irwin fracture theory involves the use of the stress intensity approach, although it is based on a potential energy concept. In the "through cracked skin" (Mode I case and also primarily considered in design) shown in Fig. 4.2.3, the stress intensity at the tip of the crack will increase as the crack propagates slowly under a static or cyclic applied stress, or as the applied stress increases at a constant crack length. At some combination of crack length and stress, a critical stress intensity level is produced and rapid crack propagation occurs.

The critical stress intensity factor for a given material is itself a material property since it is dependent on other physical and mechanical properties of the material. It is a measure of how well a material is able to transfer load in the region of a crack tip, and is related to stress and crack length.

The effect of material thickness on the critical stress intensity factor is shown in Fig. 4.2.4 for a typical material. In thin materials, the state of stress at the crack tip is two-dimensional, and is termed plane stress. As thickness increases, an additional degree of restraint is added. The stress state now becomes three-dimensional and produces strain in the plane of the plate; hence, it is termed plane strain.

After a certain thickness is reached, the critical stress intensity factor remains constant and called K_{IC} as shown in Fig. 4.2.4. Other values are listed in Fig. 4.2.5 for reference. Usually the major airplane manufacturers have their own test data for design.

The "through crack length" and stress relationship is as follows:

Stress intensity factor

$$K = S_g \sqrt{\pi a}\, \alpha \qquad (4.2.1)$$

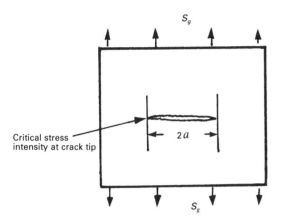

Fig. 4.2.3 *Stress intensity at crack tip for through cracked skin.*

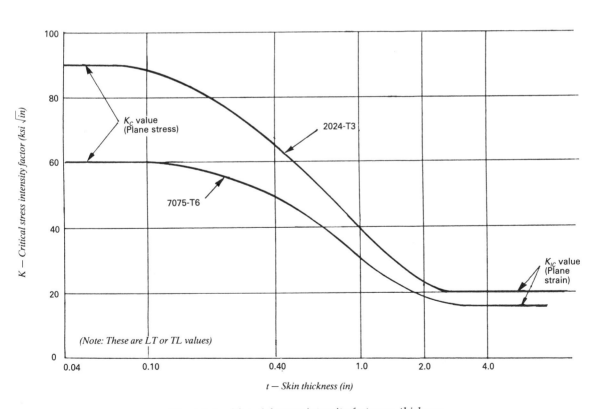

Fig. 4.2.4 *Material stress intensity factor vs. thickness.*

Airframe Structural Design 97

Material		Thickness Range	K_{IC} (ksi$\sqrt{\text{in}}$)		
			LT	TL	SL
2014-T651	Plate	0.0–1.0	22	21	18
2024-T351	Plate	0.5–2.0	31	29	22
-T851	Plate	0.5–1.5	23	20	17
7050-T7451		0.5–2.0	33	28	25
-T7452	Hand Forging	1.5	34	22	22
7075-T6	Extrusion	0.5–0.75	29	21	19
-T651	Plate	0.5–2.0	26	22	18
-T7351	Plate	0.5–2.0	30	30	23
-T7651	Plate	0.5–2.0	27	23	19
7178-T651	Plate	0.5–1.0	24	20	15
-T7651	Plate	0.5–1.0	28	24	17
300M (235 ksi)	Plate	0.5	47	—	—
H-11 (220 ksi)	Plate	0.5	34	32	—
17-4PH (175 ksi)	Plate	0.5	42	38	—
D6AC (210 ksi)	Plate	0.5–1.0	36	—	—

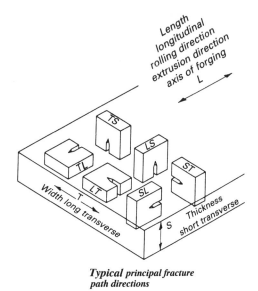

Typical principal fracture path directions

(Ref. 4.1, above data are for information only)

Fig. 4.2.5 K_{IC} *values for aluminum and steel alloys.*

Skin gross area stress

$$S_g = \frac{K}{\left(\sqrt{\frac{\pi}{2}}\right)(\sqrt{\ell})\,\alpha} \quad (4.2.2)$$

where:

K = Critical stress intensity factor in ksi $\sqrt{\text{in}}$ (from Fig. 4.2.4, Fig. 4.2.5 or other sources)

S_g = Skin gross area stress in ksi

ℓ = Crack length (or $2a$) in inch

α = Correction factor for skin width (most design cases, use $\alpha = 1.0$ because the structure has usually been designed as an infinite width skin)

(2) Crack growth characteristics of a material under cyclic stresses are a measure of its ability to contain a crack and prevent it from rapidly attaining critical length.

The "through crack growth" rate for a flat skin can be expressed as stress intensity factor as follows:

$$\Delta K = S_{\max}(1-R)^m \sqrt{2a} \quad (4.2.3)$$

where:

ΔK = Crack growth stress intensity factor in ksi $\sqrt{\text{in}}$

$S_{\max}$ = Alternating stress S_a plus mean stress S_m in ksi

R = stress ratio in $\dfrac{S_{\min}}{S_{\max}}$

m = Empirical material constant ($m = 0.5$ for aluminum)

$2a$ = Crack length ($\ell = 2a$) in inch

Note: The geometric factor such as curvature or reinforcement should also be considered in above equation.

A sample crack growth rate curve for 7075-T6 aluminum sheet ($t = 0.09$ in) is shown in Fig. 4.2.6. For other curves, refer to Ref. 4.1, Ref. 15.63 or other sources.

Corrosion and Embrittlement

Corrosion and embrittlement associated phenomena have for some time exerted considerable influence on the selection of materials. As strength levels of structural materials have increased, it has usually been found that the material's tolerance to these phenomena has been reduced. Modifications to heat treatment have, in many instances, been able to offer satisfactory compromises between strength and resistance to the particular phenomenon involved.

The major corrosion related phenomena to be considered in selecting materials include the following:

- General surface corrosion
- Aqueous stress corrosion cracking
- Corrosion fatigue

General surface corrosion must be considered when selecting magnesium alloys, aluminum alloys, and low-alloy steels. To remain in serviceable condition, magnesium alloys must rely on surface treatments and protective coatings, which are at best only stop gap measures for a metal as reactive as magnesium. For this reason, magnesium has been severely restricted in military applications, and is limited to areas that are readily accessible for inspection in commercial applications. Aluminum alloys are provided with relatively good resistance to corrosion by combinations of aluminum cladding, surface conversion treatments (as anodizing), and organic protective coatings. For low-alloy steels, cadmium plating, phosphate type conversion coatings, and organic

98 Airframe Structural Design

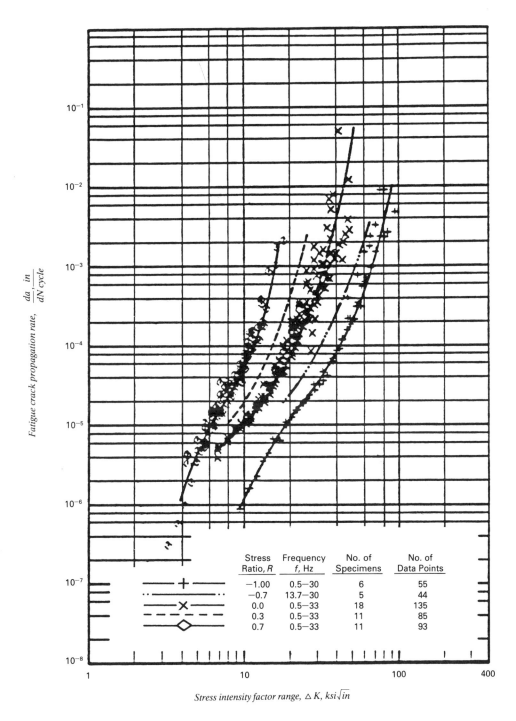

Fig. 4.2.6 Fatigue-crack-propagation data for 0.090-inch-thick 7075-T6 aluminum alloy (Ref. 4.1).

finishes are generally used to obtain the required resistance to corrosion. Of the common airframe structural materials, only titanium alloys with their excellent resistance to general surface corrosion can be used unprotected.

The resistance of an alloy to aqueous stress-corrosion cracking is not necessarily related to its resistance to general surface corrosion. Almost all alloy systems, even gold, have been found susceptible to some form of stress corrosion. Aluminum alloys are susceptible to stress corrosion cracking (SCC) in varying degrees depending on composition, heat treat condition, grain direction, stress level and nature of the corroding environment. Cracks start at the surface

Airframe Structural Design 99

and propagate slowly until fracture occurs. In general, the 7000-series alloys are more susceptible to SCC than the 2000-series alloys. Fig.4.2.7 shows the effect of grain direction on resistance to SCC 7075-T6 extrusions.

Environmental Stability
Environmental stability can be broadly defined as the ability of a material to retain its original physical and mechanical properties after exposure to the operating environment, particularly temperature and stress. In all alloy systems, there is an inherent tendency for microstructural changes to occur at elevated temperatures. In all precipitation-hardened alloys, exposure to temperatures equal to about 80 to 90 percent of the aging temperature will result in progressive overaging and loss of strength. The micro-structural changes that occur are diffusion rate controlled processes; they are dependent upon temperature, time and stress; and in addition are reasonably predictable.

Availability and Producibility
Availability and producibility imply more than just being able to obtain material on a commercial basis. The material must also be available in all forms required to fabricate the structure. The forms required may include continuous sheet, large size plate, and close tolerance extrusions. A structure could probably be fabricated with mill products that are less than optimum with respect to overall size or dimensional tolerance. However, the application may prove to be not cost effective.

Once a material has been produced commercially, lead time for material procurement must be considered for scheduling purposes, but is rarely influential in the initial selection of a material. However, the availability of a material in large quantities will depend on production capability and industry-wide demand. These affect material cost, which frequently influences material selection.

Material Cost and Fabrication Characteristics
Decisions involving the use of material in aircraft will frequently be based on whether or not it will be cost-effective in the intended application. Raw material and manufacturing costs will therefore be important considerations.

The basic method used to analyze the manufacturing costs involves the application of the selected material-to-aluminum complexity factor to aluminum parameters. This complexity factor relates the fabrication characteristics of the material to those of aluminum, and is expressed in terms of relative effort required to perform a given manufacturing operation.

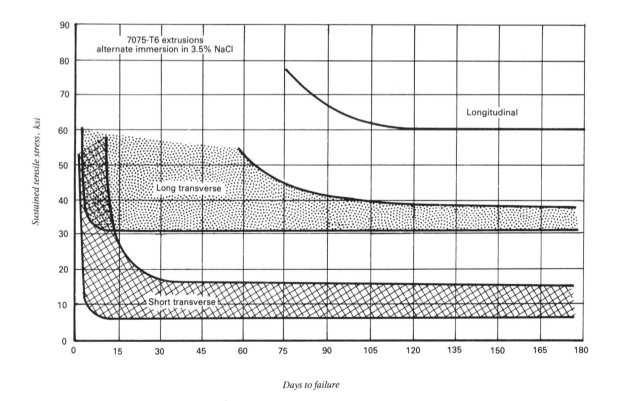

Fig. 4.2.7   Effect of grain direction on stress-corrosion cracking.

4.3 Aluminum Alloys

In commerical aviation and military transport aircraft where aluminum material counts for about 80% of the structural materials used, that material and its cost become major economic problems. Aluminum 2014 (aluminum-copper-magnesium alloy) has been used since 1920 on aircraft structures. In the 1940s, the demand for more tensile strength led to the introduction of the 7000 series alloys (Al-Zn-Mg-Cu), but the problems of stress corrosion cracking and low fatigue resistance soon become apparent.

Stress corrosion remained a problem until 1960 when the introduction of the T73 double ageing treatment produced a dramatic improvement. But the treatment reduced the tensile strength by about 10% and the additional development work was on to regain strength while retaining satisfactory stress corrosion resistance. In the early 1970s, it resulted in the appearance of 7050 and 7010 with an improved balance of properties.

The problems of fatigue became prominent since the 1950s and metallurgists have been less successful in developing fatigue resistant alloys. One of them is 2024-T3 and it still remains the yardstick for good fatigue resistance. The fatigue failure of aircraft wing spars caused several accidents in the early 1950s and the 7000 series alloys were in part to blame in that it had offered the designer higher tensile strength but no commensurate improvement in fatigue strength.

By the late 1950s the safe-life design philosophy was giving way to fail-safe design since material growing cracks could be more safely monitored in service, following a much improved understanding of crack behavior. Crack stoppers could be designed in — but usually with weight penalties — so the demand was increasing for 7000 series strength with 2024-T3 levels of fatigue crack growth resistance and toughness.

The demand for improved toughness reflects another deficiency in the 7000 series alloys. By using higher purity alloys much has been achieved and the 7075 and 7010 alloys have appreciably better strength in the presence of cracks than its predecessors.

Recently, the premier new alloys, perhaps the most important aluminum material development are aluminum-lithium alloys which is about 10% lighter than conventional aluminum alloys, and about 10% stiffer. Substitution of aluminum-lithium for conventional alloys in an existing aircraft design would reduce weight by 8 to 10%, and about 15% weight reduction could be achieved for new design. Primarily because of its superior fatigue performance, the high cost of lithium material, safety precautions in casting, the need for scrap segregation and handling, and closer control of processing parameters all continue to increase product costs more than three times above those of conventional aluminum alloys. But the advantage of using this material is that the manufacturers can use existing machinery and equipment to work and workers do not need special training. Other recently developed aluminum alloys can provide outstanding combination of strength, fracture toughness, fatigue resistance, and corrosion resistance for aircraft components. The basis for these alloy systems, called wrought PM alloys, is the rapid-solidification process. The commercially available wrought PM aluminum alloys are 7090 and 7091. See Fig. 4.3.1 for the material grain comparison between 7050 and 7091.

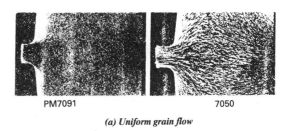

(a) Uniform grain flow

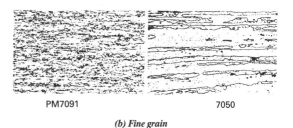

(b) Fine grain

PM = Powder Metallurgy

Fig. 4.3.1 *Microstructure improvement of PM7091.*

Design Considerations

Today, the structural designer no longer chooses a material solely on the basis of its strength qualities, but on its proven ability to withstand minor damage in service without endangering the safety of the aircraft. The residual strength after damage, described as the toughness, is now uppermost in the engineer's mind when he chooses alloys for airframes. Damage caused by fatigue is the main worry because it is difficult to detect and can disastrously weaken the strength of critical components.

So whereas about a decade ago aluminium alloys looked as if they had reached a technical plateau, engineers have now been able to clarify their needs as a result of the work done on fracture mechanics (these matters will be discussed in detail in Chapter 15), and metallurgists have changed their composition and treatment techniques to meet the new toughness requirements.

For pressurized cabins and lower wing skins — two areas particularly prone to fatigue through the long-continued application and relaxation of tension stresses — the standard material (for commercial transports, the entire fuselages of which are pressurized) is an aluminium alloy designated 2024-T3. For upper wing skins, which have to withstand mainly compression stresses as the wing flexes upwards during flight, 7075-T6 (with zinc and chromium introduced) is used. This alloy is also used extensively for military aircraft structures, which generally have stiffer wings and — except for the cockpit area — unpressurized fuselages.

Recommended application of common aluminum alloys are shown in Fig. 4.3.2.

Material	Recommended Application
2024-T3, T42, T351, T81	Use for high strength tension application; has best fracture toughness and slow crack growth rate and good fatigue life. -T42 has lower strength than -T3. Thick plate has low short transverse properties and low stress corrosion resistance. Use -T81 for high temperature applications.
2224-T3 2324-T3	8% improvement strength over 2024-T3; fatigue and toughness better than 2024-T3.
7075-T6, T651, T 7351	Has higher strength than 2024, lower fracture toughness, use for tension applications where fatigue is not critical. Thick plate has low short transverse properties and low stress corrosion resistance (-T6). -T7351 has excellent stress corrosion resistance and better fracture toughness.
7079-T6	Similar to 7075 but has better thick section ($>$ 3 in) properties than 7075. Fracture toughness, between 7075 and 2024. Thick plate has low stress corrosion resistance.
7150-T6	11% improvement strength over 7075-T6. Fatigue and toughness better than 7075-T6.
7178-T6, T651	Use for compression application. Has higher strength than 7075, lower fracture toughness and fatigue life.
Aluminum-Lithium	Compared to conventional aluminum alloys: 10% lighter, 10% stiffer, and superior fatigue performance.
PM Aluminum	Compared to conventional aluminum alloys: Higher strength, good fatigue life, good toughness, higher temperature capability and superior corrosion resistance.

Fig. 4.3.2 Recommended application of common aluminum alloys.

Designations For Alloying Elements

Aluminum 99.00% Min.	1xxx
Aluminum-Copper	2xxx
Aluminum-Manganese	3xxx
Aluminum-Silicon	4xxx
Aluminum-Magnesium	5xxx
Aluminum-Magnesium-Silicon	6xxx
Aluminum-Zinc	7xxx
Aluminum-Other Elements	8xxx
Unused	9xxx

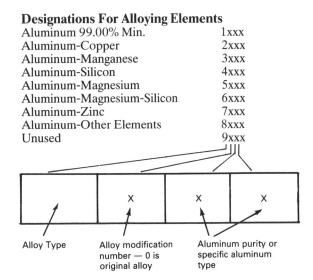

Heat Treat Conditions

F: As fabricated, usually forgings where machining will be done prior to heat treatment. Castings are also included. There is no guarantee on properties.

O: Annealed — softest condition for forming.

W: Solution heat treated, unstable condition occuring after quenching, forming possible in this condition similar to O.

T: Heat treated to stable tempers other than O or F.
 T3 — Solution heat treated, cold worked and naturally aged (usually sheet or plate)
 T4 — Solution heat treated and naturally aged
 T6 — Solution heat treated and artificially aged
 T7 — Solution heat treated and overaged
 T8 — Solution heat treated, cold worked and artificially aged

One or more numbers may appear after the heat treat number — T7xx. These show a stress relieving operation or the extent of aging or overaging.

For example, 7075 — T7351 is overaged aluminum-zinc alloy that has been stress relieved by stretching.

Fig. 4.3.3–Fig. 4.3.10 show mechanical properties of a few selected aluminum alloys.

Specification	QQ-A-250/4																					
Form	Sheet						Plate												Sheet		Plate	
Temper	T3						T351												T361			
Thickness, in.	0.008–0.009	0.010–0.128		0.129–0.249		0.250–0.499		0.500–1.000		1.001–1.500		1.501–2.000		2.001–3.000		3.000–4.000		0.020–0.062	0.063–0.249	0.250 0.500	
Basis	S	A	B	A	B	A	B	A	B	A	B	A	B	A	B	A	B	S	S	S	
Mechanical properties																					
F_{tu}, ksi:																					
L	64	64	65	65	66	64	66	63	65	62	64	62	64	60	62	57	59	68	69	67	
LT	63	63	64	64	65	64	66	63	65	62	64	62	64	60	62	57	59	67	68	66	
ST	—	—	—	—	—	—	—	—	—	—	—	—	—	52	54	49	51	—	—	—	
F_{ty}, ksi:																					
L	47	47	48	47	48	48	50	48	50	47	50	47	49	46	48	43	46	56	56	54	
LT	42	42	43	42	43	42	44	42	44	42	44	42	44	42	44	41	43	50	51	49	
ST	—	—	—	—	—	—	—	—	—	—	—	—	—	38	40	38	39	—	—	—	
F_{cy}, ksi:																					
L	39	39	40	39	40	39	41	39	41	39	40	38	40	37	39	35	37	47	48	46	
LT	45	45	46	45	46	45	47	45	47	44	46	44	46	43	45	41	43	53	54	52	
ST	—	—	—	—	—	—	—	—	—	—	—	—	—	46	48	44	47	—	—	—	
F_{su}, ksi	39	39	40	40	41	38	39	37	38	37	38	37	38	35	37	34	35	42	42	41	
F_{bru}, ksi:																					
(e/D = 1.5)	104	104	106	106	107	97	100	95	98	94	97	94	97	91	94	86	89	111	112	109	
(e/D = 2.0)	129	129	131	131	133	119	122	117	120	115	119	115	119	111	115	106	109	137	139	135	
F_{bry}, ksi:																					
(e/D = 1.5)	73	73	75	73	75	72	76	72	76	72	76	72	76	72	76	70	74	82	84	81	
(e/D = 2.0)	88	88	90	88	90	86	90	86	90	86	90	86	90	86	90	84	88	97	99	96	
e, percent (S-basis)																					
LT	10	—	—	—	—	12	—	8	—	7	—	6	—	4	—	4	—	8	9	9	
E, 10^3 ksi	10.5	10.5						10.7										10.5		10.7	
E_c, 10^3 ksi	10.7	10.7						10.9										10.7		10.9	
G, 10^3 ksi	4.0	4.0						4.0										4.0		4.0	
μ	0.33	0.33						0.33										0.33		0.33	
Physical properties																					
ω, lb/in^3								0.101													

Fig. 4.3.3   Design mechanical and physical properties of 2024-T3 aluminum alloy sheet and plate.

(Ref. 4.1)

Specification	QQ-A-250/4																
Form	Coiled sheet			Flat sheet and plate												Sheet	Plate
Temper	T4			T42				T62		T72	T81		T851			T861	
Thickness, in.	0.010-0.249			0.010-0.499	0.500-1.000	1.001-2.000	2.001-3.000	0.012-0.499	0.500-3.000	0.010-0.249	0.010-0.249		0.250-0.499		0.500-1.000	0.020-0.062	0.063-0.249	0.250-0.500
Basis	A	B	S	S	S	S	S	S	S	S	A	B	A	B	S	S	S	S
Mechanical properties																		
F_{tu}, ksi: L	62	64	62	61	60	58	64	63	—	67	68	67	68	66	71	72	70
LT	62	64	62	61	60	58	64	63	60	67	68	67	68	66	70	71	70
F_{ty}, ksi: L	40	42	38	38	38	38	50	50	—	59	61	59	61	59	63	67	64
LT	40	42	38	38	38	38	50	50	46	58	60	58	60	58	62	66	64
F_{cy}, ksi: L	40	42	38	38	38	38	—	—	—	59	61	59	61	58	63	67	64
LT	40	42	38	38	38	38	—	—	—	58	60	59	61	59	65	68	67
F_{su}, ksi	37	38	37	37	36	35	—	—	—	40	41	38	39	38	40	40	40
F_{bru}, ksi: (e/D = 1.5)	93	96	93	92	90	87	—	—	—	100	102	102	104	101	108	110	108
(e/D = 2.0)	118	122	118	116	114	110	—	—	—	127	129	131	132	129	140	142	140
F_{bry}, ksi: (e/D = 1.5)	56	59	53	53	53	53	—	—	—	83	86	87	90	87	90	96	93
(e/D = 2.0)	64	67	61	61	61	61	—	—	—	94	97	102	105	102	105	112	109
e, percent (S-basis) LT	—	—	—	8	—	4	5	5	5	5	—	5	—	5	3	4	4
E, 10^3 ksi																		
E_c, 10^3 ksi									(See Fig. 4.3.3)									
G, 10^3 ksi																		
μ																		
Physical properties ω, lb/in³									0.101									

(Ref. 4.1)

Fig. 4.3.4 Design mechanical and physical properties of 2024-T4, T6, T8 aluminum alloy sheet and plate.

Specification	QQ-A-250/12																				
Form	Sheet							Plate													
Temper	T6 and T62							T651													
Thickness, in.	0.008–0.011	0.012–0.039		0.040–0.125		0.126–0.249		0.250–0.499		0.500–1.000		1.001–2.000		2.001–2.500		2.501–3.000		3.001–3.500		3.501–4.000	
Basis	S	A	B	A	B	A	B	A	B	A	B	A	B	A	B	A	B	A	B	A	B
Mechanical properties																					
F_{tu}, ksi: L	—	76	78	78	80	78	80	77	79	77	79	76	78	75	77	71	73	70	72	66	68
LT	74	76	78	78	80	78	80	78	80	78	80	77	79	76	78	72	74	71	73	67	69
ST	—	—	—	—	—	—	—	—	—	—	—	—	—	70	71	66	68	65	67	61	63
F_{ty}, ksi: L	—	69	72	70	72	71	73	69	71	70	72	69	71	66	68	63	65	60	62	56	58
LT	61	67	70	68	70	69	71	67	69	68	70	67	69	64	66	61	61	58	60	54	56
ST	—	—	—	—	—	—	—	—	—	—	—	—	—	59	61	56	58	54	55	50	52
F_{cy}, ksi: L	—	68	71	69	71	70	72	67	69	68	70	66	68	62	64	58	60	55	57	51	52
LT	—	71	74	72	74	73	75	71	73	72	74	71	73	68	70	65	67	61	64	57	59
ST	—	—	—	—	—	—	—	—	—	—	—	—	—	67	70	64	66	61	64	57	59
F_{su}, ksi	—	46	47	47	48	47	48	41	44	44	45	44	45	44	45	42	43	42	43	39	41
F_{bru}, ksi: (e/D = 1.5)	—	118	121	121	124	121	124	117	120	117	120	116	119	114	117	108	111	107	110	101	104
(e/D = 2.0)	—	152	156	156	160	156	160	145	148	145	148	143	147	141	145	134	137	132	135	124	128
F_{bry}, ksi: (e/D = 1.5)	—	100	105	102	105	103	106	97	100	100	103	100	103	98	101	94	97	89	93	84	87
(e/D = 2.0)	—	117	122	119	122	121	124	114	118	117	120	117	120	113	117	109	112	104	108	98	103
e, percent (S-basis) LT	5	7	—	8	—	8	—	9	—	7	—	6	—	5	—	5	—	5	—	3	—
E, 10^3 ksi		10.3												10.3							
E_c, 10^3 ksi		10.5												10.6							
G, 10^3 ksi		3.9												3.9							
μ		0.33												0.33							
Physical properties ω, lb/in³									0.101												

(Ref. 4.1)

Fig. 4.3.5 *Design mechanical and physical properties of 7075 aluminum alloys sheet and plate.*

Specification	QQ-A-200/11															
Form	Extrusion (rod, bar, and shapes)															
Temper	T6, T6510, T6511, and T62															
Cross-sectional area, in²	≤ 20													> 20, ≤ 32	≤ 32	
Thickness, in	Up to 0.249		0.250-0.499		0.500-0.749		0.750-1.499		1.500-2.999		3.000-4.499				4.500-5.000	
Basis	A	B	A	B	A	B	A	B	A	B	A	B	S		A	B
Mechanical properties																
F_{tu}, ksi:																
L	78	82	81	85	81	85	81	85	81	85	81	84	78		78	81
LT	76	80	78	81	76	80	74	78	70	74	67	70	65		65	67
F_{ty}, ksi:																
L	70	74	73	77	72	76	72	76	72	76	71	74	70		68	71
LT	66	70	68	72	66	70	65	68	61	65	56	58	55		50	53
F_{cy}, ksi:																
L	70	74	73	77	72	76	72	76	72	76	71	74	70		68	71
LT	72	76	74	78	72	76	71	74	67	71	61	64	60		56	58
F_{su}, ksi	42	44	43	45	43	45	42	44	41	43	40	41	38		37	39
F_{bru}, ksi:																
($e/D = 1.5$)	112	118	117	122	117	122	116	122	115	120	109	113	105		100	104
($e/D = 2.0$)	141	148	146	153	146	153	145	152	144	151	142	147	136		135	140
F_{bry}, ksi:																
($e/D = 1.5$)	94	99	97	103	96	101	95	100	93	98	89	92	87		83	87
($e/D = 2.0$)	110	117	115	121	113	119	112	118	110	116	105	110	104		99	103
e, percent (S-basis)																
L	7	—	7	—	7	—	7	—	7	—	7	—	6		6	—
E, 10^3 ksi	10.4															
E_c, 10^3 ksi	10.7															
G, 10^3 ksi	4.0															
μ	0.33															
Physical properties																
ω, lb/in³	0.101															

Fig. 4.3.6   Design mechanical and physical properties of 7075 aluminum alloy extrusion.

(Ref. 4.1)

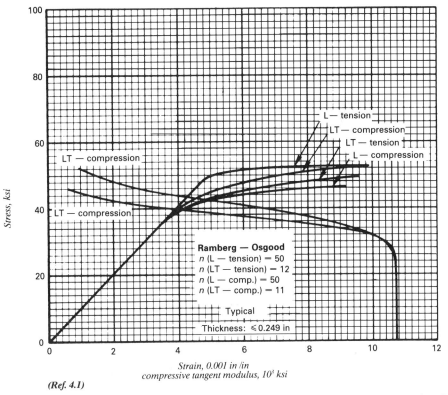

Fig. 4.3.7 Typical tensile and compressive stress-strain and compressive tangent-modulus curves for 2024-T3 aluminum alloy sheet at room temperature.

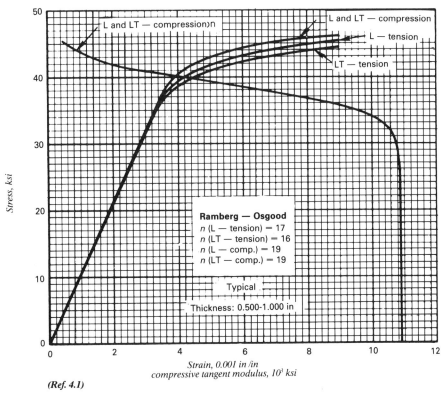

Fig. 4.3.8 Typical tensile and compressive stress-strain and compressive tangent-modulus curves for 2024-T42 aluminum alloy plate at room temperature.

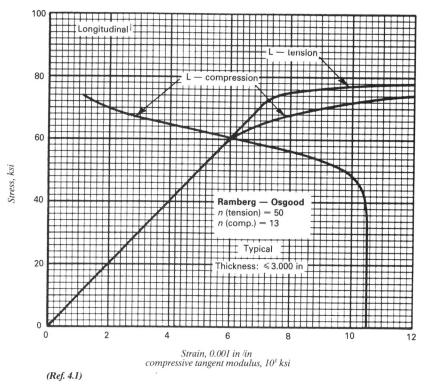

Fig. 4.3.9 Typical tensile and compressive stress-strain and compressive tangent-modulus curves for 7075-T6 and T651 aluminum alloy rolled-bar, rod, and shape at room temperature.

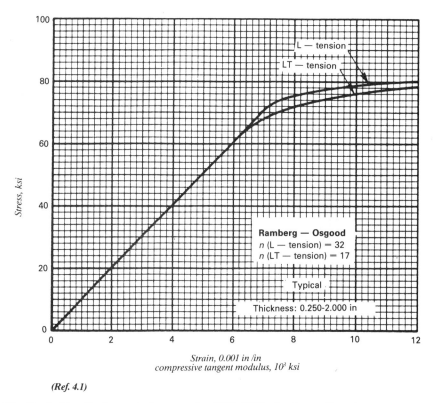

Fig. 4.3.10 Typical tensile stress-strain curves for 7075-T651 aluminum alloy plate at room temperature.

4.4 Titanium

The use of titanium as a commercial material goes back only to about 1950. Offsetting its industrially attractive properties is an extreme chemical reactivity, and there were considerable difficulties to be overcome in refining the metal from its ore and forming it into forging or castings. These problems delayed its introduction until the demands of aviation bulldozed them aside.

Several titanium alloys are used in aeronautical engineering, among them Ti-6Al-4V, and Ti-4Al-4Mo-2Sn-0.5Si. Ti-6Al-4V was developed in 1956 in America and is now probably the most widely used titanium alloy. Ti-4Al-4Mo-2Sn-0.5Si is a more specialized alloy developed in Britain and is little used elsewhere.

The first large-scale airborne application of titanium was the Lockheed SR71 and North American XB-70 in the 1960s. These projects, along with other investigations of titanium in support of aerospace programs, were responsible for solving many of the production problems. Titanium is a suitable alternative to light alloys (i.e. aluminium alloys) when prolonged operating temperatures are greater than the 150°C or so that can be withstood by aluminium without excessive deformation due to creep, or when somewhat greater strength is required without significant weight increases.

Titanium alloy bolts are now produced in quantity. For the same size as the equivalent steel bolts they have the same strength, about $F_{su} = 95000$ psi in shear strength, and adequate fatigue properties, but weight only two-thirds as much. They are, however, more expensive, though over the life of the aeroplane the weight reduction may outweigh this disadvantage. Titanium rivets are also quite extensively employed, Hi-Lok titanium fasteners for instance.

Due to the very high cost of machining of titanium parts, the economic production of titanium components therefore depends to some extent on forming them as nearly as possible to the required size, usually by forging, or by the newer precision forging so that as little as possible is machined where the cost of metal removal is paramount.

Titanium is now finding increasing use in aircraft as volume production increases. Titanium alloys may comprise up to 5% of the structural weight of commercial aircraft. This figure rises to as much as 25% on the newer military aircarft (the F-14 structure contains 24.4% by weight, that of the F-15 contains 26%) and the Rockwell B-1 contains 21% titanium. Nevertheless, there is no prospect that it will displace aluminium and its alloys because strength is not the only criterion. Rigidity is frequently essential, particularly where thin members act in compression, and a reasonable bulk such as that provided by low-density light alloys is necessary to resist buckling.

On a strength/density basis. titanium has an edge over both aluminum and steel. Also, it generally has corrosion resistance superior to both. Aluminum has usually not been considered for applications above the 350°F region. Titanium, on the other hand, is used to 1150°F, with most of its applications lying between room temperature and 1000°F. It is used in airframes as high-strength and high-toughness forgings, or where there is too little airframe space for aluminum. It is used in firewalls to isolate engines from structures, and in wing skins and support structures for military aircraft, ie. both the F-14 and B-1 use titanium carry-through wing center box structures, and the familiar SR71 reconnaissance aircraft with its extreme speed, requires an all-titanium surface. In addition, the titanium alloys have better ratios of fatigue properties to strength than do either the aluminum or steels.

Specification	MIL-I-9046, Type III, Comp.									
Form	Sheet, strip, and plate									
Condition	Annealed					Solution treated and aged				
Thickness, in	≤ 0.1875		0.1875–2.000		2.001–4.000	≤ 0.1875	0.1875–0.750	0.751–1.000	1.001–2.000	2.001–4.000
Basis	A	B	A	B	S	S	S	S	S	S
Mechanical properties										
F_{tu}, ksi:										
L	134	139	130	135	130	160	160	150	145	130
LT	134	139	130	138	130	160	160	150	145	130
F_{ty}, ksi:										
L	126	131	120	125	120	145	145	140	135	120
LT	126	131	120	131	120	145	145	140	135	120
F_{cy}, ksi:										
L	132	138	126	131	126	154	150	145	—	—
LT	132	138	126	138	126	162	—	—	—	—
F_{su}, ksi	79	81	76	79	76	100	93	87	—	—
F_{bru}, ksi:										
(e/D = 1.5)	197	204	191	198	191	236	248	233	—	—
(e/D = 2.0)	252	264	245	254	245	286	308	289	—	—
F_{bry}, ksi:										
(e/D = 1.5)	171	178	163	170	163	210	210	203	—	—
(e/D = 2.0)	208	216	198	206	198	232	243	235	—	—
e, percent (S-basis)										
L	8	—	10	—	10	5	8	6	6	6
LT	8	—	10	—	10	5	8	6	6	6
E, 10^3 ksi	16.0									
E_c, 10^3 ksi	16.4									
G, 10^3 ksi	6.2									
μ	0.31									
Physical properties										
ω, lb/in³	0.160									

(Ref. 4.1)

Fig. 4.4.1 Design mechanical and physical properties of Ti-6Al-4V.

Design mechanical and physical properties of Ti-6Al-4V is shown in Fig. 4.4.1 and typical stress-strain curve for Ti-6Al-4V is shown in Fig. 4.4.2.

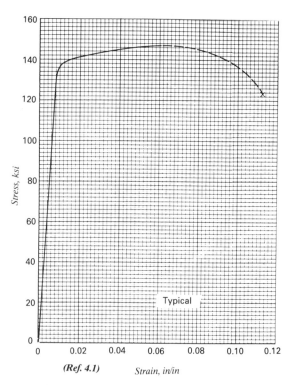

(Ref. 4.1)

Fig. 4.4.2 *Typical tensile stress-strain curve for Ti-6Al-4V.*

4.5 Steel Alloys

Where tensile strengths are required, greater than those obtainable from titanium alloys, the high strength steels are still supreme. A range of maraging, high-strength nickel-chrome, precipitation-hardening, stainless, etc. steels are available with strength up to 300000 psi (300M steel). These are used in critical areas such as landing gear units and other compact but highly loaded fittings, their use often being dictated not only by weight considerations but by the lack of space available in some areas.

Selection of high-strength alloy steel is based upon high tensile to yield ratios. These values usually are found in upper limit of heat treatment i.e. 300M steel (280000–300000 psi) and 4340 steel with tensile strength 260000 psi (260000–280000 psi). The best strength-weight ratios are achieved by heat treating steels at the maximum range that yields the desired structural reliability.

Experience shows that steels heat treated near the upper limits exhibit a tendency to fracture without appreciable deformation. Therefore, a selection of steels that considers only maximum tensile strength and hardness can result in premature failure when subjected to impact loads. Such failures can occur during attachment of fittings or rough shop handling. These loads can be either dynamic of static in origin and, although not great, be sufficient to cause failure in brittle steel. Because of this, steels are tempered to lower tensile strengths, which will increase impact values. The problem is one of selecting a steel that will give the service desired. To arrive at this selection, the engineer must determine the toughness characteristics of constructional alloys. Toughness of a material is the ability displayed to absorb energy by deformation.

Martensitic Stainless Steels
Contain 12–18% chromium and no nickel and are heat treatable by quench and temper. Maximum of 140–230 ksi attainable from 410 and 420 steels. 440C can be tempered to 275–285 ksi.
  Typical Usage: Cutlery, turbine blades, etc.
  Corrosion Resistance: Lowest of the stainless steels because of relative low chromium content and no nickel. Should be used in mild environments only. They will rust in moderate to severe corrosive conditions.

Ferritic Stainless Steels
Contain 15–30% chromium, no nickel, are not heat treatable, and have relatively low strength. They are annealed at 1200–1500°F or are slow cooled from above 1700°F.
  Typical Usage: Piping and vessels in refineries and chemical plants.
  Corrosion Resistance: Excellent resistance to high temperature sulfide corrosion. Improved resistance to other environments.

Austenitic Stainless Steels (18-8 type)
Contain 18% or more chromium and 3.5 to 22% nickel. 321 and 347 stainless steels contain titanium and columbium respectively as stabilizing alloys to reduce the susceptibility to intergranular corrosion resulting from heat sensitization. Low carbon content also reduces the susceptibility to sensitization. These alloys are non heat treatable.
  Typical Usage: Limited usage in aircraft industry. Large usage in chemical industry for piping and reaction vessels; can be used in sea water applications.
  Corrosion Resistance: Resistant to most corrosives and sea water. These alloys are nonmagnetic unless cold worked.

Precipitation Hardened Stainless Steels
Contain very little carbon, 15–17% chromium 4–7% nickel and other minor alloying elements. These alloys are solution treated and can be hardened to very high strengths.
  Typical Usage: Aircraft industry for airframe applications where high strength and excellent corrosion resistance are required. Also used in elevated temperature locations.
  Corrosion Resistance: Excellent for marine environment.

High Strength Low Alloy Steels
Principally iron base alloys, which can be hardened to very high strengths. The common steel alloy for use in the 180 to 200 ksi range is the 4130 alloy. 4340 steel has a strength range of 200 ksi up to 280 ksi and is commonly used in the 260 and 280 ksi range.
An even higher strength alloy is 300M, most commonly used for aircraft landing gear components. It can be hardened to the 240 to 290 ksi range.
  Typical Usage: Airframe and landing gear components.

Alloy	Hy-Tuf 4330V	D6AC 4335V	AISI 4340 D6AC	AISI 4340	0.40C 300M	0.42C 300M
Form	All wrought forms			Bar, forging, tubing		
Condition	Quenched and tempered			Quenched and tempered		
Basis	S	S	S	S	S	S
Mechanical properties:						
F_{tu}, ksi	220	220	260	260	270	280
F_{ty}, ksi	185	190	215	215	220	230
F_{cy}, ksi	193	198	240	240	236	247
F_{su}, ksi	132	132	156	156	162	168
F_{bru}, ksi:						
($e/D = 1.5$)	297	297	347	347	414	430
($e/D = 2.0$)	385	385	440	440	506	525
F_{bry}, ksi:						
($e/D = 1.5$)	267	274	309	309	344	360
($e/D = 2.0$)	294	302	343	343	379	396
e, percent:						
L	10	—	—	10	8	7
T	5	—	—	—	—	—
E, 10^3 ksi	29.0					
E_c, 10^3 ksi	29.0					
G, 10^3 ksi	11.0					
μ	0.32					
Physical properties:						
ω. lb/in.^{3}	0.283					

(Ref. 4.1)

Fig. 4.5.1 Design mechanical and physical properties of low-alloy steels.

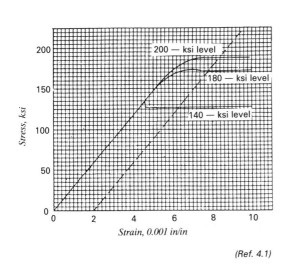

Fig. 4.5.2 Typical tensile stress-strain curves at room temperature for heat-treated AISI 4340 alloy steel (all products).

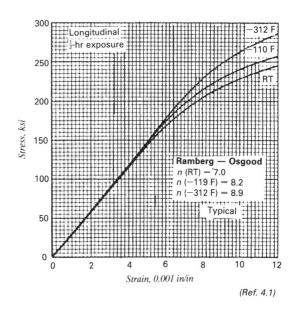

Fig. 4.5.3 Typical tensile stress-strain curves at cryogenic and room temperature for AISI 4340 alloy steel bar, $F_{tu} = 260$ ksi.

Airframe Structural Design 111

Corrosion Resistance: Poor to most environments. Components made from these alloys must be plated for corrosion protection and wear. Cadmium plating, electrodeposited below 220 ksi and vacuum deposited above 220 ksi, is most commonly used for basic corrosive protection. In addition, a primer — topcoat paint system is used for additional protection. Chromium plating is usually used for wear surface.

Fig. 4.5.1 and Fig. 4.5.2 show mechanical properties of several high strength steel alloys.

4.6 Composite Materials

Research into composite materials started in the early 1960s and at this early stage the target was a material as stiff as metal but less dense. Composite materials are now reaching the production line and aircraft designers, appreciating the weight savings available, are continually looking for further applications.

Composite or advanced composite materials will be defined as a material consisting of tiny diameter, high strength, high modulus (stiffness) fibers embedded in an essentially homogeneous matrix, see Fig. 4.6.1 for thermoset resin comparison. This results in a material that is anisotropic; having mechanical and physical properties that vary with direction and heterogeneous: consisting of dissimilar constituents that are separately identifiable. Broadly speaking, today's advanced composite materials can be broken down into two basic classes; *organic matrix* and *metal matrix* materials.

- The organic matrix systems consist of high strength fibers, such as boron or graphite, which provide the basic strength of the material and a matrix such as epoxy, polymide, or any of the thermoplastic materials which stabilize these thin or tiny fibers in compression and acts to redistribute load in shear between fibers in case of individual fiber failure or laminate transition. See Fig. 4.6.2 for reinforcement forms.
- The metal matrix composites in current use are boron/aluminum and graphite/aluminum, although some much higher temperature matrix materials such as titanium and nickel are currently being developed. The benefit of these materials is found primarily at higher temperatures. In addition, the metal matrix provides a much better foundation

Form	Advantages	Disadvantages
Tape	Maximum Structural properties, design flexibility	Drapability, possible fiber misalignment
Bi-directional reduced lay-up	Good drapability, lay-up costs	Some loss of properties due to fiber crimp, width limitations, less design flexibility
Unidirectional fabrics	Improved drapability, fiber alignment, minimal reductions in fiber strength	Slight weight penalty
Pre-plied fabrics	Reduces lay-up costs	Loss of design flexibility
Stitched fabrics, pre-forms	Provides exceptional fiber stability need for pultrusion, resin injection molding; some forms provide "Z" direction strength	Weight penalty, loss design flexibility, increased cost

Fig. 4.6.2   Reinforcement forms' comparison.

Characteristics	Thermosetting Resins				
Property	Polyester	Epoxy	Phenolic	Bismaleimide	Polymide
Processability	Good	Good	Fair	Good	Fair to difficult
Mechanical properties	Fair	Excellent	Fair	Good	Good
Heat resistance	180°F	200°F	350°F	350°F	500–600°F
Price Range	Low — Medium	Low — Medium	Low — Medium	Low — Medium	High
Delamination resistance	Fair	Good	Good	Good	Good
Toughness	Poor	Fair to Good	Poor	Fair	Fair
Remarks	Used in secondary structures, cabin interiors, primarily with fiberglass	Most widely used, best properties for primary structures: principal resin type in current graphite production use	Used in secondary structures, primarily fiberglass, good for cabin interiors	Good structural properties, intermediate temperature resistant alternative to epoxy	Specialty use for high temperature application

Fig. 4.6.1   Comparison of properties for thermoset resins.

against buckling for the filaments in compression and therefore produces much high compression strength capability, especially at higher temperatures.

During the last decade, aircraft designers have been using thermoset carbon (graphite) fiber composites as a metal replacement in aircraft industry. However, the advances in thermoplastic technology have attracted designers' attention to switch to this material. The strength-to-weight ratios, toughness, corrosion resistance, stiffness-to-weight ratio, wear resistance, producibility and etc. make it suitable contenders.

Now, designers will have to be more precise when referring to carbon or graphite fiber composites. This term can mean either thermoset or thermoplastic fiber/resin composites. A thermoset resin is one which solidifies when heated under pressure and cannot be remelted or remolded without destroying its original characteristics (ie. epoxy resin). On the other hand, thermoplastic resins soften when heated and harden when it cools down. This heating and cooling process can be repeated and the material remolded at will, see Fig. 4.6.3 for these two materials' comparison.

- Improved toughness and better impact damage
- More resistant to heat and chemical
- Do not have to be stored in refrigerators
- Resistant to attack from moisture
- Can be reheated and reformed
- Low cost of manufacturing
- Much less time of curing
- Easily repaired

Fig. 4.6.3 Advantages of thermoplastic vs. thermoset materials.

More detailed discussions regarding the advanced composite design and characteristics are in Chapter 14.

4.7 Corrosion Prevention and Control

The aircraft industry has utilized little of the vast amounts of data gained from history and research on corrosion to design out corrosion and to manufacture corrosion resistant aircraft. This Chapter relates some of the more pronounced corrosion problems faced today and the means being employed to implement corrosion prevention into design and manufacturing practices. Although stress corrosion is generally considered today as the paramount corrosion problem because of its catastrophic implications, other types of corrosion create major problems too. The problems most associated with corrosion are derived from water intrusion into structural joints from exterior surfaces, lack of adequate protective coatings, contaminated fuels, applied stresses to inadequately protected high-strength alloys in their short transverse grain direction, dissimilar metal couples and hydrogen infusion in high-strength steels. Some of the most common corrosion problems encountered today are shown in Fig. 4.7.1.

- Water intrusion into structure
- Corroded fasteners and resulting corroded structure
- Exfoliation corrosion
- Pitting of exterior skins
- Corrosion of steel and hydrogen embrittlement
- Dissimilar metal couples
- Corrosion of fuel tanks
- Fitliform corrosion
- Stress corrosion

Fig. 4.7.1 Aircraft corrosion problems.

Exfoliation corrosion is compared to stress corrosion in failure pattern with the differences being in the propagation of the failure. Proper barrier coatings have shown that this type of corrosion can be prevented or at least controlled. Pitting corrosion problems are being combated successfully with paint and in thin skin areas with anodic films. Galvanic potential differences are being resolved with closer material selection and platings.

Paint, platings and baking are being employed to diminish the threat of hydrogen embrittlement.

Stress corrosion, with ultimate catastrophic failure, is the area of most concern when considering the airworthiness of the aircraft and its systems; however, stress corrosion is not the most prevalent form of corrosion being encountered on today's aircraft and large expenditures are made each year to combat the everyday disintegration of structure by its environment.

Water, with its dissolved salts and other chemicals, constitutes the largest single corrodent known and creates by far the major portion of all corrosion. Control of corrosion by preventing water, and other corrosive fluids, from entering the aircraft structure would be the ideal solution, and great strides have been made in this direction by "environmental" sealing. The environmental sealing task consists of sealing all of the exterior joints and other points of fluid entry by applying an elastomeric compound to the faying surfaces of exterior skins.

Field experience and research and development work have defined the corrosion problems associated with aircraft. This Chapter deals only with the implementation of the corrosion prevention and control into the design and manufacture of aircraft.

Common Types of Corrosion

Corrosion may vary from superficial discoloration to deep pitting and intergranular corrosion. The severity depends upon previously mentioned factors. Several types of corrosion which have been encountered in aircraft include general surface corrosion, pitting, exfoliation, intergranular corrosion, stress corrosion, corrosion fatigue, galvanic corrosion and fretting corrosion.

(1) Surface Corrosion: This type of corrosion is visually uniform and is objectionable only from an appearance standpoint. Dulling of a bright or polished surface, etching by acid cleaners, or oxidation (discoloration) of steel are examples of surface corrosion. Corrosion resistant and stainless steels can become tarnished or oxidized in corrosive environments. Surface corrosion can

indicate a breakdown in the protective coating system, however, and should be examined closely for more advanced attack. If surface corrosion is permitted to continue, the surface may become rough and surface corrosion can lead to more serious types of corrosion.

(2) Pitting Corrosion: Pitting may occur in local patches and is generally characterized by the presence of loose corrosion products. On aluminum and magnesium parts these corrosion products appear white or gray. Most metals and their alloys are subject to pitting corrosion. A cross section through a pitted surface would show shallow to deep depressions. These can be considered as potential starting points for more serious corrosion and fatigue. Removal of corrosion products will reveal minute pits or holes in the surface. Pitting corrosion is more serious than general surface corrosion because of its localized or concentrated nature. See Fig. 4.7.2

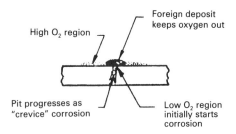

Fig. 4.7.2 Pitting corrosion.

(3) Exfoliation Corrosion: This type of corrosion occurs in the high strength aluminum alloys and is a form of intergranular corrosion. It is of special seriousness in extruded or other heavily worked shapes in which the grains of metal have been extruded in length and flattened. Corrosion products along these grain boundaries exert pressure between the grains and the result is a lifting or leafing effect. This type of corrosion usually starts at grain ends as in a machined edge, groove, chamber or machined hole and can progress through an entire section. See Fig. 4.7.3.

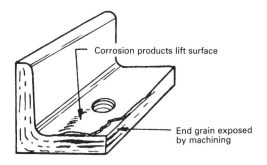

Fig. 4.7.3 Exfoliation corrosion.

(4) Intergranular Corrosion: Intergranular corrosion proceeds along grain boundaries in a metal alloy and can penetrate through a section in a short time. For example, minute particles (precipitation) are produced along grain boundaries in an aluminum or steel alloy, especially with improper heat treatment. These particles by their formation, produce less corrosion resistant areas nearby and these areas corrode rapidly in the presence of an electrolyte or conductor. These grain boundaries are anodic to the much larger adjacent cathodic grains.

(5) Stress Corrosion: Corrosion damage can result from the combined action of stress and corrosion. It can proceed either along grain boundaries or across the grains and can penetrate through a section in a short time. Stresses may be induced by working (as in bending or rolling), by non-uniform cooling during heat treatment and by machining or grinding. These stresses are internal or residual. Internal stresses can be induced during installation of a part such as *press fitting*. These stresses are especially dangerous because they are difficult to measure or predict and usually do not become evident until a part or assembly has failed by cracking. Stress corrosion cracking can occur in both aluminum and steel alloys, including the stainless steels. Stress corrosion cracking can advance more rapidly in a part or assembly which is subjected to pulsating or alternating stresses. This effect is sometimes called corrosion fatigue and can occur at stresses below the nominal fatigue strength of the metal. This type of failure can also start with pitting corrosion (stress raisers), leading to cracking. See Fig. 4.7.4.

(6) Galvanic Corrosion: Galvanic corrosion can occur where two metals or alloys in different groups in the galvanic series (shown in Fig. 4.7.5) are in contact in the presence of moisture. This type of corrosion is usually accompanied by a buildup of corrosion products in the contact area. Corrosion progresses more rapidly the farther apart the metals are in the galvanic series. Metals located toward the anodic end of the table will corrode sacrificially. For example, aluminum will corrode in contact with stainless steel, magnesium will corrode in contact with titanium. Steel which has been cadmium-plated will cause less corrosion of aluminum, since cadmium is closer to aluminum in the galvanic series.

Galvanic or dissimilar metal corrosion can be reduced by insulating the metals from each other at areas of contact and by applying protective coatings on both metals. It is important to note that the comparative areas of two metals in a dissimilar metal joint can influence the severity of corrosion in the anodic member. A large uncoated cathode area (stainless steel or titanium) will greatly accelerate corrosion at a local defect in a protected anodic (aluminum, for example) surface because of the high current flow concentrated at the defect. See Fig. 4.7.6.

(7) Fretting Corrosion: Fretting corrosion has been defined not as a true electrochemical corrosion, but as the deterioration of one or both of two contacting surfaces by the mechanical displacement of surface particles. The resulting fine par-

ticles can oxidize quickly and thereby further destroy the surfaces by their abrasive action.

Method of Prevention
The effective way to prevent or minimize corrosion is to exclude those external agents which promote or accelerate corrosion. To accomplish this, protective coatings such as plated films and cladding are used. These protect the base metal by sacrificial corrosion. On stainless steel alloys the natural occuring oxide films serve as a protection. However, in crevices, under accumulations of dirt or under fasteners, the stainless steel can become active and corrode since oxygen is excluded, thereby preventing formation of

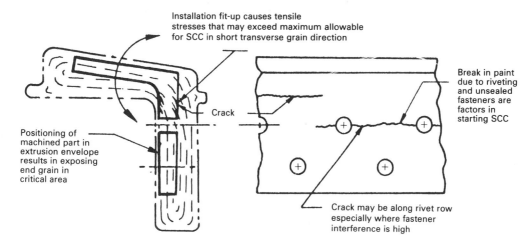

Fig. 4.7.4   Stress corrosion.

Electromotive Group	Metal or alloy
Group I	All magnesium alloys
Group II	All aluminum alloys, zinc, cadmium, and nickel-zinc alloy
Group III	Steel, cast iron, lead, tin, lead-tin solder, and 400 series stainless steel
Group IV	200, 300 and 400 series stainless steel, precipitation hardening steel (PH 13-8Mo, 15-5PH, 17-4PH, 17-7PH, AM 350, A-286), nickel, inconel, monel, titanium, chromium, silver, copper, brass, bronze, 70 Cu-30Ni and Hasteloy B
Group V	Graphite, gold and platinum

Fig. 4.7.5   Classification of dissimilar materials in the galvanic series.

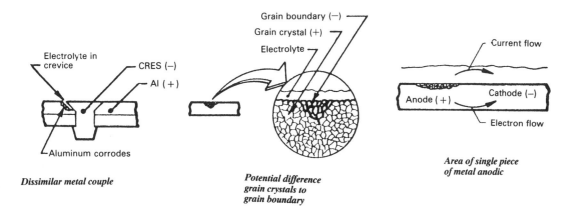

Fig. 4.7.6   Galvanic corrosion

Airframe Structural Design   115

the protective oxide.

Protective coatings such as *paint* prevent corrosion by shielding the metal surface from moisture; and with primers, as with zinc chromate primer, surface protection is given by release of corrosion inhibition.

Coatings such as *cadmium plating* on steel alloys and cladding on aluminum alloys protect the base metal by sacrificial corrosion. Cadmium is anodic (higher in the galvanic series. See Fig. 4.7.5.) to steel alloys and aluminum (essentially pure) is anodic to alloyed aluminum. These metals will provide protection as long as they are sufficiently intact.

Chromium plating applied directly to steel, as on landing gear piston surfaces, is used primarily as a mere resistant coating. When applied in sufficient thickness, it protects the base metal by excluding moisture. Cracks or other defects in chromium plating are especially serious because the underlying metal (steel) is anodic to the chromium and will corrode rapidly.

Stainless steel alloys, nickel alloys and titanium are protected by a thin, tightly adhering, complex oxide film which keeps moisture away from the active base metal. This film, as long as it remains intact or is permitted to reform after minor abrasions or scratches when oxygen is present, will protect the metals from corroding. In certain instances where oxygen is not available, as under washers or on faying surfaces, the protective film will not form and corrosion may occur.

(1) Stress Corrosion Prevention:
- Use materials which exhibit good stress corrosion resistance.
- Minimize stress concentrations.
- Do not allow sustained tensile stresses to exceed the allowables specified in Fig. 4.7.7. These values are appropriate for designs with geometrical stress concentration factor of 1.0. Sustained tensile stress allowables for other materials are dependent on the particular alloys, heat treatments, form of materials, application of structure, plus many other detail factors.
- Use stress relief heat treating and surface cold working (such as shot peening) to remove residual surface tension stresses. Removal of residual tension stresses is especially important for high heat-treat steel and aluminum.
- Use shims to further reduce or relieve fit-up stresses, (see Chapter 7.0).
- Use control of tolerances to reduce assembly fit-up stresses.
- Specify torque requirements of fasteners when uncontrolled torque could result in excessive stress in the part (such as female lugs).
- Minimize stresses in the short transverse plane.
- Avoid using interference-fit bushings or interference-fit fasteners which could produce stress in the transverse or short transverse planes of susceptible materials.

(2) Galvanic Corrosion Prevention:
- Use metal combinations close together in the galvanic series.
- Use large anodic metal areas in combination with small cathodic metal areas.
- Review proposed material combinations and platings for compatibility within groups
- Avoid the use of dissimilar metal combinations in corrosive environments such as galley and lavatory areas.
- Insulate dissimilar metals either by protective coatings or by interposing an inert material between them. Use special precaution with dissimilar metal combinations involving magnesium.
- Eliminate access of water to the metals by sealants, protective coatings, etc.

(3) Fretting Corrosion Prevention:
- Use a lubricant on the contact surfaces.
- Seal the joint to prevent entrance of a corrosive agent.
- Consider adhesive bonding in addition to mechanical attachments for vital fatigue critical parts.

(4) Intergranular Corrosion Prevention:
- Avoid exposure of short transverse grain structure.
- Use a protective film such as plating, cladding, anodizing, or a sealing compound.
- Use a cold working process on the surface grain structure such as shot peening.
- Use alloys and heat-treat conditions least susceptible to intergranular corrosion such as 7075-T73.

(5) Design for Corrosion Prevention — General:
- Provide adequate ventilation and drainage to minimize the accumulation of condensed vapors and moisture.
- Avoid depressed areas in integral fuel tanks where drainage is not provided.
- Avoid the use of absorptive materials (such as felts, asbestos, and fabrics) in contact with metallic surfaces.
- Provide permanent joints with additional

Material	Stress Allowables (Ksi)		
	L	LT	ST
2024-T3, T4 plate	35	20	8
extrusion	50	18	8
7075-T6 plate	50	45	8
extrusion	60	32	8
7075-T76 plate	49	49	25
extrusion	52	49	25
7075-T3 forging	50	48	43
plate	50	48	43
extrusion	53	48	46
7178-T6 plate	55	38	8
extrusion	65	25	8
7178-T76 plate	52	52	25

(Ref. 4.1) Stress path directions, see Fig. 4.2.5.

Fig. 4.7.7 The sustained tensile stress allowables (aluminum alloys).

protection from corrosion by assembling with faying surface sealant and with fasteners that are installed wet with sealant. The sealant provides a barrier, in this case, to prevent entrance of moisture and fluids which could promote corrosion.
- Provide easy access for the purpose of corrosion inspection and part replacement.

References

4.1 *MIL-HDBK-5D, Metallic Materials and Elements for Flight Vehicle Structures.* U.S. Government Printing Office. Washington, D.C. 1986.
4.2 *MIL-HDBK-17A, Plastics for Flight Vehicles, Part I — Reinforced Plastics.* U.S Government Printing Office. Washington, D.C. 1977.
4.3 *MIL-HDBK-17A, Plastics for Flight Vehicles, Part II — Transparent Glazing Materials.* U.S. Government Printing Office. Washington, D.C. 1977.
4.4 Anon: 'Materials: Technology Keeps Pace with Needs.' *Aviation Week & Space Technology*, (Oct. 13, 1986), 67−99.
4.5 *Buyers' Guide, Modern Plastics Encyclopedia — Engineering Data Bank.* McGraw Hill Company, New York, N.Y. 1977-78.
4.6 *MCIC-HB-01, Damage Tolerant Design Handbook (A Compilation of Fracture and Crack-Growth Data for High-Strength Alloys).* Metals and Ceramics Information Center, Battelle Co. Columbus, Ohio, Jan. 1975.
4.7 Ramberg, W. and Osgood W.R.: 'Description of Stress-Strain Curves of Three Parameters.' *NACA TN902*, (July 1943).
4.8 Wilson, M. and Boradbent, S.: 'Materials Technology.' *Flight International*, (Sept. 11, 1975), 371−376.
4.9 Steinberg, M. A.: 'Materials for Aerospace.' *Scientifcic American* (Oct. 1986).
4.10 Brown, Jr, W.F., Espey, G.B. and Jones, M.H.: 'Considerations in Selection of Wing and Fuselage Sheet Materials for Transonic Transports.' *SAE Transactions*, Vol. 70 (1970), 583.
4.11 Dietz, J. and McCreery, L.H.: 'High-Strength Steel 4340.' *Aero Digest.* (May 1956), 48−52.
4.12 Brandes, E.A.: *Smithells Metals Reference Book, 6th Edition.* Butterworth & Co. Ltd. 1983.
4.13 Tortolano, F.W.: 'Powder Metallurgy.' *Design News*, (Dec. 3, 1984), 333−364.
4.14 Anon.: *Metals Handbook, 9th Edition.* American Society for metals, Metals Park, OH 44073, 1979.
4.15 Sloan, B.G.: 'Strength-Weight Characteristics of Metals for High Speed Aircraft Structures.' *Aero Digest*, (Apr. 1956), 38.
4.16 Rittenhouse, J.B. and Singletray, J.B.: 'Space Materials Handbook.' *AFML-TR-68-205*, 3rd edition, 1968.
4.17 Hibert, C.L.: 'Factors and Prevention of Corrosion.' *Aero Digest*, (Mar. 1955).
4.18 Doyle, W.M.: 'Development of Strong Aluminium Alloys (Hiduminium − RR.58).' *Aircraft Engineering*, (July, 1964), 214−217.
4.19 Anon.: 'Aircraft Transparencies.' *Aircraft Engineering*, (Dec. 1964).
4.20 Mathes, J.C.: 'Magnesium Alloys in the Aircraft Industry.' *Aircraft Engineering*, (Nov. 1941), 323−324.
4.21 Anon.: 'Some Notes on Alclad.' *Aircraft Engineering*, (Nov. 1941), 325.
4.22 Kennedy: The Prospects for Materials.' *Aeronautical Journal of the Royal Aeronautical Soiety*, (Jan. 1969).
4.23 Graham, R. 'Materials and Structures for Future Civil Aircraft.' *Aeronautical Journal of the Royal Aeronautical Society*, (May. 1969), 411.
4.24 *Advanced Composites Design Guide, First Edition*, U.S. Air Force Materials Laboratory: 1983.
4.25 Anon.: *18% Nickel Maraging Steels.* The International Nickel Company, Inc. 67 Wall Street, New York, N.Y. 10005. 1964.
4.26 Wells, R.R.: 'New Alloys for Advanced Metallic Fighter-Wing Structures.' *Journal of Aircraft*, (July, 1975), 586−592.
4.27 Anon.: *Aerospace Structural Metals Bandbook.* Battelle's Columbus, OH 43201-2693, 1986.
4.28 Anon.: *Catalog of Literature: Metallurgy, Design, Engineering, Metalworking Appliations.* Aluminum Company of America (ALCOA), 1051 ALCOA Building, Pittsburgh, PA. 15219. Jan. 1970.
4.29 Anon.: *Forming ALCOA Aluminum.* Aluminum Company of American (ALCOA), Pittsburgh, PA. 15219. 1962.
4.30 Anon.: *Machining ALCOA Aluminum.* Aluminum Company of America (ALCOA), Pittsburgh, PA. 15219. 1967.
4.31 Anon.: *ALCOA High Quality Forgings.* Aluminum Company of America (ALCOA). Pittsburgh, PA., 15219. 1968.
4.32 Anon.: *ALCOA Aluminum Hand Forgings and Rings.* Aluminum Company of America (ALCOA). Pittsburgh, PA. 16219. 1969.
4.33 Anon.: *ALCOA Forgings.* Aluminum Company of America (ALCOA). Pittsburgh, PA. 15219.
4.34 Anon.: *Aluminum Standards & Data. (1968−1969)* Aluminum Association. 420 Lexington Ave., New York, N.Y. 10017. 1968.
4.35 Anon.: *Standards for Aluminum Sand and Permanent Mold Castings.* Aluminum Association. 420 Lexington Ave. New York, N.Y. 10017.
4.36 Anon.: *Aluminum Forging Design Manual.* Aluminum Association. 420 Lexington Ave. New York. N.Y. 10017. 1967.
4.37 Kaufman, J.G.: 'Design Mechanical Properties, Fracture Toughness, Fatigue Properties, Exfoliation, and Stress-Corrosion Resistance of 7050 Sheet, Plate, Extrusion, Hand forgings, and Die Forgings.' *NASA Contract N0019-72-C-0512*, (Feb. 1974).
4.38 Anon.: 'The World of Titanium.' *Aircraft Engineering*, (Jan. 1975).
4.39 Sendeckyj, G.P.: *Composite Materials − Vol. 2 Mechanics of Composite Materials. Academic Press*, New York. 1974.
4.40 Jones, R.M.: *Mechanics of Composite Materials.* Scripta Book Co./McGraw Hill Book Company, New York. 1975.
4.41 Wells, N.J.: 'Critical Design Factors for High-strength Steel.' *Machine Design*, (Oct. 1953), 149−157.
4.42 Green, E.A.: 'Steel in Airframes.' *Journal of Metals*, (Feb. 1960), 129−131.
4.43 Orden, J.V. and Pettit, D.E.: 'A Close Look at 7475 and 2024 Aluminum for Aircraft Structures.' *Metal Progress*, (Dec. 1977), 28−31.
4.44 Anon.: *Aircraft Engineering*, (Jan. 1975).
4.45 Loyell, D.T.: 'Structural Materials Trends in Commercial Aircraft.' *AIAA paper No. 78-1552*. (Aug. 1978).
4.46 Hyatt, M.V., Quist, W.E. and Quinlivan, J.T.: 'Improved Aluminum Alloys for Airframe Applications.' *Metal Progress*, (Mar. 1977).

CHAPTER 5.0

BUCKLING AND STABILITY OF STRUCTURES

5.1 Introduction

The three most important structural components of an aircraft, namely, the wings, fuselage and empennage, are considered from the point of view of stressing as beams or cantilevers with variable loading along their lengths. The demand for a high pay load (passengers, cargo, military weapons and so on) necessitates an economical use of the materials of construction, and hence a more accurate prediction of strength than is the case in most engineering structures. This is enhanced further by reduction of safety factors, over past years, from the general value of 2.0 to 1.5 which is used today for structural ultimate design and in the future the 1.33 factor may be considered in use.

The more or less standard design for wings, consisting of two-spar or three-spars, lends itself fairly readily to analysis, which is somewhat complicated by the addition of stressed-skin upper and lower wing surfaces. Semi-monocoque fuselage construction, which has become really popular in modern aircraft design, has presented the designer with a number of complex problems, perhaps the most important of which is the calculation of the bending strength of tubes with thin walls strengthened by longitudinal stiffeners or stringers and stabilized laterally by formers or frames.

Aircraft compression structure can be classified as columns or plates or panels and shells. The types of instability failure is different for each classification and can be described as general buckling or instability in the case of the long slender column, or local instability failure commonly regarded to as crippling for short built-up plate and panel elements.

For a perfect column, the lateral deflection is zero up to P_{cr} (column failure load) and then monotonously increases without increasing P (column axial load). When the column is not perfectly straight, it deflects before P_{cr} is reached, as shown in Fig. 5.1.1, and the load may never reach the value of P_{cr}. Aircraft structures behave as imperfect structure and the difference between perfect and imperfect structure should be understood and structural limitations recognized.

The primary structure of an airplane usually consists basically of a box beam. The main structural box of the wing or tailplane is a well-known example; a semi-monocoque fuselage is another. For any given loading condition of the aircraft, the material in the box is stressed mainly in tension, in shear, or in compression, depending on its location in the box cross-section. The aim of the designer is to make the material fulfil these three functions in the most economical manner. In tension, it is limited only by the quality of material available. In shear, this is again substantially the case, although it is well known that very light shear webs over great depths do not develop as high an effective failing stress as do more sturdy webs. This property of dependence on the intensity of loading is much more marked in the case of the compression structure, which is liable to instability in various ways. The fundamental problem of the design of such a structural element is the stabilizing of the compression surface. The ingenuity of the designer in this respect has led to various types of structure being adopted: one may mention the two-spar wing box, the thick skins reinforced by stringers and supported by ribs in the case of a wing, or by frames in the case of a fuselage, and the sandwich structure.

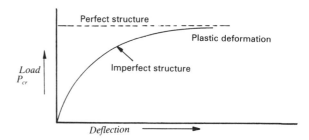

Fig. 5.1.1 Perfect and imperfect column structure

It is assumed that it is required to design a surface carrying a certain ultimate compressive end load. The surface consists of a skin stiffened by longitudinal stringers, and supported at intervals by ribs or frames. As a limiting case, this type of structure can include the two-spar wing, in which there are only two "stringers" or "spar-caps" which are stabilized by the spar webs. The trend of modern design has been towards the thick stiffened skin type of construction, the thick skin being needed for torsional stiffness in the case of the wings or tailplanes and for carrying pressurizing loads in the case of fuselages. In addition to the reinforced skin cover, there may also be spar caps which carry part of the end load.

Of the various design conditions, the one of overriding importance is that the structure should have adequate strength. The primary purpose of this design therefore is to determine the structure which has a minimum weight when its external dimensions and its strength are given. In general, the design conditions often involve other aspects besides strength. The provision of adequate stiffness may be one; ease of manufacturing, and limitation of working stresses by fatigue considerations, may be others; they are not inevitable design conditions, whereas strength considerations are.

In a subject with so broad a scope, it is inevitable that any solution involving a reasonable amount of labor is not a precise and complete one; in fact, it may be assumed to be design condition of the structure that the labor involved in design should not be prohibitive. Some reasonable limits to the range of the problem explored, and perhaps to the accuracy of the answer desired, must therefore be set. To illustrate this point, one may note that it is necessary to include many forms of stringers so as to pick the best type for the particular structure — typical shapes being Z-section, Y-section, hat-section, J-section or integrally stiffened panels and so on. A certain amount of experience is necessary in order to choose a limited range of sections which must be subjected to theoretical analysis. Even when a basic type such as the Z-section has been chosen, there are several shape parameters which can be widely varied. The various modes of instability are often quite heavily coupled so that analysis is difficult and depends on a large number of parameters: the failing stress may be appreciably different from the instability stress because of initial eccentricity (mainly due to manufacture tolerance, see Fig. 5.1.2), panel shear loading due to box torsion, or lateral air, fuel pressure and crushing loading, and so on.

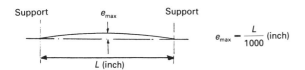

Fig. 5.1.2 *Initial eccentricity due to manufacturing tolerance.*

It is good practice to avoid eccentricities in any structure. Many times eccentricities do occur, however, and they must be accounted for in the design. For example, if a stringer is spliced from two different sections, the centroidal axes of the sections may differ in location as shown in Fig. 5.1.3. This will affect the strength of the member locally and must be considered — at the splice point and in the adjacent bays. The splice should be made at a wing rib or fuselage frame.

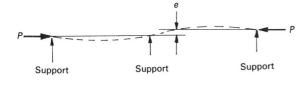

Fig. 5.1.3 *Locate eccentricity at support (i.e. rib).*

Usually the space available for compression member is dictated by fuel volume (wing box for instance) requirements and other things. The choice of skin-stringer panel types is usually decided in advance by responsible engineers after considerable investigation and study. Very often the rib spacing (wing) and frame spacing (fuselage) is established by things other than structural efficiency. The following items should be kept in mind:

- The highest possible allowable stress for a given load and material will result in lowest weight. If stringers are to be used, the shape should be carefully studied. For example, the hat section shown in Fig. 5.1.4(a) will have a higher crippling stress than the normal hat shape shown in Fig. 5.1.4(b). But it is more difficult to be formed and assembled.

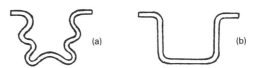

Fig. 5.1.4 *Hat-section stringers.*

- The desirable structure support spacing is one for which the stringer $\dfrac{L'}{\rho}$ (see Eq. 5.2.2) allows a column stress equal to the crippling stress. This gives the least number of supports at highest possible stress.
- The allowable panel stress for skin-stringer panel will vary with the ratio of skin to stringer thickness.
- The allowable shear stress for the skin in a skin-stringer panel may be estimated from flat panel data for diagonal tension field beam.
- The rivet spacing should be such as to prevent buckling between attachments, as discussed in Section 5.5.
- Local bending stresses on the surface structure due to pressure acting normal to the surface vary as L^2. Thus the effect can be lessened by reducing the support spacing. The allowable bending stress is usually higher than the allowable axial stress since the effect is localized over a relatively short length. This value also may be obtained by test if it is worthwhile.
- Interaction of compression, shear and bending:

$$R_s^2 + R_c^2 = 1.0$$

(for compression and shear)

$$R_s^2 + (R_c + R_b)^2 = 1.0$$

(for combined compression, bending and shear)

where $R_s = \dfrac{\text{panel shear stress}}{\text{panel allowable shear stress}}$

$R_c = \dfrac{\text{panel compression stress}}{\text{panel allowable compression stress}}$

$R_b = \dfrac{\text{panel bending stress}}{\text{panel allowable bending stress}}$

Airframe Structural Design 119

Buckling is the general term frequently used in aircraft analysis to describe the failure of a structural element when a portion of the element moves normal to the direction of primary load application. The deformation alters the mechanism by which loads are transmitted. In all instances, regardless of the complexity of the system or the nature of the primary loading (compression, shear, torsion, etc.), it is the compression stress component that forces the buckle to occur. Correspondingly, it is the compression load capability that is interrupted by the buckle formation. The general term may be applied to beam-column behavior, crippling, or any of the many complex failure modes as well as to the classical buckling behavior.

The critical equilibrium condition or buckling strength may be presented in a variety of ways. The equations commonly used to describe the equilibrium condition is as follow:

$$F_{cr} = KE \left(\frac{g_2}{g_1}\right)^2 \qquad (5.1.1)$$

where: F_{cr} = Critical stability stress such as compression, shear, bending, torsion, etc
E = Material modulus (i.e. Young's modulus, tangent modulus etc.)
g_1 = Geometry term which is usually a measure of the most significant unsupported length such as column length or panel width
g_2 = Geometry term which is usually a measure of local rigidity such as plate thickness or radius of gyration
K = A parameter which is a function of geometry, boundary conditions and loading condition

In many stability analyses, the K parameter is relatively complicated which contains all the constants of integration necessary for the equilibrium equation described in Eq. 5.1.1. However, when the boundary loading and other necessary conditions are established, the K term can then be reduced to a single coefficient for the specific case.

The general buckling equation includes an E term without indication of any limitation required when the material elastic proportional limit is exceeded. However, the use of a general modulus accounting for the decreased rigidity beyond the proportional limit is a requirement common to all buckling analyses. The general modulus E_g is a general term used to treatments of stress and strain relationship for materials beyond the proportional limit. The general modulus will describe any or all of the various modulus and shear modulus forms including the tangent and secant modulus. The fact that different forms are used with the different types of structural behaviors results from the fact that some behaviors are dependent upon total strain, whereas others are more dependent upon the slope of the stress vs. strain curve at the critical stress. (In general, the various elastic stability solutions are obtained by replacing the Young's modulus by an appropriate general modulus.) Thus, an iterative process of solving the equation

$$F_{cr} = \frac{KE_g}{\left(\frac{g_1}{g_2}\right)^2} \qquad (5.1.2)$$

for assumed value of E_g is continued until obtaining compatible F_{cr} and E_g values from a plot of the curves shown in Fig. 5.1.5, which illustrates the characteristic relationships of the stress-strain curves and corresponding general modulus curves for a given material.

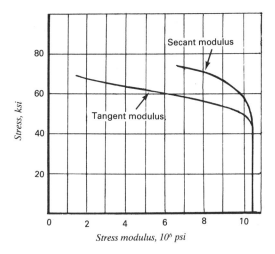

Fig. 5.1.5 Typical general modulus curves.

5.2 Columns and Beam Columns

A member subjected to compression loads (no lateral loads) is seldom of such proportions or supported in such a fashion that it can be stressed much above the yield stress of the material before it fails as a column. In general, column failures can be classed as either primary, or local. Primary failure of a column is lateral bending of the member. This lateral bending occurs about the axis of minimum moment of inertia as shown in Fig. 5.2.1. Local failure is a local collapse of the walls of the cross-section as shown in Fig. 5.2.2.

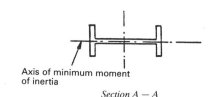

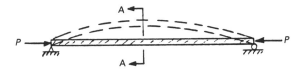

Fig. 5.2.1 Primary failure or buckling of a column member.

120 Airframe Structural Design

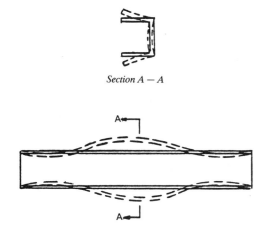

Fig. 5.2.2 *Local failure or buckling of a column member.*

A member subject to both compression loads and bending is called beam-column which is frequently used on aircraft structural analysis or design. The bending may be caused by eccentric application of the axial load, initial curvature of the member, transverse loading, or any combination of these conditions as shown in Fig. 5.2.3.

Column

Centrally loaded elements may exhibit any one of the following structural behaviors at failure:
- Material yielding
- Local crippling
- Flexural instability
- Local section torsional instability
- Column torsional instability
- Interactions between the above

Column denotes members whose cross-sectional dimensions are small compared to length and may be any combination of flat and curved elements. Fig. 5.2.4 gives an illustration of the structural behaviors and the relationship to each other. A plot of this type would normally be used only to display relationships of the different behaviors.

The modes of failure which are a result of an overall instability are column torsional instability and flexural instability. These are length dependent, whereas, the remaining modes are independent of length and the failure occurs locally. Relative positions vary for different configurations of materials.

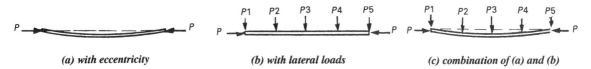

Fig. 5.2.3 *Typical beam-column cases.*

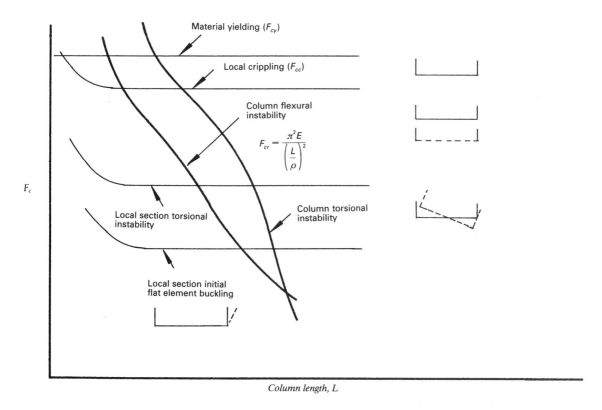

Fig. 5.2.4 *Structural modes of column failures.*

Airframe Structural Design 121

Fixity coefficients (or end fixity coefficients) as functions of the support constraints and variations in cross-section are of the form c to be used in the following general column equation as:

$$F_c = \frac{c\pi^2 E}{\left(\dfrac{L}{\rho}\right)^2} \quad (5.2.1)$$

or

$$F_c = \frac{\pi^2 E}{\left(\dfrac{L'}{\rho}\right)^2} \quad (5.2.2)$$

where: L = Column length

$L' = \dfrac{L}{\sqrt{c}}$, Effective column length

ρ = Column section radius of gyration
c = Column end fixity coefficient
E = Modulus of elasticity (if in inelastic range, use tangent modulus E_t)

For some element configurations two coefficient values are presented. One is for use in design and is prepared to facilitate determination of the required element geometry to support the design load. The other one is for analysis and is prepared to facilitate determination of the critical load (P_{cr}) for the element.

When using the fixity coefficient c for columns with varying load, stress, or cross-section the iterative process must be used to find P_{cr}. If the resulting stress is above the proportional limit, the tangent modulus is used. Using the tangent modulus E_t corresponding to the maximum stress to obtain a predicted load will give conservative predictions.

Find P_{cr} by using Young's modulus E and then compute the stress in each section. With these stresses go to the appropriate material modulus charts and find the E_t corresponding to each section. Use the tangent modulus in place of the Young's modulus and find a new P_{cr}. Continue until the new tangent modulus is approximately the same as the tangent modulus found in the previous iteration. Fig. 5.2.5 through Fig. 5.2.8 present a partial list of various column end fixity conditions usually used in aircraft design. There are a lot of other conditions which could be found in different buckling handbooks or aircraft design handbooks.

Column Shape and End Condition	End Fixity Coefficient	Column Shape and End Condition	End Fixity Coefficient
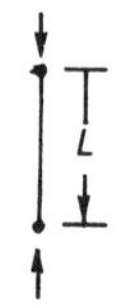 Uniform column, axially loaded, pinned ends	$c = 1$ $\dfrac{1}{\sqrt{c}} = 1$	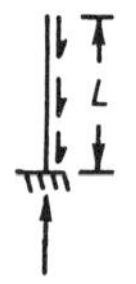 Uniform column, distributed axial load, one end fixed, one end free	$c = 0.794$ $\dfrac{1}{\sqrt{c}} = 1.12$
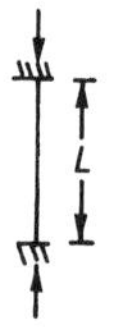 Uniform column, axially loaded, fixed ends	$c = 4$ $\dfrac{1}{\sqrt{c}} = 0.50$	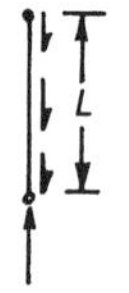 Uniform column, distribution axial load, pinned ends	$c = 1.87$ $\dfrac{1}{\sqrt{c}} = 0.732$
 Uniform column, axially loaded, one end fixed, one end pinned	$c = 2.05$ $\dfrac{1}{\sqrt{c}} = 0.70$	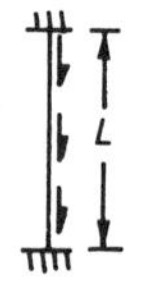 Uniform column, distributed axial load, fixed ends	$c = 7.5$ $\dfrac{1}{\sqrt{c}} = 0.365$
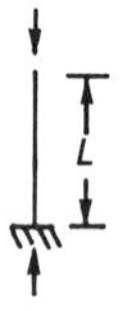 Uniform column, axially loaded, one end fixed, one end free	$c = 0.25$ $\dfrac{1}{\sqrt{c}} = 2$	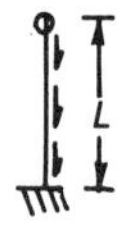 Uniform column, distributed axial load, one end fixed, one end pinned	$c = 6.08$ (approx.) $\dfrac{1}{\sqrt{c}} = 0.406$

Fig. 5.2.5 The most common use of column end fixity coefficients.

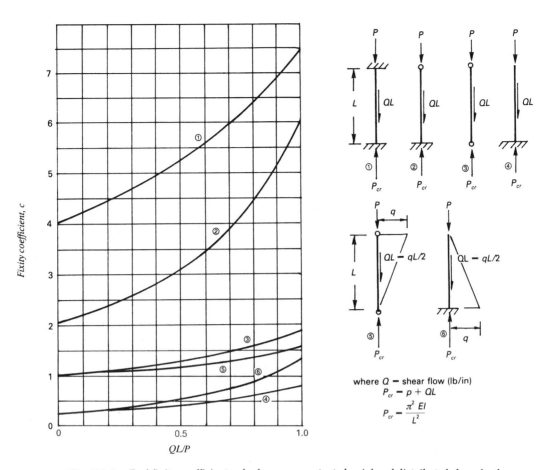

Fig. 5.2.6 *End fixity coefficients of column concentrated axial and distributed shear loads.*

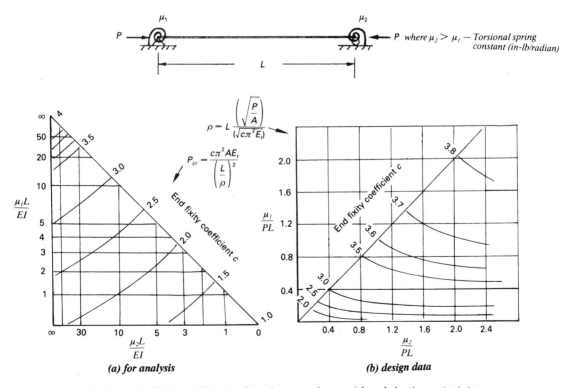

Fig. 5.2.7 *End fixity coefficients of single span column with end elastic constraints.*

Airframe Structural Design 123

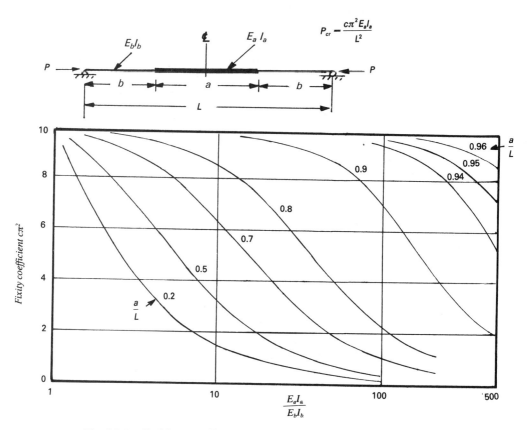

Fig. 5.2.8 End fixity coefficients of a stepped long column with pinned ends.

Example
1. Material — Aluminum alloy 7075-T6 Extrusion
   $F_{cy} = 70000$ psi and $E = 10.7 \times 10^6$ psi
2. Design the column to determine the required cross-sectional area and section properties.
3. Analyze the column as designed to determine P_{cr}.

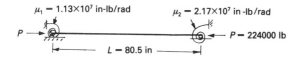

Determine area A, radius of gyration ρ and moment of inertia I:

$$\frac{\mu_1}{PL} = \frac{1.13 \times 10^7}{2.24 \times 10^5 \times 80.5} = 0.63$$

$$\frac{\mu_2}{PL} = \frac{2.17 \times 10^7}{2.24 \times 10^5 \times 80.5} = 1.2$$

From Fig. 5.2.7(b), $c = 3.5$

Assume the column is loaded in the material proportional limit, use $F_{cy} = 70000$ psi and Young's modulus = 10.7×10^6 psi

$$A = \frac{224000}{70000} = 3.2 \text{ in}^2$$

$$\rho = 80.5 \sqrt{\frac{70000}{3.5 \times \pi^2 \times 10.7 \times 10^6}} = 1.108 \text{ in}$$

$$I = A \times \rho^2 = 3.2 \times 1.108^2$$
$$= 3.927 \text{ in}^4$$

Determine P_{cr}:
$A = 3.2$ in^2; $\rho = 1.108$ in; $I = 3.927$ in^4

$$\frac{\mu_1 L}{EI} = \frac{1.13 \times 10^7 \times 80.5}{10.7 \times 10^6 \times 3.927} = 21.7$$

$$\frac{\mu_2 L}{EI} = \frac{2.17 \times 10^7 \times 80.5}{10.7 \times 10^6 \times 3.927} = 41.5$$

From Fig. 5.2.7(a), $c = 3.5$

$$\frac{F_{cr}}{E} = \frac{c\pi^2 \rho^2}{L^2}$$

$$= \frac{3.5 \, \pi^2 (1.108)^2}{80.5^2} = 0.00654$$

$F_{cr} = 0.00654 \, (10.7 \times 10^6) = 70000$ psi
$P_{cr} = F_{cr} A$
$\quad = 70000 \, (3.2) = 224000$

The Euler equation (5.2.1) is universally used to describe flexural stability of elastic or long columns. However, in the short and intermediate length range the column behavior can be considerably more complex. In this range the column strength is highly dependent upon the column cross-sectional details and the material inelastic properties. Fig. 5.2.9 illustrates two common examples of strength prediction methods that are used in various instances for short and intermediate length columns.

the parabola to the $\frac{L'}{\rho}$ value of interest. Thus, a strength prediction for a column with $\frac{L'}{\rho} = 10$ to 15, based on these curves, result in lower stresses than those obtained from crippling stresses. A Johnson-Euler family of curves for aluminum alloy with modulus of elasticity $E = 10.3 \times 10^6$ psi is plotted in Fig. 5.2.11.

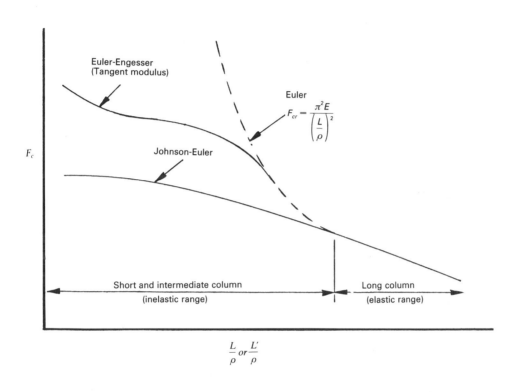

Fig. 5.2.9 Short and intermediate column failure methods.

The Johnson-Euler Column Curves

The Johnson-Euler column equation is an empirical relationship which is used to describe the flexural stability strength of columns with cross-sections having elements that are subject to local buckling or crippling. The Johnson-Euler equation will yield reasonable strength predictions for most common types of structural sections when used in conjunction with the column cross-sectional crippling strength. This method has been commonly presented as a family of curves as shown in Fig. 5.2.10. The parabolas are tangent to the Euler curve at $F_c = \frac{F_{cc}}{2}$ with the vertex, $F_c = F_{cc}$ at $\frac{L'}{\rho} = 0$. These curves are used by entering the vertical scale at $\frac{L'}{\rho} = 0$ with the crippling stress obtained from Section 5.3 and following

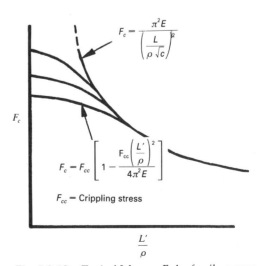

Fig. 5.2.10 Typical Johnson-Euler family curves.

Airframe Structural Design 125

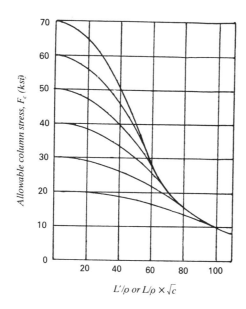

Johnson equation:
$$F_c = F_{cc} - \frac{F_{cc}^2 \left(\dfrac{L'}{\rho\sqrt{c}}\right)^2}{4\pi^2 E}$$

where:

F_{cc} — crippling stress
F_c — Stress in psi
E — 10.3×10^6 psi
c — Restraint coefficient
ρ — Radius of gyration (in)

Euler equation:
$$F_c = \frac{\pi^2 E}{\left(\dfrac{L}{\rho\sqrt{c}}\right)^2}$$

Fig. 5.2.11 *Johnson-Euler family curves for aluminum alloy with* $E = 10.3 \times 10^6$ *psi.*

Example

Given an aluminum alloy, $\dfrac{L'}{\rho} = 42$ and crippling stress $F_{cc} = 40000$ psi.

Entering the curves in Fig. 5.2.11 and following the curve for $F_{cc} = 40000$ psi and read allowable column stress $F_c = 33000$ psi.

For routine design purposes it is convenient to have column curves of allowable column stress F_c versus the effective column length $\dfrac{L'}{\rho}$ as shown in Fig. 5.2.12. These curves are plotted per equation of $F_c = \dfrac{\pi^2 E_t}{\left(\dfrac{L'}{\rho}\right)^2}$ and assume a value of F_c, then find the corresponding tangent modulus E_t from the tangent modulus curve and then solve for the term $\dfrac{L'}{\rho}$. The typical calculation is shown in example below.

7057-T6 Extrusion

F_c (assume)	E_t (from Ref. 4.1)	$\dfrac{L'}{\rho} = \pi\sqrt{\dfrac{E_t}{F_c}}$
5000	10.7×10^6	145.3
10000	10.7×10^6	102.8
20000	10.7×10^6	72.7
30000	10.7×10^6	59.3
40000	10.7×10^6	51.3
50000	10.7×10^6	46
60000	10.7×10^6	42
70000	10.4×10^6	38.3
80000	5×10^6	24.8

The above values are plotted in Fig. 5.2.12.

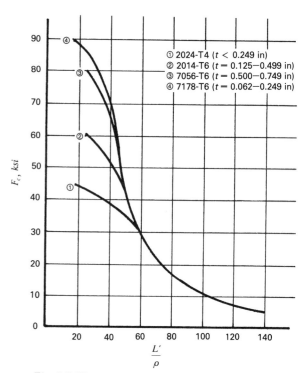

Fig. 5.2.12 *Column allowable curves for several aluminum extrusions.*

Another method to represent the column allowable stress is to use the tangent modulus Equation 4.1.4 from Chapter 4 and the Euler Equation 5.2.2 as shown below:

Tangent modulus equation: $\dfrac{E_t}{E} = \dfrac{1}{1 + \dfrac{3}{7} n \left(\dfrac{F}{F_{0.7}}\right)^{n-1}}$

126 Airframe Structural Design

Euler equation: $\dfrac{E_t}{F_c} = \dfrac{\left(\dfrac{L'}{\rho}\right)}{\pi^2}$

The problem therefore resolves itself into obtaining an expression for $\dfrac{E_t}{F_c}$ from the non-dimensional relationship. To do this, multiply both sides by $\dfrac{F_{0.7}}{F_c}$ and equate to B^2.

$$\left(\dfrac{E_t}{F_c}\right)\left(\dfrac{F_{0.7}}{E}\right) = \dfrac{1}{\dfrac{F_c}{F_{0.7}} + \dfrac{3}{7}n\left(\dfrac{F_c}{F_{0.7}}\right)^n} = B^2 \qquad (5.2.3)$$

Fig. 5.2.13 is a plot of this expression and shows $\dfrac{F_c}{F_{0.7}}$ versus B for various values of n. The shape of the stress-strain curve is given by the shape parameter n and the abscissa B incorporates the particular properties of the material $\dfrac{F_{0.7}}{E}$. These physical properties, together with the other parameters for all the commonly used materials in aircraft structural design can be found in Ref. 4.1.

Inserting a value of $F_c = \dfrac{\pi^2 E_t}{\left(\dfrac{L'}{\rho}\right)^2}$ in Equation (5.2.3)

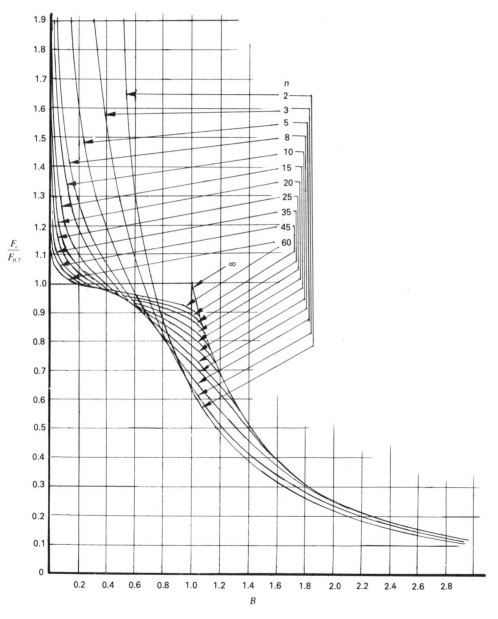

Fig. 5.2.13 $\dfrac{F_c}{F_{0.7}}$ versus B for various n values.

Airframe Structural Design 127

and then

$$B^2 = \left(\frac{F_{0.7}}{E}\right)\left(\frac{E_t}{\pi^2 E_t/(L'/\rho)^2}\right)$$

or $\quad B = \dfrac{1}{\pi}\sqrt{\dfrac{F_{0.7}}{E}}\left(\dfrac{L'}{\rho}\right) \qquad (5.2.4)$

Example

A column is made of material 7075-T6 extrusion, find column compression allowable F_c with column length = 50 in (assuming $c = 1.0$) and $\rho = 1.2$.
The column material properties are:

$$n = 26, E_c = 10.7 \times 10^3 \text{ ksi, and } F_{0.7} = 74 \text{ ksi}$$

Therefore,

$$L' = \frac{L}{\sqrt{c}} = \frac{50}{\sqrt{1}} = 50; \frac{L'}{\rho} = \frac{50}{1.2} = 41.7$$

Substituting these values into Equation 5.2.4

$$B = \frac{1}{\pi}\sqrt{\frac{74}{10.7 \times 10^3}}(41.7) = 1.1$$

Using Fig. 5.2.13 with $B = 1.1$ on bottom scale and projecting vertically upward to $n = 26$ curve and then horizontal to scale at left side of chart, read $\dfrac{F_c}{F_{0.7}} = 0.77$.

Then the $F_c = 0.77 \times F_{0.7} = 0.77\,(74) = 57$ ksi.
This $F_c = 57$ ksi can be obtained by using the column curves in Fig. 5.2.12 rather than go through the above equations. With $\dfrac{L'}{\rho} = 41.7$, using curve ③ of Fig. 5.2.12 and then read the column allowable $F_c = 57$ ksi at left side scale.

It is normally assumed that centrally loaded columns will buckle in the plane of a principal axis without rotation of the cross-sections, but experience reveals that columns having open cross-sections show a tendency to bend and twist simultaneously under axial load. The actual critical load of such columns, due to their small torsional rigidity, may be less than the critical load predicted by the generalized Euler formula as described previously.

The following sections usually are the sections subject to the torsion failure mode.

╬ I — Sections of short length and very wide flanges or webs

T — Sections of short and intermediate lengths

L — Sections with equal and unequal legs of all lengths.

It is generally not necessary to check ╬ I T sections, having relatively narrow flanges, for failure by the torsional mode.

Beam-column

A beam column is a compression member which is also subject to bending. The bending may be caused by eccentric application of the compressive load, initial curvature of the member, lateral loading, or any combination of these loadings.

A beam-column is a member subjected to transverse loads or end moments plus axial loads. The transverse loading, or end moments, produces bending moments which, in turn, produce lateral bending deflection of the member. The axial loads produce secondary bending moments due to the axial load times this lateral deflection. Compressive axial loads tend to increase the primary transverse bending moments, whereas tensile axial loads tend to decrease them. Beam-column members are quite common in airplane structures. The beams of externally braced wing and tail surfaces are typical examples, the air loads producing transverse beam loads and the struts introducing axial beam loads. In landing gears this is usually subjected to large bending and axial loads.

In general, beam-column members in aircraft structures are long and slender compared to those in buildings and bridges; thus, the secondary bending moments due to the axial loads are frequently of considerable proportion and need to be considered in the design of the members.

The principle of superposition does not apply to a beam-column, because the sum of the bending moments due to the transverse loads and the axial loads acting separately are not the same as the moments when they act simultaneously. If the axial load remains constant, however, the deflections and bending moments resulting from any system of lateral loads are proportional to the loads. The deflections or bending moments for two or more systems of lateral loads may therefore be superimposed, if the axial load is considered to act with each system. The procedure for superposition of loading conditions is shown in Fig. 5.2.14(c) are obtained as the sum of those in Fig. 5.2.14(a) and (b). The axial load P must act for each condition.

In the design of a beam-column it is necessary to assume a value of EI before the values of $\dfrac{L}{j}$ can be calculated. Where

$L = $ length of the column span (in)

$$j = \sqrt{\frac{EI}{P}}$$

$E = $ Material modulus of Elasticity (psi)
$I = $ Moment of Inertia of the cross section of column (in^4)
$P = $ Applied axial load (lb)

If the bending moment calculations show the assumed cross-section to be unsatisfactory, a new section must be assumed, and the bending moments must be recalculated. It is obviously desirable to estimate the cross-section required rather accurately before starting the bending moment calculations in order to avoid numerous trials. Fig. 5.2.15 shows one

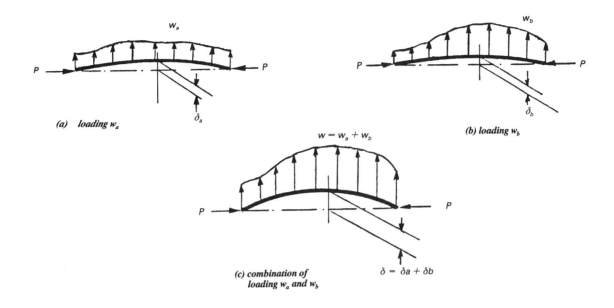

Fig. 5.2.14 Superposition of beam-column loadings.

of the approximate method of analysis, which is frequently used to design simple compression members, enabling the designer to select a cross-section without applying numerous equations.

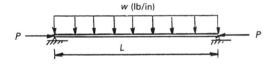

Fig. 5.2.15 Simply-support beam with applied axial loads and lateral uniform loads.

If the bending moment in a beam-column with a compressive load P is compared to the bending moment in a simple beam with no axial load, the following approximate relationship is obtained:

$$M_{max} = \frac{M_o}{1 - \left(\frac{P}{P_{cr}}\right)}$$

where M_{max} is the bending moment in the beam-column, $M_o = \left(\frac{wL^2}{8}\right)$ is the bending moment in a simple beam resisting the same lateral loading, and $P_{cr} = \frac{\pi^2 EI}{L^2}$ is the Euler column load.

Example

$I = 0.5$ in^4
$E = 10.5 \times 10^6$ psi
$P = 10000$

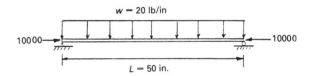

$$M_o = \frac{(20)(50^2)}{8} = 6250 \text{ in-lb}$$

$$P_{cr} = \pi^2 \, 10.5 \times 10^6 \left(\frac{0.5}{50^2}\right) = 20726 \text{ lb}$$

$$M_{max} = \frac{6250}{1 - \left(\frac{10000}{20726}\right)} = 1.93 \, (6250) = 12063 \text{ in-lb}$$

or entering in curve ③ of Fig. 5.2.16 with
$\frac{P}{P_{cr}} = \frac{10000}{20726} = 0.48$ and then read $\frac{M_{max}}{M_o} = 1.95$.

Therefore, $M_{max} = 1.95 \, (6250) = 12188$ in-lb

Fig. 5.2.16 and Fig. 5.2.17 present the beam-column magnification factors versus $\frac{P}{P_{cr}}$ for many useful loading cases. For all of the curves, EI should be considered constant on the span, and the ends of the beams are considered to be pinned.

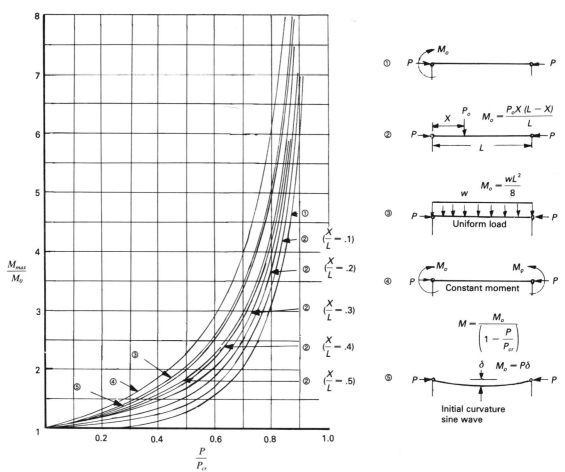

Fig. 5.2.16 Beam-column magnification factor, $\dfrac{M_{max}}{M_o}$ vs. $\dfrac{P}{P_{cr}}$.

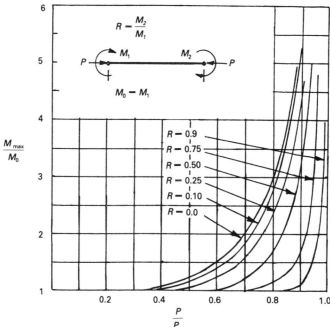

Fig. 5.2.17 Beam-column magnification factor of two different end moments, $\dfrac{M_{max}}{M_o}$ vs $\dfrac{P}{P_{cr}}$.

In the lower portion of Fig. 5.2.16, a plot of $\frac{X}{L}$ versus $\frac{P}{P_{cr}}$ is given, where $\frac{X}{L}$ defines the value of X at which the maximum bending moment M_{max} occurs for $\frac{P}{P_{cr}}$.

Example

Given conditions:

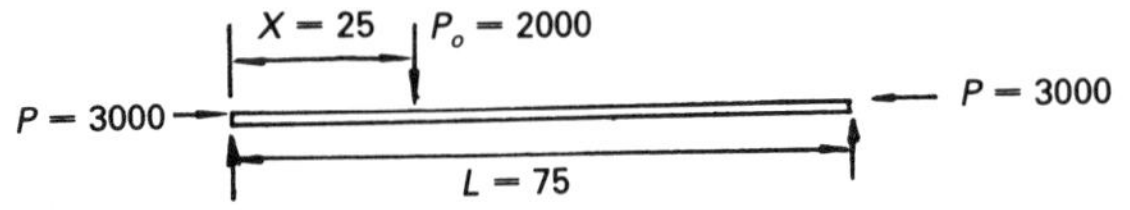

$X = 25$, $P_o = 2000$, $P = 3000$, $L = 75$, $E = 29 \times 10^6$ psi, $I = 0.15$ in^4

$$P_{cr} = \frac{\pi^2 (29 \times 10^6)(0.15)}{75^2} = 7600 \text{ lb}$$

$$\frac{P}{P_{cr}} = \frac{3000}{7600} = 0.39$$

The initial maximum bending moment is

$$M_o = \frac{P_o X(L-X)}{L} = \frac{2000(25)(75-25)}{75} = 33400 \text{ in-lb}$$

and $\frac{X}{L} = \frac{25}{75} = 0.33$

Entering Fig. 5.2.16 with $\frac{P}{P_{cr}} = 0.39$, read $\frac{M_{max}}{M_o} = 1.45$.

Thus,

$M_{max} = 1.45(33400) = 48500$ in-lb

Example

Given conditions:

$$P_{cr} = \frac{\pi^2 (29 \times 10^6)(0.10)}{30^2} = 32000 \text{ lb}$$

$$\frac{M_2}{M_1} = 0.48 \text{ and } \frac{P}{P_{cr}} = 0.78$$

From Fig. 5.2.17, $\frac{M_{max}}{M_o} = 1.65$

Hence

$M_{max} = 1.65(10000)$
$= 16500$ in-lb

There are two failure modes to be considered for beam-columns. If the axial load P is greater than the critical column load P_{cr}, then the member will fail by flexural instability. If the axial load P, combined with the transverse load w, produces a stress level greater than the material or element allowable stress, then the bending mode is critical. If neither of these modes of failure exceeds its critical value, the beam-column will reach a state of equilibrium.

If the axial load P were tension instead of compression, the moment and deflection would be decreased at any point along the beam-column.

Beam-columns under axial compressive loads are far more critical than beam-columns under axial tension loads. Axial compressive loads increase the bending moment and add the possibility of instability or buckling types of failure. An analysis of this type of loading must consider the limitations of both beams and columns. Other commonly used beam-column equations are shown in Fig. 5.2.18 through Fig. 5.2.22.

Moment M_A at support	Loading case	
$-\frac{wL^2}{2}\bar{\mu}$	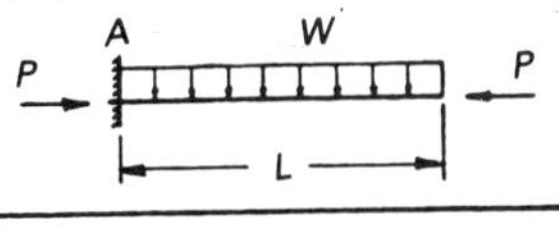 Uniform load w over length L, axial P	$\bar{\mu} = \frac{2}{(L/j)^2}\left(\frac{1}{\cos L/j} - 1\right)$
$-QL\bar{v}$	End point load Q at tip, axial P	$\bar{v} = \frac{\tan L/j}{L/j}$
$-\frac{M}{\cos L/j}$	End moment M, axial P	$\frac{1}{\cos L/j} = 1 + \frac{\bar{\mu}}{2}\left(\frac{L}{j}\right)^2$
$\frac{-Pf}{1-P/P_{cr}}$	Initial deflection $y = f \sin\frac{\pi x}{2L}$, axial P	$P_{cr} = \frac{\pi^2 \bar{E}I}{4L^2}$; $f =$ initial deflection

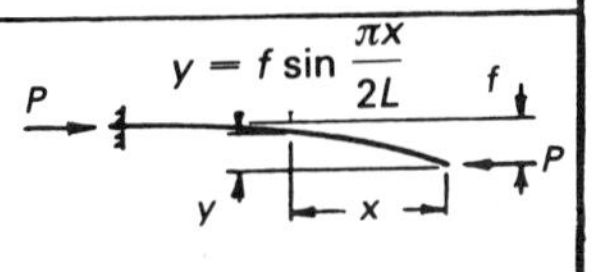

(Values of L/j, $\bar{\mu}$ and $\bar{v}$ cab be found from Fig. 5.2.21)

Fig. 5.2.18 *Beam-columns — cantilever.*

Fixed end Moment	Loading Case	Moment at Center
$-\dfrac{wL^2}{12}\beta''$	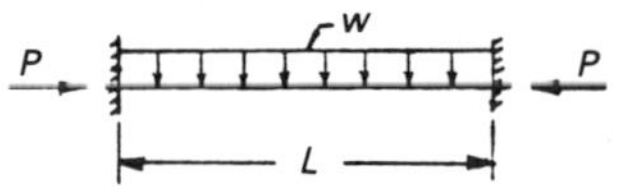 Uniform load w over length L with axial P at both ends	$\dfrac{wL^2}{24}\alpha''$
$-\dfrac{QL}{8}v''$	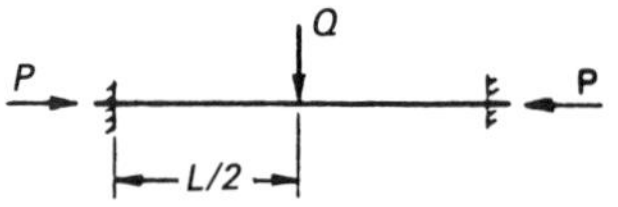 Point load Q at $L/2$ with axial P	$\dfrac{QL}{8}v''$
$\pm\dfrac{M}{2}\dfrac{\alpha''}{2\beta''}$	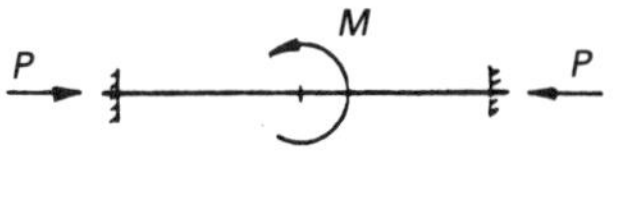 Applied moment M at center with axial P	$\pm\dfrac{M}{2}$
$\dfrac{2}{3}Pf\beta''$	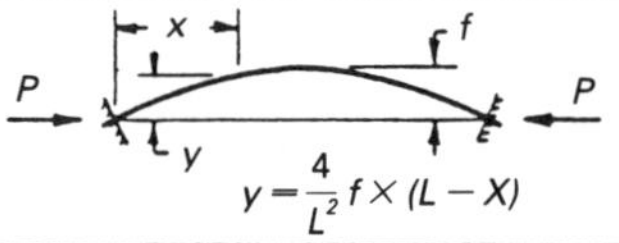 Initial parabolic deflection $y=\dfrac{4}{L^2}fX(L-X)$ with axial P	(Parabola) $-\dfrac{1}{3}Pf\alpha''$
$\dfrac{Pf}{1-\dfrac{P}{P_{cr}}}\left(\dfrac{2}{\pi}\dfrac{1}{v}\right)$	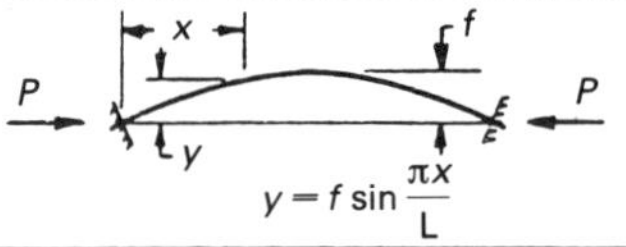 Initial sine deflection $y=f\sin\dfrac{\pi x}{L}$ with axial P	(Sine curve) $\dfrac{-Pf}{1-\dfrac{P}{P_{cr}}}\left(1-\dfrac{2}{\pi v\cos\dfrac{L}{2j}}\right)$
Moment at left end $-\dfrac{6EI\Delta}{L^2\beta''}$	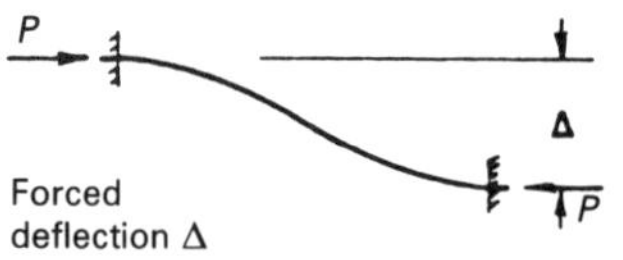 Forced deflection Δ	Moment at right end $\dfrac{6EI\Delta}{L^2\beta''}$

(Values of v, v'', L/j, α'' and β'' can be found from Fig. 5.2.21)

Fig. 5.2.19 Beam-column with both ends fixed.

Moment M_A at left support	Loading case	Beam column coefficients
$\dfrac{wL^2}{8}\dfrac{\gamma}{\beta}$	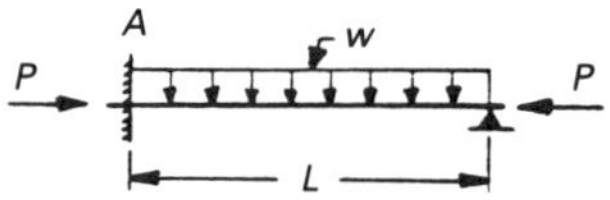 Uniform load w, fixed at A, pinned right, axial P	$\alpha=\dfrac{6\left(\dfrac{L}{j}\csc\dfrac{L}{j}-1\right)}{\left(\dfrac{L}{j}\right)^2}$
$-M\dfrac{\alpha}{2\beta}$	 Moment M at pinned right end, axial P	$\beta=\dfrac{3\left(\tan\dfrac{L}{j}-\dfrac{L}{j}\right)}{\left(\dfrac{L}{j}\right)^2\tan\dfrac{L}{j}}$
$-\dfrac{3EI\Delta}{L^2\beta}$	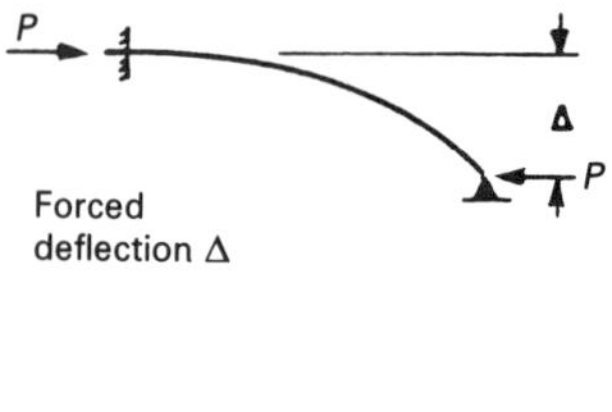 Forced deflection Δ	$\gamma=\dfrac{3\left(\tan\dfrac{L}{2j}-\dfrac{L}{2j}\right)}{\left(\dfrac{L}{2j}\right)^3}$

(Values of L/j, γ, α and β can be found from Fig. 5.2.21)

Fig. 5.2.20 Beam-column with one end fixed.

$\dfrac{L}{j}$	α	β	μ	ν	γ	β''
0	∞	1.000	1.000	1.000	1.000	1.000
0.50	1.030	1.017	1.027	1.021	1.026	1.004
1.00	1.131	1.074	1.116	1.092	1.111	1.017
1.50	1.343	1.192	1.304	1.242	1.291	1.040
$\dfrac{\pi}{2}$	1.388	1.216	1.343	1.273	1.329	1.044
2.00	1.799	1.436	1.702	1.557	1.672	1.074
2.10	1.949	1.516	1.832	1.660	1.797	1.082
2.20	2.134	1.612	1.992	1.786	1.949	1.091
2.30	2.364	1.733	2.190	1.943	2.139	1.101
2.40	2.669	1.885	2.443	2.143	2.382	1.114
2.50	3.050	2.086	2.779	2.407	2.703	1.123
2.60	3.589	2.362	3.241	2.771	3.144	1.134
2.70	4.377	2.762	3.913	3.300	3.786	1.147
2.80	5.632	3.396	4.984	4.141	4.808	1.161
2.90	7.934	4.555	6.948	5.681	6.680	1.176
3.00	13.506	7.349	11.696	9.401	11.201	1.192
3.10	45.923	23.566	39.481	31.018	37.484	1.208
π	∞	∞	∞	∞	∞	1.216
3.20	−32.706	−15.740	−27.647	−22.395	−26.245	1.227
3.40	−7.425	−3.079	−6.065	−4.527	−5.738	1.267
3.60	−4.229	−1.457	−3.331	−2.381	−3.131	1.315
3.80	−2.996	−0.813	−2.267	−1.541	−2.111	1.370
4.00	−2.357	−0.460	−1.701	−1.093	−1.569	1.436
4.20	−1.979	−0.232	−1.352	−0.814	−1.234	1.516
4.40	−1.743	−0.065	−1.115	−0.624	−1.007	1.612
4.49	−1.6680	0	−1.032	−0.557	−0.927	1.663
4.60	−1.596	0.068	−0.945	−0.487	−0.843	1.732
4.80	−1.515	0.185	−0.818	−0.382	−0.720	1.885
5.00	−1.491	0.298	−0.719	−0.299	−0.623	2.086
5.20	−1.528	0.417	−0.641	−0.231	−0.546	2.362
5.40	−1.644	0.559	−0.578	−0.175	−0.484	2.762
5.60	−1.889	0.754	−0.526	−0.127	−0.431	3.396
5.80	−2.405	1.075	−0.483	−0.085	−0.387	4.555
6.00	−3.746	1.802	−0.447	−0.048	−0.349	7.349
6.20	−11.803	5.881	−0.416	−0.013	−0.316	23.566
2π	∞	∞	−0.405	0	−0.304	∞

$\bar{\mu}$ and $\bar{\nu}$ can be determined by taking μ and ν values for $\dfrac{2L}{j}$ instead of $\dfrac{L}{j}$. α'' and β'' can be found by taking α and β values for $\dfrac{L}{2j}$ instead of $\dfrac{L}{j}$.

Fig. 5.2.21 Beam-column coefficients for Fig. 5.2.18, Fig. 5.2.19 and Fig. 5.2.20.

Case	Critical Load	Maximum Useful Load
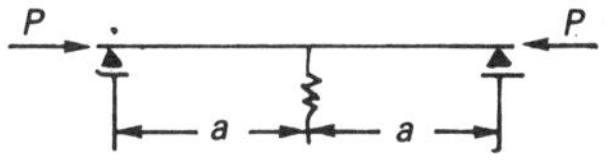 (two spans, spring at center)	Approx. $P_{cr} = \dfrac{\pi^2 EI}{4a^2} + 0.375\,\omega a$	$P = \dfrac{\pi^2 EI}{a^2}$
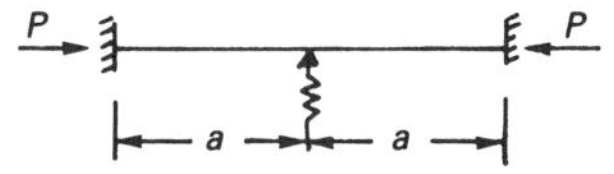 (two spans, fixed ends)	Approx. $P_{cr} = \dfrac{\pi^2 EI}{a^2} + 0.40\,\omega a$	$P = 2.046 \dfrac{\pi^2 EI}{a^2}$
More than two spans 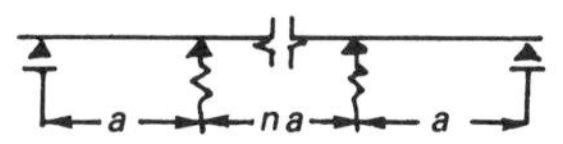	Approx. $P_{cr} = \dfrac{\pi}{2}\sqrt{\dfrac{\omega}{a} EI}$	$P = \dfrac{\pi^2 EI}{a^2}$
Uniform elastic support 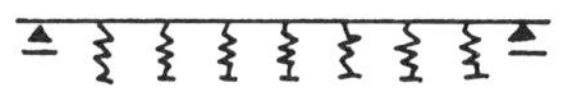 ω − Spring constant per inch	$P_{cr} = 2\sqrt{\omega EI}$ $= \dfrac{2\pi^2 EI}{\lambda^2}$	Wavelength λ $\lambda = \pi\sqrt[4]{\dfrac{EI}{\omega}}$

Fig. 5.2.22 Beam-column with elastic supports.

5.3 Crippling Stress

Extrusion, formed sheet metal, and thin-walled tubes are all subject to a local or crippling type of failure. This type of failure is characterized by a local distortion of the cross-sectional shape. The beginning of distortion usually occurs at a load appreciably less than the failing load with the more stable portions of the cross-section continuing to take additional load while supporting the already buckled portions until complete collapse occurs. Although it is possible to predict theoretically the stress at which the first buckling occurs, no satisfactory theory exists for the prediction of the average stress at failure. Thus, it is necessary to rely on test results or empirical methods which have been found to give satisfactory agreement with test results.

When the corners of a thin-walled section in compression are restrained against any lateral movement, the corner material can continue to loaded even after local buckling has occurred in the section. The remaining material is largely ineffective in supporting additional load above the local buckling load. When the stress in the corners exceeds the yield stress, the section loses its ability to support any additional load and fails. The average stress on the section at the ultimate load is called the crippling stress F_{cc}. Fig. 5.3.1 shows the cross-sectional distortion occurring over one buckle length in a typical thin-walled section. Fig. 5.3.2 shows the stress distribution over the the cross-section just before crippling.

Crippling stress is a form of instability involving inelastic axial strain of the more stable portions of a structural element resulting in permanent deformation of the section. Its longitudinal axis of the section does not necessarily deflect laterally at crippling stress as it does during column buckling.

Crippling stress is commonly assumed to be independent of component length. However, crippling stress tests indicate that the strength capability of a section is strongly affected by small changes in length at $\frac{L'}{\rho}$ values in the range of 10 or less.

Crippling stress is calculated as if the stress were uniform over the cross-section. In reality, parts of the sections buckle at a load below the critical crippling load with the result that stable areas such as intersections and corners reach a higher stress than the buckled elements. At failure, the stress in corners and intersections is always above the material yield stress F_{cy}, although the crippling stress may be considerably less than the yield stress. The use of the compression yield stress as crippling strength cut-off has been used since there is no proven analytical method for the prediction of the crippling stress.

The following is a method to analyse the crippling stress allowables for extruded, machined and formed sections. Formed and extruded or machined sections are analysed in the same manner, although different values are used for each. The section is analysed by:

- The section is broken down into individual segments as shown in Fig. 5.3.3. Each segment (or element) has a width b, a thickness t, and will have either no edge free or one edge free.
- The crippling stress calculated is as follow:

$$F_{cc} = \frac{b_1 t_1 F_{cc1} + b_2 t_2 F_{cc2} + \ldots}{b_1 t_1 + b_2 t_2 + \ldots} = \frac{\Sigma b_n t_n F_{ccn}}{\Sigma b_n t_n} \quad (5.3.1)$$

where
$b_1, b_2, \ldots$ Length of the individual segments
$t_1, t_2 \ldots$ Individual segment thickness
$F_{cc1}, F_{cc2} \ldots$ Allowable crippling stresses corresponding to computed $\frac{b}{t}$ values of the individual segments as shown in Fig. 5.3.4 and Fig. 5.3.5.

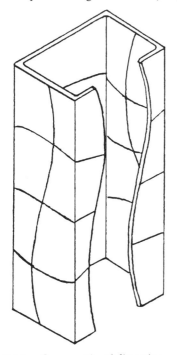

Fig. 5.3.1 Cross-sectional distortion.

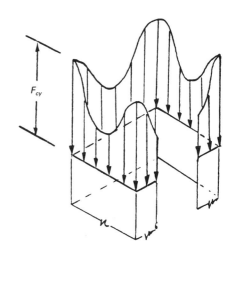

Fig. 5.3.2 Stress distribution.

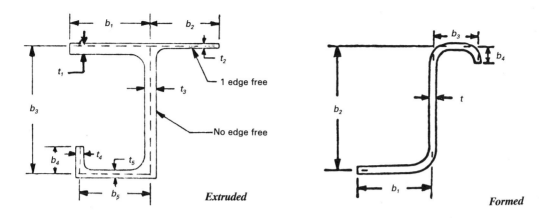

Fig. 5.3.3 Typical segment broken down of a section.

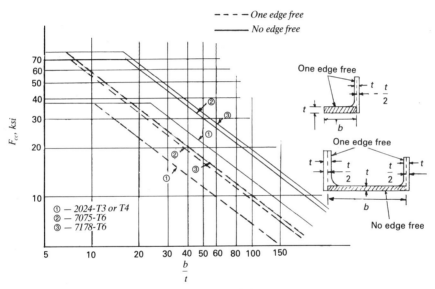

Fig. 5.3.4 Crippling stresses F_{cc} of aluminum extrusion alloys.

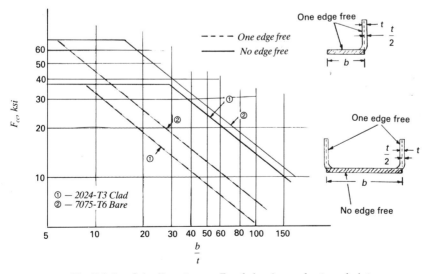

Fig. 5.3.5 Crippling stresses F_{cc} of aluminum sheets and plates.

Example

Find crippling stresses of the following two cross-sections:

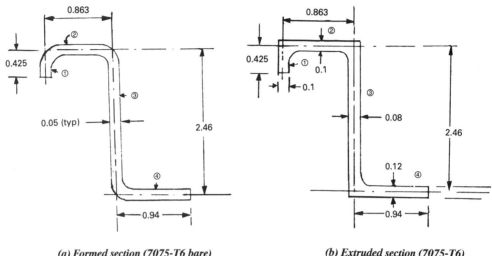

(a) Formed section (7075-T6 bare) *(b) Extruded section (7075-T6)*

(a) Formed Section (7075-T6 Bare). Using Fig. 5.3.5 curve ②,

Segment	Free Edge	b_n (in)	t_n (in)	$\dfrac{b_n}{t_n}$	$t_n b_n$ (in²)	F_{ccn} (ksi)	$t_n b_n F_{ccn}$ (kips)
①	one	0.425	0.05	8.5	0.021	52	1.09
②	no	0.863	0.05	17.3	0.043	63	2.71
③	no	2.46	0.05	49.1	0.123	27.5	3.38
④	one	0.94	0.05	18.7	0.047 Σ0.234	27	1.27 Σ8.45

$$F_{cc} = \frac{\Sigma t_n b_n F_{ccn}}{\Sigma t_n b_n} = \frac{8.45}{0.234} = 36.1 \text{ ksi}$$

(b) Extruded Section (7075-T6). Using Fig. 5.3.4 curve ②,

Segment	Free Edge	b_n (in)	t_n (in)	$\dfrac{b_n}{t_n}$	$t_n b_n$ (in²)	F_{ccn} (ksi)	$t_n b_n F_{ccn}$ (kips)
①	one	0.425	0.1	4.3	0.043	70.5	3.03
②	no	0.863	0.1	8.6	0.086	70.5	6.06
③	no	2.46	0.08	30.8	0.197	44	8.67
④	one	0.94	0.12	7.8	0.113 Σ0.439	66	7.46 Σ25.22

$$F_{cc} = \frac{\Sigma t_n b_n F_{ccn}}{\Sigma t_n b_n} = \frac{25.22}{0.439} = 57.45 \text{ ksi}$$

5.4 Buckling of Thin Sheets

In general, a structural component composed of stiffened sheet panels will not fail when buckling of the panel skin occurs since the stiffening units can usually continue to carry more loading before they fail. However, there are many design situations which require that initial buckling of panel skin satisfy certain design specifications or criteria. For example, the top surface skin on a low wing passenger airplane should not buckle under accelerations due to air gusts which occur in normal everyday flying, thus preventing passengers from observing wing skin buckling in normal flying conditions. Another example would be that no buckling of fuselage skin panels should occur while a commercial airplane is on ground with full load aboard in order to prevent the public from observing buckling of fuselage skin. In many airplanes, fuel tanks are built integral with the wing or fuselage (basically for military fighters or bombers fuselages), thus to eliminate the chances of leakage developing, it is best to design that no buckling of panel skin that bound the fuel tanks occur in 1.0–1.2g flight and landing conditions. In some cases, aerodynamic or rigidity requirements may dictate no buckling of panel skin. To ensure that buckling will not occur under certain load requirements, it is good practice to be conservative in selecting or calculating the boundary restraints of the sheet panels.

The general buckling equation 5.1.1 has been discussed in Section 5.1 and shown below:

$$F_{cr} = KE\left(\frac{g_2}{g_1}\right)^2$$

For thin plate buckling, the above equation becomes,

$$F_{cr} = \frac{k\pi^2 E t^2}{12(1-\mu^2)b^2} \quad (5.4.1)$$

or $\quad F_{cr} = KE\left(\frac{t}{b}\right)^2 \quad (5.4.2)$

where: K and k = Non-dimensional coefficients or constants that depend upon conditions of edge restraint and shape of plate (K_c = compression and K_s = shear).
E = Young's modulus in psi (use tangent modulus E_t if in inelastic range)
t = Thickness of plate in inch
b = Width of plate in inch
μ = Poisson's ratio

Assume a plate as shown in Fig. 5.4.1 is a square configuration.

When the two vertical edges are not being restrained against buckling, the stress required to cause buckling will be as

$$F_{cr} = 0.9 E \left(\frac{t}{L}\right)^2 \quad (5.4.1)$$

where $\quad K_c = 0.9$

If we let the vertical edges be supported by a hinge (actually are simply supported on all four edges conditions) so that the edges can be rotated but must remain in a straight line, this buckled plate must bend

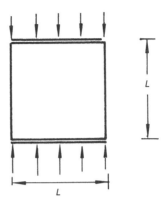

Fig. 5.4.1 A simple square plate.

in two directions and the resistance to buckling is greatly increased; in fact, it will maintain four times the load previously carried with no edge support. Therefore, the buckling equation for this all four edges simply supported is

$$F_{cr} = 3.6 E \left(\frac{t}{L}\right)^2 \quad (5.4.2)$$

where $\quad K_c = 3.6$

When the plate is lengthened in the direction of loading, the principal restraint against buckling is by bending of the plate across the minimum plate dimension, b, as shown in Fig. 5.4.2. This plate is buckled into three waves, each of them being square and acting in the same manner as the plate shown in Fig. 5.4.1.

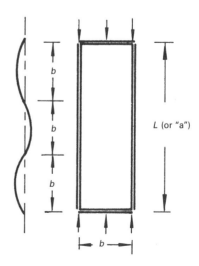

Fig. 5.4.2 A rectangular plate with four edges simply supported.

It is seen that the minimum dimension of the plate or panel is the most important in the buckling equation, so

$$F_{cr} = K_c E \left(\frac{t}{b}\right)^2$$

where $K_c = 3.6$

The dimension b in above equation is always in the perpendicular direction of loading.

Non-dimensional coefficients or constants of common design cases are shown in Fig. 5.4.3, Fig. 5.4.4, Fig. 5.4.7 and Fig. 5.4.8.

		Support conditions at edges	Sources							
			McDonnell-Douglas	Lockheed	NASA	Convair	Rockwell Int'l	Northrop	Republic	Boeing
Plate panel all edges supported	① 		3.62	3.62	4.0	4.0	3.62	3.62	3.62	3.62
	② 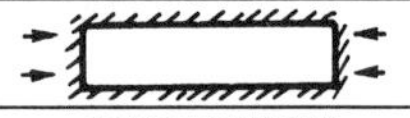		6.3	6.3	6.98	6.98	6.98	6.28	6.45	6.3
	③ 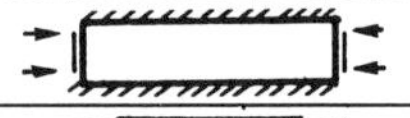		6.25	6.3	—	6.98	6.98	6.28	6.3	6.3
	④ 		3.7	6.32	—	4.0	4.0	3.62	3.7	6.32
	⑤ 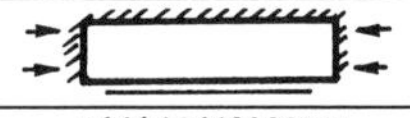		5.1	5.02	—	5.4	—	4.9	5.0	5.0
	⑥ 		5.0	5.02	—	5.4	5.41	4.9	4.9	5.0
Plate panel one edge free	⑦ 		1.2	1.16	1.28	1.28	—	1.2	1.05	1.12
	⑧ 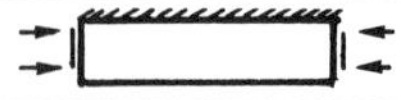		1.17	1.16	—	1.28	1.28	1.2	1.14	1.12
	⑨ 		0.45	0.378	—	0.429	—	0.367	0.4	0,4
	⑩ 		0.429	0.378	0.43	0.429	0.429	0.367	0.388	0.4

▭ Simply supported edge   ▬ Fixed edge

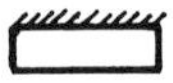

Fig. 5.4.3 Compression buckling coefficients K_c (flat plates).

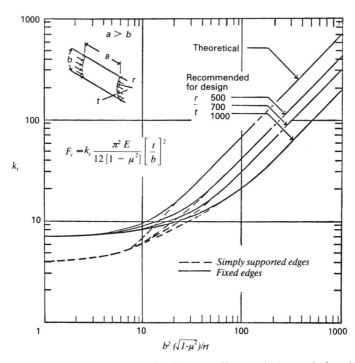

Fig. 5.4.4 Compression buckling coefficients k_c (curved plates).

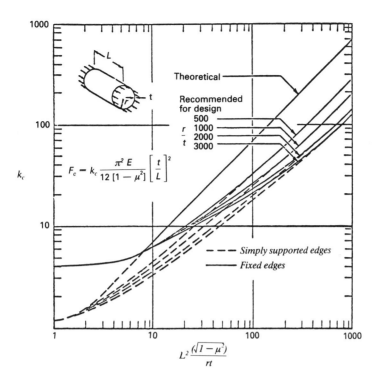

Fig. 5.4.5 Compression buckling coefficients k_c (circular cylinders).

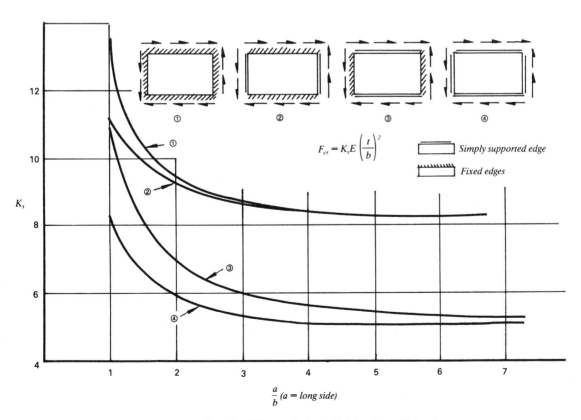

Fig. 5.4.6 Shear buckling coefficients K_s (circular cylinders).

Airframe Structural Design 139

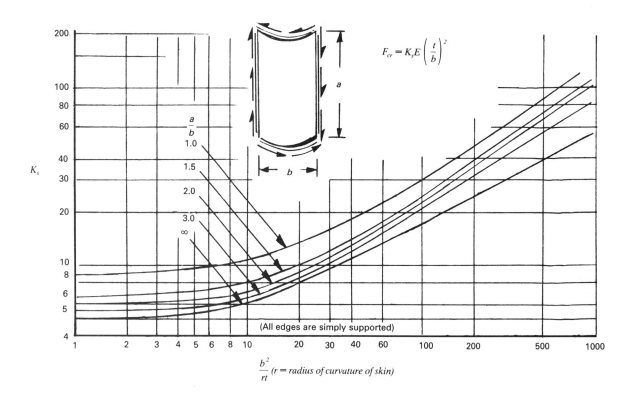

Fig. 5.4.7 Shear buckling coefficients K_s (curved plates, b at curved side).

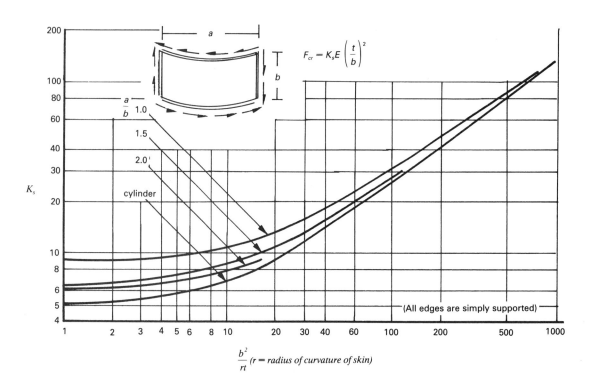

Fig. 5.4.8 Shear buckling coefficients K_s (curved plates, b at straight side).

5.5 Thin Skin-stringer Panel — Compression

This form of construction is a logical development of the necessity of providing a continuous surface for an aeroplane, combined with the requirement that the weight of structure should be as small as possible. Although it is now a commonplace of aircraft engineering, its conception and rapid development over past years has been an engineering triumph of no less importance for aviation than the more publicized developments in other fields of aeronautics.

The components of skin-stringer panel may be classified as follows:

- Longitudinal reinforcing member:
  These are the stringers and longerons of fuselage shells and the spar flanges of wings. They are capable of carrying appreciable tensile loads and, when supported, compressive loads as well. They can carry small secondary bending loads, but their bending rigidity are negligible compared with that of the sections of which they form a part and so it is customary to describe them as direct load carrying members.
- Skin:
  Like all thin shells, this is best suited to carrying load in its own surface as "membrane" stresses. Tensile, compressive and shear loads can be carried, but reinforcement (lateral support) is required for all but the first. The thin skins used in aircraft can only sustain and transmit normal pressure over very short distances by bending. Pressurization loads in a circular section fuselage can, however, be taken by "hoop" tension stresses.
- Transverse reinforcing members:
  These are the rings, frames, bulkheads or diaphragms of fuselages and the ribs of wings. In the design of these members most attention is directed to providing stiffness and strength in the plane of the member. They are therefore usually incapable of carrying much lateral load.

The stringers and rings or ribs are attached to the skins by lines of rivets, spot welds or perhaps bonding (so far this technique has been successfully applied on fuselage construction such as Caravelle, Trident, European Air Bus A-300 and so on). These joints will be called upon to transmit forces mainly along their length. Forces parallel to the skin and directed at right angles to the stringers or rings or ribs will be limited by the torsional flexibility of these members. Forces normal to the skin will be limited in magnitude by the small bending strength of the skin and stringers. The primary function of these joints is thus the transmission, by shear forces, of direct loads in the reinforcing members to the skin and vice versa. Their secondary functions are indeed essential to the working of the structure but do not give rise to such large loads.

Fig. 5.5.1 shows some typical skin-stringer constructions which have been used on existing aircraft structures. Fig. 5.5.1(a) and (b) are the most popular constructions in recent structural design especially the Fig. 5.5.1(a) configuration due to its high structural efficiency (or high Farrar's efficiency factor described in Fig. 5.6.8) and easy assembly. The Fig. 5.5.1(b) is not as efficient as (a) but it has good fail-safe characteristics (see detailed discussion in Chapter 15.0) due to the double row of fasteners attached between stringer and skin. This J-stringer is used to splice the spanwise wing skins or fuselage longitudinal skin joints as shown in Fig. 5.5.2.

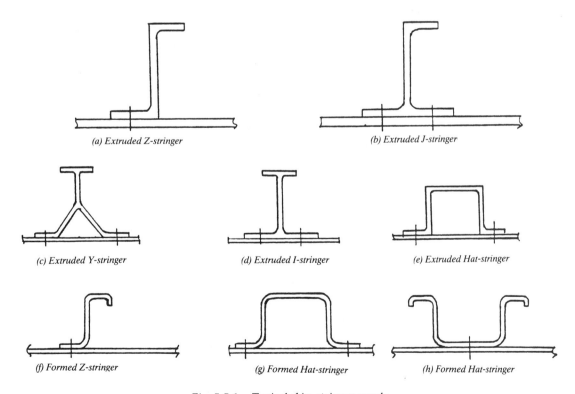

(a) Extruded Z-stringer
(b) Extruded J-stringer
(c) Extruded Y-stringer
(d) Extruded I-stringer
(e) Extruded Hat-stringer
(f) Formed Z-stringer
(g) Formed Hat-stringer
(h) Formed Hat-stringer

Fig. 5.5.1 Typical skin-stringer panels.

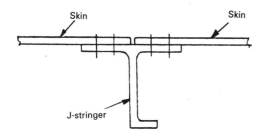

Fig. 5.5.2 The J-stringer used for skin splice as well as compression member.

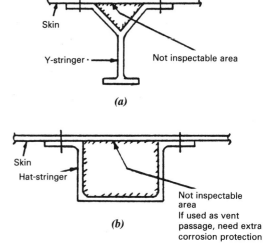

Fig. 5.5.3 The corrosion problem for the Y- and hat-stringer configuration.

Fig. 5.5.1(c) and (d) which are used on some transport aircraft are not adopted again in structural design because of its complication of the stringer attached to rib structure. The Y-stringer configuration is not accepted by commercial operators owing to the corrosion problem where inspection is very difficult as shown in Fig. 5.5.3, even though this configuration has the highest structural efficiency.

Fig. 5.5.1(e) hat-stringer configuration is generally not accepted by structural design because it has the same corrosion problem as the Y-stringer configuration but it could be used on wing upper surfaces as fuel tank vent (fuel pressure) passage as well as compression member to carry compression loads. This configuration has been widely used on modern aircraft for that purpose; however, extra corrosion protection is required in the enclosed area as shown in Fig. 5.5.3(b).

The formed stringers are usually applied on light compression load application such as fuselage skin-stringer panel. Fig. 5.5.1(f) and (h) configurations are frequently seen on existing aircraft fuselage construction, (refer to Chapter 11). The configuration shown in Fig. 5.5.1(g) has higher structural efficiency than the other two but is not acceptable for commercial transports due to the same reason as mentioned before for corrosion problems.

Effective Skin Widths

A skin-stringer panel (or skin-stiffener panel), such as shown in Fig. 5.5.4, is subjected to axial load and the skin will buckle at a stress of $F_{cr,sk} = \dfrac{K_c E}{\left(\dfrac{t}{b}\right)^2}$. Buckling of the skin does not constitute a panel failure. In fact, the panel will carry additional load up to the stress at which the stringer fails. As the stringer stress is increased beyond the skin buckling stress, the skin adjacent to the stringer will carry additional stress because of the support given by the stringers. Stress in the skin between stringers is illustrated in Fig. 5.5.4.

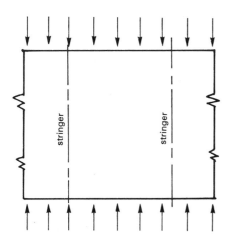
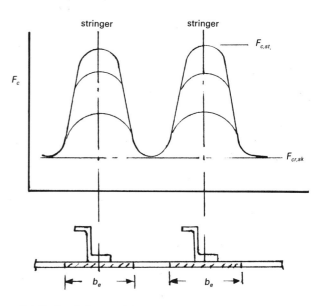

Fig. 5.5.4 Skin stress distribution between stringers.

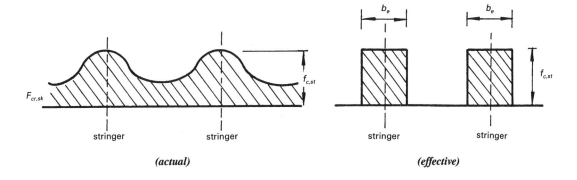

Fig. 5.5.5 *Effective width of skin stress distribution.*

It is noted that the skin stress at the center between stringers does not exceed the initial buckling stress of $F_{cr,sk}$ no matter how high the stress becomes at the stringers.

The stringer stabilizes the skin against buckling with the result that some width of skin acts as part of the stringers and carries loads beyond the general panel buckling stress up to the stringers failure stress. The width of skin which acts as part of the stringer is known as the "effective width" as shown in Fig. 5.5.5. The effective width of skin acts as part of the stringer and is incorporated in the computations of the section properties of the stringer and the load in the skin per stringer spacing as shown below,

$$P_{c,sk} = f_{c,st} \, t \, b_e$$

or $$b_e = \frac{P_{c,sk}}{t f_{c,st}} \quad (5.5.1)$$

where $P_{c,sk}$ = Load in the skin per stringer spacing
$f_{c,st}$ = Stringer compression stress
t = Skin thickness
b_e = Effective width of skin

Therefore, the effective width, at a given stress, is equal to the panel width (between stringers) at which buckling will just begin

$$F_{cr,st} = K_c E \left(\frac{t}{b_e}\right)^2$$

or $$b_e = t\left(\sqrt{\frac{K_c E}{F_{cr,st}}}\right) \quad (5.5.2)$$

where $F_{cr,st}$ = Stringer compression buckling stress
K_c = Skin compression buckling coefficient
E = Young's modulus (use E_t in inelastic range)

A buckled panel exerts a twisting force on the stringers as shown in Fig. 5.5.6. The direction of this twisting moment reverses with each change in buckle direction. These reverses occur at frequent intervals along the length of the stringer and the net twisting moment on the stringer is zero. For a large panel with thin skin, the torsional stiffness of a stringer is large in comparison to the force tending to twist it. This effect produces a fixed edge condition for the skin and the compression buckling coefficient is 6.32 (refer to Fig. 5.4.3. case ②).

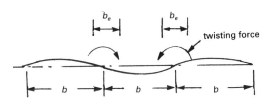

Fig. 5.5.6 *Twisting stringers due to skin compression buckling.*

A narrow panel with thicker skin produces buckling forces so great that the stringer will twist locally. This panel will act as if it had hinged edges and its compression buckling coefficient is 3.62 (refer to Fig. 5.4.3 case ①). It is seen that there are two limits for the compression buckling coefficient K_c and it has been found from tests that $K_c = 3.62$ for $\frac{b}{t} = 40$ and $K_c = 6.32$ for $\frac{b}{t} = 110$. There is a gradual transition from one K_c value to the other, as plotted in Fig. 5.5.7.

Example

Assuming allowable crippling stress of stiffeners = 25000 psi

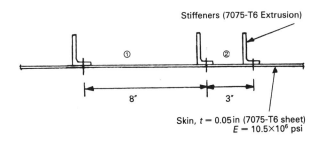

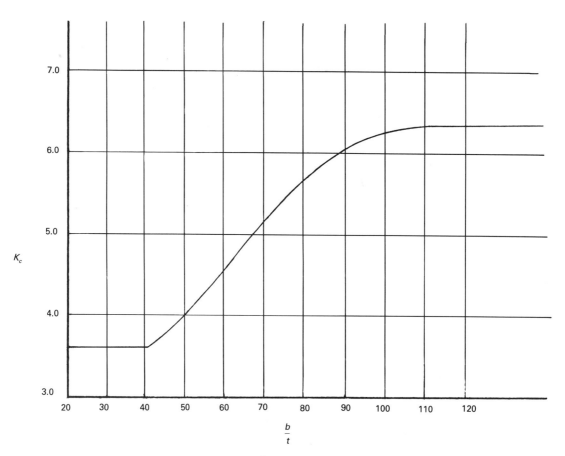

Fig. 5.5.7 K_c value vs. $\dfrac{b}{t}$ for all skin-stringer construction.

From the effective width equation, $b_e = t \sqrt{\dfrac{K_c E}{f_{st}}}$

From Fig. 5.5.7,

$\left(\dfrac{b}{t}\right)_① = 160$ then $K_c = 6.25$

$\left(\dfrac{b}{t}\right)_② = 60$ then $K_c = 4.8$

$b_{e\,①} = (0.05) \sqrt{\dfrac{6.25\,(10.5 \times 10^6)}{25000}} = 2.56$ in

$b_{e\,②} = (0.05) \sqrt{\dfrac{4.8\,(10.5 \times 10^6)}{25000}} = 2.24$ in

Inter-rivet Buckling

Inter-rivet buckling is of special importance on compression panels with skin attached by rivets. Examples of structures where the spacing of rivets (or attachments) is important are: (1) skin-stringer wing surfaces, (2) fuselage stringers, and (3) shells supporting rings. The effective width of skin acting with any compression panel is important in the interest of structural efficiency and weight economy. If the skin buckles between rivets, it can not carry the compression load and the calculated effective width will be erroneous.

The buckling equation 5.4.1 for a flat plate with free edges shown below.

$$F_{cr} = 0.9\,E\left(\dfrac{t}{L'}\right)^2$$

or $$F_{cr} = 0.9\,c\,E\left(\dfrac{t}{L}\right)^2$$

where $c =$ Column end fixity coefficient
$L =$ Column length

Therefore, for a plate with $c = 4$ (see Fig. 5.2.5), the above buckling equation becomes:

$$F_{ir} = 3.6\,E\left(\dfrac{t}{s}\right)^2 \qquad (5.5.3)$$

where $F_{ir} =$ Inter-rivet buckling stress
$s \;=$ Rivet spacing (see Fig. 5.5.8)

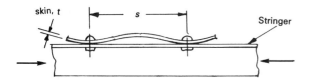

Fig. 5.5.8 *Inter-rivet buckling.*

Equation 5.5.3 has been plotted and shown in Fig. 5.5.9 which provides for inter-rivet buckling in common aluminum alloys. The curve applies to universal head rivets. Suitable corrections must be made if other than universal head rivets are used and the corrections as below:

$c = 4.0$ Universal head
$c = 3.5$ Spotwelds
$c = 3.0$ Brazier head
$c = 1.0$ Countersunk rivets

Normally the skin-stringer construction will be designed so that the rivet spacing is derived from the crippling stress of the stringer. However, when the inter-rivet buckling stress of the skin is reached before the crippling stress of the stringer the skin exhibits the ability to maintain the inter-rivet buckling stress while the stringer continues to take load.

Example A

Obtain the spacing for universal head rivet spacing for the following:

Material: 7075-T6 clad skin $t = 0.050$ in
Compression stress = 32 ksi

Enter Fig. 5.5.9 at 32 ksi and proceed horizontally to the curve ① of 7075-T6 clad material. Next, go vertically to the 0.050 thickness line. Finally, read the spacing 1.68 in. See dotted lines in Fig. 5.5.9.

If countersunk rivet heads are used

c for universal head = 4.0
c for countersunk head = 1.0

$$\text{Spacing required} = 1.68 \frac{\sqrt{1}}{\sqrt{4}} = 0.84 \text{ in}$$

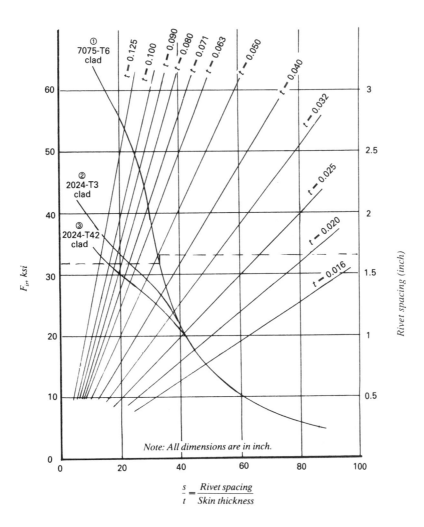

Fig. 5.5.9 *Inter-rivet buckling curves for aluminum alloys. (Euler-Engesser Equation with $c = 4.0$)*

Another method of obtaining the inter-rivet buckling spacing is to calculate effective width of skin based on the crippling stress of the stringer and it is necessary to space the rivets of skin such that $F_{cc} = F_{ir}$. Then enter the appropriate material column curve as shown in Fig. 5.2.12 and read out the corresponding $\dfrac{L'}{\rho}$ value. This value multiplied by 0.577 is the rivet spacing to skin thickness ratio $\dfrac{s}{t}$.

Example B

Use the same data given in Example A to obtain the universal head rivet spacing by using column curve method.

Enter Fig. 5.2.12 with compression stress = 32 ksi and read $\dfrac{L'}{\rho} = 57$

$$\dfrac{s}{t} = 0.577 \left(\dfrac{L'}{\rho}\right) = 0.577\,(57) = 32.9$$

Therefore, rivet spacing = 32.9 (0.050) = 1.64 in

If the stringer flange next to the skin cripples at a stress lower than the average crippling stress of the stringer, the buckling of the flange will force buckling of the skin at the reduced value. The rivet spacing should be based on the F_{cc} of the stringer attaching to the skin and the effective width of skin acting with the stringer should be reduced accordingly.

When panels are loaded in combined compression and shear or tension and shear, the rivet of skin to stringer presents a problem in net section. It is often necessary to compromise the rivet spacing in order to prevent the skin from becoming critical in net section.

5.6 Skin-stringer Panel — General

For thin skin supported by sturdy stiffeners, the initial buckling stress can usually be calculated assuming that the skin buckles between stringers with some rotational restraint by the stringer, while the failing stress can be calculated by considering the stringer together with an "effective width" of skin to fail in flexure as an Euler strut. The effective width concept accounts, in a simple way, for the interaction between the skin and the stiffener.

Where the skin is thick the initial buckling stress of the skin may be comparable to the failure stress of the stringer and both may approach the yield stress of the material. It may no longer be possible to regard the stringer as "sturdy" and it becomes necessary to take into account the flexural, torsional, and local deformations of the stiffener. These deformations govern the restraint given by the stringer to the skin, at initial buckling. They also govern the final modes of failure, where the stiffener fails as a strut with one of the above modes predominating.

The skin-stringer (or called stiffened panel) construction can develop several separate type of instability, which may be coupled to a greater or less degree.
- Initial buckling (skin buckling): This generally involves waving of the skin between stringers in a half-wavelength comparable with the stringer pitch.

There will also be a certain amount of waving of the stringer web and lateral displacement of the free flange. For some proportions these may become larger than the skin displacements, and the mode becomes more torsional or local in nature. See Fig. 5.6.1.

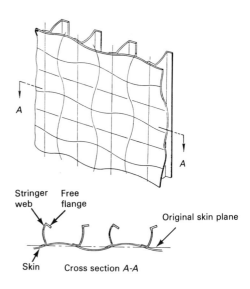

Fig. 5.6.1   Initial buckling of a skin-stringer panel.

- Local instability: A secondary short wavelength buckling may take place in which the stringer web and flange are displaced out of their planes in a half-wavelength comparable with the stringer depth. There will be smaller associated movements of the skin and lateral displacements of the stringer free flange.
- Flexural instability (or called Euler mode): Simple strut instability of the skin-stringer construction in a direction normal to the plane of the skin. There may be small associated twisting of the stringers. The half-wavelength is generally equal to the rib or frame spacing. See Fig. 5.6.2.

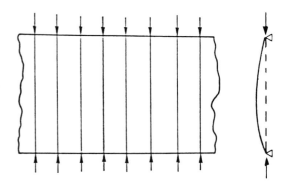

Fig. 5.6.2   Panel instability of a skin-stringer panel.

- Torsional instability: The stringer rotates as a solid body about a longitudinal axis in the plane of the skin, with associated smaller displacements of the skin normal to its plane and distortions of the stringer cross section. The half-wavelength is usually of the order of three times the stringer pitch. See Fig. 5.6.3. and Fig. 5.6.4.
- Inter-rivet buckling: Buckling of the skin as a short strut between rivets. This can be avoided by using a sufficiently close rivet pitching along the stringer. See Fig. 5.6.5.
- Wrinkling: A mode of instability similar to inter-rivet buckling in which the skin develops short-wavelength buckling as elastically supported along strut (or as column of stringer). For all practical skin-stringer constructions it can be avoided by keeping the line of attachments very close to the stringer web. See Fig. 5.6.6.

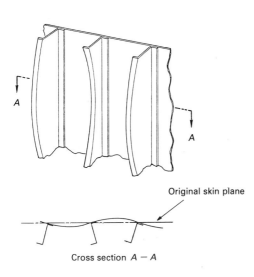

Fig. 5.6.3 Torsional buckling of a skin-stringer panel.

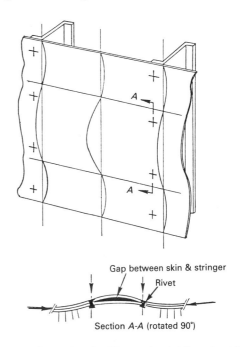

Fig. 5.6.5 Inter-rivet buckling and wrinkling of a skin-stringer panel.

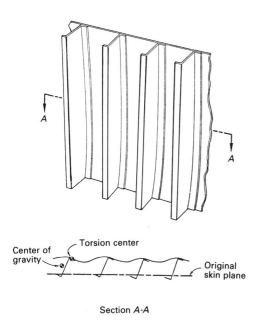

Fig. 5.6.4 Flexual and torsional buckling of a skin-stringer panel.

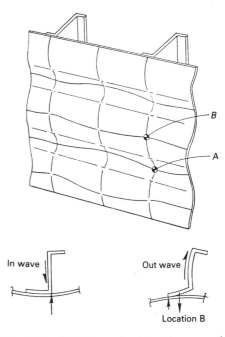

Fig. 5.6.6 Wrinkling of a skin-stringer panel.

The buckling and failure of skin-stringer panels are complex problems encountered in this kind of construction (such as wing surfaces subjected to compression loads). Because there are many dimensional parameters and different modes of buckling to be considered, the determination of the strength of panels and the selection of the most efficient panel are challenging problems, which require skill and considerable experience on the part of structural engineers. In spite of extensive studies, the design of the panels in actual practice is not based on theoretical solutions alone. Very often test data and the design charts, prepared on the basis of data, also are used.

In the following, rational analytical analysis is presented for computing the failing stresses of flat skin-stringer panels. Because of the high efficiency and advantages due to simplicity in shape and construction, and because of the need to clarify the mechanism of failures and to explain how the methods of analysis are revised, Z-section stringer panels are considered herein. It is assumed that the stringers are extruded or formed, and that the pitch and the diameters of rivets used in the construction of panels are such that they yield the potential strength of the panels.

When the skin-stringer construction approaches its Euler instability stress, development of skin buckling, local instability, torsional instability, inter-rivet buckling or wrinkling will so reduce the flexural stiffness as to cause premature collapse. If the Euler instability stress is reasonably remote, instability of skin buckling will not precipitate failure, and the structure will carry increased load, with the skin buckled, until failure occurs by the onset of instability of local torsional, inter-rivet or wrinkling. In general, an excessive margin of flexural stiffness is needed to prevent failure due to any of these four modes.

Initial Buckling

The local buckling mode which is the first to develop is a mixture of modes of skin buckling, local instability and torsional instability, the predominant type of buckling being dictated by the geometry of the particular skin-stringer construction used. The stress at which this initial buckling occurs has been determined theoretically for a wide range of skin-stringer combinations, allowance being made for all the interactions between the various modes of distortion. Typical results are shown in Fig. 5.6.7 in which the non-dimensional buckling stress $\frac{f_b}{f_o}$ is plotted against $\frac{A_s}{bt}$ for various values of $\frac{t_s}{t}$. (Note: f_o is the buckling stress of the skin if pin-edged along the stringers, and f_b the actual initial buckling stress). The upper portions of the curves correspond to a skin-buckling-and-stringer local type of instability, while the lower portions of the curves correspond to a stringer torsional-and-lateral type of instability over a longer wavelength. The change of slope in the curves takes place when these two types of initial buckling occur at the same stress.

Flexural Instability

Considering a stringer associated with a pitch b of skin, the whole cross section is fully effective until

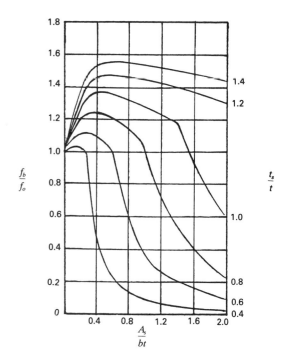

Fig. 5.6.7 Initial buckling stress of flat panel with Z-section stringer. ($\frac{d}{h} = 0.3$, see Fig. 5.6.8.)

initial local buckling occurs. In general, the stringer will not develop pure flexural instability: there will also be a certain amount of stringer twisting, the analysis of which is far from simple. However, the type of design which this analysis will show to be most efficient is one in which flexural-torsional coupling is small and it will therefore be assumed that pure flexural instability occurs. Use Farrar's efficiency factor,

$$F = f \sqrt{\frac{L}{NE_t}} \qquad (5.6.1)$$

where

f = Mean stress realized by skin and stringer at failure ($f = \frac{N}{T}$)

N = Compressive end load carried per inch width of skin-stringer panel

T = Average panel thickness which has the same cross-section area as the skin-stringer panel
$= \frac{tb + A_s}{b}$

$f_o = 3.62 E \left(\frac{t}{b}\right)^2$

f_b = Initial local buckling stress
E = Young's modulus of skin-stringer material
E_t = Tangent modulus of skin-stringer material
L = Rib or frame spacing
A_s = Cross-section area of stringer
t = Skin thickness
t_s = Stringer web thickness

b = Stringer spacing
h = Depth of stringer
d = Flange width of stringer

F is a function of $\dfrac{A_s}{bt}$ and $\dfrac{t_s}{t}$ and it is a measure of the structural efficiency of the skin-stringer construction. The quantity F ($= f \sqrt{L/NE_t}$) is plotted against $\dfrac{A_s}{bt}$ and $\dfrac{t_s}{t}$ in Fig. 5.6.8. It is seen that an optimum value of $\dfrac{A_s}{bt}$ and $\dfrac{t_s}{t}$ exists, at which for a given N, E_t and L the stress realized will be a maximum. For this optimum design with $\dfrac{A_s}{bt} = 1.5$ and $\dfrac{t_s}{t} = 1.05$ then

$$f = 0.95 \sqrt{\dfrac{NE_t}{L}}$$

which means that the maximum value of $F = 0.95$ is achieved for Z-stringer panels.

It is also noticeable that a ridge of high realized stress exists for the family of designs where the two types of local buckling occur simultaneously; if for any reason the minimum-weight design cannot be used, it is economic to use designs of this family. (The general principle seems to emerge that the most efficient designs are those in which failure occurs simultaneously in all possible buckling modes.)

In practice the full theoretical value of F is not achieved; experimental results indicate that about 90% is an average realized value. This reduction is due to the effect of initial eccentricity of the structure, and (unlike the case of a simple strut) it occurs even at low design stresses. Suppose that the Euler stress and skin buckling stress of a particular skin-stringer construction are both 30 ksi. When an actual compressive stress of 27 ksi is reached, it is likely that the bow due to initial eccentricity will give an additional stress (due to bending) of 3 ksi in the skin: thus the total stress in the skin will be 30 ksi, the skin will buckle and induce premature flexural failure. A simple method of allowing for these effects is to design for a theoretical failing load per inch N (assuming no eccentricity) somewhat greater than that actually required. A more refined method, which gives slightly more efficient structures, is to provide a slightly greater margin of flexural stress than is allowed for local buckling, the actual margin depending on the standard of straightness (typical $e = \dfrac{L}{1000}$ inch as shown in Fig. 5.6.9) of the stringers which can be achieved.

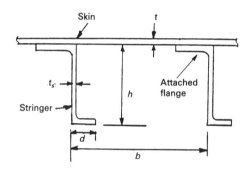

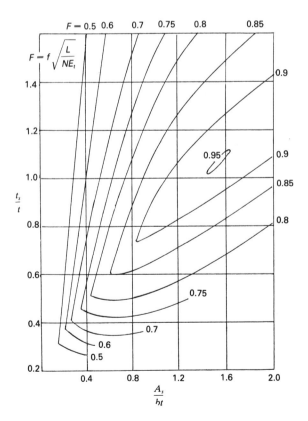

Fig. 5.6.8 *Contour of F values for Z-stringer where initial buckling coincides with failure.*

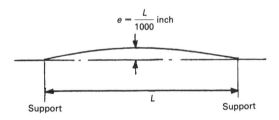

Fig. 5.6.9 *Typical straightness of stringer.*

In Fig. 5.6.10, the structural efficiency, as measured by $f \sqrt{\dfrac{L}{NE_t}}$, of the designs is plotted against the ratio of the skin buckling stress to the stress at failure. It is seen that the skin should either not be allowed to buckle at all, or should buckle at a comparatively low stress, if good structural efficiency is desired.

In Fig. 5.6.11, the best possible results using Z-section stringers are given for current light alloy material. It is assumed that the optimum design is used, and the mean stress thus achieved is plotted against the value of $\frac{N}{L}$. It is seen that the working stress, and hence the structural weight, is dictated entirely by the value of $\frac{N}{L}$ which is used, and typical values of this quantity of various aircraft components are shown.

Type of panel	Theoretical best value of F	Realized value of F
Z-section, primary buckling causing failure	0.95	0.88
Hat-section, primary buckling causing failure	0.96	0.89
Y-section, primary buckling causing failure	1.25	1.15

It shows that of the range explored, the Y-section stringer is the most efficient, although for values of $\frac{N}{L}$ less than 100 (see Fig. 5.6.11) the buckled skin and Z-section is as good and in practice is more robust.

The results show that at present aluminium alloys are the most efficient at high values of the loading coefficient $\frac{N}{L}$; at lower values of $\frac{N}{L}$, magnesium alloys are more efficient; at lower values still, wood is efficient, and for very low values of $\frac{N}{L}$ (small model airplanes) balsa wood is indicated as the best structural material.

Considerations Affecting Rib or Frame Spacing

Since the stress achieved by the skin-stringer construction varies as $\frac{N}{L}$, a light construction for a wing surface for example is achieved by placing the ribs close together. When this is done, however, the weight of ribs may be considerable, and the structure of minimum weight in fact may be associated with a wider rib spacing, the increase of weight of the skin and stringers being more than offset by the saving on ribs.

The design of ribs is a detail matter and the weight of a rib cannot be predicted in practice from pure theoretical considerations. Accordingly, a more or less empirical approach must be adopted.

Consideration of Riveting

Special NACA riveting have the manufactured head on the stringer side of panel. A comparison of NACA riveting and conventional riveting is shown in Fig. 5.6.12.

The manufactured head, either flat or brazier, is placed on the flange side in NACA riveting, and the upset head is either upset into a countersink or upset flat on the skin side. Upsetting is accomplished by squeezing the rivet in a press rather than by hammering. This riveting process tends to yield higher allowable stresses for wrinkling than is obtained with conventional riveting. The increase is large for the same rivet location, and it is ascribed to the stiffening effect of the large rivet head on the attached flange as compared to conventional riveting where the rivet head is really in contact with the flange only at the edge of the hole.

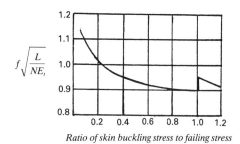

Fig. 5.6.10 Effect of skin buckling on theoretical stress realized by optimum Z-stringer-skin construction.

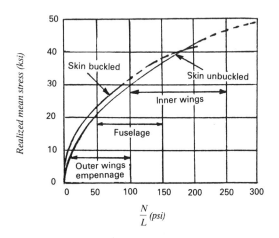

Fig. 5.6.11 Theoretical stress realized by optimum Z-stringers and skin in aluminum alloy.

Now it can be established that the structural efficiency of a skin-stringer construction can be measured by the constant F, which is $f\sqrt{\frac{L}{NE_t}}$ and which has a definite maximum value for any given type of stringer. The results of some similar calculations and test for stringer sections other than Z-stringer are given below:

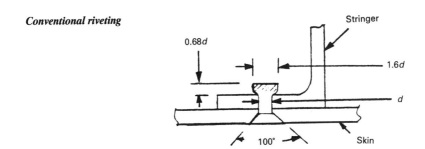
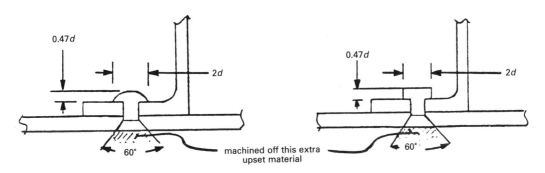

Fig. 5.6.12 Comparison of NACA riveting and conventional riveting.

Design Curves

In modern aircraft structural design, the high accuracy calculation to obtain the highest efficiency of the structure is possible through the use of computer analysis. A system of computer programs to optimize and analyze skin-stringer panel (or called stiffened panel) for fuselage and lifting surfaces (wing and empennage structures) have been developed by aircraft manufactures for their own use. These optimization programs consist of synthesis of the least weight within the input constraints that is adequate for multiple input load cases. Each input load case consists of compression, shear and surface pressure; output includes failure modes and margins of safety. All these analyses have been correlated with several hundred tests. In Fig. 5.6.13 shows one of the typical strength envelope of the stressed-skin construction of Z-stringer.

The horizontal scale shows panel compression loads N (kip/in), and the vertical scale shows panel shear flow Q (kip/in). This panel can carry any combination loads of N and Q as long as these loads are within the envelope curve. The margin of safety is calculated as below:

Assume a load case of N and Q as shown at point B in Fig. 5.6.13 and then draw a straight line between O and B and intersect envelope curve at point A. Therefore

$$\text{M.S.} = \frac{\overline{OA}}{\overline{OB}} - 1$$

Fig. 5.6.14 is a series of curves for which it can be readily determined whether or not a section is optimum. In order to develop these curves, several assumptions had to be made. Since there are so many dimensions on a typical section it would be nearly impossible to vary them all. Fig. 5.6.14 are based on the section as shown below:

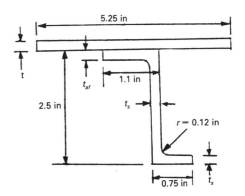

Skin (7075-T76 plate)
Stringer (7075-T6511 extrusion)
L = Length of panel = 26 in

Airframe Structural Design 151

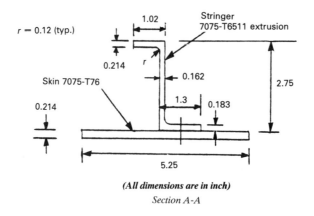

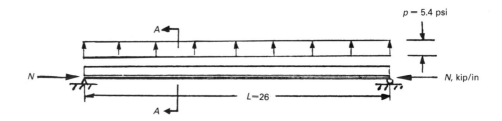

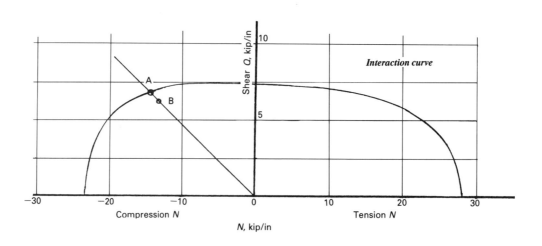

Fig. 5.6.13 Skin-stringer panel allowable strength envelope curve (interaction curve).

A constant lateral pressure of 10 psi is applied to the section assuming variable dimensions for t, t_{af} and t_s from which a series of compression allowable of N with $Q = 0$ are generated. A typical example is also shown in Fig. 5.6.14.

Similar curves are developed for the formed sheet stringers (Z-section) as shown in Fig. 5.6.15.

152 Airframe Structural Design

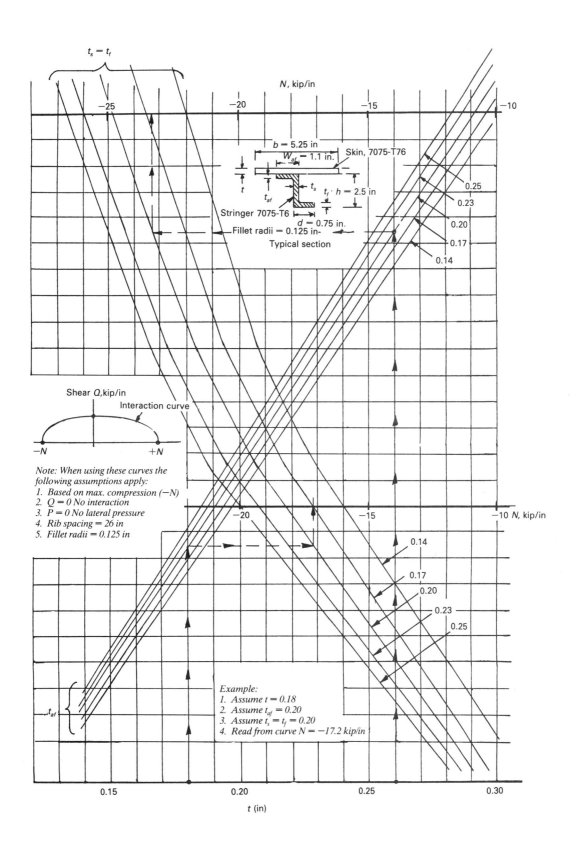

Fig. 5.6.14 *Skin-stringer (extruded) panel allowable curves.*

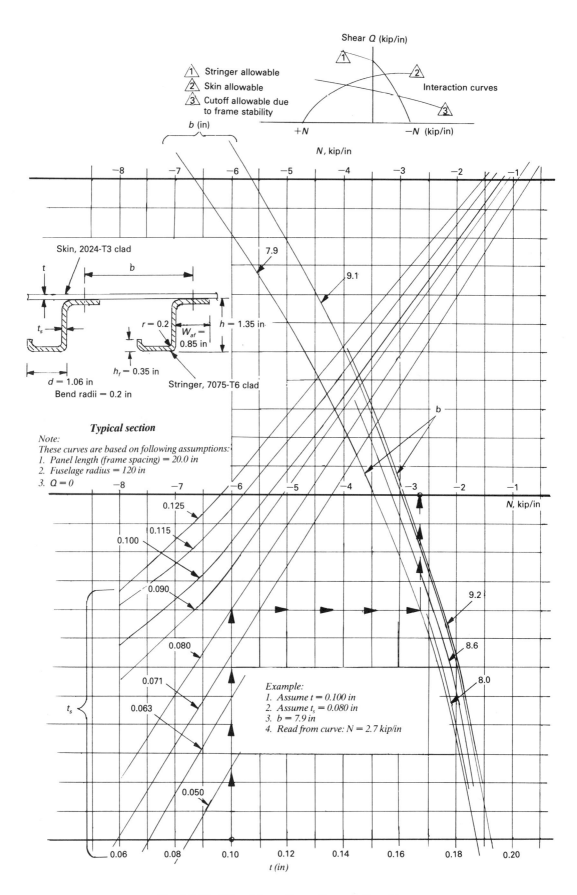

Fig. 5.6.15 Skin-stringer (formed) panel allowable curves.

154 Airframe Structural Design

5.7 Integrally Stiffened Panel

The high performance levels in machines and equipment continue to place more exacting demands on the design of structural components. In aircraft, where weight is always a critical problem, integrally stiffened structural sections as shown in Fig. 5.7.1 have proved particularly effective as a lightweight, high strength construction. Composed of skin and stiffeners formed from the same unit of raw stock, these one-piece panel sections can be produced by several different techniques. Size and load requirements are usually the important considerations in selecting the most feasible process.

For highly loaded long panels, extrusions or machined plates are most commonly employed. Since extrusion width capacities are somewhat limited, attachment of a series of extruded panels with spanwise splices may be required in some applications. Splicing can often be eliminated or reduced with billet or plate stock of sufficient width to permit machining an integral structure or, at least, to provide a structure with fewer lineal sections. Section discontinuities such as encountered in the region of cutouts can often be produced more easily from machined plate.

From a cost standpoint it is usually better to extrude integrally stiffened structures than to machine a section of the same size from a billet or plate. However, the minimum extrusion thickness is about 0.1 inch and frequently, where loads are light, considerable additional machining is needed to produce an efficient design. In some of these cases, machining from a billet seems to be more economical. Furthermore, under light loads a machined section with risers as shown in Fig. 5.7.1(a) may be sufficient. Actually, limitations on minimum thickness for both machined plates and extrusions may preclude the efficient use of these methods for lightly loaded panels. Where thin section applications are encountered, rolled and forged integrally stiffened structures may be more suitable.

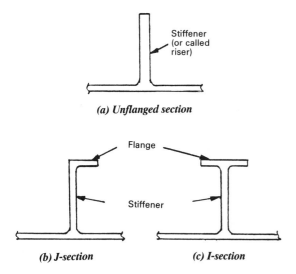

Fig. 5.7.1 Three popular sections of integrally stiffened panels.

From a structural standpoint, appreciable weight savings are possible through the integral section design which also develops high resistance to buckling loads. In addition, the reduction in the number of basic assembly attachments gives a smooth exterior skin surface. Another potential advantage of the integrally stiffened structure is the elimination of attached flanges with a corresponding redistribution of this material for optimum stiffener proportions. However, some production development work is necessary to exploit this advantage fully. In aircraft applications, the most significant advantages of integrally stiffened structures over comparable riveted panels (skin-stringer panel) have been:

- Reduction of amount of sealing material for pressurized shell structures.
- Increase in allowable stiffener compression loads by elimination of attached flanges.
- Increased joint efficiencies under tension loads through the use of integral doublers, etc.
- Improved performance through smoother exterior surfaces by reduction in number of attachments and nonbuckling characteristics of skin.

Integral fuel tanks and pressurized shells usually create sealing problems for a riveted structure. These problems are eliminated to a large extent by integral stiffeners. Integrally stiffened structures have their greatest advantage in highly loaded applications because of their minimum section size.

The integral structure is often designed so that the skin panel is unbuckled at ultimate loads. For a compression panel, it can be shown that optimum design provides either a nonbuckled skin or a skin that is completely buckled with maximum material in the stiffener. However, most aircraft structures carry shear loads in addition to axial load and from this standpoint unbuckled skin is preferable.

A method is given enabling rapid determination of the optimum cross-sectional dimensions of compression surfaces having unflanged integral stiffeners, and consideration is given to the effects of practical limitations on the design. The theoretical efficiency of the optimum integral design is found to be only 85% of that of optimum skin-stringer panel with Z-stringer design.

In view of the current interest in employing machined surfaces which are unflanged (a very popular design due to its easy machining process with less cost) for major aircraft components, it is highly desirable to have a rapid method of achieving optimum (i.e. minimum weight) design of these surfaces. Such a method, presented in this Chapter, is only for compression axial load which is the primary design load. The interaction between axial loads and panel shear loads are beyond the scope of this discussion and the detailed analysis of this interaction is complicated and computer assistance is needed. However, 80—90% of the panel sizing are dependent on the axial loads; therefore, for preliminary design purpose, it is sufficient to use the axial loads to size the structural panel.

Assumptions (see Fig. 5.7.2):

- The panels are assumed to be sufficiently wide to allow their treatment as simple column, i.e. no restraint is imposed on the longitudinal edges of the panels.

Airframe Structural Design 155

- Each panel is assumed in the analysis to be pin-ended over its bay length L; but account may be taken of end fixity by regarding L as the effective pin-ended length rather than as the actual bay length.
- The depth of the stiffeners is assumed to be small compared with the overall (aircraft wing for example) structural depth, and hence movements of the panel neutral axis have a negligible effect on the value of the load intensity
- The ratios $\dfrac{t_s}{d}$ and $\dfrac{t}{b}$ are assumed sufficiently small to allow application of thin-plate buckling theory.
- The ribs or frames are assumed to impose no restraint on the local buckling.
- The effective modulus throughout the panel in the plastic range is assumed to be the tangent modulus.
- The best designs are assumed to be those for which the initial buckling and Euler instability stresses coincide.

The initial buckling stresses of a variety of skin-stiffener combinations are presented in Fig. 5.7.3 as ratios of section buckling stress to simply-supported skin buckling stress f_b/f_o and these buckling stresses take full account of the interaction between plate and stiffener buckling, but any effects of the stiffener root fillet have been neglected.

It is noted that for a given material, load intensity and effective bay length, the achieved stress is a maximum (and hence the weight is a minimum) when F is a maximum. See Equation 5.6.1.

For a given material, for which the relationship between f and E_t is known, equation of $F = f\left(\sqrt{\dfrac{L}{NE_t}}\right)$ or $f = F\left(\sqrt{\dfrac{NE_t}{L}}\right)$ may be used to plot curves of f against the structural index $\dfrac{N}{L}$ for various values of F, and this has been done in Fig. 5.7.4 for an aluminum 2024-T6 extrusion material. Also shown in Fig. 5.7.4 is a curve for $F = 0.95$ which is the largest theoretical value achievable with Z-stringer of skin-stringer panel.

Design Procedures for An Unflanged Stiffened Panel

Notations

b = Skin width between stiffener (or riser), see Fig. 5.7.2.
d = Stiffener depth, from skin center line to tip
E_t = Tangent modulus
f = Panel applied stress
f_b = Panel initial buckling stress
f_o = Initial buckling stress of a long plate of width

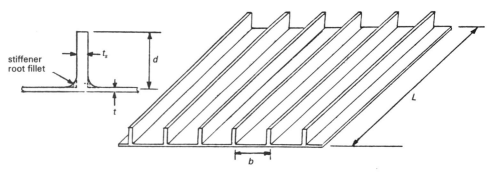

Fig. 5.7.2 Typical unflanged integrally stiffened panel.

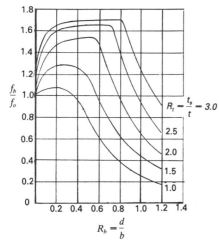

Fig. 5.7.3 Initial buckling stress of panel having unflanged integrally stiffened panel.

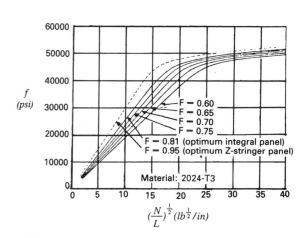

Fig. 5.7.4 Graphs of panel failing stress against structural index for 2024-T3 aluminum alloys.

b and thickness t, simply supported along its edges $= 3.62 E_t \left(\dfrac{t}{b}\right)^2$

f_e = Euler instability stress
F = Farrar's efficiency factor
ρ = Radius of gyration of skin-stiffener combination about its own neutral axis
L = Effective pin-ended bay length

$J_1 = b \left[\dfrac{E_t}{(NL^3)}\right]^{\frac{1}{4}}$

$J_2 = t_s \left[\dfrac{E_t}{(NL)}\right]^{\frac{1}{2}}$

$J_3 = d \left[\dfrac{E_t}{(NL^3)}\right]^{\frac{1}{4}}$

$J_4 = t \left[\dfrac{E_t}{NL}\right]^{\frac{1}{2}}$

N = Load per unit panel width
t = Skin thickness
t_s = Stiffener (or riser) thickness
$R_b = \dfrac{d}{b}$

$R_t = \dfrac{t_s}{t}$

For any panel section,

local buckling stress $f_b = \left(\dfrac{f_b}{f_o}\right)\left[3.62 E_t \left(\dfrac{t}{b}\right)^2\right]$ (5.7.1)

Euler instability stress $f_e = \pi^2 E_t \left(\dfrac{\rho^2}{L^2}\right)$ (5.7.2)

where $\rho^2 = \dfrac{b^2 R_b^3 R_t}{12(1 + R_b R_t)^2}(4 + R_b R_t)^2$

ρ is radius of gyration of the skin-stiffener (riser) combination about its neutral axis.

Relating the stress in the panel to the load intensity

$f = \dfrac{N}{t(1 + R_b R_t)}$ (5.7.3)

Now impose the condition,

$f = f_b = f_e$

Take Equation (5.7.1) × (5.7.2) × (5.7.3)², obtain

$f^4 = \pi^2 E_t^2 \dfrac{\rho^2 f_b (3.62)}{L^2 f_o b^2}\left[\dfrac{N^2}{(1 + R_b R_t)^2}\right]$

Hence $f = F \left(\dfrac{NE_t}{L}\right)^{\frac{1}{2}}$ (5.7.4)

where Farrar's efficiency factor,

$F = 1.314 \dfrac{[R_b^3 R_t (4 + R_b R_t)]^{\frac{1}{4}}}{1 + R_b R_t}\left(\dfrac{f_b}{f_o}\right)^{\frac{1}{4}}$ (5.7.5)

Determination of Actual Dimensions of Optimum Panel

From equation 5.7.3 and 5.7.4,

$t = \dfrac{1}{F(1 + R_b R_t)}\left(\dfrac{NL}{E_t}\right)^{\frac{1}{2}}$

$= J_4 \left(\dfrac{NL}{E_t}\right)^{\frac{1}{2}}$ (5.7.6)

For the optimum design, $R_b = 0.65$, $R_t = 2.25$ and $F = 0.81$, therefore,

$t = 0.501 \left(\dfrac{NL}{E_t}\right)^{\frac{1}{2}}$ (5.7.7)

From Equations (5.7.1), (5.7.4), (5.7.5) and (5.7.6)

$b = 1.103 \dfrac{\sqrt{F(1 + R_b R_t)}}{[R_b^3 R_t (4 + R_b R_t)]^{1/2}}\left(\dfrac{NL^3}{E_t}\right)^{\frac{1}{4}}$

$= J_1 \left(\dfrac{NL^3}{E_t}\right)^{\frac{1}{4}}$ (5.7.8)

For the optimum design,

$b = 1.33 \left(\dfrac{NL^3}{E_t}\right)^{\frac{1}{4}}$ (5.7.9)

and t_s may then be calculated from

$d = R_b b$ (5.7.10)

and

$t_s = R_t t$ (5.7.11)

For example, the optimum design,

$d = 0.65 b$ (5.7.12)

$t_s = 2.25 t$ (5.7.13)

Effect of Limitations on Stiffener Thickness and Pitch

Along the length of any one integral surface (extending over a number of bays or spans) it may be considered necessary, for ease of machining, to keep t_s and b constant, even though N and L will vary. To facilitate design in these circumstances, contours of

and $\quad J_1 = b \left(\dfrac{E_t}{NL^3}\right)^{\frac{1}{4}}$

$J_2 = t_s \left(\dfrac{E_t}{NL}\right)^{\frac{1}{2}}$

have been plotted in Fig. 5.7.5 and the use of these contours is described below:

Design an optimum panel, i.e. $F = 0.81$, $R_t = 2.25$ and $R_b = 0.65$ and to follow the following steps:

(a) From the curves of f against $\dfrac{N}{L}$ for the material such as shown in fig. 5.7.4 of 2024-T6 aluminum to find f corresponding to $\dfrac{N}{L}$ and with $F = 0.81$.

(b) From the curve of E_t against f for the material, find E_t corresponding to f.
(c) Determine t, b, d and t_s from Equations (5.7.7) to (5.7.13).

At another section on this panel,

(d) Assume a value for E_t' (e.g. as a first approximation make $E_t' = E_t$).
(e) Since let $b' = b$ and $t_s' = t_s$ and then J_1' and J_2' may be calculated.
(f) Determine F' from Fig. 5.7.5.
(g) Determine f' from Fig. 5.7.4.
(h) Find E_t' corresponding to f'.

(i) Repeat steps (d) to (h) until the value of E_t' in step (d) and (h) coincide. Repetition will be unnecessary in the elastic range.
(j) Read off R_b' and R_t' from Fig. 5.7.5, hence find

$$d' = R_b' b' \text{ and } t' = \frac{t_s'}{R_t'}$$

A variety of panel designs and panel weight may be obtained by placing the optimum panel section at various positions, but experience should enable a choice of the correct position to achieve overall minimum structural weight.

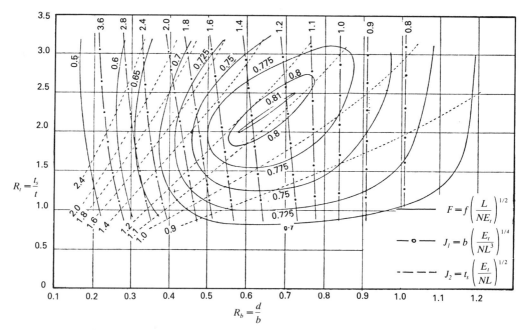

Fig. 5.7.5 Design chart for unflanged integrally stiffened panel (with limitations of stiffener thickness and pitch).

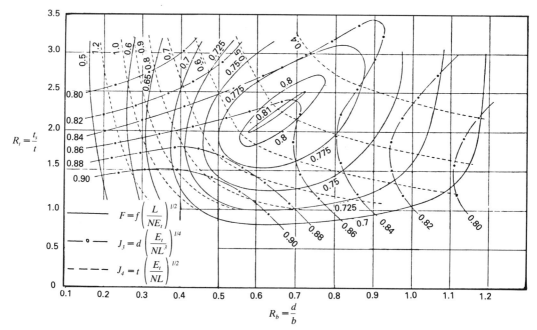

Fig. 5.7.6 Design chart for unflanged integrally stiffened panel (with limitations on skin thickness and stiffener depth).

158 Airframe Structural Design

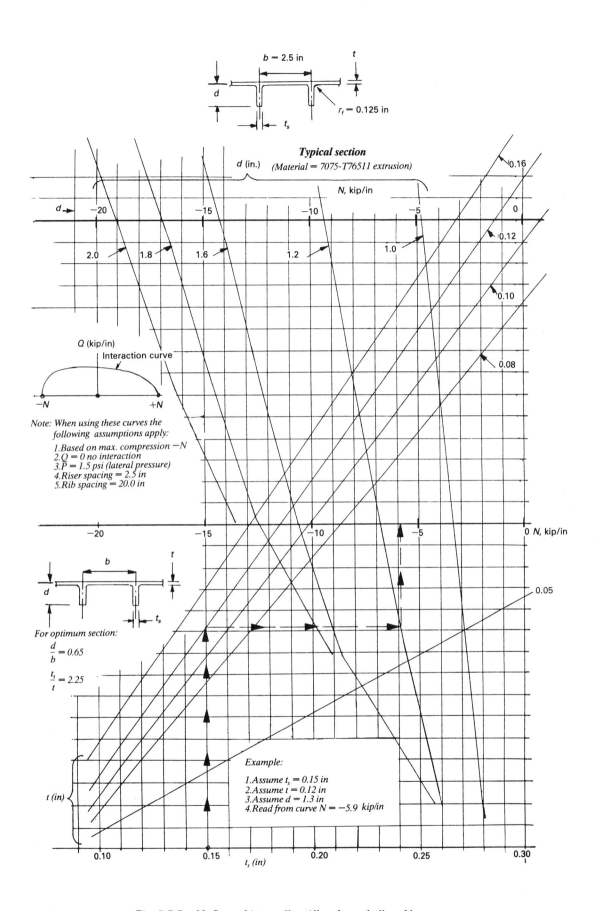

Fig. 5.7.7   Unflanged integrally stiffened panel allowable curves.

Effect of Limitations on Skin Thickness and Stiffener Depth

Additional limitations may be imposed by (a) skin thickness requirements and (b) maximum depth of stiffener. Fig. 5.7.6 which includes contours of

$$J_3 = d \left[\frac{E_t}{(NL^3)} \right]^{\frac{1}{4}}$$

and

$$J_4 = t \left[\frac{E_t}{NL} \right]^{\frac{1}{2}}$$

has been prepared to facilitate optimum design of panels subjects to such limitations.

If the integrally stiffened panel construction is to conform to the ideal compression material distribution, it will be necessary to be able to taper skin thickness, stiffener pitch and stiffener dimensions. With unflanged integrally stiffened panel, this can be done on existing machines with easy operation and appears to be an economic production. With flanged integrally stiffened panel, the problem of machining these various tapers is great and does not appear to be an economic proposition.

Fig. 5.7.7 presents a series of curves for designing the unflanged integrally stiffened panels made from aluminum 7075-T76 extrusions. Since there are so many dimensions for this panel section and using fixed values for L, b and r_f and variable dimensions for d, t_s and t from which the N allowables (with $Q = 0$) are generated.

References

5.1 Farrar, D.J.: 'The Design of Compression Structures for Minimum Weight'; *Journal of The Royal Aeronautical Society*, (Nov. 1949), 1041–1052.

5.2 Emero, D.H. and Spunt, L.: 'Wing Box Optimization under Combined Shear and Bending.' *Journal of Aircraft*, (Mar.–Apr., 1966), 130–141.

5.3 Melcon, M.A. and Ensrud, A.F.: 'Analysis of Stiffened Curved Panels under Shear and Compression'. *Journal of the Aeronautical Sciences*, (Feb. 1953), 111–126.

5.4 Catchpole, E.J.: 'the Optimum Design of Compression Surfaces having Unflanged Integral Stiffeners'. *Journal of the Royal Aeronautical Society*, (Nov. 1954), 765–768.

5.5 Rothwell, A.: 'Coupled Modes in the Buckling of Panels with Z-Section Stringers in Compression'. *The Aeronautical Journal of the Royal Aeronautical Society*, (Feb. 1968), 159–163.

5.6 Saelman, B.: 'Basic Design and Producibility Considerations for Integrally Stiffened Structures.' *Machine Design*, (Mar. 1955) 197–203.

5.7 Bijlaard, P.P.: 'On the Buckling of Stringer Panels Including Forced Crippling.' *Journal of the Aeronautical Sciences*, (July 1955), 491–501.

5.8 Cozzone, F.P. and Melcon, M.A.: 'Non-dimensional Buckling Curves — Their Development and Application.' *Journal of The Aeronautical Sciences*, (Oct. 1946), 511–517.

5.9 Yusuff, S.: 'Buckling and Failure of Flat Stiffened Panels.' *Journal of Aircraft*, (Mar 1976), 198–204.

5.10 Meyer, R.R.: Buckling of 45°Eccentric-Stiffened Waffle Cylinders. *Journal of The Royal Aeronautical Society*, (July 1967), 516–520.

5.11 Niu, C.Y. and Baker, R.W.: 'Design Curves for Compression Panels.' *Lockheed Handbook (unpublished)*, (Apr. 1975).

5.12 Ensrud, A.F.: *The Elastic Pole*. Lockheed Aircraft Corp., 1952.

5.13 Gerard, G. and Becker, H.: 'Handbook of Structural Stability (Part I — Buckling of Flat Plates)'. *NACA TN 3781*, (July 1957).

5.14 Becker, H.: 'Handbook of Structural Stability (Part II — Buckling of Composite Elements)'. *NACA TN 3782*, (July 1957).

5.15 Gerard G. and Becker, H.: Handbook of Structural Stability (Part III — Buckling of Curved Plates and Shells)'. *NACA TN 3783*, (Aug. 1957)

5.16 Gerard, G.: 'Handbook of Structural Stability (Part IV — Failure of Plates and Composite Elements)'. *NACA TN 3784*, (Aug. 1957).

5.17 Gerard, G.: 'Handbook of Structural Stability (Part V — Compressive Strength of Flat Stiffened Panels)'. *NACA TN 3785*, (Aug. 1957).

5.18 Gerard, G.: 'Handbook of Structural Stability (Part VI — Strength of Stiffened Curved Plates and Shells)'. *NACA TN 3786*, (July, 1958).

5.19 Gerard, G.: 'Handbook of Structural Stability (Part VII — Strength of Thin Wing Construction)'. *NACA TN D-162*, (Sept. 1959)

5.20 Needham, R.A.: 'The Ultimate Strength of Aluminum Alloy Formed Structural Shapes in Compression.' *Journal of the Aeronautical Sciences*, (Apr, 1954).

5.21 Shanley, F.R.: 'Inelastic Column Theory.' *Journal of the Aeronautical Sciences*, (May, 1947).

5.22 Cox, H.L.: 'The Application of the Theory of Stability in Structural Design.' *Journal of the Royal Aeronautical Society*, (July 1957)

5.23 Jones, W.R.: 'The Design of Beam-Columns.' *Aero Digest*, (June 1935), 24–28.

5.24 Legg, K.L.C.: 'Integral Construction — A survey and an Experiment.' *Journal of the Royal Aeronautical Society*, (July 1954), 485–504.

5.25 Boley, B.A.: 'The Shearing Rigidity of Buckled Sheet Panels.' *Journal of The Aeronautical Sciences*, (June 1950), 356–374.

5.26 Fischel, R.: 'Effective Widths in Stiffened Panels Under Compression.' *Journal of Aeronautical Sciences*, (Mar. 1940).

5.27 Fischel, R.: 'The Compressive Strength of Thin Aluminum Alloy Sheet in the Plastic Region.' *Journal of Aeronautical Sciences*, (Aug. 1941).

5.28 Timoshenko, S. and Gere, J.M.: *Theory of Elastic Stability*. McGraw-Hill Book Co., Inc. New York, N.Y., Second Edition, 1961.

5.29 Peterson, J.P. and Anderson, J.K.: 'Bending Tests of Large-diameter Ring-stiffened Corrugated Cylinders.' *NASA TN D-3336*, (Mar. 1966).

5.30 Dickson, J.N. and Brolliar, R.H.: 'The General Instability of Ring-stiffened Corrugated Cylinders under Axial Compression.' *NASA TN D-3089*, (Jan. 1966).

5.31 Dow, N.F., Libove, C. and Hubka, R.E.: 'Formulas for the Elastic Constants of Plates with Integral Waffle-like Stiffening.' *NACA Report 1195*, (1954).

5.32 Kempner, J. and Duberg, J.E.: 'Charts for Stress Analysis of Reinforced Circular Cylinders under Lateral Loads.' *NACA TN 1310*, (May 1947).

5.33 Ramberg, W. and Osgood, W.R.: 'Description of

Stress-Strain Curves by Three Parameters.' *NACA TN 902*, (July 1943).
5.34 Bleich, H.H.: *Buckling Strength of Metal Structures*. McGraw-Hill Book Company, Inc., New York, N.Y. 1952
5.35 Gerard, G.: 'The Crippling Strength of Compression Elements.' *Journal of Aeronautical Sciences*, (Jan. 1958).
5.36 Maddux, G.E.: *Stress Analysis Manual*. Air Force Flight Dynamics Laboratory (AFFDL-TR-69-42), 1969.
5.37 Dow, N.F, Hickman, W.A. and Rosen B.W.: 'Effect of Variation in Rivet Strength on the Average stress at Maximum Load for Aluminum-Alloy, Flat, Z-stiffened Compression Panels that Fail by Local Buckling.' *NACA TN 2963*, (June 1953).
5.38 Hickman, W.A. and Dow, N.F.: 'Data on the Compressive Strength of 75S-T6 Aluminum-Alloy Flat Panels with Longitudinal Extruded Z-Section Stiffeners.' *NACA TN 1829*, (Mar 1949).
5.39 Hickman, W.A. and Dow, N.F.: 'Data on the Compressive Strength of 75S-T6 Aluminum-Alloy Flat Panels Having Small, Thin, Widely Spaced, Longitudinal Extruded Z-Section Stiffeners.' *NACA TN 1978*, (Nov. 1949).
5.40 Rossman, C.A., Bartone, L.M. and Dobrowski, C.V.: 'Compressive Strength of Flat Panels with Z-Section Stiffeners.' *NACA RR 373*, (Feb. 1944).
5.41 Niles, A.S.: 'Tests of Flat Panels with Four Types of Stiffeners.' *NACA TN 882*, (Jan. 1943).
5.42 Schuette, E.H., Barab, S. and McCracken, H.L.: 'Compressive Strength of 24S-T Aluminum-Alloy Flat Panels With Longitudinal Formed Hat-Section Stiffeners.' *NACA TN 1157*, (Dec. 1946).
5.43 Holt, M. and Feil, G.W.: 'Comparative Tests on Extruded 14S-T and Extruded 24S-T Hat-Shape Stiffener Sections.' *NACA TN 1172*, (Mar. 1947).
5.44 Dow, N.F., Hickman, W.A. and McCracken, H.L.: 'Compressive-Strength Comparisons of Panels Having Aluminum-Alloy Sheet and Stiffeners with Panels Having Magnesium-Alloy Stiffeners.' *NACA TN 1274*, (Apr. 1947).
5.45 Dow, N.F. and Hickman, W.A.: 'Design Charts for Flat Compression Panels Having Longitudinal Extruded Y-Section Stiffeners and Comparison with Panels Having Formed Z-Section Stiffeners.' *NACA TN 1389*, (Aug. 1947).
5.46 Gallaher, G.L. and Boughan, R.B.: 'A Method of Calculating the Compressive Strength of Z-stiffened Panels that Develop Local Instability.' *NACA TN 1482*, (Nov. 1947).
5.47 Hickman, W.A. and Dow, N.F.: 'Compressive Strength of 24S-T Aluminum-Alloy Flat Panels with Longitudinal Formed Hat-Section Stiffeners Having Four Ratios of Stiffener Thickness to Skin Thickness.' *NACA TN 1553*, (Mar. 1948).
5.48 Dow, N.F., Hickman, W.A. and Rosen, B.W.: 'Data on the Compressive Strength of Skin-Stringer Panels of Various Materials.' *NACA TN 3064*, (Jan. 1954).
5.49 Fischel, J.R.: 'The Compressive Strength of Thin Aluminum Alloy Sheet in the Plastic Region.' *Journal of The Aeronautical Sciences*, (Aug. 1941).
5.50 Rothwell, A. and Liacos, N.J.: 'Stiffened Shear Webs: Post-Buckling Failure of Stiffeners of Less Than Critical Rigidity.' *Aeronautical Journal*, (Dec. 1979).
5.51 Shanley, F.R.: Engineering Aspects of Buckling — The Buckling of Simple Columns and Flat plates simply explained for the Engineer.' *Aircraft Engineering*, (Jan. 1939), 13−20.

CHAPTER 6.0

CUTOUTS

6.1 Introduction

The aircraft structure is continually faced with requirements for openings at webs and panels to provide access or to let other members such as control rods or cables, hydraulic lines, electrical wire bundles, etc., pass through. Other cutouts such as windows, doors, servicing panels, hatches, bomb-bays, inspection access holes, etc. cause a recurring headache for the structural engineer. As soon as one makes a hole in a load-bearing skin, a stronger surrounding structure must be introduced to provide alternate paths to carry the loads. Perhaps the most noticeable feature of cutouts is the rounding of the corners; sharp corners cause excessively high stress concentrations (K_t) as shown in Fig. 6.1.1.

The most efficient structure is when the load path is most direct. Cutouts in structures invariably increase the structural weight because the structure adjacent to the cutout must be increased to carry the load which would have been carried in the cutout panel, plus the forces due to the redistribution of this load. The procedure used in the analysis for the effect of cutouts may be condensed to the following two methods, depending primarily upon the geometrical relationship of the cutouts to the remainder of the structure.

- When cutouts are relatively small to medium in size the effect is localized. This means that only structure in the immediate vicinity of the cutout is appreciably involved in the redistribution; i.e. "Donut-doubler" or with standard round flanged holes. See Fig. 6.1.2.
- Sections having relatively large size cutouts must consider the effects of the cutout in computing the section properties. This means that the entire section will be affected by the cutout instead of the effect being localized, such as framing cutouts in

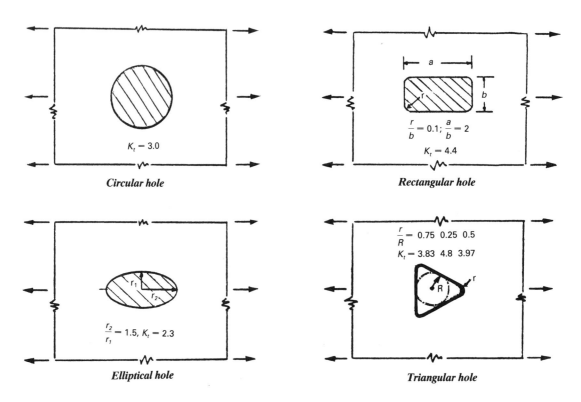

Fig. 6.1.1 Stress concentration for different cutouts (for flat plate).

webs of large rectangular opening. If the holes (round or rectangular) are so large, some shear is carried by frame action of the adjacent cap members. This should be taken into account if the caps are to be subjected to high axial loads as the ability to carry axial loads will be reduced. See Fig. 6.1.3.

Frequently it will provide an access opening or manholes on the skin-stringer panel or integrally stiffened panel for accessibility, inspection, etc., such as wing and empennage structures shown in Fig. 6.1.4. These panels usually are designed for high axial-loads (sometimes it reaches 25,000 lb/in) and careful detail design for these cutouts is very important to achieve long fatigue structural life and prevent buckling in

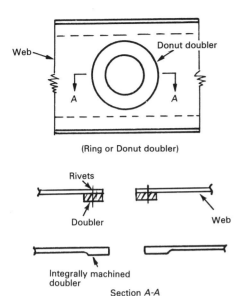

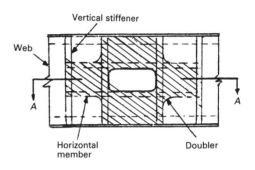

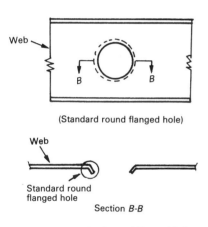

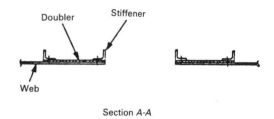

(b) *Standard round flanged holes*

Fig. 6.1.2   *Typical shear beam cutout reinforcements around relatively small holes.*

Fig. 6.1.3   *Typical shear beam with framing members and doubler around the relative large cutout.*

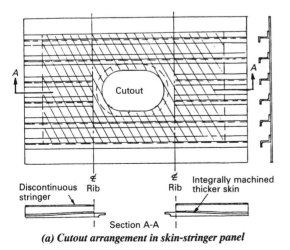

(a) *Cutout arrangement in skin-stringer panel*

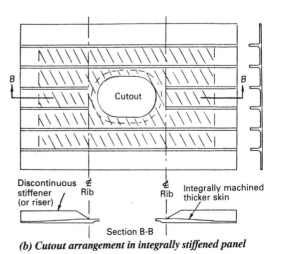

(b) *Cutout arrangement in integrally stiffened panel*

Fig. 6.1.4   *Typical cutout arrangement in skin-stringer panel and integrally stiffened panel.*

Airframe Structural Design 163

compression loads.

Transport and fighter fuselages contain numerous cutout areas of different sizes and shapes located in various regions of the fuselage body as shown in Fig. 6.1.5. However, cutouts such as those in large passenger openings, cargo openings, service openings, emergency exits, window, etc., often occur in the regions where high loads must be resisted and therefore additional structure is needed to carry the loads around the openings. Such openings require the installation of jambs as well as strengthening of the abutting internal structure as shown in Fig. 6.1.6.

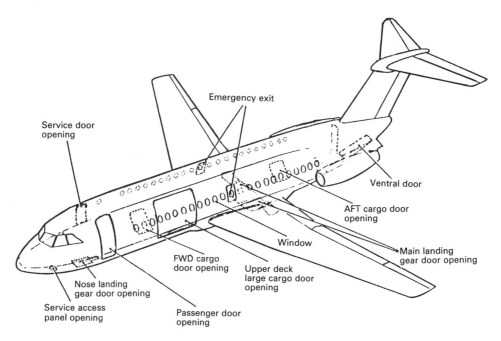

Fig. 6.1.5 General arrangement of openings of a commercial transport.

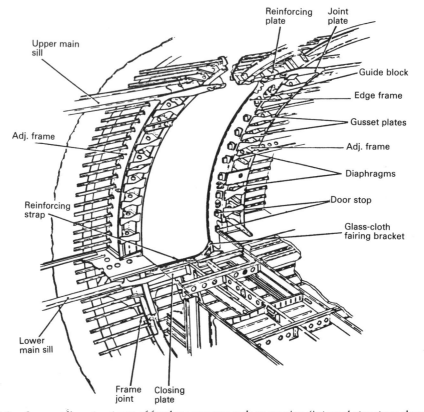

Fig. 6.1.6 Surrounding structures of fuselage passenger door opening (internal structure shown).

164 Airframe Structural Design

6.2 Lightly Loaded Beams

The ideal construction for most shear-carrying beams is a tension field (or diagonal tension beam per Ref. 6.8). However, in some cases it is advantageous, and in other cases necessary, to incorporate circular, flanged holes in the beam webs. These cases come under two main categories:
- Lightly loaded or very shallow beams. In such cases it may not be practical to construct an efficiently designed tension field beam because of minimum gage considerations and other restrictions due to the small size of the parts involved. It may then be advantageous from a weight standpoint to omit web stiffeners and, instead, introduce a series of standard flanged lightening holes, as shown in Fig. 6.2.1.
- Moderately loaded beams with access holes. Where it is necessary to introduce access holes into the web of a shear-carrying beam, a light, low cost construction is obtained by using a flanged hole with web stiffeners between the holes.

Lightly Loaded or Very Shallow Beams

The following two types of beam construction are considered. The standard flanged lightening holes as shown in fig. 6.2.2 are centered and equally spaced.
- The limiting conditions for the design curves is given in Fig. 6.2.3.

$$0.25 \leq \frac{D}{h} \leq 0.75;$$

web thickness $0.016 \leq t \leq 0.125$

$$0.3 \leq \frac{D}{b} \leq 0.7; \qquad 40 \leq \frac{h}{t} \leq 250$$

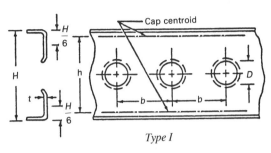

Type I

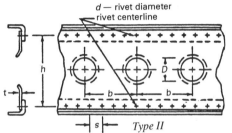

Type II

(Note: $\frac{H}{6}$ is the assumed effective depth of beam cap)

Fig. 6.2.2 Lightly loaded or very shallow beams.

$\frac{D}{(inch)}$	2.0	2.5	3.0	3.5	4.0	4.5	5.0	6.0
$\frac{f}{(inch)}$	.25	.3	.4	.45	.5	.5	.5	.55

(a) Lightening holes of typical flanged holes (45° flanged)

D_o = Outside diameter
$R = 0.155$ inch
D = Inside diameter

D_0 (Inch)	D (Inch)	α (Inch)
1.7	0.8	0.2
1.95	1.05	0.2
2.65	1.7	0.25
3.0	2.05	0.25
3.65	2.7	0.25
3.9	2.95	0.25
4.95	3.8	0.4
5.95	4.8	0.4
6.95	5.8	0.4
7.44	6.3	0.4
7.95	6.8	0.4
8.95	7.8	0.4
9.45	8.3	0.4

(b) Lightening holes with beaded flanged

Fig. 6.2.1 Common flanged lightening holes. ($t = 0.032\ in - 0.125\ in$)

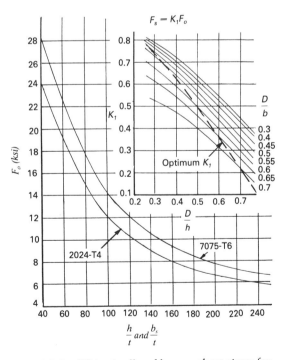

Fig. 6.2.3 Ultimate allowable gross shear stress for aluminum alloy webs with flanged holes as shown in Fig. 6.2.1(a).

- The ultimate allowable gross shear stress of the web is given by:

$$F_s = K_1 F_o$$

where K_1 and F_o are given in Fig. 6.2.3. F_o is the ultimate allowable web stress of a shallow beam without holes.

- To cover the case of large or closely spaced holes and rivets the net shear stresses should be calculated. These values must not exceed F_{su}.

At web between holes: $f_s = \dfrac{q}{t}\left(\dfrac{b}{b-D}\right)$ (6.2.1)

Web net (vertical) section at holes:

$$f_s = \dfrac{q}{t}\left(\dfrac{h}{h-D}\right) \quad (6.2.2)$$

Web net section at rivets: $f_s = \dfrac{q}{K_r t}$ (6.2.3)

where $K_r = 1.0$ for type I;

$$K_r = \dfrac{s-d}{s} \quad \text{for type II}$$

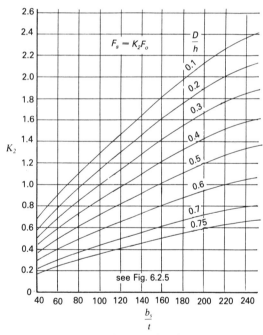

Fig. 6.2.4 *Correction factor K_2 for aluminum alloy webs with stiffeners and flanged lightening holes as shown in Fig. 6.2.1(a).*

- Because of the non-uniformity of stresses caused by the holes, the web-to-flange riveting of the type II beam should be designed by the greater of:

$$q_r = 1.25\, q \quad \text{or} \quad q_r = 0.67\left(\dfrac{b}{b-D}\right) q$$

- The equivalent-weight-web thickness of a plane web beam with lightening holes is given by:

$$t_{eq} = t\left(1 - \dfrac{0.785\, D^2}{h\, b}\right)$$

For a given web height h, web thickness t, and loading q, the equivalent-weight web thickness t_{eq} will be minimum when the hole diameter D and hole spacing b are chosen so as to lie on the curve for optimum K_1 factor in Fig. 6.2.3. This generally will be a beam where $\dfrac{D}{h}$ is approximately 0.25 and $\dfrac{D}{b}$ is approximately 0.45. It should be noted that if the web height h, hole diameter D, and loading q are given generally then the lightest beam will be obtained when the hole spacing b is such that $\dfrac{D}{b}$ is approximately 0.45.

Example

Consider the following shallow beam with lightly loaded,

Given: $q = 300$ lb/in
$h = 3.0$ in
$t = 0.04$ in
Material: 7075-T6 Aluminum bare plate,
$F_{su} = 45000$ psi

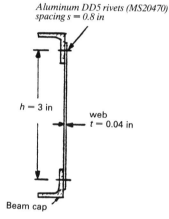

Determine the optimum hole diameter and spacing and check the riveting requirements.

$$f_s = \dfrac{q}{t} = \dfrac{300}{0.04} = 7500 \text{ psi}$$

$$\dfrac{h}{t} = \dfrac{3}{0.04} = 75$$

Therefore from Fig. 6.2.3, $F_o = 18800$ psi

From $F_s = K_1 F_o$;

or $K_1 = \dfrac{F_s}{F_o} = \dfrac{7500}{18800} = 0.4$

Note: considering $F_s = f_s = 7500$ psi for this calculation to optimize the hole diameter and spacing.

Then from Fig. 6.2.3 with optimum K_1 factor curve,

$$\dfrac{D}{h} = 0.57 \text{ and } \dfrac{D}{b} = 0.65$$

Therefore,
$$D = 0.57h = 0.57(3) = 1.71 \text{ in}$$
(or use standard hole diameter, i.e. $D = 1.5$ in)
$$b = \frac{D}{0.65} = \frac{1.71}{0.65} = 2.63 \text{ in}$$

If $D = 1.5$ in
$$\frac{D}{h} = \frac{1.5}{3} = 0.5$$
$$\frac{D}{b} = 0.7 \quad \text{(from Fig. 6.2.3 with corresponding } K_1 = 0.4)$$
$$b = \frac{D}{0.7} = \frac{1.5}{0.7} = 2.14 \quad \text{(used)}$$

Beam web allowable strength with $D = 1.5$ and $b = 2.14$
$$q_{all} = K_1 F_o t = 0.4 (18800) 0.04 = 300.8 \text{ lb/in}$$
$$M.S. = \frac{q_{all}}{q} - 1 = \frac{300.8}{300} - 1 = 0.003 \quad \text{O.K.}$$

Web-to-Beam cap flange riveting,
$$q_r = 1.25 \, q = 1.25 \, (300) = 375 \text{ lb/in}$$
or
$$q_r = 0.67 \left(\frac{b}{b-D}\right) q$$
$$= 0.67 \left(\frac{2.14}{2.14 - 1.5}\right) (300) = 672 \text{ lb/in}$$

with DD5 aluminum rivets (MS20470) at spacing $s = 0.8$ in, (watch inter-rivet buckling at beam cap with compression load).

DD5 ($d = \frac{5}{32}$ in. in diameter) rivet shear allowable = 815 lb

(from MIL-HDBK-5D per Ref. 4.1 or see Fig. 7.1.3.)

then $q_{all} = \frac{815}{0.8} = 1018 \text{ lb/in}$

$$M.S. = \frac{q_{all}}{q} - 1 = \frac{1018}{300} - 1 = \text{high} \quad \text{O.K.}$$

Net web shear check
- At web between holes;
$$f_s = \frac{q}{t}\left(\frac{b}{b-D}\right) = \frac{300}{0.04}\left(\frac{2.14}{2.14 - 1.5}\right)$$
$$= 25078 \text{ psi} < F_{su}$$
$$= 45000 \text{ psi} \quad \text{O.K.}$$

- Web net vertical section at holes;
$$f_s = \frac{q}{t}\left(\frac{h}{h-D}\right) = \frac{300}{0.04}\left(\frac{3}{3-1.5}\right)$$
$$= 25078 \text{ psi} < F_{su}$$
$$= 45000 \text{ psi} \quad \text{O.K.}$$

- Web net section at rivets;
$$f_s = \frac{q}{K_r t} = \frac{300}{0.81 (0.04)}$$
$$= 9260 \text{ psi} < F_{su} = 45000 \text{ psi} \quad \text{O.K.}$$

where $K_r = \frac{s - d}{s} = \frac{0.8 - 0.156}{0.8} = 0.81$

Moderately Loaded Beams

A sketch of the type of construction considered in the design of moderately loaded beams requiring access holes is given in Fig. 6.2.5. The access holes are centered between the single angle stiffeners and are formed to the flanged lightening holes as shown in Fig. 6.2.1. The beaded flanged lightening holes of Fig. 6.2.1(b) are recommended because of its good crack-free edge of the round hole compared with that of Fig. 6.2.1(a) which had found cracks by past experience.

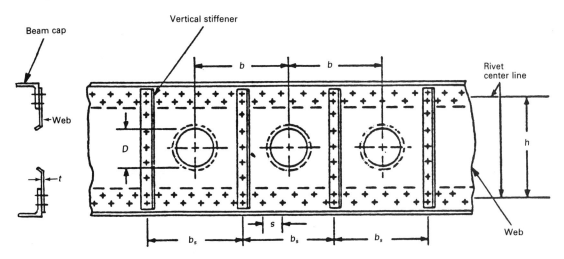

Fig. 6.2.5 Moderately loaded beam with access holes.

Airframe Structural Design 167

Shear webs of the type shown in Fig. 6.2.5 are often used in aircraft construction particularly in large, low loaded shear beams. Their function is to provide stiffness and accessibility when loads are fairly low; hence in most applications the web gauge is light, i.e. between 0.02 to 0.1 in. They are used for flap and control surface beams, ribs and formers, and floor supports. Very often beads are used in place of stiffeners where cap is formed by flanging the web (see type I of Fig. 6.2.2). The holes are always flanged as shown in Fig. 6.2.1.

No clear-cut analytical procedure is available. However, with the help of existing test data and reports, an experienced designer can select acceptable methods as described in this chapter. Sometimes, static tests are frequently conducted for the interest of weight and safety.

(1) Flanged-hole Webs with Intermediate Vertical Stiffeners-Method I
- The following limiting conditions must be satisfied when using the design curves in Fig. 6.2.4.
  (a) $0.025 \leq t \leq 0.125$

  $115 \leq \dfrac{h}{t} \leq 1500$

  $0.235 \leq \dfrac{b_s}{h} \leq 1.0$

  (b) To prevent net shear failure between holes, the maximum hole size in relation to panel rectangularity is given by:

  $$\dfrac{D}{b_s} \leq 0.85 - 0.1 \left(\dfrac{h}{b_s}\right) \quad (6.2.4)$$

  It should be noted that a staggered arrangement of holes between panels is preferred to having them in a single line.

  (c) The single angle stiffeners must satisfy the following conditions:

  $t_o \geq t$

  $$\dfrac{A_o}{b_s t} \geq 0.385 - 0.08 \left(\dfrac{b_s}{h}\right)^3 \quad (6.2.5)$$

  $$I_o \geq \dfrac{F_s t b_s h^3}{10^8 (h - D)} \quad (6.2.6)$$

  where I_o is the moment of inertia of the stiffener about its center of gravity and parallel to the skin line and the approximate equation is

$$I_o = \dfrac{t_o b_{o2}^3 (4 b_{o1} + b_{o2})}{12 (b_{o1} + b_{o2})} \quad (6.2.7)$$

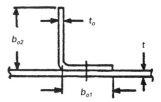

(d) The maximum width-to-thickness ratios of the single angle stiffeners are limited by Fig. 6.2.6.

- The ultimate allowable gross shear stress is given by

$$F_s = K_2 F_o \quad (6.2.8)$$

where F_o is obtained from Fig. 6.2.3 as function of $\dfrac{b_s}{t}$

K_2 is given in Fig. 6.2.4.

- To provide for the non-uniformity of stress caused by the access holes, the following riveting requirements must be satisfied.
  (a) Web-to-flange riveting:

  $$q_r = 1.25 F_s t \left(\dfrac{h}{h - D}\right) \quad (6.2.9)$$

  (b) Flange-to-stiffener rivets are designed to transmit the following load:

  $$P_{st} = \dfrac{0.0024 A_o F_s b_s}{t} \left(\dfrac{h}{h - D}\right) \quad (6.2.10)$$

  (c) Web-to-stiffener riveting should conform to Fig. 6.2.7.

For single angle stiffeners, the required tensile strength of the rivets per inch run is:

$$P_{ten} = 0.20 \, t \, F_{tu}$$

where F_{tu} is the ultimate tensile strength of the web material. This criterion is satisfied by the table in Fig. 6.2.7 for MS 20470 rivets.

t_o	0.032	0.040	0.050	0.063	0.071	0.080	0.090	0.100	0.125
$\dfrac{b_{o1}}{t_o}$	12.5	12.5	11.5	10.0	9.5	9.0	9.0	9.0	9.0
$\dfrac{b_{o2}}{t_o}$	16.0	16.0	15.0	13.0	12.0	12.0	12.0	12.0	12.0

Fig. 6.2.6   Max. width-to-thickness ratios.

Web thickness (t)	0.025	0.032	0.040	0.051	0.064	0.072	0.081	0.091	0.102	0.125
Rivet	AD-4	AD-4	AD-4	AD-5	AD-5	AD-6	AD-6	DD-6	DD-6	DD-8
Rivet diameter (d)	$\frac{1}{8}$	$\frac{1}{8}$	$\frac{1}{8}$	$\frac{5}{32}$	$\frac{5}{32}$	$\frac{3}{16}$	$\frac{3}{16}$	$\frac{3}{16}$	$\frac{3}{16}$	$\frac{1}{4}$
Rivet spacing (s)	$\frac{1}{2}$	$\frac{5}{8}$	$\frac{5}{8}$	$\frac{7}{8}$	$\frac{3}{4}$	1.0	$\frac{7}{8}$	1.0	1.0	$1\frac{3}{8}$

Fig. 6.2.7 *Web-to-stiffener riveting spacing (MS 20470 rivets).*

- The equivalent-weight web thickness t_{eq} for a beam with stiffeners and access holes is given by:

$$t_{eq} = t\left(1 - \frac{0.785 D^2}{h b_s} + \frac{A_o}{b_s t}\right) \quad (6.2.11)$$

For a given web height h, hole diameter D, and loading q, the minimum equivalent-weight beam t_{eq} will usually be obtained when the hole (and stiffener) spacing is the closest permissible value, provided the hole spacing criterion given in Eq. (6.2.4) is satisfied. The procedure to follow is outlined in the following example.

Example
Given: $q = 570$ lb/in
$h = 12$ in
$D = 5$ in
Web and Stiffener Materials: 7075-T6

Determine the web thickness, required stiffener area and spacing.

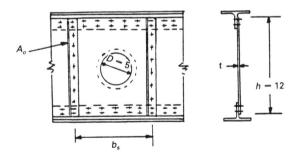

Optimizing the hole and spacing, let

$$\frac{D}{b_s} = 0.85 - 0.1\left(\frac{h}{b_s}\right) \quad \text{from Eq. (6.2.4)}$$

Therefore,

$$\frac{5}{b_s} = 0.85 - 0.1\left(\frac{12}{b_s}\right)$$

obtain $b_s = 7.3$ in

The following procedure is used to obtain the lightest web thickness:

① t (in)	② $\frac{b_s}{t}$	③ F_o (psi)
	$\frac{7.3}{①}$	Fig. 6.2.3
0.04	183	8400
0.051	143	10400
0.064	114	12600
0.072	101	14000
0.081	90	15600

④ K_2	⑤ F_s (psi)	⑥ f_s (psi)
Fig. 6.2.4	③ × ④	$\frac{570}{①}$
1.32	11088	14250
1.08	11232	11176
0.92	11592	8906
0.84	11760	7917
0.8	12480	7037

The lightest web design is such that the value of column ⑤ equals that of column ⑥. Therefore, $t = 0.051$ in is the lightest design for this given configuration.

$$M.S. = \frac{F_s}{f_s} - 1$$

$$= \frac{11232}{11176} - 1 = 0.01 \quad \text{O.K.}$$

Required stiffener area A_o is determined from Eq. (6.2.5).

$$A_{o\,(min.)} = b_s t\left[0.385 - 0.08\left(\frac{b_s}{h}\right)^3\right]$$

$$= 7.3\,(0.051)\left[0.385 - 0.08\left(\frac{7.3}{12}\right)^3\right]$$

$$= 0.137\,\text{in}^2$$

It is customary to use $t_o > t$,
let $t_o = 0.081$ in and from Fig. 6.2.6

$$\frac{b_{o1}}{t_o} = 9 \text{ and } \frac{b_{o2}}{t_o} = 12$$

$$b_{o1} = 9(0.081) = 0.73 \text{ in}$$
$$b_{o2} = 12(0.081) = 0.97 \text{ in}$$

Therefore, actual $A_o = (0.73 + 0.97)(0.081)$
$= 0.14 \text{ in}^2 > A_{o\text{ (min)}}$ O.K.

Required stiffener moment of inertia is given by Eq. (6.2.6)

$$I_{o\text{ (min)}} = \frac{F_s t b_s h^3}{10^8 (h - D)}$$

$$= \frac{11232(0.051)(7.3)(12^3)}{10^8 (12 - 5)}$$

$$= 0.01 \text{ in}^4$$

The actual stiffener moment of inertia is computed by Eq. (6.2.7)

$$I_o = \frac{t_o b_{o2}^3 (4 b_{o1} + b_{o2})}{12 (b_{o1} + b_{o2})}$$

$$= \frac{0.081 (0.97)^3 [4 (0.73) + 0.97]}{12 (0.73 + 0.97)}$$

$$= 0.014 \text{ in}^4 > I_{o\text{ (min.)}} = 0.01 \text{ in}^4 \quad \text{O.K.}$$

Web-to-flange (beam cap) riveting strength is obtained from Eq. (6.2.9).

$$q_r = 1.25 F_s t \left(\frac{h}{h - D}\right)$$

$$= 1.25 (11232)(0.051) \left(\frac{12}{12 - 5}\right)$$

$$= 1228 \text{ lb/in}$$

Using a double row of AD5 (MS 20470-AD5) rivets at spacing equal 0.8 in (rivet strength allowable, see MIL-HKBK-5D per Ref. 4.1 or Fig. 7.1.3).

$$q_{r\text{ (all)}} = \frac{596 (2)}{0.8}$$

$$= 1490 \text{ lb/in}$$

$$\text{M.S.} = \frac{q_{r\text{ (all)}}}{q_r} - 1$$

$$= \frac{1490}{1228} - 1 = 0.21 \quad \text{O.K.}$$

Web-to-stiffener riveting spacing is obtained from Fig. 6.2.7.

Using AD5 ($d = \frac{5}{32}$ in) rivets with spacing equal $\frac{7}{8}$ in.

Flange-to-stiffener rivet strength is obtained from Eq. (6.2.10)

$$P_{st} = \frac{0.0024 A_o F_s b_s}{t} \left(\frac{h}{h - D}\right)$$

$$= \frac{0.0024 (0.14)(11232)(7.3)}{0.051} \left(\frac{12}{12 - 5}\right)$$

$$= 926 \text{ lb}$$

Using 2 DD6 (MS 20470-DD6) rivets (rivet strength allowable see MIL-HDBK-5D per Ref. 4.1 or Fig. 7.1.3).

$$P_{st\text{ (all)}} = 1180 \times 2 = 2360 \text{ lb}$$

$$\text{M.S.} = \frac{P_{st\text{ (all)}}}{P_{st}} - 1$$

$$= \frac{2360}{926} - 1 = 1.55 \quad \text{O.K.}$$

The equivalent-weight web thickness t_{eq} is obtained from Eq. (6.2.11)

$$t_{eq} = t \left(1 - \frac{0.785 D^2}{h b_s} + \frac{A_o}{b_s t}\right)$$

$$= 0.051 \left[1 - \frac{0.785 (5)^2}{12 (7.3)} + \frac{0.14}{7.3 (0.051)}\right]$$

$$= 0.051 \times 1.15$$
$$= 0.059 \text{ in}^2$$

(2) Flanged-hole Webs with Intermediate Vertical Stiffeners-Method II
- Flanged Hole Web Analysis:
  Web thickness required for a given beam depth and hole diameter is based on a conservative web shear flow which is equal to:

$$q_{eff} = \frac{1.25 V}{(h - D)}$$

- Required flanged hole web thickness vs gross area shear flow q_{all} is plotted for 7075-T6 aluminum clad material in Fig. 6.2.8. Thickness ratios for 2024-T3 and 7178-T6 materials are also included in Fig. 6.2.8. The following web limitations are recommended:
  (a) Holes have standard flange as shown in Fig. 6.2.1(a).
  (b) Hole diameter is not to exceed $\frac{1}{2}$ of beam depth h.
  (c) Hole centers should not be staggered more than $\frac{1}{8}$ of the beam depth.
  (d) Minimum hole spacing b_s is 1.5 hole diameter D.
  (e) Web to stiffener attachment strength is 50% stronger in tension and shear than required for webs without hole design.
- Stiffener Analysis:
  Web thickness presented in Fig. 6.2.8 are based on a stiffener area A_o equal to

$$A_o = 0.5 (b_s - D) t \quad \text{or} \quad A_o = 0.2 t$$

The stiffener leg attached to the web must have a leg width-to-leg thickness ratio equal to or less than 12 to prevent stiffener crippling prematurely.
- Riveting Strength Requirements (Fig. 6.2.9):

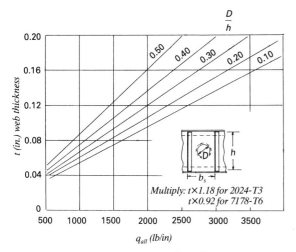

Fig. 6.2.8 Gross area shear flow, q_{all}. (7075-T6 aluminum)

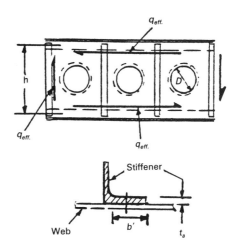

Fig. 6.2.9 Vertical stiffeners with flanged-hole webs beam.

(a) Web-to-Chord (cap flange) attachments — carrying a running load q_{att} equal to the effective net area shear flow q_{eff}.

$$q_{att} = q_{eff} = \frac{1.25\ V}{h - D}$$

(b) Stiffener to Web Attachments — should be designed for a strength 150% of that required for shear resistant webs.

$$q'_{att} = 1.5 \left(\frac{K A_o}{b_s} \right)$$

where: q'_{att} = running shear per inch for attachment of stringer to web
A_o = area of stiffener
b_s = stiffener spacing
K = 22000 for 2024 web; 25000 for 7075 web

(c) Stiffener-to-Chord Attachments — are designed to the same requirements as tension field beams. Refer to Eq. (6.2.10) of Method I.

Example
Using the same given data as in Method I.

$\frac{b_s}{D} = \frac{7.3}{5} = 1.46 \approx 1.5$ O.K.

$\frac{D}{b} = \frac{5}{12} = 0.42 < 0.5$ O.K.

with $\frac{D}{h} = 0.42$, from Fig. 6.2.8 read $q_{all} = 580$ lb/in

$M.S. = \frac{q_{all}}{q} - 1 = \frac{580}{570} - 1 = 0.02$ O.K.

(The result of this method II agrees with method I)

(3) Flanged-hole Webs with Intermediate Vertical beads

Another type of web has round holes with beaded flanges and vertical beads between holes as shown in Fig. 6.2.10. The vertical bead is as descrbed in the table in Fig. 6.2.10. The beaded flange holes used in this analysis have been described in Fig. 6.2.1(b).

Whenever using webs with formed beads as shown in Fig. 6.2.10, it is important that the beads be formed long enough to extend as close to the beam caps or flanges as assembly will allow. Short beads, ending well away from the caps, will not develop the strength indicated by the allowables given in the Fig. 6.2.11. Rivets attaching the web to the caps above a hole also need to be more closely spaced to take the higher net shear force locally (refer to the previous design as described in Method I).

The allowable shear flows shown in Fig. 6.2.11 are based on the design of $\frac{D_o}{h} = 0.6$ and the hole spacing is equal to h. The solid lines of Fig. 6.2.11 give the ultimate strength q_{all} of the web as a function of web height h and this represents the total collapsing strength of the web. The dotted lines indicate the shear flow q_{cr} at which initial buckling begins. The allowables are for pure shear only, no big lateral concentrated loads. If such loads applied, suitable vertical stiffeners must be introduced to prevent earlier collapse of the beads, see Fig. 6.2.12.

Example

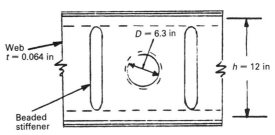

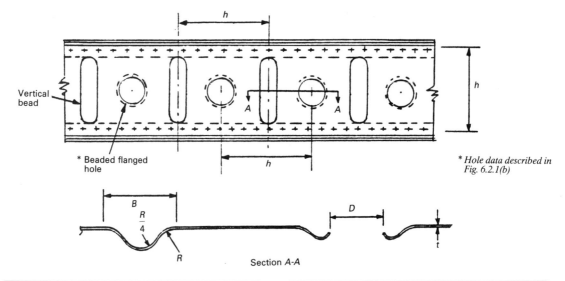

t (in)	B (in)	R (in)	t (in)	B (in)	R (in)
0.02	0.95	0.32	0.073	1.65	1.15
0.032	1.16	0.52	0.08	1.73	1.3
0.04	1.27	0.64	0.09	1.8	1.45
0.05	1.42	0.81	0.1	1.9	1.6
0.063	1.55	1.02	0.125	2.12	2.0

Fig. 6.2.10 *Beam webs with round beaded flanged lightening holes and intermediate vertical beads.*

Using the same given data as in Method I.
Check the limitation of this method before using Fig. 6.2.11.

$$\frac{D_o}{h} \approx 0.6; \text{ then } D_o \approx 0.6\,(12) = 7.2 \text{ in}$$

From Fig. 6.2.1(b)
Choose $D_0 = 7.44$ in and $D = 6.3$ in

From curves in Fig. 6.2.11 with $h = 12$ in and $q_{all} = 570$ lb/in, obtain $t = 0.064$ in

Comparison between vertical stiffeners and beads:

Vertical stiffeners	Vertical beads
$t_{eq} = 0.051$ in $D = 5$ in	$t = t_{eq} = 0.064$ in $D = 6.3$ in

From weight point of view (for this lightly loaded beam), the vertical beaded web configuration (not including stiffeners) is slightly heavier than that of vertical stiffeners.

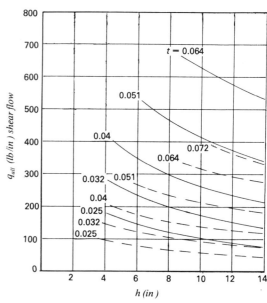

(1) Design curves for clad 2024-T4 and 7075-T6 with beaded stiffeners per Fig. 6.2.10 and flanged lightening holes per Fig. 6.2.1(b).

(2) Curves apply when:
$$\frac{D_o}{h} \approx 0.6$$
$$\frac{\text{beaded spacing}}{h} \approx 1.0$$

(3) $q_{all.}$ as shown in solid lines

(4) $q_{cr.}$ as shown in dotted lines

Fig. 6.2.11 *Beaded shear webs with lightening holes.*

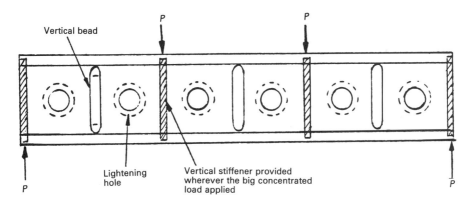

P — Lateral concentrated loads

Fig. 6.2.12 Vertical stiffeners needed wherever concentrated loads are applied.

6.3 Heavily Loaded Beams

Large holes or openings in the web as shown in Fig. 6.3.1 are undesirable, but if essential they may be compensated for by careful reinforcement. Beam webs with an access hole reinforced by either flanged or local doubler will generally be from 20 to 50% heavier than the same web without holes. Many aircraft design hours are spent solving the problem of reinforcing holes in beam shear webs. Among the many considerations a designer must resolve in designing a reinforcement are penalties imposed by weight, manufacturing costs, engineering time, clearances, accessibility, and freedom in routing auxiliary equipment.

Since holes in shear beams are unavoidable for equipment accessibility or systems routing, the following methods of analysis are presented.

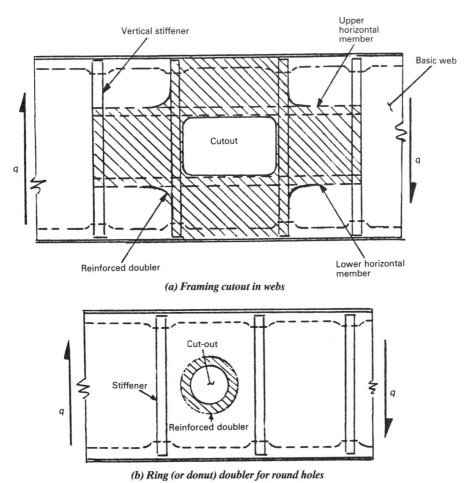

(a) Framing cutout in webs

(b) Ring (or donut) doubler for round holes

Fig. 6.3.1 Typical heavily loaded shear beam design with access hole.

Airframe Structural Design 173

Framing Cutouts in Webs

There are several methods to determine the loads in the area of framing cutouts (basically for rectangular cutouts) in beam web. One of these methods presented here is approximate and conservative. It is assumed that a shear flow q equal and opposite to that present with no cutout as shown in Fig. 6.3.2 and determine the corresponding balancing loads in the framed area. Adding this load system to the original one will give the final loads and, of course, the shear flow equal to zero in the cutout panel.

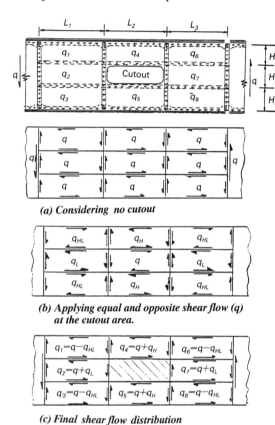

Fig. 6.3.2 *Self-balancing internal shear flow in the area of framing cutout.*

To eliminate redundancies, it is usually assumed that the same shear flow exists in the panels above and below the cutout area. It is also assumed that shear flows are the same in the panels to the left and right of the cutout. Since this represents a self-balancing load system, no external reactions outside of the framing areas are required. This is an important assumption that the designer should keep in mind.

If there were no cutouts there would be a constant shear flow q in all of the panels as shown in Fig. 6.3.2(a). Next apply a shear flow q equal and opposite to that in the center panel as shown in Fig. 6.3.2(b) and then add this load system to the original one. The final shear flow distribution around the cutout is illustrated in Fig. 6.3.2(c). The discussion is as follows.

(a) The shear flow in the panels above and below the center panel must statically balance the force due to q

$$q_H = \frac{q(H_2)}{(H_1 + H_3)}$$

(b) The shear flows in the panels to the left and right of the center panel must also statically balance the force due to q

$$q_L = \frac{q(L_2)}{(L_1 + L_3)}$$

(c) The shear flows in the corner panels must also balance the force due to the shear flow in the panel between them.

$$q_{HL} = q_H \left(\frac{L_2}{L_1 + L_3} \right)$$

The final shear flows are obtained by adding the values in Fig. 6.3.2 (a) and (b) together algebraically. The shear flow in the center panel (the cutout) is zero as it should be. The shear flows above and below and to the left and right of the cutout give greater shear flow q_H and q_L than the original q. The shear flows q_{HL} in the corner panels are smaller than the original q.

Finally there are axial loads developed in all of the framing members due to this cutout effect. These will add or substract, depending upon their directions, to any loads present before the cutout was made. These are illustrated in the following example.

Example
$R_{A1} = 30000$ lb
$Q \;\; = 12000$ lb
$R_{A4} = 30000$ lb

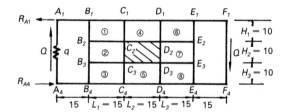

For the panels at each end of this beam,

$$q = \frac{12000}{10 + 10 + 10} = 400 \text{ lb/in}$$

For panels at ④ and ⑤,

$$q_4 = q_5 = q + q\left(\frac{H_2}{H_1 + H_3}\right)$$

$$= 400 + 400\left(\frac{10}{10 + 10}\right) = 600 \text{ lb/in}$$

For panels at ② and ⑦,

$$q_2 = q_7 = q + q\left(\frac{L_2}{L_1 + L_3}\right)$$

$$= 400 + 400\left(\frac{15}{15 + 15}\right) = 600 \text{ lb/in}$$

For panels at ①, ③, ⑥ and ⑧,

$$q_1 = q_3 = q_6 = q_8 = q - q_4\left(\frac{L_2}{L_1 + L_3}\right)$$

$$= 400 - 200\left(\frac{15}{15+15}\right) = 300 \text{ lb/in}$$

With all the shear flows known, the variations in axial load in the stiffeners and longitudinals are easily determined. These are plotted below,

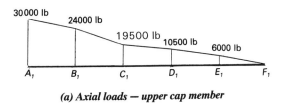

(a) Axial loads — upper cap member

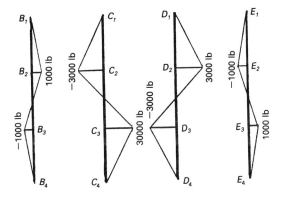

(b) Axial laods — vertical stiffeners

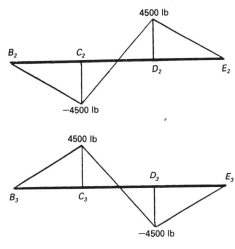

(c) Axial loads — horizontal longeron

Ring (or Donut) Doubler for Round Holes

The data presented herein consists of round holes with diameters up to 50% (or $\frac{h}{2}$) of the beam depth and ring reinforcement $\frac{W}{D}$ ratios of 0.35 to 0.50. The theoretical approach is described below (refer to Fig. 6.3.3).

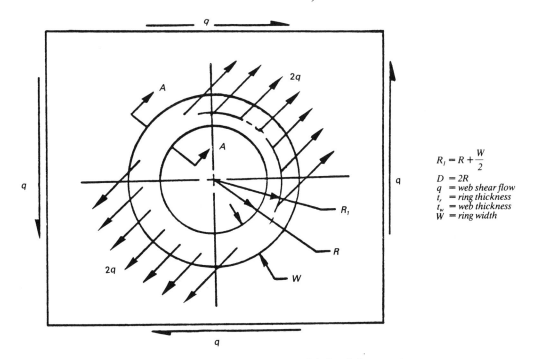

$R_1 = R + \frac{W}{2}$
$D = 2R$
q = web shear flow
t_r = ring thickness
t_w = web thickness
W = ring width

Fig. 6.3.3 Rings used for shear web hole reinforcement.

Airframe Structural Design 175

(1) Bending and tension stresses at Sect. A — A:

$$\text{Bending stress, } f_b = \frac{\text{Moment}}{\text{Section Modulus}}$$

$$= \frac{2q(0.25R)(R_1)}{t \frac{W^2}{6}}$$

$$\text{Tension stress, } f_t = \frac{2qR}{Wt_r} = \frac{qD}{Wt_r}$$

(2) Stress interaction at failure:
Ultimate tensile stress

$$F_{tu} = f_b + f_t$$

$$F_{tu} = \frac{0.75qD(D+W)}{t_r W^2} + \frac{qD}{Wt_r}$$

and

$$t_r = \frac{0.75q\left(\frac{D}{W}\right)^2 + 1.75q\left(\frac{D}{W}\right)}{F_{tu}}$$

(3) Rivet strength requirement at ring doubler:

Rivet pattern between tangent lines as shown in Fig. 6.3.4 develops a running load strength/inch.

$$2q\left(\frac{t_r}{t_r + 0.8t_w}\right)$$

Fig. 6.3.5 presents the total reinforced thickness $(t_w + t_r)$ required for gross area shear flow at various $\frac{W}{D}$ ratios.

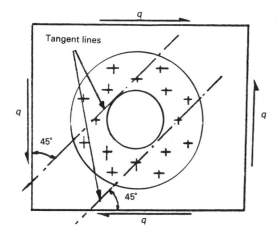

Fig. 6.3.4 Rivet pattern around hole.

Example

Given: Material 2024-T3 Clad for both web and ring. Assume the allowable shear flow $q = 750$ lb/in (buckled web without cut-out).

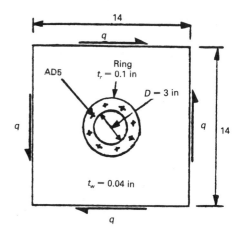

Determine the size of ring with $D = 3$ in and rivet pattern.

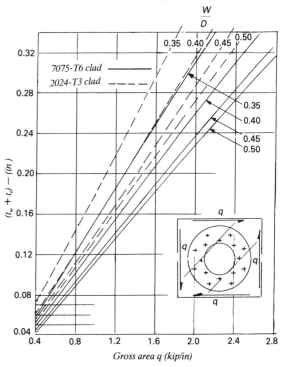

Fig. 6.3.5 Shear web hole reinforcement for 2024-T3 clad and 7075-T6 clad aluminum alloys.

From Fig. 6.3.5, $t_w + t_r = 0.14$ with $\frac{W}{D} = 0.35$.

Therefore, $t_r = 0.14 - 0.04 = 0.1$ in
and $W = 0.35(3) = 1.05$ in

$$\text{Rivet pattern} = 2q\left(\frac{t_r}{t_r + 0.8t_w}\right)$$

$$= 2(750)\left(\frac{0.1}{0.1 + 0.8(0.04)}\right)$$

$$= 1154 \text{ lb/in}$$

Use AD5 (MS20470-AD5), 596 lb/rivet per MIL-HDBK-5D (Ref. 4.1) with rivet spacing approximately 0.8 in as shown in the sketch.

6.4 Cutouts in Skin-stringer Panel (Wing and empennage)

Cutouts in aircraft wing and empennage structures constitute one of the most troublesome problems confronting the aircraft designer. Because the stress concentrations caused by cutouts are localized, a number of valuable partial solutions of the problem can be obtained by analyzing the behavior, under load (axial or shear load) of simple skin-stringer panels or integrally stiffened panels. In order to reduce the labor of analyzing such panels, simplifying assumptions and special devices may be introduced. The most important device of this nature used in this Chapter is a reduction of the number of stringers, which is effected by combining a number of stringers into a substitute single stringer as shown in Fig. 6.4.1 and Fig. 6.4.2.

If this cutout is designed for axial compression load, the local buckling problems has to be investigated carefully. In practice, the coaming stringer is designed to have a higher moment of inertia to increase its buckling strength and with smaller cross-sectional area to reduce and absorb local axial loads because of the cutouts. However, stringers next to the coaming stringer in the cutout region should have large cross-sectional area to absorb a portion of the coaming stringer load. In addition, the local skin thickness also should be beefed up such that it can redistribute both axial and panel shear loads.

Simplified Solution of Cutout Analysis — Axial Load

This Section describes the application to a panel with a cutout of a simplified three-stringer method of analysis. As indicated in Fig. 6.4.2, these substitute stringers are:

- Substitute Stringer No. 1: A substitute stringer having for its area all the effective area of the fully continuous member to one side of substitute stringer No. 2 "coaming stringer" (the stringer bordering the cutout) and placed at the centroid of the area of material for which it substitutes. The stress which this stringer develops is then the average stress for the material it replaces.
- The coaming stringer (called substitute stringer No. 2): It is the main continuous stringer bordering the cutout.
- Substitute Stringer No. 3: This one replaces all of the effective material made discontinuous by the cutout. It is located at the centroid of the material it replaces, and its stress is the average stress for the material it replaces.
- The panel skin thickness used in this analysis are the same as those of the actual structure.

It is assumed that the panel is symmetrical about both axes; the analysis can then be confined to one quadrant. It is also assumed that the cross-sectional areas of the stringers and of the skin do not vary in the longitudinal direction, that the panel is very long, and that the stringer stresses are uniform at large longitudinal distances from the cutout. In a panel with a cutout, the most important action takes places around the main stringer (coaming stringer) bounding the cutout. In accordance with the three-stringer method, it retains the main stringer as an individual stringer in the substitute structure. The Fig. 6.4.2 shows the actual structure, the substitute structure, and the distribution of the stresses in the actual structure.

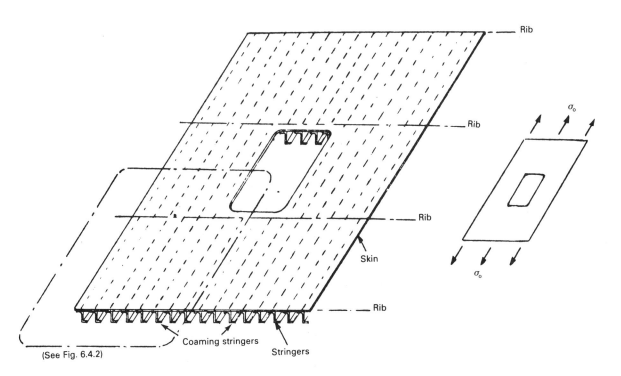

Fig. 6.4.1 Skin-stringer panel with rectangular cutout.

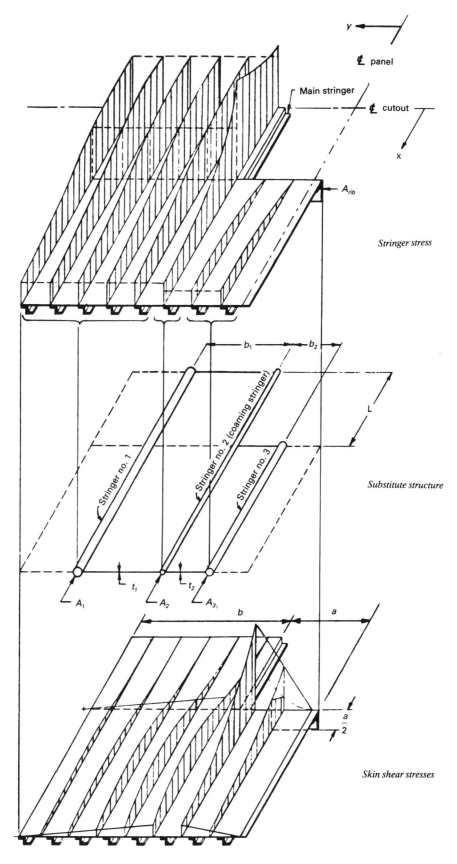

Fig. 6.4.2 *Schematic representation of actual structure and of substitute structure with stress distributions. (Axially loaded skin-stringer panel with cutout by three-stringer method)*

Symbols and Sign Conventions

A_1 Effective cross-sectional area of all continuous stringers, exclusive of main stringer (coaming stringer) bordering cutout (in^2)

A_2 Effective cross-sectional area of main continuous stringer (coaming stringer) bordering cutout (in^2)

A_3 Effective cross-sectional area of all discontinuous stringers (stringer No. 3) (in^2)

A_{rib} Cross-sectional area of rib at edge of cutout (in^2)

$$B = \sqrt{\frac{K_1^2 + K_2^2 + 2\bar{K}}{K_1^2 + K_2^2 + 2\bar{K} - \frac{K_3 K_4}{K_1^2}}}$$

C_0 Stress-excess factor for cutout of zero length, see Fig. 6.4.3.

$D = \sqrt{K_1^2 + K_2^2 + 2\bar{K}}$

E Young's Modulus of Elasticity (ksi)

G Shear Modulus (ksi)

$$K_1^2 = \frac{G t_1}{E b_1}\left(\frac{1}{A_1} + \frac{1}{A_2}\right)$$

$$K_2^2 = \frac{G t_2}{E b_2}\left(\frac{1}{A_2} + \frac{1}{A_3}\right)$$

$$K_3 = \frac{G t_2}{E b_1 A_2}$$

$$K_4 = \frac{G t_1}{E b_2 A_2}$$

$\bar{K} = \sqrt{K_1^2 K_2^2 - K_3 K_4}$

L Half-length of cutout (in)

R Stress-reduction factor to take care of change in length of cutout, see Fig. 6.4.4

b_1 Distance from A_2 to centroid of A_1, see Fig. 6.4.2

b_2 Distance from A_2 to centroid of A_3, see Fig. 6.4.2

$$r_1 = \frac{\tau_{2R} t_2}{A_3 \sigma_0}$$

$$r_2 = \frac{\tau_{2R} t_2 - \tau_{1R} t_1}{A_2(\sigma_{2R} - \sigma_0)}$$

$$r_3 = \frac{G \sigma_{2R}}{E b_2 \tau_{2R}}$$

t_1 Thickness of continuous skin (in), see Fig. 6.4.2

t_2 Thickness of discontinuous skin (in), see Fig. 6.4.2

x Spanwise (or longitudinal) distances (in), see Fig. 6.4.2

y Chordwise distances (in), see Fig. 6.4.2

σ_0 Average stress in the gross section (ksi)

σ_1 Stress in continuous substitute stringer No. 1 (ksi)

σ_2 Stress in main continuous stringer No. 2 (coaming stringer (ksi)

σ_3 Stress in discontinuous substitute stringer No. 3 (ksi)

σ_{rib} Stress in rib (chordwise), (ksi)

$\bar{\sigma}$ Average stress in net section (ksi)

τ_1 Shear stress in continuous substitute skin in t_1 (ksi)

τ_2 Shear stress in discontinuous substitute skin in t_2 (ksi)

(1) Stresses at the rib station in the substitute structure

The stringer stresses at the rib station are

$$\sigma_{1R} = \bar{\sigma}\left(1 - \frac{R C_0 A_2}{A_1}\right) \quad (6.4.1)$$

$$\sigma_{2R} = \bar{\sigma}(1 + R C_0) \quad (6.4.2)$$

where the factor C_0 is obtained from Fig. 6.4.3 and the factor R is obtained from Fig. 6.4.4.

The running shear in the continuous panel at the rib station is

$$q_{1R} = \tau_{1R} t_1 = \bar{\sigma} R C_0 A_2 K_1 \tanh K_1 L \quad (6.4.3)$$

The running shear in the discontinuous panel at the rib station is

$$q_{2R} = \tau_{2R} t_2$$
$$= \bar{\sigma} A_2 \frac{K_4}{D}\left(1 + R C_0 + \frac{K_1^2}{K}\right) \quad (6.4.4)$$

in which the factor D may be obtained from the Fig. 6.4.5.

The stresses σ_{1R} and σ_{2R} are the maximum values of σ_1 and σ_2 respectively, and are the maximum stresses in the panel. The stress σ_1 reaches its maximum at the center line of the cutout. The stress τ_1 reaches its maximum in the gross section at the section where $\sigma_1 = \sigma_2 \neq \sigma_0$. See Fig. 6.4.6.

(2) Stresses in the net section of the substitute structure

At a distance x from the center line of the cutout, the stresses in the stringer No. 1 and No. 2 are

$$\sigma_1 = \bar{\sigma}\left(1 - \frac{R C_0 A_2 \cosh K_1 x}{A_1 \cosh K_1 L}\right) \quad (6.4.5)$$

$$\sigma_2 = \bar{\sigma}\left(1 + R C_0 \frac{\cosh K_1 x}{\cosh K_1 L}\right) \quad (6.4.6)$$

The running shear in the net section is

$$\tau_1 t_1 = \bar{\sigma} R C_0 A_2 K_1 \left(\frac{\sinh K_1 x}{\cosh K_1 L}\right) \quad (6.4.7)$$

and decreases rapidly to zero at the center line of the cutout.

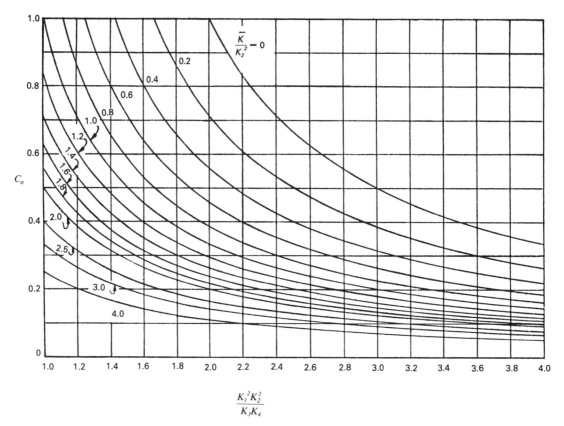

Fig. 6.4.3　Graph for factor — C_o.

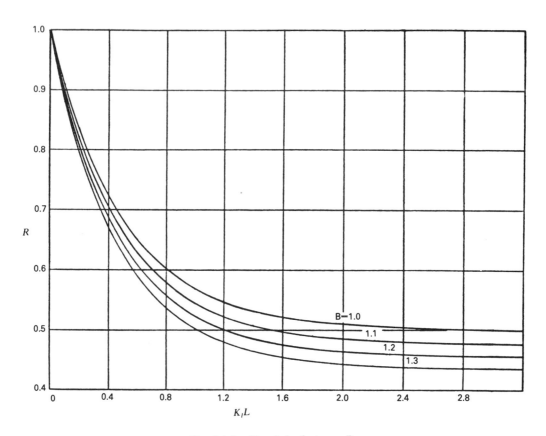

Fig. 6.4.4　Graph for factor — R.

(3) Stresses in the gross section of the substitute structure

It assumes the differences between the stresses at the rib station and the average stresses in the gross section to decay exponentially with rate-of-decay factors. The stress in the cut stringer (stringer No. 3) by the solution is

$$\sigma_3 = \sigma_0 (1 - e^{-r_1 x}) \qquad (6.4.8)$$

The stress in the main stringer (coaming stringer) is

$$\sigma_2 = \sigma_0 + (\sigma_{2R} - \sigma_0) e^{-r_2 x} \qquad (6.4.9)$$

The stress in the stringer No. 1 is

$$\sigma_1 = \sigma_0 + \frac{A_2}{A_1}(\sigma_0 - \sigma_2) + \frac{A_3}{A_1}(\sigma_0 - \sigma_3) \qquad (6.4.10)$$

The running shears in the skins are

$$\tau_1 t_1 = \tau_{2R} t_2 e^{-r_1 x} - (\tau_{2R} t_2 - \tau_{1R} t_1) e^{-r_2 x} \qquad (6.4.11)$$

$$\tau_2 t_2 = \tau_{2R} t_2 e^{-r_3 x} \qquad (6.4.12)$$

(4) Stresses in the actual structure

By the basic principles of the substitute structure, the stresses in the main continuous stringer (coaming stringer) of the actual structure are identical with the stresses in stringer No. 2 of the substitute structure. The total force in the remaining continuous stringers of the actual structure is equal to the force in stringer No. 1 of the substitute structure, and the total force in the cut stringers of the actual structure is equal to the force in stringer No. 3. In structural shear-lag theory, the force acting on a substitute stringer is distributed over the corresponding actual stringers on the assumption that the chordwise (y-direction) distribution follows the hyperbolic cosine law. Again, by the principles of the substitute structure, the shear stresses τ_1 in the substitute structure equal the shear stresses in the first continuous skin panel adjacent to the main stringer. In order to be consistent with the assumption that the chordwise distribution of the stringer stresses is uniform, the chordwise distribution of the shear stresses should be assumed to taper linearly from τ_1 to zero at the edge of the panel as shown in Fig. 6.4.2. Similarly, the chordwise distribution of the shear stresses in the cut skin panels should be assumed to vary linearly from τ_2 adjacent to the main stringer (coaming stringer) to zero at the center line of the panel.

By experimental values, it was found that the law of chordwise distribution of the shear stress τ_2 at the rib station could be approximated quite well by a cubic parabola. If the stress τ_2 is distributed according to cubic law, then the stress in the rib caused by the shear in the skin is

$$\sigma_{rib} = \frac{\tau_{2R} t_2 b}{4 A_{rib}} \left[1 - \left(\frac{y}{b}\right)^4 \right] \qquad (6.4.13)$$

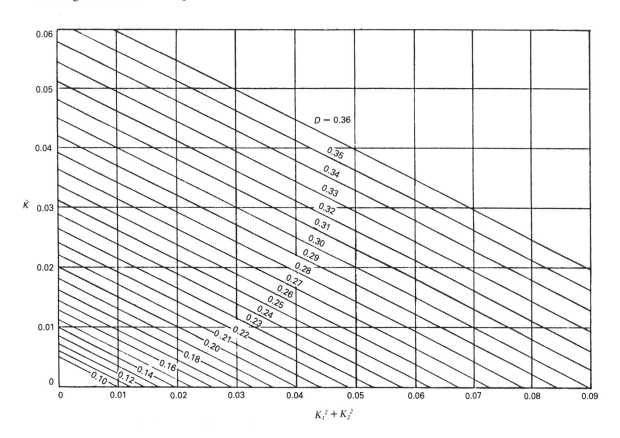

Fig. 6.4.5 Graph for factor — D.

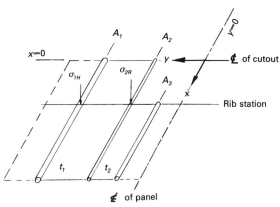

Fig. 6.4.6 Max. axial stresses at cutout.

Example

The structure for this numerical example is a 16-stringer panel with eight stringers cut and with a length of cut out equal to 30 in. The basic data is:

$A_1 = 0.703$ in^2 $t_1 = 0.0331$ in
$A_2 = 0.212$ in^2 $t_2 = 0.0331$ in
$A_3 = 1.045$ in$_2$
$b_1 = 5.96$ in $L = 15.0$ in
$b_2 = 7.56$ in

Materials for both skin and stringers are aluminum with:

$E = 10.5 \times 10^3$ ksi and $G = 4.0 \times 10^3$ ksi

This data yields the following values:

$K_1^2 = 0.01295$ $K_3 = 0.00995$ $\overline{K} = 0.00664$
$K_2^2 = 0.00944$ $K_4 = 0.00785$

From Fig. 6.4.3, obtain $C_0 = 0.6$

For $K_1 L = \sqrt{0.01295}\,(15) = 1.707$ and the exact value of

$$B = \sqrt{\frac{K_1^2 + K_2^2 + 2\overline{K}}{K_1^2 + K_2^2 + 2\overline{K} - \dfrac{K_3 K_4}{K_1^2}}} =$$

$$\sqrt{\frac{0.01295 + 0.00944 + 2(0.00664)}{0.01295 + 0.00944 + 2(0.00664) - \dfrac{0.00995\,(0.0078)}{0.01295}}}$$

$= 1.097$

From Fig. 6.4.4, obtain $R = 0.492$.

The stresses in the continuous stringers at the rib [refer to Eq. (6.4.1) and (6.4.2)] are

$$\sigma_{1R} = 8.21 \left[1 - (0.492)(0.6)\left(\frac{0.212}{0.703}\right) \right]$$

$= 7.48$ ksi

$\sigma_{2R} = 8.21\,[1 + (0.492)(0.6)] = 10.63$ ksi

The running shear in the continuous panel at the rib [Eq. (6.4.3)] is

$q_{1R} = \tau_{1R} t_1$
$\qquad = -8.21\,(0.492)(0.6)(0.212)(0.1138) \tanh 1.707$
$\qquad = -0.0547$ kip/in

The maximum running shear in the cut panel is

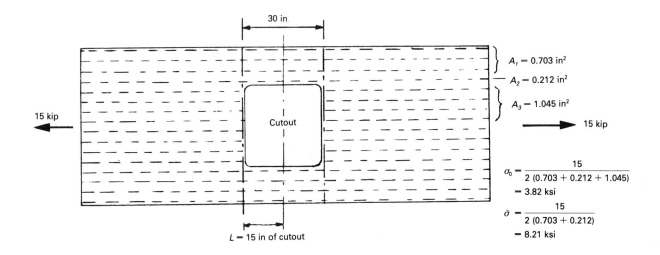

Compute $\dfrac{K_1^2 K_2^2}{K_3 K_4} = \dfrac{0.01295\,(0.00944)}{0.00995\,(0.00785)} = 1.565$

$\dfrac{\overline{K}}{K_2^2} = \dfrac{0.00664}{0.00944} = 0.704$

computed by Eq. (6.4.4). The value of D is obtained from Fig. 6.4.5 with

$\overline{K} = 0.00664$

$K_1^2 + K_2^2 = 0.01295 + 0.00944 = 0.02239$

$D = 0.189$ and

$$q_{2R} = \tau_{2R} t_2$$
$$= -8.21\,(0.212)\left(\frac{0.00785}{0.189}\right) \times$$
$$\left[1 + (0.492)(0.6) + \frac{0.01295}{0.00664}\right]$$
$$= -0.234 \text{ kip/in}$$

or the shear stress as

$$\tau_{2R} = -\frac{0.234}{0.0331} = -7.08 \text{ ksi}$$

The stresses in the cut section are computed as follows;

The rate-of-decay factors for the stresses in the gross section can now be computed

$$r_1 = \frac{\tau_{2R} t_2}{A_3 \sigma_0} = -\frac{-0.234}{1.045\,(3.82)} = 0.0587$$

$$r_2 = \frac{\tau_{2R} t_2 - \tau_{1R} t_1}{A_2(\sigma_{2R} - \sigma_0)} = \frac{-0.234 + 0.055}{0.212\,(10.63 - 3.82)}$$

$$= -0.1236$$

$$r_3 = \frac{G\,\sigma_{2R}}{E\,b_2\,\tau_{2R}} = \frac{(4.0 \times 10^3)(10.63)}{(10.5 \times 10^3)(7.56)(-7.08)}$$

$$= -0.0755$$

The stress in the cut stringers is obtained from Eq. (6.4.8)

$$\sigma_3 = 3.82\,(1 - e^{0.0587 x})$$

and in the main stringer from Eq. (6.4.9)

$$\sigma_2 = 3.82 + 6.93\,e^{-0.1236 x}$$

The stress in the continuous stringer can be found by Eq. (6.4.10).

The running shears are obtained by Eq (6.4.11) and (6.4.12)

$$q_1 = \tau_1 t_1 = -0.234\,e^{0.0587 x} + 0.179\,e^{-0.1218 x}$$

$$q_2 = \tau_2 t_2 = -0.234\,e^{-0.0755 x}$$

At $x = 1.5$ in, the point of maximum observed shear flow

$$q_{x=1.5\text{ in}} = -0.234\,e^{-0.0755\,(1.5)}$$
$$= -0.234\,(0.893) = -0.209 \text{ kip/in}$$

and shear stress is,

$$\tau_2 = -\frac{0.209}{0.0331} = -6.31 \text{ ksi}$$

Simplified Solution of Cutout Analysis — Shear Load

A method for the estimation of the maximum stresses around a small rectangular cutout in a skin-stringer panel loaded in pure shear is presented in this section. This method is based on a simplified application of the shearing theory. This method of analysis is valid only when either the length or the width of the cutout is less than the half-width of the panel.

The structure considered for the analysis is a skin-stringer panel with a rectangular cutout bounded on two sides by stringers and on the other two sides by ribs that lie at right angles to the stringers. The stringers and ribs are assumed to be of constant section and the skin is assumed to be of constant thickness. Under the simplifying assumptions on which the analysis is based, the cutout need not be centrally located in the panel but it must not be closed to an edge of the panel if the formulas are to be valid. The simplifying assumptions also imply that the results may be applied to a panel with a moderate amount of curvature.

In Fig. 6.4.7 the actual loading case investigated may be resolved into case I and case II. In case I the basic shear stress τ_0 is assumed to act along the outer edges of the panel and also at the edges of the cutout. For such a loading, the shear stresses are equal to the basic shear stress τ_0, and there are no stringer stresses. In case II shear stresses, which are assumed to act only along the edges of the cutout, are equal in magnitude but opposite in direction to those acting at the cutout as in case I. The shear stresses of case II will be called liquidating stresses because, when superposed on the stresses of case I, they neutralize the stresses at the edge of the cutout and thus produce the condition of zero shear at the edge of the cutout that exists in the actual case.

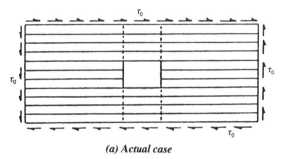

(a) Actual case

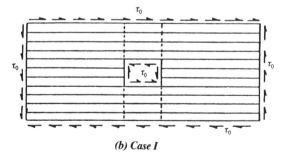

(b) Case I

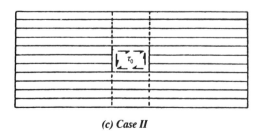

(c) Case II

Fig. 6.4.7 *Shear panel loading cases.*

Symbols

A_R Effective area of rib (in^2)
A_s Effective area of stringer (in^2)
E Young's Modulus of elasticity (ksi)
G Shear modulus of elasticity (ksi)
K_R Shear-lag parameter for stresses in rib-skin
K_s Shear-lag parameter for stresses in stringer-skin
L Half length of cutout (in)
a Chordwise distance from the edge of cutout to the edge of the panel (in)
b Half-width of cutout (in)
t Skin thickness (in)
x Spanwise distance from rib bounding cutout (in)
y Chordwise distance from stringer bounding cutout (in)
σ_R Rib stress (ksi)
σ_s Stringer stress (ksi)
τ_0 Basic shear stress in panel (ksi)
τ_R Shear stress in rib-skin caused by liquidating forces (ksi)
τ_s Shear stress in stringer-skin caused by liquidating forces (ksi)
τ' Final shear stress in stringer-skin panel (ksi)
τ'' Final shear stress in rib-skin panel (ksi)

Experimental results indicate that the stresses caused by the liquidating forces acting along the length of the panel are confined mainly to the stringers (coaming stringers) bounding the cutout and to the skin lying between these stringers. Similarly, the stresses caused by the liquidating forces acting across the width of the panel are confined mainly to the ribs bounding the cutout and to the skin lying between these ribs. Therefore, in order to simplify the problem of analysis, the actual structure is replaced by a simplified structure consisting of the aforementioned stringers and skins of the actual structure which carry the major share of the stresses caused by the cutout as shown in Fig. 6.4.8.

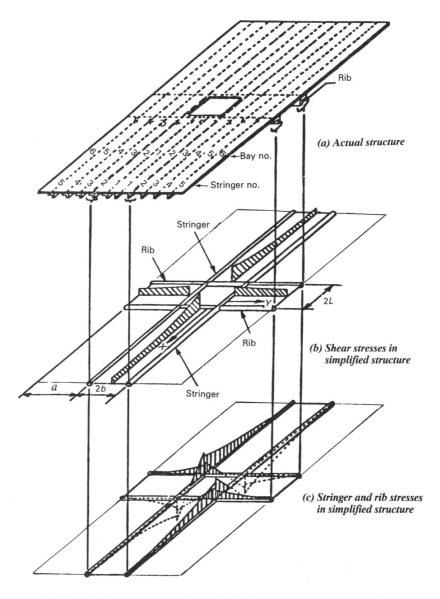

Fig. 6.4.8 Schematic representation of actual structure and stress distributions.

(1) Stringer stresses
The uniformly distributed, liquidating shear stresses, applied to the edge of the cutout parallel to the stringers, produce stresses which, because of the antisymmetry of the stringer-stress distribution in the vicinity of the cutout, are zero at the center line of the cutout and increase linearly to a maximum at the ribs. See Fig. 6.4.8. The maximum stringer stress is

$$\sigma_{s\,(max)} = \pm \frac{\tau_o t L}{A_s} \quad (6.4.14)$$

The stringer stress beyond the cutout can be expressed as

$$\sigma_s = \sigma_{s\,(max)} (e^{-K_s x}) \quad (6.4.15)$$

The shear-lag parameter K_s is defined by the equation below:

$$K_s^2 = \frac{G t}{E b A_s}$$

(2) Shear Stresses along stringer
The shear stresses caused by the liquidating forces

$$\tau_s = \tau_o K_s L\, e^{-K_s x}$$

The final shear stresses in the skin is obtained by adding the basic shear stress to the shear stresses caused by the liquidating forces as

$$\tau' = \tau_o (1 + K_s L\, e^{-K_s x}) \quad (6.4.16)$$

(3) Rib Stresses
The liquidating forces applied to the ribs produce stresses in the ribs that are zero at the center line of the cutout and increase linearly to maximum at the stringers. At the stringers the maximum stresses in the ribs are given by

$$\sigma_{R\,(max)} = \pm \frac{\tau_o t b}{A_R} \quad (6.4.17)$$

The rib stress away from the cutout is

$$\sigma_R = \sigma_{R\,(max)}\, e^{-K_R y} \quad (6.4.18)$$

in which the shear-lag parameter K_R is defined by

$$K_R^2 = \frac{G t}{E L A_R}$$

(4) Shear Stresses along rib
The shear stresses in the skin between ribs decrease exponentially

$$\tau_R = \tau_o K_R b\, e^{-K_R y}$$

The final shear stresses in the skin are the sum of the stresses caused by the liquidating forces and the basic shear stress

$$\tau' = \tau_o (1 + K_R b\, e^{-K_R y})$$

In many practical cases, it is sufficiently accurate to assume that the shear stresses are uniformly distributed over a length of a, then the above equation can be written as

$$\tau' = \tau_o \left(1 + \frac{b}{a}\right) \quad (6.4.19)$$

Example
Given:

$\tau_o = 20$ ksi
$t = 0.05$ in
$A_s = 0.5$ in^2
$A_R = 1.0$ in^2

Materials:

$E = 10.5 \times 10^3$ ksi
$G = 4.0 \times 10^3$ ksi

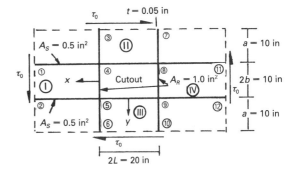

Stringer stresses:

$$K_s^2 = \frac{G t}{E b A_s}$$

$$= \frac{(4.0 \times 10^3)(0.05)}{(10.5 \times 10^3)(5)(0.5)} = 0.00762$$

or $K_s = 0.0873$

The maximum stringer stress obtained from Eq. (6.4.14)

$$\sigma_{s\,(max)} = \pm \frac{\tau_o t L}{A_s} = \pm \frac{20\,(0.05)\,(10)}{0.5}$$

$$= \pm 20 \text{ ksi}$$

The stress beyond the cutout obtained from Eq. (6.4.15)

$$\sigma_s = \sigma_{s\,(max)} (e^{-K_s x}) = \pm 20\, (e^{-0.0873 x})$$

if $x = 20$ in

$$\sigma_s = \pm 20\, [e^{-0.0873\,(20)}] = \pm 20\,(1.745)$$
$$= \pm 3.49 \text{ ksi}$$

Shear stresses along stringer:
from Eq. (6.4.16)

$$\tau' = \tau_o (1 + K_s L\, e^{-K_s x})$$
$$= 20\,[1 + 0.0873\,(10)\, e^{-0.0873 x}]$$

if $x = 0$

$$\tau' = 20\,(1 + 0.873) = 37.46 \text{ ksi};$$

if $x = 20$

$$\tau' = 20\,(1 + 0.152) = 23.05 \text{ ksi}$$

Rib stresses:

Airframe Structural Design 185

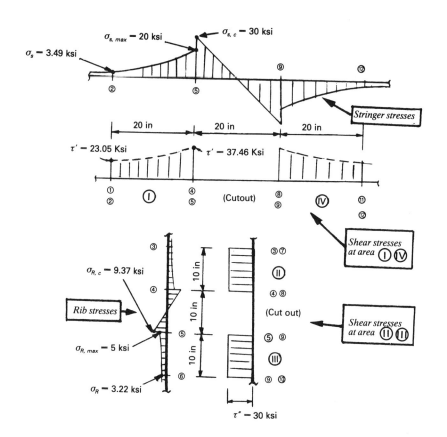

$$K_R^2 = \frac{Gt}{ELA_R}$$

$$\frac{(4.0 \times 10^3)(0.05)}{(10.5 \times 10^3)(10)(1.0)} = 0.0019$$

or $K_R = 0.044$

The maximum rib stress obtained from Eq. (6.4.17)

$$\sigma_{R\,(max)} = \frac{\tau_0\, t\, b}{A_R}$$

$$= \frac{20\,(0.05)(5)}{1.0} = 5\text{ ksi}$$

The rib stress away from cutout obtained from Eq. (6.4.18) is

$$\sigma_R = \sigma_{R\,(max)}\, e^{-K_R\, y} = 5\,(e^{-0.044\, y})$$

if $y = 0$

$\sigma_R = 5$ ksi;

if $y = 10$ in

$$\sigma_R = 5\,[e^{-0.044\,(10)}] = 5\,(0.64) = 3.22\text{ ksi}$$

Shear stress along rib:
from Eq. (6.4.19)

$$\tau'' = \tau_0\left(1 + \frac{b}{a}\right)$$

$$= 20\left(1 + \frac{5}{10}\right) = 30\text{ ksi}$$

Stresses at cutout regions ⑤:

Stringer stress $\sigma_{s,c} = \dfrac{\tau''\, t\, L}{A_s}$

$$= \frac{30\,(0.05)(10)}{0.5} = 30\text{ ksi}$$

Rib stress $\sigma_{R,c} = \dfrac{\tau'\, t\, b}{A_R}$

$$= \frac{37.46\,(0.05)(5)}{1.0} = 9.37\text{ ksi}$$

6.5 Curved Skin-stringer Panels (Fuselage)

The following method of analysis is considered applicable for establishing preliminary designed plug-type door cutouts (generally for cutout aspect ratios of approximately $\dfrac{\ell}{h} = 0.6$). Since the main loading systems are to be redistributed over a much smaller region, as shown in Fig. 6.5.1, than actually would occur. In fact, the redistribution loads should be, in general, somewhat conservative. The methods of analysis reflected herein will be modified if required for final design based on more rigorous analytical

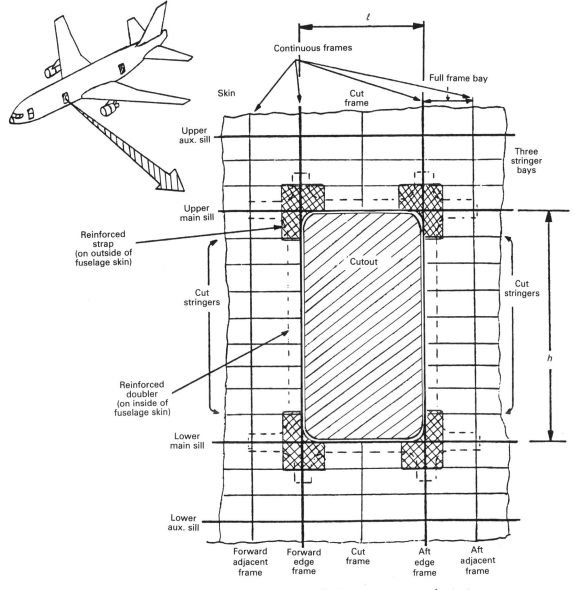

Fig. 6.5.1 Effective fuselage redistribution structure around cutout

methods such as finite element modeling.

Cutout Sizing Considerations

The following is a guide for preliminary sizing of members around a transport plug door cutout (the most common type design) of aluminum structure.

(1) Skin panels: Skin panel thickness should be based on the average panel shear flow. The panel is considered to be made up of skin or skin and doublers without inclusion of strap materials locally over frames and sills. The panel thickness should be such that the gross area shear stress should be lower i.e. 25 ksi for aluminum skin.
(2) Straps: Straps as shown in Fig. 6.5.1 are included over the main sill and edge frame members around the cutout to provide required outer chord (Fig. 6.5.5) material for redistribution of design ultimate loads in addition to providing a fail-safe load path.

Overall strap widths and thickness requirements should be first based on the critical combination of axial and bending loads in the outer chords. In addition, the areas of the strap, skin, and doublers, in the immediate vicinity of the main sill and edge frame junctions, as indicated by the crosshatching in Fig. 6.5.1 shall keep a gross area shear stress low for long fatigue life when subjected to a shear flow of $\frac{5}{3} \times q_{av}$. This is intended to account for local load concentration at the sill-frame junctions.

(3) Frame and sill outer chord members: The frame and sill outer chords consists of the frame or sill cap, strap and effective skin. For preliminary sizing of outer chord members, consider the following (see Fig. 6.5.5):
(a) Frame and sill axial loads act on the outer chord only.

(b) Bending loads $\left(\dfrac{M}{h'}\right)$ combined with the axial loads when they both produce the same type of loading (tension or compression load) in the outer chord.
(c) Consider 50% of the bending moment loads (ie. $\dfrac{0.5M}{h'}$) combined with the axial loads when the bending moment loads relieve and are of smaller magnitude than the axial loads.
(d) Consider the total bending moment loads $\left(\dfrac{M}{h'}\right)$ combined with the axial loads when the bending loads oppose and are of large magnitude than the axial loads.

Note: The above loading combinations will be different with different sizes of cutouts as well as different airplanes.

(4) Edge frame and main sill inner chord members:
  (a) The inner frame and sill chords are considered to carry bending moment loads only.
  (b) If the inner chord ultimate tensile stress is very high, the chord shall be given fail-safe design considerations.
  (c) The $\dfrac{b}{t}$ ratio for the inner chord outstanding flange should be maintained at less than 12 to prevent compression crippling buckling.
(5) Panel attachments (fasteners): The panel attachments shall be good for 1.25 times q_{max} (ultimate panel failing shear flow).
(6) Redistribution geometry: effective fuselage redistribution structure.
  (a) One frame bay forward and aft of the cutout edge frame members
  (b) Three stringer bays above and below the cutout
(7) Design conditions (ultimate load):
  (a) Flight Loads (acting alone when critical)
  (b) Flight Loads + 2.0 Load Factors × Cabin Pressure ($2 \times p$)
  (c) 3.0 Load Factors × Cabin Pressure ($3 \times p$) — (Acting along)
  (d) 2.5 Load Factors × Cabin Pressure ($2.5 \times p$) — (Use for shear and stringers in compression case)
  where p = Fuselage cabin pressure differential
(8) Fail-safe design loads use ultimate (use limit loads per FAR 25):
  Each of the primary structural members, namely the edge frames and main sills shall be individually designed as fail-safe based on the following design load conditions.
  (a) Fail-safe flight loads acting alone
    use limit loads
  (b) Fail-safe pressure loads acting alone
    Fail-safe pressure loads shall be 1.5 times a maximum operating pressure differential.
  (c) Fail-safe pressure loads combined with fail-safe flight loads
    This condition shall be (maximum operating pressure differential) times a 1.25 factor to account for dynamic effects combined with the fail-safe flight loads.
(9) Internal load redistribution cases:
  To more clearly define the redistribution loads for the four loading cases (shown below), each case should be considered independently and then combined to formulate the most critical design condition on each of the structural elements. The following designs reflect the meth-

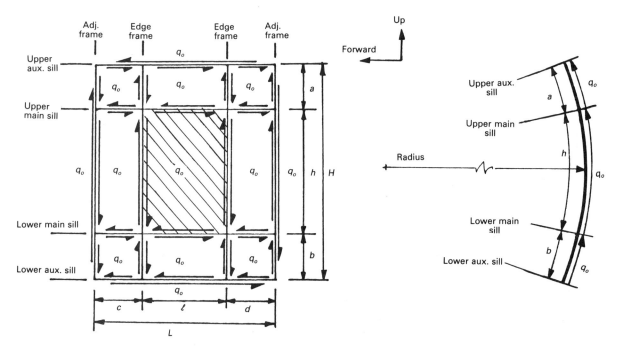

Fig. 6.5.2 Panel shear flows assuming no cutout.

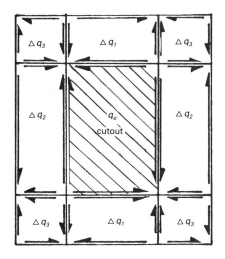
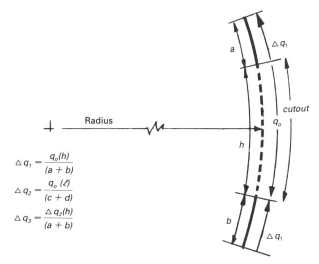

Fig. 6.5.3 Apply an equal but opposite shear flow (q_o) to the cutout and redistribute as $\triangle$ shear flows to the surrounding panels.

ods of internal load redistribution for each of the four loading cases with considerations for constant and varying shear flows.

Case I   Fuselage skin shears — flight conditions
Case II   Cut stringer loads — flight conditions
Caee III Longitudinal and circumferential tension loads — cabin pressurization conditions
Case IV   Plug pressure and door stop redistribution effects

Case I. Fuselage Skin Shears — Flight Conditions

Redistribution of constant shear flows in the vicinity of the cutout and assume that cutout in an area of nearly constant fuselage radius.

Combining the shear flows from Fig. 6.5.2 and Fig. 6.5.3, the total redistributed shear flows (average) is obtained around the cutout as shown in Fig. 6.5.4.

on the average shear flows and are determined by the above final redistributed panel shear flows.

Axial loads are considered acting in the frame or sill outer chords only since the loads are introduced from the skins. The outer chord consists of frame or sill cap plus local effective skin plus local doubler or fail-safe strap if any, as indicated in Fig. 6.5.5.

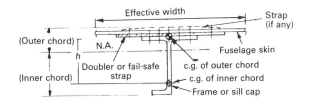

Fig. 6.5.5 Total cross-section of typical frame or sill elements.

Redistribution of variable shear flows in the vicinity of the cutout (Fig. 6.5.6) shows the shear flow variation in fuselage panels.

Then apply an equal and opposite shear flow to the cutout panel q_b and q_c and redistribute as $\triangle$ shear flows to the surrounding panels as shown in Fig. 6.5.7(a). Utilize the distribution lengths shown in Fig. 6.5.7(b) in redistributing the cutout shear flows to the upper and lower panels.

Combining the shear flows from Fig. 6.5.6 and Fig. 6.5.7(a) the total redistributed shear flows (average) are obtained around the cutout as shown in Fig. 6.5.8.

$$q_{1b} = q_a + \triangle q_{1b}$$
$$q_{1c} = q_d + \triangle q_{1c}$$
$$q_{2b} = q_b + \triangle q_{2b}$$
$$q_{2c} = q_c + \triangle q_{2c}$$
$$q_{3b} = \triangle q_{3b} - q_a$$
$$q_{3c} = \triangle q_{3c} - q_d$$

Fig. 6.5.4 Final redistributed panel shear flows around cutout.

where q_1, q_2 and q_3 are average shear flows q_{av} in each individual panel.

Axial loads in the sills and frames shall be based

Airframe Structural Design 189

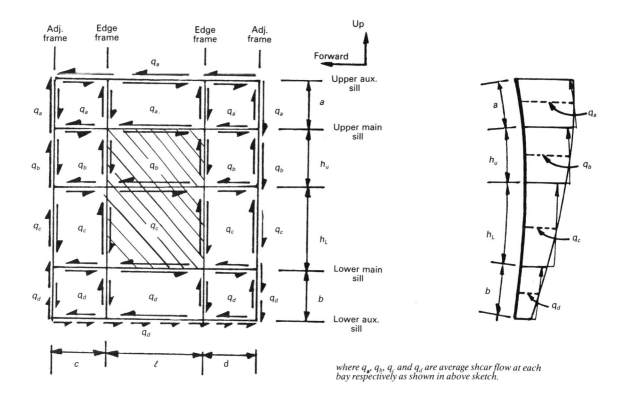

where q_a, q_b, q_c and q_d are average shear flow at each bay respectively as shown in above sketch.

Fig. 6.5.6 *Average panel shear flows assuming no cutout.*

(a) △ Shear flow distribution   **(b) Shear at cutout**

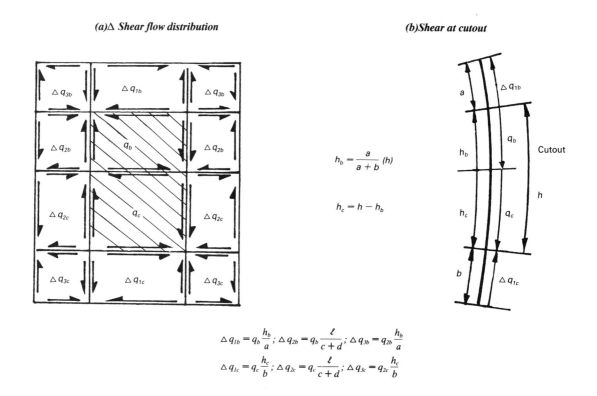

$$h_b = \frac{a}{a+b}(h)$$

$$h_c = h - h_b$$

$$\triangle q_{1b} = q_b \frac{h_b}{a} \,;\; \triangle q_{2b} = q_b \frac{\ell}{c+d} \,;\; \triangle q_{3b} = q_{2b} \frac{h_b}{a}$$

$$\triangle q_{1c} = q_c \frac{h_c}{b} \,;\; \triangle q_{2c} = q_c \frac{\ell}{c+d} \,;\; \triangle q_{3c} = q_{2c} \frac{h_c}{b}$$

Fig. 6.5.7 *△ shear flow distribution to surrounding panels.*

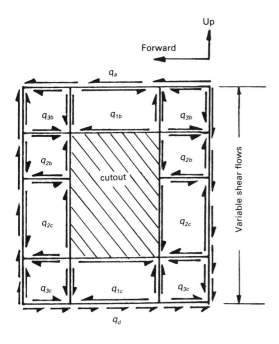

Fig. 6.5.8 *Final redistributed panel shear flows around a cutout due to circumferential variable shear flows.*

Next we consider frame shear flows, sill reactions, and frame and sill bending moments. To reflect a method of obtaining loads, the forward edge frame of a typical cutout is considered.

To minimize conservatisms in establishing frame bending moments due to curvature effects, the redistributed panel shears reflected to Fig. 6.5.4 (for constant fuselage shear flow application) and Fig. 6.5.8 (for fuselage variable shear flow application) are modified to introduce higher shears at the main sill and edge frame junctions thus minimizing the sill reactions required to put the frame in equilibrium.

- Consider a 3:1 distribution of frame shear flows when two stringer bays exist between the main and auxiliary sills.
- Consider a 5:3:1 distribution of frame shear flows when three stringer bays exist between the main and auxiliary sills. The 5:3:1 implies that the applied frame shears (average) in the upper, center, and lower portions of the frame are proportioned $\frac{5}{3}(q), \frac{3}{3}(q),$ and $\frac{1}{3}(q)$ as shown Fig. 6.5.9.

When the sills are radial (as assumed in this method), moment equilibrium is satisfied, therefore, only vertical and horizontal force equilibrium equations are available to establish sill reactions. Since four unknowns exist and only two equations, the following assumptions will be considered:

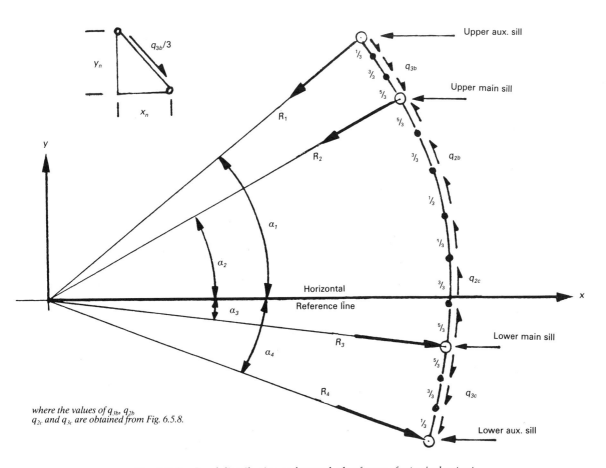

where the values of q_{3b}, q_{2b}, q_{2c} and q_{3c} are obtained from Fig. 6.5.8.

Fig. 6.5.9 *Load distribution at forward edge frame of a typical cutout.*

Airframe Structural Design 191

$R_2 = 2R_1$ at upper sills
$R_3 = 2R_4$ at lower sills

The sill force reaction can now be obtained from the following equations.

Summation vertical forces = 0

$R_1 \sin \alpha_1 + 2R_1 \sin \alpha_2 + 2R_4 \sin \alpha_3 + R_4 \sin \alpha_4$

$+ \left(\dfrac{1}{3} q_{3b}\right) y_1 + (q_{3b}) y_2 + \left(\dfrac{5}{3} q_{3b}\right) y_3 - \left(\dfrac{5}{3} q_{2b}\right) y_4$

$- (q_{2b}) y_5 - \left(\dfrac{1}{3} q_{2b}\right) y_6 - \left(\dfrac{1}{3} q_{2c}\right) y_7 - (q_{2c}) y_8$

$- \left(\dfrac{5}{3} q_{2c}\right) y_9 + \left(\dfrac{5}{3} q_{3c}\right) y_{10} + (q_{3c}) y_{11} + \left(\dfrac{1}{3} q_{3c}\right) y_{12}$

$= 0$

Summation horizontal forces = 0

$R_1 \cos \alpha_1 + 2R_1 \cos \alpha_2 - 2R_4 \cos \alpha_3 - R_4 \cos \alpha_4$

$- \left(\dfrac{1}{3} q_{3b}\right) x_1 - (q_{3b}) x_2 - \left(\dfrac{5}{3} q_{3b}\right) x_3 + \left(\dfrac{5}{3} q_{2b}\right) x_4$

$+ (q_{2b}) x_5 + \left(\dfrac{1}{3} q_{2b}\right) x_6 + \left(\dfrac{1}{3} q_{2c}\right) x_7 + (q_{2c}) x_8$

$- \left(\dfrac{5}{3} q_{2c}\right) x_9 + \left(\dfrac{5}{3} q_{3c}\right) x_{10} + (q_{3c}) x_{11} + \left(\dfrac{1}{3} q_{3c}\right) x_{12}$

$= 0$

With the sill reactions at the forward edge frame known, as determined from the above equations, the bending moments at any point on the frame may be determined by conventional methods. Sill reactions and frame moments for the aft edge frame are obtained using the same technique as for the forward edge frame.

With sill reactions known at both edge frames, sill balancing loads at adjacent frame are obtained as indicated in Fig. 6.5.10.

Fig. 6.5.10 shows a balanced sill from which shear and bending moment diagrams may be obtained. Bending moments (B.M.) in the adjacent frames are obtained by direct proportion of radial loads as follows:

B.M. (Fwd Adj. Frame) =

$\dfrac{R_{\text{Fwd Adj.}}}{R_{\text{Fwd Edge}}}$ [B.M. (Fwd Edge Frame)]

For practical structural considerations, it assumes a sill extending two bays either side of cutout as shown in Fig. 6.5.11.

The simplified assumptions above are based on adjacent frames of equal stiffness and do not reflect variable stiffnesses of frames due to induced loading effects from fuselage floor beams, etc.

**Case II. Cut Stringer Loads
— Flight Conditions**

Additional panel shear flows resulting from cut-stringer at both forward and aft edge of cutout must be considered.

Assume zero fixity of corners as shown in Fig. 6.5.12 and cut-stringer loads are assumed to be diffused to the main sills generally over one frame bay on either side of the cutout. This approach results in somewhat higher panel shears and frame axial loads than actually occur in practice. However, this conservative approach is required to consider fatigue design for less tension stresses around cutout.

If the cut-stringer loads are not the same on either side of the cutout, use average stringer loads at center line of cutout for preliminary analysis. However, large

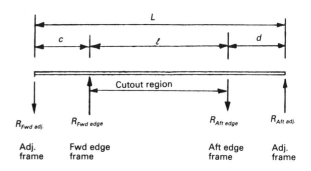

Fig. 6.5.10 Sill extending one bay either side of cutout.

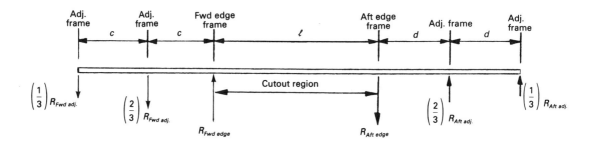

Fig. 6.5.11 Sill extending two bays either side of cutout.

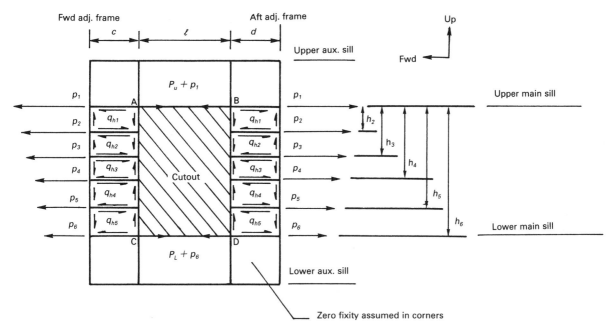

Fig. 6.5.12 Additional panel shear flows from cut-stringer loads.

door cut-outs such as passenger, service, cargo doors, etc. will necessitate considering additional Δ axial stringer loads between forward and aft adjacent frames.

Obtain P_u and P_L by conventional moment and force balance as shown below.

$$P_L = \frac{p_2 \times h_2 + p_3 \times h_3 + p_4 \times h_4 + p_5 \times h_5}{h_6}$$

and $P_u = (p_2 + p_3 + p_4 + p_5) - P_L$

Panel shear flows (ie. forward panels)

$$q_{h1} = \frac{P_u}{c}$$

$$q_{h2} = \left(q_{h1} - \frac{p_2}{c} \right)$$

$$q_{h3} = \left(q_{h2} - \frac{p_3}{c} \right)$$

.
.
.

Etc.

Note: Aft panel shears are obtained in the same manner and care should be used in establishing shear flow directions.

Case III Longitudinal and Circumferential Tension Loads — Cabin Pressurization Conditions

Hoop tension loads above and below the cutout are considered to be redistributed to the edge frames via shears in the panels between the main and auxiliary sills as shown in Fig. 6.5.13.

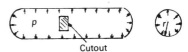

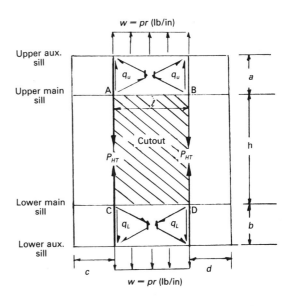

Fig. 6.5.13 Cutout panel shear flows due to cabin pressure-hoop tension.

Airframe Structural Design 193

$w = pr$ (hoop tension running load in lb/in)

$q_u = 0$ at panel center varies to

$$q_u = \frac{pr\ell}{2a} \text{ at edge frames}$$

$q_L = 0$ at panel center varies to

$$q_L = \frac{pr\ell}{2b} \text{ at edge frames}$$

Hoop tension load in edge frames

$$P_{HT} = \frac{pr\ell}{2}$$

Longitudinal tension loads are assumed to be redistributed to the upper and lower main sills in the same manner as for hoop tension, see Fig. 6.5.14.

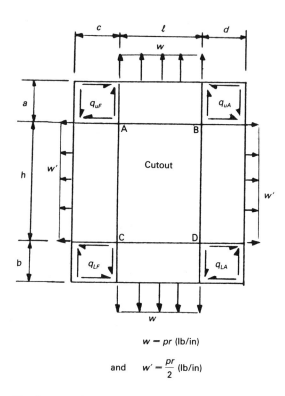

$w = pr$ (lb/in)

and $w' = \dfrac{pr}{2}$ (lb/in)

Fig. 6.5.15 *Cutout corner shear flows due to cabin pressure.*

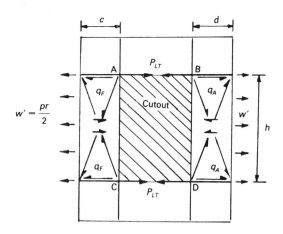

Fig. 6.5.14 *Cutout panel shear flows due to cabin pressure-longitudinal tension.*

$w' = \dfrac{pr}{2}$ (hoop tension running load in lb/in)

$q_F = 0$ at panel center varies to

$$q_F = \frac{w'h}{2c} \text{ at main sills}$$

$q_A = 0$ at panel center varies to

$$q_A = \frac{w'h}{2d} \text{ at main sills}$$

Longitudinal tension load in main sills

$$P_{LT} = \frac{w'h}{2}$$

The following reflects a method of obtaining corner panel shear flows (see Fig. 6.5.15) and incremental axial loads in the frame and sill members from corner fixity considerations. First, consider the corner as bent under combined pressure effects and establish adjusted fixed end moments (F.E.M.) considering relative stiffnesses of upper, lower, and side panels.

Then assume deformations such that a minimum of 50% fixity is provided in the corners.

Distribution factor (see Fig. 6.5.16) based on constant moment of inertia (it is usually designed such that these values of a, b, c and d are approximately equal) around corner.

$$K_1 = \frac{h}{h + \ell} \text{ (upper and lower panels)}$$

$$K_2 = \frac{\ell}{h + \ell} \text{ (forward and aft side panels)}$$

Fixed end moments:
upper and lower panels

$$M_{UL} = \frac{w\ell^2}{12} \text{ (in-lb)}$$

forward and aft side panels

$$M_{FA} = -\frac{w'h^2}{12} \text{ (lb-in)}$$

Considering 50% corner fixity based on adjusted F.E.M., the shear flows in the corner panels are obtained as reflected below.

Shear flow in upper forward corner panel:

$$q_{uF} = \frac{50\% \text{ (Adjusted F.E.M.)}}{ac}$$

$$= \frac{50\% \left[\frac{w\ell^2}{12} - K_1 (M_{UL} + M_{FA}) \right]}{ac} \text{ (lb/in)}$$

Shear flows in the other corners are generated in the same manner based on their respective geometry. The axial loads in the sills and frames are obtained by summation of q's.

The following reflects method of obtaining sill reactions add frame and sill bending moments resulting from pressurization effects. Fig. 6.5.17 shows a section above the cutout reflecting the balancing loads per unit width required to react pure hoop tension load applied at the auxiliary sill.

Integrating the unit reaction system shown in Fig. 6.5.17 over half the door width $\left(\frac{\ell}{2}\right)$, the total loading on the sills (shown in Fig. 6.5.18) is obtained.

Where the shear flow q_u (lb/in) is the same as shown in Fig. 6.5.13, loads on edge frames between lower main sill and lower auxiliary sill are obtained in the same manner as indicated above.

Induced frame and sill loads resulting from corner panel fixity considerations are shown in Fig. 6.5.15. Fixing shear flows in the corners result in a $\triangle P_{HT}$ load in the frames (tension in edge frames and compression in adjacent frames) plus local frame and sill bending due to frame curvature. Fig. 6.5.19 are examples showing loads necessary to put the forward edge frame and upper auxiliary sill in balance.

Case IV. Plug Pressure and Door Stop Redistribution Effects

Plug pressure loads on the door are redistributed to the forward and aft edge frames surrounding the door by means of stop fittings located at several points along the door edges as shown in Fig. 6.5.20. Since the door stops are somewhat eccentric from the edge frame, $\triangle$ moments are produced at the edge frame.

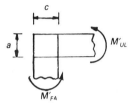

Distribution Factors	K_1	K_2
F.E.M. (+Moment)	$\frac{w\ell^2}{12}$	$-\frac{w'h^2}{12}$
Correction Values	$-K_1 (M_{UL} + M_{FA})$	$-K_2 (M_{UL} + M_{FA})$
Adjusted F.E.M.	$M'_{UL} = \frac{w\ell^2}{12} - K_1 (M_{UL} + M_{FA})$	$M'_{FA} = -\frac{w'h^2}{12} - K_2 (M_{UL} + M_{FA})$

Fig. 6.5.16 *Adjusted fixed end moments (F.E.M.).*

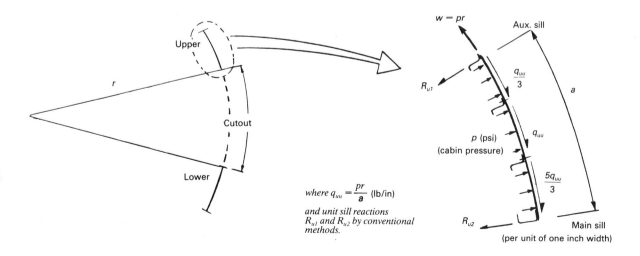

Fig. 6.5.17 *Upper cutout balancing loads.*

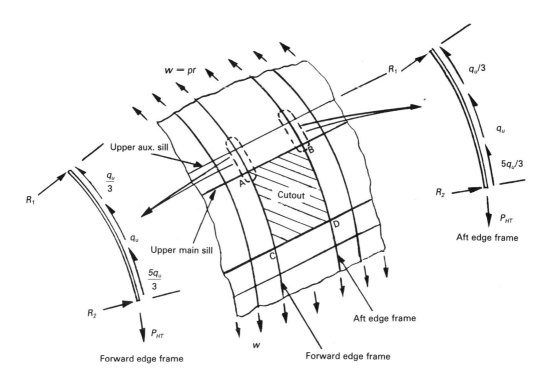

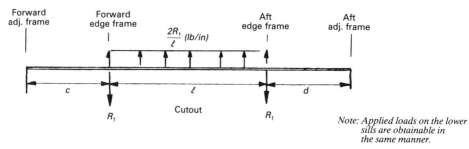

Where the shear flow q_u (lb/in) is the same as shown in Fig. 6.5.13 and loads on edge frames between lower main sill and lower auxiliary sill are obtainable in the same manner as indicated above.

Applied sill loads — upper aux. sill:

Note: Applied loads on the lower sills are obtainable in the same manner.

Applied sill loads — upper main sill:

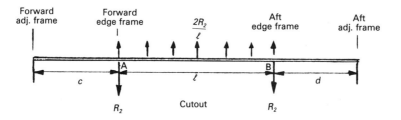

Fig. 6.5.18 Sill loads at cutout.

These Δ moments are generally reacted by providing intercostals between edge frames and reversed loads on the adjacent frames.

For preliminary analysis the 50% fixity considerations should encompass the fixity requirements for stop overhang considerations. Also, for preliminary analysis, assume the point loads introduced at the stops are uniformly distributed on the frames, thus introducing no localized bending on the frames. In general, frame bending from stop load effects will be small compared to bending from flight load considerations.

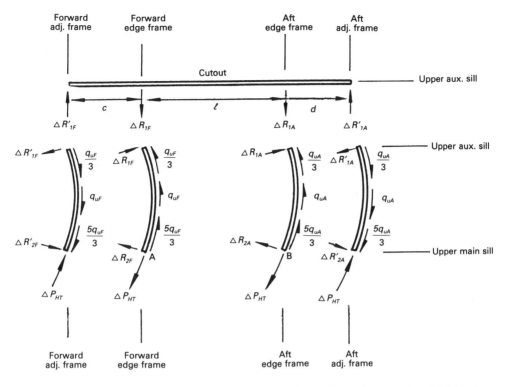

Note: These loading effects on the frames and sills when combined with the loads developed in Fig. 6.5.18 constitute the total loading due to pressurization and fixity effects.

Fig. 6.5.19 Induced frame and sill loads resulting from corner panel fixity.

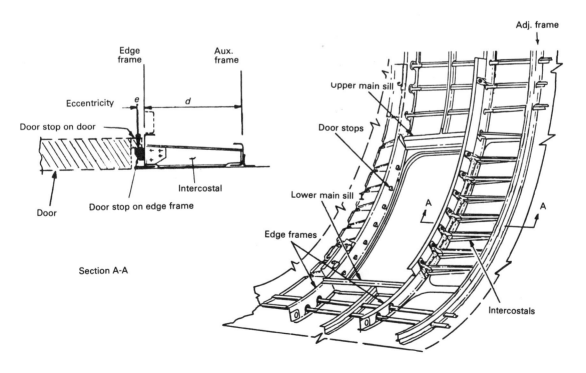

Fig. 6.5.20 Surrounding structures of fuselage cargo entrance cutout.

Airframe Structural Design 197

Example
Given data as shown below:

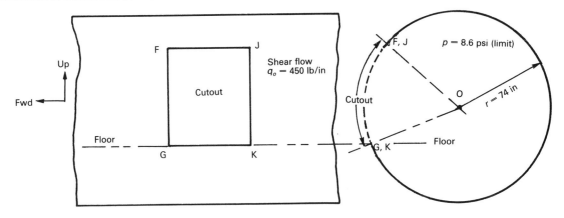

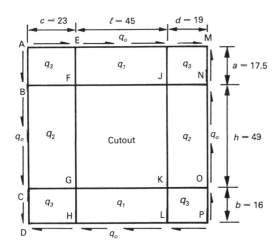

In general, the surround structure for a cutout shall consist of two main and two auxiliary sills (upper and lower), two edge frames (one forward, one aft), and two adjacent frames, door stops and corresponding intercostals, reinforced straps and various skin doublers. This structure shall absorb fuselage body shear, body bending arising from aircraft flight conditions, hoop tension and longitudinal tension due to cabin pressure.

Design conditions:
(a) 3.0 factors pressure, which is 3 (8.6) psi = 25.8 psi acting alone and to be used for tension and bending members
(b) 2.5 factors pressure, which is 2.5 (8.6) = 21.5 psi acting alone, and to be used for compression and shear structures
(c) Flight loads, plus 2.0 factors pressure which is 2 (8.6) = 17.2 psi
(d) Flight loads, without pressure

Assumed effective fuselage body structure:
(a) One frame pitch forward and aft of the edge frames
(b) Two stringer pitches above and below the main sills for the calculation.

For cutout on or near the aircraft fuselage horizontal bending axis (neutral axis), the vertical shear flow along the boundaries of the redistribution structure may be assumed uniform. The magnitude of this shear flow will be equal to the value that exists at the vertical center line of the cutout in the uncut structure. However, for cutouts which are appreciably off-set from the horizontal bending axis, for example, under floor cargo cutouts, the shear flow will not be uniform and this must be reflected in establishing the shear flows when "free-bodying" the structure. Further, the average change in shears due to the cutout maybe calculated and distributed between the top and bottom stringer panels (above and below the main sills) in a manner consistent with the loading and stiffness etc.

(i) Fuselage body shears due to flight conditions:
For simplicity, thus, cutout is assumed near the neutral axis and therefore the normal average body shear flow is considered uniform shear as $q_0 = 450$ lb/in.
(Refer to Section 6.3 framing cutout in web.)

$$q_1 = \left(1 + \frac{h}{a+b}\right) q_0 = \left(1 + \frac{49}{17.5+16}\right)(450)$$
$$= 1108.2 \text{ lb/in}$$

$$q_2 = \left(1 + \frac{\ell}{c+d}\right) q_0 = \left(1 + \frac{45}{23+19}\right)(450)$$
$$= 932.1 \text{ lb/in}$$

$$q_3 = \left[-1 + \left(\frac{\ell}{c+d}\right)\left(\frac{h}{a+b}\right)\right] q_0$$
$$= \left[-1 + \left(\frac{45}{23+19}\right)\left(\frac{49}{17.5+16}\right)\right](450)$$
$$= 255.2 \text{ lb/in}$$

(See shear redistribution around cutout sketch on next page.)

Note: Frame and sill axial loads are determined by summation of shears, and axial loads are assumed to be carried in the frame or sill outer chord, straps, skin and doublers as defined in Fig. 6.5.5.

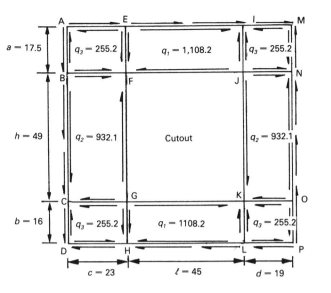

(ii) Cut stringer loads from fuselage body bending moment — flight conditions

It will be assumed that zero fixity exist at the junction of frames to main sills. The cut stringer axial loads will be diffused to the main sills over one frame pitch.

Given stringer axial loads as shown below.

Taking moments about point K to obtain the upper main sill reaction load S_9:

$$(-760)(8.5) + (760)(17) + (2200)(25)$$
$$+ (3630)(33) + (5060)(41) + (6500)(49)$$
$$= (49)(S_9)$$

$S_9 = 14480$ lb in tension

Taking moments about point J to obtain the lower main sill reaction load S_{15}:

and $S_{15} = 630$ lb in tension

Shear flow below stringer No. 9 due to diffusion of axial loads is given by

$$\frac{14480 - 6500}{19} = \frac{7980}{19}$$
$$= 420 \text{ lb/in}$$
at aft edge frame

$$\frac{7980}{23} = 347 \text{ lb/in} \quad \text{at forward edge frame}$$

Shear flow above stringer No. 15

$$\frac{630 - (-2280)}{19} = 153 \text{ lb/in} \quad \text{at aft edge frame}$$

$$\frac{630 - (-2280)}{23} = 126.5 \text{ lb/in}$$
at forward edge frame

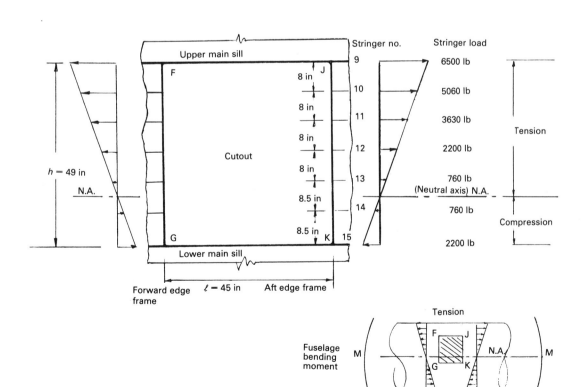

Airframe Structural Design 199

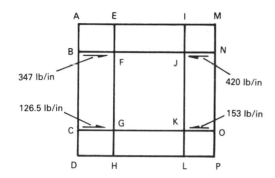

Panel shear flows, due to fuselage bending moment of flight loads, around cutout structure are shown below.

Forward edge frame shear flows:

Consider the shear between the upper main and auxiliary sills

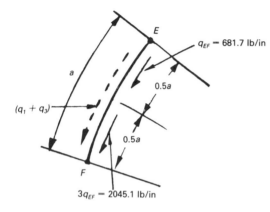

$(0.5a)(q_{EF}) + (0.5a)(3q_{EF}) = (a)(q_1 + q_3)$

$q_{EF} = \frac{1}{2}(q_1 + q_3)$

$= \frac{1}{2}(1108.2 + 255.2) = 681.7 \text{ lb/in}$

$3q_{EF} = 3(681.7) = 2045.1 \text{ lb/in}$

Similarly for the shear flows between lower main and auxiliary sills.

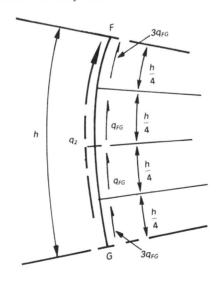

Consider the shear flows between the upper and lower main sills, thus

$\left(\frac{h}{4}\right)(3q_{FG} + q_{FG} + q_{FG} + 3q_{FG}) = q_2(h)$

$q_{FG} = \frac{1}{2}(932.1) = 466.05 \text{ lb/in}$

$3q_{FG} = 3(466.05) = 1398.15 \text{ lb/in}$

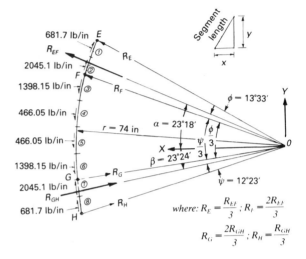

where: $R_E = \frac{R_{EF}}{3}$; $R_F = \frac{2R_{EF}}{3}$

$R_G = \frac{2R_{GH}}{3}$; $R_H = \frac{R_{GH}}{3}$

The sill reactions are assumed also to be the components of a resultant load (R_{EF} and R_{GH} as shown above) located at one-third the distance between the main and auxiliary sills and measured from the main sill.

Segment length in x and y directions:

Segment No.	Shear (lb/in)	x (in)	y (in)
①	681.7	4.2	7.6
②	2045.1	3.3	8.1
③	1398.15	3.1	11.9
④	466.05	1.0	12.2
⑤	466.05	1.0	12.2
⑥	1398.15	3.1	11.8
⑦	2045.1	3.0	7.5
⑧	681.7	3.8	7.0

Consider the vertical equilibrium of the frame:

$R_{EF} \sin \alpha + R_{GH} \sin \beta + 681.7 \, y_1 + 2045.1 \, y_2$
$- 1398.15 \, y_3 - 466.05 \, y_4 - 466.05 \, y_5$
$- 1398.15 \, y_6 + 2045.1 \, y_7 + 681.7 \, y_8 = 0$

Rearrange above equation and obtain

$R_{EF} \sin \alpha + R_{GH} \sin \beta + 681.7 \, (y_1 + y_8)$
$+ 2045.1 \, (y_2 + y_7) - 1398.15 \, (y_3 + y_6)$
$- 466.05 \, (y_4 + y_5) = 0$

$R_{EF} \sin 23°18' + R_{GH} \sin 23°24'$
$+ 681.7 (7.6 + 7.0) + 2045.1 (8.1 + 7.5)$
$- 1398.15 (11.9 + 11.8) - 466.05 (12.2 + 12.2)$
$= 0$

$\qquad 0.396 R_{EF} + 0.397 R_{GH} = 2651.4 \qquad$ (Eq. A)

Consider the horizontal equilibrium of the frame:
$- R_{EF} \cos \alpha + R_{GH} \cos \beta + 681.7 x_1 + 2045.1 x_2$
$- 1398.15 x_3 - 466.05 x_4 - 466.05 x_5$
$+ 1398.15 x_6 - 2045.1 x_7 - 681.7 x_8 = 0$

Rearrange above equation and obtain
$- R_{EF} \cos \alpha + R_{GH} \cos \beta + 681.7 (x_1 - x_8)$
$+ 2045.1 (x_2 - x_7) - 1398.15 (x_3 - x_6)$
$- 466.05 (x_4 + x_5) = 0$
$- R_{EF} \cos 23°18' + R_{GH} \cos 23°24'$
$+ 681.7 (4.2 - 3.8) + 2045.1 (3.3 - 3.0)$
$- 1398.15 (3.1 - 3.1) - 466.05 (1.0 + 1.0) = 0$
$\qquad - 0.918 R_{EF} + 0.918 R_{GH} = 45.9 \qquad$ (Eq. B)

Solving Eq. (A) and (B) yields
$\quad R_{EF} = 3322.73 \text{ lb} \quad$ and $\quad R_{GH} = 3372.73 \text{ lb}$
and, therefore,

$$R_E = \frac{R_{EF}}{3} = \frac{3322.73}{3} = 1107.58 \text{ lb}$$

$$R_F = \frac{2R_{EF}}{3} = 2215.15 \text{ lb}$$

$$R_G = \frac{2R_{GH}}{3} = \frac{2 \times 3372.73}{3} = 2248.49 \text{ lb}$$

$$R_H = \frac{R_{GH}}{3} = 1124.24 \text{ lb}$$

Note: For the case of radial sill loads and a body of constant radius, moment equilibrium is satisfied by the above procedures. If either condition is violated, special considerations of rotational equilibrium must be utilized.

In a similar manner, establish the loads on the remaining frames (forward adjacent, aft edge and aft adjacent) and then draw shear force and bending moment diagrams after finding the reactions by conventional methods of analysis for the main and auxiliary sills. Also bending moment and end loads in the frames. These load diagrams are due to flight cases only. To these have to be added the effects arising from pressurization, which will be found in the next calculation.

(iii) Hoop tension due to pressurization condition
Pure hoop tension is assumed to exist above and below the upper and lower auxiliary sills respectively. Until discrete stops are located, it is assumed that the door applies a uniform radial loading to the main frames and will be magnified by the overhang (see sketch below). This magnified loading must be reflected in higher loads in the main frames implying high hoop tension, and a reverse loading to the adjacent frames, that is a hoop compression, and this effect is achieved where corner fixity is available.

Door and skin overhang pressure load redistribution on one inch wide strip with unit load of 1.0 psi.

$$R_F (38) = 35 (18.5) \rightarrow R_F = 17.05 \text{ lb/in}$$
$$R_A (38) = 35 (19.5) \rightarrow R_A = 17.95 \text{ lb/in}$$

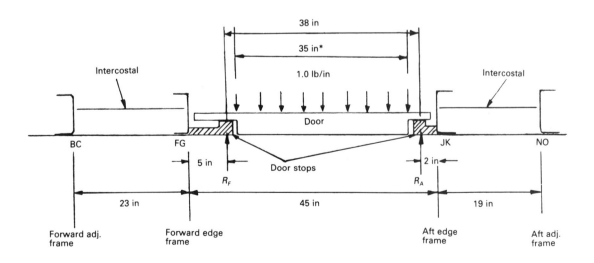

This is the pressure width of door.

Reaction loads on forward edge and adjacent frames,

taking moments about FG

$$R_{BC} = \frac{17.05\,(5)}{23} = 3.71 \text{ lb/in}$$

taking moments about BC

$$R_{FG} = \frac{17.05\,(23 + 5)}{23} = 20.76 \text{ lb/in}$$

Reaction loads on aft edge and adjacent frames

taking moment about NO

$$R_{JK} = \frac{17.95\,(21)}{19} = 19.85 \text{ lb/in}$$

taking moment about JK

$$R_{NO} = \frac{17.95\,(2)}{19} = 1.9 \text{ lb/in}$$

Forward edge frame at cabin pressure 17.2 psi (2.0 × 8.6 psi, see previously mentioned design conditions requirement):

$$P_{HT} = (357)(74) = 26418 \text{ lb}$$

$$A_1 = \frac{74^2}{2}\left[(6°46')\frac{\pi}{180°} - \sin 6°46'\right] = 0.752 \text{ in}^2$$

$$A_2 = \frac{74^2}{2}\left[(13°33')\frac{\pi}{180°} - \sin 13°33'\right]$$
$$- 0.752 = 5.267 \text{ in}^2$$

$$\left(q_u + \frac{q_u}{3}\right)\frac{17.5}{2} = 26418 \text{ lb}$$

$$\therefore q_u = 2264.4 \text{ lb/in}$$

$$\therefore \frac{q_u}{3} = \frac{2264.4}{3} = 754.8 \text{ lb/in}$$

$r\,(1 - \cos 13°33') = 74\,(1 - 0.972) = 2.06 \text{ in}$

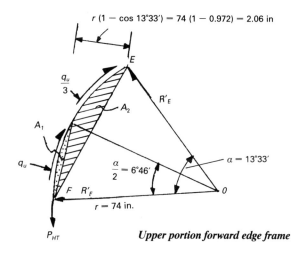

Upper portion forward edge frame

Forward edge frame

Taking moments about F,

$$2 A_1 (q_u) + 2 A_2 \left(\frac{q_u}{3}\right) - 17.38 \left(\cos\frac{\phi}{2}\right) R_E' = 0$$

$$2(0.739)(2264.4) + 2(5.285)(754.8)$$
$$- 17.26 R_E' = 0$$
$$R_E' = 656.14 \text{ lb}$$

Taking moments about E

$$2 A_1 \left(\frac{q_u}{3}\right) + 2 A_2 (q_u) - P_{HT} (2.06)$$
$$+ 17.38 (\cos 6°46') R_F' = 0$$
$$2(0.739)(754.8) + 2(5.285)(2264.4)$$
$$- 26418(2.06) + 17.26 R_F' = 0$$
$$R_F' = 1701.67 \text{ lb}$$

And similarly the shear flows and reaction R_G' and R_H' may be established. The forces at the other frames, ie. forward adjacent, aft edge and aft adjacent frames may be computed by the same method.

Panel Shears — due to cabin pressure [2.0 (8.6) psi = 17.2 psi]:

Pressure $p = 17.2$ psi

$$w = pr = 17.2 (74) = 1272.8 \text{ lb/in}$$

$$w' = \frac{pr}{2} = 636.4 \text{ lb/in}$$

Distribution factors based on constant moment of inertia around cut-out, upper and lower panels

$$K_1 = \frac{h}{h + \ell} = \frac{49}{49 + 45} = 0.521$$

forward and aft side panels

$$K_2 = \frac{\ell}{h + \ell} = \frac{45}{49 + 45} = 0.479$$

Fixed end moments (F.E.M.), see Fig. 6.5.16,

for long sides ($h = 49$ in)

$$\frac{w'(h)^2}{12} = \frac{636.4 (49)^2}{12} = -127,333 \text{ in-lb}$$

for short sides ($\ell = 45$ in)

$$\frac{w(\ell)^2}{12} = \frac{1272.8 (45)^2}{12} = 214,785 \text{ in-lb}$$

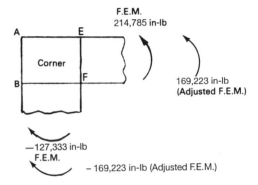

Consider one corner (by moment distribution method):

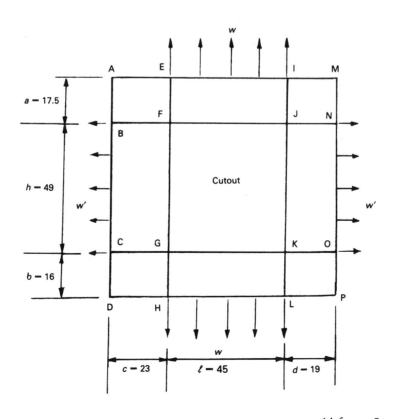

Distribution factors	K_2 0.479	K_1 0.521
F.E.M.	−127333	214785
Correction values	−41890	−45562
Adjusted F.E.M.	−169223	169223

Assume that the deformation of the corner panel will permit only 50% fixity. Thus, for the forward top corner (ABFE) the fixing shear flow is given by

$$\frac{169223\,(50\%)}{17.5\,(23)} = 210.2 \text{ lb/in}$$

and the shear flows for all other corners may be found in a similar manner.

Consider the shear flows in the structure bounded by forward and aft adjacent frames and main and auxiliary sills (A-B-F-J-N-M-I-E-A):

H-L-P-O-K-G-C).

Hence, panel shear flows due to 2.0 pressure factor can be drawn. If 3.0 pressure factor is the critical case, these shears may be obtained by simple ratio (3:2) for shear structures. However, if 2.0 pressure factor plus flight loads is the design case, all loading may be obtained by simple superposition of the flight and pressure cases.

It is now possible to draw shear force, bending moment, and axial load diagrams for all sills and frames. And finally the cutout structure can be sized.

6.6 Fuselage Cutout for Big Cargo Doors — Shear Type Door

These doors have hinges and latches along upper and lower sills with or without plugs.

It is not economic to design big cargo door cutouts, generally for cutout aspect ratio $\frac{\ell}{h} > 1$, which is very critical for pressurized cabin conditions. A method of analysis is presented here suitable for the preliminary

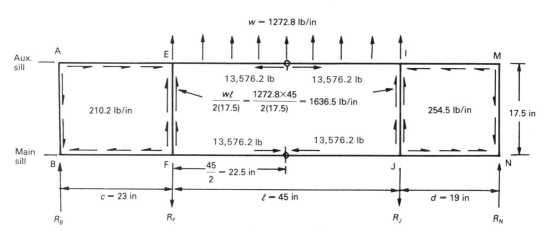

Reactions at edge frames,

$$R_F = (1636.5 + 210.2)(17.5) = 32317 \text{ lb}$$
$$R_J = (1636.5 + 254.4)(17.5) = 33093 \text{ lb}$$

Reactions at adjacent frames,

$$R_B = (210.2)(17.5) = 3678.5 \text{ lb}$$
$$R_N = (254.5)(17.5) = 4453.8 \text{ lb}$$

Axial loads in main and auxiliary sills at center of cutout

$$= \frac{32{,}317(\frac{45}{2}) - 3{,}678.5(23 + \frac{45}{2}) - 1{,}272.8(\frac{22.5^2}{2})}{17.5}$$
$$= 13{,}576.2 \text{ lb}$$

Note: The similar calculation of reaction load (R_f) is required for the running load of $w' = 636.4$ lb/in due to longitudinal pressure will be sheared to the main sills (top and bottom) to give additional loads.

And a similar procedure can be adopted for the lower structure, that is between main and auxiliary sills and forward and aft frames (C-D-

design of cargo door cutouts as shown in Fig. 6.6.1. Each opening will be framed by four or more horizontal and vertical members respectively. In order to simulate actual loading conditions several assumptions will be made, and these are outlined below based on experience gained from past aircraft designs.

Design Conditions:
1. 3.0 pressure factor acting alone — use for tension and member tension bending stresses due to pressure.
2. 2.5 pressure factor acting alone — use for shear and compression stresses.
3. Flight loads plus 2.0 pressure factor where critical.
4. Flight loads acting alone.
5. 50% of the average shear load is carried across the door; another 50% is carried by the cutout surround structure (see Fig. 6.6.2 & Fig. 6.6.3).
6. With door unlatched (on ground), limit loads will be distributed around the door cutout as for a "door removed" case. The method of cutout analysis is same as Case I and II of Section 6.5 except only ground load conditions are used.

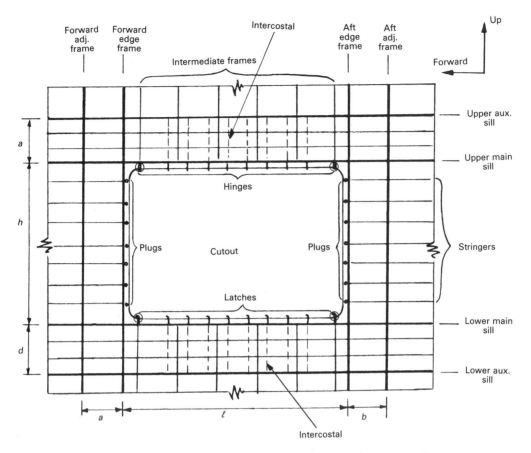

Fig. 6.6.1 Typical shear type door cutout structural arrangement.

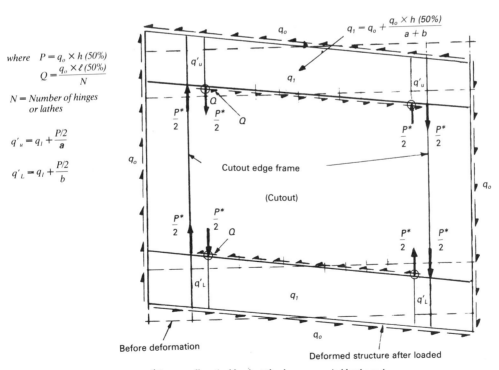

where $P = q_o \times h \ (50\%)$
$Q = \dfrac{q_o \times \ell \ (50\%)}{N}$

N = Number of hinges or lathes

$q'_u = q_1 + \dfrac{P/2}{a}$

$q'_L = q_1 + \dfrac{P/2}{b}$

$q_1 = q_o + \dfrac{q_o \times h \ (50\%)}{a+b}$

(*Assume all vertical loads at the door are carried by the end hinges and lathes: ⊕ denotes end hinges and lathes)

Fig. 6.6.2 Loads around cutout for shear type door.

Airframe Structural Design 205

7. The fuselage cutout with assumed 50% average shear load carried will be analyzed by Case I through Case IV of Section 6.5.
8. Assume 100% of hoop tension loads is carried across the door through those hinges and latches; longitudinal tension loads are considered to be redistributed to the upper and lower main sills in the same manner as shown in Fig. 6.5.14.

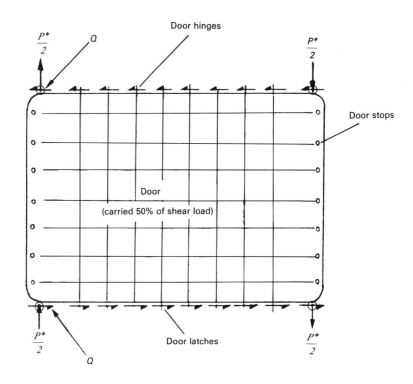

*Note: Assume door stops carry no shear load

Fig. 6.6.3 *Loads around shear type door (carried partial shear load).*

References

6.1 King, K.M.: 'Rings Used for Shear Web Hole Reinforcement.' *Aero Digest*, (Aug. 1955).
6.2 Kuhn, P., Duberg, J.E. and Diskin, J.H.: 'Stress Around Rectangular Cutouts in Skin-Stringer Panels Under Axial Load — II.' *NACA ARR L368 (ARR 3J02)*, (Oct. 1943).
6.3 Rosecrans, R.: 'A Method for Calculating Stresses in Torsion-Box Covers with Cutouts.' *NACA TN 2290*, (Feb. 1951).
6.4 Kuhn, P. and Moggio, E.M.: 'Stresses Around Large Cutout in Torsion Boxes.' *NACA TN 1066*, (1946).
6.5 Golologov, M.M.: 'Shear Distribution Due to Twist in a Cylindrical Fuselage with a Cutout.' *Journal of The Aeronautical Sciences*, (Apr. 1947).
6.6 Cicala, P.: 'Effects of Cutouts in Seminomocoque Structures.' *Journal of The Aeronautical Sciences*, (Mar. 1948).
6.7 Kuhn, P.: 'The Strength and Stiffness of Shear Webs with Round Lightening holes Having 45° Flanges.' *NACA ARR WR L-323*, (Dec. 1942).
6.8 Kuhn, P., Peterson, J.P. and Levin, L.R.: 'A Summary of Diagonal Tension.' *NACA TN 2661*.
6.9 Anevi, G.: 'Experimental Investigation of Shear Strength and Shear Deformation of Unstiffened Beams of 24ST Alclad with and without Flanged Lightening Holes.' *Sweden, SAAB TN-29*, (Oct. 1954).
6.10 Gurney, C.: 'An Analysis of the Stresses in a Flat Plate with a Reinforced Circular Hole under Edge Forces.' *R. & M. No. 1834, Aeronautical Research Committee Reports and Memoranda, London*, (1938).
6.11 Moggio, E.M. and Brilmyer, H.G.: 'A Method for Estimation of Maximum Stresses Around a Small Rectangular Cutout in a Sheet-Stringer Panel in Shear.' *NACA ARR No, L4D27*, (Apr. 1944).
6.12 Kuhn, P.: *Stresses in Aircraft and Shell Structures*. McGraw-Hill Book Company, Inc., New York, 1956.
6.13 Bruhn, E.F.: *Analysis and Design of Flight Vehicle Structures*. Tri-State Offset Company, Cincinnati, Ohio, U.S.A. 1965.
6.14 Kuhn, P. and Peterson, J.P.: 'Stresses Around Rectangular Cutouts in Torsion Boxes.' *NACA TN 3061*, (Dec. 1953).
6.15 Kuhn, P., Rafel, N. and Griffith, G.E.: 'Stresses Around Rectangular Cutouts with Reinforced, Coaming Stringers.' *NACA TN 1176*, (Jan. 1947).
6.16 Anon.: 'Airworthiness Standards: Transport Category.' *Federal Aviation Regulations (FAR)*, Vol. III, Part 25.
6.17 Anon.: 'Aeroplanes.' *British Civil Airworthiness Requirements, Section D*.

CHAPTER 7.0

FASTENERS AND STRUCTURAL JOINTS

7.1 Introduction

A complete airplane structure is manufactured from many parts. These parts are made from sheets, extruded sections, forgings, castings, tubes, or machined shapes, which must be joined together to form subassemblies. The subassemblies must then be joined together to form larger assemblies and then finally assembled into a completed airplane. Many parts of the completed airplane must be arranged so that they can be disassembled for shipping, inspection, repair, or replacement, and are usually joined by bolts or rivets. In order to facilitate the assembly and disassembly of the airplane, it is desirable for such bolted or riveted connections to contain as few fasteners as possible. For example, a semimonocoque metal wing usually resists bending stresses in numerous stringers and sheet elements distributed around the periphery of the wing cross section. The wing cannot be made as one continuous riveted assembly from tip to tip, but must usually be spliced at two or more cross sections. These splices are often designed so that four bolts, for example, transfer all the loads across the splice. These bolts connect members called *fittings*, which are designed to resist the high concentrated loads and to transfer them to the spars, from which the loads are distributed to the sheet and stringers. The entire structure for transferring the distributed loads from the sheet and stringers outboard of the splice into a concentrated load at the fitting and then distribute this load to the sheet and stringers inboard of the splice is considerably heavier than the continuous structure which would be required if there were no splice. Many uncertainties exist concerning the stress distribution in fittings. Manufacturing tolerances are such that bolts never fit the holes perfectly, and small variations in dimensions may affect the stress distribution. An additional margin of safety of 15% is used in the design of fittings. This fitting factor must be used in designing the entire fitting, including the riveted, bolted, or welded joint attaching the fitting to the structural members. The fitting factor need not be used in designing a continuous riveted joint.

The ideal aircraft structure would be a single complete unit of the same material involving one manufacturing operation. Unfortunately the present day types of materials and their method of working dictates a composite structure. Furthermore, general requirements of repair, maintenance and stowage dictate a structure of several main units held to other units by main or primary fittings or connections, with each unit incorporating many primary and secondary connections involving fittings, bolts, rivets, welding, etc. No doubt main or primary fittings involve more weight and cost per unit volume than any other part of the aircraft structure, and, therefore, fitting and joint design plays an important part in aircraft structural design.

A blanket factor of safety for all types of fittings or load conditions is not logical. The manner in which a load is applied to a joint often involves a dynamic or shock load, for example, joints or fittings in landing gear. Single pin connections often undergo rotation or movement between adjacent parts, thus producing faster wearing away of material in operation. Repeated loads often present a fatigue problem. In an airplane there are certain main fittings which, if they failed, would definitely cause the loss of the aircraft. Thus, the design fitting requirements of the military and civil aviation agencies involve many special or higher factors of safety. This is particularly so in designs involving castings.

General Design Considerations

Joints are perhaps the most common source of failure in aircraft structure and therefore it is most important that all aspects of the design are given consideration when making the structural analysis. Failures may occur for various reasons but generally because of some factor, such as secondary stresses due to eccentricities, stress concentrations, slippage of connectors, excessive deflections, etc., or some combination of conditions, all of which are difficult to evaluate to an exact degree. These factors not only affect the static strength but have a great influence on the fatigue life of the joint and the adjacent structure.

(1) Eccentricities and their effect on the part of the joint and the surrounding structures.
  - If eccentricities exist in a joint, the moment they produce must be resisted by the adjacent structure.
  - When a joint contains a dihedral or anhedral angle (such as wing structure), a rib should be provided at the vertex of the angle to eliminate the eccentricity that would exist.
  - A joint of a truss structure containing an eccentricity produces secondary stresses which would be accounted for.
  - Eccentrically loaded bolt and rivet patterns may produce excessively loaded connectors if eccentricity is not considered.

(2) Fatigue considerations.

(3) Mixed fasteners: It is not good practice to employ both rivets and bolts in combination in

a joint. Due to a better fit for the rivets, the bolts will not pick up their proportionate share of the load until the rivets have deflected enough to take up the clearance of the bolts in their bolt holes.

This tends to overload the rivets and may induce premature failure. If such a combination is absolutely necessary, it is advisable to use close tolerance bolts in reamed holes.

(4) Overall efficiency: Both sides of the joint should be considered. It is possible to design a perfect joint on one side by paying a heavy weight and production penalty on the other. A compromise should be made to make the joint and surrounding structure the most efficient. It is the overall efficiency which is the prime consideration.

(5) Splices in discontinuous members, which act in conjunction with a part or parts which are continuous past the splice, should be made as rigid as possible using generous splice members and close fitting attachments, thereby minimizing slippage which might overload the continuous material and cause premature failure.

(6) Insufficient rigidity of surrounding structure may cause excessive deflections and consequent changes in direction and magnitude of loads on certain joints, such as those in a landing gear installation.

(7) Do not use spot welds on either side at the joggled area of a joggled member; use rivets at the joggle.

(8) Do not use a long string of fasteners in a splice. In such cases, the end fasteners will load up first and yield early. Three, or at most four, fasteners per side is the upper limit unless a carefully tapered, thoroughly analyzed splice is used.

(9) Carefully insure against feather edges in all joint designs (see Fig. 7.1.1). The thickness of countersunk sheet shall be equal to or greater than 1.5 times the depth of the countersunk head of the fastener at the fatigue critical areas. For other applications, $t \geq h + 0.020$ inch shall be considered.

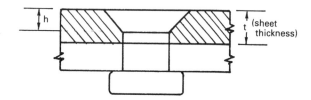

Fig. 7.1.1 Feather-edge in countersunk sheet.

(10) When possible use a double shear splice.
(11) Maintain a fastener spacing approximately four times the fastener diameter or more.
(12) Probably the most single important item regarding detail structural design is the matter of equilibrium. If the engineer will show the load in equilibrium for every part of the assembly, most errors will be prevented.

(13) Carefully select interference-fit fasteners which produce sustained tensile stress (stress corrosion crack in fastened material).

(14) In fastener installation, a minimum edge distance (see Fig. 7.1.2) of two fastener diameters plus 0.03 inch and a spacing of an approximately four diameter shall be considered. Edge distance is measured from the center of the fastener hole to the closest edge of the sheet and a shorter edge distance of $1.5d$ may possibly be considered in a few applications.

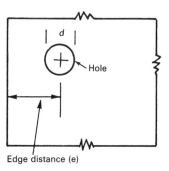

Fig. 7.1.2 Fastener minimum edge distance.

(15) The use of rivets involving tension only is poor engineering practice and should be held to a minimum.

When secondary tension loads are imposed on a standard aluminum rivet (such as the attachment of a diagonal tension web to a stiffener), use the tension allowable of rivet and sheet combination test allowable data.

Basic Criteria of Fastener Strength Allowable

The allowable loads are based on the lowest values of the following criteria:

(1) Bearing load (protruding head only) = $F_{br}dt$
where
F_{br} — Allowable ultimate bearing allowable stress of sheet material is based on either MIL-HDBK-5D "B" value per Ref. 7.6 or other sources
d — Nominal shank diameter
t — Nominal sheet thickness

(2) Shear-off load = $F_{su}\left(\dfrac{\pi d^2}{4}\right)$
where
F_{su} — Ultimate shear allowable stress of fastener material is based on MIL-HDBK-5D per Ref. 7.6 or other sources
d — Nominal shank diameter
(See Fig. 7.1.3)

(3) Countersunk fastener and sheet combinations — the allowable ultimate and yield loads are established from actual test data.

(4) Yield strength to satisfy permanent set requirements at limit load (limit load = $\dfrac{\text{ultimate load}}{1.5}$).

Fastener Diameter		3/32	1/8	5/32	3/16	1/4	5/16	3/8	7/16	1/2	9/16
Fastener Material	F_{su} (Ksi)	\multicolumn{10}{c}{Ultimate Single Shear Load (lb/fastener)}									
1100F	9										
5056(B)	28	203	363	556	802	1450	2290	3280			
2117-T3(AD)	30	217	388	596	862	1550	2460	3510			
2017-T31(D)	34	247	442	675	977	1760	2790	3970			
2017-T3(D)	38	275	494	755	1090	1970	3110	4450			
2024-T31(DD)	41	296	531	815	1180	2120	3360	4800			
Monel	49	355	635	973	1400	2540	4020	5730			
7075-H75	38	275	494	755	1090	1970	3110	4450			
A-286 CRES	90	—	—	1726	2485	4415	6906	9938			
Ti-6A1-4V & Alloy Steel	95	—	—	1822	2623	4660	7290	10490	14280	18650	23610
Alloy Steel	108	—	—	2071	2982	5300	8280	11930	16240	21210	26840
Alloy Steel	125	—	—	2397	3452	6140	9590	13810	18790	24540	31060
H-11 Steel	132	—	—	2531	3645	6480	10120	14580	19840	25920	32800

(Refer to MIL-HDBK-5D per Ref. 7.6)

Fig. 7.1.3 Fastener shear-off allowable loads.

Fittings

For structural economy, engineers in the initial layout of the aircraft should strive to use a minimum number of fittings, particularly those fittings connecting units which carry large loads. Thus, in wing structure, splicing the main beam flanges or introducing fittings near the centerline of the airplane are far more costly than splices or fittings placed farther outboard where member sizes and loads are considerably smaller. Avoid changes in direction of heavy members such as wing beams and fuselage longerons as these involve heavy fittings. If joints are necessary in continuous beams place them near points of inflection in order that the bending moments to be transferred through the joint to be kept to a small magnitude. In column design with end fittings, avoid introducing eccentricities on the beam; on the other hand make use of the fitting to increase column end fixity, thus compensating some of the weight increase due to fitting weight by saving in the weight of the beam. For economy of fabrication, the engineer should have good knowledge of shop processes and operations. The cost of fitting fabrication and assembly varies greatly with the type of fitting, shape, and the required tolerances. Poor layout of major fitting arrangement may require very expensive tools and jigs for shop fabrication and assembly. Fittings, likewise, add considerably to the cost of inspection and rejections of costly fittings because of faulty workmanship or materials. The stress analysis of most connections or fittings is more complicated than for the primary structural members due to such things as combined stresses, stress concentrations, bolt-hole fit, etc., thus an additional factor of safety is necessary to give a similar degree of strength reliability for connections as provided in the strength design of the members being connected.

(1) Fitting Factor
Under the conditions outlined below, the U.S. FAA requires an ultimate fitting factor of 1.15 to be used in the structural analysis (military aircraft may not require the use of a fitting factor).
- A fitting factor is required for joints which contain fittings when the strength of the fitting is not proven by limit and ultimate load tests.
- This factor shall apply to all portions of the fittings, the means of attachment (connections), and the bearing on the members joined.
- In the case of integral fittings, the part shall be treated as fitting up to the point where the section properties become typical of the member.

The fitting factor need not be employed in the following cases:
- When a type of joint made in accordance with approved practice is based on comprehensive test data.
- With respect to the bearing surface of a part, if the bearing factor used is of greater magnitude than the fitting factor.
- If a casting factor (for casting materials) has been used which is of greater magnitude than the fitting factor.

(2) Hinge Factor
The U.S. FAA requires that control surface hinges, except ball and roller bearings, incorporate a special factor of not less than 6.67 with respect to the ultimate bearing strength of the softest material used as a bearing, and that control system joints subjected to angular motion in push-pull systems, except ball and roller bearing systems, shall incorporate a special factor

Airframe Structural Design 209

of not less than 3.33 with respect to the ultimate bearing strength of the softest material used as a bearing.

7.2 Rivets (permanent fasteners)

Rivets are low cost, permanent fasteners well suited to automatic assembly operations. The primary reason for riveting is low in-place cost, the sum of initial rivet cost and costs of labor and machine time to set the rivets in the parts. Initial cost of rivets is substantially lower than that of threaded fasteners because rivets are made in large volumes on high-speed heading machines, with little scrap loss. Assembly costs are low. Rivets can be clinched in place by high-speed automatic machinery.

(1) Advantages:
- Dissimilar materials, metallic or nonmetallic, in various thicknesses can be joined. Any material that can be cold worked makes a suitable rivet.
- Rivets may have a variety of finishes such as plating, parkerizing, or paint.
- Parts can be fastened by a rivet, if parallel surfaces exist for both the rivet clinch and there is adequate space for the rivet driver during clinching.
- Rivets can serve as fasteners, pivot shafts, spacers, electric contacts, stops, or inserts.
- Parts that are painted or have received other finishes can be fastened by rivets.

(2) Disadvantages:
- Tensile and fatigue strengths of rivets are lower than for comparable bolts or screws. High tensile loads may pull out the clinch, or severe vibrations may loosen the fastening. Riveted joints are normally neither watertight nor airtight; however, such joints may be attained, at added cost, by using a sealing compound.
- Riveted parts cannot be disassembled for maintenance or replacement without destroying the rivet.
- Rivets produced in volume are not normally made with the same precision as screw-machine parts.

Use flush-head rivets only where:
- High aerodynamic efficiency is required, or
- Head clearances are inadequate; i.e. where a protruding head might interfere with a mating part or the operation of adjacent mechanisms.

Solid rivets and threaded collar rivets such as Hi-Loks shall not be used for the attachment of control surface hinges, hinge brackets, or similar parts which must be readily removable. Blind rivets shall be used only where access is limited to one side. Blind rivets shall not be used on control surface hinge brackets, wing attachment fittings, landing gear fittings, fixed tail surface attachment fittings or similar heavily stressed locations. Also, do not use them in the design of hulls, floats, or tanks where a fluid tight joint is required. In structural applications, the upset head must be visible for inspection. Fig. 7.2.1 illustrates the conventional rivet material characteristics and usage.

Rivet Material	Driving Characteristics	Used in Material	Recommended Use and Limitations
1100-F	Easy-soft	Soft aluminum, fiber, plastics	Low strength — not recommended in large dia. (1/16–1/8 inch)
5056(B)	Medium	Aluminum alloys	Used if the strength of 5056 is adequate and lower driving force is desirable (3/32–3/16 inch)
2117-T3 (AD)	Medium-hard	Aluminum alloys	For use in sealing application in integral fuel tanks (1/16–5/32 inch)
2017-T3(D)	Hard	Aluminum alloys	For general use
2024-T31 (DD)	Hard	Aluminum alloys	High strength, limited use due to refrigerated cond. (3/16–1/4 inch)
Monel(M)	Very hard	Copper alloys, steel and CRES	Driving force too high for large dia.
A-286	Very hard	Titanium, steel and CRES	Up to 1/4 inch dia. — Used in pylon, hot areas and corrosion environment
Titanium B120(K)	Very hard	Steel and CRES	Up to 3/16 inch dia. — Used in hot areas and corrosion environment
7075-H75	Medium-hard	Aluminum alloys	Do not use in fatigue critical areas

Fig. 7.2.1 Conventional rivet material characteristics and usage.

Semi-tubular rivets

Semi-tubular rivets shall not be used in any structural application. They are normally used in soft materials such as rubber, leather, fabric, etc., with the upset head against a harder surface, such as a washer or back-up strip. (See Fig. 7.2.2).

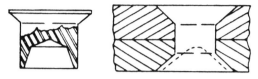

Fig. 7.2.2   Semi tubular rivet.

Blind rivets

Blind rivets, or fasteners, are designed to be installed where access to both sides of a sheet assembly or structure is not possible or practical. The blind rivet usually consists of a tubular sleeve in which a stem having an enlarged end is installed. The heads of such rivets are made in standard configurations such as brazed, universal, and flush. The rivet and stem are inserted in a correct size hole, and the stem is drawn into the sleeve by means of a special tool. The bulb or other enlargement on the end of the stem expands the end of the rivet and locks it into the hole. A typical blind rivet is the Cherrylock shown in Fig. 7.2.3.

Fig. 7.2.4 illustrates a COMP-TITE blind fastener, which is a large-bearing blind fastener developed for use with advanced composite materials. Existing blind fasteners do not provide sufficient blind head expansion and they may damage the composite structure during installation. This damage is caused by either excessive clamp load for the available bearing area or by radial fastener expansion within the fastener hole, delaminating the composite plies on the blind-side surface. The COMP-TITE blind fastener overcomes this problem by forming a large and uniform bearing surface that will not crush or delaminate the composite material during installation. As illustrated in Fig. 7.2.4, a coiled washer element is driven over the tapered end of the nut, expanding to its final diameter. The formed washer is then seated against the joint surface by the continued advance of the sleeve and corebolt. This fastener form a flat washer surface without damaging the mating surface.

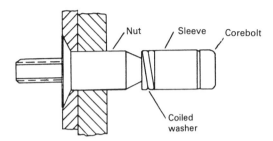

(1) Using NAS 1675-type installation tooling, the nut is restrained from turning while the corebolt is driven.

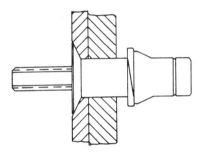

(2) The advance of the corebolt forces the washer and sleeve over the taper, expanding and uncoiling the washer to its maximum diameter.

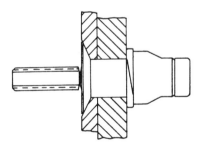

(3) Continued advance of the corebolt draws the washer and sleeve against the joint surface, preloading the structure.

At a torque level controlled by the break groove, the slabbed portion of the corebolt separates, and installation is complete.

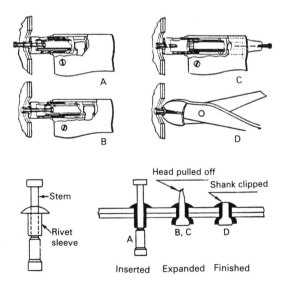

(Installation of Cherry blind rivets)

Fig. 7.2.3   Cherry blind rivets and huck blind rivets.

Fig. 7.2.4   COMP-TITE blind fastener installation sequence.

Blind Bolts

A blind bolt, like a blind rivet, is one that can be completely installed from only one side of a structure or assembly. The blind bolt is used in place of a blind rivet where it is necessary to provide high shear strength. The bolt is usually made of alloy steel, titanium, or other high-strength material. A typical blind bolt is shown in Fig. 7.2.5.

Hi-shear fasteners

Hi-shear rivets are designed for quick, permanent installations where it is desired to reduce weight and installation time. Such rivets or bolts can be used only where they can provide adequate strength. The hi-shear rivet is made of steel and employs a swaged aluminum collar to hold it in place. The collar is driven on to the end of the rivet by means of a special tool in conventional pneumatic rivet gun. A hi-shear rivet is shown in Fig. 7.2.6.

Hi-Lok fasteners

The Hi-Lok fastener is basically a high strength fastener which combines the best features of a rivet with those of a bolt and nut. It consists of two parts, a threaded pin and a threaded collar. The pin is a straight shank, precision threaded pin which is in-

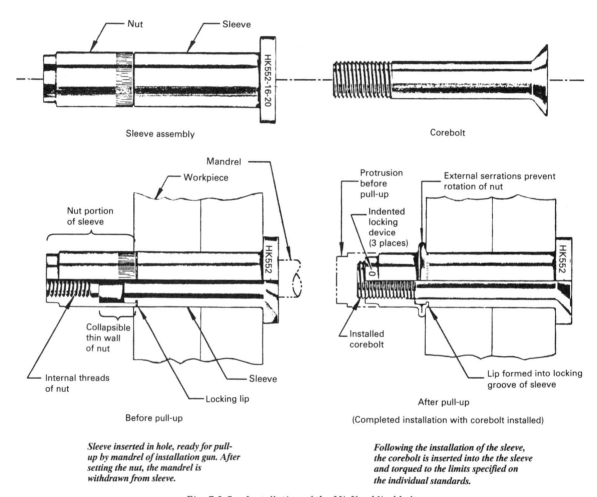

Fig. 7.2.5 *Installation of the Hi-Kor blind bolt.*

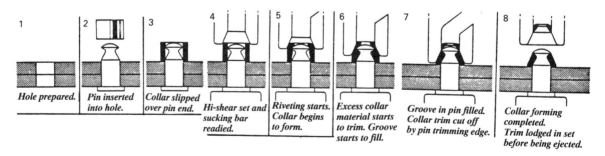

Fig. 7.2.6 *Hi-shear rivet.*

212 Airframe Structural Design

stalled in a straight walled hole, drilled normally, at 0.002 to 0.004 diametral interference. The installation of the Hi-Lok fastener is completed on one side of the assembly after the bolt has been inserted through the hole from the other side. The hexagonal wrench tip of the installing tool is inserted into a recess in the bolt, which holds the pin while the tool turns the collar. As the collar is tightened to the design torque level built into the collar, the hex portion of the collar is sheared off automatically by the driving tool. This leaves the installation with the correct amount of torque and preload. They are normally used where:
- High shear strength is required, or
- Impact riveting of solid rivets is prohibited and they cannot be squeezed, or
- Expansion of the shank would cause undesirable effects, or
- High clamp-up is desired for sheet pull-up or faying surface sealing requirements dictate.

The Hi-Lok fastener is illustrated in Fig. 7.2.7.

Taper-Lok fasteners

Taper-Lok fastener is an interference-fit, lightweight fastening system with self-sealing feature. The system comprises a close-tolerance tapered-shank bolt and a companion washernut assembly which is both self-centering and self-locking. The washernut assembly, as the name implies, consists of a nut with a free-spinning captive washer. The Taper-Lok system and installation sequence are shown in Fig. 7.2.8.

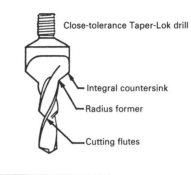

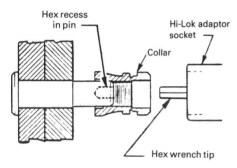

Insert the hex wrench tip of the power driver into the pin's hex recess.

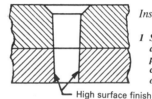

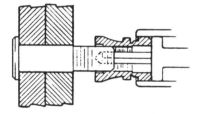

Firmly press the power driver against the collar, operate the power driver until the collar's wrenching device has been torqued off.

Installation sequence

1 Special, close-tolerance tapered drills are used for Taper-Lok hole preparation; drilling, reaming and countersinking are accomplished in one operation.

2 The tapered shank bolt is then inserted in the hole, and seated firmly in place by hand pressure. Head protrusion (in thousandths of an inch) above the structural material, divided by 0.048 gives the interference value between the bolt and the hole.

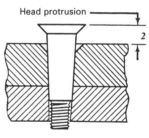

3 Full contact along the entire shank of the bolt and the hole prevents rotation of bolt while tightening the washernut. During tightening, the nut spins freely to the locking point, but the washer remains stationary and provides a bearing surface against the structure.

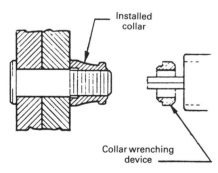

This completes the installation of the Hi-Lok fastener assembly.

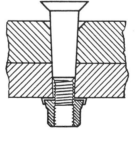

4 Torquing of washernut by conventional wrenching methods produces a controlled interference fit, seats the bolt head, and creates an evenly balanced pre-stress condition within the bearing area of the structural joint.

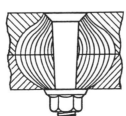

Fig. 7.2.7 Installation of the Hi-Lok fastener.

Fig. 7.2.8 Installation of tapered fastener.

Fig. 7.2.9 illustrates a LGP fastener which is a specially designed version of the HUCK lightweight titanium lockbolt system. It is specifically for graphite composite applications. It is an all titanium system with a Ti-6A1-4V pin and a flanged commercially pure titanium swage on collar. The fastener comes in both a pull stem version for installation with common pull tools or automatic drill/rivet machines such as a DRIVMATIC machine. Another LGP fastener is called LGP sleeved fastener which is designed to be installed into interference fits in graphite composite structure without causing any installation damage while gaining the improved structural fatigue, electrical continuity, and water/fuel tightness with an sealant (see Fig. 7.2.10).

7.3 Bolts and Screws (Removable Fasteners)

A bolt is an externally threaded fastener designed for insertion through holes in assembled parts, and is normally intended to be tightened or released by torquing a nut. The aircraft high-strength steel bolts and screws are shown in Fig. 7.3.1 and high-strength steel nuts are shown in Fig. 7.3.2.

To ensure satisfactory service life of bolts and nuts used in primary structure in addition to required design strength is the type of thread form and thread relief, plus material grain structure of thread forms. The basic form of thread used in aircraft design is the 60° American National Form of thread as specified in

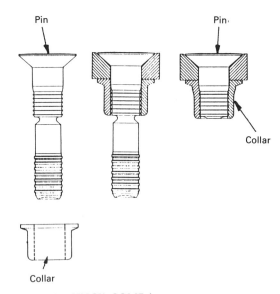

Fig. 7.2.9 HUCK-COMP fastener.
(Courtesy of Huck Manufacturing Co.)

Fig. 7.3.1 Standard bolts and screws.
(Courtesy of Standard Pressed Steel Co.)

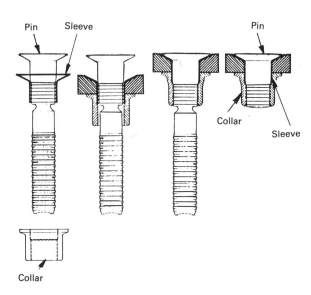

Fig. 7.2.10 HUCK-TITE interference fit fastener.
(Courtesy of Huck Manufacturing Co.)

Fig. 7.3.2 Standard nuts.
(Courtesy of Standard Pressed Steel Co.)

MIL-S-7742 which does not require a specific radius in the root of the thread. In MIL-B-7838 the minimum root radius is specified and thread is rolled after heat treat and produces a grain flow in the threads which is continuous and follows the thread contour.

Other rolled thread improvement over the MIL-B-7838 thread form is the MIL-S-8879 thread form. Its root radius is a smooth uniform radius which blends to the thread flanks with an uninterrupted surface and is a larger radius than found in any other standard form. Fig. 7.3.3. shows the three thread forms and root stress concentration. Additional fatigue strength improvements available in the rolled thread fasteners pertain to the bolt head to shank fillet radius. This radius is coldworked to the desired finish requirement after heat treat of the bolt and is free of seams or inclusions.

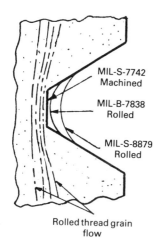

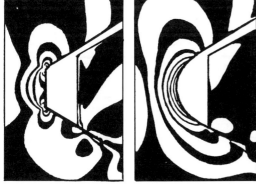

Photo-stress comparison

Fig. 7.3.3 *Comparison of thread forms and root stress concentration.*

In addition to standard bolts, screws and nuts, there are many special fasteners that have been developed to join parts or structure where the more common fasteners are not usable or do not provide adequate strength. Other special fasteners have been designed to provide for quick removal of inspection plates and cowlings. The use of special fasteners is not discussed here.

Bolts and Screws

Aircraft bolts are used primarily to transfer relatively large shear or tension loads from one structural member to another. In designing or strength checking a multiple bolt fitting, the question arises as to what proportion of the total fitting static load does each bolt transfer. This distribution could be affected by many things such as bolt fit or bolt tightness in the hole, bearing deformation or elongation of the bolt hole, shear deformation of the bolt, tension or compressive axial deformation of the fitting members and the member being connected, and a number or other minor influences. Since aircraft materials such as widely used aluminum alloys have a considerable degree of ductility, if the fitting is properly designed, the loads on the bolts approach their maximum value and will tend to be in proportion to the shear strength of the bolt. That is, if the shear strength of the bolts is not the critical strength (bolt shear strength is stronger than the local material bearing strength), the yielding of the fitting material in bearing, shear and tension will tend to equalize the load on the bolts in proportion to their shear strengths. For stresses below the *elastic limit* of the fitting material the bolt load distribution no doubt is more closely proportional to the bearing area of each bolt; on the other hand, the load distribution on each bolt is the function of fitting material modulus, cross-section, fastener pattern (spacing, rows and diameter), fastener deflections, and materials (see later section of this chapter). Since the primary interest is failing strength (based on design ultimate loads), the bolt load distribution in proportion to the bolt shear strength is usually assumed. Fig. 7.3.4 shows several high strength tension bolts.

Nuts

Fig. 7.3.5 illustrates different standard steel nuts. Nut material should be more ductile than bolt material; thus when the nut is tightened the threads will deflect to seat on the bolt threads. Therefore, it is important to select the proper nuts to match the tension bolt or, otherwise, it could decrease the capability of the tension bolt load. Usually refer to design handbook or other documents from fastener manufacturers.

Self locking nuts are widely used in the aircraft industry. The use of the self locking nut reduces assembly costs as it eliminates the bothersome cotter pin which takes extra operations and is very difficult to install on the nut in the many joints and corners of an aircraft. However, the castellated hexagonal nuts with cotter pins are mandatory on installations subject to bolt or nut rotation on any rotational hinge design.

There are two basic types of nuts: tension nuts such as 12 point nuts for tension bolt applications and shear nuts for shear bolt applications which are primarily loaded in shear.

Aluminum alloy nuts are not used on bolts designed for tension. Bolts used in places where the nut should be aluminum alloy are not allowed. Aluminum nuts are not recommended.

Since honeycomb panels generally have thin facings, the shear and tension loads in particular should be transmitted to the entire honeycomb structure wherever possible. For this reason, most fasteners go through the entire honeycomb or are bonded to the opposite facing and core by means of a potting adhesive as shown in Fig. 7.3.6. These fasteners will provide the best structural strength since the adhesive bonds the fastener to both facings and core. They add weight to the panel and are more time consuming to install.

Detail design considerations
- Avoid joints incorporating bolts or screws in conjunction with upset shank rivets or interference fit fasteners. Due to the better fit of the other fasteners, the bolts or screws will not pick up their share of the load until the other fasteners have deflected enough to take up the clearance between the bolt and its hole as previously mentioned. When this type of joint cannot be avoided, the other fasteners should be able to support the entire shear load.
- Cadmium-plated bolts and screws shall not be used in composite structure and areas with temperatures above 450°F.
- No threads shall be in bearing in a fitting as shown in Fig. 7.3.7.
- Critical applications requiring the control of torque of bolts and screws shall be indicated on the drawing such as preloaded tension bolts for tension fittings.
- Bolts smaller than 0.25 in diameter shall not be used in any single bolted structural connection including primary control systems or any application where failure would adversely affect safety of flight.
- Shear bolts with thin heads or shear nuts shall not be used where bolts are subjected to bending loads.

Tensile (ksi)	Shear (ksi)	Style and Part Number	Description	Material	Companion Locknut	Temperature Application To
160	96	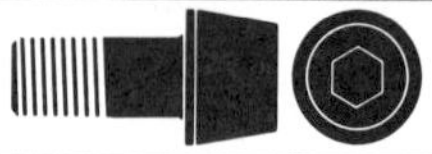 MS 20004 thru MS 20024	Internal Wrenching	Alloy Steel	42FW 42FLW	450°F
160	96	 MS 21262	Self-Locking, Cap	Alloy Steel	—	250°F
160	96	 MS 24675 thru MS 24678	Drilled Head, Cap	Alloy Steel	32FL	450°F
160	 —	NAS 563 thru NAS 572	Hex, Full Thread	Alloy Steel	42FW 42FLW	450°F
160	96	 NAS 1223 thru NAS 1235	Hex Head, Self Locking	Alloy Steel	—	250°F
160	95	 NAS 1266 thru NAS 1270	Hex Head	6Al-4V Titanium	48FLW FN T20	500°F
300	180	 EWB 30	12-Point	Alloy Steel	FN 30	450°F
300	180	 EWB 930	12-Point	Alloy Steel	FN 930	900°F
260	156	 EWB 26	12-Point	Alloy Steel	FN 26 FNF 26	450°F
260	156	 EWB 926	12-Point	Alloy Steel	FN 926	900°F
220	132	 EWB 22	12-Point	Alloy Steel	EWN 22 FNF 22 FNH 22	450°F
220	132	 LWB 22	12-Point	Alloy Steel	FN 22	450°F
220	132	 EWB TM9	12-Point	Alloy Steel	EWN TM9	900°F
220	132	 LWB 922	12-Point	Alloy Steel	FN 922	900°F

Fig. 7.3.4 Several tension and shear bolts.
(Courtesy of Standard Pressed Steel Co.)

Style and Part No.		Description	Gov't/Ind Equipment	Material	Size Range	Tensile (ksi)
450°F and 500°F						
	EWSN 26	Shear Nut		Alloy Steel	# 10 thru 1.0 inch	200-220
	EWN 22	High Tensile		Alloy Steel	1/4 inch thru 1 1/2 inch	200-220
	FN 22	Featherweight		Alloy Steel	# 10 thru 1 1/2 inch	200-220
	FNF 22 FNH 22	Featherweight		Alloy Steel	# 10 thru 1 1/2 inch	200-220
	FN 26	Featherweight		Alloy Steel	# 10 thru 7/8 inch	260
	FNF 26	Featherweight		Alloy Steel	# 10 thru 1 inch	260
	FN 30	Featherweight		Alloy Steel	# 10 thru 7/8 inch	330
750°F and 800°F						
	59FT	Thin Height	NAS 1022C	CRES Steel	# 6 thru 1 1/4 inch	60
	97	Lightweight	MS 21040 NAS 679	CRES Steel	# 4 thru 5/16 inch	125
	59FH	Regular Height	MS 21046C AN 363C NAS 1021C	CRES Steel	# 4 thru 1 1/4 inch	125

Fig. 7.3.5 Several standard steel nuts.
(Courtesy of Standard Pressed Steel Co.)

- Bolts of 180 ksi heat treat and above shall be procured to specifications which meet or exceed the requirements of MIL-B-8831 (180 ksi), MIL-B-8906 (220 ksi) and MIL-B-8907 (260 ksi).
- Bolts highly loaded in tension shall be assembled with washers under both bolt head and the nut.

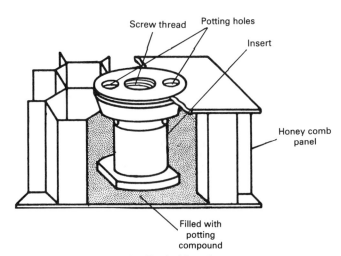

Fig. 7.3.6 Typical insert.

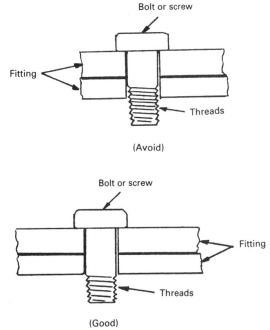

Fig. 7.3.7 No threads in bearing.

Airframe Structural Design 217

7.4 Fastener Selection

In making a fastener selection, the designer must write down all the conditions to be encountered by the overall design. These are not specific requirements which are determined for each part but are the general ranges over which the entire aircraft is expected to operate.

Fig. 7.4.1 shows the fastener characteristics influencing fastener selection.

Fastener	Advantages	Limitations
Solid rivets (Fig. 7.2.1)	Good clamp-up High rigidity Lowest cost Lowest weight Good static & fatigue in shear joint	Non-removable Limited static shear Low tension & shear Low tension fatigue High noise level during installation
Lockbolt (Fig. 7.2.10)	High clamp-up High shear strength High reliability Low noise level during installation	Non-removable Moderate cost Limited tensions Low fatigue, Moderate weight
Hi-Lok (Fig. 7.2.7)	Same as Lockbolt but somewhat lower cost and weight	Same as Lockbolt but somewhat lower cost and weight
Taper-Lok (Fig. 7.2.8)	Same as Lockbolt except high joint shear fatigue capability High regidity	Same as Lockbolt except high cost
High shear rivets (Fig. 7.2.6)	High shear strength Low cost Low weight	Non-removable Low clamp-up Low tension Poor fatigue High noise level during installation
Blind rivets (Fig. 7.2.3)	Locked spindle Moderate rigidity Low cost Low weight Blind installation	Low tension & shear Poor fatigue Moderate clamp-up Moderate reliability Non-removable
Blind bolts (Fig. 7.2.5)	Locked spindle High shear strength Blind installation	Low tension Poor fatigue Moderate clamp-up Moderate reliability High cost Higher weight Non-removable
12 point tension bolt (Fig. 7.3.4)	High static tension High tension fatigue High torque High reliability Removable	High cost High weight No flush surface
Hexagon bolt	High tension (long thread) High static shear Low cost Removable	Low tension (short thread) Low fatigue Moderate weight Moderate torque No flush surface
100° Flush bolt	High static shear Low weight Removable Flush surface	Low tension Low fatigue Low torque Moderate cost

Fig. 7.4.1 Fastener characteristics comparison.

218 Airframe Structural Design

After the designer determines the type of joint to be used, he then determines whether the fastener is loaded axially or in shear. If tension fatigue is the principal load condition, items such as thread form rolled threads (see Fig. 7.3.3) and rolled head-to-shank fillets with special design tension heads are used.

For static tension loading, the same enlarged root radius threads are not essential. For joints with incidental tension loadings, nearly any fastener can be considered, including rivets, blind rivets, and shear head fasteners.

Shear loading requires consideration of fasteners such as bolts or swaged collar fasteners with conventional or shear heads. Shear heads offer the higher strength-weight ratios. For shear fatigue loadings, the designer selects fasteners capable of interference fit or of developing high residual compression stresses in the structural material around the holes.

The process of fastener selection is usually accomplished in the following steps:
- Static strength is determined by material, diameter, head size, and thread size and length.
- Corrosion resistance — Fastener materials with good corrosion resistance are not acceptable for many applications because of high cost or galvanic corrosion they may cause in the dissimilar metals used in structure. Cadmium plating is the most commonly used finish on steel fasteners, but it leaves much to be desired. In blind rivets, alloy steel stems have been replaced by corrosion resisting steel stems to avoid unsightly rust or expensive paint touch-up.
- Material compatibility — Fastener materials, finishes, and lubricants must be selected so as to be compatible with other materials which they contact and with the functions and life of the assembly or equipment in which they are installed.
- Fatigue (Tension) — Tension fatigue failures are complex problems; therefore, very careful selection of materials, configuration, manufacturing processes, finishes, lubricants, mating parts, and installation procedures are required. Vacuum melt material is generally used to improve life of fasteners. Threads per MIL-S-8879 (with enlarged root radius) reduce stress concentrations and improve fatigue life. Rolling of threads after heat treatment and cold working of head-to-shank fillet induce compressive stresses which prolong fatigue life.
- Fatigue (Shear) — Many fasteners are used in applications where the primary loading is shear; however, fastener fatigue failures are induced by secondary loads in tension or bending. In such cases, many shear fasteners are used in tight or interference-fit holes.
- Elevated temperature strength is required and this is a function of the material as to strength at temperature and strength after exposure to temperature.
- Magnetic permeability — In or near certain types of equipment (compasses or directional control devices), magnetic permeability of fasteners is a critical characteristic.
- Fluid compatibility — The finish or lubricant on a fastener or the fastener material must be carefully considered when the fastener will come into contact with any of the many fluids used in aircraft.
- Availability — Is there more than one source for the fastener? Competition is good because of its influence on price and delivery schedules. Procurement costs are also important in fastener selection.
- Storage — Do the manufacture already have stocks of needed fasteners in the stores? Shelf life may also be a problem; for example, 2024-T31 (DD) rivets have a short shelf life after removal from the refrigerator.
- Fastener installation — It should be considered whether there is equipment for installation, in plant, availability for purchase, or whether it must be designed. Is the installation manual or automatic? How is reliability, fastener-fit, applied torque on a nut and bolt combination, etc?

In the design process, the designer considers the relationship between cost, weight, and function in selecting from the previously chosen fasteners, the optimum combination of these factors should be carefully considered. In some military applications, ground support equipment and facilities may be required. In such cases, great consideration to cost and little or no attention to weight is required. When removeability is required, the design case limits the choice to bolts or screws. Inaccessibility, when complete, requires the use of blind fasteners unless previously installed nut plates can be used with bolts or screws.

7.5 Lug Design and Analysis (Shear Type)

A method of analysis (Ref. 7.5) for the design of lugs made of aluminum or steel alloys, with static load applied axially, transversely, or obliquely (as shown in Fig. 7.5.1) is presented herein.

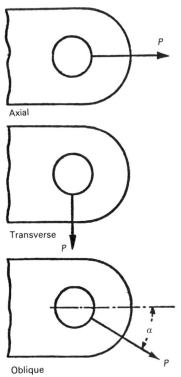

Fig. 7.5.1 *Three different applied load cases on lug.*

A tension efficiency factor is introduced to account for stress concentration effects, and curves are given for predicting such failures in lugs of steel and aluminum alloys.

The method, as described in this Section, had a theoretical basis, and its validity has been verified by comparison with many test results. This method also contains the evaluation of pin bending moments.

Lugs Loaded Axially — Case I

Three modes of failure: (a) net section tension; (b) shear tearout, assuming all the load to be transmitted on "40 degree planes"; and (c) bearing as shown in Fig. 7.5.2. It is indicated that modes (b) and (c) could be regarded as a single mode of failure and that for this mode, the allowable ultimate load for shear-bearing failure is

$$P_{bru} = K_{br} A_{br} F_{tu} \tag{7.5.1}$$

where:
- P_{bru} = Allowable ultimate load for shear-bearing failure
- K_{br} = Shear-bearing efficiency factor, see Fig. 7.5.3
- A_{br} = Projected bearing area ($A_{br} = Dt$)
- F_{tu} = Ultimate tensile strength of lug material with grain (refer to Ref. 7.6)

Over a range of lug shapes covering all lugs found in practice, K_{br} depends only on the ratio $\frac{a}{D}$.

The ratio $\frac{D}{t}$ also affects the shear-bearing efficiency factor K_{br}, but only for very large values of $\frac{D}{t}$ or $\frac{a}{D}$, and can usually be ignored.

The allowable ultimate load for tension failures

$$P_{tu} = K_t A_t F_{tu} \tag{7.5.2}$$

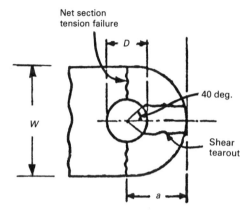

Fig. 7.5.2 *Lug net section tension and shear tearout failure under axial load.*

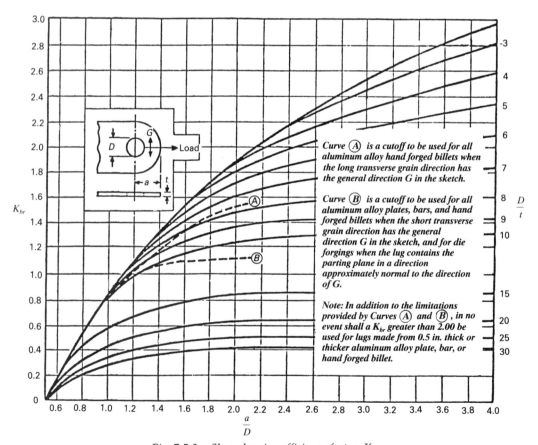

Fig. 7.5.3 *Shear-bearing efficiency factor, K_{br}.*

where
- P_{tu} = Allowable ultimate load for tension failure
- K_t = Efficiency factor for tension, see Fig. 7.5.4.
- A_t = Minimum net section for tension, $A_t = (W-D)t$
- F_{tu} = Ultimate tensile strength of lug material, refer to Ref. 7.6

Yield axial load attributable to shear-bearing is given by

$$P_y = C \frac{F_{tyx}}{F_{tux}} (P_u)_{\min} \qquad (7.5.3)$$

where
- P_y = Allowable yield load on lug
- C = Yield factor, see Fig. 7.5.5.
- F_{tyx} = Tensile yield stress of lug material across grain
- F_{tux} = Ultimate tensile strength of lug material across grain
- $(P_u)_{\min}$ = The smaller of P_{bru} or P_{tu}

Determine allowable yield bearing load on bushing,

$$P_{bry} = 1.85 F_{cy} A_{brb} \qquad (7.5.4)$$

where
- P_{bry} = Allowance yield bearing load on bushing
- F_{cy} = Compressive yield stress of bushing material
- A_{brb} = The smaller of the bearing areas of bushing on pin or bushing on lug. (The latter may be the smaller as a result of external chamfer on the bushing).

Finally, calculate margins of safety.

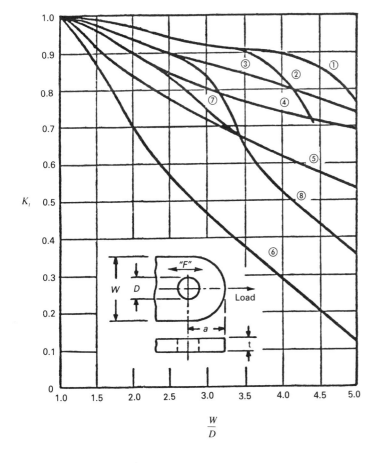

Legend L, LT and ST indicate grain in direction F in sketch:
Aluminum designation:
14S-2014
24S-2024
75S-7075

Curve ①
  4130 steel
  14S-T6 and 75S.T6 plate ≤ 0.5 in (L, LT)
  75S-T6 bar and extrusion (L)
  14S-T6 hand forged billet ≤ 144 sq in (L)
  14S-T6 and 75S-T6 die forgings (L)

Curve ②
  14S-T6 and 75S-T6 plate > 0.5 in, ≤ 1in
  75S-T6 extrusion (LT, ST)
  75S-T6 hand forged billet ≤ 36 sq in (L)
  14S-T6 hand forged billet > 144 sq in (L)
  14S-T6 hand forged billet ≤ 36 sq in (LT)
  14S-T6 and 75S-T6 die forgings (LT)

Curve ③
  24S-T6 plate (L, LT)
  24S-T4 and 24S-T42 extrusion (L, LT)

Curve ④
  24S-T4 plate (L, LT)
  24S-T3 plate (L, LT)
  14S-T6 and 75S-T6 plate > 1 in (L, LT)
  24S-T4 bar (L, LT)
  75S-T6 hand forged billet > 36 sq in (L)
  75S-T6 hand forged billet ≤ 16 sq in (LT)

Curve ⑤
  75S-T6 hand forged billet > 16 sq in (LT)
  14S-T6 hand forged billet > 36 sq in (LT)

Curve ⑥
  Aluminum alloy plate, bar, hand forged billet, and die forging (ST)
  75S-T6 bar (LT)

Curve ⑦
  18-8 stainless steel, annealed

Curve ⑧
  18-8 stainless steel, full hard. Note: for $\frac{1}{4}, \frac{1}{2}$ and $\frac{3}{4}$ hard, interpolate between Curves ⑦ and ⑧.

Fig. 7.5.4 *Efficiency factor for tension, K_t.*

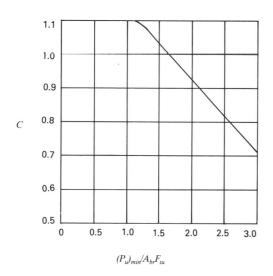

Fig. 7.5.5 Yield factor, C.

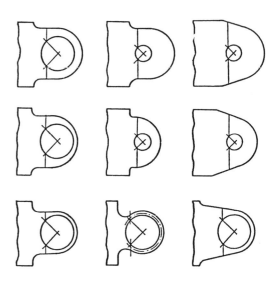

Fig. 7.5.7 Lug shapes.

Lug Loaded Transversely — Case II
(1) Obtain the areas A_1, A_2, A_3 and A_4 as follows:
   (a) A_1, A_2 and A_4 are measured on planes indicated in Fig. 7.5.6(a) perpendicular to the axial centerline, except that in a necked lug A_1 and A_4 as shown in Fig. 7.5.6(b) should be measured perpendicular to the local centerline.

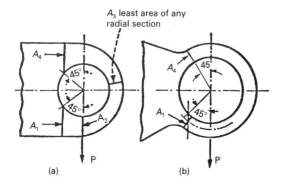

Fig. 7.5.6 Locations of cross-sectional areas of A_1, A_2, A_3 and A_4.

   (b) A_3 is the least area on any radial section around the hole.
   (c) Since the choice of areas and the method of averaging has been substantiated only for lugs of the shapes, as shown in Fig. 7.5.7, thought should always be given to assure that the areas A_1, A_2, A_3 and A_4 adequately reflect the strength of the lug. For lugs of unusual shapes (for example, with sudden changes of cross section), an equivalent lug should be sketched as shown in Fig. 7.5.8 and used in the analysis.

   Obtain the average of these areas using

$$A_{av} = \frac{6}{\left(\dfrac{3}{A_1}\right) + \left(\dfrac{1}{A_2}\right) + \left(\dfrac{1}{A_3}\right) + \left(\dfrac{1}{A_4}\right)}$$

(7.5.5)

   (d) Compute:
$$A_{br} = Dt \quad \text{and} \quad \frac{A_{av}}{A_{br}}$$

   (e) Determine allowable ultimate load
$$P_{tru} = K_{tru} A_{br} F_{tux} \qquad (7.5.6)$$

   where
   P_{tru} = Allowable ultimate load as determined for transverse loading
   K_{tru} = Efficiency factor for transverse load (ultimate), (see Fig. 7.5.9)
   A_{br} = Projected bearing area
   F_{tux} = Ultimate tensile strength of lug material across grain.

   In no case should the allowable transverse load be taken as less than that which could be carried by cantilever beam action of the portion of the lug under the load (see Fig. 7.5.10). The load that can be carried by cantilever beam action is indicated very approximately by curve A in Fig. 7.5.9; should K_{tru} be below curve A, a separate calculation as a cantilever beam is warranted.

   (f) Determine allowable yield load of lug
$$P_y = K_{try} A_{br} F_{tyx} \qquad (7.5.7)$$

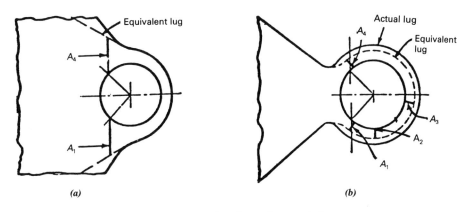

Fig. 7.5.8 Equivalent lugs of unusual shapes.

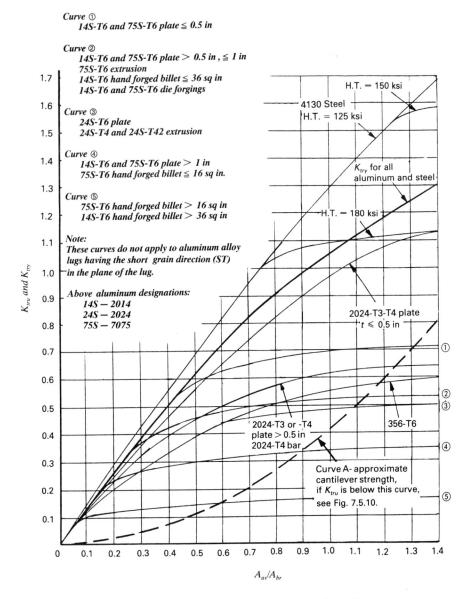

Fig. 7.5.9 Efficiency factor for transverse load, K_{tru}.

Airframe Structural Design 223

where
- P_y = Allowable yield load on lug
- K'_{try} = Efficiency factor for transverse load (yield), see Fig. 7.5.9
- A_{br} = Projected bearing area
- F_{tyx} = Tensile yield stress of lug material across grain

(g) Determine allowable yield bearing load on bushing by using Eq. (7.5.4).
(h) Calculate margins of safety.

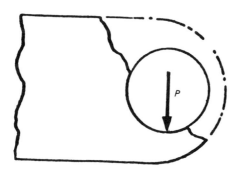

Fig. 7.5.10   Cantilever beam action of the portion of the lug under load.

Lug Loaded Obliquely — Case III

(a) Resolve the applied load into axial and transverse components and obtain the lug ultimate and yield margins of safety from the interaction equation:

$$M.S. = \frac{1}{(R_a^{1.6} + R_{tr}^{1.6})^{0.625}} - 1 \qquad (7.5.8)$$

where
for ultimate load,
- R_a = [Axial component of applied ultimate load] divided by [Smaller of P_{bru} (Eq. (7.5.1) and P_{tu} (Eq. (7.5.2)]
- R_{tr} = [Transverse component of applied ultimate load] divided by [P_{tru} (Eq. (7.5.6)]

for yield load,
- R_a = [Axial component of applied yield load] divided by [P_y (Eq. (7.5.3)]
- R_{tr} = [Transverse component of applied yield load] divided by [P_y Eq. (7.5.7)]

(b) Determine allowable yield bearing load P_{bry} on bushing by Eq. (7.5.4).

Pin Design

In general, static tests of single bolt fittings will not show a failure due to bolt bending failure. However, it is important that sufficient bending strength be provided to prevent permanent bending deformation of the fitting bolt under the limit loads so that bolts can be readily removed in maintenance operations. Furthermore, bolt bending weakness can cause peaking up a non-uniform bearing loads on the lugs, thus influencing the lug tension and shear strength. The unknown factor in bolt bending is the true value of the bending moment on the bolt because the moment arm to the resultant bearing forces is difficult to define. An approximate method for determining the arm to use in calculating the bending moment on bolt is given below:

$$\text{moment arm, } b = 0.5t_1 + 0.25t_2 + \delta \qquad (7.5.9)$$

If it is no gap ($\delta = 0$) between the lugs, the moment arm is taken as

$$b = 0.5t_1 + 0.25t_2 \qquad (7.5.10)$$

If it is desired to take account of the reduction of pin bending that results from peaking as shown in Fig. 7.5.11, the moment arm is obtained as follows:

(a) Compute the inner lug r — use D_b instead of D, if bushing is used

$$r = \frac{\left[\left(\frac{a}{D}\right) - \frac{1}{2}\right]D}{t_2}$$

(b) Take smaller of P_{bru} and P_{tu} for inner lug as $(P_u)_{min}$ and compute $\dfrac{(P_u)_{min}}{A_{br} F_{tux}}$

(c) Enter Fig. 7.5.12 with $\dfrac{(P_u)_{min}}{A_{br} F_{tux}}$ and r to obtain the reduction factor for peaking, and calculate the moment arm

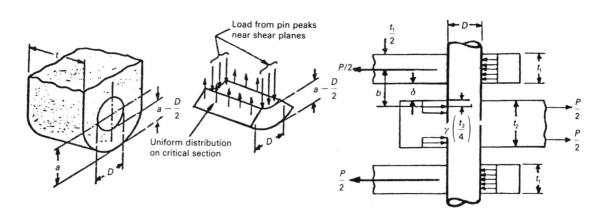

Fig. 7.5.11   Pin moment arm for determination of bending moment.

$$b = \left(\frac{t_1}{2}\right) + \delta + \gamma\left(\frac{t_2}{4}\right) \quad (7.5.11)$$

The maximum bending moment in pin from the equation as

$$M = P_u\left(\frac{b}{2}\right) \quad (7.5.12)$$

and calculate the bending stress in the pin that results from M, assuming an $\frac{MD}{2I}$ distribution (where I = moment of inertia of pin) and its margin of safety as

$$M.S. = \frac{\frac{MD}{2I}}{F_b} - 1$$

where F_b = the modulus of rupture deter-

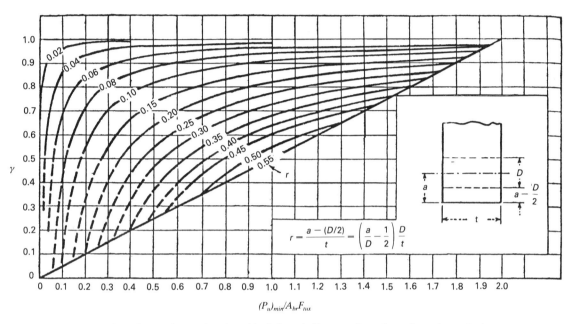

Fig. 7.5.12 Peaking factors for pin bending (dash lines indicate region where these theoretical curves are not substantiated by test data).

Bolt Dia. (in)	Bolt Area (in²)	Moment of Inertia (in⁴)	Single Shear (lb)	Tension (lb)	Bending (in-lb)
1/4	0.0491	0.000192	3680	4080	276
5/16	0.0767	0.000468	5750	6500	540
3/8	0.1105	0.00097	8280	10100	932
7/16	0.1503	0.001797	11250	13600	1480
1/2	0.1963	0.00307	14700	18500	2210
9/16	0.2485	0.0049	18700	23600	3140
5/8	0.3068	0.00749	23000	30100	4320
3/4	0.4418	0.01553	33150	44000	7450
7/8	0.6013	0.02878	45000	60000	11850
1.0	0.7854	0.0491	58900	80700	17670

Note: AN steel bolts, F_{tu} = 125 ksi, F_{su} = 75 ksi

Fig. 7.5.13 AN steel bolts allowable strength (F_{tu} = 125 ksi).

mined from Ref. 7.6 or other sources

or

$$\text{M.S.} = \frac{M}{\text{Ultimate bending moment (allowable)}} - 1$$

The ultimate allowable bending moment for $F_{tu} = 125$ ksi AN steel bolt can be obtained from Fig. 7.5.13.

Bushing and Bearing

It is customary to provide bushings in lugs with single bolts or pin fittings subjected to reversal of stresses or to slight rotation. Thus, when wear and tear takes place, a new bushing can be inserted in the lug fitting. Steel bushings are commonly used in aluminum alloy single bolt fitting lugs to increase the allowable bearing stress on the lug since the bushing increases the bearing diameter. If bushings are not used on single bolt connections, sufficient edge distance should be provided to ream hole for the next size bolt in case of excessive wear of the unbushed hole. If considerable rotation occurs a lubricator should be provided for a plain bushing or an oil impregnated bushing should be used.

Plain spherical bearings usually consist of two pieces: an outer race with a spherical inside diameter swaged over a spherical inner ball, as shown in Fig. 7.5.14. Joint rotation takes place between these two pieces with a ball-and-socket feature providing a small amount of unrestricted alignment.

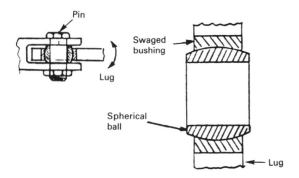

Fig. 7.5.14 *Plain spherical bearing.*

The advantage of a plain spherical bearing over an anti-friction bearing lies in the much greater load carrying capability of its area contact as opposed to the point contact or line contact of the ball. Disadvantages of the plain spherical bearings include:
- Very much higher friction
- The necessity of frequent lubrication
- Poor reliability compared to plain bushing
- Usually higher cost.

Because of these factors, mechanism designs should specify the plain spherical bearing for joints where the load is high (such as powered controls, landing gear joints, etc.); anti-friction bearings should be used elsewhere. Highly loaded joints where the aligning feature of spherical bearings is not required should utilize plain bushings.

Example

Given: Materials —
  Lug: Aluminum extrusion 7075-T6 (see Fig. 4.3.6)
  Bushing: Steel with H.T. = 150–180 ksi
  Pin: Steel with min. H.T. = 200 ksi

$a = 1.875$ in
$W = 3.75$ in
$D_b = 1.25$ in
t (or t_2) = 0.75 in
$t_1 = 0.375$ in
Pin diameter = $D = 0.75$ in

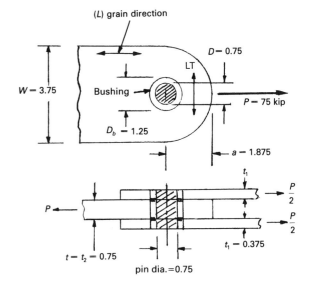

(a) Compute:

$$\frac{a}{D_b} = \frac{1.875}{1.25} = 1.5$$

$$\frac{W}{D_b} = \frac{3.75}{1.25} = 3.0$$

$$\frac{D_b}{t} = \frac{1.25}{0.75} = 1.67$$

$A_{br} = D_b t = 1.25 \,(0.75) = 0.938$
$A_t = (W - D_b)\,t = (3.75 - 1.25)\,0.75 = 1.875$

(b) Allowable shear-bearing failure load [Ref. Eq. (7.5.1)].

Enter Fig. 7.5.3 with $\frac{a}{D_b} = 1.5$; $\frac{D_b}{t} = 1.67$,
obtain $K_{br} = 1.26$

$P_{bru} = K_{br}(A_{br})(F_{tux}) = 1.26\,(0.938)\,(74)$
$\quad\quad = 87.5$ kip

(c) Allowable tension failure load: [Ref. Eq. (7.5.2)]

Enter Fig. 7.5.4 (curve ①) with $\frac{W}{D_b} = 3.0$,
obtain $K_t = 0.922$

$P_{tu} = K_t(A_t)(F_{tu}) = 0.922\,(1.875)\,(81)$
$\quad\quad = 140.03$ kip

(d) Allowable yield load: [Ref. Eq. (7.5.3)].

1. $(P_u)_{min} = P_{bru} = 87.5$ kip
   Enter Fig. 7.5.5 with
   $$\frac{(P_u)_{min}}{A_{br}(F_{tux})} = \frac{87.5}{0.938(81)} = 1.15, \text{ obtain } C = 1.1$$

2. $P_y = C\dfrac{F_{tyx}}{F_{tux}}(P_u)_{min}$
   $$= \frac{1.1\,(72)\,(87.5)}{81} = 85.6 \text{ kip}$$

(e) Allowable yield bearing load on bushing:
   [Ref. Eq. (7.5.4)]
   $$P_{bry} = 1.85\,F_{cy} \times A_{brb}$$
   $$= 1.85\,(145)\,(0.563) = 151.02 \text{ kip}$$
   where $A_{brb} = Dt = 0.75\,(0.75) = 0.563$ in^2
   $F_{cy} = 145$ ksi (From Ref. 7.6)

(f) Allowable pin bending load [Ref. Eq. (7.5.9)]
   Assume there is no gap between t_1 and t_2, therefore $\delta = 0$
   $$b = 0.5\,t_1 + 0.25\,t_2 + \delta$$
   $$= 0.5\,(0.375) + 0.25\,(0.75) + 0 = 0.375$$
   pin moment, $M = (P_u)_{min}\left(\dfrac{b}{2}\right)$
   $$= 87.5\left(\frac{0.375}{2}\right) = 16.4 \text{ in-kip}$$

   From Fig. 7.5.13, the allowable ultimate bending moment is 7.45 in-kip ($D = 0.75$ in), and for H.T. = 200 ksi steel bolt and allowable pin bending load is
   $$M' = 2\,(7.45)\left(\frac{200}{125}\right) = 23.84 \text{ in-kip}$$
   $$P_b = (P_u)_{min}\left(\frac{M'}{M}\right) = 87.5\left(\frac{23.84}{16.4}\right)$$
   $$= 127.2 \text{ kip}$$

(g) Allowable pin shear load

   From Fig. 7.5.13, the allowable ultimate shear load is 33.15 kip ($D = 0.75$ in), and for H.T. = 200 ksi ($F_{su} = 125$ ksi, from Ref. 7.6) steel bolt and total allowable pin shear load is
   $$P_s = 2\,(33.15)\left(\frac{125}{75}\right) = 110.5 \text{ kip}$$

(h) Margins of safety

   1. Lug:

      Ultimate margin in shear bearing
      $$M.S. = \frac{P_{bru}}{1.15P} - 1$$
      $$= \frac{87.5}{1.15\,(75)} - 1 = 0.01$$

      Ultimate margin in tension
      $$M.S. = \frac{P_{tu}}{1.15P} - 1$$
      $$= \frac{140.03}{1.15\,(75)} - 1 = 0.62$$

      Yield margin $M.S. = \dfrac{P_y}{\text{limit load}} - 1$
      $$= \frac{85.6}{1.15\left(\dfrac{75}{1.5}\right)} - 1 = 0.49$$

   2. Pin:
      Ultimate margin in bending
      $$M.S. = \frac{P_b}{1.15P} - 1$$
      $$= \frac{127.16}{1.15\,(75)} - 1 = 0.47$$

      Ultimate margin in shear
      $$M.S. = \frac{P_s}{1.15P} - 1$$
      $$= \frac{110.5}{1.15\,(75)} - 1 = 0.28$$

   3. Bushing:
      Yield margin $M.S. = \dfrac{P_{bry}}{\text{limit load}} - 1$
      $$= \frac{151.02}{\dfrac{75}{1.5}} - 1 = 2.02$$

Conclusion: The least margin of safety as shown above is 0.27, which is greater than requirement of a minimum margin of safety of 0.20 (recommended).

7.6 Welded and Adhesive Bonding Joints

Welded Joint

Since the overall structure of an airplane cannot be fabricated as a single continuous unit, many structural parts must be fastened together. For certain materials and types of structural units, welding plays an important role in joining or connecting structural units. Research is constantly going on to develop better welding machines and welding techniques and also to develop new materials that can be welded without producing a detrimental strength influence on the base or unwelded material.

Welding is used extensively for steel tube truss structures, such as engine mounts, general aviation airplane fuselages, and for steel landing gears and fittings. The strength of welded joints depends greatly on the skill of the welder. The stress conditions are usually uncertain, and it is customary to design welded joints with liberal margins of safety. It is preferable to design joints so that the weld is in shear or compression rather than tension, but it is frequently necessary to have welds in tension. Tubes in tension are usually spliced by "fish-mouth" joints, as shown in Fig. 7.6.1, which are designed so that most of the weld is in shear and so that the local heating of the tube at the weld is not confined to one cross section.

To weld or use fasteners: here are a few points to

consider when deciding whether welding or a mechanical fastener would be best for a particular design.

- Is the joint to be permanent or temporary? If it is temporary, welding is obviously out of the question.
- Is air tightness or leak tightness required? If so, welding may be the easiest and least expensive way to achieve this.
- Is access to both sides of the part available? This is often essential with mechanical connection, but not generally required with welding.
- What equipment is needed? What does it cost? Sometimes expensive drilling and assembly systems are needed to give a bolted or riveted part the same production rate as welding. This is especially true if there are long seams, or if many holes must be drilled.
- Is it practical to design the joints with overlapping edges? This is necessary with bolts and rivets, but not usually essential with welding.
- Does available labor have the necessary skills to use the assembly method properly? Some welding methods require highly trained personnel.

Tests show that plain carbon and chromemolybdenum steels suffer very little in loss of tensile strength due to welding. For cold rolled sheet or tubing the refinement in grain due to cold working is lost in the material adjacent to the weld which lowers the strength to a small degree. Welding, however, does produce a more brittle material which has lower resistance to shock, vibration and reversal of stress. Thus, it is customary to assume an efficiency of weld joints less than 100%. Ref. 7.6 gives the allowable ultimate tensile stress for alloy steels for materials adjacent to the weld when the structure is welded after heat treatment. For welding members subjected to bending, the allowable modulus of rupture for alloy steels when welded after heat-treatment should not exceed that as specified in Ref. 7.6. For weldable materials heat-treated after welding, the allowable stresses in the parent material near a welded joint may equal the allowable stress for the heat-treated material; in other words, no reduction for welding. However, it is good design practice to be conservative on welded joints; thus a reduction of 10% of the heat-treated properties is often used in calculating the tensile or bending strength in the member adjacent to the weld, or properly test the designed component without using the 10% reduction.

When the structural engineer considers the possibility of welding or brazing as a method to satisfy the joining requirements of a aircraft, several questions immediately arise:

- Are the alloy compositions, conditions, and forms being used readily weldable?
- To what degree is the design depending on the development of the full design strength in the joint?
- To what degree are the joints repetitious, and in what total quantity?
- Can heat treatment and/or machining be performed on the welded assembly?
- Can the weld be formed manually, by machine with operator, or automatically?

The designer of welded structures in steel can greatly help the welder obtain good joints or connections by adhering to the following general rules:

- It is much easier to obtain a good weld when the parts being welded together are of equal thickness. It is general design practice to try and keep thickness ratio between the two welded parts less than 3 to 1. Some designers try to keep within a 2 to 1 ratio in order to eliminate possibilities of welders burning the thinner sheet.
- Steel tubes often have walls as thin as 0.035 in, and the welder must control the temperature to keep from overheating the thin walls and burning holes in them.
- A weld should not encircle a tube in a plane perpendicular to the tube length.
- In general, avoid welds in tension since they produce a weakening effect. In some connections it is difficult to avoid all tension loads on welds; thus weld stresses should be kept low and if possible incorporate a fishmouth joint or finger patch to put part of weld in shear.
- Tapered gusset plates should be incorporated in all important welded joints to insure gradual change in stress intensity in members. These gussets lessen the danger of fatigue failure by reducing stress intensity.

Welding consists of joining two or more pieces of metal by applying heat, pressure, electron beam, etc., with or without filler material, to produce a localized union through fusion or recrystallization across the joint interface. Examples of common welding processes include: fusion, resistance, flash, pressure, and friction. Several terms used in describing various

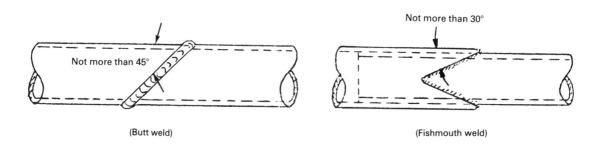

Fig. 7.6.1 Welded joints of butt weld and fishmouth weld.

sections of a welded joint are illustrated in Fig. 7.6.2.

(1) Fusion Welding:
Fusion joining is accomplished by heating the joint zone to a molten state and weld is formed when adjacent molten surfaces coalesce and freeze. A filler metal may be added to the welded joints. All fusion welds must be classified as follows:
- Class A — a vital joint (failure would jeopardize the aircraft safety.)
- Class B1 — a secondary structure (failure would not jeopardize the aircraft safety)
- Class B2 — nonstructure

The fusion weld applications are summarized in Fig. 7.6.3.

(2) Resistance Welding (spot, roll and seam welding):
Resistance spot welding is produced by the resistance to the flow of electrical current through the plates being jointed. The roll and seam welding are basically spot welding process.

In general, aluminum alloy spot welded joints should not be used in primary or critical structures without the specific approval of the military or civil aeronautic authorities. The following are a few types of structural connections in aircraft where spot welding should not be used.
- Attachment of flanges to shear webs in stiffened cellular construction in wings.
- Sonic fatigue critical areas such as flaps, slats, wing fixed trailing and leading panels, etc., if the jet engines are mounted underneath the wing.
- Attachment of shear web flanges to wing sheet covering.
- Attachment of wing ribs to beam shear webs.
- Attachment of hinges, brackets and fittings to supporting structure.
- At joints in trussed structures.
- At juncture points of stringers with ribs unless a stop rivet is used.
- At ends of stiffeners or stringers unless a stop rivet is used.
- On each side of a joggle, or wherever there is a possibility of tension load component, unless stop rivets are used.
- In general most aluminum and aluminum alloy material combinations can be spot welded.

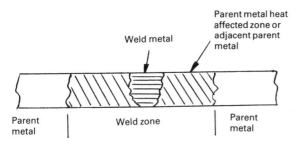

Fig. 7.6.2 *Schematic diagram of weld and parent material.*

Type of Welding	Characteristics
Gas tungsten-arc (or called T10 or heliarc)	Tungsten electrode with inert gas shielded. The most popular welding process in aerospace.
Hot wire gas and tungsten-arc	A faster welding and improved distortion.
Pulsed current gas and tungsten-arc	Better depth-to-width ratio, narrower heat affected zone and lower porosity content.
Plasma-arc	Permits higher welding speeds and improved process control.
Gas metal-arc	For thick materials.
Vacuum electron beam (E.B.)	Welding heat is generated by impingement on the metal joint by high velocity electrons. Vacuum chamber size and depth of welding are the limiting factors. It is a relatively costly process and requires close tolerance machining and straight line joint surfaces. This welding process is widely applied on titanium material (some for tension joint applications).
Non-vacuum electron beam	Least costly of the E.B. welding process and high speed welding of sheet gages with constant thickness and straight line joints.
Pulsed laser	Small function spotwelds by a concentrated coherent light beam. Generally for electronic parts and instrumentation.

Fig. 7.6.3 *Summary of fusion welding.*

(3) Flash Welding:
Flash welding is an electric current passing through a joint causing a flashing action which heats the metal to the fusion point and the end pressure is required to complete the process. This welding process is usually applied to joining tubular and solid cross-sections.

(4) Pressure Welding:
Pressure welding is produced by a radially oscillating oxyacetylene ring torch to heat the butting joint under pressure. This process is similar to a flash welding.

(5) Friction Welding:
Friction welding joint is created through heat generated by forging under rotation force, see Fig. 7.6.7. This method is also referred to as inertia welding and offers a wide range in joining dissimilar metals. This process is more economical than either flash or pressure welding.

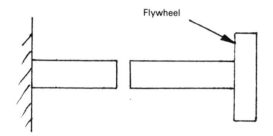

Kinetic energy of rotating flywheel is predetermined by selection of moment of inertia.

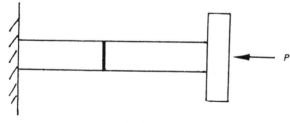

Axial force is applied and energy stored in flywheel becomes converted to heat at the joint faces.

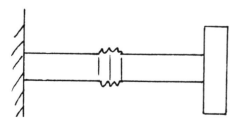

Welding and upset occurs just before rotation ceases.

Fig. 7.6.4 Principle of inertia welding.

Brazing

Brazing consists of joining metals by the application of heat causing the flow of a thin layer, capillary thickness, of nonferrous filler metal into the space between pieces. Bonding results from the intimate contact produced by the dissolution of a small amount of base metal in the molten filler metal without fusion of the base metal.

Soldering

Soldering joints are accomplished by applying a low melting point filler metal between joined surfaces. Soldering is not to be recommended as a means of structural attachment due to its inherent low resistance to creep.

Adhesive Bonded Joints

Light weight materials can often be used with adhesive bonding rather than with conventional fastening, simply because the uniform stress distribution in the joint permits full utilization of the strength and rigidity of the adherends. Payloads in aircraft can be increased.

The following are possible advantages of properly designed adhesive bonded joints:
- Adhesive bonds provide airtight joints.
- Electrochemical corrosion is reduced or eliminated. In a bonded joint adhesives are electrically insulated from each other; there are no holes to expose base metal, and cladding, anodize and other corrosion protection surfaces are not destroyed.
- Higher fatigue life of joints may possibly allow reduced sheet gages.
- Residual strength of damaged structure which is adhesively bonded can be large. In a laminated panel, cracks may grow for some time in only one layer; in stiffened panels, cracks may grow more slowly across a bonded area than across a rivet line.
- Aerodynamic surface smoothness of adhesively bonded structure is excellent.

Adhesively bonded joint may suffer from the following limitations:
- Assembly may be more expensive than for conventional joints.
- Very extensive process control over the entire bonding procedure is required.
- Curing temperatures of some adhesives may degrade other components.
- Service degradation of bonds is difficult to check.

Bonded joints shall not be used in any application in which a complete bond failure or obvious partial failure could cause loss of the aircraft. Bonded joints shall be classified for structural application as follows as a means for designating levels and types of inspection and peel strength requirements. The appropriate class shall be designated by the engineering department.

7.7 Fatigue Design Considerations (Mechanical fastened joints)

The purpose of this section is to introduce a rational method of analysis; the stress severity factor concept (Ref. 7.19) that emphasizes the fatigue characteristics of the structural joint rather than it's static strength. The stress severity concept requires a detailed analysis of the load distribution on each critical fatigue fastener within the structure. The fatigue performance of a joint is largely influenced by hole preparation, fastener type, and installation technique. Residual stresses from cold working the hole and from interference between the fastener and hole must be accounted for. The load distribution can be calculated by structural finite element model analysis.

Once the load distribution has been found, the life estimates of the structural joints may be directly obtained by combined use of the severity factors, fatigue load spectrum, and material S-N data.

(1) Basic intent of the design criterion
The basic intent of this joint criterion is to realize an equivalent fatigue life within the joint equal to or greater than the basic panel life. The joint configuration is designed such that the peak stresses within the joint are equal to or less than the peak limitation stress as defined by the basic panel. There are two essential requirements necessary before the analysis can be made: (a) the basic panel design allowable stress immediately adjacent to the joint as defined by the fatigue analysis and (b) the fastener load distribution along the length of the splice or doubler being analyzed. Once these two requirements are satisfied, the stress concentration factor can then be determined to define the joint peak stresses by using curves derived from conventional stress concentration factor data.

(2) Fastener load distribution method of analysis
The imporance of defining the point of maximum stress concentration in fatigue design warrants a detailed definition of joint fastener load distribution. The net stress level at the fastener hole coupled with the fastener load dictates the stress concentration to be expected. The load distribution along the splice or doubler length is a function of the area distribution, fastener pattern (spacing, rows and diameter), fastener deflections and materials. Excessive unconservatism can be expected in assuming equal distribution of load on a given pattern. This method of analysis is limited to stress levels within the elastic limit. Since all fatigue damage should be within the elastic range, this method is readily adaptable.

Comparison of Doublers and Splices

It is helpful to keep in mind that there are two basic differences between doublers and splices.
(1) Different purposes
- A splice's function is to transfer a given load. It is kept as short as possible to accomplish this.
- A doubler's function is to pick up load (and relieve another member). In order to do this efficiently, it must have some considerable length, although this is kept to a minimum. Therefore, doublers are, by nature, relatively long members compared to splices.

(2) General guide for doubler design
  (a) To increase the load picked up by the doubler
   - Increase the doubler planform width
   - Increase the doubler thickness
   - Increase the length of the doubler
   - Increase the number of fasteners
   - Increase the size of fasteners
   - Use stiffer fasteners (material change)
   - Use stiffer doubler material
  (b) To reduce the "peaking effect", that is the large fastener loads developed at the ends of the doubler
   - Taper the doubler thickness
   - Use tapered width at both ends
   - use more flexible (or smaller) fasteners at the ends
  (c) In order to insure all fasteners loading up efficiently, and also more consistent results, the doubler should be installed using close tolerance or reamed holes when interference fit fasteners are used.

(3) General guide for splice design
The main effort is to keep the length of the splice as short as possible. Within this limit the "peaking effect" can be dealt with as outlined in the foregoing items (b) and (c) which also apply to splices.

(4) The fastener load distribution
  (a) The fastener load in splices can be made to approach a somewhat uniform distribution efficiently since they are all acting in one direction (unless unusual intermediate applied loads are present).
  (b) In a doubler, however, the fastener loads form two groups acting in opposite directions to load and unload the doubler. Thus, the fastener loads will be larger at the ends and vanish at the center where the relative displacement between members is zero. This will not, efficiently, approach uniformity as in the case of the splice. These facts are illustrated in Fig. 7.7.1.

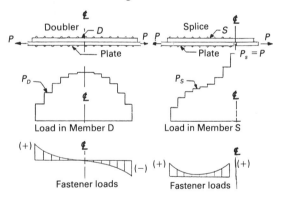

Fig. 7.7.1 Comparison of internal loads in typical doubler and splice.

  (c) Fastener fit
    The fatigue life of a joint loaded in fluctuating shear is greatly influenced by the fastener fit in the hole. A close fit or an interference fit hole must provide long term fatigue life.
  (d) Fastener hole preparation
    Drilled and punched holes have equal fatigue characteristics whereas reaming produces slight improvement. Coining process (Ref. 7.33) is the addition of a concentric groove around a hole that induces a high compres-

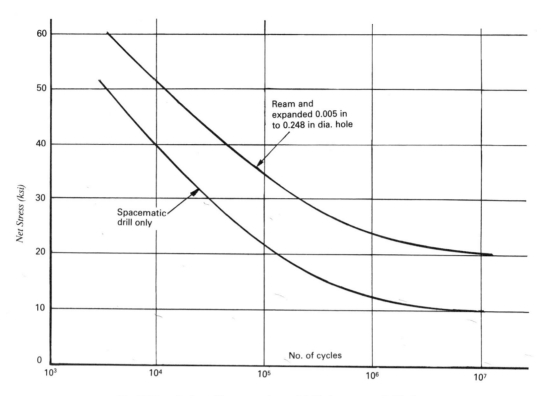

Fig. 7.7.2 Fatigue life comparison of drilled vs. expanded holes.

sive circumferential stress in the material around the hole edge. Under cyclic loading of a structure, the residual compressive stress lowers the applied mean stress at the hole and hence the fatigue life is increased. Fig. 7.7.2 illustrates the improved fatigue characteristics obtained by the use of mechanical cold work such as by expanding the hole diameter slightly with a mandrel.

Reducing Stress Concentration by Careful Detail Design

Serious fatigue troubles have, in the majority of cases, been the result of unusually high stress concentration factors existing in vital tension members or joints. The relatively simple changes in detail design may improve service life by a factor of from between 10 and 100. When designing for fatigue, one should not be too concerned about testing scatter in the order of 2 because far more potent factors are under the designer's control in detail design. However, the detail design is the most important single factor affecting the fatigue life of a structural joint and an airplane as well (see Chapter 15.0 for further discussion).

General rules and information:
- Symmetrical joints should be used wherever possible.
- Joints should be designed to distribute the loading evenly between the parts.
- Local eccentricity should be kept to a minimum in joints.
- Provide adequate increased thickness on machined members at all riveted and bolted joints.
- The thickness of machined countersunk sheets must be 1.5 times the depth of the fastener countersunk head (see Fig. 7.7.3).

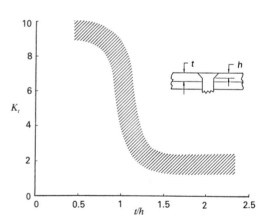

Fig. 7.7.3 Variation of stress concentration factor with countersunk fastener hole.

- The fastener edge distance should not be less than 2 times the fastener diameter (metal structures).
- Avoid single rows of fasteners at all splices; use double or three rows of fasteners whenever possible (Fig. 7.7.4).
- Use double shear joints wherever possible.
- Shotpeening the plate surfaces.
- Structural shims applied properly may reduce the

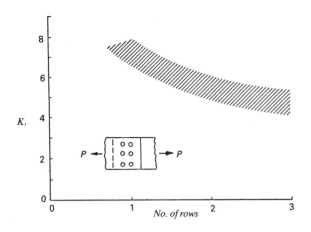

Fig. 7.7.4 *Variation of stress concentration factor with no. of rows of fasteners.*

local fastener load transfer to improve joint life
- Avoid abrupt change in cross section.
- Avoid superimposing stress concentrations.
- Avoid feather edges and sharp corners when spot-facing.
- Design with sufficient fastener clamped force (fastener torque) see Fig. 7.7.5.
- Design with sufficient fastener spacing, usually between 4 and 8 times the fastener diameter (see Fig. 7.7.6).
- Machined countersunk joints (see Fig. 7.7.7).
- Dimple countersunk joints (see Fig. 7.7.8).
- Avoid dimple countersinking in sections subject to bending (see Fig. 7.7.9).
- First fastener row is relieved with a long splice or doubler by either tapering the splice or doubler plate, or by using thin auxiliary doubler (see Fig. 7.7.10).
- Wing spanwise joint design should follow Fig. 7.7.11.

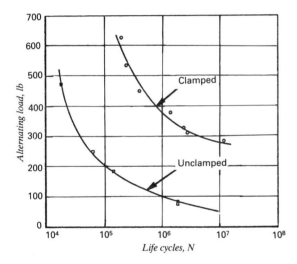

Fig. 7.7.5 *Joint life comparison between clamped and unclamped fasteners.*

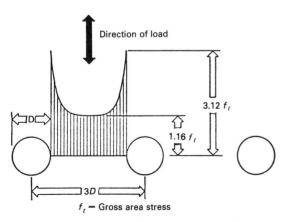

Fig. 7.7.6 *Stress distribution between two holes at 3D-spacing.*

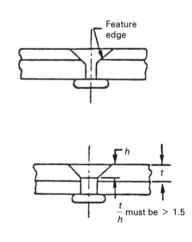

Fig. 7.7.7 *Machined countersunk joints.*

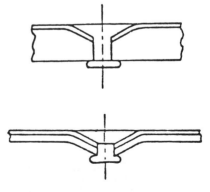

Not recommended in fatigue critical joints

Fig. 7.7.8 *Dimple countersunk joints.*

Airframe Structural Design 233

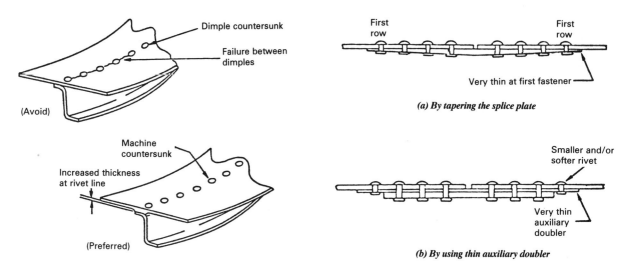

Fig. 7.7.9 Dimple countersinking subject to bending.

Fig. 7.7.10 Two typical designs of splice and doubler.

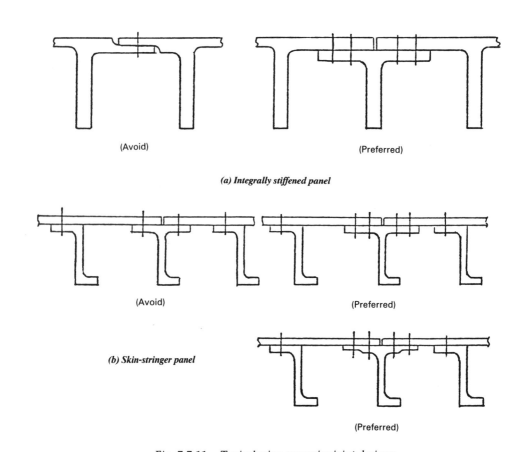

Fig. 7.7.11 Typical wing spanwise joint designs.

Fastener Load Distribution and By-pass Load

It is considered that most fatigue damage occurs under loading conditions for which the fasteners behave as linear elastic members. Even for very simple structures the treatment of the fasteners as flexible members result in rather elaborate calculations justifying the use of the finite element modeling analysis. Before simulating the structural modeling, the fastener spring constant which can be converted to equivalent structural beam member is obtained by the following hypothetical equations (see Ref. 7.8).

234 Airframe Structural Design

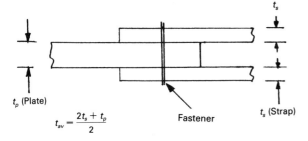

where C — Fastener constant (in/lb)
K_f — Fastener spring constant ($K_f = 1/C$)
E_{bb} — Fastener material modulus of elasticity

Fig. 7.7.12 Average thickness of plate and straps.

Case I: For steel plates and fasteners (see Fig. 7.7.12)
$$C = \frac{8}{t_{av}E_{bb}}\left\{0.13\left(\frac{t_{av}}{d}\right)^2\left[2.12+\left(\frac{t_{av}}{d}\right)^2\right]+1.0\right\} \tag{7.7.1}$$

Case II: For aluminum plate and straps and aluminum fasteners — Same as Case I.

Case III: For aluminum plates and steel fasteners (see Fig. 7.7.12)
$$C = \frac{8}{t_{av}E_{bb}}\left\{0.13\left(\frac{t_{av}}{d}\right)^2\left[2.12+\left(\frac{t_{av}}{d}\right)^2\right]+1.87\right\} \tag{7.7.2}$$

Case IV: For aluminum plate, steel straps and steel fasteners (see Fig. 7.7.12)
$$C = \frac{8}{t_{av}E_{bb}}\left\{0.13\left(\frac{t_{av}}{d}\right)^2\left[2.12+\left(\frac{t_{av}}{d}\right)^2\right]+1.43\right\} \tag{7.7.3}$$

Case V: For aluminum plate, steel straps and aluminum fasteners (see Fig. 7.7.12)
$$C = \frac{8}{t_{av}E_{bb}}\left\{0.13\left(\frac{t_{av}}{d}\right)^2\left[2.12+\left(\frac{t_{av}}{d}\right)^2\right]+0.84\right\} \tag{7.7.4}$$

Case VI: For aluminum plates and titanium fasteners (see Fig. 7.7.12)
$$C = \frac{8}{t_{av}E_{bb}}\left\{0.133\left(\frac{t_{av}}{d}\right)^2\left[2.06+\left(\frac{t_{av}}{d}\right)^2\right]+1.242\right\} \tag{7.7.5}$$

Case VII: For aluminum plates, titanium straps and titanium fasteners (see Fig. 7.7.12)
$$C = \frac{8}{t_{av}E_{bb}}\left\{0.1325\left(\frac{t_{av}}{d}\right)^2\left[2.06+\left(\frac{t_{av}}{d}\right)^2\right]+1.1125\right\} \tag{7.7.6}$$

It is convenient to simulate the structural finite element modeling by a simplified model as demonstrated by the following joint illustrations as shown in Fig. 7.7.13.

When there is more than one fastener in a row (normal to the loading, or in the axial direction) the spring constants of the individual fasteners in the row can be simply added together and considered as one fastener.

Frequently, however, in the case of doubler installations there are too many rows of fasteners for any analysis to include all of them, and it is necessary to group, or "lump", two or more rows together as one row, or one fastener. Since the end fasteners are the most highly loaded it is best to do the least grouping at the ends and the most at the middle. Fig. 7.7.14 illustrates how this is carried out.

Fig. 7.7.15 through Fig. 7.7.22 illustrate fastener load distribution of several typical joints designs and can be used as a guide for preliminary sizing purposes

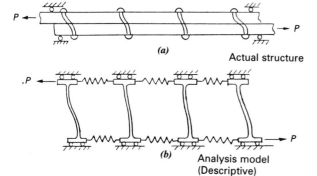

Fig. 7.7.13 Structural model of four fastener lap joint.

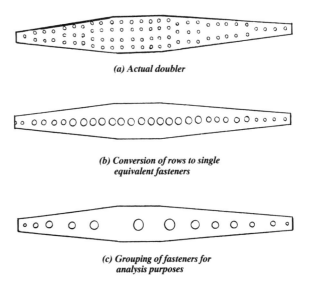

Fig. 7.7.14 Grouping of fasteners.

Airframe Structural Design 235

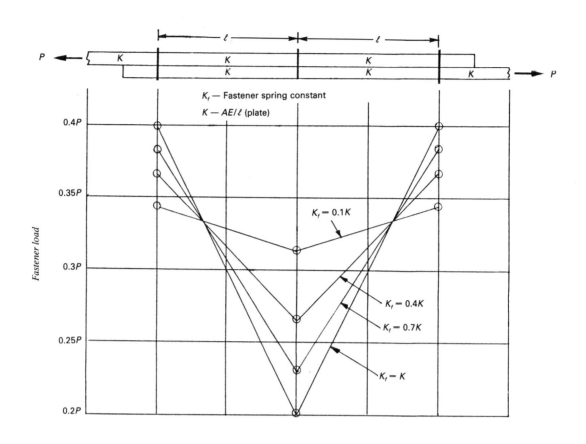

Fig. 7.7.15 Fastener load distribution of three rows of fasteners.

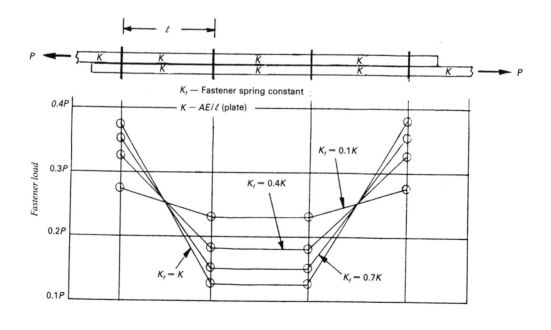

Fig. 7.7.16 Fastener load distribution of four rows of fasteners.

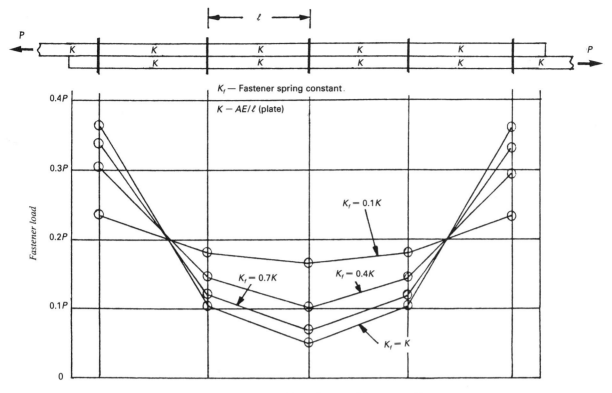

Fig. 7.7.17 Fastener load distribution of five rows of fasteners.

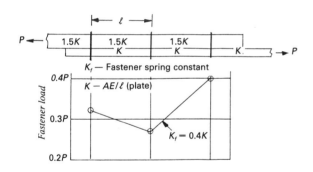

Fig. 7.7.18 Fastener load distribution of three rows of fasteners with different plate thickness.

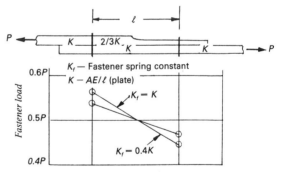

Fig. 7.7.20 Fastener load distribution of two rows of fasteners with tapered plate thickness.

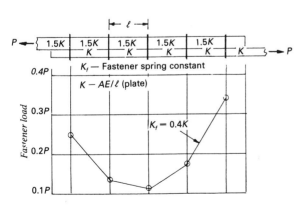

Fig. 7.7.19 Fastener load distribution of five rows of fasteners with different plate thickness.

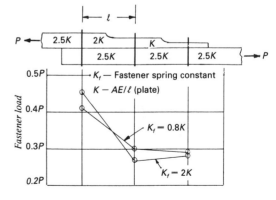

Fig. 7.7.21 Fastener load distribution of three rows of fasteners with

Airframe Structural Design 237

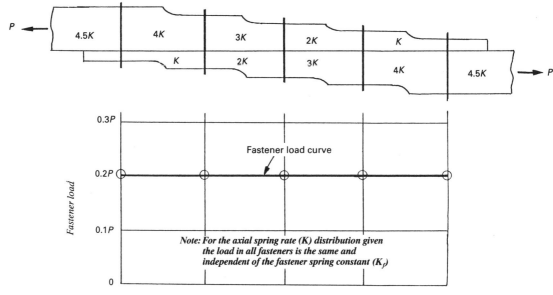

Fig. 7.7.22 Fastener load distribution of five rows of fasteners with stepped plate thickness.

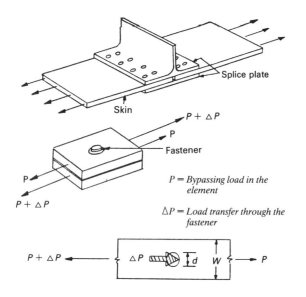

Fig. 7.7.23 Load fastener load transfer and bypass load.

Severity Factor (SF) Concept

(1) Concept (Ref. 7.7)

The severity factor SF, which is a fatigue factor that accounts for:
- Fastener type, method of installation, interference, hole preparation, etc.
- Detail design
- Fastener load distribution to avoid "peaking effect"
- Minimization of the stress concentration caused by both local load transfer at a fastener and bypass load (see Fig. 7.7.23).

(2) Local Stresses

The maximum local stress in the considered element,

$$\sigma_{max} = \sigma_1 + \sigma_2 = K_{tb}\frac{\Delta P}{dt}\theta + K_{tg}\frac{P}{wt}$$

where σ_1 = Local stresses caused by load transfer ΔP, see Fig. 7.7.24(a)
σ_2 = Local stresses caused by pypass load P, see Fig. 7.7.24(b)

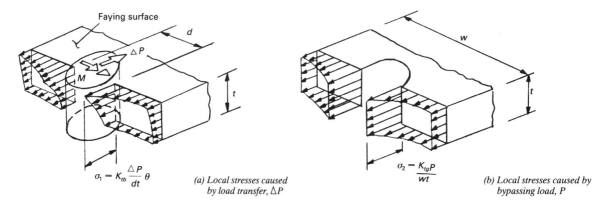

Fig. 7.7.24 Local peak stresses caused by load transfer and bypass load.

K_{tb} = Stress concentration factor, bearing stress, see Fig. 7.7.25
K_{tg} = Stress concentration factor, bypass gross area stress, see Fig. 7.7.26
θ = Bearing distribution factor, see Fig. 7.7.27

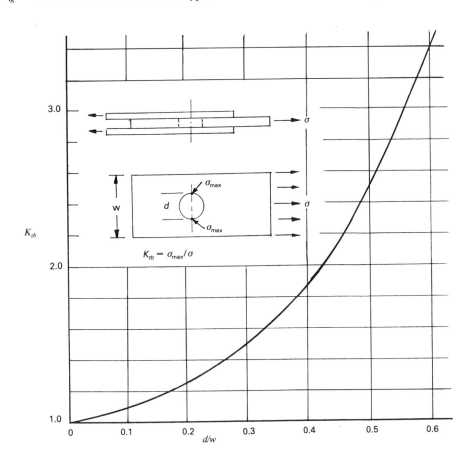

Fig. 7.7.25 *Bearing stress concentration factor — K_{tb}.*

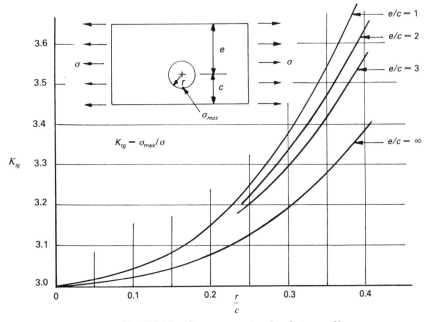

Fig. 7.7.26 *Stress concentration factor — K_{tg}.*

Airframe Structural Design 239

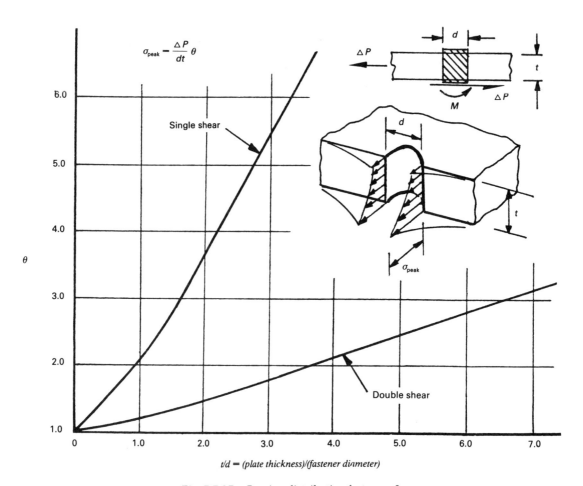

Fig. 7.7.27 Bearing distribution factor — θ

Dividing the maximum local stress by the reference gross area gives a stress concentration factor SCF for the considered element:

$$SCF = \frac{\sigma_{max}}{\sigma_{ref}}$$

As mentioned previously, the fatigue performance of fastener joint is largely influenced by fastener type and method of installation. These effects are expressed by a factor which is applied to the previously derived SCF in order to get a dimensionless factor expressing the fatigue design quality:

$$SF = \frac{\alpha\beta}{\sigma_{ref}}\left(K_{tb}\,\theta\,\frac{P}{d\,t} + K_{tg}\,\frac{P}{w\,t}\right) \quad (7.7.7)$$

where α = Hole or surface condition factor, see Fig. 7.7.28
β = Hole filling factor, see Fig. 7.7.29
(α and β are empirical factors derived from fatigue tests)

A condition of an unreinforced open hole (not cold work) in a wide sheet, the severity factor $SF = 3.0$. For a hole filled by a well driven fastener, with no load transfer, $SF < 3.0$. In case of a load transfer, usually, $SF > 3.0$.

	α
Standard hole drilled	1.0
Broached or reamed	0.9
Cold worked holes	0.7–0.8

Fig. 7.7.28 Hole condition factor — α.

	β
Open holes	1.0
Lock bolt (steel)	0.75
Rivets	0.75
Threaded bolts	0.75–0.9
Taper-Lok	0.5
Hi-Lok	0.75

Fig. 7.7.29 Hole filling factor — β.

(3) Application of Severity Factors Procedures:
- Perform a load transfer analysis.
- Evaluate α, β, θ, K_{tb} and K_{tg}.

- Calculate the severity factory *SF*.

$$SF = \frac{\alpha\beta}{\sigma_{ref.}}(\sigma_1 + \sigma_2)$$

Example I:

Consider a plate with an open hole subjected to a tensile load

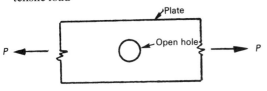

$\alpha = 1.0$ (see Fig. 7.7.28)
$\beta = 1.0$ (see Fig. 7.7.29)
$\sigma_1 = 0$ (no load transfer)

$$\sigma_2 = \frac{P}{A_{bypass}} K_{tg} = 3.0 \frac{P}{A_{bypass}}$$

Thus,

$$SF = \frac{\alpha\beta(\sigma_1 + \sigma_2)}{\sigma_{ref}}$$

$$= \frac{(1.0)(1.0)\left(0 + 3.0\frac{P}{A_{bypass}}\right)}{\frac{P}{A_{bypass}}} = 3.0$$

The result, of course, is simply the stress concentration factor for an open hole. The *SF*, therefore, may be imagined as a stress concentration factor with reference to some hypothetical stress which would exist if the entire local area of interest were replaced with a single uniform piece of structure. The *SF* is not referenced to the nominal stress at the particular hole location.

Example II:

Consider a simple lap joint

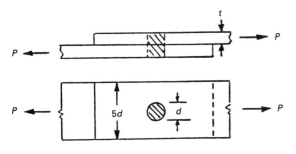

$\alpha = 0.9$ (see Fig. 7.7.28)
$\beta = 0.75$ (see Fig. 7.7.29)
$\theta = 2.0$ (see Fig. 7.7.27)
$K_{tb} = 1.26$ (see Fig. 7.7.25)

Thus

$$SF = \frac{(0.9)(0.75)}{\frac{P}{5dt}}\left[\frac{P}{dt}(2.0)(1.26) + 0\right] = 8.5$$

Because the Severity Factor is, after all, a direct measure of the severity of the state of stress at a particular location in a structural joint, relative values of Severity Factors at all fastener hole intersections in a joint give an excellent indication of which locations are the most critical from a fatigue standpoint. Fatigue tests have shown that specimens almost always fail at the location of the highest Severity Factor.

A typical joint will have many fastener hole intersections. The overriding problem in Severity Factor analysis is, therefore, that of obtaining a detailed stress distribution of the entire joint region. Once the stress distribution has been found, the Severity Factors may then be determined as per Eq. (7.7.7).

Structural Joint Life Prediction

The basic formula for predicting the fatigue life of a structural joint is that of the severity factor *SF*. In this equation θ, α and β vary with the joint material, fastener type, and installation process. For preliminary design, Fig. 7.7.27, Fig. 7.7.28 and Fig. 7.7.29 can be used to estimate the values of θ, α and β respectively. However, for detail design, coupon test specimen should be required to achieve more accurate results.

In addition, the digital computer is strongly recommended as a useful tool for analyzing member stresses and fastener load distribution in a joint. Ideally, it is desired to achieve evenly distributed loads on each fastener. This is especially important on the end fasteners as they are most sensitive in joint life analysis. Once the load distribution is obtained and θ, α and β are determined, the severity factor *SF* can be calculated. Then, knowing the defined constant amplitude loading, the predicted fatigue life of the proposed (baseline) joint can be obtained using the *SF* as calculated above in conjunction with material constant-life diagrams (or S-N curves).

Example

Given a critical support structural joint:

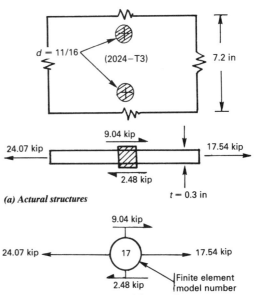

(a) Actural structures

(b) Equivalent finite element model output data

Material— 2024-T3 plate with $t = 0.3$ inch

Fastener Diameter = $\frac{11}{16}$ (or 0.6875) inch

— Threaded bolts

All loads at this joint come from the finite element model,

$$\sigma_{ref} = \frac{24.07}{7.2\,(0.3)} = 11.14 \text{ ksi}$$

$\Delta P = 9.01 - 2.48 = 6.53$ kip
$\alpha = 1.0$ (from Fig. 7.7.28)
$\beta = 0.855$ (from Fig. 7.7.29)
$K_{tb} = 1.25$ (from Fig. 7.7.25)
$K_{tg} = 3.0$ (from Fig. 7.7.26)
$\theta = 1.4$ (from Fig. 7.7.27)

From Eq. (7.7.7)

$$SF = \frac{\alpha\beta}{\sigma_{ref}}\left(K_{tb}\,\theta\frac{\Delta P}{dt} + K_{tg}\frac{P}{wt}\right)$$

$$= \frac{1.0\,(0.855)}{11.14}\left(1.25\,(1.4)\frac{6.53}{2\,(0.6875)\,(0.3)}\right.$$

$$\left. + 3.0\frac{17.54}{7.2\,(0.3)}\right)$$

$$= 0.0769\,(27.7 + 24.36)$$
$$= 4.0$$

Given structural joint fatigue loading was defined as a constant amplitude loading of 24.07 ± 24.07 kip, then

$$S_{max} = \frac{2\,(24.07)}{7.2\,(0.3)} = 22.29 \text{ ksi}$$

$$S_{min} = 0$$

$$S_a = \frac{(S_{max} - S_{min})}{2} = \frac{(22.29 - 0)}{2} = 11.145 \text{ ksi}$$

$$S_{mean} = \frac{(S_{max} + S_{min})}{2} = \frac{(22.29 + 0)}{2} = 11.145 \text{ ksi}$$

(For definitions of S_{max}, S_{min}, S_a and S_{mean}, see Chapter 15.0).

From Fig. 15.4.6 of Chapter 15.0, obtain the predicted life for this joint as 46,000 cycles.

After the predicted life of the baseline configuration is complete, a small component test specimen should be built to measure the strain gage reading at the most critical location. Each specimen should represent the typical as well as the critical part of the entire joint structure. In addition to a strain gage survey at some critical cross section, it is highly desirable to perform a stress coat, or photo stress measurement of the test specimen. A structural finite element modeling analysis should also be performed, taking into account the method of loading and supporting the test specimen.

By correlating the final analysis with the measured strain data, a high level of confidence will be obtained that an accurate stress distribution has been achieved. When the fatigue test of the specimen is completed and the test life is significantly different from the prediction, then the SF value should be revised by the following equation (also see Fig. 7.7.30).

$$K = a + b\,(SF) \quad (7.7.8)$$

where K = Fatigue quality index (see Chapter 15.0)
a and b = constants obtained from test results,
$a = 0 \sim 1.5$ and $b = 1.0 \sim 1.5$
$K = SF$, if $a = 0$ and $b = 1.0$

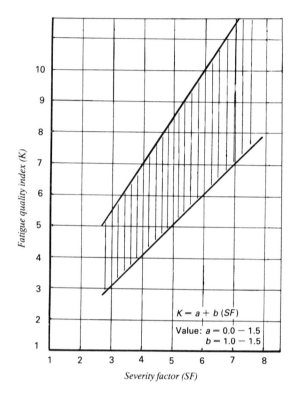

Fig. 7.7.30 Severity factor (SF) vs fatigue quality index (K).

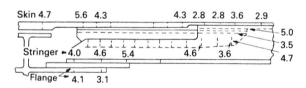

(a) Baseline design

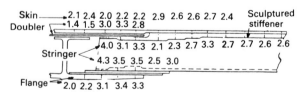

(b) Improved design

Fig. 7.7.31 SF comparison between baseline and improved design.

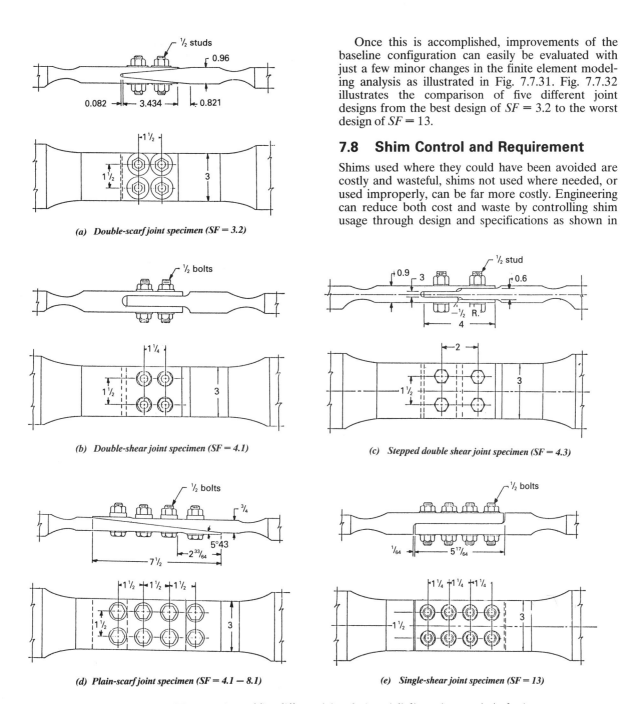

Once this is accomplished, improvements of the baseline configuration can easily be evaluated with just a few minor changes in the finite element modeling analysis as illustrated in Fig. 7.7.31. Fig. 7.7.32 illustrates the comparison of five different joint designs from the best design of $SF = 3.2$ to the worst design of $SF = 13$.

7.8 Shim Control and Requirement

Shims used where they could have been avoided are costly and wasteful, shims not used where needed, or used improperly, can be far more costly. Engineering can reduce both cost and waste by controlling shim usage through design and specifications as shown in

Fig. 7.7.32 SF comparison of five different joint designs (all dimensions are in inches).

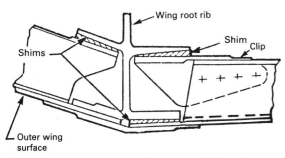

Fig. 7.8.1 Wing joint design (with shims).

Fig. 7.8.1. Another improvement design is that of the wing joint which eliminates the need for shimming as illustrated in Fig. 7.8.2.

Shims are used in aircraft production to control structural fit-up, to maintain contour or alignment, and for aesthetic purposes. In attempting to control shim usage by specification, the engineer now depends on the maximum shim thickness in the parts list and the note "shim if and as required". The mechanic must then decide when to shim, what the shim taper and thickness should be, what gap to allow, and whether the gap should be shimmed or pulled up with fasteners.

An examination of improper shimming practices reveals some unsatisfactory results. These include a great number of unnecessary shims, waste of manufacturing manhours, and increased production costs.

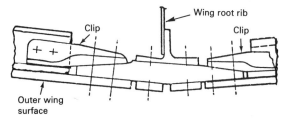

Fig. 7.8.2 *Wing joint design (without shims).*

The responsibility for shims belongs to the engineer who must evaluate each of the designs to determine proper shimming requirements. It must be specified when, where, what, and how to shim. Above all, the engineer must not delegate the responsibilities to the mechanic.

Estimates of Tension Stress from Pull-up

Each problem is unique and requires a study of deflection. Some example analyses are included in Fig. 7.8.3.

After the tension residual stress (or sustained tensile stress) is estimated, do not allow this stress to exceed the allowable (as shown in Fig. 4.7.7).

Example of pull-up

(a) Sequence of assembly may cause gap as shown

Clamping stresses (of cross-section ▨ ↕ t, 1 inch)

Case	Loading condition	Stress
(1) 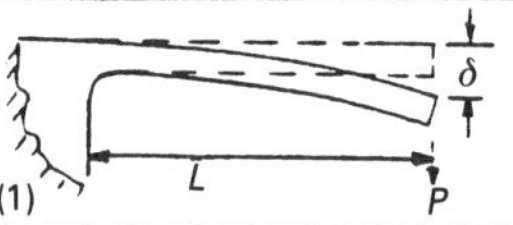	$P = \dfrac{E\delta}{4}\left(\dfrac{t}{L}\right)^3$	$f_b = \dfrac{1.5\,E\,\delta\,t}{L^2}$
(2) 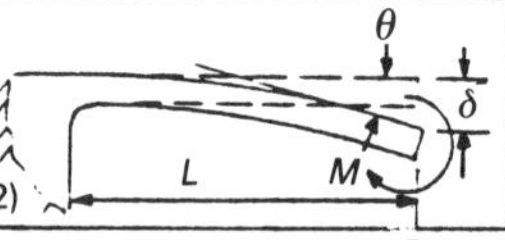	$M = \dfrac{E\,\delta\,t^3}{6L^2} - \dfrac{E\,\theta\,t^3}{12L}$	$f_b = \dfrac{E\,\delta\,t}{L^2} - \dfrac{E\,\theta\,t}{2L}$
(3) 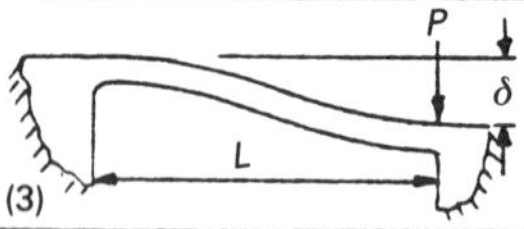	$P = E\,\delta\left(\dfrac{t}{L}\right)^3$	$f_b = \dfrac{3E\,\delta\,t}{L^2}$
(4) 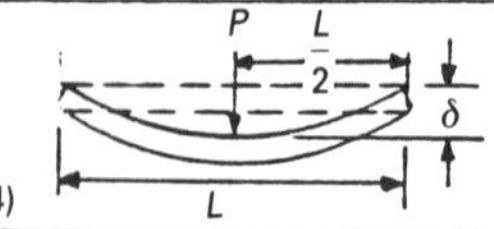	$P = 4E\,\delta\left(\dfrac{t}{L}\right)^3$	$f_b = \dfrac{6E\,\delta\,t}{L^2}$
(5) 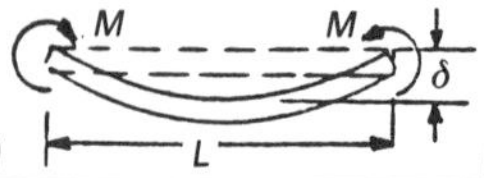	$M = \dfrac{E\,\delta\,t^3}{1.5L^2}$	$f_b = \dfrac{4E\,\delta\,t}{L^2}$
(6) 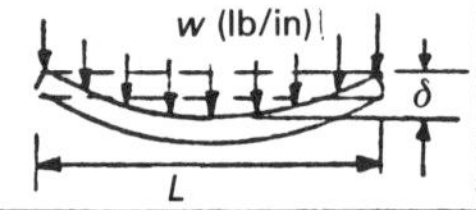	$w = \dfrac{6.4E\,\delta\,t^3}{L^4}$	$f_b = \dfrac{4.8E\,\delta\,t}{L^2}$
(7) 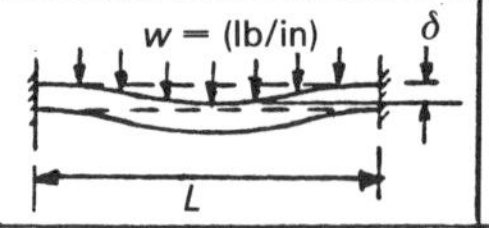	$w = \dfrac{32E\,\delta\,t^3}{L^4}$	$f_b = \dfrac{16E\,\delta\,t}{L^2}$
(8) 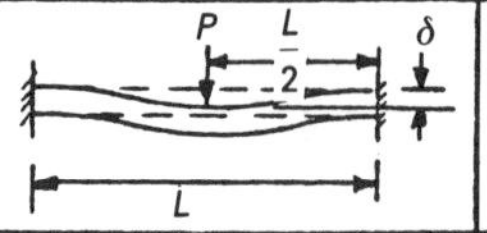	$P = 16E\,\delta\left(\dfrac{t}{L}\right)^3$	$f_b = \dfrac{12E\,\delta\,t}{L^2}$
(9) 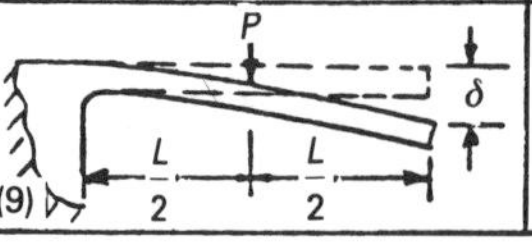	$P = \dfrac{4E\,\delta}{5}\left(\dfrac{t}{L}\right)^3$	$f_b = \dfrac{2.4E\,\delta\,t}{L^2}$

Fig. 7.8.3 *Pull-up (or clamp-up) stresses.*

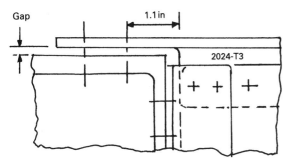

(b) Define maximum gap for pull-up
- Maximum allowable tensile sustained stress F_t = 50 ksi (refer to Fig. 4.7.7)
- Maximum pull-up gap [see case (3) of Fig. 7.8.3]

$$\delta = \frac{F_t L^2}{3\,Et} = \frac{50\,(1.1)^2}{3\,(10.4)\,(10^3)\,(0.1)} = 0.02 \text{ in (gap)}$$

Fatigue Consideration

Fatigue performance depends on alternating stresses and mean stresses. While pull up rarely affects the alternating stress in a given part, it will increase the mean stress. The significance of this increase is illustrated in Fig. 7.8.4. The tension residual stress due to pull-up is 60,000 psi at point A. The normal operating mean stress has been established at 45,000 psi from design condition. When the residual stress is added to the design mean stress and evaluated on the material S-N curve, the part life is significantly reduced at all alternating stress levels as illustrated in Fig. 7.8.5.

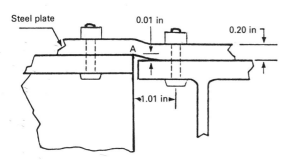

Fig. 7.8.4 Pull-up stress due to the mismatch between two structures.

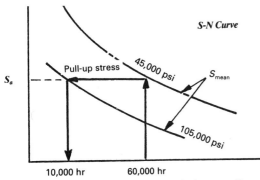

Fig. 7.8.5 Reduction of structural life due to pull-up stress.

How to Minimize Shims

To minimize shims the engineer shall:
(a) Calculate part tolerances
(b) Check sequence of assembly
(c) Define max. gap for pull-up
(d) minimize stress concentration
(e) Use materials which exhibit good stress corrosion resistance
(f) Do not allow residual tensile stresses to exceed the allowables specified
(g) Use stress relief heat treating and surface cold working (such as shot peening) to remove residual surface stresses. Removal of residual tension stresses is especially important for high heat treat steel and aluminium alloys
(h) Use control of tolerances to reduce assembly fit-up stresses
(i) Mimimize stresses in the short-transverse plane
(j) Necessary shims:
- Specify shim material and size.
- If taper is required, minimum shim thickness should not be less than 0.03 inch.
- Specify structural shim or non-structural shim, see Fig. 7.8.6.

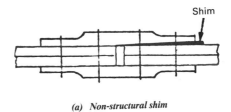

(a) Non-structural shim

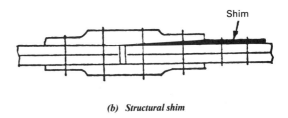

(b) Structural shim

Fig. 7.8.6 Differences between structural and non-structural shims.

References

7.1. Ritchie, Oscar: 'Design Trade-offs that Determine Fastener Selection.' *SAE paper No. 670886. SAE Transactions*, (Vol. 76, 1967), 2801-2807.

7.2. Ramsey, C.L. and Ingram, Jr., J.C.: 'Structural Fast-

7.3. Hills, J.F.: 'Advantages of Some New Fastener Systems.' *SAE paper No. 680206.*
7.4. Smith, C.R.: 'Effective Stress Concentrations for Fillets in Landed Structures.' *Experimental Mechanics*, (Apr. 1971), 167–171.
7.5. Melcon, M.A. and Hoblit, F.M.: 'Developments in the Analysis of Lugs and Shear Pins.' *Product Engineering*, (June, 1953), pp. 160–170.
7.6. MIL-HDBK-5D: *Metallic Materials and Elements for Flight Vehicle Structures.* U.S. Government Printing Office, Washington, D.C. 1983.
7.7. Jarfall, L.E.: *Optimum Design of Joints: The Stress Severity Factor Concept.* The Aeronautical Research Institute of Sweden, 1967.
7.8. Tate, M.B.: 'Preliminary Investigation of The Loads Carried by Individual Bolted Joints.' *NACA TN 1051*, (1946).
7.9. McCombs, W.F.: 'Analytical Design Methods for Aircraft Structural Joints.' *AFFDL – TR-67-184*, (1968).
7.10. Peterson, R.E.: *Stress Concentration Design Factors.* John Wiley and Sons, 1953.
7.11. Hartmann, C.C.: 'Static and Fatigue Strengths of High-Strength Aluminum-Alloy Bolted Joints.' *NACA TN 2276*, (1951).
7.12. Rosenfeld, S.J.: 'Analytical and Experimental Investigation of Bolted Joints.' *NACA TN 1458*, (1947).
7.13. Grover, H.J.: 'Fatigue of Aircraft Structures.' *NAVAIR 01-1A-13*, (1966).
7.14. Anon.: *Fatigue and Stress Corrosion Manual for Designers.* Lockheed-California Company.
7.15. Deneff, G.V.: 'Fatigue Prediction Study.' *WADD TR 61-153*, (1962).
7.16. Niu, Michael Chun-yung (牛春匀): 'L-1011 Fastener Handbook.' *Lockheed-California Company. Report No. CER 51-013*, (Jan. 24, 1973).
7.17. Anon.: 'Fastener Technology Catches Up.' *Aviation Week & Space Technology*, (Dec. 14, 1987), 89-93.
7.18. Bruhn, E.F.: *Analysis and Design of Flight Vehicle Structures.* Tri-State Offset Company, Cincinnati, Ohio 45202. 1965.
7.19. Niu, Michael Chun-Yung (牛春匀): *Structural Joint Analysis Handbook (Fatigue).* Lockheed-California company, 1973. (unpublished).
7.20. HANDBOOK H28: 'Screw-Thread Standards for Federal Services (Part I, II, & III).' *U.S. Department of Commerce, National Bureau of Standards (1957).*
7.21. Sines, G. and Waisman, J.L.: *Metal Fatigue.* McGraw-Hill Book Company, Inc., New York, N.Y. 1959.
7.22. Kimball, D.W. and Barr, J.H.: *Elements of Machine Design.* John Wiley and Sons, Inc. new York, N.Y. 1935.
7.23. Kuenzi, E.W.: 'Determination of Mechanical Properties of Adhesives for Use in Design of Bonded Joints.' *U.S. Dept. Agriculture Forest Service, Prod. Lab., Report No. 1851*, (1957).
7.24. MIL-A-5090B: *Adhesive: Airframe Structural; Metal to Metal.* Government Printing Office, Washington, D.C. 1955.
7.25. McClure, J.G.: 'The Prestressed Bolt.' *Machine Design* (Sept. 15, 1960).
7.26. Stewart, W.C.: *Bolted Joints – ASME Handbook.* McGraw-Hill Book Company, N.Y.
7.27. Anon.: 'Fasteners Reference Issue.' *Machine Design*, (1987).
7.28. Parmley, R.O.: *Standard Handbook of Fastening and Joining.* McGraw-Hill Book Company.
7.29. Anon.: *Assembly Engineering.* A Hitchcock Publication, Wheaton, Ill.60187.
7.30. Cole, R.T.: 'Fasteners for Composite Structures.' *Composite*, (July, 1982).
7.31. Phillips, J.L.: 'Fastening Composite Structures with HUCK Fasteners.' *Technical Paper of HUCK Manufacturing Co.* (1984).
7.32. Phillips, J.L.: 'Fatigue Improvement by Sleeve Coldworking.' *SAE Paper No. 730905. SAE Transactions*, (Vol. 82. 1973), 2995–3011.
7.33. Speakman, E.R.: 'Fatigue Life Improvement Through Stress Coninning Methods.' *Douglas Paper No. 5506. Douglas Aircraft Company*, (1969).
7.34. Khol. R.: 'Fasteners That Fight Fatigue.' *Machine Design*, (Feb. 20, 1975).
7.35. Boucher, R.C.: 'Table Speeds Calculation of Strength of Threads.' *Product Engineering*, (Nov. 27, 1961).
7.36. Baumgartner, T.C. and Kull, F.R.: 'Determining Preload in a Bolted Joint.' *Machine Design*, (Feb. 13, 1964).
7.37. Anon.: *SPS Threaded Fasteners – Section I, Reference Guide to Bolts and Screws.* Standard Pressed Steel Co., Santa Ana, Calif. 92702. 1966.
7.38. Anon.: *SPS Threaded Fasteners – Section II, Reference Guide to Self-Locking Nuts.* Standard Pressed Steel Co., Santa Ana, Calif. 92702. 1966.
7.39. Hopper, A.G. and Thompson, G.V.: 'How to calculate and Design for Stress in Preloaded Bolts.' *Product Engineering*, (Sept. 14, 1964).
7.40. Cobb, B.J.: 'Torque and Strength Requirements for Preloading of Bolts.' *Product Engineering*, (Aug. 19, 1963).
7.41. Leftheris, B.P.: *Stress Wave Riveting.* Grumman Aerospace Corp., Bethpage, new York 11714. 1972.
7.42. Cornford, A.S.: 'Bolt Preload – How can you be sure it's right?' *Machine Design*, (Mar. 6, 1975).
7.43. Dann, R.T.: 'How Much Preload for Fasteners;.' *Machine Design*, (Aug. 21, 1975).
7.44. Whaley, R.E.: 'Stress-Concentration Factors for Countersunk Holes.' *Experimental Mechanics*, (Aug. 1965).
7.45. Harris, H.G. and Ojalvo, I.U.: *Simplified Three-Dimensional Analysis of Mechanically Fastened Joints.* Grumann Aerospace Corp., Bethpage, New York 11714. 1974.
7.46. Thrall, Jr. E.W.: 'Fatigue Life of Thick-Skinned Tension Joints.' *Aeronautical Engineering Review*, (Nov., 1953), 37–45.
7.47. Brilmyer, H.G.: 'Fatigue Analysis of Aircraft Bolts.' *Aeronautical Engineering Review*, (July, 1955), 48–54.
7.48. Anon.: *Adhesive Bonding ALCOA Aluminum.* Aluminum Company of America, Pittsburgh, Pa. 1967.
7.49. Trembley, W.H.: 'Fastening Analysis to Optimized Design.' *SAE Paper No. 730309. SAE Transactions*, (Vol. 82, 1973), 1066–1074.
7.50. Silha, C.W. and Schwenk, W.: 'Fasteners with a Memory.' *SAE Paper No. 730900. SAE Transactions*, (Vol. 82, 1973), 2982–2987.
7.51. Gill, F.L.: 'A Corrosion Inhibiting Coating for Structural Airframe Fasteners.' *SAE Paper No. 730902. SAE Transactions*, (Vol. 82, 1973), 2988–2994.
7.52. Robinson. G.I.: 'Influences of Grain Flow on the Strength of Lugs.' *Aircraft Engineering*, (Sept., 1951), 257–260.
7.53. Anon.: *Shur-Lok Specialty Fasteners for Industry, (9th edition)* Shur-Lok Corp., P.O. Box 19584, Irvine, CA. 92713.

CHAPTER 8.0

WING BOX STRUCTURE

8.1 Introduction

The purpose of this chapter is to explain the basic principles of wing design that can be applied to any conventional airplane such as shown in Fig. 8.1.1

It will be noted that any wing requires longitudinal (lengthwise with the wing) members to withstand the bending moment which are greatest during flight and upon landing. This is particularly true of the cantilever wings, which are normally employed for high-performance aircraft. Light aircraft often have external struts for wing bracing, and these do not require the type of structure needed for the cantilever wing as shown in Fig. 8.1.2.

The outline of the wing, both in planform and in the cross-sectional shape, must be suitable for housing a structure which is capable of doing its job. As soon as the basic wing shape has been decided, a preliminary layout of the wing structure must be indicated to a sufficient strength, stiffness, and light weight structure with a minimum of manufacturing problems.

There are several types of wing structure for modern high speed airplanes; thick box beam structure (usually built up with two or three spars for high aspect-ratio wings as shown in Fig. 8.1.1(a)), multi-spar box structure for lower aspect-ratio wings with thin wing airfoil as shown in Fig. 8.1.1(b), and delta wing box as shown in Fig. 8.1.3.

The wing is essentially a beam which transmits and gathers all of the applied airload to the central attachment to the fuselage. For preliminary structural sizing

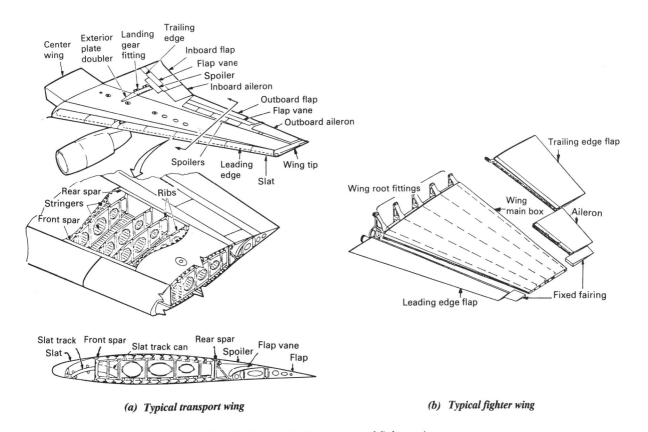

Fig. 8.1.1 Typical transport and fighter wing.

and load purposes it is generally assumed that the total wing load equals the weight of the aircraft times the limit load factor times a safety factor of 1.5. In addition to this applied load, other loads that may also be applied to the wing may include:
- Internal fuel pressure (static & dynamic) which may influence the structure design
- Landing gear attachment loads

Fig. 8.1.2   Wing with external struts.

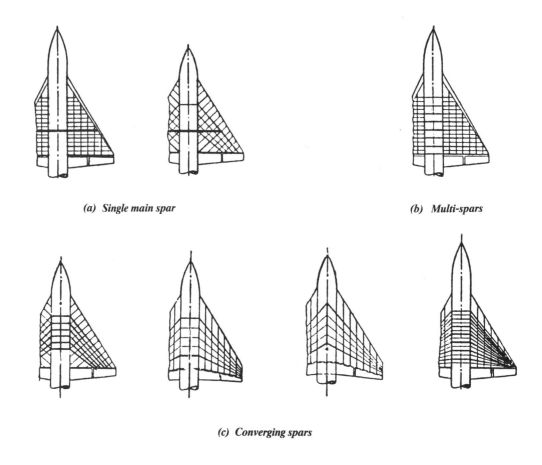

(a) Single main spar

(b) Multi-spars

(c) Converging spars

Fig. 8.1.3   Several structural arrangements for delta wing box.

248   Airframe Structural Design

- Wing leading and trailing loads

These are generally secondary loads in wing design, the primary loads resulting from the applied airload. The local concentration of these loads may however require a rib to distribute the load to the overall structure. The applied airloads result in increasing shear and bending moments toward the wing root with the shear carried by the wing spars and the bending moment by the wing covers. Rather than referring to bending moment what is generally defined as cover load N_x, the load per inch measured along the chord line. If this load is divided by the thickness of the cover skin the result is the average stress of the cover at that point.

Since the covers typically represent fifty to seventy percent of the structural weight of the wing, it is imperative that the covers be designed as efficiently as possible.

Since the lower cover is loaded primarily in tension, its design is fairly straight forward. It requires careful material selection in order to assure fairly high tensile strength to density ratio combined with good fracture toughness and fatigue life. Certain aluminum alloys such as 2024-T3 and the newer alloys such as 7475-T7351 are excellent candidates along with most of the titanium alloys such as Ti-6A1-4V.

An additional consideration of tension cover design is improving the fatigue strength by utilizing interference fit fasteners. In this process a fastener is installed in a hole that is several thousandths of an inch, typically 0.003 inch, smaller than the fastener diameter. This produces radial compression and tangential tension stresses at the edge of the hole. Since the tangential tension stresses are larger than the stress produced by most of the applied loads, the edge of the hole sees less stress cycling and therefore a lower effective stress concentration resulting in increased fatigue life.

The upper cover optimum design is far more complex and configuration dependent. Since the upper cover is loaded primarily in compression, its design efficiency is dictated primarily by how well it can be stabilized, that is, prevented from buckling.

In order to enforce a mode requires that the cover be supported and restrained from moving up or down at the particular location. Many techniques are available to accomplish this and will be discussed in later sections of this chapter. The selection of the optimum cover stabilization technique is very configuration dependent. For thin wings, multi spar and full depth honeycomb tend to be the lowest weight construction. For deeper wings, wing cover with skin-stringer panel become attractive but ribs have to be spaced closely enough to prevent the stiffeners from failing as a column.

Variable Swept Wing
In choosing a variable swept wing, the designers endeavour to make the aircraft as adaptable as possible to the varying flow conditions encountered during the operation of supersonic flight. The aerodynamic concept of an aircraft of this type has to take into account both the requirements for economic cruise at Mach 2−3 and those for slow flight, particularly during the take-off and landing phases. In order to obtain reasonably satisfactory slow flight characteristics, the wing structure could be incorporated by adding various high lift devices and airbrakes or spoilers. Fig. 8.1.4 shows the outer swept wings are mated to fuselage by titanium pins (one per wing). Pins are first cold soaked in liquid nitrogen at −320°F for 3 hours to provide a shrink fit.

Fig. 8.1.4 B-1 bomber titanium pivot pin (38 in. long & 18 in. dia.)

From the viewpoint of structural strength, variable swept wing has disadvantages. The transmission of forces and moment at the stub wings, as shown in Fig. 8.1.5, requires care in designing such elements as the sweeping mechanism, and gives a weight and space penalty which adversely affects the fuel capacity. An absolutely reliable and synchronized operation of the sweeping mechanism is another prerequisite for the maintenance of stability over the whole speed range, especially for safety during landing. The system is for this reason required to be fail-safe. Certain difficulties also arise in routing the linkages for the flight controls and fuel supply lines, as it is indispensable to utilize the great capacity of the outer wing for fuel stowage. While, with a backwards rotation of the wing, the aircraft's center of gravity moves towards the rear, the same applies to the aerodynamic center, so that a part of the change of longitudinal moment produced by the movement of the center of gravity is compensated. Still more pronounced is the backwards movement of the aerodynamic center with increase of Mach number, which acts in the same direction. It is, therefore, only necessary for the resulting residual moment, which in every case is smaller than the largest of the above mentioned additional moments, to be trimmed.

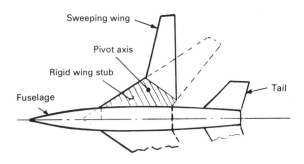

Fig. 8.1.5 *Sweeping wing location.*

One of the most demanding tasks is the design of the wing pivots. These have to take the entire loads imposed on them by the outer wing panels: drag, bending and torsion. See Section 8.7.

As with strength calculations, a pivot does not introduce new factors which are so unusual as to render existing methods of calculating stiffness inapplicable. Unlike the localized effects of the discontinuity at the bearing on strength, its stiffness — like that of any root attachment — does have an overall effect on deflections and this has to be accommodated in static and dynamic aeroelastic calculations. The stiffness of "massive" attachments is not always easy to calculate reliably, and early tests are therefore desirable. It is normal practice in flutter calculations to vary important parameters, and the stiffness contribution of the pivot is no exception.

It is also current practice to check at an early stage not only flutter speeds but also modal damping, which is more indicative of conditions applying within the flight envelope rather than beyond it. As soon as a design is sufficiently defined, the response of the whole aircraft is evaluated over a practical range of frequencies to check the level of accelerations induced by turbulent air. This procedure not only checks crew comfort, but confirms the vibration spectra for equipment and fatigue life of structure.

In order to achieve versatile performance, wing drag may be minimized for high-speed flight by having a small wing, a high wing loading being met by high-lift devices as mentioned previously to compensate and give good take-off, approach and landing characteristics. At intermediate speeds, however, where the wing is still unswivelled and very high normal acceleration required, the high-lift devices may be developed for practical setting.

This introduces new loading cases, and flutter calculations will have to take account of attachment stiffness and derivatives appropriate to the extended position at higher speeds than usual for flaps. The mode of usage will have to be taken in the fatigue calculations on associated local structure.

A broad study of the structural problems associated with variable swept wings for aircraft has been discussed in Section 8.7 of this chapter.

Advanced Airfoils

Advanced transport technology studies show that supercritical airfoils can provide greater gains by increasing airfoil thickness and/or decreasing wing sweep at the same cruise Mach number, rather than by increasing cruise speed. Although increases in wing depth alone of approximately 30% can be used to reduce wing weight, it has been determined that the greatest benefit is achieved by a combination of increased depth, reduced sweep, and increased aspect ratio. The airfoil difference between a conventional and supercritical wing is shown in Fig. 8.1.6. The disadvantages of supercritical wing are:
- The incompatibility of the sharply "undercut" trailing edge with extensive flaps.
- The extremely close tolerances needed to maintain laminar flow.

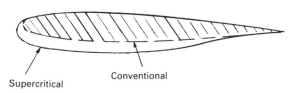

Fig. 8.1.6 *Conventional and supercritical airfoil comparison.*

Another possibility also is under study at NASA. This is tailoring the flexibility of the wing so that aerodynamic loads can be used to flex it to the proper size and shape for the performance envelope in which it is operating. The wing might be structured, for example, so that it normally assumes the proper aerodynamic shape for take-off and landing. As the aircraft becomes airborne, the aerodynamic forces on the wing would be sufficient to flex it to a shape best suited for cruise. Such a wing could eliminate, at least in part, the use of mechanical flaps, slats, spoilers or other high-lift devices in aircraft. This peculiar wing structural design is used on the air superiority fighter.

Brief Summary of Wing Loads (reference only)
(1) General
- Positive high angle of attack (+HAA)
- Negative high angle attack (−HAA)
- Positive low angle of attack (+LAA)
- Negative low angle of attack (−LAA)
- Dive maneuvers
- Flaps down maneuver — takeoff configuration
- Flaps down maneuver — landing configuration
- Taxiing
- Jacking
- Maneuver with certain wing fuel tanks empty
- Flutter
- Control surface reversal
- Roll initiation
- Unsymmetrical spanwise lift distribution
- Fatigue
- Fail-safe
- Fuel vapor or refueling pressures

- Thermal gradients
- Lightning strike

(2) Spar conditions
  - Fuel slosh
  - Fuel head — crash conditions
  - Concentrated shear loads
  - Fuselage pressure in center section

(3) Rib conditions
  - Rib crushing
  - Concentrated load redistribution
  - Fuel slosh
  - Fuel head
  - Wing cover stabilization
  - Sonic fatigue

(4) Leading edge conditions
  - Hail strike
  - Thermal anti-icing
  - Duct rupture
  - Sonic fatigue — engine reverse thrust

(5) Trailing edge and fairing load conditions
  - Sonic fatigue
  - Buffet
  - Slosh and gravel impact
  - Minimum gage
  - Positive and negative normal force pressures

(6) Other special conditions for military aircraft (see Ref. 3.4).

8.2 Wing Box Design

It appears that the primary structural design problem is one of general structural layout — first, whether a large percentage of the wing bending shall be carried by the spars, or whether the cover should be utilized to a large extent; and, second, in which direction should be primary wing ribs run — along the flight path, or normal to the rear spar in the wing?

Regarding the first, it is fairly obvious that the cover should be utilized for a large percentage of the bending material. This is true, since it appears that torsional rigidity is required and, since it is, this same torsion material may as well be used for both primary bending and torsion material. Spanwise stiffeners spaces fairly close together are, as a consequence, required to keep the buckling of the bending material down to a minimum.

In consideration of the direction of wing ribs, Fig. 8.2.1(a) shows the somewhat conventional structure; Fig. 8.2.1(b) shows the wing ribs parallel to the flight path. It may be noted here that some opinions hold it necessary to have the wing ribs parallel to the flight path in order to insure a smooth aerodynamic shape between the spars (assuming a two-spar wing). This latter arrangement seems to have too many disadvantages to be structurally sound and, further, if spanwise stringers are utilized between the spars, then the rib riveting will not particularly further aggravate the aerodynamic contour because a large amount of riveting is already required for the spanwise stiffeners. For the sample illustration chosen, the total rib length is 28% longer for the wing with the ribs parallel to the flight path, with corresponding weight loss.

Some of the manufacturing problems that exist with the sweptback wing are:
- Bending the spar caps is difficult.
- The skin gages required are extremely thick because the skin is of a necessity a large part of the bending material. These thick gages will probably require multiple brake operations, rather than hydropresses, because of huge capacity presses needed.
- Angles of 90° in jigs, fixtures, bulkheads, and spar webs are important to the workman. Any variance to 90° imposes a hardship.

The triangular section A, shown in Fig. 8.2.2, is indeterminate. It is noted that the torsion, shear, and bending are perfectly stable without the skin indicated area as A. Further, it is perfectly stable to cut out all attachments and utilize only the skin to take out the torsion. Therefore, if both forms of structure are present, a consistent deformation analysis is required to determine the percentage of torque-carrying for each structure. For the center-of-pressure aft, the primary torque and the primary bending produce additive torsions which make this condition critical for torque in the wing root rib bulkhead.

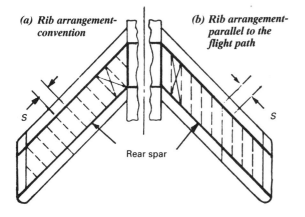

Fig. 8.2.1 Comparison of rib direction (rectangular box).

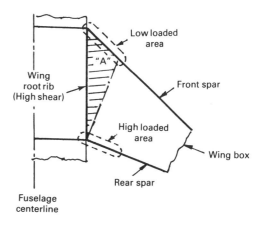

Fig. 8.2.2 Wing root load distribution problem of swept wing (a high indeterminate structure).

In the initial weight estimate of the aircraft, the weight of the wing root rib bulkhead and fuselage must not be neglected.

Another design method is by using a single main beam approach as illustrated in Fig. 8.2.3, which is practical to apply on very high swept wings with a relatively thick wing box. It is simple and it works.

The fore and aft spar locations as shown in Fig. 8.2.4 are approximate locations early in the design during layout of high lift devices. The design of the wing body joint, and development and sizing of the hydraulic components, control components, and electrical systems may require changing spar locations as design progresses. However, firm spar locations must be established very early in the design and preferably by the time the final mathematically defined loft are available. In any case, both are required before final layouts and drawings can be started.

The rear spar must be located at a suitable chordwise station, leaving sufficient space for the flaps and for housing the controls to operate the flaps, ailerons and spoilers. A rearward shift of this spar increases the cross-sectional area of the torsion box (and incidentally the fuel storage space) but the reduction in the sectional height will make it less efficient in bending. Similar criteria apply to the front spar when it is moved forward. It is noted that the best flap chord for simple plain and split flaps is about 25% of the wing chord, but highly efficient slotted flap systems are more effective with flap chords of up to 35% or even 40% of the wing chord. In general, the front spar is located at about 15% chord, the rear spar at 55 to 60%. About 5 to 10% chord should be available between the nested flap and the rear spar for control system elements. The central part of the wing, bounded by the front and rear spars, takes the loads from the nose and rear sections and carries them to the fuselage, together with its own loads. Primary wing structure of transport aircraft is in effect a leak-proof, integral fuel tank, the arrangement of which in the spanwise direction is dictated by considerations of balancing the aircraft for various fuel loads. Center tanks should be avoided from the outset, although for long-range aircraft they are more or less essential.

Fig. 8.2.5 shows a preliminary view of spars and maximum wing thicknesses. In conjunction with a preliminary cubic mathematically defined loft, these are used for fuel quantities and management design (locations of end ribs), and to establish the torsional and bending material of the wing box.

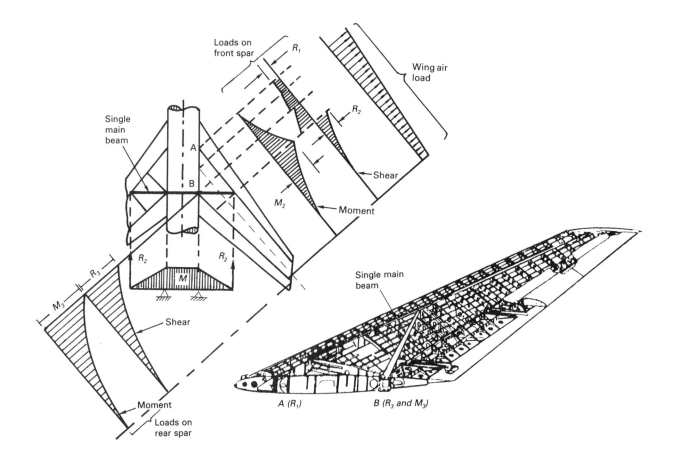

Fig. 8.2.3 Single beam design for very high swept wing.

252 Airframe Structural Design

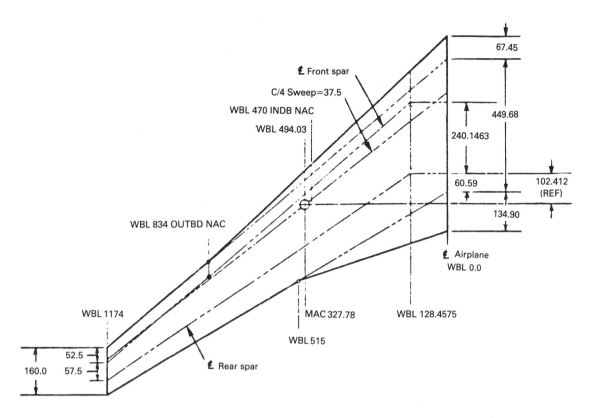

Fig. 8.2.4 Wing plan view layout of a transport.

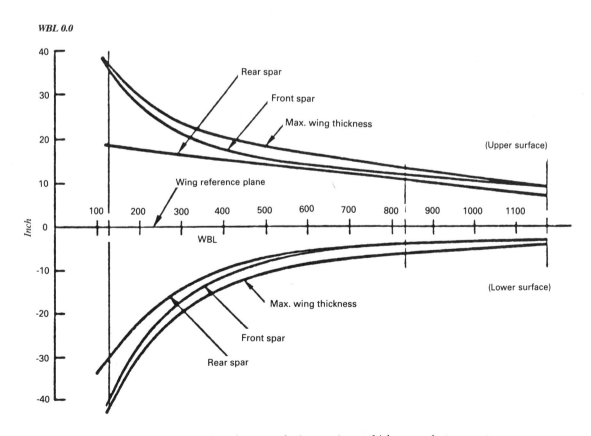

Fig. 8.2.5 Preliminary view of spars and wing maximum thicknesses of a transport.

Airframe Structural Design 253

The final wing loft is established in two stages. The high speed loft is finalized first. The lines from this loft are then examined and modified for producibility, particularly in the leading edge and trailing edge. As an example, the contour in the vicinity of the wing spoilers could be modified so that the spoilers are interchangeable with one another. Also, the left hand spoilers are interchangeable with the right hand spoilers. The inboard spoilers and the leading edge flaps were developed the same way.

Wing Layout

Since the wing design has to allow for so many factors — planform, spar and stringer location, landing-gear attachment and retraction, power plant, ailerons, flaps, and a host of others — it is desirable to make preliminary studies to make sure that every design feature has been properly incorporated. The following recommendations are for taking full advantage. The instructions are for a two-spar wing and may be modified for any other type of design.

- Draw planform of wing with the necessary dimensions, to scale, to satisfy aspect ratio, area, and sweepback.
- Determine the mean geometric chord and check that the relation of the wing to the fuselage is such that the center of gravity lies in the lateral plane perpendicular to the mean geometric chord at the mean aerodynamic center.
- Locate the front spar at a constant percentage of the chord, from root to tip. The front spar is located at between 12 to 17% of the chord. Note that the constant percentage line of the chord may not parallel the leading edge of the wing. Indicate the spar location by its center line.
- Locate the rear spar similarly. The rear spar is located from 55 to 60% of the chord — usually 60% to accommodate a 30% chord aileron. Neither the front nor rear spar needs to be extended to the extreme wing tip, since the extreme wing-tip structure is inherently rigid and capable of transmitting tip loads to the spars and adjacent structure.
- Mark out the aileron. The leading edge of the aileron may be parallel to the rear spar. If the rear spar is located at 60% of the chord, then the aileron chord should not exceed about 30% of the chord, since some allowance must be made for rear spar cap width, aileron gap, space for control systems, etc.
- If a flap is used for a high lift device, it may extend the entire flap distance inboard of the aileron. Here, some additional study may be necessary if a considerable flap area is desired. In that case, the aileron chord might be increased, even if such an aileron has some adverse characteristics. Increasing the aileron chord may necessitate moving the rear spar slightly forward to give sufficient clearances. If the flap chord is less than the aileron chord, an auxiliary spar is needed to support the flap hinges, and may have to be added to the wing structure.
- The wing rib spacing may be spotted in next. Ribs are likely to be located at each aileron and flap hinge. The rib spacing is determined from panel-size considerations, to which reference should be made. Some adjustments in the rib spacing may be desirable to get hinge-rib locations to coincide with the rib stations. Reinforced ribs are also called for engine-mount attachments, landing-gear attachments, and fuel-tank or external store supports. Also, such ribs may suggest relocations of the other ribs in order to obtain a more pleasing pattern.
- Spanwise stringers may be located. These may be placed parallel to each other or at constant percentages of the wing chord. These spanwise stringers are not normally carried out to the tip, but are rather discontinued at intervals inboard of the tip so that fewer and fewer stringers are left from the mid-span outboard.
- The main elements of the wing structure have now been located. If the layout is considered satisfactory, other details may now be added. Some of them may cause reconsideration of the structure locally. For example, the wheel well for the retraction of the landing gear may necessitate some re-designing.

The layout outlined above is shown primarily in outline form by means of centerlines, with subsidiary sketches showing structural details. Until the stress analysis is completed, such dimensions as thicknesses and limiting dimensions should be omitted. After the stress analysis is completed, a more detailed drawing is possible.

Lightning Strike Consideration

Airplanes flying in and around thunderstorms are often subjected to direct lightning strikes, and also to nearby lightning discharges that produce corona and streamer formations. The possibility that lightning may ignite fuel vapors has been strongly suspected in at least several transport airplane accidents. The surveys show a lightning strike frequency of one strike each 2000 hours of flights operation.

(1) Lightning Strike Zone

It is convenient, for the purpose of the discussion that follows, to define three zones (on the external surface of the airplane), each having a different lightning strike probability (see Fig. 8.2.6).

(a) Zone 1:
Surfaces of the aircraft for which there is a high probability of direct stroke attachment. These areas are:
- All surfaces of the wing tips located approximately within 18 inches of the tip (measured parallel to the lateral axis of the aircraft)
- Forward projections such as engine nacelles, external fuel tanks, propeller disks
- Any other projecting part that might constitute a point of direct stroke attachment

(b) Zone 2:
Surfaces for which there is a probability of strokes being swept rearward from a Zone 1 point of direct stroke attachment (such as engine nacelles, propellers, etc). Zone 2 extends approximately 18 inches laterally to each side of fore and aft lines passing through the Zone 1 forward projections points of stroke attachment.

(c) Zone 3:
Surfaces for which there is a low probability

of either direct or swept strokes. Ignition sources in these areas would exist only in the event of streamering. This zone includes all surfaces of the aircraft not coming under the definitions for Zones 1 and 2.

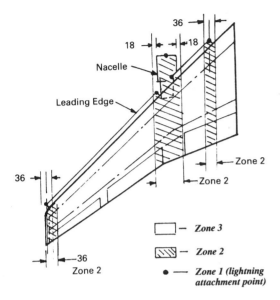

Fig. 8.2.6 *Lightning zones of a transport wing. (Dimensions in inch)*

Lightning strike damage is generally confined to the tip areas and will result in burn through damage if the cover thicknesses are inadequate or other preventions such as diverter strips are not used. Since a wing box generally contains fuel (i.e. transports and advanced fighters) and should consider:
(a) Flammable mixtures are likely to exist in any part of the tank and venting system
(b) Certain types of vent outlet designs may be susceptable to ignition of vapors by either streamering or direct strokes
(c) Stroke attachment to semi-insulated parts of the fuel tank can cause sparking on the inside of the tank with sufficient energy to ignite flammable vapors.

The final item (c) is the primary concern in the integral fuel tank. Components such as access doors and filler necks must be designed with care. Paint films, anodizing, and other corrosion protection treatments usually serve as electrical insulators. Two basic solutions to prevent internal arcing and sparking have been used:
(a) to provide a continuous electrical contact around the entire periphery of the part; or
(b) to design the part in such a way that any arcing or sparking that might occur would take place on the outside of the fuel-containing wing box rather than on the inside.

The (a) approach is the best. Any other approach is guesswork and requires extensive testing. Do not hesitate to investigate by testing if any doubt at all exists. A review of fastener patterns, method of installation (that is, with wet primer as sealant), and joints between structural members should be a continuing procedure to ensure against inadvertent insulation and possible arcing.

Resistance to Lightning Strike Skin Penetration

The forward motion of an airplane causes lightning strokes to bounce across a surface from fore to aft. A stroke hitting a leading edge tip, the corner of a leading edge slat, or the leading edge of the engine cowl will bounce aft in Zones 1 and 2, hitting the wing several times. Enough conductivity must be available to carry off the charge and keep resistance heating below the ignition point of potential flammable mixtures in the integral fuel tank area. A 0.080 inch gage aluminum alloy or equivalent is adequate to carry off the charge without heating to a potential fuel vapor ignition point.

Composites are being used extensively in limited leading edge and trailing edge minimum skin thickness areas. This presents a problem to electrical wiring installation in these areas. An electromagnetic field around a structure results, as shown in Fig. 8.2.7, from the flow of current from a lightning strike. If the leading edge and trailing edge are all-aluminum alloy, no problem exists.

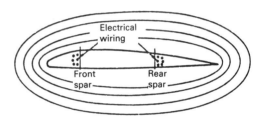

(a) Aluminum alloy leading edge & trailing edge

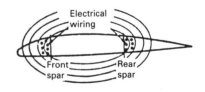

(b) Composite leading edge & trailing edge

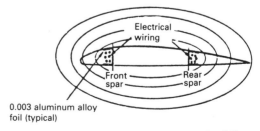

(c) Composite with aluminum alloy foil

Fig. 8.2.7 *Composite edge panels affect electromagnetic field.*

Airframe Structural Design 255

Lightning strike inducing higher voltage in the wiring may result in damage to engine instruments. Also, overloading wiring to fuel quantity probes in the integral tank may result in arcing to the structure. These problems could be solved by:
- Running wiring in conduits
- Putting fuel quantity probe wiring inside the integral tank
- Protecting the engine gage wiring with 0.003 self-bonding aluminum alloy foil on the fiberglass skin panels to increase electromagnetic field as shown in Fig. 8.2.7(c).

8.3 Wing Covers

In the consideration of bending material it is convenient to classify wing structure according to the disposition of the bending-load resistant material: (a) all bending material is concentrated in the spar caps; (b) the bending material is distributed around the periphery of the profile; (c) skin is primarily bending material. Typical wing cross-section in which the bending material is concentrated in the spar caps is shown in Fig. 8.3.1.

Fig. 8.3.1 Three spar wing — all bending materials concentrated at the spar caps.

(1) Advantages of the concentrated spar cap type:
- Simplicity of construction (mostly used on general aviation aircraft).
- Because of the concentration of material, the spar caps can be so designed that buckling occurs near the ultimate stress of the material; this allows the use of higher allowable stresses.

(2) Disadvantages of the concentrated spar cap type:
- Skin will buckle at a very low load. The load-carrying ability of the skin, in so far as bending is concerned, is therefore negligible, which means that it has a certain amount of material which is not being utilized.
- Skin can be in a wave state having relatively large amplitudes which disturbs the airflow over the wing profile and causes an increase in drag.
- Fatigue failures due to the local bending stress in the buckled sheet.

Typical wing cross sections in which the bending material is distributed around the periphery of the profile. The distributed bending material consists of stiffening elements running in a spanwise direction. In high speed airplanes, the wing structure is usually made of multiple spars which are primarily shear material and carry vertical shear. Very little bending material is contributed by the spars. They may be built-up shear webs or channel sections, as shown in Fig. 8.3.2.

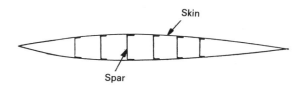

Fig. 8.3.2 Multi-spar skin bending material.

The wing bending loads which cause compression at the upper surface of the wing are generally somewhat higher than those causing compression at the lower surface. This requires that the stiffening elements along the upper surface be more efficient and also more closely spaced than those on the lower surface.

The torsional moments are primarily resisted by the skin and the front and rear spars. The portion of the wing aft of the rear spar is usually over the greater portion of the chord for control surfaces which does not resist any of the torsional loads (see Fig. 8.3.3).

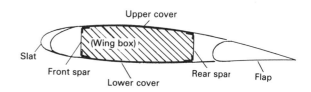

Fig. 8.3.3 Typical wing torque box enclosed area.

Fig. 8.3.4 shows five panels on the wing lower surface to meet the Federal Aviation Regulations (FAR), Volume III, part 25 that require fail-safe structure. This structure shall be able to carry 80% of limit load times 1.15 dynamic factor after a structural failure. Another approach is allowed if testing and analysis can establish safe life structure, that is, if the manufacturer can convince the Federal Aviation Agency (FAA) that failure will not occur. Obviously, this is rather hard to accomplish. Each spanwise splice between panels is a tear-stopper which tends to stop the failed panel to continuously crack to the adjacent panels. The rivet patterns and shear strength shall be designed such that it is strong enough to transfer a failed panel's load (fail-safe load) into the adjacent panels.

The upper panels on a wing structure are also designed to be fail-safe, but since the only structural separation that can occur is during ground operations where the tension loads are small and the FAR 25 requirements can easily be satisfied. The skin panels are designed as wide as possible to minimize the weight and expense of spanwise joint.

Since positive flight design load factors are always higher than for negative flight, the wing upper surface is usually critical for compression loads. When large weights (such as tip tanks) are concentrated at the wing tip, the upper surface near the tip may be critical in tension for positive flight conditions. The following

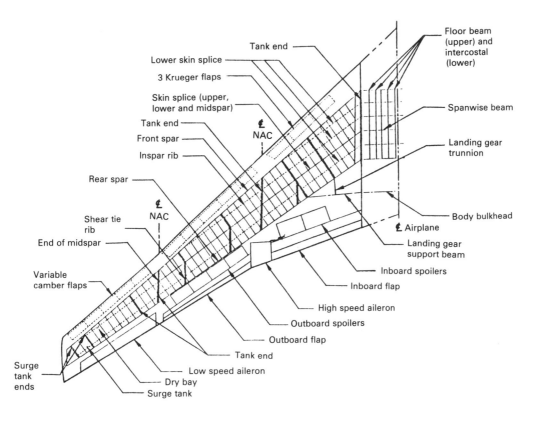

Fig. 8.3.4 *Wing lower surface have five panels.*

loads must be considered in the design of a compression surface:
(a) Direct compression induced by bending of the entire section.
(b) Shear flows — Maximum panel shear flows induced by wing box torsion loads.
(c) Combination of maximum compression panel load with corresponding local shear flow, or maximum shear flow with corresponding local compression load to optimize the least weight structure.
(d) Local bending effects caused by surface aerodynamic pressure load — consists of air loads normal to the surface of the wings. The summation of the components of these pressures normal to the airplane reference plane over the entire wing surface is equal to the airplane weight times the design load factor plus or minus the effects of tail load. For the conditions critical for the wing upper surface (usually +HAA, +LAA), the air loads normal to the upper surface are negative, i.e. they are suction pressures and act upward. When wing fuel tanks are pressurized, this pressure adds to the external pressures. Inertia loads due to fuel, structure and articles of equipment usually act opposite to the above and must be considered. Fig. 8.3.5 shows the critical wing cover axial loads at different locations.
(e) Local bending effects caused by wing tank fuel loads which includes fuel vapor pressure, refueling pressure, inertia, etc.
(f) Local bending effects caused by wing bending crushing loads — crushing load is a radial loading caused by curvature of the wing cover as it bends. As a wing is loaded it naturally deflects and this load is reacted by ribs as discussed in Section 8.5. This load always acts inward to compress the ribs. Generally, the crushing and inertia loading are less than the air loading on the compression cover and act in the opposite direction. The effect of these loads on the design of the cover structure is generally small and depends to a good extent on the rib spacing. Frequently, the effect of air load is included while the relieving crushing and inertia load effects are conservatively neglected.

When corrugated panels are used, the shear flow is distributed between the skin and corrugation in proportion to their respective thicknesses. If skin-stringer panels are used, the entire shear flow is carried by the skin. To properly take these normal loads into account the longitudinal members are treated as beam-columns or panel compression instability as specified previously in Chapter 5.

It is good practice to avoid eccentricities in any structure. Many times eccentricities do occur and they must be accounted for in the design. For example, if a stringer is spliced from two different sections, the centroidal axes of the section may differ in location as shown in Fig. 8.3.6. This will weaken the strength of the member locally and must be considered at the splice point and in the adjacent bays; therefore, this splice should be made at a rib location.

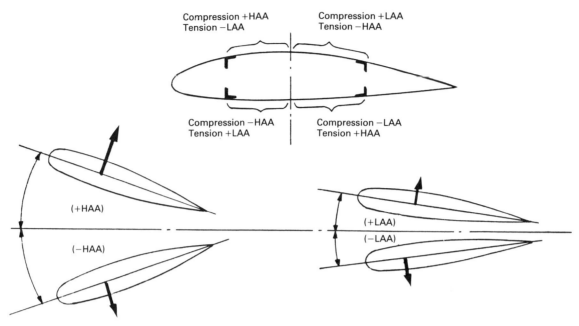

Fig. 8.3.5 Critical wing cover axial loads at wing surfaces.

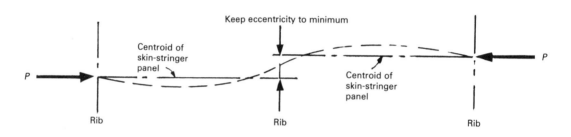

Fig. 8.3.6 The wing panel affected by eccentricity.

The principal source of eccentricity occurs where stringers end. To properly provide for this eccentricity these stringers should be ended only at ribs where the shear load due to surface pressures and eccentricity or loading can be resisted without over-straining the skin. The stringer should be tapered at the end to prevent a sharp change in section. The stringer will tend to carry the same stress as the skin since they are both tied together. A sharp change of section can overload the rivets near the end and may cause failure. It is good practice to space the rivets near the end reasonably close together and also taper the stringer thickness near the end to reduce relative deflection between the stringer and skin (refer to Chapter 7 for detailed discussion).

Skin-Stringer Panels

The most common wing covers of transports are skin-stringer panels as shown in Fig. 8.3.7. Wing skins are mostly machined from a thick plate to obtain the required thickness at different locations and then required pads can be integral; otherwise the pads or doublers have to be riveted or bonded on the basic skin around cutouts. The machined skins combining with machined stringers are the most efficient structures to save weight. This machined skin process has been adopted by modern aircraft structures.

There are many advantages in using integral machined skins. The skins can be taperad spanwise and chordwise, thickened around holes and to produce rib lands as shown in Fig. 8.3.8. With integral skins, when designing a sweep wing, with its associated problems at the root, the ability to place end load-carrying material where it is required is a great advantage.

Optimum distributions of area between skin and stiffener for minimum weight exist. Various studies show that the optimum ratio of stiffener area to skin area is approximately 1.4, assuming equal buckling stress in the skin and stiffener. The optimum design of unflanged integrally stiffened panels and from results obtained therein, the ratio of stiffener area to skin area is 1.7. Based on the equality of Euler column failure stress of the composite section with the initial buckling stress of the skin, the optimum ratio of stiffener area to skin area for Z section stiffeners is obtained to be approximately 1.5; and the ratio of stiffener thickness to skin thickness for minimum weight is 1.05. It should be noted that in a practical design, the skin area will be a higher fraction of the

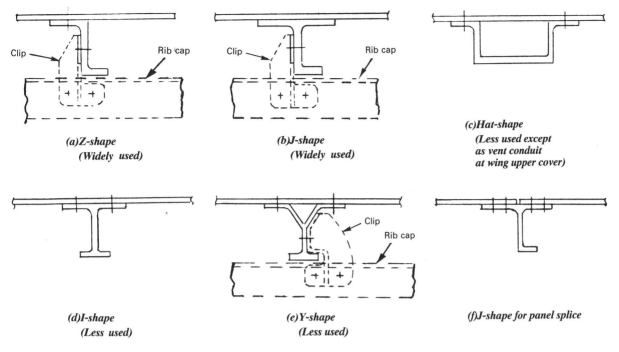

Fig. 8.3.7 *Typical wing skin-stringer panels.*

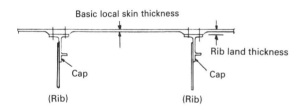

Fig. 8.3.8 *Rib lands in integral wing skin-panel.*

total weight than indicated in this discussion because of interacting shear loads, fatigue, and stiffness considerations. It is recognized that the upper as well as the lower wing cover must be designed to fatigue criteria.

An example will illustrate the importance of considering the effect of interacting loads in the effort to obtain minimum weight design. It is common practice to optimize a cover panel for compression loading by maintaining a 10–15% allowance for shear loading; however, the resulting geometry has to be revised later since the skin thickness is found to be inadequate to carry the shear loads. Presently only a trial and error procedure is available to obtain a panel geometry that is nearly optimum for the combined loading. Considering an aluminum alloy skin and skin-stringer panel with a compression loading of 21000 lb/in and shear loading of 3400 lb/in, the compression optimization analysis yielded the following dimensions: stringer spacing $b = 2.7$ in, skin thickness $t_s = 0.15$ in, and effective thickness $t_e = 0.376$ in. The combined margin of safety was a negative 10%. In a subsequent optimization analysis, the stringer pitch was maintained at $b = 2 (2.7) = 5.4$ in. The resulting skin thickness was 0.22 in and the effective thickness obtained was 0.396 in, representing a 5% weight increase. The margin of safety for combined compression and shear was a positive 1%. The increase in stringer pitch has the additional beneficial effect on the so-called non-optimum weight; for this case, it permits the elimination of 40 000 fasteners and 1000 stringer-to-rib clips, and a reduction in the amount of material for tank sealing. The combined effect of the increase in stringer pitch amounted to a 400 lb weight reduction and a cost reduction. The increase in skin thickness will also yield increases in torsional stiffness and fatigue life due to reduced stress levels.

In the design of current transports, the allowable tension stress, based on fatigue considerations on the lower cover, is somewhat lower than the maximum compression stress on the upper cover; however, the average working stresses are of the same order of magnitude, resulting in approximately equal weight for the upper and lower covers. Centroids of sections should be as close to the skin as possible for maximum centroidal depth of the wing box and minimum panel eccentricity.

Fig. 8.3.9 shows three most advanced wing structures of their maximum upper compression and lower tension loads or stresses. These maximum compression allowable stresses reach as high as 60 000 psi and, to obtain such high efficient structure, good detailed structural design is necessary to achieve this goal. One of the most important detail designs is the structural cut-out such as fuel probe holes, inspection doors, etc. For detailed discussion of how to design structural cut-outs, refer to Chapter 6.

Airframe Structural Design 259

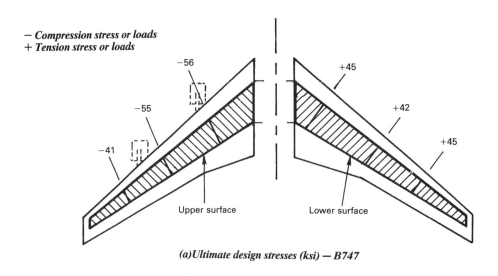

(a) Ultimate design stresses (ksi) — B747

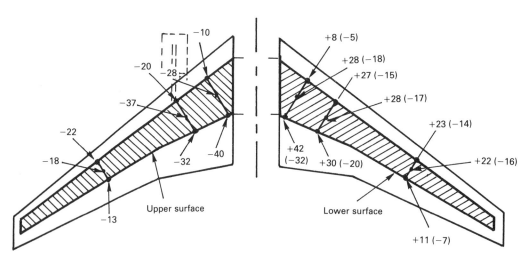

(b) Ultimate design loads (kip/in) — DC-10

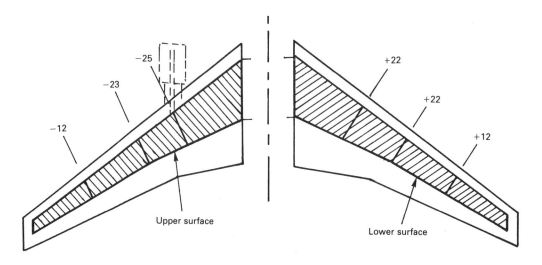

(c) Ultimate design loads (kip/in) — L-1011

Fig. 8.3.9   Ultimate design stresses or loads (for reference only) — wing covers.

Integrally Stiffened Panels

Present trends toward higher performance levels in machines and equipment continue to place more exacting demands on the design of structural components. In aircraft, where weight is always a critical problem, integrally stiffened structural sections (see Fig. 8.3.10) have proved particularly effective as a light weight, high-strength construction. Composed of skin and stiffeners formed from the same unit of raw stock, these one-piece panel sections can be produced by several different techniques. Size and load requirements are usually the important considerations in selecting the most feasible process.

Integrally stiffened structures have their greatest advantage in highly loaded applications because of their minimum section size. Investigations have indicated that an integrally stiffened section such as shown in Fig. 8.3.11 can attain an exceptionally high degree of structural efficiency. A weight reduction of approximately 10–15% was realized by the use of an integrally stiffened structure.

A study on a typical transport's upper surfaces was made, comparing integrally stiffened and builtup skin-stiffener types of construction with a rib spacing of 26 inches. The analysis has been made under the simplifying assumption that an optimum design is attain-

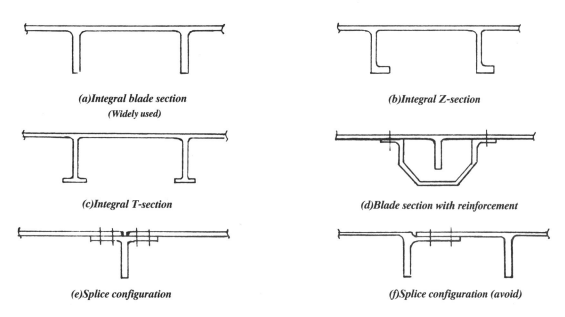

(a)Integral blade section
(Widely used)

(b)Integral Z-section

(c)Integral T-section

(d)Blade section with reinforcement

(e)Splice configuration

(f)Splice configuration (avoid)

Fig. 8.3.10 *Typical integral stiffened panels (planks).*

For highly loaded long panels, extrusions or machined plates are most commonly employed. Section discontinuities such as encountered in the region of cutouts can often be produced more easily from machined plate. From a cost standpoint it is usually better to machine a section from the extruded integrally stiffened structures than to machine a section of the same size from a billet or plate.

From a structural standpoint, appreciable weight savings are possible through the integral-section design which also develops high resistance to buckling loads. In addition, the reduction in the number of basic assembly attachments gives a smooth exterior skin surface. In aircraft applications, the most significant advantages of integrally stiffened structures over comparable riveted panels has been:
- Reduction of amount of sealing material for pressurized fuel tank structures
- Increase in allowable stiffener compression loads by elimination of attachment flanges
- Increased joint efficiencies under tension loads through the use of integral doublers, etc.
- Improved performance through smoother exterior surfaces by reduction in number of attachments and nonbuckling characteristics of skin
- Light weight structures

able for all stations along the wing span for both designs. In this analysis, all non-optimum (such as joints, cutouts, etc.) factors are ignored. The integrally stiffened skin and the stringers are manufactured from 7075-T6 aluminum alloy extrusions and the skin is 7075-T6 bare plate. In both cases a 10% margin for shear-bending interaction was maintained. The resulting weight of the integrally stiffened upper surface is 6000 lb and that of the skin-stringer surface is 6600 lb, indicating the builtup configuration to be approximately 10% heavier than the integral construction. It should be noted that in order to obtain the true weight difference, all non-optimum factors must be taken into account. The integrally stiffened design will have a relatively low weight for the so-called non-optimum features. This is attributed to the machined local padding and reinforcing material and permitted by integral cover construction. In contrast, the builtup type of design generally requires a relatively large non-optimum weight because of the many chordwise splices for ease of tank sealing and fabrication, discrete doublers, etc.

From the foregoing one concludes that the lightest cover panel design can be obtained with an integrally stiffened cover structure supported by sheet metal ribs with a preference for a large spacing. If the use of integral skin is prohibited for such reasons as exfoliation, etc, then special attention must be given to the non-optimum factor for using builtup skin-stringer panel. End grain exposure and residual stresses (threshold stresses) due to fabrication "pull-up" presented a stress corrosion risk. Airplane operator objections, due to experience with integrally stiffened panels on some competitors' airplane, led to a decision not to use them on some of these commercial airplanes.

Cover Panel Splice Design

It is practical to use double or stagger row of fasteners to splice panels together as shown in Fig. 8.3.12. Avoid joints incorporating bolts or screws in conjunction with upset shank rivets or interference fit fasteners. The extensive machining to sculpture the wing plank splice resulted in exposure of short transverse grain flow in the extrusion web junction. Basically, the aluminum extrusion alloys, especially 7075-T6510 and 7178-T6510 materials, have a high susceptibility toward stress cracking in the short transverse direc-

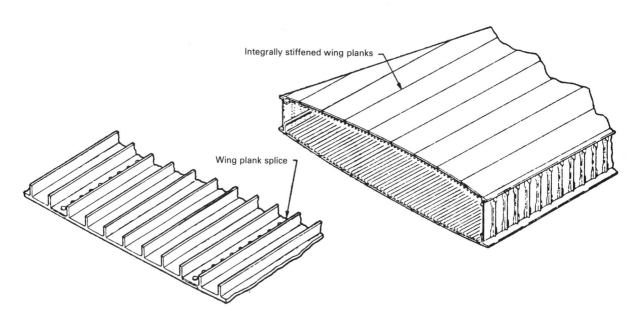

Fig. 8.3.11 Integrally stiffened panels — wing covers.

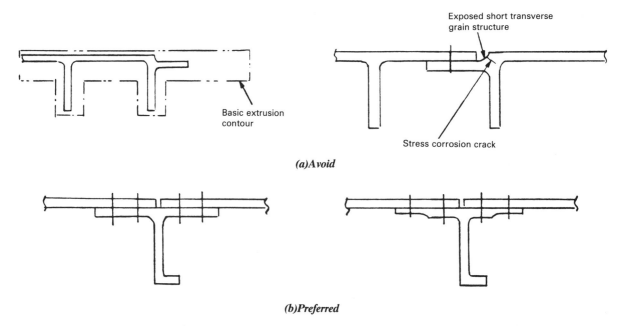

Fig. 8.3.12 Wing cover panel splice configurations.

262 Airframe Structural Design

tion. The combination of a susceptible alloy and exposed residual tensile stresses in the transverse grain resulted in the high incidence of stress corrosion cracking failures.

Stringer Run-out Design

Fig. 8.3.13 shows a wing cover stringer run-out which has given many fatigue problems because of unanticipated patterns of load transfer at end fasteners. This problem is analyzed by the stress severity factor (SF) method (refer to Chapter 7). This method determines the elastic load transfer and bypass load for each fastener, and specifies a severity factor for each location. The effects of fastener size, fastener tightness, materials of fasteners, stringers and skin are all reflected in that solution. The SF method is useful as a comparison of several designs of determining the best way to reduce the severity factor and thus to give the best fatigue performance.

Stringers are parallel to the rear spar or front spar, and because of wing taper, some of them must ultimately terminate on a special edge member stringer near the front spar (for aft sweep wing design). The splice fitting shown in Fig. 8.3.14 is designed for reactions of both stringer axial loads, thus minimizing the possibility of skin cracks at the end of stringers. Other than designing to minimize these cracks, the stringer ends and skins are padded to reduce bearing stresses and tension stress around the fastener holes. Adequate ramps from the basic stringer thickness to the cover skin padded thickness are a necessity. However, careful attention should be given to this location due to past experience of fatigue cracks.

Access Holes

The techniques selected for installation of the various access holes contained in the wing box structure presents one of the more critical design challenges. Structural integrity and aircraft maintainability are prime considerations. The penetrations have been carefully tailored so that load eccentricities and stress concentrations are held to a minimum. Design of the sealing method and the finish of the sealing surfaces have been considered as important factors. Lightning protection has increasingly influenced the design of the modern jet aircraft and tests should be conducted to assure that the design will not result in a safety hazard in fuel tanks. There are two major designs of access holes, with stressed door (carried loads) or non-stressed door (not carried loads)

For access hole with stressed door (Fig. 8.3.15):
- Increased stiffness
- Reduced fretting corrosion between door and door landing
- Improved tank seal
- Improved electrical bonding between door and fixed structure, thus minimizing the possibility of in-tank arcing following a lightning strike
- Lighter structure

For access hole with non-stressed door (see Fig. 8.3.16):
- All doors are designed as clamped in equipment with none of the installation fasteners penetrating wing surface structure, thus minimizing fatigue crack problems.
- Due to clamped-in door, it eliminates the installation problems of using close tolerance fastener holes and precise location of fastener holes between door and support structure.

Fig. 8.3.17 shows a doubler around a skin access door cutout. This is a typical example of a failure at end fasteners. The answer is to machine all doublers into the tension critical skins or spar webs, remembering to use a 7° maximum/ramp of the machine-in doubler.

Leading Edge and Trailing Edge Panel Attached to Wing Panels

Any notch at the wing skin such as at the front and rear spar where the leading edge, trailing edge, control surfaces, etc, are attached to will result in a stress concentration. The notch in the skin for a spoiler hinge, for example, is unacceptable as shown in Fig. 8.3.18. The best design is no notches, but if no other better design is possible, then attention should be paid to the detailed design, adding thicker pads around notch corners and, if it is possible, conducting tests.

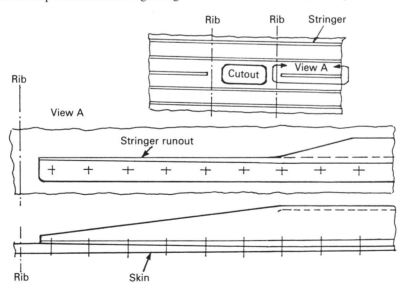

Fig. 8.3.13 A typical stringer run-out of skin-stringer panel.

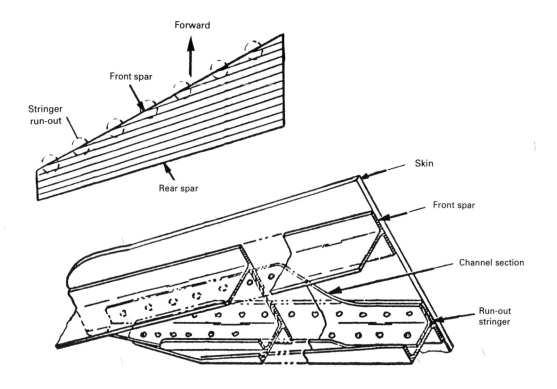

(a)Stringer run-out of a skin-stringer panel

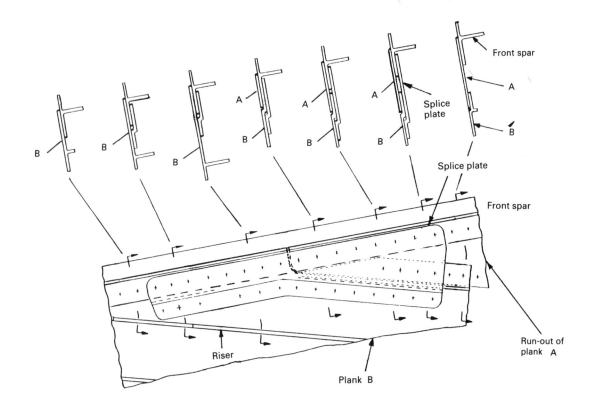

(b)Run-out of an integrally stiffened panel (plank)

Fig. 8.3.14   Typical stringer run-out joint at front spar.

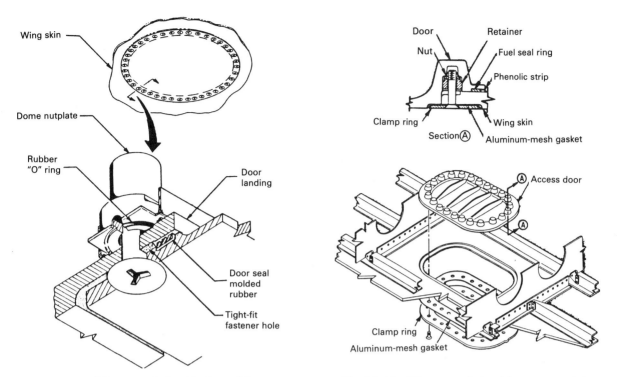

Fig. 8.3.15 Wing access hole with stressed door.

Fig. 8.3.16 Wing access hole with non-stressed door.

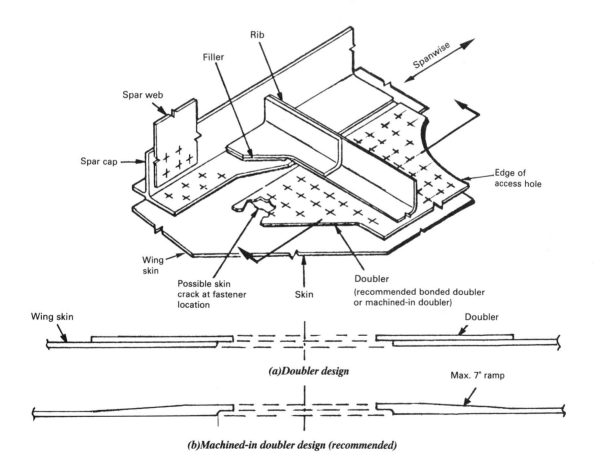

Fig. 8.3.17 Reinforcement around access hole.

Airframe Structural Design 265

Stress concentrations in wing box structures will result at end fasteners where trailing edge and leading edge skins, such as fixed leading and trailing edge panels, rivet to the wing spar caps or wing skins. Several approaches are introduced as follows:

(1) A corrugated splice strap shown in Fig. 8.3.19 called a "wiggle plate" is developed to support and splice the edge panels to the wing box. The wiggle plate acts like an accordion. Using this design restricts "wiggle plate" to pick up spanwise loads.

(2) A second approach is to use a sacrificial doubler to attach interchangeable leading or trailing edge panel as shown in Fig. 8.3.20. The design allows the use of interference attachments in the heavier spar flanges and avoids degradation in the spar cap if replacement of the edge panel becomes necessary.

(3) A third approach is used where aerodynamic loads are low and minimum skin gages are used. This approach involves using either laminated or honeycomb fiberglass panels (see Fig. 8.3.21). These panels may be fastened directly to the spar flange with hole filling rivets, since the modulus of elasticity of fiberglass can be designed to be as little as 25% of aluminum alloy, thus reducing induced loads to an acceptable level. Induced loads are a direct function of the modulus of elasticity of the panels being elongated.

(4) Another approach is to use gooseneck hinges, particularly if the leading edge or trailing edge panel is to be used as an access door (see Fig. 8.3.22). The maximum spanwise length of panel is governed by bending stresses in the gooseneck hinges as the spar elongates.

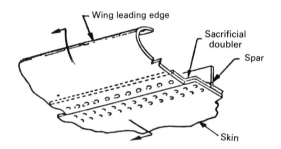

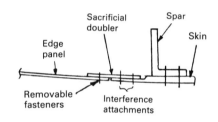

Fig. 8.3.20 Sacrificial doubler design.

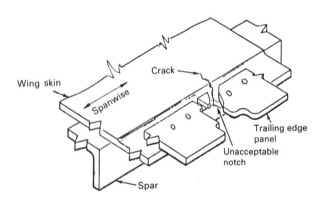

Fig. 8.3.18 A notch at wing skin for spoiler hinge.

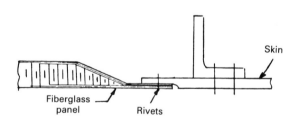

(a) Attached to wing skin

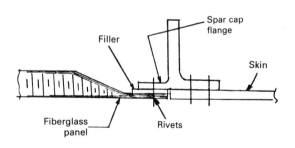

(b) Attached to wing spar cap flange

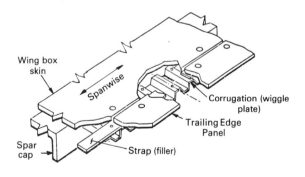

Fig. 8.3.19 A corrugated splice strap (wiggle plate) design.

Fig. 8.3.21 Fiberglass panel attached to wing box design.

Fighter Wing Covers

High performance modern fighters require not only high speed with very thin wing box but also higher design loads as high as 6.0g × 1.5 = 9.0g (ultimate load), which produce very high cover axial stresses that are a function of wing box cross-sectional moment of inertia. Fighter wing covers generally are machined from a thick plate to various required thicknesses, and, from a structural stand point, plate covers are more efficient than either skin-stringer panel or integrally stiffened panel as shown in Fig. 8.3.23.

Design criteria of a fighter wing is different than a transport such as removable wing cover for maintenance, multi-root fittings for removable wings, integral fuel tank seal problem (no access holes and sealed externally, see Section 8.8) etc.

Various Wing Box Data

Typical wing box data of various modern fighters and transports are illustrated in Fig. 8.3.24 and Fig. 8.3.25 respectively.

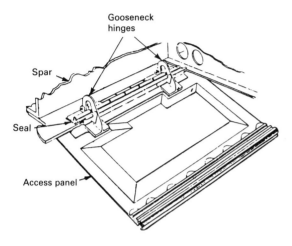

Fig. 8.3.22 Access panel attached to wing box by gooseneck hinges.

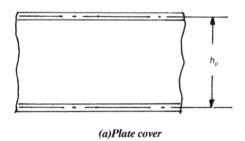

(a) Plate cover

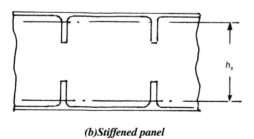

(b) Stiffened panel

R_i = Moment of inertia's ratio = I_p/I_s = $(h_p/h_s)^2$
Assume $h_p = 10''$ and $h_s = 9''$
then $R_i = 23.4\%$

Fig. 8.3.23 Moment of inertia comparison between a stiffened panel and plate cover (equal cross-sectional areas).

Fighter Name	Wing Skin	
	Upper	Lower
F-4	7075-T651 (Plate 3 in)	7075-T651 (Plate 3 in)
F-5E	7075-T651	7075-T7351
F-8 (A-7)	7079-T6 (Plate 1/4 in) 7049-T7xxx 7050-T7xxx	7075-T6 (Plate 1/4 in)
F-14	Ti-6Al-4V (Ann. plate 0.5–0.75 in)	Ti-6Al-4V (Ann. plate 0.5–0.75 in)
	Leading edge: 2219-T6 (bare) Control surfaces: 2024-T6, -T8 (Honeycomb)	
F-15	2124-T851 (Plate, 0.5–1.5 in)	Ti-6Al-4V (Ann. plate 1.5 in) 2124-T851 (Plate)
F-16	2024-T851 (Plate, 0.5–1.0 in)	7475-T7351 (Plate, 0.5–1.0 in)
F-18	7050-T76 Plate, 0.5 in)	7475-T7351 (Plate, 0.75 in)
	Graphite/Epoxy Composite	

Fig. 8.3.24 Typical fighter wing box material data comparison.

Airplane Name	Cover	Type of Construction	Material Skin/Str.	Panel Shape	No. of Spars; Type of Ribs; Rib Spacing
1649	Upper Lower	Integral Integral	7075 7075	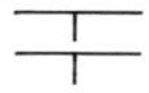	2 spars; rib-truss;
L-188	Upper Lower	Integral Integral	7178-T6 7075-T6	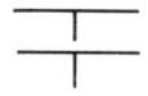	2 spars; rib-truss; 18 in.
B707	Upper Lower	Skin-str. Skin-str.	7178/7178 2024-T3/7075-T6	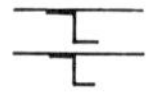	2 spars; rib-web; 28.5 in.
DC-8	Upper Lower	Skin-str. Skin-str.	7075/— 7075/—	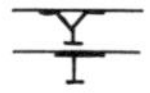	2 spars; rib-web; 32 in.
DC-9	Upper Lower	Skin-str. Skin-str.	7075-T6/— 2024-T4/—	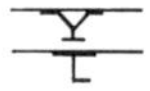	2 spars; rib-web;
CV880	Upper Lower	Skin-str. Skin-str.	7075/— 2024/—		3 spars; rib-web; 22 in.
CV990	Upper Lower	Skin-str. Skin-str.	7179/— 2024/—		3 spars; rib-web; 22 in.
BAC 111	Upper Lower	Integral Integral	7075 2024-T4	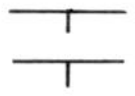	3 spars; rib-web; 23 in.
C130	Upper Lower	Integral Integral	7178/7075 7178/7075	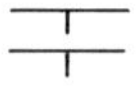	2 spars; rib-truss; 18 in.
P-3	Upper Lower	Integral Integral	7075-T6		2 spars; rib-truss; 18 in.
C141	Upper Lower	Integral Integral	7075-T6 7075-T6	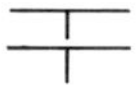	2 spars; rib-truss; 23 in.
C-5A	Upper Lower	Integral Integral	7075-T6 7075-T6	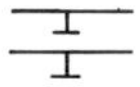	3 spars; rib-truss; 30 in.
B747	Upper Lower	Skin-str. Skin-str.	7075-T6/7075-T6 2024/2024	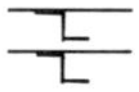	3 spars; rib-web; 25 in.
L-1011	Upper Lower	Skin-str. Skin-str.	7075-T76/7075-T6 7075-T76/7075-T6	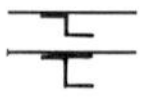	2 spars; rib-web; 21 in.
DC-10	Upper Lower	Skin-str. Skin-str.	7075-T651/7075-T6 2024-T351/7075-T6		2 spars; rib-web; 18—34 in.
A-300	Upper Lower	Skin-str. Skin-str.	7075-T6/— 2024-T3/—	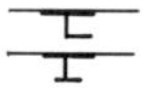	3 spars; rib-web;
B-1	Upper Lower	Integral Integral	2219-T851/— 2219-T851/—	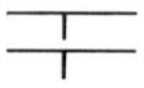	2 spars; rib-web;
Trident	Upper Lower	Skin-str. Skin-str.	7075-T6/7075-T6 2024/2024	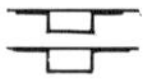	3 spars; rib-web;
B727	Upper Lower	Skin-str. Skin-str.	7178-T651/ 7178-T6511 2024-T351/ 2024-T3511	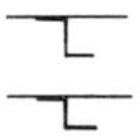	2 spars; rib-web;
B737	Upper Lower	Skin-str. Skin-str.	7178-T651/— 2024-T351/—	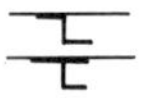	2 spars; rib-web;
VC-10	Upper Lower	Integral Integral	7075-T6 2024-T4	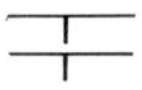	4 spars; rib-web;
F.28	Upper Lower	Skin-str. Skin-str.	7075-T6/7178-T6 2024-T3/7075	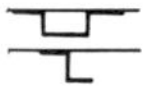	2 spars; rib-truss;
KC-135	Upper Lower	Skin-str. Skin-str.	7175-T6/7178-T6 7178-T6/7178-T6	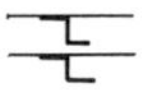	2 spars; rib-web; 28.5 in.
Concorde	Upper Lower	Integral Integral	Hiduminium RR.58 Hiduminium RR.58	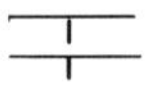	Multi-spars; rib-truss;
B757 and B767	Upper Lower	Skin-str. Skin.str.	7150-T6/7150-T6 2324-T3/2224-T3	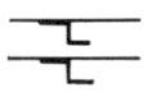	2 spars; rib-web;

Fig. 8.3.25 Typical transport wing box data comparison.

8.4 Spars

For strength/weight efficiency, the beam (or spar) cap should be designed to make the radius of gyration of the beam section as large as possible and at the same time maintain a cap section which will have a high local crippling stress. The cap sections for large cantilever beams which are frequently used in wing design should be of such shape as to permit efficient tapering or reducing of the section as the beam extends ouboard. Fig. 8.4.1 shows typical beam cap sections for cantilever metal wing cover construction where additional stringers and skins are also used to provide bending resistance. These cap sections are generally of the extruded type although such sections as (c) is made from sheet stock. These cap sections are almost always used with a beam web composed of flat sheet, which is stiffened by vertical stiffeners riveted to the web as shown in Fig. 8.4.2.

The air loads act directly on the wing cover which transmits the loads to the ribs. The ribs transmit the loads in shear to the spar webs and distribute the load between them in proportion to the web stiffnesses. In the past it has been customary to design wings with three or more spars. The use of several spars permit a reduction in rib stresses and also provides a better support for the spanwise bending material. Another important purpose is so designed for structural fail-safe feature. Refer to discussion in Chapter 15.0.

Space requirements for the housing of fuel tanks and landing gears (when retracted) is the main reason for the at least two spar wing box construction. A two-spar wing construction usually consists of a front and rear spar, the front spar located that the wing leading edge slats can be attached to it and the rear spar located that the control surface such as these hinge brackets of flaps, aileron, spoiler, etc, can be attached to it. Furthermore, the front and rear spars combined with wing skin panels form as the closing member of the torsion-resistant box and also serves as integral fuel tank.

Different types of spar beam construction are shown in Fig. 8.4.3 and spars can be divided into two basic types; shear web type and truss type. The shear web type is widely adapted to design the modern wing spar for its structural efficiency as described later.

The aerospace structures engineer is constantly searching for types of structures and methods of

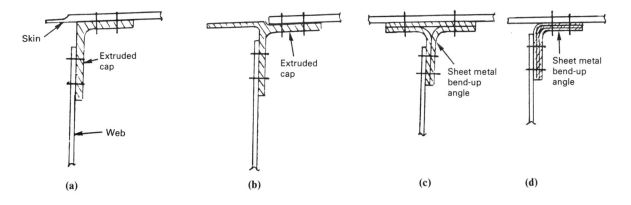

Fig. 8.4.1 Typical spar cap sections.

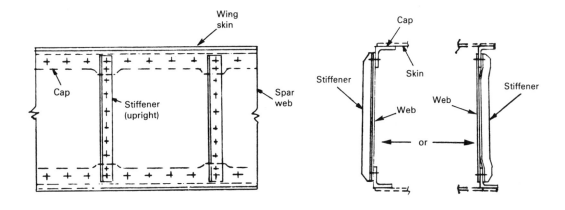

Fig. 8.4.2 Typical spar construction.

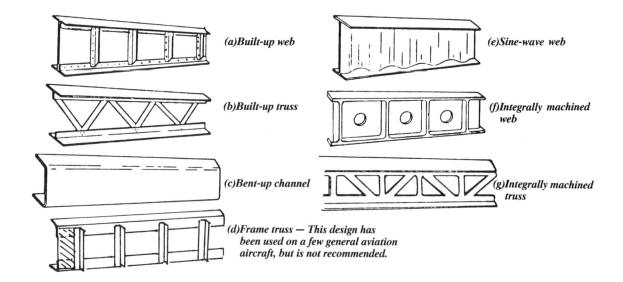

Fig. 8.4.3 *Typical spar configurations.*

structural analysis and design which will save structural weight and still provide a structure which is satisfactory from a fabrication and economic standpoint. The development of a structure in which buckling of the webs is permitted with the shear loads being carried by diagonal tension stresses in the web is a striking example of the design of aerospace structures from the standard structural design methods into other fields of structures.

The design of a metal beam composed of cap members riveted or spotwelded to web members is a common airplane structural design. In this section, the two basic types of shear beam construction are discussed, i.e. shear resistant (non-buckling) type and diagonal-tension field (buckling) type. However, the more common case where the beam web wrinkles and goes over into a semi-diagonal-tension will be introduced later.

A shear resistant beam is one that carries its design load without buckling of the web, or, in other words, it remains in its initially flat condition. The design shear stress is not greater than the buckling shear stress for the individual web panels and the web stiffeners have sufficient stiffness to keep the web from buckling as a whole. In general, a thin web beam with web stiffeners designed for non-buckling is not used widely in airplane structures as its strength-weight ratio is relatively poor. In built-in or integral fuel tanks, it is often desirable to have the beam webs undergo no buckling or wrinkling under the buckling criteria of 1.0−1.5g of level flight loads in order to give better insurance against leaking along riveted web panel boundaries. It is realized that the buckling web stress is not a failing stress as the web will take more before collapse of the web takes place, thus in general the web is not loaded to its full capacity for taking load and the web stiffeners are only designed for sufficient stiffness to prevent web buckling and sufficient strength to take the full failing strength of the web.

Investigations have indicated that stiffened non-buckling web-stiffener (shear web resistant type) constructions are lighter than the diagonal-tension type for certain regions of high structural load index.

A large majority of the beams in aircraft wing and tail surfaces have sloping spar caps because of the taper of the structure in both planform and box depth. This sloping of the spar caps relieves the beam web of considerable shear load and should not be neglected as illustrated in Fig. 8.4.4.

The two primary conditions which determine the overall efficiency of a spar are its construction cost and its efficiency as a load-carrying member. The incomplete tension field beam is particularly adaptable to mass production because of its component parts. Webs require a simple cutting operation, and for the spar caps and vertical stiffeners extrusions or bend-up sections are used. Because of the high degree of redundancy present in an incomplete tension field beam, it will carry load even when severely damaged. Semi-tension type beams have a better strength-weight ratio and are much stiffer than the truss-type beam. Shear stiffness of the spars is important because large deflections will cause deep shear wrinkles, which may become permanent in the leading edge of the wing. Construction of a truss-type spar requires considerable more time because of the large number of individual parts and because more elaborate tooling is necessary. A truss-type spar has either none or a small degree of redundancy; which means that if any one member in the spar fails, its load carrying ability will be destroyed. For modern airplane design, the wing box is considered an important fuel tank that the truss-type spars can not serve this purpose.

Fig. 8.4.5 shows the behavior of spar web construction in semi-tension type. The effect of the vertical component of web tension is to compress the vertical stiffeners (upright), bend the caps in the plane of the webs, introduce vertical shear loads on the web-to-cap rivets. The horizontal component of web tension compresses both beam caps, bends the end

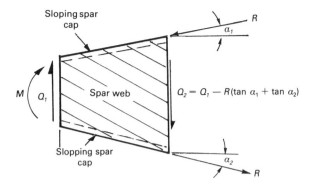

Fig. 8.4.4 *Sloping spars relieve the spar web shear load.*

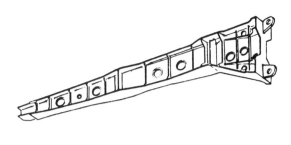

Fig. 8.4.6 *Integrally machined spar.*

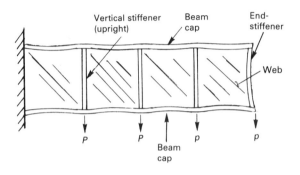

Fig. 8.4.5 *The behavior of spar web construction.*

stiffeners inward, produces horizontal shear forces on the web-to-end stiffener rivets and on any vertical web splices that may exist. The horizontal web tension, however, has no tendency to bend interior stiffeners or to load the web-to-upright rivets. Web wrinkles tend to produce sinusoidal deformations of web stiffener flanges and to a lesser extent out of the plane of the web. The buckling of the web also loads the web-to-upright rivets in tensions, an effect which sometimes is responsible for panel failure.

The effect of material utilization factor on cost for an integrally stiffened (machined one piece including cap, web and stiffeners as shown in Fig. 8.4.6) spar which is fabricated from aluminum, titanium or steel materials, and the cost of the finished spar installed in the wing, (i.e. fighter thin airfoil construction) is far less than the cost of a built-up assembly of individual caps, web and stiffeners riveted together. For this special reason, the structural engineer would like to adapt the integrally machined spars for their design. On the other hand, although this method of construction gives a weight saving, the fail-safe feature is far less than the built-up spar and, therefore, other fail-safe methods should be carefully considered such as adding a titanium strap or crack stopper on the spar web to retard or stop the crack propagation (refer to spar design as shown in Fig. 8.4.17); another design uses three or more spars concept to compensate it.

General Rules of Spar Design
1. Machine pads or add doublers to the web around spar web cutout to reduce local stresses as shown in Fig. 8.4.7.
2. It is strongly recommended to use double rows (or stagger rows) of fasteners between spar caps and webs, and also between spar caps and wing box skin as shown in Fig. 8.4.8.
3. Spar web splice doublers should be designed such that it is strong enough to carry not only the vertical shear force but also the spar axial force at the spar cap where the tapered doubler along spanwise is recommended (see Fig. 8.4.9).
4. The tension fitting is required wherever appreciable concentrated loads exist, such as engine pylon fittings, main landing gear support fittings (if gear is supported at wing box), flap track fittings, aileron support fittings, slat track fittings, wing jack fittings, etc. At these locations, the local material thicknesses of spar cap, web as well as skin should be made thicker to reduce local principal stresses as shown in Fig. 8.4.10.
5. Do not allow any fixed leading or trailing edge panel to be directly riveted to the spar cap to avoid potential fatigue cracks. If fiber glass ($E = 3.5 \times 10^6$ psi) honeycomb panels are used, the panels may be riveted to spar caps with soft rivets (i.e. aluminum rivets), providing careful stress analysis has been done. Some typical configurations are shown in Fig. 8.4.11.
6. In the area of the wing sweepback break, the spar-cap horizontal flange and local wing skin can be easily spliced by double shear splice plates and an additional tension fitting should be provided to take care of the remaining part such as spar-cap vertical flange and local web as shown in Fig. 8.4.12. Careful detailed design should be given to this crtical area.
7. Clips, provided for the support of wires, hydraulic tubes, control rods, ducts, etc, should be fastened to spar vertical stiffener only. See Fig. 8.4.13.
8. Fastener spacing along vertical stiffeners should not be too close to make the local web net area shear critical. In addition, the fasteners going through the spar cap and stiffener should be at least two fasteners with diameter of one size bigger than adjacent attachments as shown in Fig. 8.4.14.

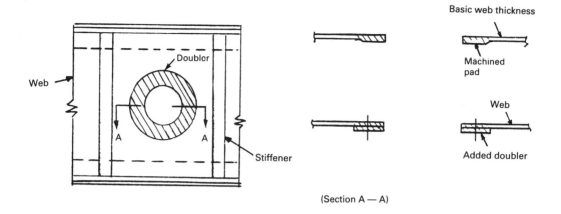

Fig. 8.4.7 Spar web reinforcement around cutout.

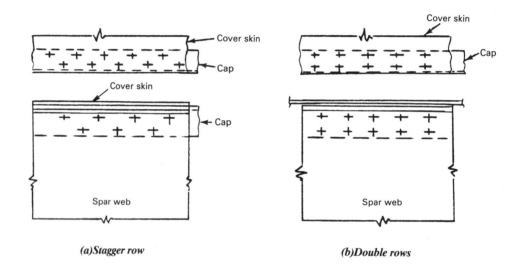

(a) Stagger row   (b) Double rows

Fig. 8.4.8 Spar cap fastener pattern (avoid using single rows of fasteners).

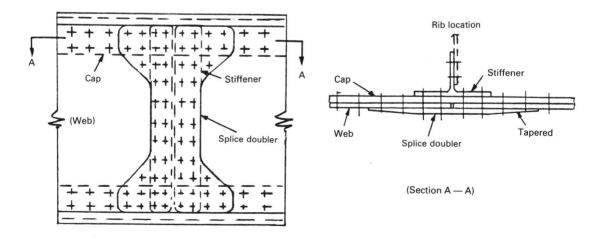

Fig. 8.4.9 Typical spar web splice design.

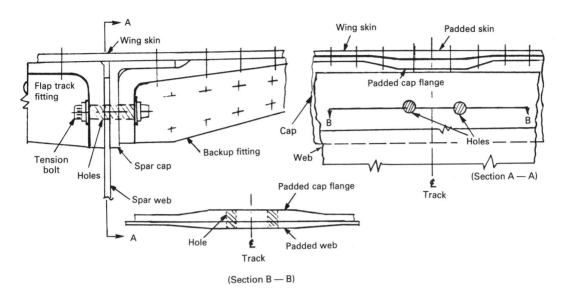

Fig. 8.4.10 *Local padded thickness where the concentrated load applied.*

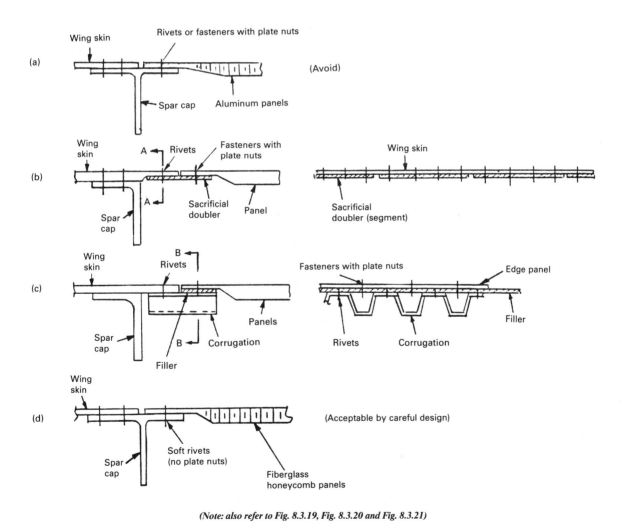

(Note: also refer to Fig. 8.3.19, Fig. 8.3.20 and Fig. 8.3.21)

Fig. 8.4.11 *Typical installations of edge panels to wing box.*

Airframe Structural Design 273

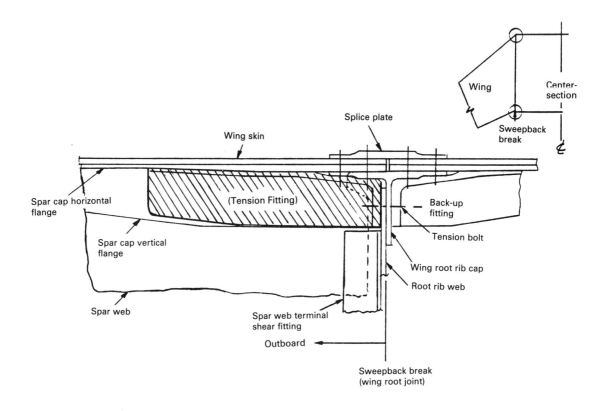

Fig. 8.4.12　Wing root joint at spar location.

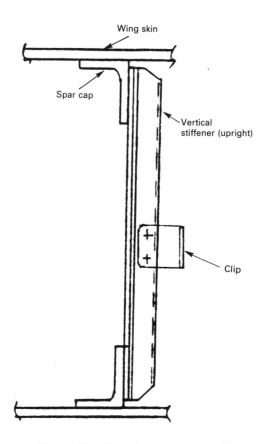

Fig. 8.4.13　Control system support clips.

274　Airframe Structural Design

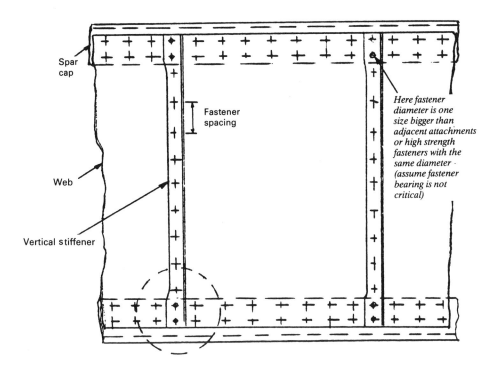

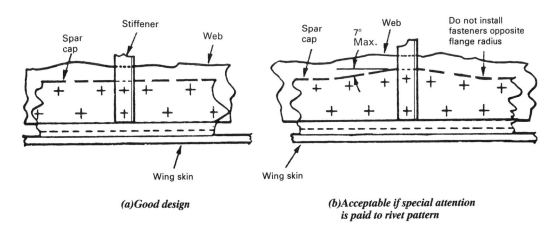

(a)Good design

(b)Acceptable if special attention is paid to rivet pattern

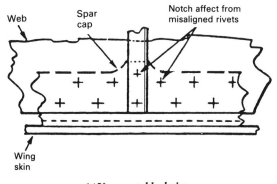

(c)Unacceptable design

Fig. 8.4.14 Fastener pattern at the location of spar cap and vertical stiffener.

Airframe Structural Design 275

Fail-safe Design

A truly fail-safe structure must not fail when a shear beam is damaged for any reason. Here, multiple beam construction has the advantage of supplying alternate load paths for tension in case any single beam web should fail. Assuming that a fatigue crack would start as usual at the lower wing cover, it would probably run upward through the web unless stopped in some manner.

Fig. 8.4.15 shows two types of "crack stopper" that are formed basically by the creation of an artificial joint. The design consists of a spanwise web splice and stiffener located about one-third of the distance above the lower surface. The remaining two-thirds of the shear web would be able to sustain considerable load should the first one-third be broken. Even though the scheme has satisfactorily stopped fatigue cracks, its use involves considerable extra weight, parts, and assembly labor.

Another design, which avoids the artificial splice, is shown in Fig. 8.4.16. Here the web is thickened at the attachment to the spar cap (tee extrusion). Steel attachments (high strength steel) are made equally critical in bearings in the web and cap; vertical stiffeners are added as required on the web. Upon the fracture of a spar cap, the load to be transferred divides between the thickened web and the adjacent wing skin structure. Bearing deformation in the holes will allow the load to be distributed over a number of attachments. The thickened web is sized to carry the additional load.

Another design feature is that the wing front and rear spars are integrally machined from aluminium alloy rolled or forged plate. This method of construction gives weight savings at reasonable cost compared with fabricated spar construction. Each section of spar has a continuous horizontal stringer crack stopper introduced approximately a third of the way up the shear web from the lower predominantly tension flange as shown in Fig. 8.4.17.

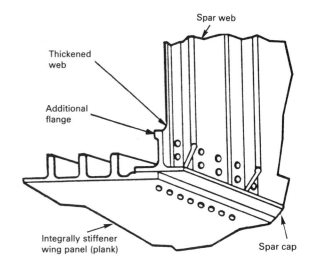

Fig. 8.4.16   Fail-safe spar shear web design.

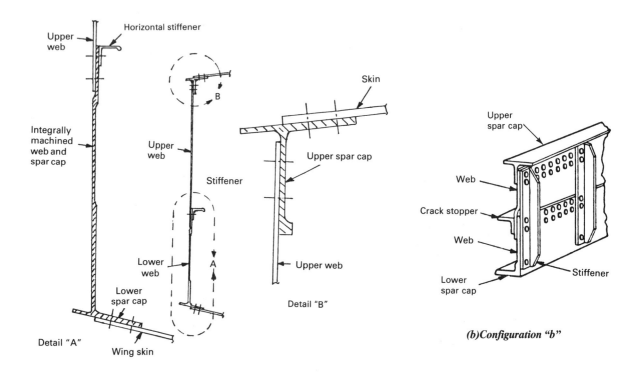

Fig. 8.4.15   Spar shear web crack stopper for fail-safe feature.

the web are large enough for a man to crawl through and provide access from bay to bay. Forged clips are used for attachment of stringers to ribs in lieu of bolting to stringer and rib cap flanges. Rib bulkeads are also provided for such purposes as flap, aileron, pylon and landing gear support, tank ends and redistribution of loads at the sweep and dihedral break.

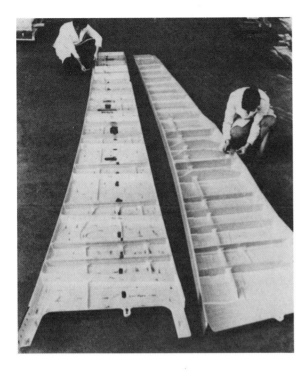

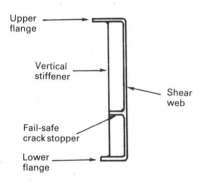

Fig. 8.4.17 Integrally machined A310 front and rear spars.

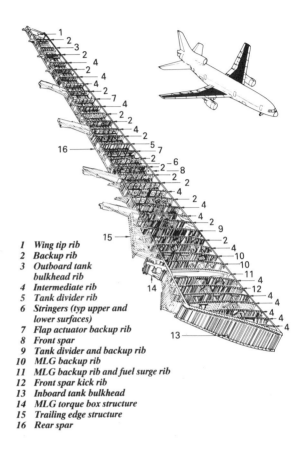

1 Wing tip rib
2 Backup rib
3 Outboard tank bulkhead rib
4 Intermediate rib
5 Tank divider rib
6 Stringers (typ upper and lower surfaces)
7 Flap actuator backup rib
8 Front spar
9 Tank divider and backup rib
10 MLG backup rib
11 MLG backup rib and fuel surge rib
12 Front spar kick rib
13 Inboard tank bulkhead
14 MLG torque box structure
15 Trailing edge structure
16 Rear spar

Fig. 8.5.1 Two-spar wing box construction of Lockheed L-1011.
(Courtesy of Lockheed Aeronautical Systems Co.)

8.5 Ribs and Bulkheads

For aerodynamic reasons the wing contour in the chord direction must be maintained without appreciable distortion. Therefore, ribs are used to hold the cover panel to contour shape and also to limit the length of skin-stringer or integrally stiffened panels to an efficient column compressive strength. The rib also has another major purpose, to act as a transfer or distribution of loads. The applied loads may be only distributed surface air and/or fuel loads which require relatively light internal ribs to carry through or transfer these loads to main spar structures. Fig. 8.5.1 illustrates a typical wing box with different ribs and their functions.

A typical wing rib, illustrated in Fig. 8.5.2, is composed of caps, stiffeners and webs. Lightening holes in

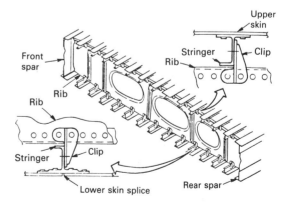

Fig. 8.5.2 Typical rib construction.

Airframe Structural Design 277

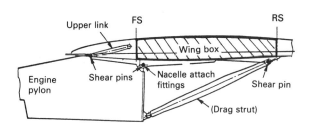
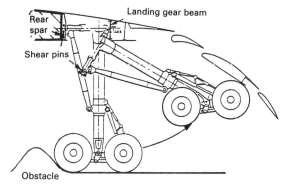

(a) Engine nacelle support fuses (shear pins) *(b) Main landing gear support fuses (shear pin)*

Fig. 8.5.3 Break-away design of engine pylon and main landing gear support.

Basically, there are many types of rib construction similar to the spar shown in Fig. 8.4.3. The aircraft industry generally uses shear web rib design due to a number of advantages. Its web acts as a fuel slosh inhibiter. The web gage should be a minimum 0.04 inch thick of aluminum alloy to withstand fuel slosh. The rib cap members and shear web inherently require gradual cross sectional change, eliminating load concentrations. The web provides continuous support for the wing cover panels for internal integral fuel tank pressures of up to 20–25 psi at the tip of the wing box.

The shear web rib is somewhat forgiving for small changes in load criteria or analysis, and is a distinct original release schedule and eventually has a "growth airplane" advantage. A truss type rib has none of the above advantages. In addition, it generally will be heavier, particularly on deep ribs where column lengths in compression are a problem. Truss member end design for fixity and concentrated loads of truss members are a distinct disadvantage, particularly where they attach to tension members. The one advantage of truss type rib is where tubing runs or control components require cutouts that do not leave room for a shear web.

The rib structure in the torque box should be put to double use wherever possible. Flap tracks supported from the same ribs that support the landing gear is the best example. When the airplane is on the ground, the flap has no load and vice versa. Another example is putting fuel management, internal tank ends at the nacelle support or flap track support ribs to take advantage of the heavier structure required for a tank end rib. In addition, the shear ties between stringers can readily be sealed.

All concentrated load points on the wing box such as engine nacelle support, landing gear support, and flap track support are designed to ensure that the wing box integral fuel tank will not break open and spill fuel in a crash condition. Fig. 8.5.3 shows examples of nacelle support and landing gear support fuses. Flap track fuses are similar. Shear pins are generally used since shear pin failures are most readily predictable. Shear failure testing, of course, is a necessity. The local backup structure in the wing box is designed approximately 15% stronger than the shear pin in

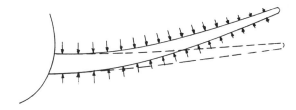

Fig. 8.5.4 Wing crushing loads due to flexure bending.

ultimate load condition.

This type of approach has proved successful in a number of instances in service, where violent training flight maneuvers have resulted in engines breaking away from the wing and where horizontal loads on the gear from landing short of the runway or where excessive sink rates have resulted in gear support failure.

Function of wing ribs
(1) Wing bulkheads are frequently constructed as solid webs, although webs with access holes or trusses may be used.
(2) Wing ribs carry the following loads:
  - The primary loads — acting on a rib are the external air loads and the transfer of them to the spars.
  - Inertia loads — fuel, structure, equipment, external stores (missiles, rockets, etc.).
  - Crushing loads due to flexure bending — when a wing box is subjected to bending loads, the bending of the box as a whole tends to produce inward acting loads on the wing ribs as shown in Fig. 8.5.4 Since the inward acting loads are oppositely directed on the tension and compression side they tend to compress the ribs.
  - Redistributes concentrated loads — such as nacelle and landing gear loads to wing spars and cover panels.
  - Supports members — such as skin-stringer panels in compression and shear.
  - Diagonal tension loads from skin — when the wing skin wrinkles in a diagonal tension field

the ribs act as compression members.
(3) The manner in which the rib structure resists external loads and reaction forces acting on the ribs depends on the type of construction.
- In the truss-type ribs the distributed external loads and reaction forces are applied as concentrated loads at the joints and the structure is analyzed as a simple truss. The outer members on which the distributed loads are relied upon to transfer these loads, in shear, to the points where they can then be considered as concentrated loads. These outer members are therefore subjected to combined bending and compression or bending and tension. Fig. 8.5.5 illustrates the structural stability requirement of truss rib cap.
- Shear web type ribs are usually employed either to distribute the concentrated loads, such as the nacelle and engine or landing gear to the shear beams.
- Webs with lightening holes and stiffeners are applied to resist bending moments by the rib cap members and shear by the web.
(4) Ribs must effect a redistribution of shear flows in a wing where concentrated loads are applied or where there is a change in cross section such as cutouts, dihedral change or taper change, etc.
(5) The analysis of rib is usually similar to that of a simple beam. The items to check are:
- Shear in web, or axial loads in truss members
- Rib caps due to bending loading on ribs
- Shear attachment of rib to spar and wing covers
- Tension attachment of wing covers to rib (usually gives a combined shear and tension loading)
- Effects of crushing loads on rib
- Effects of shear flow distribution on rib if it borders a cutout.
- Effects of loads normal to the plane of the rib from such items as fuel pressure, slosh etc.

Rib Spacing

The spacing of the wing ribs usually has to be established early in the design phase. Since the weight of the ribs is a significant amount of the total box structure, it is important to include the ribs in the overall optimization consideration of the structure. This is illustrated in Fig. 8.5.6 where the relative weight of ribs and cover panels is presented for a specific spanwise of the wing. It is advantageous to select a larger rib spacing; for equal structural weight it leads to cost savings and less fatigue hazards.

Wing rib spacing will increase with the depth of the wing box. Thus, considering the typical wing which is tapered in planform and in depth, the optimum wing structure would have a variable rib spacing with the maximum spacing at the inboard end. Of course, practical considerations such as alignment of a control surface support structure with box beam ribs and heavy skin for stiffness or fatigue reasons may dictate a compromise in box beam rib spacing.

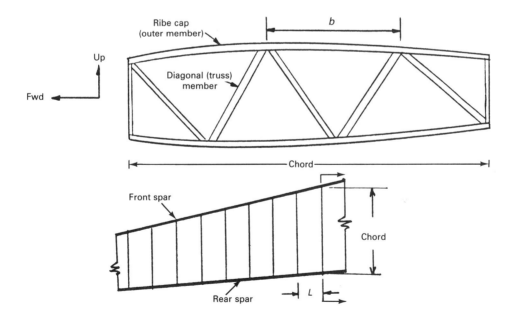

Rib cap stiffness required as in following equation:

$$\left(\frac{\pi^2}{4}\right)\left(\frac{E I_c}{L D}\right)\left(\frac{L}{b}\right)^4 \geq 1.0$$

Where: E = Modulus of elasticity of rib cap material
I_c = Moment of inertia of rib cap
L = Rib spacing
b = Distance between diagonals
D = Spanwise bending stiffness of the cover panel per inch of chord (E I cover /in)

Fig. 8.5.5 Structural stability requirement of truss rib cap.

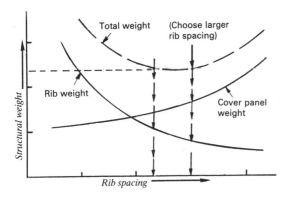

Fig. 8.5.6 Determination of rib spacing by structural weight comparison.

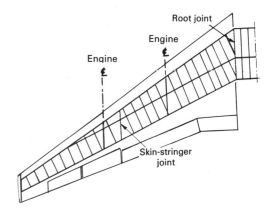

(a) DC-8

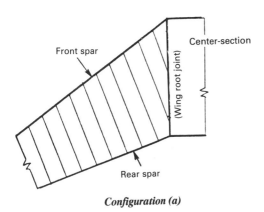

Configuration (a)

Configuration (b)

Fig. 8.5.7 Transport wing rib arrangments outside of wing root joint.

Wing rib arrangements outside of wing root joint is critical for designing the compression structural stability, especially for the wing upper surface. The rib spacing here is considered as important as the root joint design and the two basic arrangements are shown in Fig. 8.5.7.

Rib Arrangement of Modern Transports

Some typical wing rib arrangements (including spars) of modern transports are illustrated in Fig. 8.5.8.

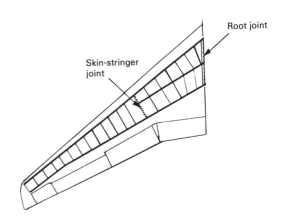

(b) Trident

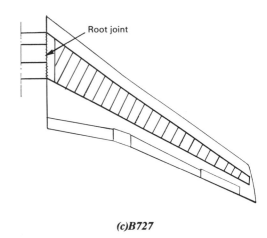

(c) B727

Fig. 8.5.8 Typical wing spar and rib arrangement of transports.

280 Airframe Structural Design

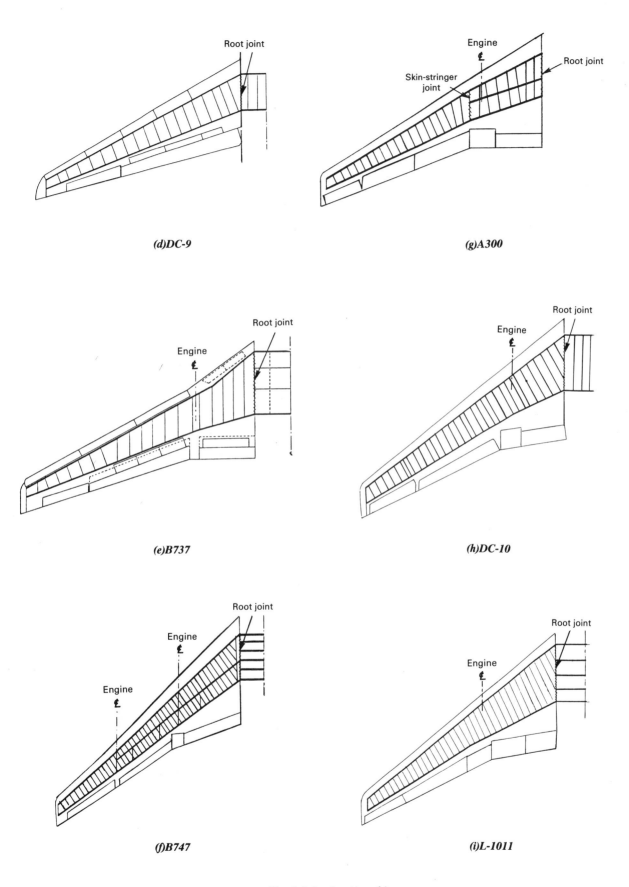

Fig. 8.5.8 (continued.)

Airframe Structural Design 281

8.6 Wing Root Joints

Wing joint design is one of the most critical areas in aircraft structures, especially for fatigue consideration of long life structure. The detail design and analysis of the joint has been given in Chapter 7.0.

There are basically two types of wing joint design, i.e. fixed joint and rotary joint, discussed in Fig. 8.6.1 and Fig. 8.6.2.

The best fatigue design, of course, is one with no joints or splices. This is accomplished on the modern transports, which have no joints across the load path except at the side of the fuselage. Wing sweep plus dihedral and manufacturing joint requirements make the joint at the side of fuselage necessary. Remarkable double shear, minimum eccentricity joints are shown in Fig. 8.6.3–Fig. 8.6.6. Note that the centroid of the skin and stringer segment is lined up with the neutral axis of the splice plate. It is important to keep the joint short. A long joint tends to pull load in from adjoining areas, therefore use two-row fastener joints (see Fig. 7.7.4) wherever possible.

At the stringer ends, the local skins are padded to reduce bearing stresses and tension stress around the fastener holes. Adequate ramps from the basic skin thickness to the padded thickness are a necessity. Do not use a variety of fits for the fasteners. Hole sizes should be held as tight as practical and close ream holes are used in those joints. For further detail design, refer to Section 7.7 of Chapter 7.0.

As customary in the design of any new airplane, much attention is paid to the design of joints. Many configurations were studied and the more promising one is subjected to small component tests in the development stages. The analysis of joints in the past proved to be very unreliable and tests were the only reliable methods for selecting a design. With the advantage of high speed computer techniques and improved structural simulation methods, it has become apparent that test and analysis plus complementary tools lead to a fast selection of proper configurations. Correlation of small component test results with analytical techniques will increase the probability of successful selection.

Joint	Ref. Fig.	Advantages	Disadvantages
Spliced plates	8.6.3	Widely used due to its light weight and more reliable and inherent fail-safe feature.	Slightly higher cost, manufactural fitness
Tension bolts	8.6.4	Less manufactural fitness required, easy to assemble or remove. More economic for military flighter with thin airfoil.	Heavy weight penalty
Lug (shear type)	8.6.5	(Same as above)	(Same as above)
Combination of spliced plates and tension bolts	8.6.6	Reliable and inherent fail-safe feature, and less manufactural fitness required.	Heavy weight penalty

Fig. 8.6.1 Wing root fixed joints.

Joint	Ref. Fig.	Advantages	Disadvantages
Swivelling pivot* (horizontal movement)	8.7.1 – 8.7.8	Required by variable (swivelling) geometry wing design for the best aerodynamic efficiency to accomplish missions required.	Structural weight penalty, high cost, complicate design, close tolerance requirements make manufacturing maintenance, repair, and replacement difficult and expensive
Folding joints	–	Mainly designed for the aircraft to fit on a carrier hangar deck. (with limited space)	Structural weight penalty, complicate joint control mechanism

*More detailed discussion can be found in Section 8.7.

Fig. 8.6.2 Wing root rotary joints.

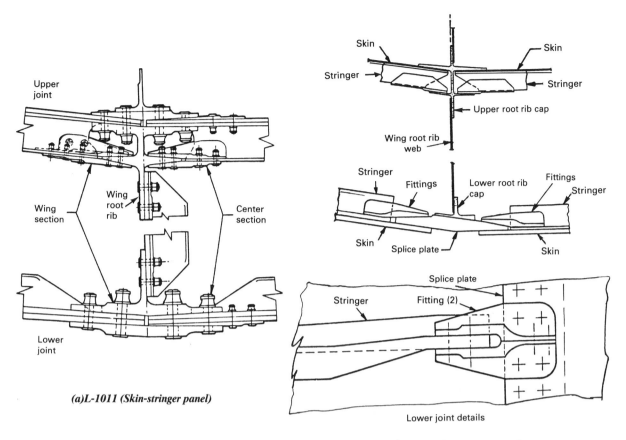

(a)L-1011 (Skin-stringer panel)

(b)B727 (Skin-stringer panel)

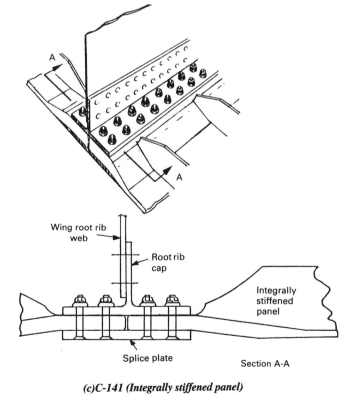

(c)C-141 (Integrally stiffened panel)

Fig. 8.6.3 *Wing root joint-spliced plates.*

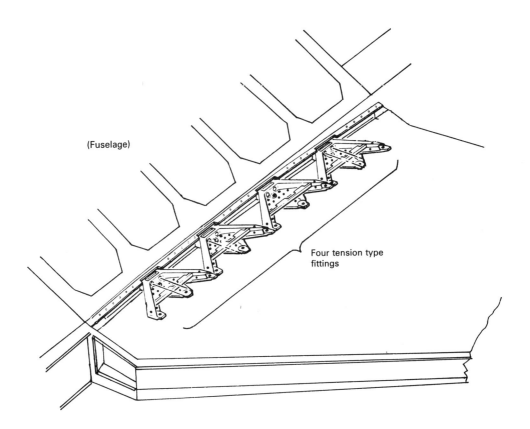

(a) F-16 fighter

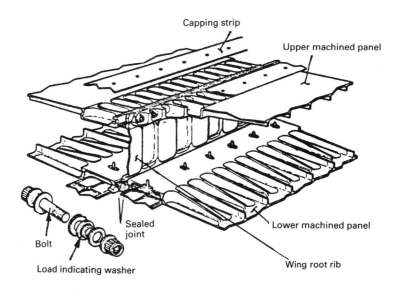

(b) Concord supersonic transport
(Courtesy of Aerospatiale)

Fig. 8.6.4 Wing root joint — tension bolts.

Fig. 8.6.5 Wing root joint — lug (shear type).

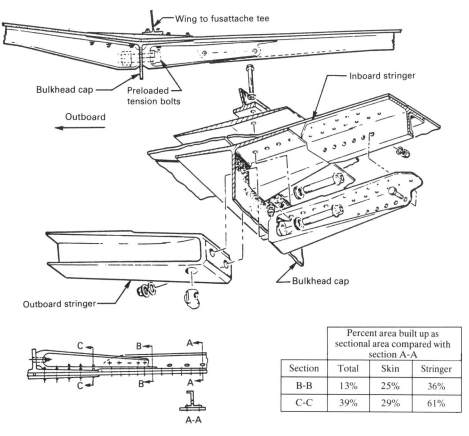

Section	Percent area built up as sectional area compared with section A-A		
	Total	Skin	Stringer
B-B	13%	25%	36%
C-C	39%	29%	61%

Fig. 8.6.6 Wing root joint — combination of spliced plates and tension bolts.

Airframe Structural Design 285

One of the peculiar designs is in the area of the sweepback and dihedral break of the transport wing box; spar caps are fabricated from machined forgings as shown in Fig. 8.6.7(a). The front spar forged cap extends from the airplane centerline to outboard of the fuselage; the rear spar forged cap extends from the airplane centerline to the aerodynamic break where there is a slight change in sweep angle of the rear spar. Spar webs are continuous in the high load transfer region of the sweep break. Continuous spar caps and webs across the dihedral and sweep break are a major factor in achieving good fatigue life.

The forged spar caps are spliced to machined extruded caps at the outboard ends. A typical spar cap splice is illustrated in Fig. 8.6.7(b). The caps and splice members are tapered to minimize stress concentrations. Spar cap areas are increased at the start of the joint to reduce the stress levels for increased resistance to fatigue.

Most of the lightly loaded wings for general aircraft adapt a single main front spar and an auxiliary rear spar construction. Therefore, the wing root joint usually is a triple point lug joint as illustrated in Fig. 8.6.8. The upper and lower lugs at the front spar pickup wing bending loads, vertical shear loads and wing torque; the single lug at the auxiliary rear spar takes wing vertical shear loads and torque only.

If it is a high swept fighter wing as shown in Fig. 8.2.3, the rear spar should be the main spar and the front spar is the auxiliary spar. To ensure the structural integrity and life, these lugs generally are machined from forging materials which is not only to reduce weight but also to minimize the local stress concentration. Double shear design as shown in Fig. 8.6.8 have to apply to all lug construction to obtain the most efficient joint; Fig. 8.6.9 illustrates highly loaded wing root joint designs for swept wing fighters.

A fail-safe design feature of "bolt within a hollow tube" may be required (multi-spar lug design may waive) in case the failure of hollow tube and then the bolt will take the fail-safe load as illustrated in Fig. 8.6.8.

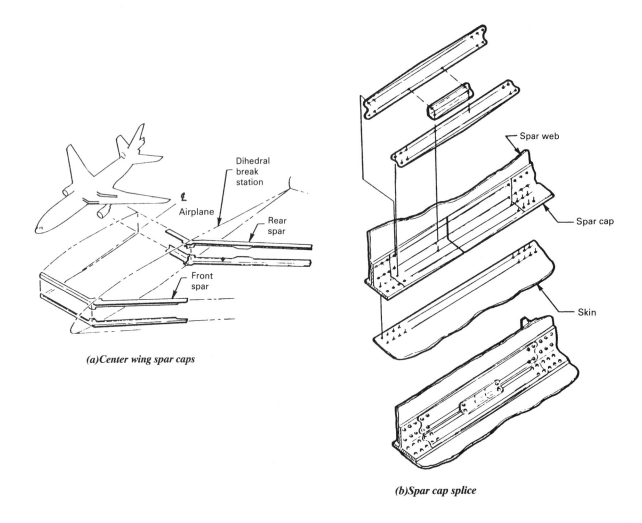

(a)Center wing spar caps

(b)Spar cap splice

Fig. 8.6.7 *Continuous spar caps across the wing diherdral and sweep break.*

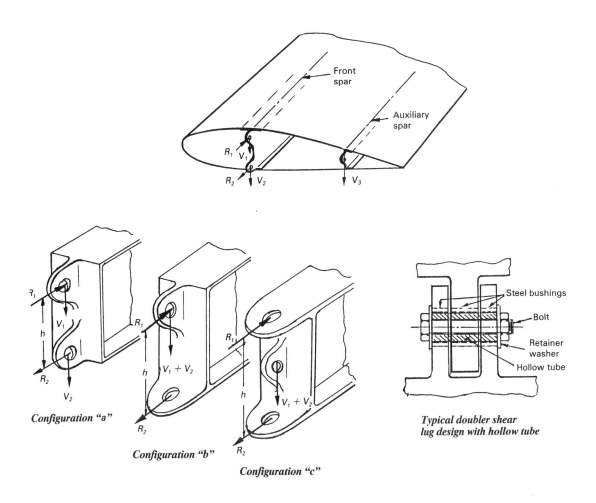

Configuration		Advantages	Disadvantages
"a"	Upper and lower lugs take axial load and vertical load shared by these two lugs	Stronger fittings Less machining cost	Smallest moment arm (h) and produce the highest lug axial loads Difficult to install due to its close tolerance holes requirement Vertical load distribution is difficult to predict
"b"	Upper and lower lugs take axial load and vertical load taken by upper lug only	Easy to install Load distribution is clear Longer moment arm (h) and then produce moderate lug load	Not enough space to install lower lug, if beam depth is too shallow
"c"	Upper and lower lugs take axial load and vertical load taken by center lug only	Easy to install Load distribution is clear Longest moment arm (h) and then produce small lug load	Heavy weight because of the third lug Highest machining cost Not enough space for the center lug, if beam depth is too shallow
Note: Always use double shear design with steel bushings to ensure fatigue life.			

Fig. 8.6.8 Wing lug design and comparison.

Airframe Structural Design 287

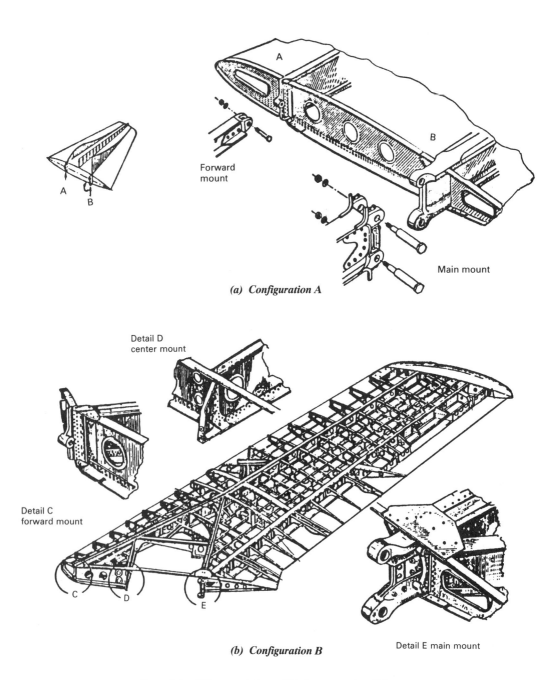

Fig. 8.6.9 Wing root joints of high swept wing fighters.

8.7 Variable Swept Wings

This section presents a broad survey of the structural problems associated with variable swept wing aircraft as shown in Fig. 8.7.1. Variable sweep allows an aircraft to fly throughout a broad regime of speed and altitude efficiently and without excessive power requirements. Tailored lift drag, improved ride quality, lessening of fatigue damage, and reasonable control sensitivities are advantages. Structural problems fall into two general categories: (a) Because of the number of wing positions, the equivalent of many fixed-wing aircraft must be investigated, analyzed and tested; (b) there are many unusual problems which have not be considered in conventional design.

Category (a) presents the problem of managing and assimilating large amounts of data. Computer programs and a family of cross-plots assist greatly in design. Category (b) presents new fail-safe criteria, a large number of possible flutter-critical configurations, unavoidable free play in mechanisms which affect flutter speeds. dynamic loads, pivot mechanism bearing life, and requires high reliability in materials. Analyses and wind-tunnel tests have shown that free play in mechanical joints may or may not cause significant service problems depending upon the mechanical arrangement selected and the actual degree of free play under service conditions.

One of the obstacles which has delayed development of the variable-sweep aircraft is the change in stability, control characteristics and structural stiffness as wing sweep is varied. Nevertheless, it has taken two decades of engineering developmemt to achieve this sweep wing ideal, because of the magnitude of the associated problems of structural and systems engineering and aerodynamic stability and control.

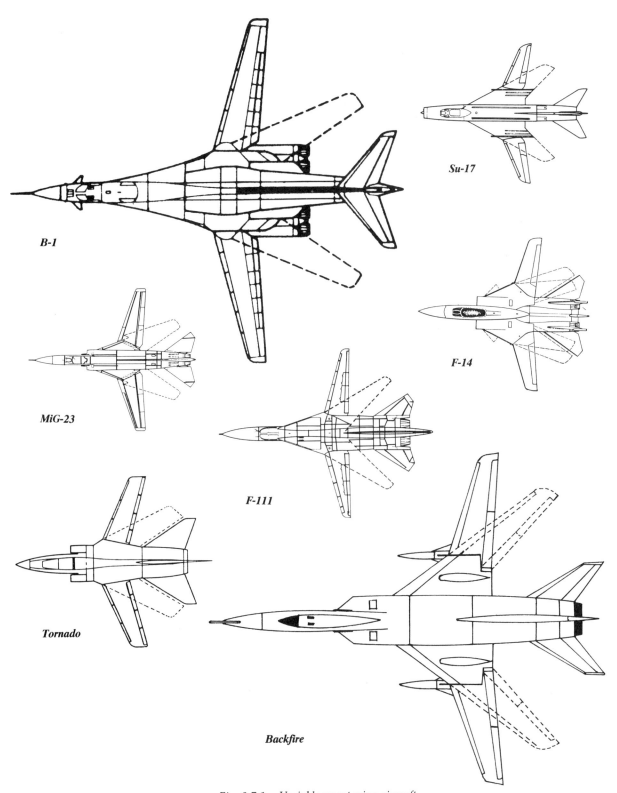

Fig. 8.7.1 *Variable swept wing aircraft.*

Airframe Structural Design 289

Pivot Mechanism

For the problems in which all, or a major portion, of the exposed wing surface is movable, there are many candidate arrangements in the pivot mechanism. Four examples have been selected for brief qualitative discussion.

(1) Concept I (Fig. 8.7.2)

The "shoe-in-track" type offers the advantages of low wing thickness requirements, a degree of redundancy of load paths, and an efficient load reaction means on the wing-box side of the joint. Based upon evaluations of this concept, it has been found several distinct disadvantages. An efficient strength design in this type has less rigidity than some other concepts. The arm-like beams "wipe" through a significant volume as the wing sweeps and, thereby, (a) reduce the available space for route electrical, mechanical, and fluid lines and, (b) require a relatively great expanse of fuselage structure which must be designed to support high static loads at many discrete points.

(2) Concept II (Fig. 8.7.3)

Employing a large ring of inclined roller bearings, offers low friction and requires only nominal depth of wing cross section. On the other hand, this concept has highly redundant load paths, has the same routing provisions problem as does Concept I, and requires a large number of parts fitting precisely with close tolerances.

(3) Concept III (Fig. 8.7.4)

A track-type concept with roller bearing contacts, offers a workable system requiring a minimum of wing thickness, distributed reaction loads, and low friction. Disadvantages evident in this design include: (a) the requirement for a large number of close tolerance parts; (b) the same routing difficulties characteristic of Concepts I and II; (c) many detail features which adversely affect functional reliability, and (d) the design is highly redundant and complicated in essentially all respects, particularly the actuation mechanism.

The Concepts I, II, and III share some common disadvantages — ice, debris, etc., could become major maintenance and reliability problems.

(4) Concept IV (Fig. 8.7.5)

A single vertical pin-clevis arrangement appears to be a design which is in many ways most ideal from a structural point of view. It offers the utmost in structural simplicity. Load paths are determined with confidence, a minimum of volume is occupied by the "hinge", actuation mechanism arrangement is simple, very few moving parts are involved, and a minimum weight design is possible. Disadvantages include: (a) relatively high journal bearing operating stresses, (b) somewhat greater local wing thickness is required in order to accomplish the necessary load transfers; and (c) great reliance is placed upon the integrity of single-load paths.

Many studies made for alternate designs have generally concluded that an arrangement similar to

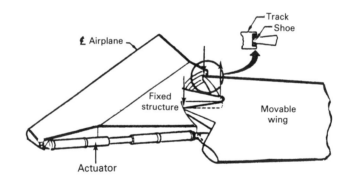

Fig. 8.7.2 Track and shoe design.

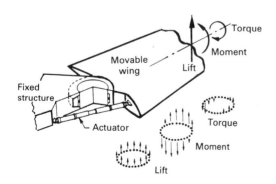

Fig. 8.7.3 Moment bearing design.

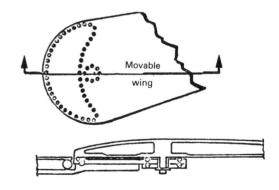

Fig. 8.7.4 Track with roller bearing design.

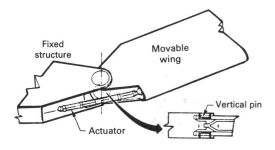

Fig. 8.7.5 *Vertical pin design.*

The method of transferring wing vertical shear from the movable to the fixed wing is one of the primary considerations in the vertical pin joint concept selection. A unique method is to "tilt" the planes of the bearings so that they intersect at the location of the net center of pressure of the movable wing.

Concept IV should be the choice. Some engineers would modify this concept to the extent of providing additional fail-safe features rather than relying so completely upon achieving a fool-proof safe-life design. This concept has been applied on the modern variable sweep wings.

Variations of these primary concepts are as numerous as the number of designers who toil away at the problem. The vertical pin design usually provides the lightest structural arrangement because it provides the least interruption to the wing bending load path. For extremely thin wings, the vertial pin design may not be feasible, and track and shoe or moment bearing designs will offer the best solution. Studies indicate, however, that the vertical pin concept is feasible for wing thickness-to-chord ratios down to 7–8%. Any further discussion of pivot points throughout the remainder of this section will be limited to the vertical pin concept.

In a vertical pin wing joint, wing bending moment is taken across the joint as a couple consisting of equal and opposite loads acting approximately parallel to and in the plane of the upper and lower wing skins. A vertical pin through the pivot axis transfers this couple force from the movable outer wing to the fixed center section. The two basic vertical pin joint designs, the single shear and the double shear joints, are described schematically in Fig. 8.7.6.

In the single shear joint, a full-depth vertical pin is required to balance the moments due to lug offset and this typical design is shown in Fig. 8.7.7.

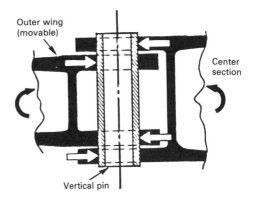

(a)Single shear

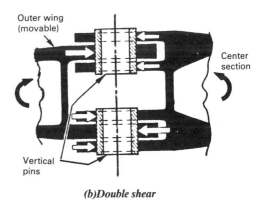

(b)Double shear

Fig. 8.7.6 *Typical vertical pin wing pivot design.*

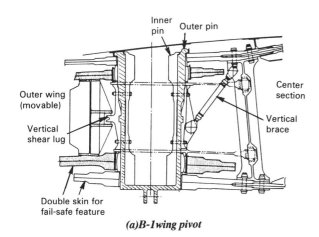

(a)B-1 wing pivot

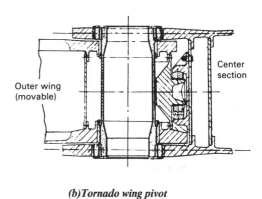

(b)Tornado wing pivot

Fig. 8.7.7 *Single shear pivot design (vertical pin design).*

Airframe Structural Design 291

This "tilted bearing" concept is described schematically in Fig. 8.7.8. Shear and bending load components are resolved into loads in the plane of the lugs and pivot bearings. Self-aligning spherical bearings are required for this joint design because of the slight misalignment which occurs between inboard and outboard lugs when the wing sweeps away from the optimum points. Separate shear transfer structure

and hold. The fixed glove area increases in size and weight as the pivot point moves outboard. As mentioned before, the incorporation of flow control devices in the fixed glove may be necessary if the glove becomes too large and this, too, means a weight increment. Total wing structural weight is seen to decrease as the pivot moves outboard over the practical range considered.

(a) A truss

These bearings were tilted and aimed toward the airload center-of-pressure airloads and bending moments are carried as bearing radial pressure

(b) Added pivots

This arrangement eliminates the need for vertical shear webs and thrust bearings in the pivot

(c) Applied on swept wing

Fig. 8.7.8 *Wing pivot truss concept.*

and separate thrust bearings are eliminated in this design.

The structural weight of the wing is of course sensitive to pivot point location. In general, the optimum spanwise location of the pivot point for minimum wing weight will be further outboard than the aerodynamic considerations will allow. A trade-off must be made to arrive at an overall optimum. As shown in Fig. 8.7.9, the major wing items that are subject to weight variation with pivot location are as follows:
(a) Outer wing box
(b) Fixed carry-through box (center section)
(c) Pivot lug area
(d) Sweep actuator
(e) Fixed glove

As the pivot point moves further outboard, it can be seen that the outer wing box weight will decrease because its span is shortened. On the other hand, the fixed carry-through box weight goes up as it becomes longer. The pivot lug area will show a decrease in weight because the wing bending moment is decreasing faster than the wing thickness; consequently, the spanwise axial load that must be carried by the lugs decreases. The sweep actuator will show a weight decrease because it has a smaller outer panel to sweep

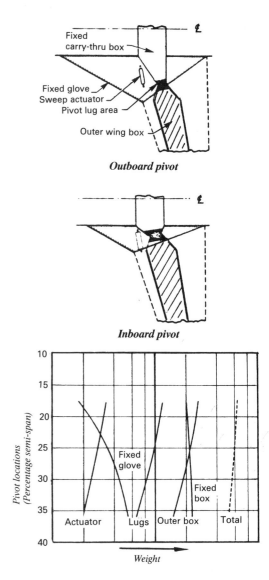

Fig. 8.7.9 *Swept wing component weight vs. pivot location.*

Structural Dynamics

As in the case of static loads for design, variable swept wings present a multiplicity of possible dynamic load and flutter considerations. Dynamic loads overshoot and the dynamic influences on repeated loads are very possibly affected by the stiffness and free play in the pivot mechanism. Flutter-free flight must be assured over a broad flight regime for many wing positions, and quite often for a multiplicity of possible wing mass configurations — such as a variety of wing-mounted weapons, external fuel tank, and internal fuel variations. If external stores must be retained throughout the range of wing sweep positions, another variable is added in the form of the articulating pylon supports which are required to align the stores streamwise.

The mechanism such as that shown in Concept IV (Fig. 8.7.5) can be thought of as an interruption, or discontinuity, in the structure of the wing box as compared to the straight through uninterrupted wing box of the same typical fixed wing aircraft. If the torsional, bending, and shear stiffnesses of the beam structure through this region do not differ significantly from those expected in a fixed wing, and, if free play (see Fig. 8.7.10) in the joints is negligible, unusual flutter problems would be unlikely.

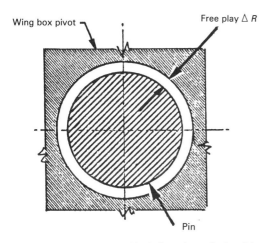

Fig. 8.7.10 A pivot pin and hole free play relationship.

In any rotating or moving joint similar to a wing pivot mechanism, free play to some degree is almost certainly unavoidable — at least without some elaborate, complex preload scheme. Interference fits in manufacturing are not compatible with low friction and minimum wear journal, ball, roller, or sliding shoe bearings. Some small freedom for alignment and uniform pressure distribution is necessary. On the other hand, excessive free play is definitely undesirable. Close tolerance requirements make manufacturing, maintenance, repair, and replacement difficult and expensive. So, the structure engineers have the problem of deciding how much free play he can permit. It is too generous in allowing clearances and tolerances to simplify and economize in manufacturing, it may very well reduce the serviceability of the article. Free play in the joint or joints of the pivot mechanism potentially affect flutter speeds, dynamic wing loads, and can "pound out" bearings.

For several years there has been the necessity to consider free play effects in irreversible control surfaces such as tabs, elevators, rudders, etc. Tests on all movable "slab type" horizontal tail surfaces which are controlled in pitch by linkages and actuators have shown a significant flutter sensitivity to free play in the pitch direction. Relatively small amounts of free play can lower flutter speeds by 15–20%.

Pivot materials

As mentioned previously in discussions of the various pivot mechanical concepts, loads are large and highly concentrated and space or volume in the vicinity of the pivot is at a premium. These facts almost certainly dictate the use of high strength and dense materials, such as steel. To avoid excessive weight, non-optimum material must be eliminated to the maximum degree practicable in the design. Experience has shown that the reliability of these high-strength materials is quite sensitive to variations in manufacturing processes. This means that precise process standards must be established and rigidly followed during manufacture. Generally speaking, bearing material and lubrication performance demands are greater than those of "off-the-shelf" bearings For the case of the journal bearing type of pivot joint, one might desire to have a secondary or dual rotating bearing surface as a fail-safe feature.

Structural Fatigue

A benefit of paramount importance to the structural engineer is reduced fatigue damage. For the same reasons that a variable swept wing offers the option of a "smoother ride" in a given environment, so does this geometric feature offer the opportunity to reduce or minimize structural and material fatigue damage. The less a vehicle responds to a given level of turbulence, the less the magnitude of cyclic or repeated loads and stresses. These facts offer the opportunity to achieve long service life with a minimum of structural weight penalty for military bombers and commercial transports, and, in some cases, make the gust sensitivity of structure of high manoeuvre-factor fighters non-existent.

Fail-safe Considerations

There also arises the question as to whether fail-safe features must be incorporated in this vital mechanical pivoting system. Very few, if any, landing gears have alternate load paths to carry load in case of major member failure. One may think of the wing pivot mechanism as just another mechanism like that of the landing gear. However, there are differences in the results of failure. Many landing gear failures do not result in loss of the vehicle nor loss of life or serious injury to occupants. The landing gears are required to function only during taxi, take-off, and landing and have little influence on safety during most of the flight. On the other hand, at all times during aircraft operation, with the possible exception of low taxi speeds, the loss of a wing panel will almost certainly result in loss of the vehicle and occupant lives. The reliability of a wing pivot mechanism, therefore, should be considered even more vital to flight safety.

Incorporation of alternate load paths in pivot mechanisms such as that previously shown in Concept IV (see Fig. 8.7.5) leads to problems. Generally speaking, more premium space or volume is required, more structural weight will be involved, and it is conceivable that the added complexity of design can adversely affect maintenance operations, safe-life type of material reliability, and probability of mechanical malfunction. For example, consider a fighter designed for high-maneuvre-load factor such as a $7.33g$ limit or $1.5 (7.33) = 11.0g$ ultimate. A minimum weight design might have two elements in parallel resisting the ultimate load, and, in such a case, the failure of one element would leave fifty per cent of the ultimate strength available. The remaining strength would normally correspond to that required for 5.5g manoeuvres. The aircraft safety is now entirely a function of the probability of encountering 5.5g loads before failure is detected and repaired. This probability, according to manoeuvre statistics, is high for fighter and attack aircraft. Thus, essentially no measure of safety has been gained. For the fail-safe features to be of value, additional load paths in parallel, or more than 50% ultimate strength in the remaining member is required. Providing the additional strength or a multiplicity of load paths can be significant problems in thin, space limited, high performance fighter wings.

In the case of passenger transports, fail-safe features more easily add to the reliability with less penalty, since these aircraft normally experience incremental load factors which are much smaller percentages of their ultimate design capability. Adequate and thorough consideration of fail-safe design philosophy most certainly must be greater in depth than the above criteria philosophy. These expressed philosophies, however, should serve as a reminder that the structural designer has a new important task.

Carry-through (Center Wing) Structure

Design of a carry through structure which will satisfy all operational requirements is a complex interdisciplinary task. It requires detailed coordination to ensure that all design criteria and interfaces are properly considered. Geometric constraints, design and operational loadings, weight goals, material characteristics, service environment, service life, and damage tolerant criteria must be considered.

(1) Space Truss

A space truss design which would use uniaxially loaded members that enable individual members to be placed in a manner which would effectively transmit the design loadings. Good damage tolerance performance could be obtained by use of multi-element members. This design concept has a high structural efficiency for wing-fuselage loads. However, the truss became highly redundant and inefficient for point (or concentrated) loadings at landing gear supports, etc. This factor, coupled with the nonstructural assembly required to contain fuel caused the total weight to be higher (see Fig. 8.7.11).

(2) Constant depth box

This design is much simpler in geometry and generally easier to fabricate. It eliminates the discontinuities in the upper cover and minimizes the size of spars and ribs (see Fig. 8.7.12).

Reduction of the depth of the box, however, requires heavier covers to carry the same design loadings. In addition, weight must be included to provide fuel tanks. Deflection compatibility between these tanks and the wing carry-through structure produced some unusual design problems. The total weight of this design generally is found to be heavier but has been adopted on many fighters and one of these examples is shown in Fig. 8.7.13.

(3) Conical box

A conical design is highly efficient in carrying torsional loadings and typical of those applied to the wing carry-through structure when the wings are in the swept position. However, a review of the design loading distribution shows that a large portion of the shear loadings is removed at the side of body. Also, this configuration is inefficient in bending and for fuel containment, nevertheless, it is the highest manufacturing cost design (see Fig. 8.7.14).

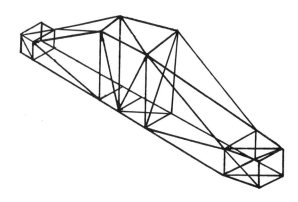

- *Utilize uniaxially loaded truss members to transmit carry-through structure load*
- *Payoff would be high structural efficiency and low cost*
- *Damage tolerance would be through multi-element members*
- *Fuel containment would be by bladder cells or other low modulus shell*

Fig. 8.7.11 Carry-through structure — space truss.

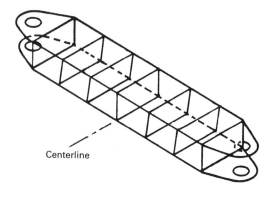

- *Simple geometry for welding*
- *Lighter support structure*
- *Heavier cover panels*
- *Less fuel capacity*

Fig. 8.7.12 Carry-through structure — constant depth box.

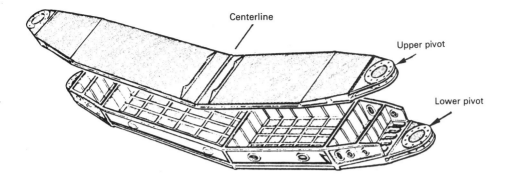

Fig. 8.7.13 F-14 wing carry-through box.
(Courtesy of Grumman Corp.)

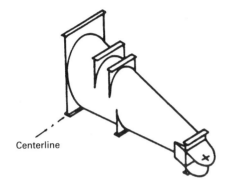

- *The center section is a truncated cone shape with pivot lug welded at the outboard end*
- *Spokes or rods are used to stabilize the cone skin*
- *Fuel volume is substantially decreased*

Fig. 8.7.14 Carry-through structure — conical box.

(4) Minimum rib and splice design
The design can be simplified by minimizing the number of ribs and splices. This approach generally increases the efficiency and reduces the manufacturing costs. However, the elimination of these items reduces the damage containment capabilities of the structure. Significant material (weight) must be added to provide the required level of damage tolerance (see Fig. 8.7.15)

(5) Built-up variable depth box
This configuration is so designed to fit the "blended" wing and body aerodynamic design of the specific aircraft involved. The forward and aft of the carry-through box may be more regarded as fuselage structures. Thus, they are characterized by provisions for longitudinal bending strength rather than wing spanwise bending strength. The wing carry-through structure inboard of the pivot pin is the principal spanwise bending component, and it should be designed by multi-spar elements for fail-safe consideration. A substantial proportion of the fuselage bending strength is contributed by the principal longerons as indicated in Fig. 8.7.16.

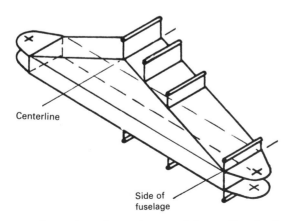

Fig. 8.7.15 Carry-through structure — minimum rib and splice design.

Airframe Structural Design 295

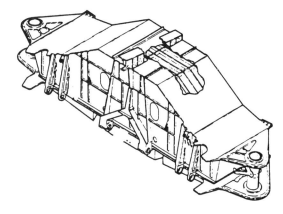

- *Higher structural stiffness and lighter structure*
- *Good fuel capacity*
- *Highly efficient for concentrated load supports*
- *Relative improvement in damage tolerance, inspectability, repairability, etc.*
- *Not practical for passenger transports*
- *Best fit for "blended" wing carrythrough structure*

(a) Built-up variable depth box

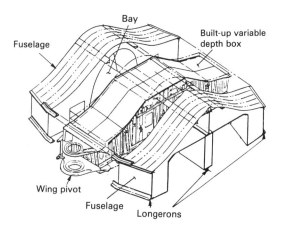

(b) Built-up variable depth box and fuselage interfaces

Fig. 8.7.16 *Carry-through structure — built-up variable depth box.*

8.8 Wing Fuel Tank Design

Fuel tanks for aircraft may be constructed of aluminum alloy, fuel-resistant materials, or stainless steel. Tanks that are an integral part of the wing are of the same material as the wing and have the seams sealed with fuelproof sealing compound. The material selected for the construction of a particular fuel tank depends upon the type of airplane in which the tank will be installed and the service for which the airplane is designed. Fuel tanks and the fuel system, in general, must be made of materials that will not react chemically with any fuels that may be used. Aluminum alloy, because of its light weight, strength, and the ease with which it can be shaped and welded, is widely used in fuel tank construction.

In the earlier day, many aircraft used synthetic-rubber bladders for fuel cells in the wing box and even today many fighters still use bladders in fuselage fuel tanks for combat bullet-proof purposes.

Metal fuel tanks generally are required to withstand an internal test pressure of 3.5 psi without failure or leakage. Furthermore, they must withstand without failure any vibration, inertia loads, and fluid and structural loads to which they may be subjected during operation of the aircraft.

Fuel tanks located within the fuselage of a transport aircraft must be capable of withstanding rupture and retaining the fuel underneath the inertia forces that may be encountered in an emergency landing. These forces are specified as $4.5g$ downward, $2.0g$ upward, $9.0g$ forward, and $1.5g$ sideward. Such tanks must be located in a protected position so that exposure to scraping action with the ground will be unlikely.

A number of special requirements are provided for the installation of fuel tanks in aircraft. These requirements are established principally to provide for safety, reliability, and durability of the fuel tank.

The method for supporting tanks(separate metal tanks inside of the wing box structure) must be such that the fuel loads are not concentrated at any particular point of the tank. Non-absorbent padding must be provided to prevent chafing and wear between the tank and its supports. Synthetic-rubber bladders or flexible fuel-tank liners must be supported by the structure so that they are not required to withstand the fuel load. A positive pressure must be maintained within the vapor space of all bladder cells under all conditions of operation, including the critical condition of low airspeed and rate of descent likely to be encountered in normal operation. Pressure is maintained by means of a tank vent tube, the open end of which is faced into the wind to provide continuous ram air pressure.

The testing program should be carried out in connection with an integral tank program which should consist of the following parts:
(a) tests to prove the suitability and physical properties of sealing materials
(b) tests to prove the application of sealants to detail structural design
(c) test of the final integral tank as designed.

As a consequence, the following criteria has been established.
- Compression surfaces should not have compression wrinkles at limit load (some designers prefer $1.5g$).
- Shear panels should not wrinkle at limit load (some prefer $1.5g$).

In designing any kind of a tank, the engineer has to depend on quality of workmanship. If workmanship is known to be questionable or poor, the designer must go to unusual lengths in his design to be sure that the production line will fulfill its requirements.

Transport Wing Fuel Tanks

The definition of transport wing box is a thicker airfoil with bigger wing planform which the interior can be accessible through access holes for tank sealing and repairing. The following items must be considered in the design of a transport wing fuel tank:

(1) Doublers, or other flat metal strengthening members, are to be so designed that sealing material can be applied where these members intersect or paralled other structure. Flange spar caps, ribs and bulkheads, should be kept away from the tank cavity. The tank cavity must be accessible for the replacement of all rivets or other attachments. It should be possible to make any spot repair in the tank within one hour.

(2) The structure must be arranged to permit complete drainage to one point and corner seals must be flexible and large enough in size to isolate fittings and other complicated structures in the tank corners. Avoid locating end bulkheads at places where the structure is complicated. Use additional bulkheads, if necessary, to accomplish this.

(3) Avoid the use of countersunk rivets in the tank area, if possible. All rivets must be tight and special fabrication precautions should be made to insure that such tightness is obtained. Access doors on lower surfaces should seal on the inside of the structure.

(4) Minimum sheet thickness should be 0.064 inch with exceptions in certain areas as determined from the amount of bearing and the rivet configuration. The absolute minimum sheet thickness should be 0.05 in. Do not permit gaps between structural members if there is any way to avoid them. The maximum gaps in the structure must not be any greater than 0.03 in.

(5) All parts, accessories, brackets, clips, fittings, and attachments that require service removal or replacement should be so mounted that removal will not disturb cavity integrity.

(6) All tank walls should be accessible on both the interior and exterior surfaces so that leakage detection and repair can be readily accomplished.

(7) The number of parts comprising a tank wall should be kept to a minimum. Design for the combining of parts wherever possible.

(8) Fewer parts mean fewer seams to leak, fewer fasteners penetrating the seal plane, fewer fillets to apply, and less chance of channeling in case of a leak. Avoid abrupt section changes and sharp corners in the vicinity of a seal. Tank wall intersections of less than 90° increase the difficulty of cleaning, sealing, and repairing seals and therefore decrease the seal plane reliability.

(9) It is of utmost importance to design integral fuel tank subassemblies (that is , skin panels, spars, tank end ribs) so that they may be as structurally complete on the seal plane as possible prior to major assembly. This will generally allow the use of fasteners or bolts to obtain the primary seal gaps, and will greatly facilitate sealing operations.

(10) Domenuts as shown in Fig. 8.8.1 are not recommended for primary use on the tank seal plane. Not only do nutplates make additional seal, but are also easily failed because of overtorquing by too long a grip length of screw.

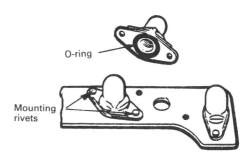

Fig. 8.8.1 Domenut (sealing nut).

(11) Wherever possible, design the parts so that the fillet seal is on the unloaded edge of the joint as illustrated in Fig. 8.8.2. The relative importance of this requirement depends primarily on the structure gages and fastener pattern. When heavy gages are involved and/or locating the fillet seal on the unloaded seam causes an increase in seal plane complexity (that is additional parts and/or fasteners, preassembly seals, etc.), consideration can be given to applying the seal on the loaded edge.

(12) Avoid placing the seal where there is a difference in rigidity of the parts being sealed. The sealed parts should have similar rigidities.

(13) All fillet seals should be readily accessible for proper application of the fillet material, for inspection after the seal is applied, and for repair in service.

(14) Avoid conditions where a seam is not in the direct line of vision of the sealer. See Fig. 8.8.3.

(15) All portions of tank should be readily accessible by convenient access doors. Having to crawl through one compartment to reach another may be considered.

(16) Tubing and equipment should be kept away from tank walls to increase accessibility to sealed walls.

(17) Stringer termination at tank end ribs generally improves sealing reliability between the two adjacent cavities. Continuous stringers (through tank end ribs) can reduce tank-to-tank sealing reliability, but their use improves the sealing that prevents external leakage. This is because the continuous stringer concept has a more consistent structural section (no hard spots) and fewer potential leak paths to the exterior when compared to the spliced stringer design.

(18) Closed section stringers (excluding vents) should not be used because their interior is inaccessible.

(19) Stringer ends should be designed without splice angles, splice tees, or splice fittings. Machined

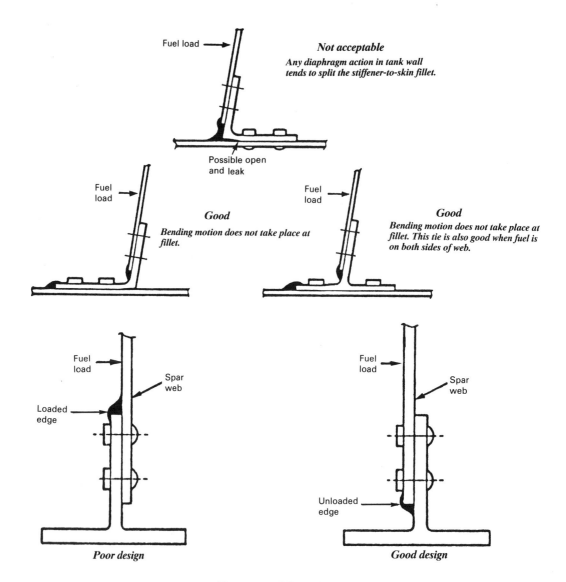

Fig. 8.8.2 Fillet seals.

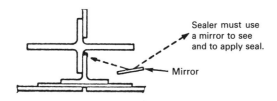

Fig. 8.8.3 Seal seam.

stringer ends are preferred because they eliminate the preassembly seal associated with splice angles and reduce the number of parts on the seal plane.

(20) Wherever possible, the stiffeners shall be placed on the dry side of the tank wall to avoid sealing around the stiffener. This also provides a convenient attachment point for brackets, etc. without penetration of the tank wall.

(21) All items inside and outside of the tank should be attached to stiffeners or intermediate ribs wherever possible to avoid loading or penetrating the tank wall.

(22) Chordwise skin joints are not desirable because this design requires splice structure across the primary load paths. This in turn would cause additional leak paths in an area that is highly susceptible to working.

(23) Fuel tank ventilation —
- Fuel tank must be suitably vented from the top portion of the air space. Such air vents should be arranged so as to minimize the possibility of stoppage by ice formation.
- Where large fuel tanks are used, the size of the vent tubes or other means should be so proportioned as to permit rapid changes in internal air pressure, thereby preventing collapse of the tank in a steep glide dive.

(24) Install all fasteners wet.

Transport Fuel Tank Geometry vs Structural Wing Relief

Careful fuel management can result in lighter weight wing structures. Fuel management is an important consideration in the structural design of an aircraft. Each of the three cases shown in Fig. 8.8.4 represents a transport wing (fuel in military fighter aircraft is generally stowed in fuselage and external fuel tanks, but in modern fighters, fuel has also been stowed in the wing). The c.g. fuel management is important with internal fuel tanks. The weight of the fuel supply acts down at its c.g. This creates a counterclockwise bending moment at the root. These moments are subtracted to obtain the final root bending moment.

In Case A there are two fuel tanks. By feeding first from the inboard tank and then from the outboard, a counterclockwise moment corresponding to trace A is obtained.

In Case B, there are two fuel tanks; however, the inboard tank is much longer than in Case A. Therefore, its c.g. remains further outboard and a counterclockwise moment corresponding to Case B results.

In case C, there are three tanks and by feeding first from the root tank, next the middle, and finally the outboard, a moment corresponding to trace C is obtained. Since these values are the highest in magnitude, it is advantageous to use them. However, it is not always possible to position fuel tanks in the optimum locations.

If, for example, the engines are placed on the wing, fuel tanks may not be placed adjacent to them. Tanks may not be placed in areas where their mass might induce flutter.

Furthermore, fuel management procedures must not be too complex or the airline personnel will object. In short, considerable coordination is required to come up with a satisfactory method which results in an optimum structure.

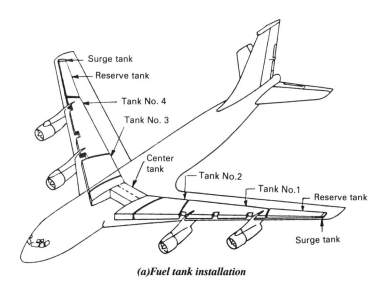

(a) Fuel tank installation

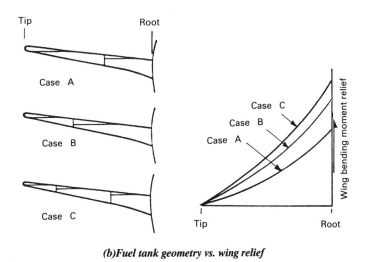

(b) Fuel tank geometry vs. wing relief

Fig. 8.8.4 *Transport fuel tank geometry vs. wing load relief.*

Fighter Wing

High speed fighters usually require thin wings which, from structural stand point and internal space, the wing covers are designed without access holes. One of the requirements is to repair the tank seal externally without removing wing covers. For those restrictions, the groove injection sealing (non-curing sealant) is the only method to seal the fuel tank rather than the common fillet seal for transports with thicker wing sections. The grooves must form a continuous interconnected network completely enclosing the fuel on all sides as illustrated in Fig. 8.8.5.

The non-curing sealant are injected by high pressure gun through a special insert. A series of inserts are installed permanently on the wing upper and lower convers for the horizontal grooves and on the spar web for the vertical grooves. The grooves can be located either along fastener rows or between fastener rows as shown in Fig. 8.8.6.

The groove could be machined either from substructure or wing covers (not recommended for certain cases) and the final selection should be based on the structural strength requirement. Fig. 8.8.7 illustrates the design cases, and the mismatch gap between the two contact surfaces should keep flat and under 0.003 in. The most critical loactions are at the corners where the spar and enclosed rib are joined. Their depth mismatch should also be kept to the minimum, otherwise a specially designed tapered shim should be installed to smooth the mismatch problem.

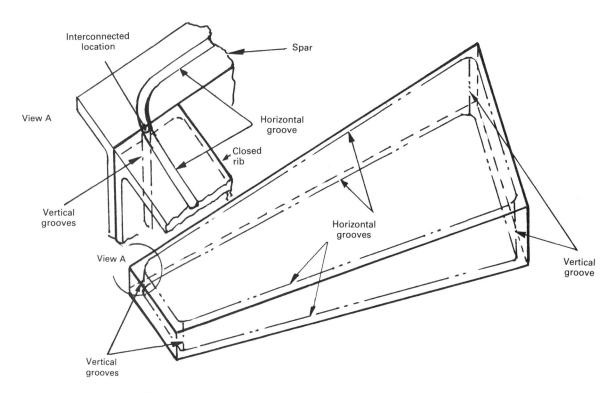

Fig. 8.8.5 Interconnected network groove sealing of a wing box.

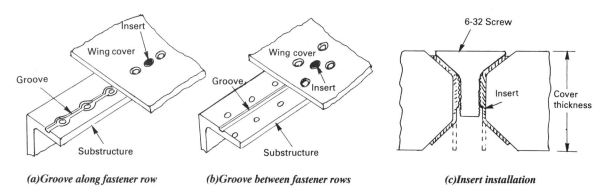

(a) Groove along fastener row *(b) Groove between fastener rows* *(c) Insert installation*

Fig. 8.8.6 Groove locations and insert.

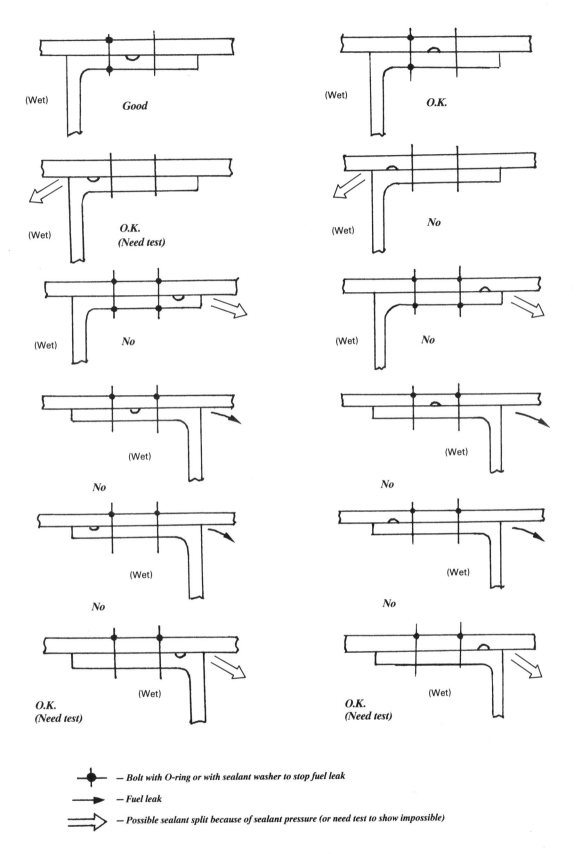

Fig. 8.8.7 *Fuel groove design cases.*

References

8.1. Catchople, E.J.: 'The Optimum Design of Compression Surfaces having Unflanged Integral Stiffeners:' *Journal of The Royal Aeronautical Society*, (Nov. 1954), 765−768.
8.2. Saelmen, B.: 'Basic design and Producibility Considerations for Integrally Stiffened Structures.' *Machine Design*, (Mar. 1955)
8.3. Cozzone, F.P. and Melcon, M.A.: 'Nondimensional Buckling Curves — Their Development and Application.' *Journal of The Aeronautical Sciences*, (Oct. 1946).
8.4. Farrar, D.J.: 'The Design of Compression Structures for Minumum Weight.' *Journal of The Royal Aeronautical Society*, (Nov. 1949).
8.5. Anon: 'AFFDL-TR-79-3047.' *Aircraft Integral Fuel Tank Design Handbook*, (1983)
8.6. Osgood, Carl C.: *Spacecraft Structures*. Prentice-Hall, Inc., New Jersey, 1966.
8.7. Asbroek, J.G.: *Contributions to the Theory of Aircraft Structures*. Delft University Press, Groningen, The Netherland, 1972.
8.8. Emero, D.H. and Spunt, L.: 'Wing Box Optimization under Combined Shear and Bending'. *J. Aircraft*, (Mar.−Apr. 1966).
8.9. Becker, H.: Handbook of Structural Stability, Part II — Buckling of Composite Elements, *NACA TN 3782*, (July 1957).
8.10. Baker, Allen S.: 'Integral Fuel Tank Design.' *Aero Digest*.
8.11. Barfoot, J.E.: 'Design of Sweptback Wings.' *Aero Digest*, (June 1947).
8.12. Williams, W.W.: 'An Advanced Extensible Wing Flap System for Modern Aeroplanes.' *AIAA Paper No. 70-911*, 1970.
8.13. Pazmany, L., Prentice, H. and Waterman, C.: 'Potential Structural Materials and Design Concepts for Light Airplanes.' *NASA CR-73258*, (Oct. 1968).
8.14. Smetana, F.O.: 'A Study of NACA and NASA Published Information of Pertinence in the Design of Light Aircraft.' *NASA CR-1484*, (Feb. 1970).
8.15. Torenbeek, E.: *Synthesis of Subsonic Airplane Design*. Delft University Press, The Netherland, 1976.
8.16. Anon: Lockheed-California Company, *Fatigue and Stress Corrosion Manual for Designers*, (Nov. 1968)
8.17. Goff, Wilfred E.: 'A-300B Wings Production.' *Flight International*, (Dec. 9, 1971).
8.18. Staff: 'Supercritical Wing.' *Flight International* (Jul. 1, 1971)
8.19. Staff: 'Fixed or Variable Geometry Wings?'. *Interavia*, (Mar 1964).
8.20. Heath, B.O.: 'Problems of Variable Geometry Aircraft.' *Flight International*, (Oct. 12, 1967).
8.21. Special Report: 'European Airbus — Full Details and Cutaway Drawing of The Airbus Industrie A-300.' *Flight International*, (Apr. 6, 1972).
8.22. Burns, B.R.A.: 'The Ins and Outs of Swing Wings.' *Flight International*, (Jul. 13, 1972).
8.23. O'Malley, Jr. J.A. and Woods, R.J.: 'First Revelation of Details of World's First Variable-Sweep Research Airplane, Bell X-5.' *Aero Digest*, (Aug. 1953).
8.24. Bodet, P.: 'Contribution to the Structural Analysis of Swept Wings.' *Aircraft Engineering*, (Jul 1954).
8.25. Anon.: 'The DeHavilland DH 121 Trident.' *Aircraft Engineering*, (May, 1962).
8.26. Anon.: 'Vickes VC-10.' *Aircraft Engineering*, (Jun 1962).
8.27. Anon.: 'BAC One-Eleven — Structural Design.' *Aircraft Engineering*, (May, 1963).
8.28. Anon.: 'Short Belfast: Structural Design.' *Aircraft Engineering*, (Sept. 1963).
8.29. Anon.: 'Engineering Problems Associated with Supersonic Transport Aircraft.' *Aircraft Engineering*, (Nov., 1963).
8.30. Anon.: 'Trident: Structural Design.' *Aircraft Engineering*, (Jun 1964).
8.31. Swihart, J.M.: 'Variable Sweep Today.' *Aircraft Engineering*, (Oct. 1965), 300−308.
8.32. Alexander, Jr., M.M.: 'Structural Problems Associated with Variable Geometry.' *Aircraft Engineering*, (May, 1966), 34−39.
8.33. Heldenfels, R.R.: 'Structural Prospects for Hypersonic Air Vehicles.' *Aircraft Engineering*, (Nov. 1966), 18−20.
8.34. Bradshaw P.: 'Design Features of the Hawker Siddeley 748.' *Aircraft Engineering*, (May 1967), 45−48.
8.35. Anon.: 'Fokker F.28 Fellowship.' *Aircraft Engineering*, (Jun 1967), 10−45.
8.36. Landgral S.K., and Herring, R.N.: 'Aerodynamic Design Considerations of Variable Geometry Aircraft.' *SAE No. 670880, Society of Automotive Engineers, Inc.*, (1967)
8.37. Swihart, J.M.: 'Variable Geometry in a Supersonic Transport Aircraft.' *SAE No. 670878, Society of Automotive Engineers, Inc.*, (1967).
8.38. Harvey, J.W.: 'Structural Considerations for Variable Sweep Wings.' *SAE No. 670881, Society of Automotive Engineers, Inc.*, (1967).
8.39. Paget, F.M.: 'Mechanical Aspects of Variable Sweep Wings.' *SAE No. 670882, Society of Automotive Engineers, Inc.*, (1967).
8.40. Newell, G.C. and Marsh, F.E.: 'Reliability of the SST Wing-Sweep Actuation System.' *SAE No. 670884, Society of Automotive Engineers, Inc.*, (1967).
8.41. Foreman, C.R.: 'The Design and Testing of a Wing Pivot Joint for Variable Geometry Aircraft.' *AIAA/ASME 11th Structures, Structural Dynamics, and Materials Conference*, (1970).
8.42. Brown, D.A.: 'Advanced Airfoils Studied for Transports.' *Aviation Week & Space Technology*, (Jun. 22, 1970), 55−60.
8.43. Anon.: 'Assessment of Lift Augmentation Devices.' *AGARD-LS-43-41*, (1971).

CHAPTER 9.0

WING LEADING AND TRAILING EDGES

9.1 Introduction

The purpose of wing leading and trailing edges (high-lift devices) are to increase the aerodynamic (C_{Lmax}, refer to Chapter 3.0) of the airplane and reduce the stall speed, especially the take-off and landing speed.

There are many different types of wing leading and trailing edges used to increase the maximum lift at low speed flight as shown in Fig. 9.1.1 below. (*Caution: Some of the high-lift devices are still under patent protection*).

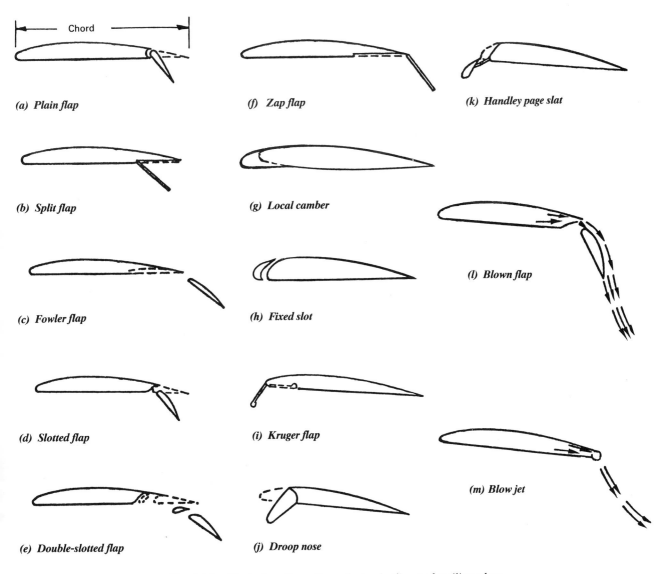

Fig. 9.1.1 Typical configurations of wing leading and trailing edges.

Airframe Structural Design 303

The leading and trailing edges applied to a wing section is usually 8–15% of the chord for leading edges and 65–75% for trailing edges. The leading edge consists of a fixed slot, drooped nose, Handley page slat and Kruger flap and small amounts of local camber (used on Learjet business aircraft). The fixed slot in a wing conducts flow of high energy air into the boundary layer on the upper surface and delays airflow separation to some higher angle of attack and lift. Slats and slots can produce significant lift which are usually used in conjunction with flaps. The use of a slot has two advantages:

- There is only a negligible change in the pitch moment;
- No significant change in drag at low angles of attack.

The trailing edge consists of more configurations than the leading edges as shown in Fig. 9.1.1. The plain flap is a simple hinged portion of the trailing edge and the effect of the camber added well aft on the chord causes a significant increase lift. The split flap consists of a plate deflected from the lower surface of the wing section and produces a slightly greater change in lift than the plain flap. The slotted flap is similar to the plain flap but the gap between the main wing section and flap leading edge is given specific contours which can cause much greater increases in lift than the plain or split flap. The fowler flap arrangement is similar to the slotted flap. The difference is that the deflected flap segment is moved aft along a set of tracks which increases the chord and effects an increase in wing area. This device is characterized by a large increase in lift with minimum changes in drag and that is why this design is so popular in transport aircraft.

The deflection of a flap causes large nose down moments which create pitching moments that must be balanced by the horizontal tail. This factor along with other mechanical complexity of the installation and maintainability may complicate the choice of a leading edge or trailing edge or the combination of both. Fig. 9.1.2 illustrates the comparison of five different combinations of wing leading edges and trailing edges.

A typical transport wing leading and trailing edge structures is shown in Fig. 9.1.3. General aviation aircraft only have trailing edge devices for simplicity (usually, plain flap and ailerons with simple hinge design). Military fighters' edges are similar to that of a transport but using a simple hinge design (see Fig. 8.1.1(b)) except for very high performance advanced fighters.

The leading edge of the wing consists of a fixed leading edge structure and slotted slats, Kruger flaps, drooped nose, or hinged flap. Construction of the slats is considerably more complex due to the slat shape and anti-icing requirements (usually required on outboard slats). The fixed leading edge structure which is attached to the wing box at the front spar consists mainly of track rib assemblies for supporting slats and intermediate frame supports. The lower surface of the fixed leading edge consists of access panels which provide access for inspection and maintenance into the entire fixed leading edge structure as well as all the services installed within this area. Each panel is secured to surrounding structure with quick action quad-lead type fasteners. All of the doors have

	Wing Leading and Trailing Configuration		Advantages	Disadvantages
	Leading Edge	Trailing Edge		
A	2 position slat (retracted and landing)	Double slot	Simpler leading edge than variable camber Kruger (VCK).	Retains double slotted trailing edge flap. No take-off L/D improvement.
B	3 position slat (retracted, landing, and take-off)	Double slot	Substantial L/D improvement for high altitude airport take-off.	Retains double slotted trailing edge. Leading edge slat only marginally simpler than VCK.
C	3 position slat	Single slot	Substantial L/D improvement. Simpler trailing edge flap.	$C_{L\max}$ is difficult to achieve due to limited design freedom in designing slat and wing surface.
D	Slotted variable camber Kruger	Single slot	Sufficient $C_{L\max}$ capability. Simpler trailing edge flap.	No take-off L/D improvement. Excessive flap separation in landing position, retains complex VCK.
E	Sealed variable camber Kruger	Simple slot	Sufficient $C_{L\max}$ capability. Improvement in take-off L/D at both sea level and high altitude airports.	Retains complex VCK.

Fig. 9.1.2 Comparison of five configurations of wing leading edges and trailing edges.

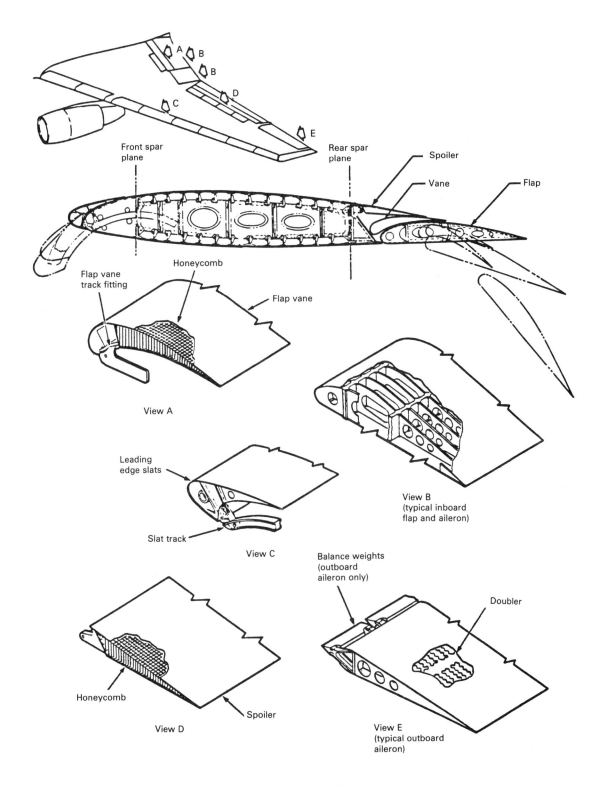

Fig. 9.1.3 Typical leading and trailing edges of a transport.

non-structural suspension hinges for their retention to structure when open. The entire fixed leading edge area will be vented by circulating air through openings in the lower surface. The trailing edge of the wing consists of flaps, vanes, aileron, spoilers, fixed trailing edge panels and all supporting structures. The extension of the slats forward and flaps aft to increase the wing chord and thus provide additional wing lift during take-off and landing. The fixed trailing edge structure primarily provides structural and functional support for the wing spoilers, flaps, and ailerons. The flap structure consists of spars, chordwise ribs and

Airframe Structural Design 305

skin panels. The leading edge and trailing edge portions of the flaps are removable in segments for inspection and maintenance. The removable trailing edge section directly behind the main landing gear (if any) is designed with sufficient strength to resist damage from slush and foreign objects deflected by the gear.

The lower surface skin of the flap is designed mainly for sonic environment from the jet engine exhaust as well as the entire flap if the jet engine is mounted underneath the wing. The flaps are weather sealed with drain holes provided on the lower surface to prevent the accumulation of moisture. Replaceable chafing strips are installed on the flap and vane assemblies to protect the structure from wear in all areas where chafing could occur, such as along the flap leading edge seal and along the spoiler trailing edges. An aileron is provided on each side of the airplane (some military airplanes use tailerons to eliminate ailerons) and ailerons may be interchangeable units supported by A-frame or simple multihinges attached to rear spar of the wing. The design is such that any single hinge may fail without loss of the control surface or endangering the safe flight of the airplane (this is very important). The construction of the ailerons is the same as previously described for flap structure. The purpose of spoilers is to kill wing lift during landing or rapid drop altitude in flight, etc., and spoilers are located forward of flaps only. The spoilers are designed to be used interchangeable at any spoiler locations if it is possible. Each spoiler segment is supported by three hinges or two hinges with fail-safe design to permit any one hinge to fail without causing loss of the entire surface.

Aerodynamic tolerance requirements are prescribed such that the leading edge slats and trailing edge flaps and vanes do not deflect adversely from their nominal positions at the design loads specified for certain flight conditions. The values shown in Fig. 9.1.4 should be determined, which are the maximum acceptable deviations from nominal settings for specific landings, take-offs and cruise flight conditions. These values are used to determine the minimum acceptable structural bending stiffness (EI), torsional stiffness (GJ) and support structural stiffness to achieve that deviation.

For the interest of designers, several plan views of wing and high-lift geometry for transports are illustrated in Fig. 9.1.5—9.1.8.

Drive Systems

There are presently two common ways (see Fig. 9.1.9) of activating flaps: (a) use an actuator to extend and retract it using hydraulic pressure; or (b) use an elec-

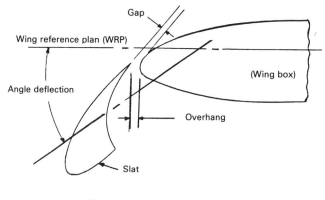

(a) Leading edge slat

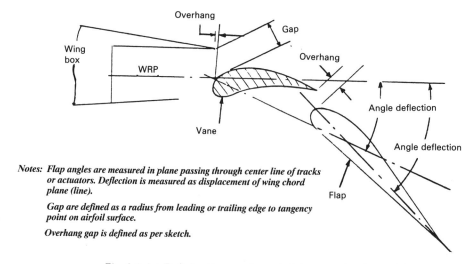

Notes: Flap angles are measured in plane passing through center line of tracks or actuators. Deflection is measured as displacement of wing chord plane (line).

Gap are defined as a radius from leading or trailing edge to tangency point on airfoil surface.

Overhang gap is defined as per sketch.

Fig. 9.1.4   Defining the aerodynamic tolerances for slat and flap.

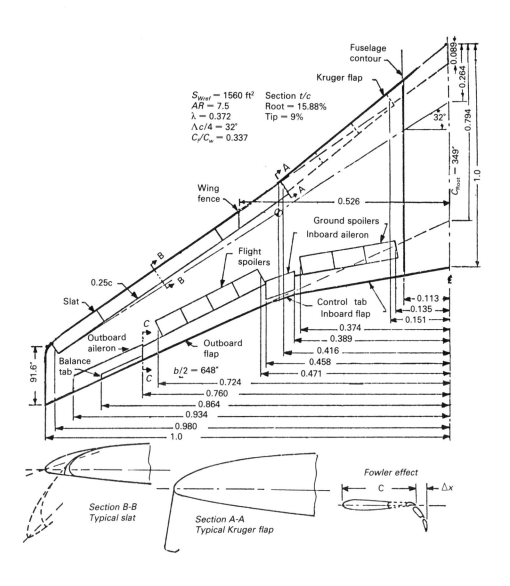
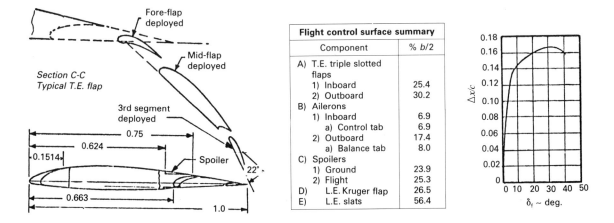

Fig. 9.1.5 Wing and high-lift geometry — B727-200 (reference only).

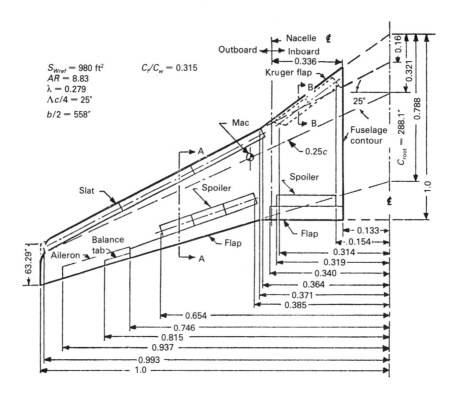

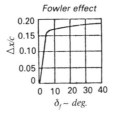

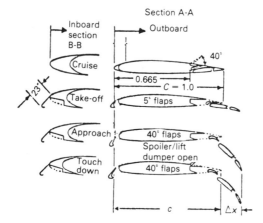

Flight control surface summary		
Component	% $b/2$	
	Trailing edge	
	Inboard	Outboard
Flaps	20.7	37.5
Spoilers	16.5	26.9
Ailerons		19.1
Balance tab		6.9
	Leading edge	
Kruger flaps	16.0	
Slats		62.9

Fig. 9.1.6 Wing and high-lift geometry — B737-200 (reference only)

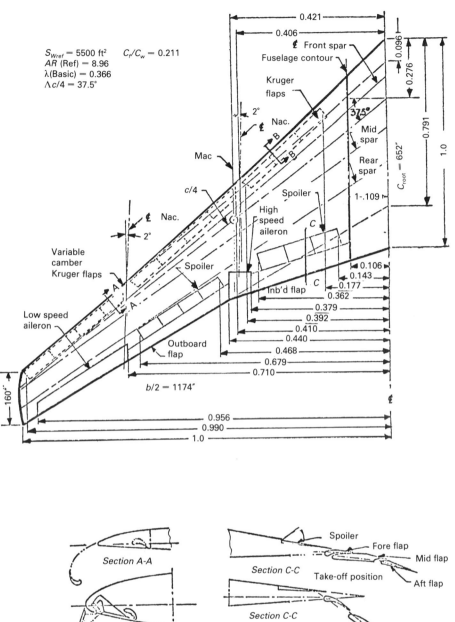

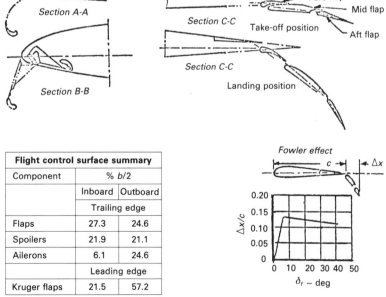

Flight control surface summary		
Component	% b/2	
	Inboard	Outboard
	Trailing edge	
Flaps	27.3	24.6
Spoilers	21.9	21.1
Ailerons	6.1	24.6
	Leading edge	
Kruger flaps	21.5	57.2

Fig. 9.1.7 Wing and high-lift geometry — B747 (reference only).

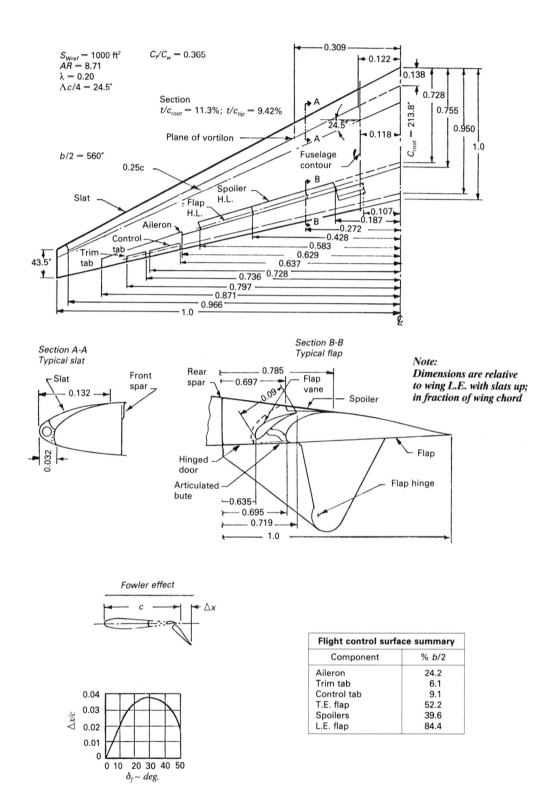

Fig. 9.1.8 Wing and high-lift geometry — DC9-30 (reference only).

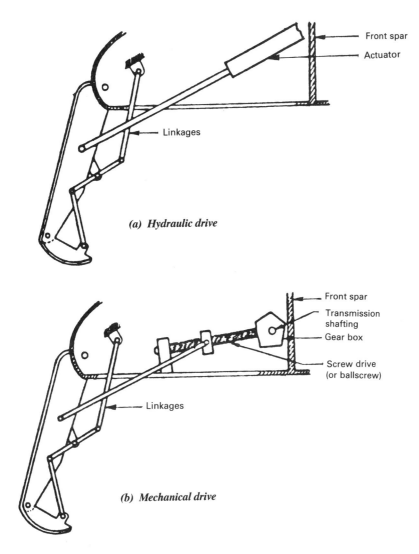

Fig. 9.1.9 Drive system, hydraulic vs. mechanical drive.

tric motor which through a series of drive shafts, gear boxes, and mechanical linkages produces the required extention and retraction of the flap.
(1) Hydraulic drive
  (a) Advantages:
- Simple to install, less components involved
- Overall hydraulic plant efficiency increased in as much as hydraulic system utilization is increased
- Less weight is involved since only tubing, fluid, actuator and support installation have to be dealt with

  (b) Disadvantages:
- Possibility of leaking in tubing and total hydraulic failure makes the flap inoperable
- Increased possibility of fire hazard along the wing area

(2) Mechanical drives
  (a) Advantages:
- A very positive control of the flap achieved since actuation or retraction is done by any electric motor through transmission shafting, gear boxes, and screw drive.
- Operation of the system completely independent since powered by a separate and independent electric motor drive
- Do not create the possibility of fire hazard in the wing area of the plane.

  (b) Disadvantages:
- A considerable weight involved due to the electric motor, transmission shafting, gear boxed and screw drive components
- Numerous wearing parts like bearings, thread portion of the drive screw, and gears which require frequent changing
- Requires a lot of maintenance, changing lubrication oil in the gear boxes; greasing on outside wearing parts

(3) Cable drive
This drive system has been applied on the leading edge of F28, DC9, DC10, etc, for its light weight

and simplicity, but cable pre-loading will be required to prevent the slack due to the applied loads, temperature changes and wing box deflections.

Fail-safe Considerations

The wing leading and trailing edges fail-safe philosophy must be selected so as to comply with the safety conditions and criteria such as FAA and military specifications. The aircraft must be shown by analysis, tests, or both to be capable of continued safe flight and landing within the normal flight envelope after any of the failure or jamming in the control systems. They are:

(1) Any single failure or disconnection or failure of a mechanical or structural element, hydraulic components;
(2) Any probable combination of failures such as dual hydraulic system failure, any single failure in combination with any probable hydraulic or electrical failure, etc;
(3) Any jam in a control position;
(4) Physical loss of retraction of more than one slat or the outboard flap on one side of the aiplane may result in loss of roll control;
(5) Physical loss of retraction of all slats or the outboard flaps on both sides of the airplane may result in pitch-up;
(6) Physical loss of an inboard flap (if there is any) could result the potential catastrophic damage to horizontal tail;
(7) Inadvertent extension of any one flap or any combination of flaps at high speeds;
(8) Inadvertent extension under airloads at high angles of attack of more than one slat on one side of the airplane;
(9) Flap asymmetry during extension or retraction could result loss of roll control and fail-safe back-up system of the asymmetry detection and lock-out, as shown in Fig. 9.1.10, should be provided;
(10) Flap load alleviating system malfunctions which results in failure of the system to properly control the flap angle of resulting in excess of design loads;
(11) All moving leading edges and trailing edge flaps shall remain affectively locked in the retracted position during the design high speed flight;
(12) Physical separation of an inboard aileron (if there is any) or outboard aileron from the airplane results in possible unacceptable loss of roll control capability;
(13) Multi-hinge design (usually more than three hinges) for outboard aileron is a common application to meet fail-safe requirements and also give the stiffness requirement by flutter;
(14) Balance weight or other means is required on outboard aileron to prevent flutter;
(15) Inadvertent actuation of the spoilers could result in pitch-up;
(16) Spoiler free to deploy at all speeds up to V_D may result in flutter or excessive dynamic loading of the wing structure.

Patent Summary

The mechanism in an aircraft wing leading and trailing flaps extending and retracting is a unique and sophisticated design work and will not be discussed in detail in this chapter. However, some of these designs, shown in Fig. 9.1.11, have U.S.A. patent rights and can be used as design references to those who are interested. Due to the limited space of this book, it is impossible to describe in detail each design; instead, certain original sketches are selected for the reader's information shown in Fig. 9.1.12.

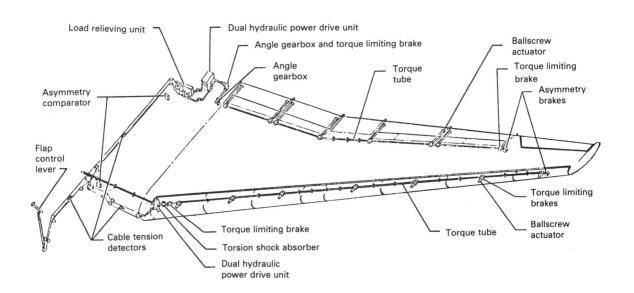

Fig. 9.1.10 Fail-safe back-up system of wing slat and flap systems — L-1011 (reference only).

Title	U.S.A. Patent No. (Dated)	Application	Refer Fig. 9.1.12
1. Airplane Wing	2,381,678 (Aug. 7, 1945)	L.E.	(1)
2. Airplane and Control Device Thereof	2,381,681 (Aug. 7, 1945)	L.E. and T.E.	
3. Flap and Jet Device for Airplane Wings	2,466,466 (April 5, 1949)	T.E.	
4. Wing Lift Modification Means	2,500,512 (Mar. 14, 1950)	L.E.	
5. Wing with Interconnected Flap and Nose Slat	2,538,224 (Jan. 16, 1951)	L.E. and T.E.	
6. Wing Flap Mechanism	2,608,364 (Aug. 26, 1952)	T.E.	(2)
7. Airplane Flap Control Mechanism	2,620,147 (Dec. 2, 1952)	T.E.	(3)
8. Aircraft Flap Mechanism	2,661,166 (Dec. 1, 1953)	T.E.	(4)
9. Aircraft Flap Control	2,677,512 (May 4, 1954)	T.E.	
10. Retractable Surface Support	2,688,455 (Sept. 7, 1954)	T.E.	
11. Slat Mechanism for Airplanes with Sweptback Wings	2,702,676 (Feb. 22, 1955)	L.E.	
12. Aircraft Wing Leading Edge and Slot	2,755,039 (July 17, 1956)	L.E.	
13. Aircraft Wing with Means to Increase Lift Through Control of Air Flow	2,772,058 (Nov. 27, 1956)	T.E.	
14. Wing Flap Actuating Mechanism	2,779,555 (Jan. 29, 1957)	T.E.	
15. Airplane Wing with Slotted Flap, Cove Lip Door, and Spoiler	2,836,380 (May 27, 1958)	T.E.	(5)
16. External-Flow Jet Flap	2,891,740 (June 23, 1959)	T.E.	
17. Aircraft Wing	2,908,454 (Oct. 13, 1959)	T.E.	(6)
18. Jet Propelled Aircraft with Jet Flaps	2,912,189 (Nov. 10, 1959)	T.E.	
19. Aircraft Flap Structure	2,924,399 (Feb. 9, 1960)	T.E.	

L.E. — Leading Edge; T.E. — Trailing Edge

Fig. 9.1.11 Wing high-lift devices — U.S.A. patent summary.

Title	U.S.A. Patent No. (Dated)	Application	Refer Fig. 9.1.12
20. Multiple Position Airfoil Slat	2,938,680 (May 31, 1960)	L.E.	(7)
21. Power Flap for Aircraft	2,964,264 (Dec. 13, 1960)	T.E.	
22. Aerodynamically Automatic Airfoil Slat Mechanism	2,973,925 (March 7, 1961)	L.E.	(8)
23. Trailing Edge Flaps for Airplane Wings	3,061,244 (Oct. 30, 1962)	T.E.	
24. Airplane Wing Flaps	3,112,089 (Nov. 26, 1963)	T.E.	
25. Area Increasing Slotted Flaps and Apparatus Thereof	3,126,173 (March 24, 1964)	T.E.	
26. High Lift System for Aircraft Wings	3,128,966 (April 14, 1964)	L.E.	
27. Airplane Wing Flap	3,129,907 (April 21, 1964)	T.E.	
28. High Lift Slotted Flap	3,195,836 (July 20, 1965)	L.E.	(9)
29. High Lift Flaps for Aircraft Wings	3,203,647 (Aug. 31, 1965)	T.E.	
30. High Lift Device Actuating Mechanism	3,244,384 (April 5, 1966)	L.E.	
31. Means for Positioning a Rotating Wing Slat Device	3,212,458 (Sept. 13, 1966)	L.E.	(10)
32. Variable Airfoil High-Lift Slat and Slot for Aircraft	3,273,826 (Sept. 20, 1966)	L.E.	
33. Vertical or Short Take-off and Landing Aircraft	3,361,386 (Jan. 2, 1968)	L.E. and T.E.	(11)
34. Aircraft	3,363,859 (Jan. 2, 1968)	L.E.	(12)
35. Leading Edge Flap and Apparatus Thereof	3,375,998 (April 2, 1968)	L.E.	
36. Extendible Wing Flap Arrangement for Airplanes	3,438,598 (April, 15, 1969)	T.E.	(13)
37. Flap Systems for Aircraft	3,447,763 (June 3, 1969)	T.E.	
38. Jet Flaps	3,478,987 (Nov. 18, 1969)	T.E.	

L.E. — Leading Edge; T.E. — Trailing Edge

Fig. 9.1.11   (continued)

Title	U.S.A. Patent No. (Dated)	Application	Refer Fig. 9.1.12
39. Multiple Slotted Airfoil System for Aircraft	3,480,235 (Nov. 25, 1969)	T.E.	(14)
40. Inverting Flap Shapes and Mechanisms	3,481,561 (Dec. 2, 1969)	T.E.	
41. Continuous Slot Forming Leading Edge Slats for Cranked Wings	3,486,720 (Dec. 30, 1969)	L.E.	
42. Aircraft Wing Variable Camber Leading Edge Flap	3,504,870 (April 7, 1970)	L.E.	(15)
43. Leading Edge Flap of Variable Camber and Thickness	3,524,610 (Aug. 18, 1970)	L.E.	
44. High Lift Flaps for Aircraft	3,528,632 (Sept. 15, 1970)	T.E.	
45. Methods and High Lift Systems for Making an Aircraft Wing More Efficient for Take offs and Landings	3,556,439 (Jan. 19, 1971)	L.E.	(16)
46. Long Structural Column Support	3,568,957 (Mar. 9, 1971)	T.E.	(17)
47. Jet Flap Controlling Means	3,595,500 (July 27, 1971)	T.E.	
48. Folding Flap	3,617,018 (Nov. 2, 1971)	L.E.	(18)
49. Wing with Slotted Flap Mounted at the Leading Edge and/or at the Trailing Edge	3,638,886 (Feb. 1, 1972)	L.E.	
50. Aircraft	3,698,664 (Oct. 17, 1972)	T.E.	(19)
51. Auxiliary Flap Actuator for Aircraft	3,706,431 (Dec. 19, 1972)	T.E.	(20)
52. Pneumatic Leading Edge Flap for an Aircraft Wing	3,711,039 (Jan. 16, 1973)	L.E.	
53. Fluid Dynamic Lift Generating or Control Force Generating Structures	3,716,209 (Feb. 13, 1973)	L.E.	
54. High Lift Leading Edge Device	3,743,219 (July 3, 1973)	L.E.	(21)
55. Leading Edge Flap Mechanism	3,743,220 (July 3, 1973)	L.E.	
56. Airplane Flaps	3,767,140 (Oct. 23, 1973)	T.E.	(22)
57. Aerodynamic Surfaces	3,778,009 (Dec. 11, 1973)	T.E.	

L.E. — Leading Edge; T.E. — Trailing Edge

Fig. 9.1.11   (continued)

Title	U.S.A. Patent No. (Dated)	Application	Refer Fig. 9.1.12
58. Aircraft Flap System	3,785,594 (Jan. 15, 1974)	T.E.	(23)
59. Flap System	3,790,106 (Feb. 5, 1974)	T.E.	
60. Bi-directional Deflectible Control Flap for Airfoils	3,799,474 (Mar. 26, 1974)	T.E.	
61. Aircraft Flap System	3,819,133 (June 25, 1974)	T.E.	
62. Flap Arrangement for Thrust Deflection in Aircraft	3,827,657 (Aug. 6, 1974)	T.E.	(24)
63. Airfoil with Extendible and Retractable Leading Edge	3,831,886 (Aug. 27, 1974)	L.E.	(25)
64. Airfoil Camber Change System	3,836,099 (Sept. 17, 1974)	L.E.	(26)
65. Trailing Edge Flap and Actuating Mechanism Thereof	3,853,289 (Dec. 10, 1974)	T.E.	(27)
66. STOL Flaps	3,874,617 (April 1, 1975)	T.E.	(28)
67. Variable Camber Multi-Slotted flaps	3,897,029 (July 29, 1975)	L.E.	
68. Leading Edge Flap	3,910,530 (Oct. 7, 1975)	L.E.	(29)
69. Flap Mechanisms and Apparatus	3,917,192 (Nov. 4, 1975)	L.E. and T.E.	
70. Variable Camber Airfoil	3,941,334 (Mar. 2, 1976)	L.E.	(30)
71. Parametrically Shaped Leading Edge Flaps	3,949,956 (April 13, 1976)	L.E.	
72. Actuating System for Wing Leading-Edge Slats	3,949,957 (April 13, 1976)	L.E.	(31)
73. STOL Aircraft	3,977,630 (Aug. 31. 1976)	T.E.	(32)
74. Trailing Edge Flaps Having Spanwise Aerodynamic Slot Opening and Closing Mechanism	3,987,983 (Oct. 26, 1976)	T.E.	(33)
75. Upwardly Extendible Wing Flap System	4,007,896 (Feb. 15, 1977)	T.E.	(34)
76. Leading Edge Flap for a Airfoil	4,398,688 (Aug. 16, 1983)	L.E.	(35)

L.E. — Leading Edge; T.E. — Trailing Edge

Fig. 9.1.11 (continued)

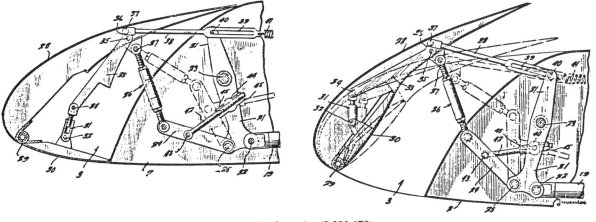

(1) Airplane wing (2,381,678)

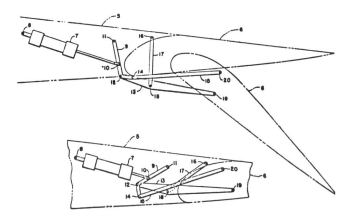

(2) Wing flap mechanism (2,608,364)

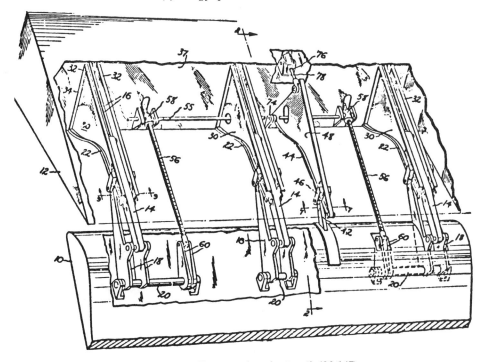

(3) Airplane flap control mechanism (2,620,147)

Fig. 9.1.12 U.S.A. patent high-lift devices sketches.

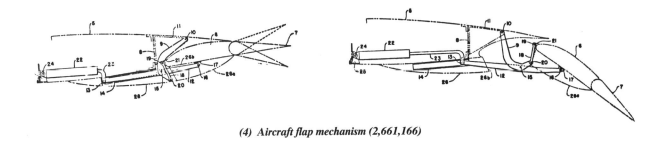

(4) Aircraft flap mechanism (2,661,166)

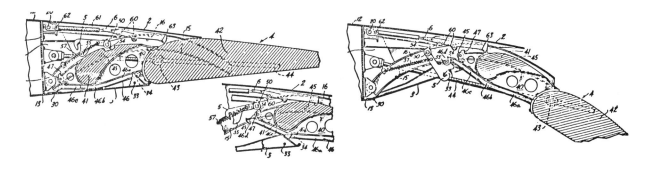

(5) Airplane wing with slotted flap, cove lip door, and spoiler (2,836,380)

(6) Aircraft wing (2,908,454)

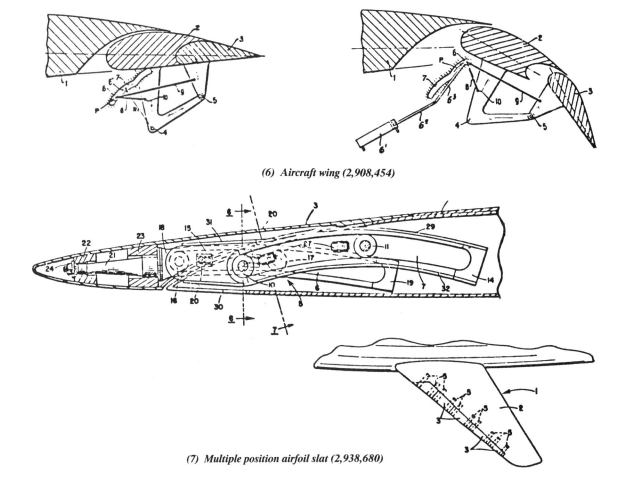

(7) Multiple position airfoil slat (2,938,680)

Fig. 9.1.12   (continued)

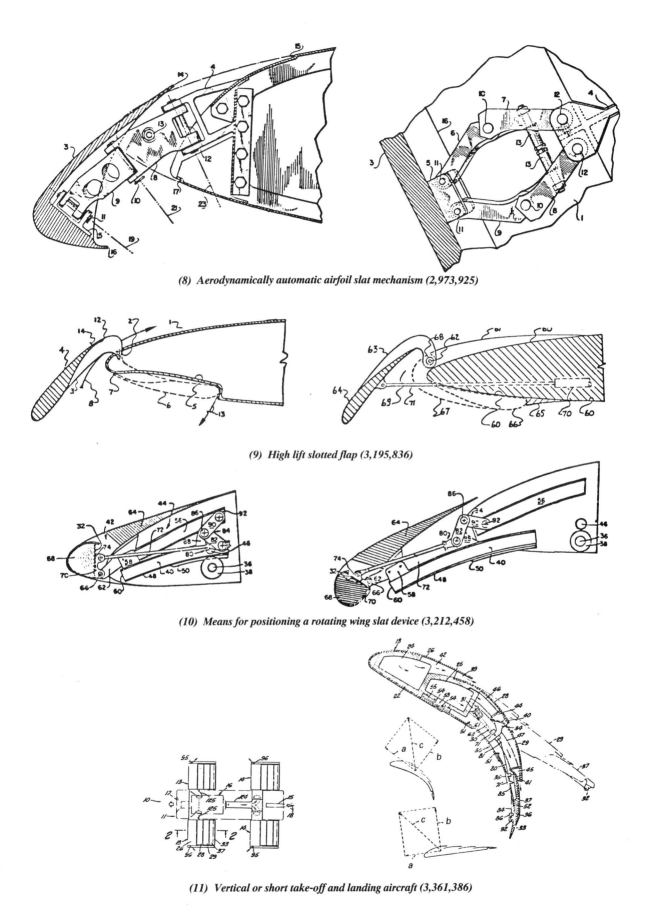

(8) Aerodynamically automatic airfoil slat mechanism (2,973,925)

(9) High lift slotted flap (3,195,836)

(10) Means for positioning a rotating wing slat device (3,212,458)

(11) Vertical or short take-off and landing aircraft (3,361,386)

Fig. 9.1.12   (continued)

Airframe Structural Design 319

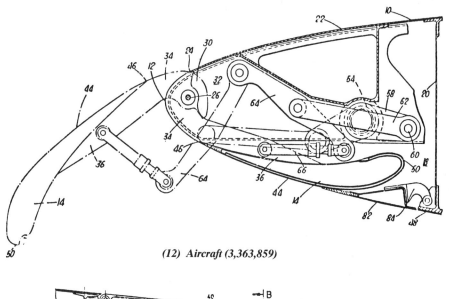

(12) Aircraft (3,363,859)

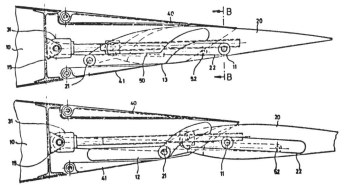

(13) Extendible wing flap arrangement for airplanes (3,438,598)

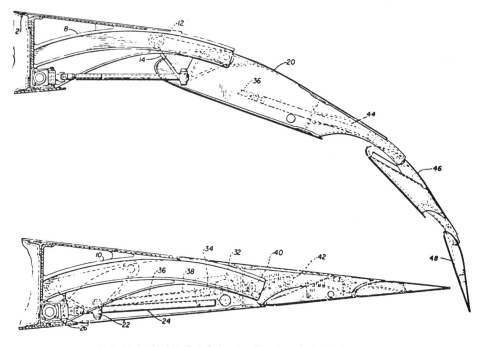

(14) Multiple slotted airfoil system for aircraft (3,480,235)

Fig. 9.1.12   (continued)

320 Airframe Structural Design

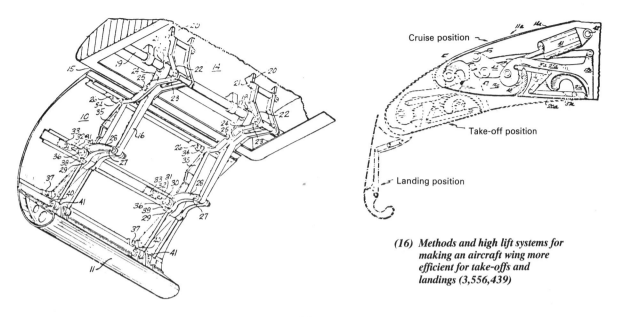

(15) Aircraft wing variable camber leading edge flap (3,504,870)

(16) Methods and high lift systems for making an aircraft wing more efficient for take-offs and landings (3,556,439)

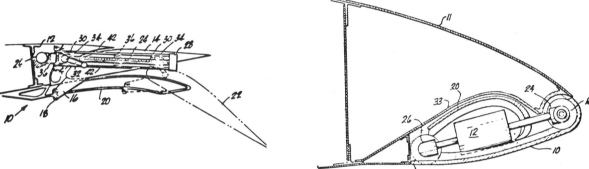

(17) Long structural column support (3,568,957)

(18) Folding flap (3,617,018)

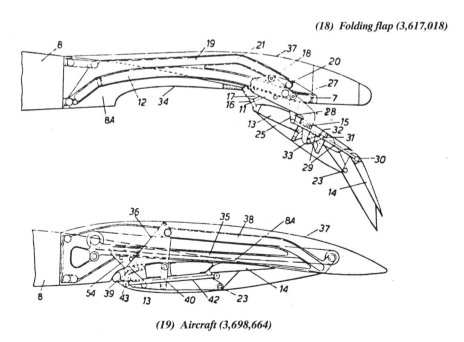

(19) Aircraft (3,698,664)

Fig. 9.1.12 (continued)

Airframe Structural Design 321

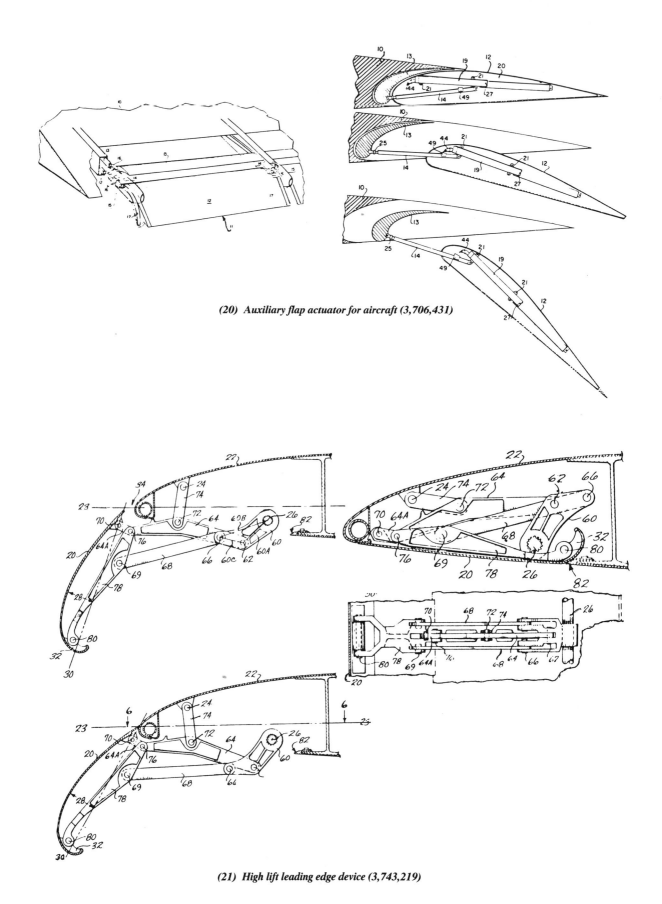

(20) Auxiliary flap actuator for aircraft (3,706,431)

(21) High lift leading edge device (3,743,219)

Fig. 9.1.12 (continued)

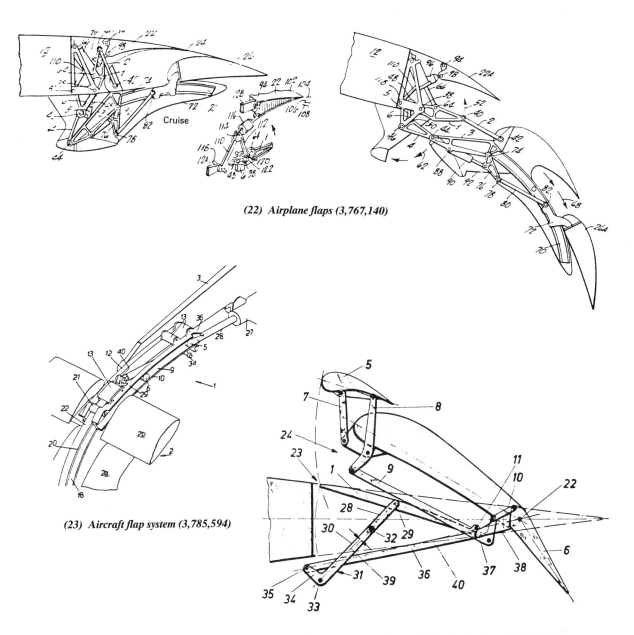

(22) Airplane flaps (3,767,140)

(23) Aircraft flap system (3,785,594)

(24) Flap arrangement for trust deflection in aircraft (3,827,657)

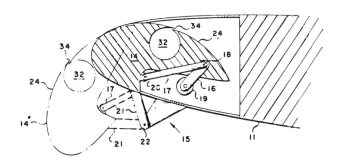

(25) Airfoil with extendible and retractable leading edge (3,831,886)

Fig. 9.1.12   (continued)

Airframe Structural Design 323

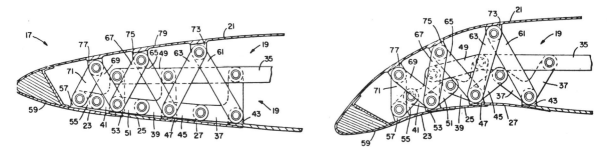

(26) Airfoil camber change system (3,836,099)

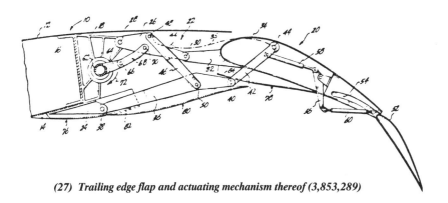

(27) Trailing edge flap and actuating mechanism thereof (3,853,289)

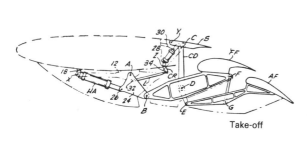

(28) STOL flaps (3,874,617)

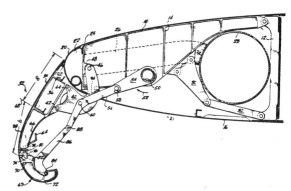

(29) Leading edge flap (3,910,530)

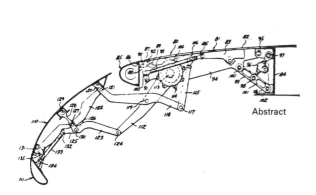

(30) Variable camber airfoil (3,941,334)

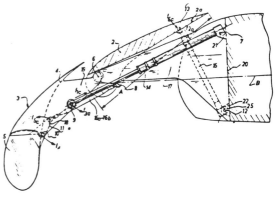

(31) Actuating system for wing leading edge slats (3,949,957)

Fig. 9.1.12 (continued)

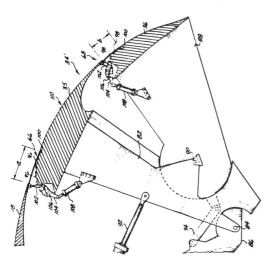

(32) STOL aircraft (3,977,630)

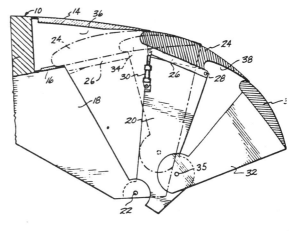

(33) Trailing edge flaps having spanwise aerodynamic slot opening and closing mechanism (3,987,983)

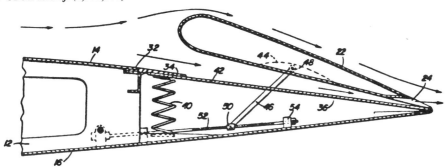

(34) Upwardly extendible wing flap system (4,007,896)

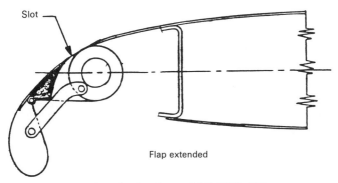

(35) Leading edge flap for an airfoil (4,398,688)

Fig. 9.1.12 (continued)

Airframe Structural Design 325

9.2 Leading Edges

The increased circulation associated with the deflection of an effective trailing edge-device induces an upwash at the nose. The local suction peak increases on airfoils which are liable to leading edge stall; the flow will separate at an angle of attack which is below that of the basic wing. Leading-edge (L.E.), high-lift devices (Fig. 9.1.1) are intended primarily to delay the stalling to higher angles of attack and the requirements are:

- Must delay flow separations to large angles of attack;
- Must stow in the most forward portion of the wing section (within 10—15% of the wing chord) and prefer not to have the control mechanism penetrate the fuel tank;
- Either in retracted or extended positions, the devices should not be deflected beyond the required gaps under air load or wing bending.

Wing leading edges consist of fixed leading edge structure, with or without slats. If slats are used, when retracted, they form the front profile of the wing ahead of the fixed edge, which is contoured to form a secondary profile behind the slats when they are fully extended (see Fig. 9.2.1). The outboard slats are usually designed to provide hot air anti-icing or de-icing systems either utilizing engine bleed air or electric heating systems; the inboard slats etc. generally do not require anti-icing or de-icing systems because they are not aerodynamic sensitive as compared with the outboard.

Slats

Slat are small, highly cambered airfoils forward of the wing leading edge, which experience large suction forces per unit of area. The associated profile drag increment and pitching moment changes are small and for optimized configurations, a C_{Lmax} increment due to a full-span slat between 0.5 and 0.9 can be obtained without any appreciable increase in tail load and trim drag. In view of the large stalling angle of attack, the angles of pitch during take-off and landing are large and proper attention must therefore be paid to the visibility from the cockpit, particularly if the angle of sweep is large and has a low aspect ratio. Slats may be used to advantage by varying their effectiveness in the spanwise direction in order to delay the onset of the tip stall and the associated pitch-up of sweptback wings. This may be achieved by varying the gap width, the slat deflection angle or the relative slat chord in the spanwise direction.

Because of its flexible and longer span structure, each slat has two or more snubbers or other similar

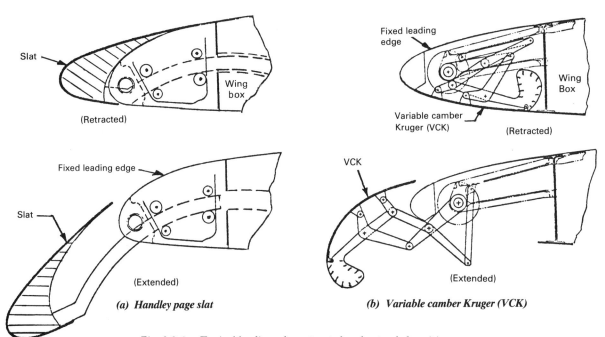

(a) *Handley page slat*

(b) *Variable camber Kruger (VCK)*

Fig. 9.2.1 *Typical leading edge retracted and extended positions.*

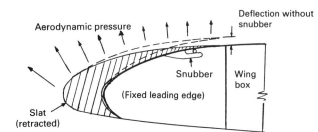

Fig. 9.2.2 *The function of slat snubbers.*

devices to ensure that the trailing edge of the slat is snugged down to wing contour when the slats are fully retracted (see Fig. 9.2.2). This type of slat is one of the most popular types of leading edge devices for several decades and many of the high subsonic jet transports have used it.

Slats consist of three sections — the nose section, center section, and trailing edge (refer to Fig. 9.2.3) connected by a front and rear spar. Close rib spacing (light former ribs are generally used) is generally required along the length of the slat to carry extreme high aerodynamic pressure (i.e. more than 10 psi) at

326 Airframe Structural Design

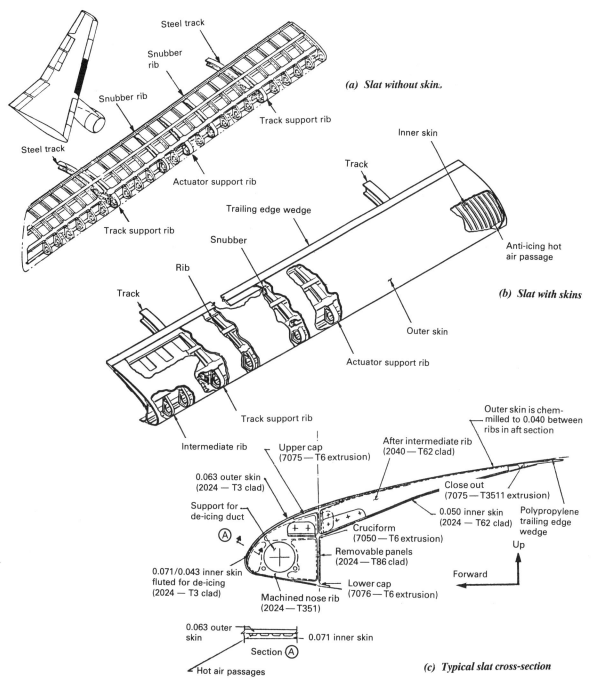

Fig. 9.2.3 Typical built-up slat arrangement — L-1011.
(Courtesy of Lockheed Aeronautical Systems Co.)

the slat surface. The main support ribs where the support track and actuator are attached are required to react to the concentrated loads.

To facilitate inspection of the slat interior, the inside skin and the vertical web of the front beam are removable panels. The minimum skin thickness (aluminum) at the nose section of the slats should not be less than 0.063 inch because this area is to protect the wing as well as leading edge control system from rain erosion and hail damage. Furthermore, heat resistant materials, such as 2024-T3 aluminum used on slats as shown in Fig. 9.2.3(c) are carefully selected for the nose section so that it can withstand the anti-icing hot air with temperature of 400°F or higher.

Each slat is generally supported by two curved tracks, which are made of either high strength steel or titanium to react very high track bending moments during the slats in extended cases. The principle track ribs are as shown in Fig. 9.2.4 and are actuated by a single telescoping ball screw actuator. The hydraulic power is designed strong enough to overcome jam situations in case one of the ball screws jams. Obviously the local support structure is also designed for this jam condition. On some airplanes, the cables and pulleys are used to control the moving of the slats. If cables are used, the cable circuit pre-tension

Airframe Structural Design 327

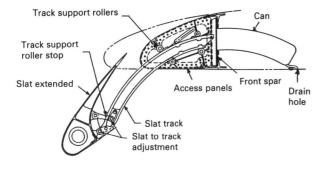

Fig. 9.2.4 *Relationship of slat, slat track and track support.*

levels will be considered and allow for pre-loading the slats against the extended and retracted stops provided at each drive track. This pre-loading will be required to locate the retracted and extended position of the slats accurately under operational conditions. An example of a cable system is illustrated in Fig. 9.2.5.

As mentioned previously, each slat is generally supported by two tracks and each use four rollers to carry vertical loads; the lower pair being adjustable so that the slat-to-leading edge gap can be set to suit the tracks.

A small adjustable linkage is designed at the end of track where the slat is attached to, see Fig. 9.2.6. Side

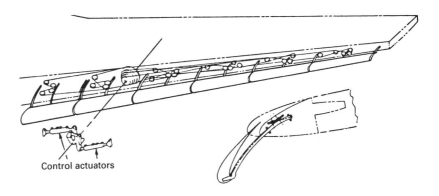

(a) McDonnell-Douglas DC-10 wing (DC-9 also uses a similar system)
(Courtesy of Douglas Aircraft Co.)

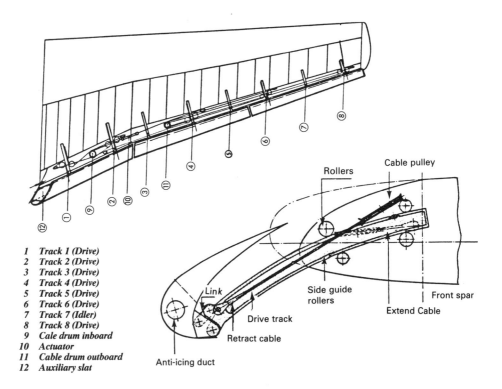

1 Track 1 (Drive)
2 Track 2 (Drive)
3 Track 3 (Drive)
4 Track 4 (Drive)
5 Track 5 (Drive)
6 Track 6 (Drive)
7 Track 7 (Idler)
8 Track 8 (Drive)
9 Cale drum inboard
10 Actuator
11 Cable drum outboard
12 Auxiliary slat

(b) Fokker F28 Fellowship transport
(Courtesy of Fokker Corp.)

Fig. 9.2.5 *Wing slat with cable drive system.*

328 Airframe Structural Design

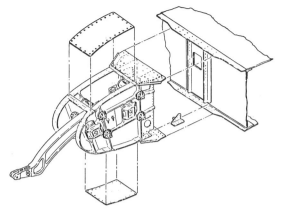

(a) Slat track assembly

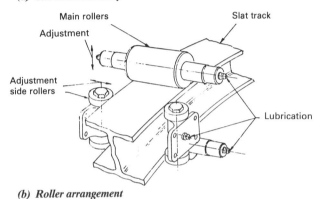

(b) Roller arrangement

Fig. 9.2.6 *Typical assembly of slat track.*

loads are carried by rollers running on the track side flanges and mounted on the pair of slat track ribs. One track per section carries side loads. Suitable linkages are being introduced to absorb expansion due to thermal de-icing of the slats or wing expansion due to wing bending if three or more slat tracks are used to support the leading slat structure. Another example is shown in Fig. 9.2.5(b) of the slats supported by three tracks, a spanwise sliding movement will be possible between tracks No. 1, 3, 5, 6 and 8 and the appropriate slats. Datum track No. 2, 4 and 7 control the spanwise position of the slats and therefore provide no spanwise degree of freedom. These features are mainly introduced for the retracted position of the slats to allow for differential expansion and wing deflection. Slats, supported by three or more tracks, are light weight structures and also have failsafe features by its nature; but the high vertical track loads due to wing deflection are a big problem and cannot be neglected.

Special bumper type fittings are installed at the ends of each slat to maintain slat symmetry during the extension and retraction cycle.

Slat track ends and hydraulic actuator rear ends (if not using telescoping ball screw) shown in Fig. 9.2.4 are projected into a fuel-tight recess in the front spar web. Slat-track fuel-tank cans are a must to stop fuel flow. These cans are designed to withstand as high as 20 psi or more crushing pressure and the most efficient can structure is a cylindric member with circumferential stiffener reinforcements to stablize the thin skin buckling for light weight design.

Kruger Flaps

Kruger flaps perform in the same way as slats; but they are thinner and more suitable for installation on thin wings. Kruger flaps are often used on the inboard part of wings such as Boeing's commercial airplanes, in combinating with outboard slats, to obtain positive longitudinal stability in the stall.

So far basically two kinds of Kruger flaps are used on transport airplanes, i.e. fixed Kruger flap and variable camber Kruger (VCK) flap as shown in Fig. 9.2.7. The Kruger flap is a simple and light weight structure compared with slat design but its high lift capability is not as efficient as the slat. The fixed Kruger flap generally is made of one piece of aluminum, or magnesium, or advanced composite material, etc. As shown in Fig. 9.2.8, the Boeing 727 airplane uses the fixed Kruger flap with one piece of magnesium casting hung on three hinges. Fig. 9.2.9 illustrates the curved Kruger flap arrangement.

The Boeing 747 wing variable camber Kruger flap shown in Fig. 9.2.10 is a most intriguing use of glass fibre (or composite material) in the leading edge flaps, which are flat when they form the under surface of the leading edge, but are bent on extension.

The advantages of using Kruger flaps rather than slats are as follows:
* Simple structure, easy to manufacture with cost savings as well as weight savings.

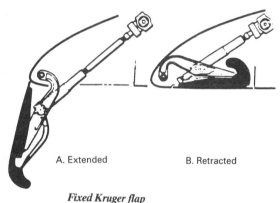

Fixed Kruger flap

Variable camber Kruger flap

Fig. 9.2.7 *Kruger flaps.*

Airframe Structural Design 329

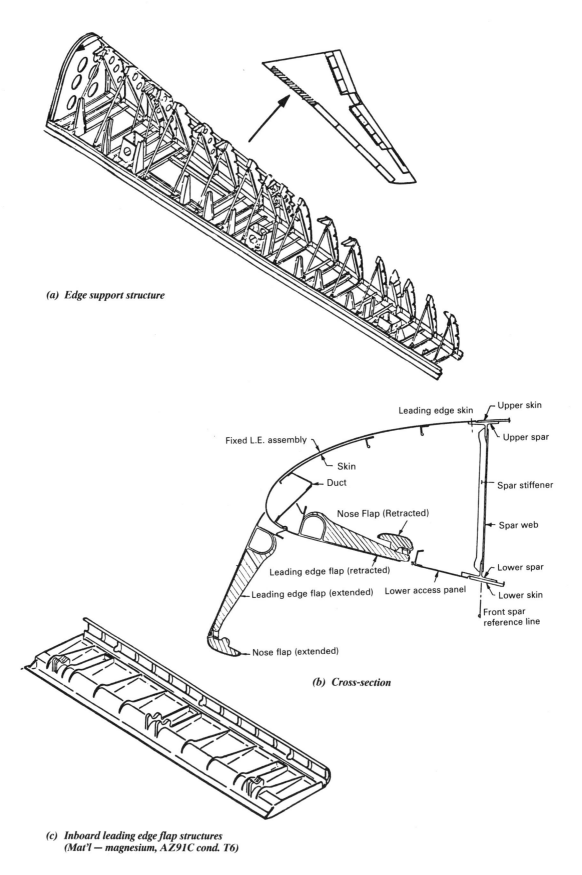

(a) Edge support structure

(b) Cross-section

(c) Inboard leading edge flap structures (Mat'l — magnesium, AZ91C cond. T6)

Fig. 9.2.8 *Fixed Kruger flap — B727.*
(Courtesy of The Boeing Co.)

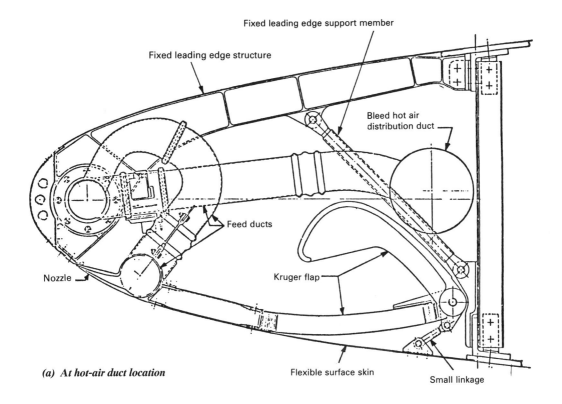

(a) At hot-air duct location

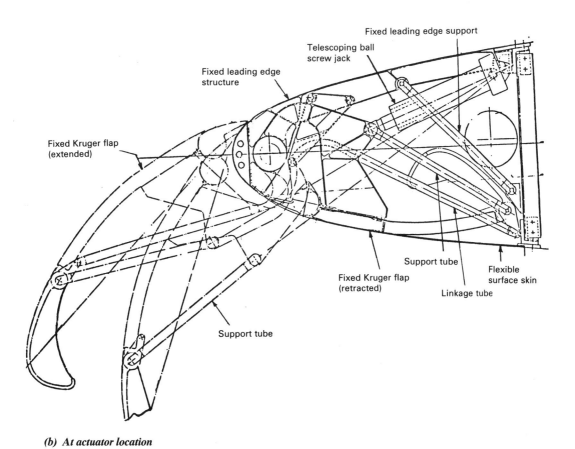

(b) At actuator location

Fig. 9.2.9 Curved fixed Kruger flap arrangement.

Airframe Structural Design 331

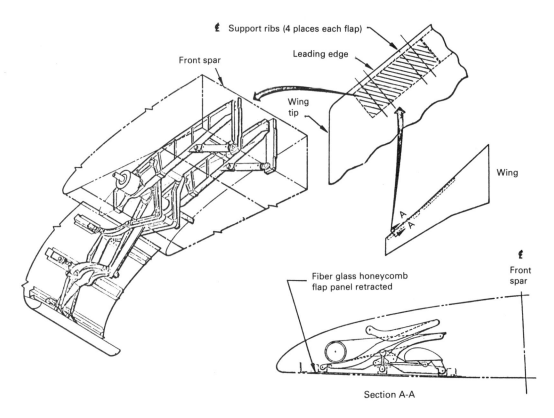

Fig. 9.2.10   Variable camber Kruger flap — B747.
(Courtesy of The Boeing Co.)

- No flap tracks are required to support the flap structure and, therefore, the recess into a fuel tank can be avoided.
- When retracted, the Kruger flaps conform to the fixed leading edge profile, which is aerodynamically the most efficient means at cruise speeds to avoid the air leakage that occurs on slat designs. This is a most important reason for choosing Kruger flap designs (see Fig. 9.2.11).

The disadvantages are:

- It is not practial to use Kruger flaps on thin wing airfoils such as military fighter wings.
- Aerodynamic high lift capability is not as efficient as slats.
- The control mechanism of moving Kruger flaps is complicated as well as the design of de-icing or anti-icing.
- As the Kruger flap mechanism is exposed, when the flap is extended, to dust, dirt, water, ice, snow, etc., kicked up by the reverse thrust blast, it is necessary to quickly close the leading edge in order to keep out contaminants.

Choosing a proper flap system for certain proposed airplanes is not an easy job. Some ground rules are required in the early stage in the specification, and different trade-off studies should be made before reaching a final decision.

Droop Nose

Droop nose is less effective than a slat, but it is mechanically simple and rigid and particularly suitable for thin airfoil sections. The droop nose has become very popular with supersonic aircraft with a sharp or relatively sharp leading edge on thin wings, as shown in Fig. 9.2.12.

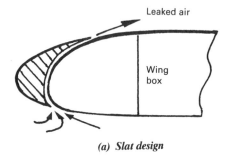
(a) Slat design

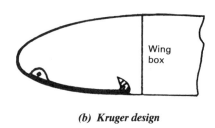
(b) Kruger design

Fig. 9.2.11   Increased drag due to air leakage.

332   Airframe Structural Design

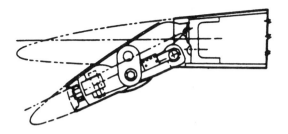

(a) F-5 fighter
(Courtesy of Northrop Corp.)

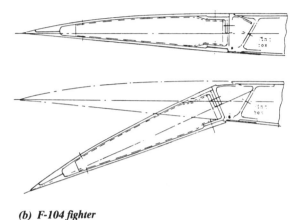

(b) F-104 fighter

Fig. 9.2.12 *Droop nose of a very thin fighter wing.*

Droop nose not only has the advantage of being able to droop at low speed and lift at high speeds, but it is sometimes used to protect the leading edge air flow from early separation while pulling high g loads at combat maneuvering speeds. Since this design has an abrupt change in chord direction, the air flow has difficulty turning the corner without separating. It may need a cove behind the nose to allow accelerated air from below to be ejected at the aft end of the cove. To solve this problem, a variable camber leading edge concept is adopted (see Fig. 9.2.13) in which the upper and lower flexible skins are designed to permit smooth contour deflection of the leading edge from the supersonic cruise position to the fully deflected high-lift position. It has no sliding joints or gaps on the leading edge nose or upper surface. A faired sliding joint exists on the lower surface at the front spar in the region of positive pressures.

One of the droop nose designs for thicker wing airfoil is shown in Fig. 9.2.14, which is typical for subsonic transports. Statistical data shows that very few modern transports use this except fighters.

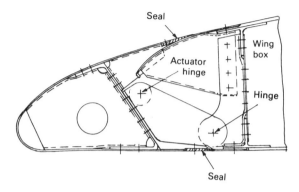

Fig. 9.2.14 *A typical droop nose design for transport wing.*

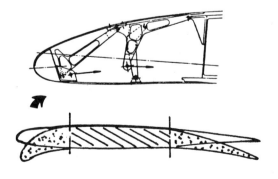

Fig. 9.2.13 *Flexible droop nose design.*

Besides the kinematics of the device of moving the droop nose, the other important device is obviously the actuation devices such as conventional hydraulic cylinders, rotary actuators, jack screws (or called ball screw), eccentric actuator, etc. Before choosing any type of these actuation devices, attention should be given to their stiffness, mechanical advantage, backlash and failure mode characteristics. Further, cost, weight and maintainability comparison on a particular application are also investigated. Fig. 9.2.15 illustrates the actuation devices for leading edge droop nose (also applicable to the trailing edge flaps).

The Vought Eccentuator is a structural hinge that can support the large hinge moments inherent in the advanced aerodynamic concepts of thin airfoils employing maneuvering variable camber. The basic structural concept of the eccentuator is a one piece structural tapered circular beam that connects a wing leading or trailing edge to its support structure while it changes its beam geometry on command to provide the required up and down deflections as shown in Fig. 9.2.16. The unique aspect is the kinematics of the device which features equal and opposite rotation of a tapered bent beam inside a carrier to produce planar motion of the leading or trailing edge of the beam.

Fixed Slots
Fixed slots of leading edges as shown in Fig. 9.2.17 are the simplest devices for postponing leading edge stall, but their profile drag penalty is generally prohibitive for effective cruising, except on some low speed STOL aircraft or general aviation aircraft.

To reduce the profile drag penalty during cruise, a device could be designed to close the slot. Fig. 9.2.18 illustrates one of these devices. But the complexity of the device may prevent the designer from choosing this approach.

Airframe Structural Design 333

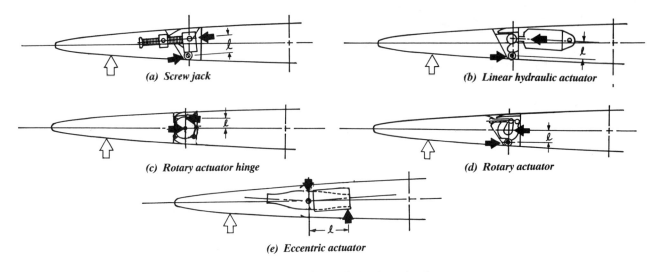

Fig. 9.2.15 Actuation devices for leading edge droop nose.
(Courtesy of LTV Aircraft Products Corp.)

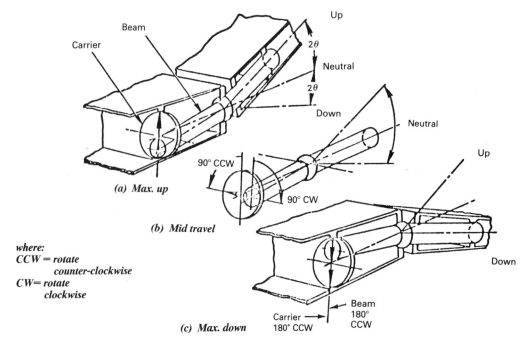

Fig. 9.2.16 Vought eccentuator motion (U.S.A. Patent No. 3,944,170)
(Courtesy of LTV Aircraft Products Group)

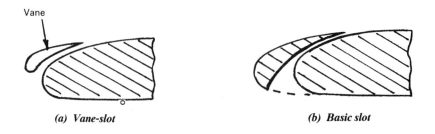

Fig. 9.2.17 Typical fixed slot of wing leading edge.

334 Airframe Structural Design

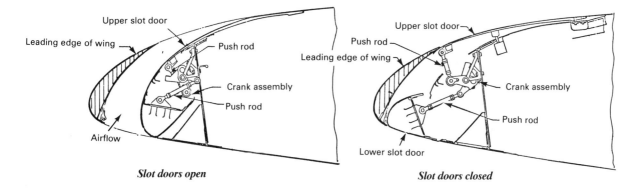

Fig. 9.2.18   Wing open-and-close fixed slot design.

Comments

Leading edge devices are beneficial if there is any possibility of leading edge stall occurring, but with the comparatively thick sections used on light aircraft and small propeller transports, slats or leading edge flaps are not normally considered necessary. Stall prevention on the outboard wing should be obtained by a proper choice of the sectional shape, moderate taper ratio and washout. If necessary, an increased leading edge radius, drooped nose or a part-span slot may be considered.

9.3 Trailing Edges

There are many types of trailing edge (T.E.) flaps used to increase the maximum lift coefficient to shorten airplane take-off and landing distance. The design of flap system is more complicated than the previous leading edge and poses very challenging work for designers. The flap applied to the trailing edge of a section consists of a wing section usually from 25—35% of the chord length; some special design for mission requirement may reach as high as nearly 40%. The determination of the flap chord length is also a function of wing box structural stiffness and strength requirement as well as the volume required for fuel taken to achieve the airplane's performance. Therefore, a trade-off study should be carefully investigated before freezing the final configuration. Fig. 9.3.1 illustrates a typical trailing edge arrangement for a transport.

Split Flaps

Split flaps consist of a stiffened plate on the wing lower surface, hinged just aft of the rear spar, by means of a piano hinge (Fig. 9.3.2). The drag due to flap deflection is large. Although its structural simplicity and low weight are attractive this type of flap has not been used because of its low efficiency.

Plain Flaps

Plain flaps are hardly more than a hinged part of the trailing edge (Fig. 9.3.3). The best performance is obtained with a sealed gap. If the deflection exceeds

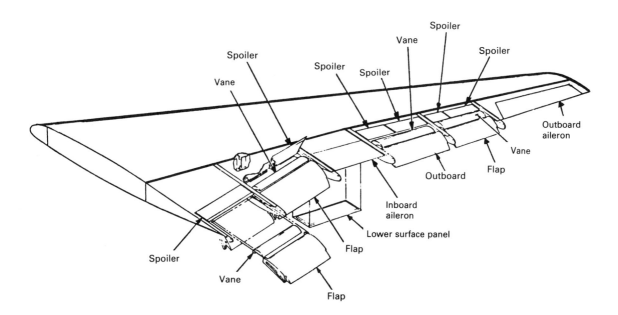

Fig. 9.3.1   A typical trailing edge arrangement — transport.

Airframe Structural Design 335

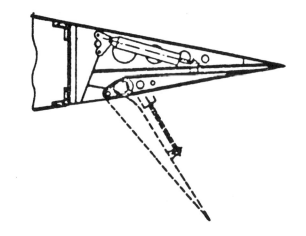

Fig. 9.3.2 Split flap.

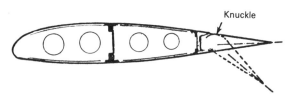

Fig. 9.3.3 Plain flap.

Single Slotted Flaps

Single slotted flaps derive their favorable action from a specially contoured slot through which air is admitted from below the wing. The upper surface boundary layer is stabilized by the suction on the flap's leading edge and diverted at the trailing edge of the basic wing. A new boundary layer is formed over the flap which permits an effective deflection of up to 40 degrees. The performance is sensitive to the shape of the slot, which is determined by the kinematics of deflection. A simple fixed hinge is not very effective, but the achievement of an optimum slot shape requires a system which can only be realized by means of a track and flap carriage assembly. Single slotted flaps with fixed hinge are commonly used on light aircraft and a few transports (see Fig. 9.3.4).

10 to 15 degrees, the flow separates immediately after the knuckle, the lift effectiveness drops progressively and the drag increases to values comparable to those of split flap. Most fighters and general aviation aircraft still use this design because it is very efficient for mounting on very thin wings by rotary hinge or power hinge, eccentuator, or torque tube rotated by the rotary power at the side of the fuselage which is commonly used on general aviation aircraft and earlier day fighter aircraft.

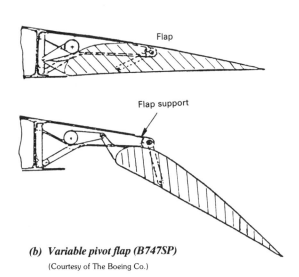

(b) *Variable pivot flap (B747SP)*
(Courtesy of The Boeing Co.)

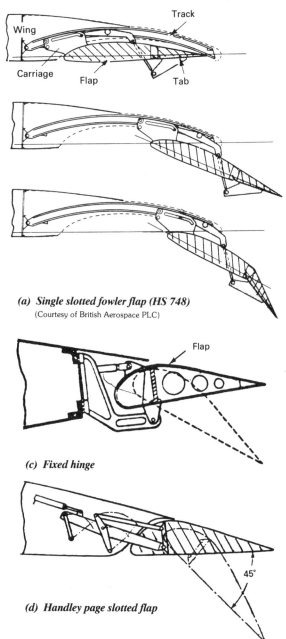

(a) *Single slotted fowler flap (HS 748)*
(Courtesy of British Aerospace PLC)

(c) *Fixed hinge*

(d) *Handley page slotted flap*

Fig. 9.3.4 Single slotted flaps.

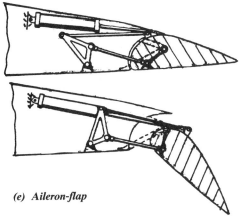

(e) Aileron-flap

Fig. 9.3.4 *(continued)*

If the two flap elements are independently adjustable, the maximum deflection may be increased up to higher degrees. This rather complex system is occasionally used on STOL aircraft, such as the DHC-7. Fig. 9.3.6(b) shows the system used on the GAF N-22 Nomad full span flaps with aileron control at high speeds, augmented by spoiler-ailerons at large flap deflections.

(3) Using tracks as shown in Fig. 9.3.7 to support double slotted flaps is also popular to increase the wing area as well as the lift coefficient. This design could be applied on the smaller flap chord (move wing rear spar more aft to obtain bigger wing fuel tank volume) but still meet the same performance of take-off and landing or just increase the wing area for improving the airplane take-off and landing performance.

Double Slotted Flaps

Double slotted flaps are markedly superior to the previous type at large deflections, because separation of the flow over the flap is postponed by the more favorable pressure distribution. Double slotted flaps are most popular for transports and various degrees of mechanical sophistication are possible (see Fig. 9.3.5–9.3.7):

(1) Flaps with a fixed hinge and a fixed vane of relatively small dimensions [Fig. 9.3.5 (a)] are structurally simple, but may have high profile drag during take-off. If the vane is made retractable for small flap deflections, the flap is effectively single-slotted during the take-off, which improves the drag and hence the climb performance. The external flap supports cause a noticeable parasite drag.

(2) Double slotted flaps may be supported on a four-bar mechanism (Fig. 9.3.6). During extension the slot shape closely approximates the aerodynamic optimum, but the flap supports cause a disturbance in the flow through the slot. This configuration is probably most suitable for application on long-range transports.

Triple Slotted Flaps

Triple slotted flaps are used on several transport aircraft with very high wing loadings. In combination with leading edge devices, this system represents almost the ultimate achievement in passive high-lift technology, but Fig. 9.3.8 shows that complicated flap supports and controls are required.

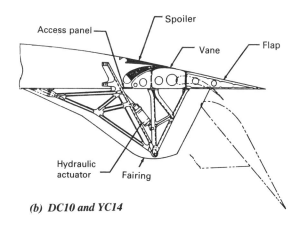

(b) DC10 and YC14

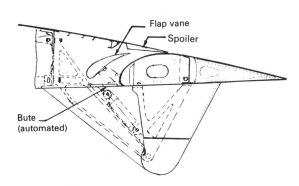

(a) DC9 flap (fixed vane)

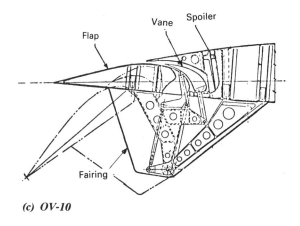

(c) OV-10

Fig. 9.3.5 *Double slotted flaps — hinge type*
(Courtesy of Douglas Aircraft Co.)

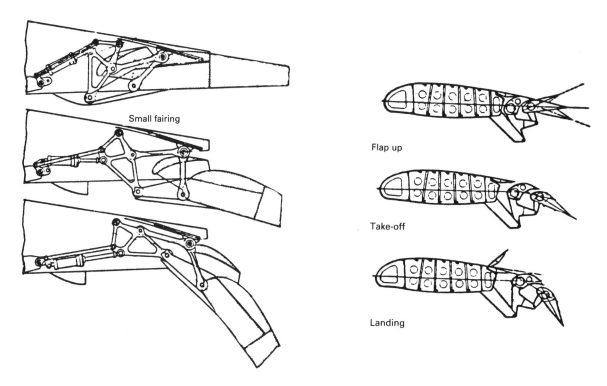

(a) DC-8 (four bar linkage design with very small fairing)
(Courtesy of Douglas Aircraft Co.)

(b) GAF N-22 Nomad (with individual adjustment of flap segment and drooped aileron)

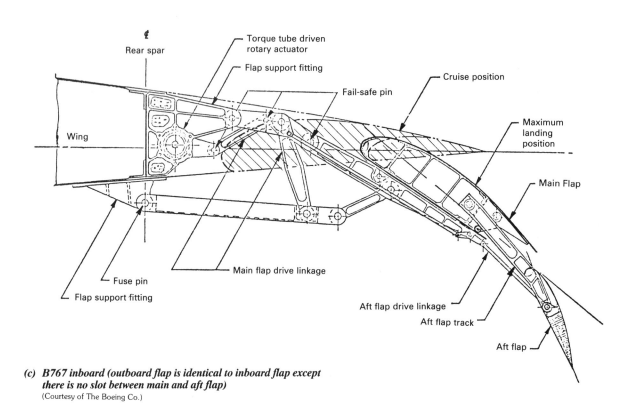

(c) B767 inboard (outboard flap is identical to inboard flap except there is no slot between main and aft flap)
(Courtesy of The Boeing Co.)

Fig. 9.3.6 *Double slotted flaps — linkage type.*

338 Airframe Structural Design

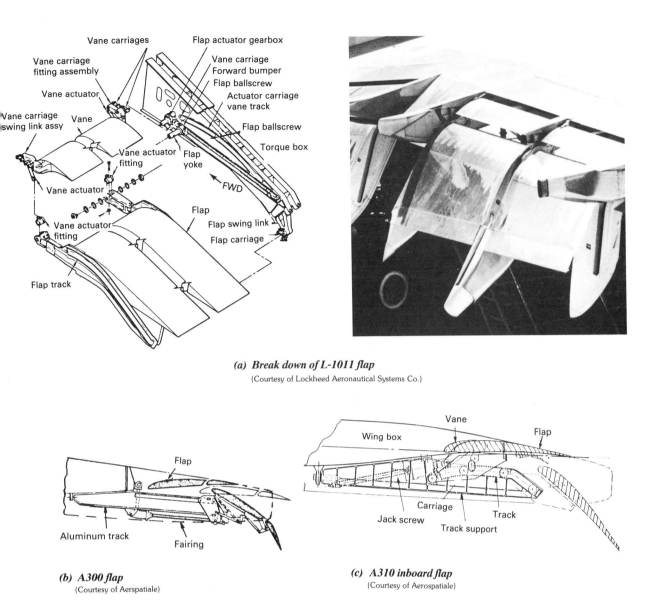

(a) Break down of L-1011 flap
(Courtesy of Lockheed Aeronautical Systems Co.)

(b) A300 flap
(Courtesy of Aerspatiale)

(c) A310 inboard flap
(Courtesy of Aerospatiale)

Fig. 9.3.7 *Double slotted flaps — track type.*

The YC-15 externally-blown flap configuration as shown in Fig. 9.3.9 is an extension of high-lift flap technology which uses engine exhaust flow to produce increased lift. It requires no internal ducting or auxiliary thrust deflection devices. A slotted flap is lowered into the engine exhaust stream which cause the exhaust stream to be spread laterally and then deflected downward so that a relatively thin jet sheet leaves the flap trailing edge at approximately the same angle as the flap.

Fowler Flaps
Fowler flaps are slotted flaps traveling aft on tracks over almost their entire chord and subsequently deflecting to their maximum angle (Fig. 9.3.10). The favorable performance is derived from the effective wing area extension, yielding a gain in lift for very little extra drag. Configurations have been devised with one, two and even three slots, depending upon the magnitude of the required lift. These systems not only have considerable chord extension due to the basic flap motion, but the flap sections move relative to each other and the extended length of the chord is therefore in excess of the nested length. The structural complications of multi-track supports and their weight penalty is usually the limiting factor. The shape of the slots must be very carefully optimized to obtain good performance.

Flap Track Supports
The aerodynamic engineer, however, expresses concern over the risk in potential degradation of performance of the flap system due to blocking of the airflow by the flap tracks. The general consensus is that the outboard flap has to be fail-safe since loss of an outboard flap is unacceptable from a flight safety standpoint. Six different track configurations are discussed; one two-track arrangement, three three-track arrangements, and two four-track arrangements. These track configurations are shown in Fig. 9.3.11.

Airframe Structural Design 339

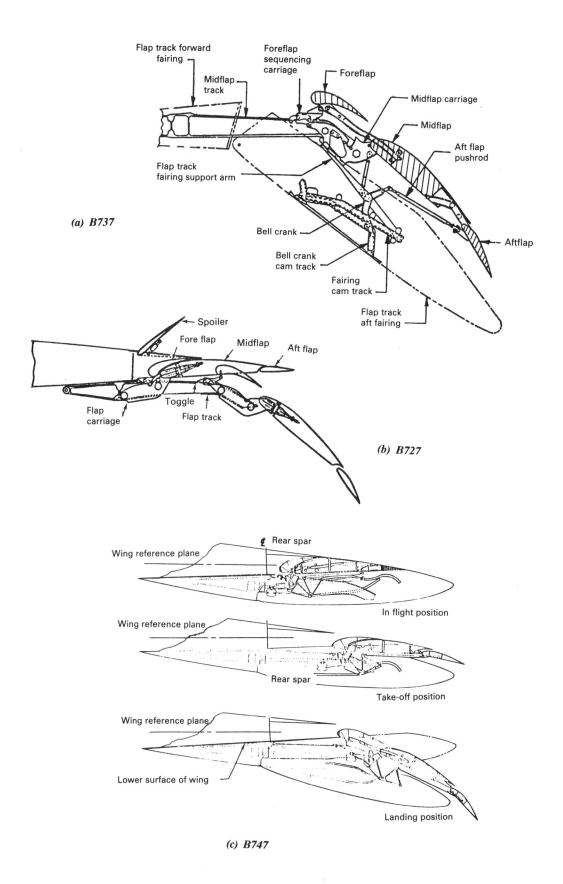

Fig. 9.3.8 *Triple slotted flaps.*
(Courtesy of The Boeing Co.)

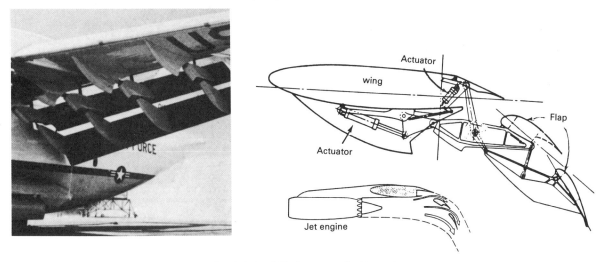

Fig. 9.3.9 YC-15 externally blown flap.
(Courtesy of Douglas Aircraft co.)

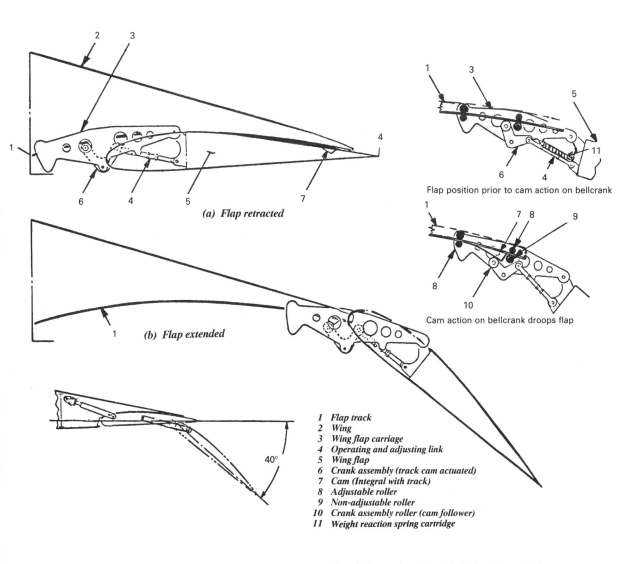

1. Flap track
2. Wing
3. Wing flap carriage
4. Operating and adjusting link
5. Wing flap
6. Crank assembly (track cam actuated)
7. Cam (Integral with track)
8. Adjustable roller
9. Non-adjustable roller
10. Crank assembly roller (cam follower)
11. Weight reaction spring cartridge

Fig. 9.3.10 Typical fowler flap — Lockheed Electra (see Fig. 9.3.14 for flap track.)

Airframe Structural Design 341

Flap Track Arrangement	Track Configuration	Flap Track Arrangement	Track Configuration
A	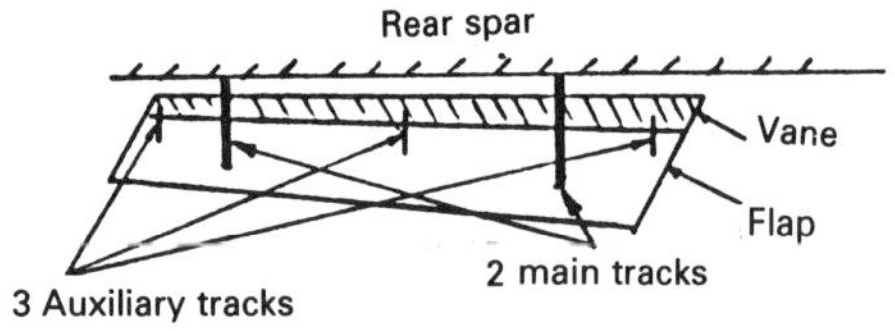 Rear spar / Vane / Flap / 3 Auxiliary tracks / 2 main tracks	D	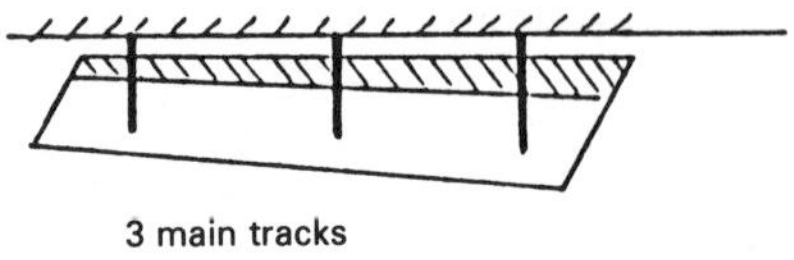 3 main tracks
B	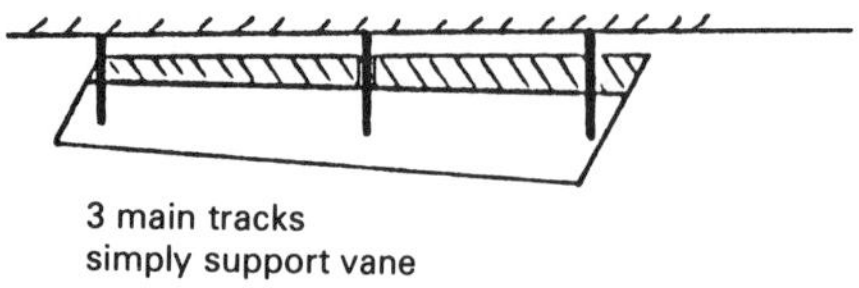 3 main tracks simply support vane	E	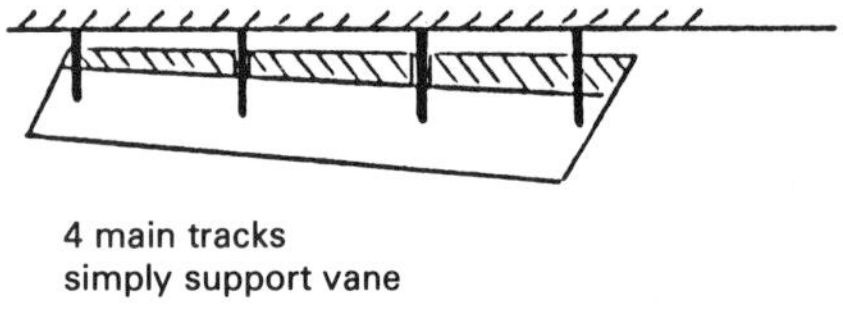 4 main tracks simply support vane
C	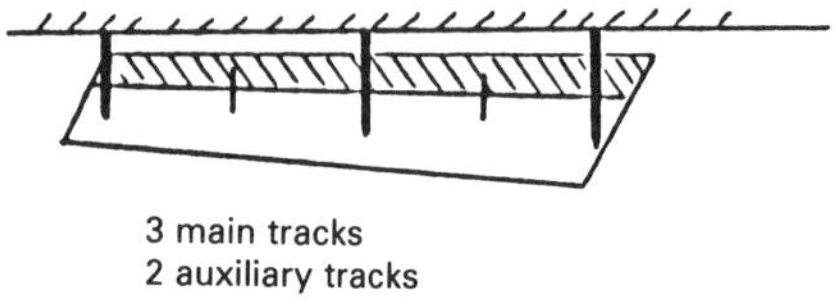 3 main tracks 2 auxiliary tracks	F	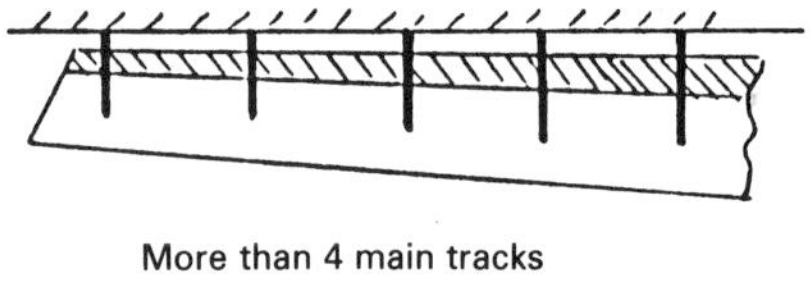 More than 4 main tracks

Flap Track Arrangement	Advantages	Disadvantage
A	1. Minimum fairing area. 2. Minimum flow blockage between vane and T.E. of spoilers.	1. Heaviest structure, primarily because of fail-safe requirements and limited depth of tracks due to threading through vane. 2. Additional latches required at ends of flap structure to force the retracted flap to conform with wing deflection.
B	1. No threading of vane through tracks required. 2. No latches at ends of flap structure required. 3. Less flow blockage between main flap and vane than in "A" arrangement.	1. Special trimming of flap and required to accommodate fairings. 2. Large deflections under failed track conditions.
C	1. Less weight than "A" arrangement. 2. No threading of vane through tracks required.	1. Fairing at flap ends present problems. 2. Large deflections under failed track conditions.
D	1. Low blockage of flow by tracks. 2. Considerable less weight than "A" arrangement.	1. Large deflections under failed track conditions. 2. Threading of tracks through vane required.
E	1. No threading of tracks through vane required. 2. Considerable less weight than "A" arrangement.	1. More flow blockage between vane and spoiler. 2. Fairings at ends of flap difficult to install.
F	1. Excellent control of aerodynamic gap. 2. Even distribution of load over tracks.	1. More flow blockage between vane and spoiler. 2. Threading of tracks through vane required.

Fig. 9.3.11 Flap track support arrangements.

Arrangement D indicates the least amount of flow blockage; however, threading the tracks through the vane is required. Deflection of flap with a failed track is probably unacceptable for aero-elastic performance. Arrangement E does not require threading but indicates some more flow blockage and presents a problem with respect to installation of the fairings at the ends of the flap structure. Arrangement F does not have the fairing problem, but the tracks have to be threaded through the vane. Slot blockage is about the same as arrangement E. However, the relative deflection between vane and main flap is much smoother. Since the arrangement of the wing box structure is to a large part dependent on the location of the flap support structure, therefore attention should be given to this design effort at the preliminary design.

Aerodynamic tolerance requirements are prescribed such that the trailing edge flaps do not deflect adversely from their nominal positions at loads specified for certain flight conditions. The values should be listed by aerodynamic group for the maximum acceptable deviations from nominal settings for a specific landing, take-off and cruise flight condition. Slat and flap sketches defining the aerodynamic tolerances are shown in Fig. 9.1.4.

The flap tracks, which present particular structural problems because of the long rearward extension, are schemed in steel, titanium and light aluminum alloy.

- Steel tracks: It is not practical to use heat treat higher than 200,000 psi due to reduced ductility. From a weight point of view, higher than 200 ksi H.T. steel may be considered but great care must be taken. Steel tracks are widely used on modern transports with very good record. Typical fowler flap track designs are shown in Fig. 9.3.12—9.3.15.

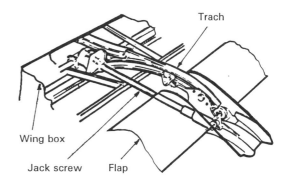

Fig. 9.3.12 *Convair 880 inboard flap track arrangement.*
(Courtesy of General Dynamics Corp.-Convair)

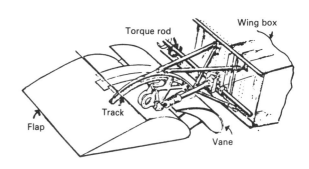

Fig. 9.3.13 *Mechanism of the double-slotted fowler flaps Transall Aircraft).*

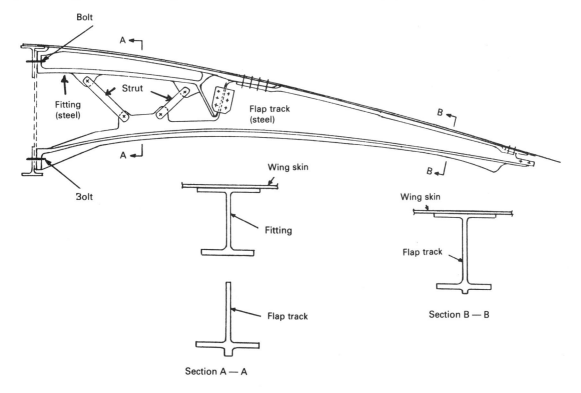

Fig. 9.3.14 *Flap track construction — Lockheed Electra.*

Airframe Structural Design 343

- Titanium tracks: These have high strength to weight ratio than other materials and very good corrosion resistance, but they are not widely used because of high cost (very high machining costs).
- Aluminum alloy with steel inserts: This design has been used on several transports. Flap support boxes with steel roller tracks along the the roller path are used to increase the bearing strength (Fig. 9.3.15).

The flap tracks are designed individually for fail-safe, or long fatigue-life steel with lower heat treat material if designed for single load path. Sometimes flap structure uses more than three track supports and each track may not be required for fail-safe. The whole track system is enclosed in a hinged aerodynamic fairing constructed of built-up or honeycomb structure. these tracks and fairings are aligned streamwise, and deflect downward during flap operation.

Flap Construction

The construction of flap structure is depended on where the jet engines are mounted, i.e. underneath the wing surfaces, or rear fuselage or other places. If the jet engines are mounted underneath the wing surface, the criteria of design flap structure is sonic fatigue and extra attention should be given to where the attachments are used to join the lower skin to either caps or rib caps. (Refer to Chapter 15.0 for detailed design to avoid the cracks due to the sonic fatigue environment.)

Flap structure design is essentially similar to that of designing elevators, rudders, and ailerons. There are nose section, middle section and trailing section as illustrated in Fig. 9.3.16.

The two or three spars with multi-close-ribs construction (as shown in Fig. 9.3.17–9.3.19) are widely used on transports with jet engines mounted underneath wing surfaces because this construction has better sonic fatigue resistance. The bulkhead ribs are always required where the flap tracks and actuators are attached. The intermediate ribs are usually made of aluminum sheet formers with lightening holes for weight saving purposes or truss ribs as shown in Fig. 9.3.18.

The surface skin of the nose and middle section are machined or chem-milled from aluminum sheets with a local thicker pad, as shown in Fig. 9.3.19, where the attachments (countersunk fasteners) will go through.

The chem-milled process is cheaper than machined skins but the chem-milled skin is recommended where the sonic fatigue is not critical; however, use chem-milled process for cost standpoint, the special treatment at the edge of the chem-milled areas is required. Another cost saving design is that the skins are bonded doublers for reinforcement in the area of spar and rib attachments to improve sonic fatigue (see Chapter 15.0). The lower skins are designed to be removable for maintenance by unscrewing fasteners.

The spars are either built-up beams with extruded caps and webs or formers bent from sheet metal. Lightening holes are generally required at front spar web for inspection (maintenance purpose), if the nose skins are permanently riveted to the flap structure. Wedge structures are generally tapered full-depth honeycomb panels and easy to remove from flap rear spar due to its vulnerability to foreign object (see Fig. 9.3.20).

The light weight flaps structure can be achieved without considering the sonic environment condition, if the jet engines are not mounted near wing surfaces such as military fighter engines inside the fuselage or as engines at rear fuselage of transport airplanes; or if the propeller engines are mounted foward of the wing leading edge. Those types of engine mounts have

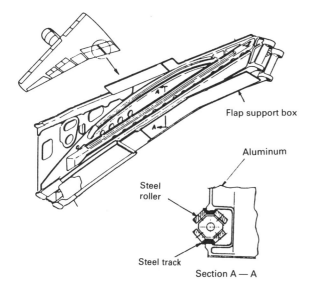
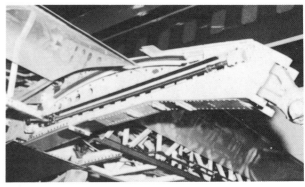

Fig. 9.3.15 Flap track construction — Lockheed L-1011 (also see Fig. 9.3.7(a)).
(Courtesy of Lockheed Aeronautical Systems Co.)

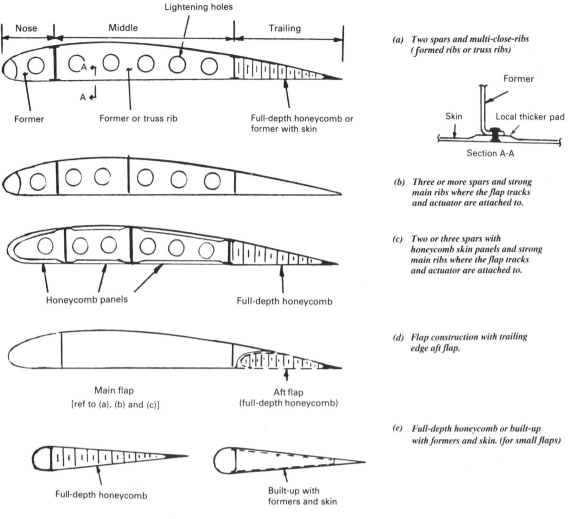

(a) *Two spars and multi-close-ribs (formed ribs or truss ribs)*

(b) *Three or more spars and strong main ribs where the flap tracks and actuator are attached to.*

(c) *Two or three spars with honeycomb skin panels and strong main ribs where the flap tracks and actuator are attached to.*

(d) *Flap construction with trailing edge aft flap.*

(e) *Full-depth honeycomb or built-up with formers and skin. (for small flaps)*

Fig. 9.3.16 *Typical flap constructions.*

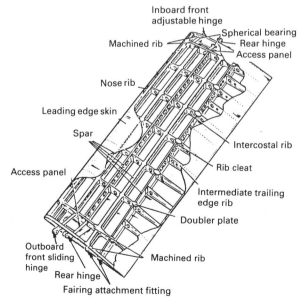

Fig. 9.3.17 *Flap construction of BAC One-eleven transport.*
(Courtesy of The British Aerospace PLC)

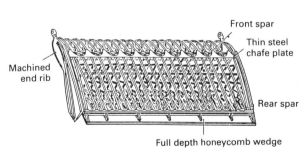

Fig. 9.3.18 *Flap construction (truss ribs) of L-1011.*
(Courtesy of Lockheed Aeronautical Systems Co.)

Airframe Structural Design 345

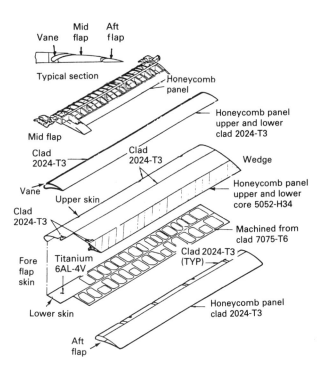

Fig. 9.3.19 Flap construction of B737.
(Courtesy of The Boeing Co.)

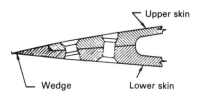

(a) For L.E. flap

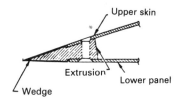

(b) For L.E. flap

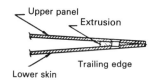

(c) For T.E.

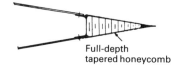

(d) For T.E.

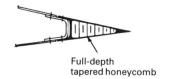

(e) For T.E.

Fig. 9.3.20 Typical wedge construction.

much less sonic fatigue problems to the trailing edge flap structure.

Honeycomb panel structure is one of the best structures to resist sonic fatigue cracks, therefore, some transports (Fig. 9.3.21) use aluminum honeycomb panels at upper and lower surfaces of the middle section and light weight fiber glass honeycomb panels at nose and trailing sections. This design configuration is similar to that as illustrated in Fig. 9.3.16(c) and (d). Recently, high strength composite materials have been used on military airplanes and the civil transports have gradually tried to use them on wing edges, control surfaces, etc.

The full-depth tapered honeycomb panel structure is adopted on military high-speed fighter flaps and also small general aviation flaps due to the fact that they are small and have thin sections [Fig. 9.3.16(e)].

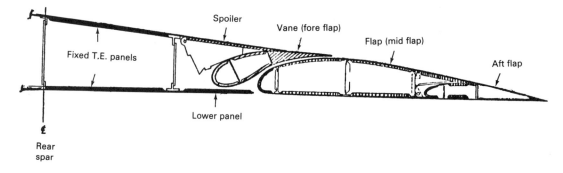

Fig. 9.3.21 Honeycomb construction of B747 flap.
(Courtesy of The Boeing Co.)

However, the back-up ribs which are bonded into the honeycomb inside the flap box or externally attached to the flap box structure are required to transfer the flap loads to the main wing structure.

9.4 Wing Control Surfaces

Wing spoilers and ailerons are considered as wing control surfaces, which combining with elevators and rudders, can be maneuvered by the pilot to control the motion of the airplane. The spoilers are used for many purposes. In flight, the spoilers can contribute to roll control and can be deployed as speed brakes or on the ground to kill lift as well as an air brake.

Ailerons should be located toward the tip of the wing span and its function to the roll motion of the airplane. Most transports have two ailerons per side of the wing, namely inboard aileron (or called high-speed aileron) and outboard aileron (low-speed aileron). Some of the military aircraft, e.g. F-14, F-111, etc. have no aileron at all and the roll is controlled by differential tailplane (Taileron) operation. The purpose of doing this is to obtain the maximum lift on the wing to slow down the landing speed or to shorten the take-off distance.

Spoilers

Spoilers may be fitted for various reasons; they often combine more than one function and usually occupy a substantial part of the flap span, just behind the rear spar. As shown in Fig. 9.4.1, the spoiler can be deflected in an almost upright position. Immediately after landing, or in the event of an aborted take-off, they may be activated either by the pilot or automatically. As a result of the interrupted airflow over the flaps, the wing loses a large parts of its lift, which increases the normal force on the tires and makes braking more effective. In addition, they create considerable drag and these combined effects increase the deceleration by some 20%. Outboard spoilers may be used in flight when an appreciable increment in drag is required to obtain a high rate of descent or improved speed stability with a constant angle of descent.

Inboard spoilers are not deflected in this case to avoid disturbing the flow over the empennage and prevent buffeting. For this reason the inboard spoilers are only installed to decrease the lift on the ground and are referred to as ground spoilers (liftdumpers), while the others are known as inflight spoilers. When acting as drag-producing devices, these spoilers are referred to as airbrakes (speedbrakes) which may be installed both on the upper and lower surfaces of the wing.

The location of actuator hinges should be close to one of the spoiler hinges to prevent structural deflection which will interfere with the adjacent structure (see Fig. 9.4.2–9.4.4).

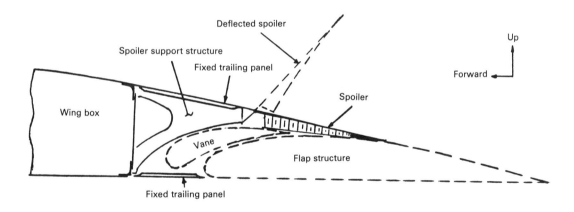

Fig. 9.4.1 Spoiler relative location at typical transport wing structure.

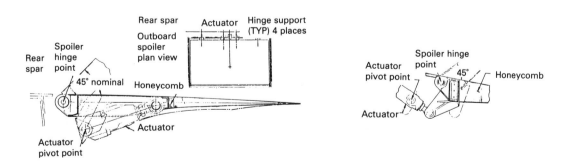

Fig. 9.4.2 Spoiler — B747.
(Courtesy of The Boeing Co.)

Airframe Structural Design 347

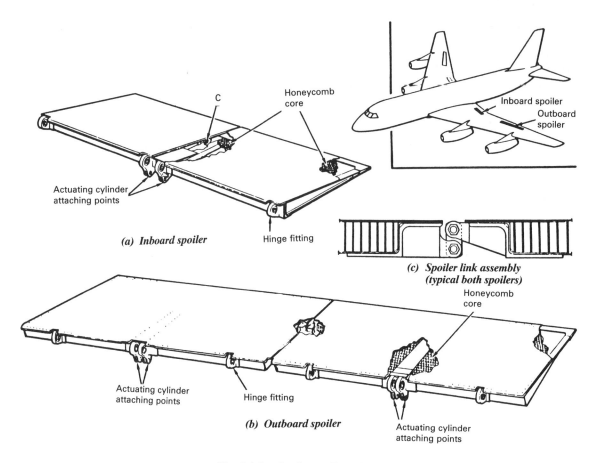

Fig. 9.4.3 Spoiler — Convair 880.
(Courtesy of General Dynamics Corp.-Convair)

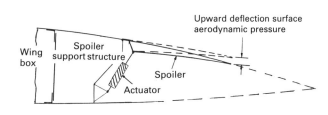

Fig. 9.4.5 Spoiler and support structure arrangement.

In order to prevent wing surface spoiler flutter during high speed flight, the stiffness of the spoiler system, made up of the spoiler surfaces, actuators and actuator support structure (as shown in Fig. 9.4.5) is required. These are:
- Spoiler bending and torsional stiffness (EI&GJ) — reflects minimum acceptable stiffness to meet flutter requirements.
- Actuator and support structure stiffness — reflects minimum acceptable stiffness.

Several of the current jet airplanes have had spoiler control problems that resulted from structural deflections caused by the landing gear during the take-off run (upward landing gear loads and then the wing bending down). This deflection has resulted in the spoilers lifting and thereby killing the wing lift during

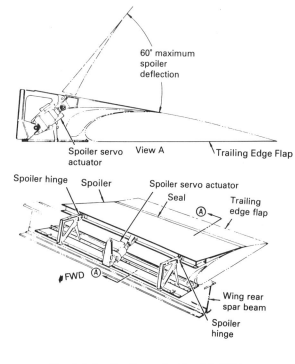

Fig. 9.4.4 Spoiler — Lockheed L-1011.
(Courtesy of Lockheed Aeronautical Systems Co.)

the most critical portion of the take-off run (Fig. 9.4.6). Control system design is another must to be established to accommodate the structural deflections. Attention should be given to the spoiler hinge due to the wing forced bending (wing deflections) for three-hinge design which is a typical design for most spoilers as illustrated in Fig. 9.4.7.

Ailerons

The aileron span is chosen to be as small as possible

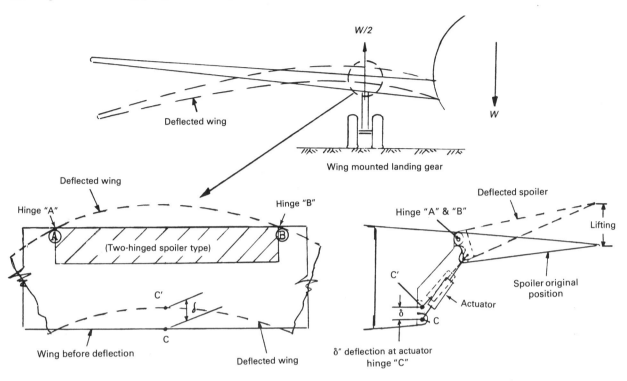

Fig. 9.4.6 *Spoiler lifting caused by upward landing gear loads (two-hinged spoiler type design).*

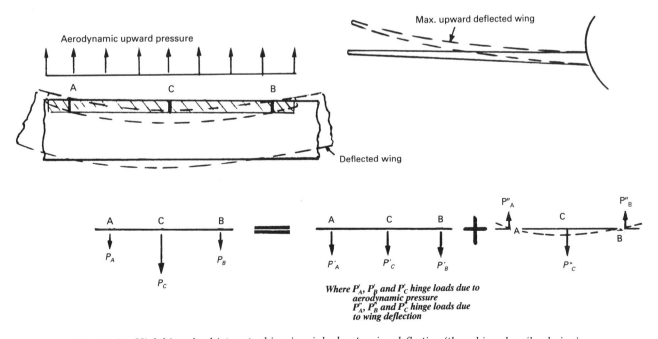

Fig. 9.4.7 *High hinge load (at center hinge) mainly due to wing deflection (three-hinged spoiler design).*

Airframe Structural Design 349

in order to obtain the largest possible flap span, within the limits imposed by roll control requirements. If the aileron of a thin wing is deflected at a high speed, the load on it causes the wing to twist in the opposite direction. This could, in some cases, offset the load produced by the aileron, resulting in a roll in the wrong direction (aileron reversal). For this reason some high-speed aircraft employ ailerons located closer to the wing root, where adequate torsional rigidity is present. This is one of the important reasons why wing box torsional rigidity (GJ value) should be considered in preliminary design phase for high-speed airplanes today. The outboard ailerons operate only at low speeds, while the small-span "high-speed" ailerons are operative both at low and high speeds. A suitable location for the high-speed aileron is behind the (inboard) engines (wing mounted jet engines case), since the flap will have to be interrupted there to keep it free from the jet efflux. In addition, to reduce the size of the ailerons on high-speed aircraft, these are frequently assisted by spoiler-ailerons.

The aileron must, first of all, be statically or mass-balanced to minimize the possibility of flutter. Static balance is obtained by equating the moments of mass forward of the aileron hinge line to those aft of the hinge line. Because the hinge line is usually well forward, statics balance requires the addition of weights to the aileron nose. The amount of weight used can be lessened by the use of fabric covering, composite materals or honeycomb structure aft of the hinge line as shown in Fig. 9.4.8.

Mass balance is obtained by locating the center of mass close to the hinge line, either directly on it or slightly forward of it. The alternative method is to use hydraulic dampers instead of mass balance weight. The advantage of using dampers is mainly to save weight but it cost more and this method has been applied on outboard aileron of Lockheed L-1011 [see Fig. 9.4.9 (b)].

The outboard aileron (low speed aileron) requires multi-hinge support (three or more hinges) structures of flutter prevention in case of failure of one or two hinges, another reason is that the actuator hinge is designed at the very inboard end aileron hinge to control the up and down rotation to reduce the effects of aero-elastic distortion of the wings as shown in Fig. 9.4.7. Multi-hinges are also required between aileron and wing to make aileron structure more effective. Due to the multi-hinge supports, the higher hinge loads owing to wing deflection, similar to that of spoiler design as shown in Fig. 9.4.7 and Fig. 9.4.10, should be carefully analyzed. Another very important consideration is the structural torsional stiffness (GJ) that the minimum acceptable stiffness is required to prevent aileron flutter.

In case balance weights are required, provisions will be made to facilitate increasing or decreasing the mass of each balance weight to compensate for effects of changes and repairs. The following inertia load factors (military aircraft may require higher load factors) will be used to design the balance weight attachments:
- A limit load factor of ±100g in a direction normal to the plane of the control surface

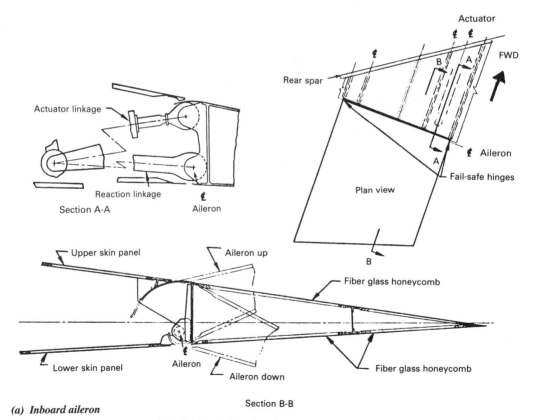

Fig. 9.4.8 Inboard and outboard ailerons — B747.
(Courtesy of The Boeing Co.)

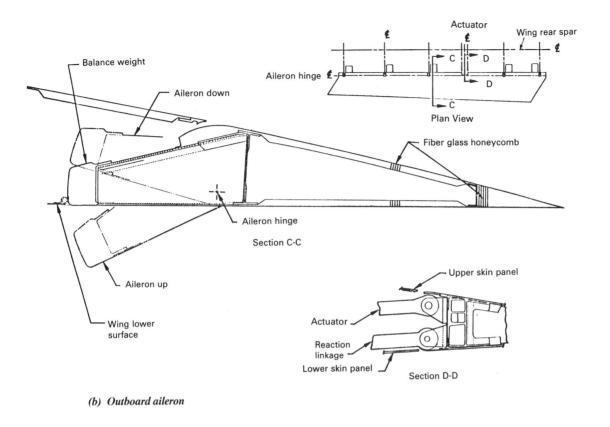

(b) Outboard aileron

Fig. 9.4.8 (continued).

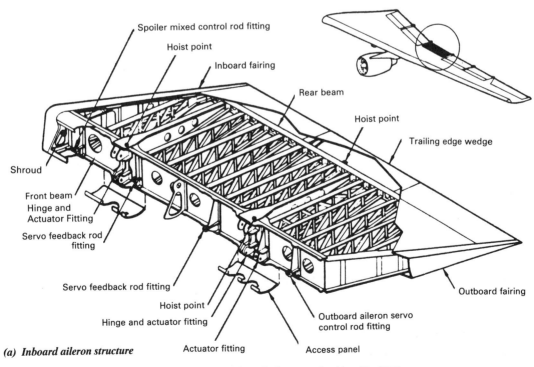

(a) Inboard aileron structure

Fig. 9.4.9 Inboard and outboard aileron — Lockheed L-1011.
(Courtesy of Lockheed Aeronautical Systems Co.)

Airframe Structural Design 351

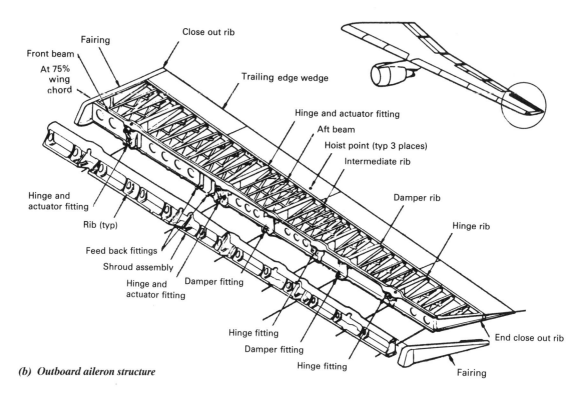

(b) Outboard aileron structure

Fig. 9.4.9 (continued).

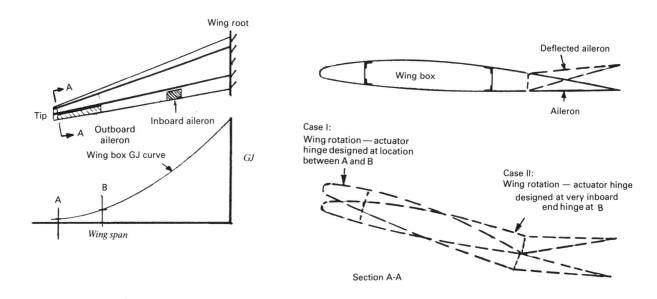

Fig. 9.4.10 Actuator location design of outboard aileron structure.

- A limit load factor of ±30g in the other two mutually perpendicular directions.

9.5 Fixed Leading and Trailing Edges

Fixed Leading Edges (Transport Airplanes)

The transport wing leading edge consists of fixed structure and slats. When slats are retracted, the slats form the front profile of the wing ahead of the fixed leading edge, which is contoured to form a secondary profile behind the slats when they are fully extended. Attached to the wing front box at the front spar, the fixed leading edge structure consists of surface panels (usually made from fiber-glass honeycomb panels), slat track ribs or other high-lift device support structures, intermediate support frames, and access doors on the lower surface. Each of the rib assemblies is attached between an extension of the upper and lower wing box skin panels or spar caps and to the web stiffeners on the front spar. The rib assemblies and intermediate support frames are permanently attached to the wing box structure, but surface panels and lower surface doors are screwed on the wing box with specially designed plate-nuts as shown in Fig. 9.5.1 to protect any damage around the screw holes which leads to fatigue cracks. Other typical fastener installations of edge panels to spar structure have been discussed in Chapter 8.0.

The only design load on fixed leading edge structures is aerodynamic surface pressure and the surface panels are generally designed not to carry any load from wing box bending and torsion loads because the surface panels are generally designed to be removable for inspection (maintenance reasons) on modern transports.

Leading edge support structures are generally light weight structure and either use close truss ribs with thick skin covers (Fig. 9.2.8) for the Kruger flap installation or use heavy ribs with stiffened cover panels (Fig. 9.5.2). Some designs use honeycomb panels to save weight, but attention should be given to the outer skin minimum thickness or toughness to prevent the damage from hail. In addition, track ribs with roller assembly (see Fig. 9.2.6) should be provided to support the concentrated loads from leading edge (slats). These ribs are usually machined from forgings and attached to the wing front spar where the wing ribs are located.

For structural simplicity, cost and weight saving, some airplanes are designed without any moving leading edges along its leading edge of the wing structure such as earlier day transport planes (see Fig. 9.5.3). Because of these reasons, even today's commuter aircraft are designed without any moving leading edges. However, after careful trade-off studies, the modern high efficient airplanes require high-lift devices to achieve the accomplishments with today's technology advancement.

Fixed Trailing Edge

The wing fixed trailing edge section completes the aerodynamic contour of the wing and provides the supporting structure for the wing control surfaces. Honeycomb skin panels, control surface hinge support ribs, intermediate supports, flap tracks and spanwise auxiliary beams (used mostly at inboard wing for swept wing of transport airplanes) constitute the basic fixed trailing edge structure (see Fig. 9.5.4).

To facilitate hydraulic lines, control cables or rods continuity, an opening should be provided at each support ribs, track and other support structure immediately aft of the rear spar.

The lower surface of the trailing edge section, forward of the flaps and aileron, is hinged for ready access to interior structure and equipment. These panels near jet engine blast area are designed for sonic fatigue conditions as well as aerodynamic pressure loads.

The only load on fixed trailing edge surface panels is usually designed to take aerodynamic pressure and will not react any other loads from wing bending and torsion loads. Refer to Fig. 8.3.21 and Fig. 8.4.11 of Chapter 8.0 for typical installations of edge panels to wing spar structure to prevent premature fatigue cracks. There have been a lot of such crack experiences on different airplanes, especially at wing lower surfaces owing to its frequent tension loads.

Obviously the critical load to design the flap track structure is during the flap extended conditions combining with flap side loads (aerodynamic loads plus flap inertia loads). These side loads will be reacted either by flap track side braces or trailing edge surface panels. Another flap support structure is designed as a small torque box (as shown in Fig. 8.4.58) which can carry loads from any directions.

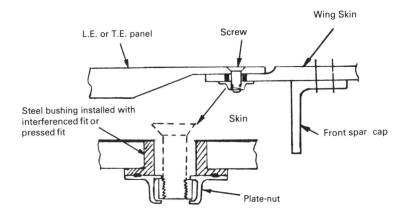

Fig. 9.5.1 Fixed leading edge panels attached to wing box structure.

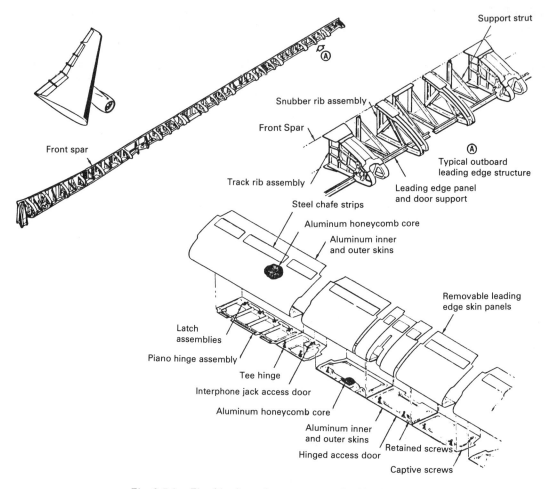

Fig. 9.5.2 Fixed leading edge structure — Lockheed L-1011.
(Courtesy of Lockheed Aeronautical Systems Co.)

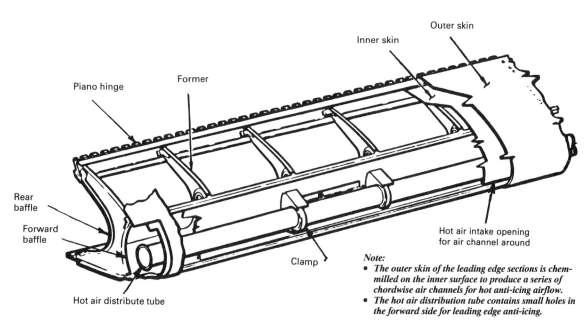

Fig. 9.5.3 Wing leading edge construction — Convair-880.
(Courtesy of General Dynamics Corp. — Convair)

354 Airframe Structural Design

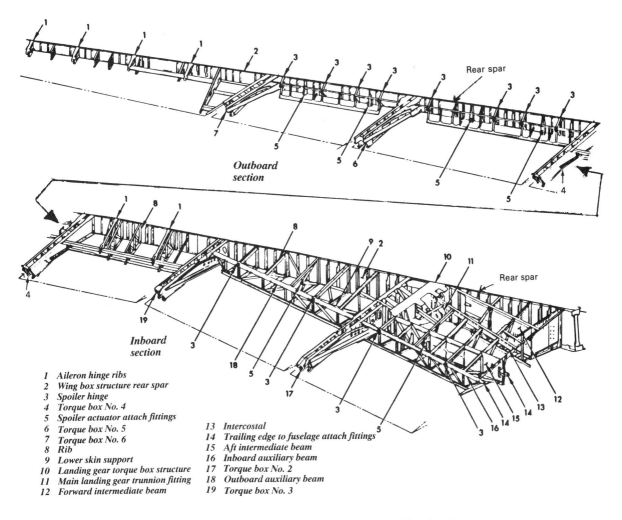

1. Aileron hinge ribs
2. Wing box structure rear spar
3. Spoiler hinge
4. Torque box No. 4
5. Spoiler actuator attach fittings
6. Torque box No. 5
7. Torque box No. 6
8. Rib
9. Lower skin support
10. Landing gear torque box structure
11. Main landing gear trunnion fitting
12. Forward intermediate beam
13. Intercostal
14. Trailing edge to fuselage attach fittings
15. Aft intermediate beam
16. Inboard auxiliary beam
17. Torque box No. 2
18. Outboard auxiliary beam
19. Torque box No. 3

Fig. 9.5.4 Wing trailing edge structure — Lockheed L-1011.
(Courtesy of Lockheed Aeronautical Systems Co.)

9.6 Design Considerations

General

- A fiber glass (or equivalent) overlay should be used on all trailing edge structures exposed to debris from the runway generally right behind the wing mount main landing gears.
- Provide quick access to all areas and equipment requiring frequent maintenance.
- Interchangeable parts should be designed to be interchanged without removing adjacent structure.
- Use same size and grid length of fasteners in all removable panels.
- Drain holes should be sufficient to eliminate all moisture traps. The minimum drain hole diameter should be $\frac{3}{8}$ inch.
- Honeycomb, either aluminum or fiber glass depending on strength requirements, should be a prime candidate for leading edge and trailing edge skins.
  - Fiberglass will generally be the lightest in minimum gage areas.
  - Hydropressed inner skins should be used only on flat, aluminum-alloy honeycomb panels or where contour can be simulated with a radius. If a varying contour exists, costs are prohibitive in developing matched hydroform dies. Manufacturing requires a minimum gage of 0.016 inch for the hydropress pans in order to handle them without damages.
- Pressure relief doors are required to protect the wing fixed leading structure, in case of hot bleed air duct rupture.
- Bearings on control surfaces should be replaceable without removing the control surface from the airplane.
- All major single pin bushed joints should have provisions for lubrication and should be readily accessible.

Aerodynamic Seals

Aerodynamic seals, a major drag reducer, are used to seal control surfaces and high-lift devices to the fixed structure to eliminate air entering into wing cavities at one point and exiting at another, which creates turbulence drag. On the wing it is most practical to pay more attention to seal the lower surface to restrict air flow from the lower surface to the upper surface.

Airframe Structural Design 355

- Seals like those for electrical bonding jumpers should be included in the total design problem and should not be designed as afterthought.
- Dacron-covered type seals should be used where scrubbing action is present. Skydrol resistant material (silicone non-dacron covered) is satisfactory for static seals.
- Seals should be retained in extruded channels for quick replacement. Consideration should be given to extruding the retaining channel as part of the structural member it is fastened to and counted as part of the structural strength. Occasionally space will not allow use of a retainer. In such cases attach with screws (never rivet).
- The vertical gaps between the upper and lower fixed structure should be sealed, joining the seals between the lower surface fixed structure and the movable surfaces. An example would be between an aileron and a flap.

Hail and Rain Protection

The area between points A and B on the leading edge in Fig. 9.6.1, where the angle of impact is less than 15°, is most susceptible to hail damage. 2024-T3 aluminum alloy of 0.063 inch is the minimum thickness and material necessary to prevent indentations from hailstones. Either a clad or unpainted anodized exterior surface is satisfactory. Clad can be used only if stretch forming to the leading edge contour does not destroy the corrosion-prevention characteristics of the clad.

Hail damage can not only occur on wing leading edge but flaps, spoilers and ailerons. Fig. 9.6.2 shows control surfaces that were damaged by hailstorm and in some areas the skin was actually punctured and in others the bond between the skin and the honeycomb core was broken.

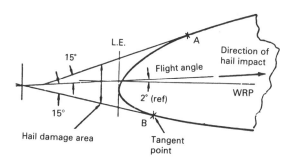

Fig. 9.6.1 Possible hail damage area on the leading side between A and B.

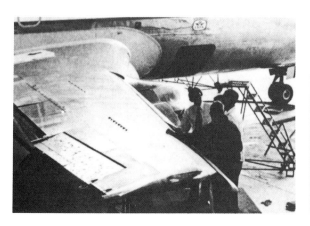
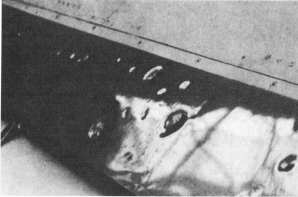

Fig. 9.6.2 Wing control surfaces damaged by hailstorm.

References

9.1 Fowler, H.D.: *Fowler Flaps for Airplanes — An Engineering Handbook.* Wetzel Publishing Co., Inc., 1948.

9.2 Nonweiler, T.R.: 'Flaps, Slots and Other High Lift Aids.' *Aircraft Engineering*, (Sept. 1955), 274–286.

9.3 Anon.: 'Take-off and Landing.' *AGARD-CP-160*, (1974).

9.4 MacLellen, A.: 'The Mechanism, Causes and Cures of Transonic Control-Surface Buzz.' *Aero Digest*, (Dec. 1956).

9.5 Anon.: 'Control Surface Flutter Considerations.' *Aero Digest*, (June 1944).

9.6 Williams, W.W.: 'An Advanced Extensible Wing Flap System for Modern Aeroplanes.' *AIAA Paper No, 70-911*, (1970).

9.7 Cotta, R.: 'The Case of the Extended Slat (What caused a TWA B-727 to take a nose dive over Michigan?)' *Aviation Week &Space Technology,* (June 1979).

9.8 Fowler, H.D.: 'Determination of the Influence of Propeller Thrust on C_L and C_D of an Airplane Using A Triple-slotted Flap at Various Deflections.' *SAE Paper No. 690724,* (1969).

9.9 Cornish, J.J. and Tanner, R.F.: 'High Lift Techniques for STOL Aircraft.' *SAE Paper No. 670245,* (1967).

9.10 Chacksfield, J.E.: 'Variable Camber Airfoils.' *Aeronautical Journal,* (May 1980), 131–139.

9.11 Anon.: 'Quiet Aircraft Flight Tests Schedules.' *Aviation Week &Space Technology,* (Apr. 17, 1978).

9.12 Velupillai, D.: 'Boeing Flexes Its Wings.' *Flight International,* (Nov. 24, 1979).

9.13 Gilbert, W.W.: 'Mission Adaptive Wing System for Tactical Aircraft.' *J. Aircraft,* (July 1981), 597–602.

9.14 Anon.: 'Installing the Heater Cable Directly in the Redesigned Leading Edge.' *Aircraft Engineering,* (May, 1973).

9.15 Williams, A.L.: 'A New and Less Complex Alternative to the Handley Page Slat.' *Technical Paper, Northrop Corp,* (1985).

9.16 Harvey, S.T. and Norton, D.A.: 'Development of the Model 727 Airplane High Lift System.' *SAE paper No. S408,* (Apr. 1964).

9.17 McRae, D.M.: 'The Aerodynamics of High-lift Devices on Conventional Aircraft.' *Journal of The Royal Aeronautical Society,* (June, 1969).

9.18 Stevens, J.H.: 'An Assessment of the Jet Flap.' *Interavia,* (Sept. 1958).

9.19 Anon.: 'Variable Camber Wing Being Developed.' *Aviation Week &Space Technology,* (Apr. 30, 1979).

9.20 Schaetzel, S.S.: 'The Icing Problem.' *Flight International,* (Aug. 1951).

9.21 Anon.: 'High-lift Ressearch (Youngman-Baynes Aircraft With Full-span Slotted Flaps).' *Flight International,* (Sept. 30, 1948).

9.22 Liebe. W.: 'The Boundary Layer Fence.' *Interavia,* (1952). 215–217.

9.23 Furlong, G.C. and McHugh, J.G.: 'A Summary and Analysis of the Low-speed Longitudinal Characteristics of Swept Wings at High Reynolds Number.' *NACA Report No. 1339,* (1957).

9.24 Anon.: 'Vought Presents the Eccentuator: (A Structural Hinge Concept For Variable Camber).' *Brochure Published by Vought Corp.*

9.25 Smith, A.M.O.: 'High-lift Aerodynamics.' *J. Aircraft,* (June 1975), 501–530.

9.26 Hoerner, S.F. and Borst, H.V.: *Fluid-Dynamic Lift.* P.O. Box 342, Brick Town, NJ 08723, 1975.

CHAPTER 10.0

EMPENNAGE (TAIL) STRUCTURE

10.1 Introduction

Empennage structure evolves essentially as does the wing (see Chapter 8 and Chapter 9). The aspect ratio of either a vertical surface or a horizontal surface usually tends to be smaller than a wing aspect ratio. The low aspect ratio, of course, means less bending moment because of less span. To date, the aerodynamist has not been able to make a case for a leading-edge device on an empennage surface. Reliability of the controls, plus controls and structural weight trade-offs, favor the use of larger surfaces rather than the complication of leading edge devices. With no devices, the leading edge can be included in the torque box torsional calculations, since it is not all cut up with a slat or flap. Fig. 10.1.1 illustrates typical empennage configurations.

The type of construction employed in the fixed control surfaces, stabilizer, and fin is usually similar to the types of wing construction discussed previously in Chapter 8.

- Single spar construction with auxiliary rear spar and all bending material concentrated in the spar cap only. This design is popular in light aircraft construction
- Single spar with pivot at root which can be rotated as flying wing or used as differential tailplane operation (taileron). This design has been applied on modern high speed fighters such as the F-14, F-111, F-15, F-18, etc. for their horizontal tail and the SP-71 vertical tail.
- Two-spar construction with all bending materials concentrated in the spar caps or most bending materials concentrated in the built-up surface panels, i.e. L-1011 Horizontal stabilizer.
- Multi-spar construction with spars resisting all the bending loads such as DC-10 fin structure.

The only concentrated loads which must be taken into account are those at the hinges supporting the movable control surfaces, i.e. elevators or rudders. However, control surfaces are not designed purely on a strength basis, stiffness of structure to prevent flutter is an important item in the design. Large deflections of the fixed surface (main box) will impose severe loads on the movable surface attached to it and may also cause bending at the hinge brackets. Fig. 10.1.2 illustrates a general arrangement of the conventional transport tail.

For specific detail design to resist sonic fatigue, see Chapter 15.

Conventional Tail

The conventional horizontal stabilizer assembly consists of left and right outboard sections attached to a center section, or torque box, within the aft fuselage, see Fig. 10.1.2. In large transport planes, the stabilizer is designed to pivot on two self-aligning bushing type hinge joints attached to a heavy bulkhead in the fuselage, and the angle of attack is adjusted by means of an electrically driven or manually operated ball nut and jackscrew, which is attached to the forward side of the center section. All vertical load distributions on the stabilizer are reacted at these three above mentioned attachment points. The removable leading edge is provided with an deicing system (usually required). The tip is removable as in the wing design. A rubber aerodynamic seal fills the gap between the stabilizer left and right outboard sections and the aft fuselage.

The Lockheed Jetstar has an unique design, that is the all-moving tailplane as shown in Fig. 10.1.3, which

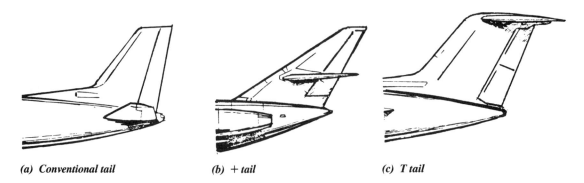

(a) Conventional tail   *(b) + tail*   *(c) T tail*

Fig. 10.1.1   *Typical arrangement of the transport tail.*

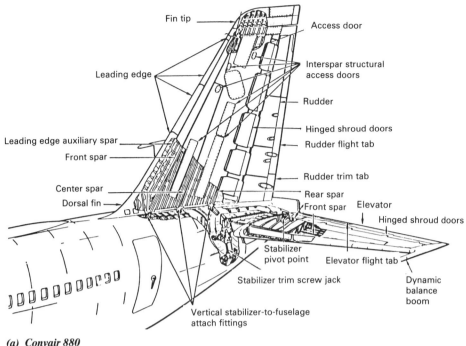

(a) *Convair 880*
(Courtesy of General Dynamics Corp. — Convair.)

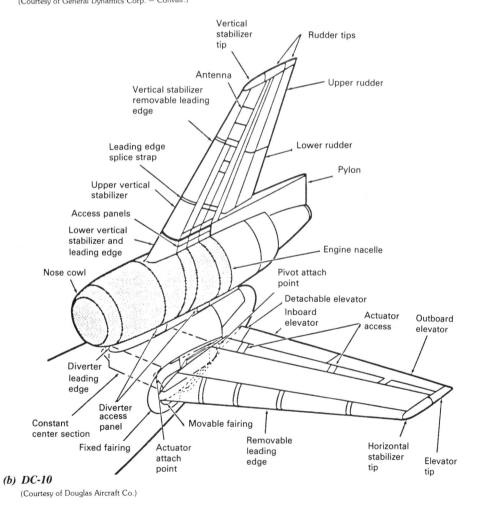

(b) *DC-10*
(Courtesy of Douglas Aircraft Co.)

Fig. 10.1.2 *Structural arrangement of transport tails.*

Airframe Structural Design 359

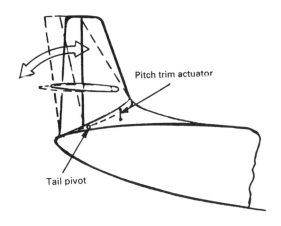

Fig. 10.1.3 One piece moving tailplane of Lockheed Jetstar empennage.

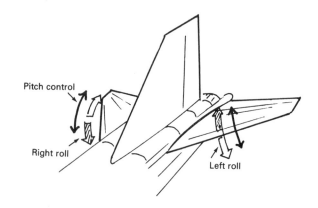

Fig. 10.1.5 Taileron operation.

moves in one piece with the fin. The fulcrum is at the base of the fin rear spar, and the whole assembly is actuated by scissor links attached to the front spar.

Fig. 10.1.4 shows the all-movable horizontal stabilizer (flying tail) which was used on the L-1011 transport; the British Trident and Lockheed F-104 also use this similar design. The horizontal tail is moved by hydraulic actuators in response to control-column pitch inputs.

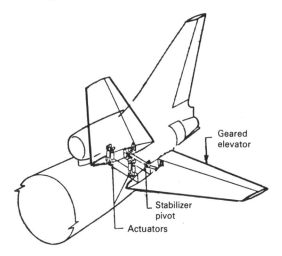

Fig. 10.1.4 All-movable horizontal stabilizer of L-1011.
(Courtesy of Lockheed Aeronautical Systems Co.)

Modern air-superiority fighter demands adequate rolling power under all conditions by a taileron (horizontal stabilizer surfaces are moved either together, as pitch control, or independently for additional control in roll) and plenty of pitching moment to take account of the quite considerable out-of-balance forces which especially occur when the wing sweeps for the variable sweep wings. Taileron operation is shown in Fig. 10.1.5.

T-Tail

From a static loads standpoint, the design of a T tail is as straightforward as a fuselage-mounted arrangement. However, because of flutter considerations, it is necessary that the vertical fin and the attachment of the horizontal tail to be principally designed to stiffness requirements. The primary parameter for T-tail flutter is, of course, the fin torsional stiffness, and with this arrangement, the vertical fin stiffness required is heavily dependent on the mass of the horizontal stabilizer. Because of this T-tail characteristic, it is very important to design for minimum horizontal tail size in order to minimize the fin stiffness requirement. A comparison of the vertical fin stiffness requirement between a low tail and a T-tail design is shown in Fig. 10.1.6. This data shows that the T-tail requires about 1.5 times the stiffness at the vertical fin root and about 40 times the stiffness at the tip than does the low tail arrangement. Obviously, this results in a higher structural weight for the vertical fin for the T-tail.

Another important parameter for a T-tail design is the effect of stabilizer dihedral on the flutter speed, which directly affects the required fin stiffness. The flutter speed can be increased appreciably by incorporating negative dihedral into the horizontal stabi-

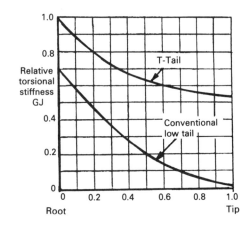

Fig. 10.1.6 Comparison of fin torsional stiffnesses for conventional tail and T-tail design.

360 Airframe Structural Design

lizer design. For instance, 5° of horizontal stabilizer negative dihedral can result in a reduction of approximately 30% in the required fin stiffness for flutter reduction.

The T-tail as shown in Fig. 10.1.7 includes a variable incidence horizontal stabilizer and fixed vertical stabilizer (fin), together with their associated movable surfaces, i.e., elevator and rudder as shown in Fig. 10.1.7(a). Fig. 10.1.7(b) uses an all-movable horizontal stabilizer. The fin built into the aft fuselage structure is same as previously mentioned for conventional fin construction. The horizontal stabilizer pivots in bearings housed in the support structure at the top of the fin rear spar. The jack arm in the center of the horizontal stabilizer is connected to the screw of a variable incidence jackscrew bolted to mounting plates at the top of the fin from spar.

elevators and rudders. Fig. 10.1.8 illustrates some typical design cases such as beaded skins, skin dimpling for stiffness, chordwise flutes, etc.

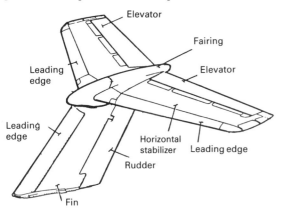

(a) Transport tail

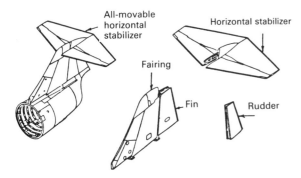

(b) Fighter airplane tail

Fig. 10.1.7 Typical T-tail arrangements.

Another configuration of horizontal stabilizer T-tail design is a fixed stabilizer which is similar to that of the previously mentioned conventional tail. The only moving parts are elevators and tab (if any).

The leading edges of the fin and horizontal stabilizer are usually designed for anti-icing and also are detachable for maintenance.

Tail of General Aviation Airplanes
The airload on tail surfaces of general aviation airplanes are fairly low and single main spar with light stiffened skins are generally the typical construction of the horizontal stabilizer box, vertical stabilizer box,

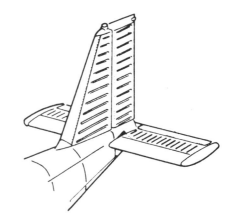

Case (a) Beaded skin or skin dimpling

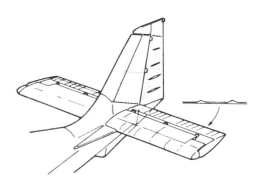

Case (b) Elevators and rudder-beaded skin or skin dimpling; main boxes — spar and ribs

Case (c) Spar and ribs design

Fig. 10.1.8 Typical tail design cases of general aviation airplanes.

Various Transport Tail Box Data
Typical tail box data of modern transports are illustrated in Fig. 10.1.9 for reference.

Airplane Name	Vertical/ Horizontal Tail	Type of Construction	Material Skin/Str.	Panel Shape	No. of Spars; Type of Ribs
L-1011	Vert.	Skin-Str.	7075-T6/7075-T6	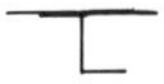	2 spars; Truss ribs
	Horiz.	Integral	7075-T76		2 spars; Truss ribs
DC-10	Vert.	Spars	7075-T6 panels		4 spars; Ribs-web
	Horiz.	Integral	7075-T651		2 spars; Ribs-web
B747	Vert.	Skin-Str.	7075-T6/		2 spars; Ribs-web
	Horiz.	Skin-Str.	7075-T6/		2 spars; Ribs-web
B727	Vert.	Skin-Str.	2024-/7075-T6	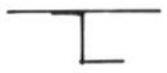	2 spars; Ribs-web
	Horiz.	Skin	2024-T3		2 spars; Ribs-web
Trident	Vert.	Skin-Str.	2024-/2024-		2 spars; Ribs-web
	Horiz.	Skin-Str.	2024-/2014-		2 spars; Ribs-web
DC-8	Vert.	Beaded panels			3 spars; Arch ribs
	Horiz.	Integral			3 spars; Arch ribs
A300	Vert.	Skin-Str.			3 spars; Truss ribs
	Horiz.	Skin-Str.		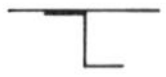	2 spars; Ribs-web
F.28	Vert.	Honeycomb panels			4 spars; Ribs-web
	Horiz.	Skin-Str.			2 spars; Ribs-web
C-141	Vert.	Skin-Str.			3 spars; Truss rib
	Horiz.	Skin-Str.		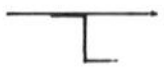	2 spars; Truss ribs
B707 B737	Vert.	Skin	2024-T3		2 spars; Ribs-web
	Horiz.	Skin	2024-T3		2 spars; Ribs-web
DC-9	Vert.	Skin/Str.		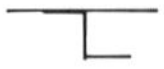	2 spars; Ribs-web
	Horiz.	Integral		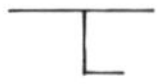	2 spars; Ribs-web
YC-15	Vert.	Skin-Str.	7050-T76/ 7050-T76	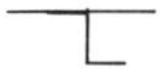	2 spars; Ribs-web
	Horiz.	Skin-Str.	7050-T76 7050-T76		2 spars; Ribs-web

Fig. 10.1.9 Comparison of transport tail box structures.

Brief Summary of Tail Loads (reference only)
(1) General
- Sonic fatigue
- Fatigue
- Fail-safe
- Control surface hinge load due to the stabilizer forced bending (refer to Fig. 9.4.7)
- Hoisting
- Flutter
- Control surface reversal
- Control surface effectivity
- Rib crushing
- Control surface support
- Actuator support
- Concentrated load redistribution

(2) Leading edge
- Hail strike

(3) Stabilizer
- Instantaneous elevator
- Fin gust (on T tail)
- Positive maneuver
- Negative maneuver
- Unsymmetrical spanwise load distribution

(4) Fin
- Instantaneous rudder — yaw initiation
- Dynamic overyaw
- Check maneuver
- Engine out
- Fin gust

(5) Aft Fuselage
- Redistribute vertical and horizontal stabilizer concentrated loads
- Tail skid load

10.2 Horizontal Stabilizer

The conventional horizontal tail consists of a fixed tail box, or an adjustable incidence box or an all-movable box and elevators. The horizontal stabilizer is usually a two-spar structure consisting of a center structural box section and two outer sections. The stabilizer assembly is interchangeable (usually adapt symmetrical airfoil section) as a unit at the fuselage attach points, and the outer sections are interchangeable at the attachment to the center box.

A pivot bulkhead is located at the juncture of the center box and outer section at each side of the fuselage. Each bulkhead contains a pivot bearing at the aft end and an actuator attach point at the forward end. This provides a four-point, fail-safe support arrangement for the stabilizer assembly. In the event of any single pivot or actuation point failure, the remaining three points will continue to support the stabilizer. Another design concept is a single jackscrew at the middle of the front spar and two pivot bearings at each side of the rear spar. Of course, each of these fittings has to be designed for fail-safe capability.

The center box and the main box structure of the outer section are designed with the primary bending material distributed in spar caps and cover panels or in spars only. Fail-safe design criteria of the wing box should be applied on horizontal stabilizers to ensure the structural integrity.

The leading edge structure of the horizontal stabilizer outer sections is composed of several segments, and each segment is removable without disturbing adjacent segments. The construction of skin and formed ribs is designed to resist hail and foreign object damage.

Access doors are provided in the leading edge structure, front spar web, and aft closing shear web for inspection and maintenance of internal structure. The stabilizer assemblies are weather sealed with drain holes provided on the lower surface to prevent entrapment of moisture.

Transports

There are two basic horizontal stabilizer box constructions for modern transport:

(1) Box construction with spars, closer light rib spacing (usually less than 10 inches) and surface (may be tapered skins) without stringer reinforcements. The feature of this design is the low manufacturing cost and high torsional stiffness (due to thicker skins) required by flutter analysis. This design has been used on some horizontal stabilizers as shown in Fig. 10.2.1.

In Fig. 10.2.1, the front and rear spars and the skin form a box beam of the horizontal stabilizer which is the main structural member of the stabilizer. Airfoil contour and torsional strength are maintained by ribs at close spacing. The bending moment and beam shear are carried by the spars and local effective skin. The stabilizer torsion is reacted by torsional shear in the skin and spar web. Attachment of the outboard sections to the center section is at the front and rear spars only, with no other structural tie between the outboard and center section skin.

Provisions are included in the aft fuselage structure that allows the stabilizer center section to be withdrawn through the left or right side of the aft fuselage should repair be required in service.

Fig. 10.2.2 shows a peculiar design of the built-up tailplane center-section (without covers), which is better for maintenance than the conventional box-type structure. The variable-incidence screw-jack is applied at the front, while the pivots are attached to the rear spar.

(2) Box construction with spars, stronger ribs and surface skins with stringer reinforcements (skin-stringer or integrally stiffened panels) is a lighter weight structure. Typical configurations of this design are shown in Fig. 10.2.3 and Fig. 10.2.4.

Fig. 10.2.3 illustrates a horizontal stabilizer box structure that consists of two spars and surface covers, which are integrally stiffened and machined from 7075-T76 aged stabilized alloy for superior exfoliation and stress corrosion characteristics. The stabilizer pivots about its rear spar and is actuated by four independent hydraulic actuators attached to the front beam. Elevators are mechanically positioned by cable linkage in accordance with a fixed gearing relationship to the horizontal stabilizer position. The horizontal stabilizer is supported at either side of the fuselage by fail-safe bearing as shown in Fig. 10.2.5.

Flying Tail of Fighter Planes

The main advantage of a flying tail is the reduction in tail size and consequent reduction in drag during cruise. But, for high speed modern fighters, the flying

Airframe Structural Design 363

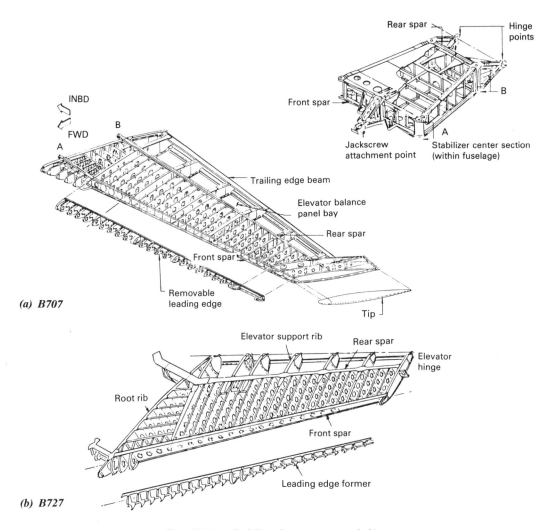

Fig. 10.2.1 Stabilizer box — spars and skins construction.

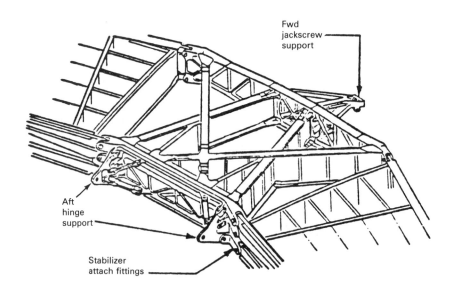

Fig. 10.2.2 Horizontal stabilizer center section — B737.
(Courtesy of The Boeing Co.)

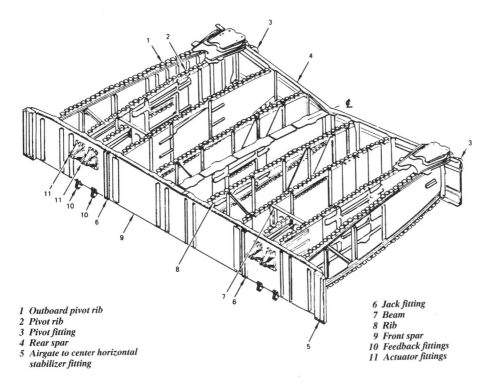

1 *Outboard pivot rib*
2 *Pivot rib*
3 *Pivot fitting*
4 *Rear spar*
5 *Airgate to center horizontal stabilizer fitting*
6 *Jack fitting*
7 *Beam*
8 *Rib*
9 *Front spar*
10 *Feedback fittings*
11 *Actuator fittings*

(a) Center-section

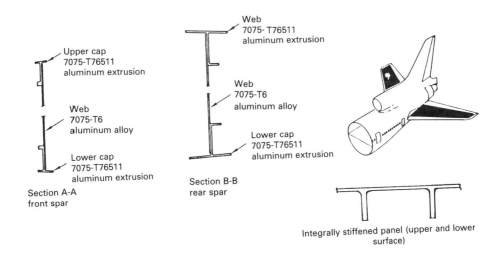

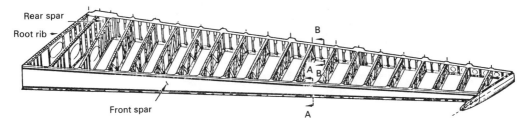

(b) Outboard section

Fig. 10.2.3   Horizontal stabilizer structure — L-1011.
(Courtesy of Lockheed Aeronautical Systems Co.)

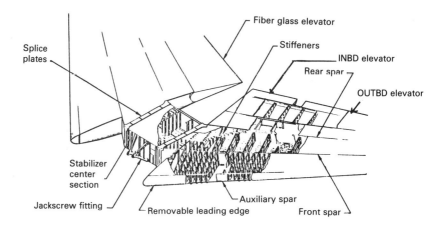

(a) Structural arrangement

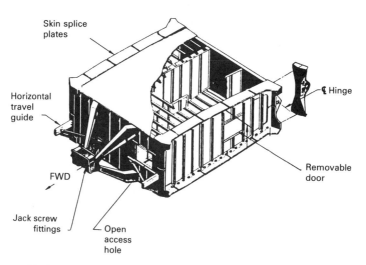

(b) Center section

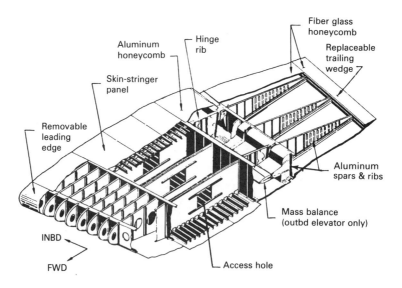

(c) Outboard section

Fig. 10.2.4 *Horizontal stabilizer structure — B747.*
(Courtesy of The Boeing Co.)

366 Airframe Structural Design

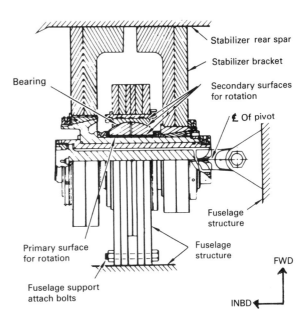

Fig. 10.2.5 L-1011 horizontal stabilizer pivot and bearing.
(Courtesy of Lockheed Aeronautical Systems Co.)

tail is almost the only design choice because of its thin and small section and conventional design cannot meet the requirements.

In Fig. 10.2.6 shows a typical flying tail suitable for a supersonic fighter aircraft. The entire surface pivots about a shaft which projects from the aft fuselage, with actuation accomplished by means of a control horn.

The leading edge, trailing edge and tip are usually designed for full depth honeycomb structure with surface skins. The box beam construction also has a full depth honeycomb core as shown in Fig. 10.2.7, except in the pivot region, where the intercostal beams and the root and intermediate ribs are required to carry the local high concentrated loads. The basic materials of this construction are aluminum alloy; however, composite materials, such as graphite/epoxy, boron, etc, are widely used on modern air superiority fighters due to its light weight, high strength and stiffness.

Some examples of horizontal stabilizer designs of existing fighter planes are briefly described below.

- Grumman F-14: Basically the tail box is a multi-spar construction with composite material skin application. The leading and trailing edge structures are built-up with full depth honeycomb

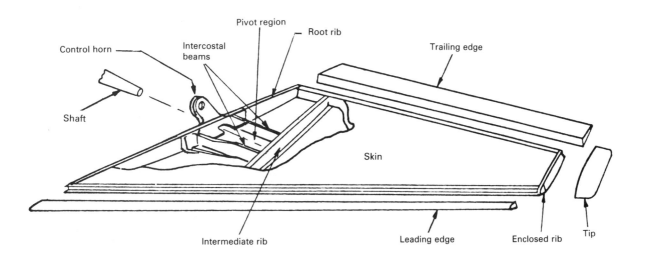

Fig. 10.2.6 Typical structural components of flying tail.

Airframe Structural Design 367

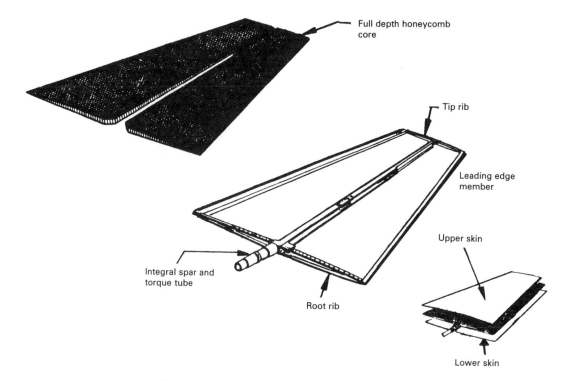

Fig. 10.2.7   Typical full depth honeycomb construction.

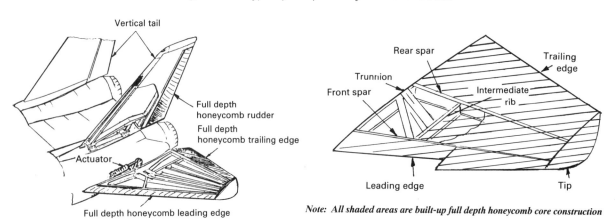

Fig. 10.2.8   Flying tail construction — F-14.

Fig. 10.2.9   Flying tail construction — F-15.

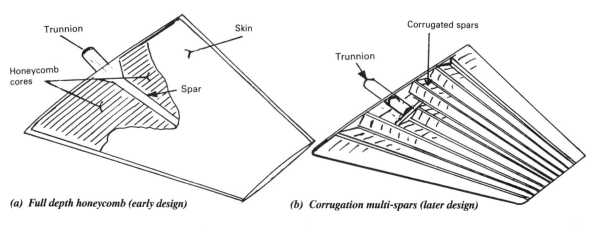

(a) Full depth honeycomb (early design)

(b) Corrugation multi-spars (later design)

Fig. 10.2.10   Flying tail construction — F-16.

design (see Fig. 10.2.8).
- McDonnell-Douglas F-15: It is a two-spar construction with full depth aluminum-honeycomb core and boron-composite skin except in the root area which provides the machined tailplane-spar attachment fitting for mounting trunnion. The complete trailing edge panels and outboard leading edge panels are built-up with full depth Nomex honeycomb core and aluminum skin (see Fig. 10.2.9).
- General Dynamic F-16: The flying tail is made up of an integral main spar (one spar construction) and pivot fitting, with titanium root and reinforced plastic tip ribs enclosed by the upper and lower skins. Aluminum honeycomb-cores fill the spaces between the upper and lower skins and are bonded to these components. Upper and lower skins are graphite epoxy composite material to provide stiffness and light weight. The horizontal tail is interchangeable from left to right and vice versa (see Fig. 10.2.10).
- Russian Mikoyan MiG 21: The flying tail is an all-movable tailplane design and a built-up structure with one main spar and ribs covered with aluminum skin to form a complete torque box structure including leading and trailing edge as one integral unit. The metal fitting is provided at the trunnion location to transfer the tail load to pivot shaft (see Fig. 10.2.11).

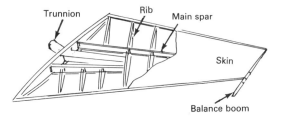

Fig. 10.2.11 Flying tail construction (conventional design) — MiG-21.

10.3 Vertical Stabilizer (Fin)

Structural design of vertical stabilizers is essentially the same as for horizontal stabilizers. The vertical stabilizer box is a two- or multi-spar structure (general aviation airplanes usually use single spar design) with cover panels (with or without ribs). The root of the box is terminated at the aft fuselage juncture with fittings or splices or the box spars terminate on bulkheads in the aft fuselage that are canted (for swept fin) into the plane of the spars as shown in Fig. 10.3.1, thus transmitting the fin loads directly into the fuselage structure and avoiding the fatigue-critical structural splices.

The T-tail arrangement places the horizontal stabilizer in a favorable flow field during low-speed, high angle-of-attack operations. Mounting the horizontal stabilizer on top of the fin has significant effect on the torsional frequency of the fin; the inertia of the stabilizer reduces the natural frequency of the combination by a large factor. This means that flutter coupling between the rudder and the fin is reduced drastically so that the rudder need not have as much

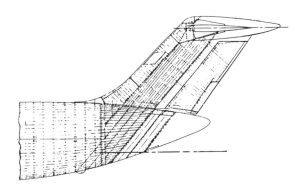

Fig. 10.3.1 Vertical stabilizer structure arrangement.

balance weight. The thicker symmetric airfoil usually is used to design T-tail fin box structure that will give both higher bending and torsional stiffnesses which are badly needed for the T-tail flutter problems. For economic reasons, the current two AMST (Advanced Medium STOL Transport) prototypes (as shown in Fig. 10.3.2) select a constant vertical fin section which permits the use of all-identical fin rib assemblies as well as all-identical leading edge rib assemblies, and also all-identical rudder rib assemblies. This permits a cost saving of approximately 41% on rib assemblies and 34% on rudder assemblies on these airplanes. However, the trade-off study between cost and weight penalty has to be carefully evaluated.

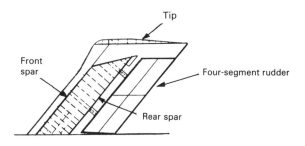

Fig. 10.3.2 Constant vertical tail section.

The span of the T-tail fin is approximately one-third shorter than conventional tails, but, on the other hand, the higher structural stiffness is required. Therefore, typical skin-stringers or integrally stiffened panels are suitable for this requirement.

Navy air superiority fighters, such as the F-14, use two vertical fins because of limited vertical space as carriers were required (as well as the need for good lateral control for aerodynamic reasons).

It is worthwhile to mention that the Lockheed strategic reconnaissance SR-71 (blackbird) uses all-movable vertical stabilizers sized for a minimum acceptable level of directional stability at design cruise Mach number. Rudders are discarded as inadequate because extremely large deflections are required to control the engine-out condition. Also, because of the

Airframe Structural Design 369

high sustained operating temperatures encountered at high cruise Mach numbers, the rudder-hinge stagnation temperature would cause local heating problems. All-movable vertical stabilizers increased control effectiveness two-and-a-half times; therefore, engine-out control required smaller deflections and gave lower trim drag. (One of the earliest examples of an all-movable vertical stabilizer on a supersonic aeroplane was the Sud-Quest S.Q. 9000 Trident fighter prototype of 1953.)

One of the main advantages of twin (on the An-22) or triple vertical tail surfaces for propeller-driven airplanes (as shown in Fig. 10.3.3) is that it is possible to place them directly behind the propellers of twin-engine airplanes. This allows the slipstream to strike the surfaces with full force, giving good rudder control at low air-speeds. However, the horizontal tail must be sufficiently rigid to prevent buffeting and flutter.

This end-plate effect increases the horizontal tail efficiency as much as 22%, depending on the ratio of the vertical tail span to the horizontal tail span. Similarly, tests have shown an increase of elevator efficiency which is dependent also on the foregoing ratio. The efficiencies of the vertical surfaces and the rudder are less for the twin or triple than for the single surface by approximately 50%.

Where additional stability with minimum area is of prime importance, the single tail outshines the double and triple fins. In fact, it was for this reason that the twin surfaces of the B-24 were redesigned into a single unit, as the double tail provided insufficient directional stability at low airspeeds.

The horizontal surfaces should be situated sufficiently high on the fuselage to clear the wing wake, especially when the flaps are deflected. It is recommended that the ratio of tip to root chords be not less than 0.5 and preferably greater.

The use of twin or triple vertical tail surfaces introduces additional structural problems to the design of the stabilizer (same phenomenon on winglet). It is possible in certain flight conditions to have fin and rudder loads acting inboard, which may be represented by the force P acting at the center of pressure of the surface. Assuming no load on the horizontal tail, the vertical loads produce a moment $P \times \delta$ at the

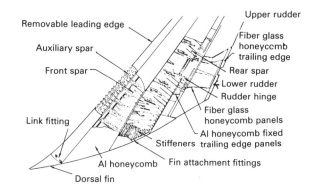

(a) Vertical stabilizer arrangement

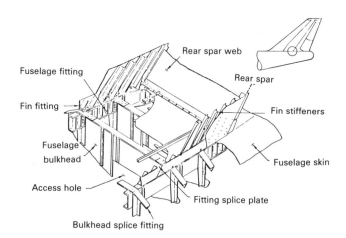

(b) Root joint

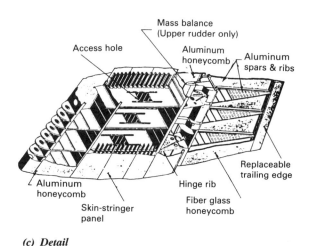

(c) Detail

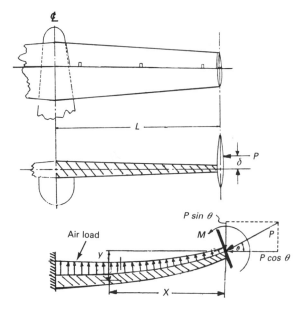

Fig. 10.3.3 Aerodynamic loads on twin-vertical fin design.

Fig. 10.3.4 Vertical stabilizer structure — B747.
(Courtesy of The Boeing Co.)

370 Airframe Structural Design

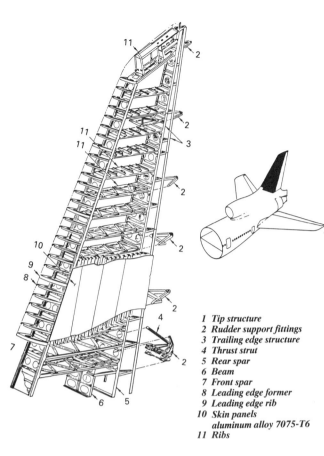

Fig. 10.3.5 Vertical stabilizer structure — L-1011.
(Courtesy of Lockheed Aeronautical Systems Co.)

1. Tip structure
2. Rudder support fittings
3. Trailing edge structure
4. Thrust strut
5. Rear spar
6. Beam
7. Front spar
8. Leading edge former
9. Leading edge rib
10. Skin panels aluminum alloy 7075-T6
11. Ribs

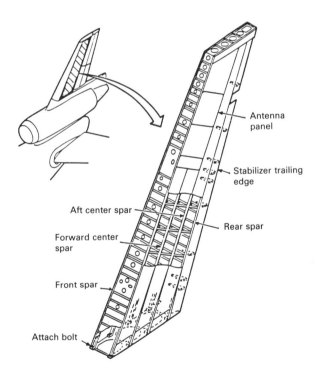

Fig. 10.3.6 Vertical stabilizer structure — DC-10.
(Courtesy of Douglas Aircraft Co.)

horizontal stabilizer-fin juncture, plus an axial compression in the stabilizer beams equal to P. This primary moment causes the horizontal stabilizer to rotate, which in turn causes the axial compression to induce secondary bending moments equal to the product of the horizontal component of the force and the distance y.

When apportioning the vertical tail area, the moment $P \times \delta$ may be eliminated by dividing the area equally and symmetrically above and below the horizontal tail. The axial compression still remains, but the magnitude of the secondary moments will decrease.

Transports

There are basically three types of fin construction: (a) two spars with skin-stringer panels as shown in Fig. 10.3.4 and Fig. 10.3.5 respectively; (b) multi-spars design (Fig. 10.3.6); and (c) two spars, closer light rib spacing and surface skins without stringer reinforcements which is similar to that of the horizontal stabilizer box design (Fig. 10.2.1).

Fighter Fin

The vertical stabilizer construction of fighters is similar to that of horizontal stabilizer as previously discussed in Section 10.2. Basically, the vertical stabilizer has a rudder which is attached to the fin box by fittings. The all-movable vertical stabilizers used on the SR-71 have not been seen on modern fighters because, after trade-off studies, this design does not pay-off except under special conditions such as the SR-71.

10.4 Elevator and Rudder

The FAA (Federal Aviation Administration of the United States) sets forth certain requirements for the design and construction of the tail and control surfaces for an airplane; and because the technician must perform repair and maintenance in conformity with such requirements.

Movable tail surfaces should be so installed that there is no interference between the surfaces or their bracing when any one is held in its extreme position and any other is operated through its full angular movement. When an adjustable stabilizer is used, stops must be provided at the stabilizer to limit its movement, in the event of failure of the adjusting mechanism, to a range equal to the maximum required to balance the airplane.

Elevator trailing-edge tab systems must be equipped with stops that limit the tab travel to values not in excess of those provided for in the structural report. This range of tab movement must be sufficient to balance the airplane under all speeds of normal flight except those specified by FAR (Federal Aviation Regulations) that are in excess of cruising speed or near stalling speed.

When separate elevators are used, they must be rigidly interconnected so that they cannot operate independently of each other. All control surfaces must be dynamically and statically balanced to the degree necessary to prevent flutter at all speeds up to 1.2 times the design dive speed. The installation of trim and balancing tabs must be such as to prevent any

Airframe Structural Design 371

free movement of the tab. When trailing-edge tabs are used to assist in moving the main surface (balance tabs), the areas and relative movements must be so proportioned that the main surface is not over-balanced at any time.

Basically the design of both the elevator and rudder is similar to that of aileron construction (refer to Chapter 9 Section 4) and the wedge design (refer to Chapter 9 Section 3)

In Fig. 10.4.1, the skins are bonded scalloped doublers for reinforcement in the area of spar and rib attachments. The addition of bonded scalloped doublers has shown by test and by service experience to provide increased fatigue life for skin panels subjected to aerodynamic turbulence or sonic fatigue. Hinge bearings are designed for long-life, anti-friction roller types which can be easily lubricated in service and may be replaced without removing the surface from the airplane.

All holes in the hinge support fittings for hinge pin bolts are bushed with cadmium plated stainless steel bushings installed with wet primer to guard against corrosion from dissimilar metals. Sufficient material is provided around the bushings to permit installation of oversize bushings for service replacement. The hinge pin bolts are plated with dense chrome to increase wear life and guard against corrosion.

The surfaces are mass balanced and the balance weights (the design inertia load factors identical to that of ailerons as discussed in Section 4 of Chapter 9) are easily accessible through hinged access panels. The balance limits permit weight variations resulting from minor repairs and some repainting.

The surface assemblies are weather sealed with drain holes provided in the lower surfaces to prevent entrapment of moisture. Access doors and panels are provided for inspection and service of all internal structures and mechanisms.

The structural design for both the elevator and rudder are similar in construction, front and rear spar (some designs have omitted the rear spar) and the skin form a box beam which is the structural member of the elevator or rudder. Close spaced ribs are provided to stabilize skin strength due to the high torsional load and local aerodynamic pressure (as shown in Fig. 10.4.2). The control actuator hinge is located underneath the very inboard end hinge of the elevator or located at the side of the bottom hinge of the rudder structure and, therefore, all the torsional loads are reacted at this actuator hinge. The multi-hinge supports are necessary to react the local aero-dynamic loads and transfer to the stabilizer main box. It is also for fail-safe reason in case of failure hinges to prevent flutter problems. Another important require-

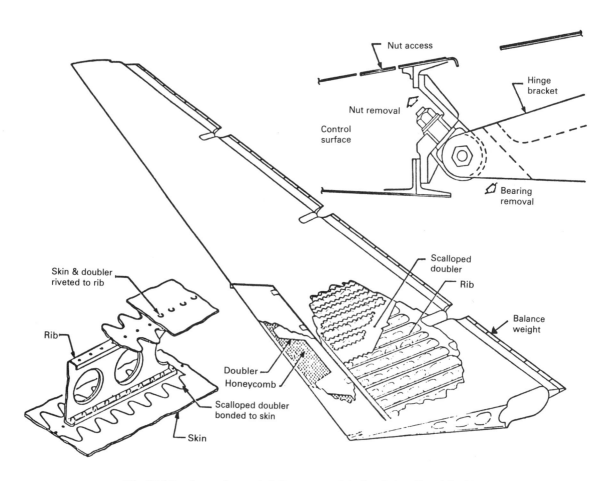

Fig. 10.4.1   Increasing sonic fatigue strength by bonded scalloped doublers.

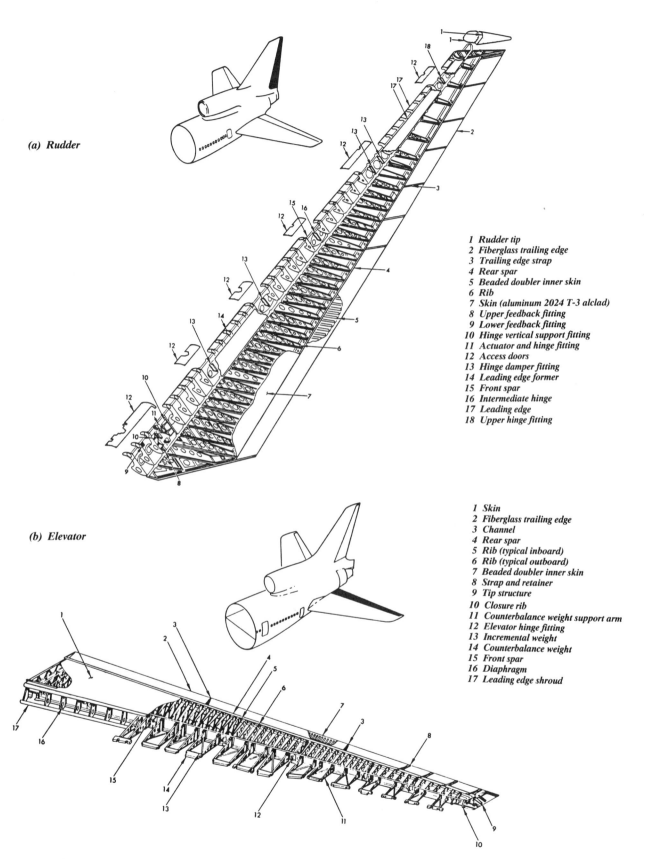

Fig. 10.4.2 Rudder and elevator construction — L-1011.
(Courtesy of Lockheed Aeronautical Systems Co.)

ment is the minimum acceptable torsional stiffness (GJ), and actuator and support structure stiffness to meet high speed flutter requirements.

Owing to the multi-hinge design, the extreme high hinge loads on the elevator or rudder are caused by the attached main stabilizer box bending deflection which is similar to that of the spoiler case as shown in Fig. 9.4.7 of Chapter 9. This condition should not be neglected. From a structural standpoint, some transport elevators are divided into two segments, i.e. inboard elevator and outboard elevator as shown in Fig. 10.2.4. At low speeds all the elevators are used, but at high speeds the inboard elevator (or so called high speed elevator) is allowed to operate because of the structural effectiveness (to prevent the so called elevator reversal). The same design is also applied on rudder structures, namely upper and lower rudders as shown in Fig. 10.3.4. The advantages of this design is to save weight but will pay off for a more complex control system. Fig. 10.4.3 illustrates a four-segment rudder to augment the rudder aerodynamic efficiency under the condition of smaller vertical tail area and, obviously, the design makes the control system more complex (military transports such as the YC-15 use this design, see Fig. 10.3.2).

The honeycomb panel structure is commonly applied on the elevator and rudder surface (Fig. 10.2.4, 10.3.4 and 10.4.4) in lieu of solid skin to strengthen the surface buckling and also to meet the important torsional stiffness (GJ).

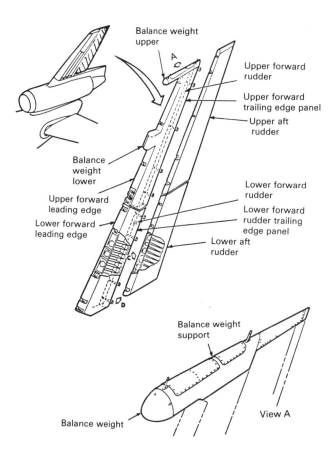

Fig. 10.4.3 Four-segments rudder design — DC-10.
(Courtesy of Douglas Aircraft Co.)

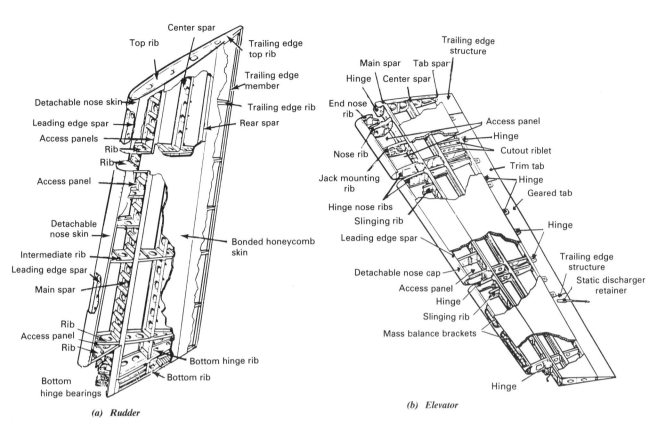

Fig. 10.4.4 Honeycomb panel structure design — BAC 111 rudder and elevator.
(Courtesy of British Aerospace PLC)

References

10.1 Byrnes, A.L., Hensleigh, W.E. and Tolve, L.A.: 'Effect of Horizontal Stabilizer Vertical Location on the Design of Large Transport Aircraft.' *Journal of Aircraft*, (Mar.–Apr., 1966).

10.2 Multhopp, H.: 'The Case for T-Tails.' *Aero Digest*, (May, 1955), 32–35.

10.3 Cleveland, F.A.: 'Lockheed-Georgia's T-Tail Concept.' *Lockheed-Georgia Quarterly*, (Summer 1965), 25–26.

10.4 Lipsick, L.P.: 'Twin Tail Surfaces.' *Aero Digest*, (Apr. 1947), 70.

10.5 Harmon, S.M.: 'Comparison of Fixed-Stabilizer, Adjustable-Stabilizer and All-Movable Horizontal Tails.' *NACA ACR No. L5H04*, (Oct., 1945).

10.6 Anon.: 'Advanced Development of Conceptual Hardware Horizontal Stabilizer, Vol. I.' *AFFDL-TR-78-45*, Vol. 1.

10.7 Velupillai, D.: 'T-tails and Top Technology.' *Flight International*, (Oct. 13, 1979).

CHAPTER 11.0

FUSELAGE

11.1 Introduction

The fuselage of a modern aircraft is a stiffened shell commonly referred to as *semi-monocoque construction*. A pure monocoque shell is a simple unstiffened tube of thin skins, and as such is inefficient since unsupported thin sheets are unstable in compression and shear. In order to support the skin, it is necessary to provide stiffening members, frames, bulkheads, stringers and longerons.

The stiffened shell semi-monocoque type of fuselage construction as shown in Fig. 11.1.1 is similar to wing construction with distributed bending material. The fuselage as a beam contains longitudinal elements (longerons and stringers) transverse elements (frames and bulkheads) and its external skin. The longerons carry the major portion of the fuselage bending moment, loaded by axial forces resulting from the bending moment. The fuselage skin carries the shear from the applied external transverse and torsional forces, and cabin pressure.

In addition to stabilizing the external skin, stringers also carry axial loads induced by the bending moment. Frames primarily serve to maintain the shape of the fuselage and to reduce the column length of the stringers to prevent general instability of the structure. Frame loads are generally small and often tend to balance each other, and as a result, frames are generally of light construction.

Bulkheads are provided at points of introduction of concentrated forces such as those from the wings, tail surfaces and landing gear. Unlike frames, the bulkhead structure is quite substantial and serves to distribute the applied load into the fuselage skins.

It should be apparent that there are some similarities and some differences between the structural components of a fuselage and a wing.

- The function of the stringers and skins of the

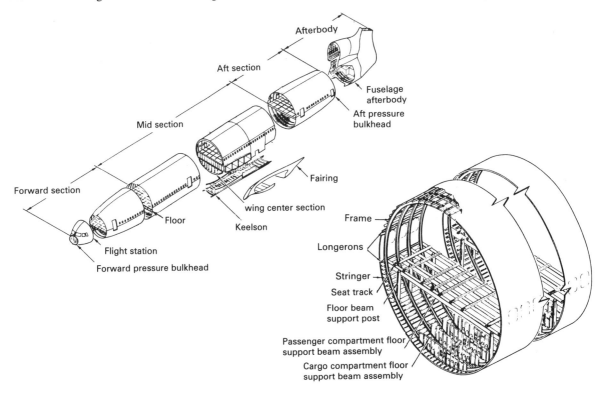

Fig. 11.1.1   Typical semi-monocoque stiffened shell — L-1011.
(Courtesy of Lockheed Aeronautical Systems Co.)

376 Airframe Structural Design

fuselage and wing are equivalent. By virtue of their greater curvature, fuselage skins, under compression and shear loads, are more stable. Additionally, external pressure loads are much lower on the fuselage than on the wing. As a result, the skin thickness required on a fuselage generally will be found to be thinner than on wing skins.
- Fuselage longerons and stringers and wing beam caps serve similar functions of carrying axial loads induced by bending.
- In the fuselage, transverse shear loads are carried by the skin while in the wing these loads are predominately resisted by the spar webs.
- Fuselage frames are equivalent in function to wing ribs, except that local airloads will have a large influence on the design of wing ribs while the design of fuselage frames may be influenced by loads resulting from equipment mounted in the fuselage.

A structural system comprised of a thin-skinned shell stiffened by longitudinal stringers which, in turn, are supported by transverse frames forms a semi-monocoque structure. Thin sheets are very efficient in resisting shear and tension loads in the plane of the sheet, but must be stiffened by longitudinal and transverse members to withstand compression loads and loads normal to the plane of the skin. The design of a semi-monocoque system requires an understanding of sheet stability, panel buckling and semi-diagonal tension field action.

Semi-monocoque structure is very efficient, i.e., it has a high strength to weight ratio, and it is well suited for unusual load combinations and locations. It has design flexibility and can withstand local failure without total failure through load redistribution.

Ultimate Strength of Stiffened Cylindrical Structure

The design of a semi-monocoque structure involves the solution of two major problems: (a) the stress distribution in the structure under all load conditions; and (b) a check of the structure to determine if the resulting stresses can be efficiently sustained by the individual components of the structure as well as the total assembled structure.

In the early design stage, rough analysis is adequate, and redundancies should be removed by assumptions. "The most elaborate analysis possible cannot make a poor design into a good one — any time available at this stage might better be spent on improving the design." This is also true for the whole field of structural design.

The first step towards an understanding of how to design a semi-monocoque structure is a general understanding of how this type of system will fail under load. There are three types of instability failure of semi-monocoque structures:

(1) Skin Instability

Thin curved sheet skins buckle under relatively low compressive stress and shear stress and if the design requirements specify no buckling of the sheet under load, the sheet has to be relatively thick or the section must have closely spaced stringers. This results in an inefficient design and the structure will be unsatisfactory from a strength to weight standpoint. Internal pressure tends to alleviate the condition.

Longitudinal stringers provide efficient resistance to compressive stresses and buckled skins can transfer shear loads by diagonal tension field action. Therefore, buckling of the skin is not an important factor in limiting the ultimate strength of the structural system. Stress redistribution takes place over the entire structure when buckling of the skin panels occurs and therefore it is important to ascertain when this buckling starts. In some cases, the design criteria wil specify that buckling is not allowed below a certain percent of limit load or design load.

(2) Panel Instability

The internal rings or frames in a semi-monocoque structure, such as fuselage, divide the longitudinal stringers and their attached skin into lengths called panels. If these frames are sufficiently rigid, a semi-monocoque structure if subjected to bending will fail on the compression side as shown in Fig. 11.1.2. The longitudinal stringers act as columns with an effective length equal to that of the frame spacing which is the panel length. Initial failure thus occurs in a single panel and is referred to as *panel instability failure*.

In general, this type of failure occurs in semi-monocoque structure because the rings are sufficiently stiff to promote this type of failure. Since the inside of a fuselage carries various loads such as passengers, electronic equipment, engines, armament, etc., and the frames transfer the loads to the skin, this requires considerable strength and stiffness in the frames. Even minor lightly loaded formers must be several inches deep to provide space for wiring, tubing and control lines to pass through the web of the former which again provides relative stiffness for supporting the longitudinal stringers in column action.

When the skins buckle under shear and compressive stresses, the skin panels transfer additional shear forces by semi-diagonal tension field action which produces again higher axial loads and bending in the stringers which must be considered in arriving at the panel failing strength.

(3) General Instability

Failure by general instability extends over a distance of two or more frames as shown in Fig. 11.1.2(a) and is not confined between two adjacent frames as is the case with panel instability failure. In panel instability, the transverse stiffness provided by the frames is sufficient to enforce nodes in the stringers at the frame support points [Fig. 11.1.2(b)]. Therefore general instability failures occur when the transverse frames are not rigid enough to enforce longitudinal stringer node points at each frame.

Since general instability failure occurs as a result of transverse frame failure, adding longitudinal stringers will not contribute to the buckling strength.

The goal in the design of a semi-monocoque structure is to insure that the system will fail as a result of panel instability and not through failure of the transverse frames, which is general instability.

The expression for the required frames stiffness to preclude general instability failure of a stiffened shell in bending is as follows:

$$(EI)_f = \frac{C_f M D^2}{L}$$

In a study of available test data C_f was found to be $\frac{1}{16000}$. Therefore,

$$(EI)_f = \frac{MD^2}{16000 L}$$

where E = modulus of elasticity
I = moment of inertia of frame
D = diameter of stiffened fuselage
L = frame spacing
M = bending moment on fuselage

This equation provides a preliminary minimum sizing requirement for fuselage frames.

Brief Summary of Fuselage Loads — Transport
(Reference only)
(1) Ultimate design conditions
  • Flight loads (acting alone)
  • Flight loads + cabin pressure (p = maximum

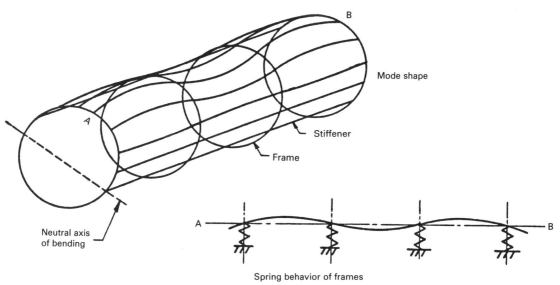

(a) Panel instability

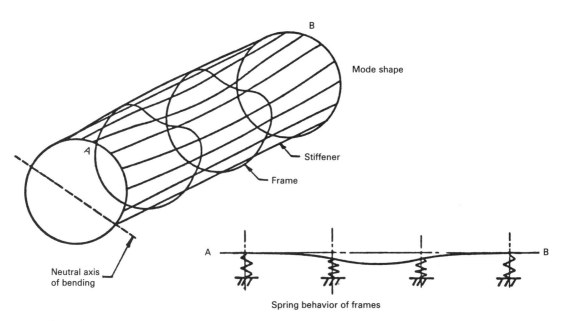

(b) General instability

Fig. 11.1.2 Mode shapes for panel and general instability of stiffened cylinders in bending.

differential pressure loads)
- Cabin pressure only ($1.33 \times p$)
- Landing and ground loads
(2) Fail-safe design conditions
- Fail-safe flight loads (acting alone)
- Fail-safe flight loads + cabin pressure
- Cabin pressure only
(3) Fatigue (Fig. 11.1.3)
- Fatigue loads based on flight profiles developed by manufacturer to encompass anticipated airplane usage
- Fatigue objective — design flight hours of service life without modification of primary structure
(4) Special area conditions
- Depressurization of one compartment
- Bird strike
- Hail strike
- Cargo and passenger loads on floors
- Crash load (emergency landing)

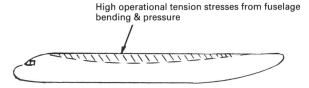

Fig. 11.1.3 *Entire airframe potentially fatigue critical area.*

11.2 Fuselage Configuration

This section discusses some of the factors that influence the configuration of the airplane fuselage and how the externally applied loads, fatigue, and fail-safe philosophies are applied in the detail structure design. The transport structure will be discussed in detail since it is representative of design practices. Some comparisions will be made with more recent models to show how processing development and operating environment influence detail design.

Ground rules:
- Analyze the differences in various structural arrangements in terms of:
  — Producibility
  — Structural Efficiency
  — Weight
  — Provisions for ducting and control cables
  — Effect on fuselage outside diameter
  — Noise attenuation (for commercial airplanes)
- Establish the impact of a future version on the structural arrangement.
- Compare the fuselage structural arrangement with other aircraft that are applicable.
- To obtain a minimum structural wall depth

General Requirements
Assume that the requirements for a particular airplane have been established; namely, to transport a payload, people, at subsonic speeds in a comfortable environment. This dictates that the fuselage must be pressurized to provide a comfortable environment for the passengers. As shown in Fig. 11.2.1, the most efficient pressure-carrying structure has a cylindrical

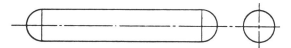

Fig. 11.2.1 *Ideal pressure shell.*

cross-section with spherical end caps. Here is where the compromising design begins.

The efficient pressure structure is compromised (Fig. 11.2.2) to satisfy the aerodynamicist. The aerodynamicist is compromised by operational requirements for crew visibility. It must be obvious by now that compromise is the name of the game. From a structural standpoint, passengers are the worst possible payload. Every cutout or opening in a pressurized structure is a compromise. Fig. 11.2.3 shows the effect of passenger requirements. They must have doors to get in, windows to look out, more doors to load food for their comfort, more doors to load their luggage, and still more doors so they can get out in a hurry in emergencies.

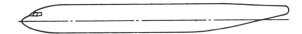

Fig. 11.2.2 *Aerodynamic smoothness.*

Fig. 11.2.3 *Passenger requirements (cutouts).*

Finally, after providing space for a weather radar antenna and large cavities for the wing and landing gear, the fuselage configuration is established, not completely but in profile anyway. As can be seen in Fig. 11.2.4, the only detail remaining of the original efficient structural pressure vessel is the closure bulkhead at the aft end. And space considerations have compromised this to a curvature flatter than a true spherical shape.

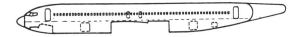

Fig. 11.2.4 *Final configuration of a commercial transport.*

With configuration established it is possible to determine the primary structural requirements. The primary flight loads applied to the fuselage are shown in Fig. 11.2.5. They are lift, thrust, and pitching moment applied by the wing, and maneuvering tail loads from the empennage. These loads should combine along with pressure in the proper combinations. It should be pointed out here that these loads represent only flight and environment conditions. Secondary loads, or those loadings associated with the func-

Airframe Structural Design 379

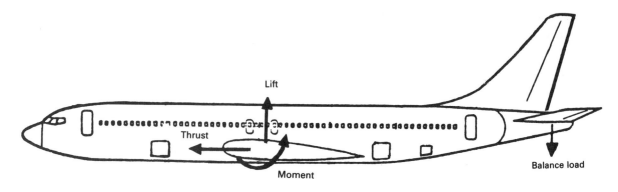

Fig. 11.2.5 Primary flight loads on fuselage body.

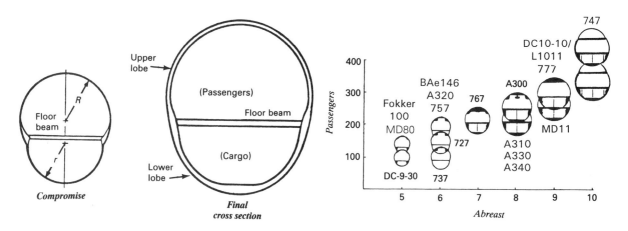

Fig. 11.2.6 Typical double lobe cross-section.

Fig. 11.2.7 Seat abreast influence cross-section.

tion or utility of the airplane, impose additional structural requirements on the fuselage structure. Some of the more significant of those secondary loads come from large equipment such as galleys, passengers and their seats, baggage, and cargo, which must be restrained not only for normal flight conditions, but in most cases for extreme loadings imposed under crash conditions.

The final consideration in configuration design is the cross-section for passenger transport which is predominantly influenced by passengers. Fig. 11.2.6 shows a double lobe cross-section to satisfy passengers on the upper lobe and gives enough room for cargo volume at the lower lobe. This cross-section is a common one for narrow body transports of five to seven abreast seats; more than eight abreast usually go to a single circular cross-section as shown in Fig. 11.2.7.

11.3 Fuselage Detail Design

Skin and Stringers

The largest single item of the fuselage structure is the skin and its stiffeners. It is also the most critical structure since it carries all of the primary loads due to fuselage bending, shear, torsion and cabin pressure. These primary loads are carried by the fuselage skin and stiffeners with frames spaced at regular intervals to prevent buckling and maintain cross-section. Fig. 11.3.1 shows typical skin-stiffener panels.

Note that there has been very little change in the basic structural concept since the earliest metal stressed skin airplanes. These skin/stiffener combinations have proved over the years to be lightweight, strong structures that are relatively easy to produce and maintain.

The most efficient structure is the one with the least number of joints or splices; therefore, the skin panels are as large as possible, limited only by available mill sizes. Stringers, being rolled from strip stock, are limited in length by manufacturing techniques.

Single lap splices shown in Fig. 11.3.2(a) are typically used for the longitudinal skin joints. This is the lightest design and does not impose a severe aerodynamic penalty on a subsonic airplane. The transverse splices, those normal to the air stream, are flush butt splices [Fig. 11.3.2(b)] because a lap or step here would have an appreciable effect on boundary layer turbulence and drag.

Fig. 11.3.3 illustrates the design of a typical longitudinal and circumferential splice. The circumferential splice consists of a strap doubler with a scalloped doubler placed between the strap and the skin break.

Stringer splice locations are established by another set of rules. Since the skin and stringers are working together, they should both be spliced at the same

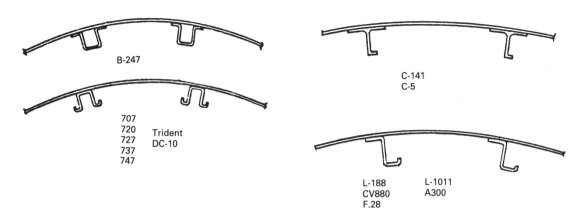

Fig. 11.3.1 Typical transport skin-stringer panels.

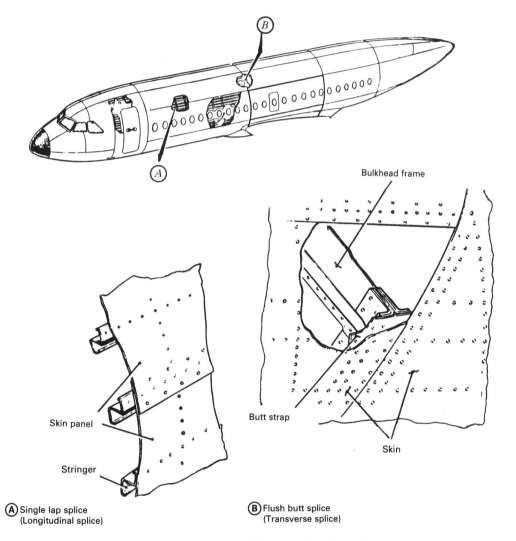

Fig. 11.3.2 Typical transport fuselage splices.

location. This maintains the relative stiffness of the skin/stringer combination, which is desirable from a fatigue standpoint (see Fig. 11.3.4).

The fuselage cross-section in the ideal shape is of a true cylinder such as the L-1011, DC-10 and A300 transports, and the cabin pressure loads are carried by hoop tension in the skin with no tendency to change shape or induce frame bending. The repeated tension loading is a critical fatigue condition; therefore, it must have a fail-safe design where an individual part

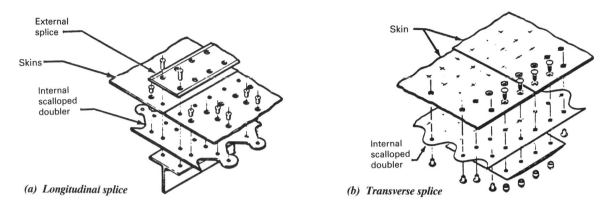

Fig. 11.3.3 Fuselage splices with scalloped doubler.

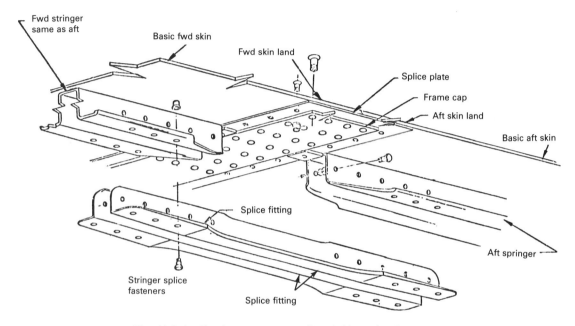

Fig. 11.3.4 Fuselage transverse splice of skin and stringer.

failure can be sustained or retained until it is found and repaired before it results in a catastrophic failure.

The two most common fail-safe design concepts are breaking the component down into several small overlapping pieces where, if one fails, its load can be carried by adjacent parts, or utilizing a restrainer or fail-safe strap that will contain a failure within controllable limits. The latter method is applied in the skin of transport design where the critical design loading is pressure.

Fig. 11.3.5 shows a typical skin, stringer, fail-safe strap, and frame attachment. The tear strap, which is riveted, spotwelded or bonded to the skin, is sized such that a skin crack will stop when it reaches the strap and the strap can carry the load that the skin has given up. Tear straps are located at or between each frame station (see Fig. 11.3.6). It has been proved by test that a 20 to 40 inch crack can be sustained without a catastrophic failure. In the areas where the skin thickness is determined by bending loads, the stress level from hoop tension is low enough that fatigue is not critical (see Fig. 11.3.7 for design tension allowables).

The discussion of skin, stringer, and fail-safe pressure design has been somewhat short compared to its importance. Most thought, study, and testing has gone into this phase of the structure than any other because poor design details in this area are unforgiving.

Utilization of the airplane over shorter route segments means that fatigue is a primary design consideration. The lower operating pressure differential means that minimum gage skins could be used in the hoop tension areas if a satisfactory design for fail-safe crack length control could be developed.

Shown in Fig. 11.3.8 are fuselage skins with doublers which are bonded to the skin to extend fatigue life. Minimum thickness (0.036 inch) skins have a waffle pattern doubler bonded on them, which acts as a fail-safe strap in the station plane and a bearing plate for the rivets that attach the stringers. A by-product of this design is the elimination of the

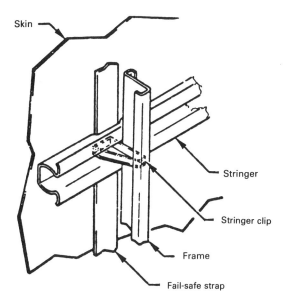

Fig. 11.3.5 Fail-safe strap located between skin and frame.

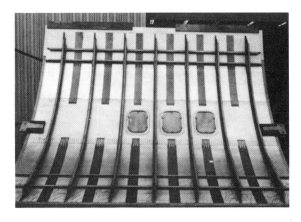

Fig. 11.3.6 Fail-safe strap located between frames.
(Courtesy of Lockheed Aeronautical Systems Co.).

Airplane	Fail-safe Straps (Tear Straps)			Hoop $\left(f = \dfrac{PR}{t}\right)$ Tension Stress
	Location	Material	Thickness	
L-1011	Between frames	Ti	0.020	14000
DC-10	At frames	Ti	0.025	14350
B707	At frames	2024-T3	0.063	12700
B737	At frames	2024-T3	0.04	15,000
B747	At frames	2024-T3	0.071	16200
DC-8	At frames	Ti	0.025	12900
DC-9	At frames	Ti	—	10100
CV880	—	—	—	8500
C130	—	—	—	12000
C141	Between frames	Ti	0.020	14000
C-5	At frames	Ti	0.020	18000
Electra	—	—	—	11000

(Note: The above are approximate values)

Fig. 11.3.7 Fuselage design hoop tension stresses and fail-safe straps.

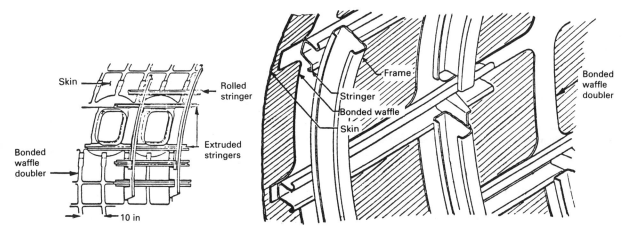

Fig. 11.3.8 Use waffle doubler design instead of fail-safe straps.

Airframe Structural Design 383

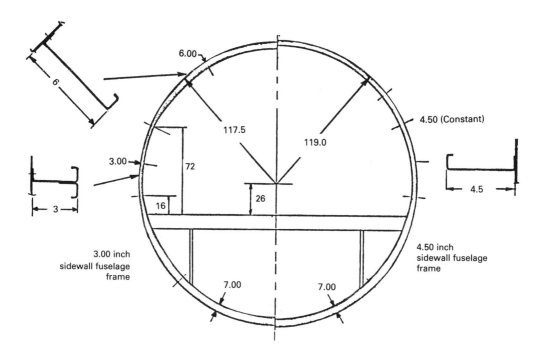

(Note: Lower fuselage frames are identical for both configurations)

Fig. 11.3.9 *Effect of sidewall frame depth study of L-1011 fuselage diameter.*

stringer joggle where the stringers have to step over individual fail-safe straps (or tear straps).

Examination of the various typical fuselage configurations of commercial airplanes reveal that they are basically similar combinations of skin-stringer-ring (frame) structures, with the interior trim line (one inch greater than the frame depth) providing an overall cabin wall thickness. Ring or frame spacing is in the order of 20 inches and stringer spacing varies between 6 to 10 inches. Commonly, the passenger transport fuselage sidewall (window and door area) design replaces stringers with heavier thickness skins so that a quieter cabin can be obtained and the skin fatigue stress can be reduced because of cabin pressurization cycles. In the sidewall region, the frame depth can be kept to a minimum (provided that adequate working space is not a problem) because no significant concentrated loads are involved. Above and below the sidewall region, ample space is available to profile for increased frame depth as required. Another advantage is to reduce fuselage diameter to save structural weight and less fuselage frontal area to reduce aerodynamic drag with the same internal width across the cabin (see Fig. 11.3.9).

Fig. 11.3.10 illustrates several examples of transport fuselage skin, stringer and frame arrangements; other airplane fuselage information is tabulated in Fig. 11.3.11.

Frame and Floor Beam

Fuselage frames perform many diverse functions such as:

- Support shell (fuselage skin-stringer panels) compression/shear
- Distribute concentrated loads
- Fail-safe (crack stoppers)

They hold the fuselage cross-section to contour shape and limit the column length of longerons or stringers. Frames also act as circumferential tear strips to ensure fail-safe design against skin crack propagation. In addition, they distribute externally and internally applied loads onto the shell, redistribute shear around structural discontinuities, and transfer loads at major joints. The construction of a typical fuselage frame is shown in Fig. 11.3.10. The frame cap is usually a Z-section which runs around the periphery inside the stringers. In the shell area between stringers the cap is attached to the skin by means of an angle or clip.

As pointed out previously, frame spacing can have a substantial impact on compressive skin panel design. The weight of bulkheads and flooring are also affected by frame spacing as indicated in Fig. 11.3.12.

Heavy cabin frames and bulkheads represent extremes in the matter of radial constraint on the expansion of the cabin shell. Owing to their flimsiness, conventional cabin frames, whose main function is the

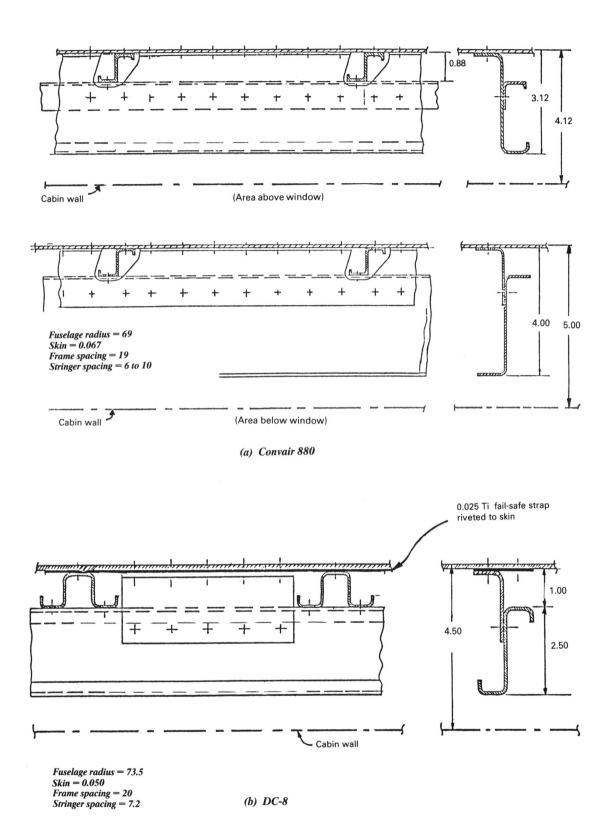

Fig. 11.3.10 Transport fuselage structural arrangement. (Dimensions in inch and materials are aluminum except as noted.)

Airframe Structural Design 385

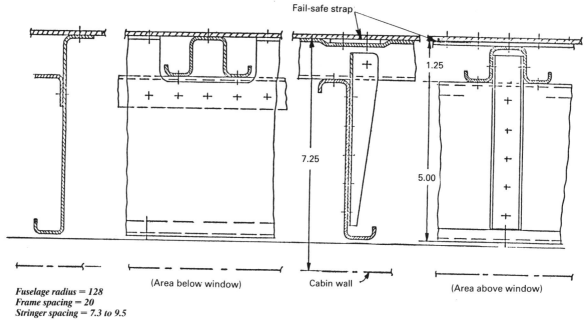

(c) B747

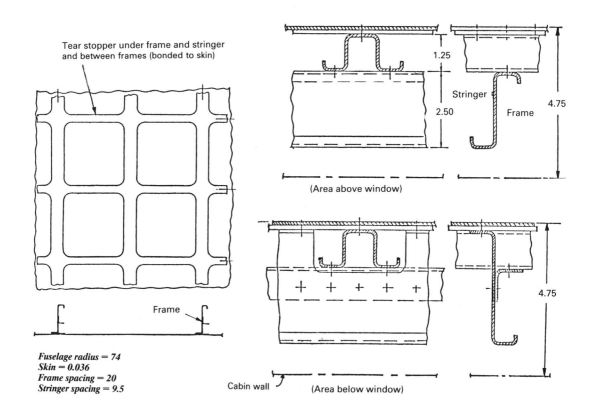

(d) B737

Fig. 11.3.10 (continued)

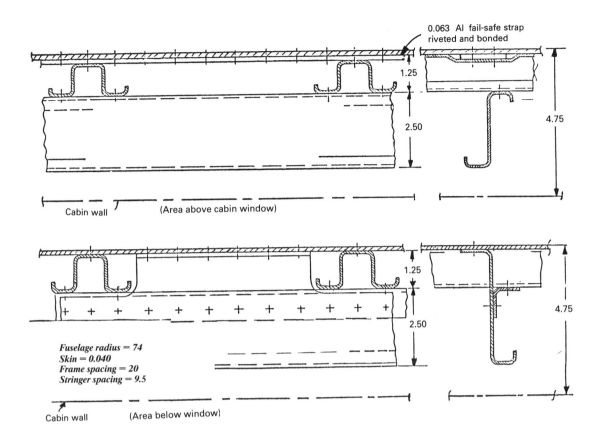

(e) B707 and B727

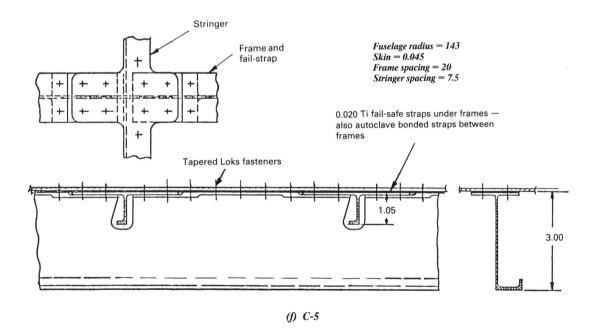

(f) C-5

Fig. 11.3.10 (continued)

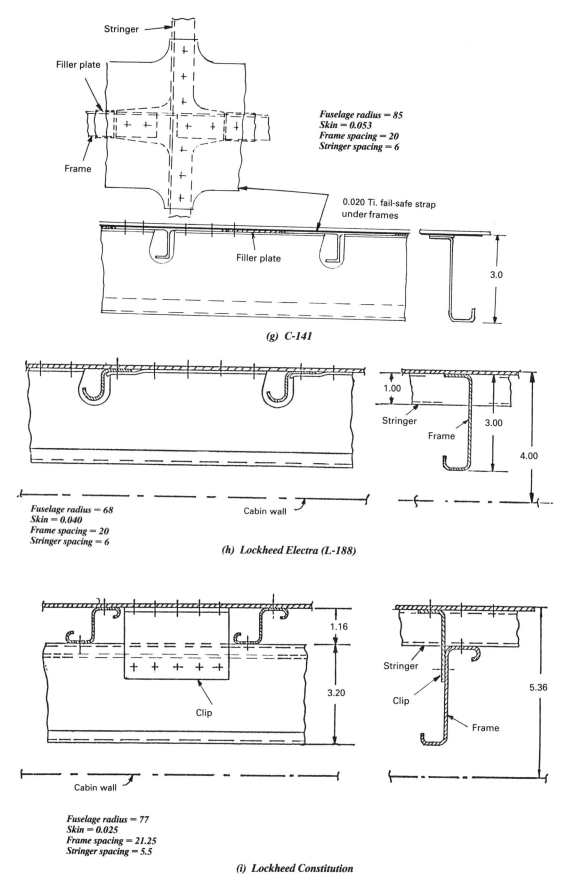

(g) C-141

(h) Lockheed Electra (L-188)

(i) Lockheed Constitution

Fig. 11.3.10 (continued)

Aircraft	Radius R_{max} (inches)	$T_{min.}$ (inches)	Material Skin/Str.	Str. Spacing (inches)	Str. Shape	Frame Depth/ Spacing (inches)
B707	74 *	0.04	2024-T3/7075-T6	9	 ⊐⊏	3.8/20
B727	74 *	0.04	2024-T3/7075-T6		 ⊐⊏	3.8/20
B737	74 *	0.036	2024-T3/7075-T6		 ⊐⊏	3.8/20
B757	74 *	0.04			 ⊐⊏	3.8/20
B747	128 *	0.071	2024-T3/7075-T6	7.5–9.5	 ⊓ ⊏	6.3/20
B767	99 *				 ⊐⊏	/22
1649	69.6	0.032			 ⊏	4.4/21.3
L-188	68	0.04	2024-T3/7075-T6	6	 ⊏	3.75/19
C-130	85	0.032	2024-T3/7075-T6	—	Longerons	
C-141	85	0.053	7079-T6/7075-T6	6	 ⊏	3/20
C-5	143 *	0.045	7079-T6/7075-T6	7.5	 ⊏	8/20
L-1011	117.5	0.075	2024-T3/7075-T6	9	 ⊏	3–6/20
DC-8	73.5 *	0.05	2024-T6/7075-T6	7.2	 ⊐⊏	3.5/20
DC-9	65.8 *		2014-T6/7075-T6	7	 ⊐⊏	/19
DC-10	118.5	0.071	2024-T3/7075-T6	6.5–8	 ⊐⊏	4.5–5.5/20
YC-15	108	0.05	2024-T3/7075-T76		 ⊏	6/24
CV880	69 *	0.067	2024-T3/7075-T6	6–10	 ⊐⊏	4/19
CV990	69 *	0.067	2024-T3/7075-T6		 ⊐⊏	4/19
A300	111	0.063			 ⊐⊏	/20
A310	111	0.063			 ⊐⊏	/20
A320	77 *				 ⊐⊏	/20
BAC111	77	0.060	2024-T3/		 ⊏	/20
Trident	72.75	0.048	2014-T/2014-T	6.5	 ⊐⊏	/20
Concord	56.5 *	0.055	RR58/RR58		 ⊏ ⊤	/21.5

* Double lobe cross-section

(a) Transports

Aircraft	Fuselage	
	Skin	Frames
F-4	7178-T6 (Bare) 7178-T76 (Bare)	7079-T652 (Forgings) 7079-T7352 (Forgings)
F-8 (A-7)	7075-T6 (Clad)	7075-T65x (Forgings) 7049-T7xxx 7050-T7xxx
F-14	2024-T6, -T8 (Sheet, clad and bare) Ti-6Al-4V (Ann. at engine inlet)	2024-T851 (Plate, 1.0–3.0 in) Ti-6Al-4V (Ann./RA, 2.0-3.0 in) D6C (220–240 ksi, 4.0 in)
F-15	7075-T6, -T76 (Clad) 2024-T62, -T72, -T81 (Clad)	2124-T851 (Plate) 7075-T7352 (Forgings) Ti-6Al-4V, (Ann. plate & Forgings)
F-16	2024-T62, -T81 (Sheet)	2124-T851 (Plate, 3.0–6.0 in) 7475-T7351 (Plate, 4.0 in)
F-18	7075-T6 (Clad, 0.08 in) 7075-T76 (Clad, 0.08 in)	7075-T7xxx (Plate, 1.0–4.0 in) 7050-T73651 (Plate, 2.0–6.0 in)

(b) Fighters

Fig. 11.3.11 *Fuselage information.*

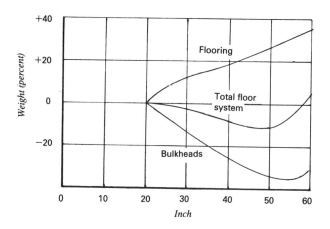

Fig. 11.3.12 *Effect of frame spacing on floor and bulkhead weight.*

preservation of the circular shape against elastic instability under the compressive longitudinal loads, have little constraint on the radial expansion of the shell. Bulkheads, on the other hand, are usually so stiff that the cabin shell is locally allowed only a very small fraction of its radial expansion.

Between these two extremes there are frames of various degrees of stiffness, all of them causing local disturbances in the otherwise smooth stress distribution. One needs to consider two basic types of transverse frames — ordinary formers and frames (including bulkheads) specially stiffened for various purposes (see Fig. 11.3.13). It is necessary to consider

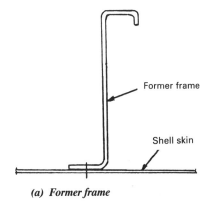

(a) *Former frame*

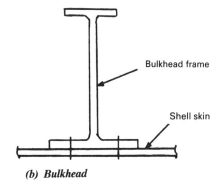

(b) *Bulkhead*

Fig. 11.3.13 *Two basic types of frames.*

not only the bending stresses induced in the shell but also the hoop and other stresses induced in the frames themselves.

The former-frames at 20-inch pitch (the pitch between frames has the conventional value of 20 inches for transport airplanes) can at best only be 38% effective in helping to reduce the maximum hoop stress in the skin. Considering their inefficient cross-sectional shape for resisting radial loads applied to their outer flange, and due to the small amount of material they contain in relation to the skin, their contribution to the reduction of skin hoop-stress is, in practice, inappreciable. The actual fuselage skin stress of modern transports are approximately 85% of the calculated hoop stresses ($f = \dfrac{PR}{t}$). The connection between the frame and stringer is attached by stringer clip. the typical clip and its functions are shown in Fig. 11.3.14.

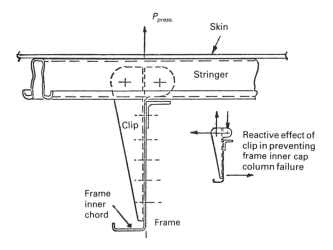

Fig. 11.3.14 *Fuselage stringer clip usage.*

The purposes of stringer clip are:
- Transfers skin panel normal pressure loads to frame
- Helps break up excessive column length of stringers
- Provides some degree of compressive strength of frame inner chord (cap)
- Acts as frame web panel stiffener

Bulkhead weight decreases as frames move farther apart, until the additional stoppers to prevent skin crack propagation overcomes the advantage. On the other hand, flooring weight usually increases as the frames move farther apart because the span increases between the lateral floor beams attached to the frames. This relationship of combined floor and bulkhead weight variation is shown in Fig. 11.3.12.

Acoustical loads created by propeller tips have an effect on fuselage panel fatigue and, as a consequence, directly affect frameweight.

An interesting example of fail-safe design is shown in Fig. 11.3.15. Every frame in the fuselage is attached to the horizontal floor-beam which acts as tension ties across the fuselage to resist the cabin pressure loads. As an additional path for distributing these loads, a

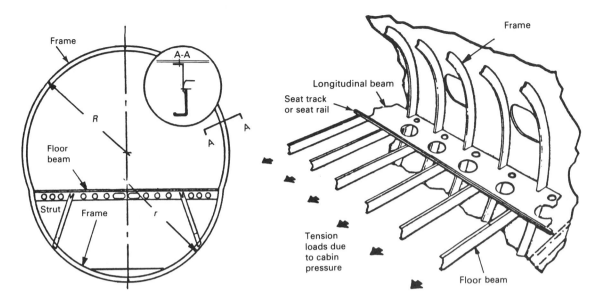

Fig. 11.3.15   Fail-safe design by using longitudinal beam along side of fuselage.

longitudinal beam along the side of the fuselage is provided. A longitudinal beam at the cusp constitutes one cap of the beam and the seat rail constitutes the other. Thus, should a transverse floor beam become damaged its share of the tension load is transferred by means of the longitudinal side beam to the adjacent cross ties.

Frames have been mentioned in previous discussion. Now we take a closer look at their requirements and design. In the crown areas where the contour is round, pressure loads are carried in hoop tension; the frames are secondary, or stabilizing members to maintain the shape and to keep the skin and stiffeners from buckling under bending loads. The essence of the frame requirements shows up in the nonradial contour areas where pressure loads combine with the stabilizing loads to produce large bending moments. Fig. 11.3.16 shows a typical skin, stringer, frame, and shear tie arrangement. The shear tie attachment to the skin between stringers is the shear load path between frame and fuselage skin. Due to the pressure load, the skin wants to assume a radial shape. The shear tie holds the skin in the desired shape locally between stringers, and the frame holds the several shear ties and stringers to the desired shape. The frame is, therefore, experiencing a bending load, i.e., the skin and shear ties are trying to bend the frame from the desired contour into the more natural radial shape for carrying pressure.

The detail buildup of parts shown in Fig. 11.3.16 is typical of the lower lobe frames. The frame web inner cap is a flange, and several shear tie segments are combined in one hydroformed part with the reinforced cap, usually an extruded angle, paralleling the skin and attached to the frame web. Fig. 11.3.17 illustrates an economic process of forming fuselage frames.

A peculiar deep frame design of using flanged cutouts with beaded frame web near the cutout to stabilize web buckling under loads instead of using conventional reinforced cap is shown in Fig. 11.3.18.

The fuselage cross-section view presented in Fig. 11.3.19 indicates the general design and arrangement of cabin and belly cargo flooring. The main deck consists of floor panels, longitudinal stiffeners, lateral floor beams and vertical struts which reduce bending moments in the floor beams. Seat tracks (or seat rails) such as shown in Fig. 11.3.20 also act as stiffening members and provide 9.0g foward crash load restraint for the passenger seats. Typical design loads for cabin floors are shown in Fig. 11.3.21, and an additional consideration is to design the passenger floor to withstand, without collapse, an in-flight depressurization caused by the sudden opening of a large hole in the below cargo compartments as required for all wide-body fuselages by FAA. The best way to solve this design condition is to use a sidewall venting system as shown in Fig. 11.3.22. Similar venting systems have been applied on all wide-body fuselages, such as the L-1011, DC-10, B747 and also A300 series transports.

If the passenger floor is designed for both passenger and cargo loads, the type of construction should be reinforced for the heavy cargo loads as well as belly cargo floors and vary in accordance with specific design requirements. All cargo floors and belly cargo floors should provide rollers, ball mates, or similar means to freely move containers and pallets as shown in Fig. 11.3.23. (Rollers in bulk cargo areas are not required.)

A method for graphically depicting the relationship between floor weight and floor area, categorized by floor function, is shown in Fig. 11.3.24. Considerable variation can occur, depending upon many design considerations. In this example, the floor designed for containers only does not have floor panels except for walkways along the sides of the fuselage. Bulk flooring is somewhat heavier, as it is constructed of full floor panels with rollers added for containers and pallets. Floors capable of transporting military vehicles, such

Airframe Structural Design   391

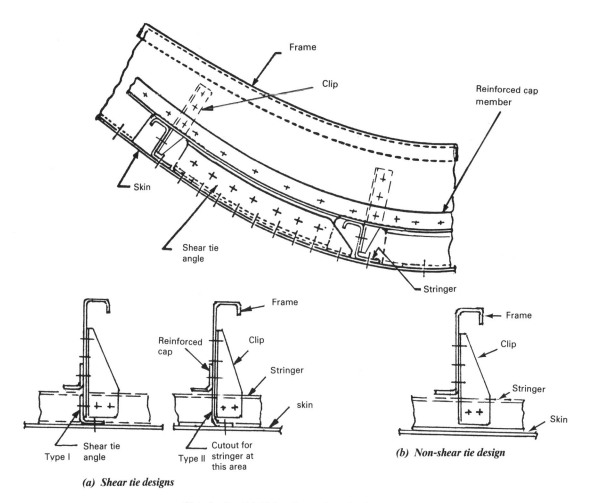

(Note: See Fig. 11.3.10 for other configurations)

Fig. 11.3.16   Typical shear tie configurations.

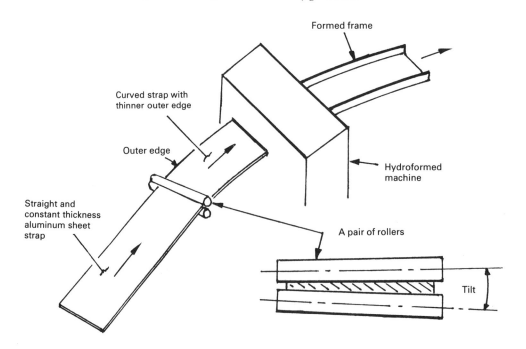

Fig. 11.3.17   A common process of forming fuselage frame.

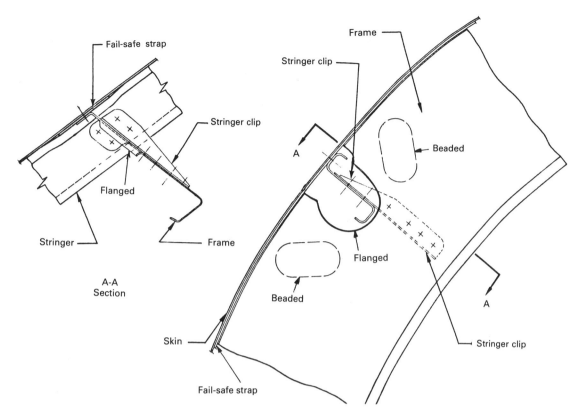

Fig. 11.3.18 Frame design — franged cutouts with beaded frame web.

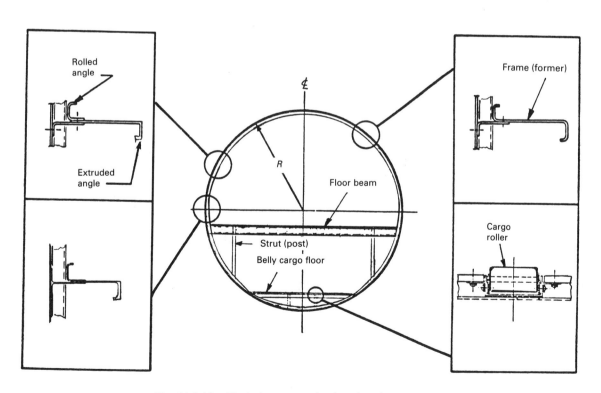

Fig. 11.3.19 Typical transport fuselage interior arrangement.

Airframe Structural Design 393

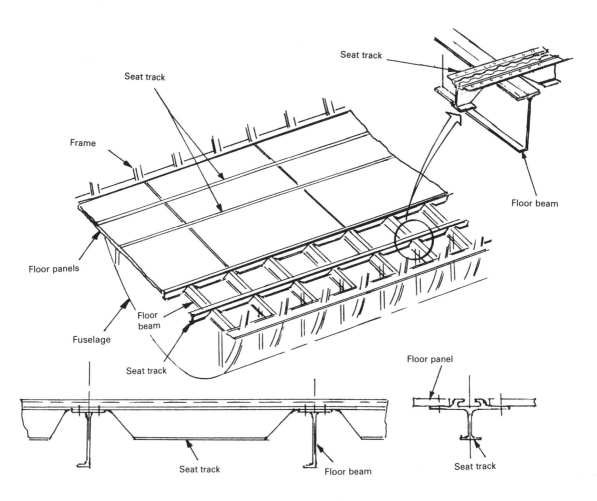

Fig. 11.3.20 Transport passenger flooring.

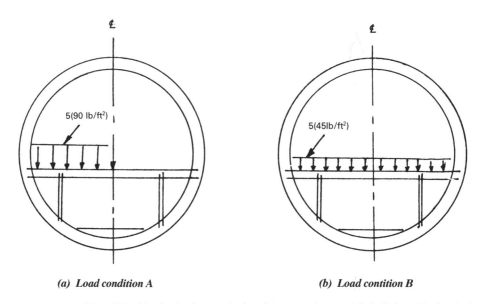

(a) Load condition A

(b) Load contition B

Note: Use 5×750 lb/lineal ft load besides the above two load conditions in conjunction with the flight and landing loads, whichever is critical.
The above loads are much higher for cargo floor.

Fig. 11.3.21 Design loads for transport passenger floors.

Fig. 11.3.22 Fuselage sidewall venting system.

(a) Upper deck cargo floor

(b) Lower deck belly cargo floor

Fig. 11.3.23 Cargo floors.

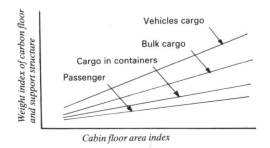

Fig. 11.3.24 Passenger floor type and structural weight relationship.

as the C-5 are the heaviest. On the average, the penalty for passenger floors is approximately 2 lb/ft² and for all-vehicle operation, approximately 5 lb/ft².

Fig. 11.3.25 shows the fuselage center section floor beams (at the intersection of fuselage and wing) which is typical for passenger transports. Because of the wing box, all floor beams are running longitudinally, which not only support floor loads but also stabilize the wing upper cover panels. Seat tracks are installed right on top of these beams and also act as beam caps. One of the center section floor beam configurations of a wide-body transport is shown in Fig. 11.3.26.

Military cargo transports floor design is completely different from passenger transports because they will

Airframe Structural Design 395

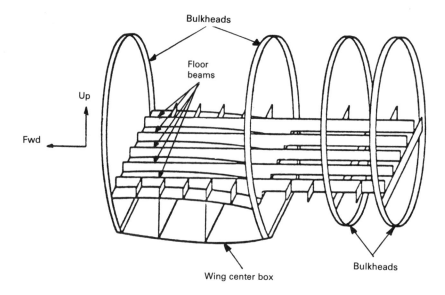

Fig. 11.3.25 Typical transport fuselage center section floor beams arrangement.

Fig. 11.3.26   Center section floor beams — L-1011.
(Courtesy of Lockheed Aeronautical Systems Co.)

carry heavy military equipment. Another important requirement is that the floor surface should be close to the ground, then the equipment can be driven in and out of the fuselage. Fig. 11.3.27 shows a typical military transport fuselage cross-section.

Pressure Bulkhead

The cylindrical shell of a pressure-cabin is closed at the rear by some kind of dome [Fig. 11.3.28(a)] in preference to a flat bulkhead [Fig. 11.3.28(b), except for supporting rear fuselage engine mount as shown in Fig. 13.5.3], which would have to be heavily braced to withstand the pressure. However, some smaller planes and executive transports would rather use it for simplicity. From a structural point of view, a hemispherical shell provides an ideal rear dome because the membrane stresses for a given amount of material are the least. On the basis of this argument, the best shape for the rear dome of a pressure cabin is the hemisphere. Assuming this to be correct, the problem of choosing the most efficient method of joining the hemispherical dome to the forward cabin shell and to the rear fuselage is challenging work.

This design is guided by two basic considerations:
(1) Owing to the comparatively heavy membrane force involved, it is desirable to avoid any radial offset between the shell and the dome skins.
(2) There must not, in the neighborhood of the joint, be any reduction in the longitudinal bending stiffness of the fuselage wall, the maintenance of which the elastic stability of the wall depends on.

It is not easy to outline a scheme of designing that satisfies both these requirements and a scheme which entails local stiffening of the dome around its base, both on its inner and outer surfaces. The joint itself is made by sandwiching together the three skins — aft section fuselage, dome and aft body fuselage — to form a single lap-joint. The dome and the aft body fuselage wall are further directly connected by fastening the latter to the outer stub-stringers that form the external reinforcement of the dome. The problem of determining the forces and moments introduced by the differential free radial expansion of aft section fuselage, dome and aft body fuselage wall is discussed in this section.

In considering the design of the rear dome of a pressure cabin, the objective is to achieve a minimum weight for the dome itself and a minimum amount of interference stresses at the junction of dome and fuselage walls. As a matter of academic interest it may be proved that a circular opening with completely rigid edges is most economically closed by a 60° spherical cap as shown in Fig. 11.3.29.

The large transport fuselage pressure dome is located at the rear of the fuselage compartment and usually the top of the dome rim is canted slightly forward in the vertical plane to accommodate the taper of the aft bottom fuselage. Fig. 11.3.30 illustrates an aluminum dome construction which consists of stretch-formed gores of sheet aluminum bonded together at the edges. Bonded to the outside surface

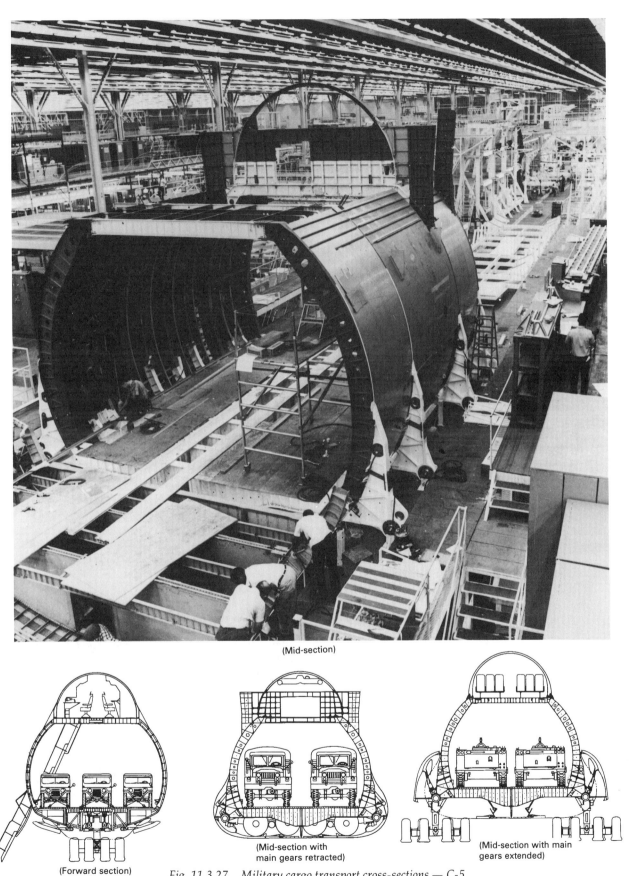

Fig. 11.3.27 *Military cargo transport cross-sections — C-5*
(Courtesy of Lockheed Aeronautical Systems Co.)

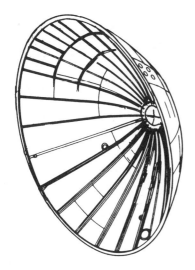

(a) Dome

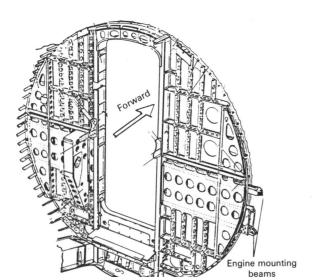

(b) Flat bulkhead

Fig. 11.3.28 *Typical pressure bulkheads.*

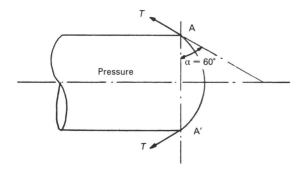

Fig. 11.3.29 *Rear dome with 60° spherical cap.*

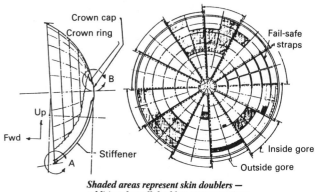

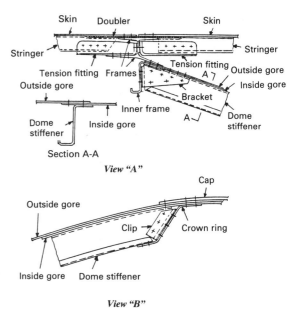

Fig. 11.3.30 *Structural configuration of L-1011 fuselage dome.*

of the shell are doublers at locations of shell penetrations.

In addition, radial and circumferential fail-safe straps are bonded to the outside surface to inhibit the propagation of cracks. Reinforcement of the shell is accomplished with radial stiffeners made of aluminum sheet. Twelve stiffeners are riveted to the shell where the bonded gores overlap.

At the periphery of the bulkhead, the shell and stiffeners attach to a flange of a fuselage frame. At the apex, the shell and stiffeners attach to a crown ring which is covered with a crown cap. The joint to fuselage consists of tension type fittings splicing the stringers and fuselage skin butt type splice.

11.4 Forward Fuselage

The need for better visibility for the pilot of an aircraft has occupied the attention of designers since man first began protecting himself against the windstream by sitting inside the airframe. In addition to problems on structural considerations, streamlining requirements and the necessity of providing comfort for the occupants, the engineer has been plagued by some adverse characteristics of those transparent materials that best lend themselves for use in cabin enclosures and canopies.

Flight Station Design for Transports

The cockpit is that portion of the airplane occupied by the pilot. From the cockpit radiates all controls used in flying and landing the airplane. On a propeller-driven airplane the seats for the pilot and co-pilot and the primary control units for the airplane, excluding cables and control rods, must be so located with respect to the propellers that no portion of the pilots or controls lie in the region between the plane of rotation of any propeller and the surface generated by a line passing through the center of the propeller hub and making an angle of 5° forward or aft of the plane of rotation of the propeller. This requirement is necessary to protect the pilot and the airplane in case a propeller blade should be thrown from the engine shaft and pass through the fuselage.

The windows and windshield sections for the pilot's compartment must be installed in such a manner that there will be no glare or reflections that will interfere with the vision of the pilot, particularly while flying at night. The windshield panes that the pilots will be directly behind in the normal conduct of their duties and the supporting structures for such panes must have sufficient strength to withstand, without penetration, the impact of a 4-lb bird when the velocity of the airplane relative to the bird along the flight path of the airplane is equal to cruising speed at sea level.

In the DC-3 type of transport planes (Fig. 11.4.1), where the pilots are placed relatively close to the flat V-shaped windshields, vision is considered adequate by the majority of pilots. The development of planes with pressurized cabins introduced complications in that the demand for increased mechanical strength to resist the cabin pressures resulted in decreased window areas and in curved surfaces to obtain strength without excessive weight. The curved surfaces (as shown in Fig. 11.4.2) resulted in a tendency

Fig. 11.4.2 Boeing 307 stratoliner (first of the pressurized cabin skyliner).
(Courtesy of The Boeing Co.)

for the pilots to pull to the left in landing, although not to a serious extent.

In older types of aircraft, blind areas were created by the wings; in modern transports, the protruding areas of the fuselage at the side and in front of the cockpit are the more important obstructions. In aircraft with conventional landing gears (two main gears and a tail wheel), difficulty is encountered in taxiing because the nose of the plane projects above the pilot's head; this structural obstruction has eliminated much or even all the direct forward vision. The development of tricycle landing gear has greatly improved forward visibility while the plane is on the ground, but in some models the fuselage is so high off the ground and the nose projects so far forward that considerable restriction in the immediate field of view is still encountered. In aircraft with pressurized cabins there is a tendency to have a heavier and more elaborate framework to support the stresses to which the windshield is subjected. Both Fig. 11.4.3 and Fig. 11.4.4 illustrate the cockpit framework for the modern transports.

To give a clean aerodynamic shape of supersonic transports during transonic and supersonic flight, a visor is raised to cover the windshield. This visor also provides protection from the effects of kinetic heating. The visor as shown in Fig. 11.4.5 consists of a single glazed unit with four panes in front and one on each side. It is carried on two rails at the rear and a swinging frame link at the front and is raised and lowered by a hydraulic jack which drives the swinging front frame.

In addition, the entire nose fairing forward of the windshield, including the visor and its mechanism, is dropped to improve view on take-off and landing. The nose has three positions (up, intermediate and down) and is raised or lowered by a hydraulic droop nose actuator. The visor and droop nose controls are interlocked in such a manner that the intermediate and down droop selections for the nose fairing can only be made after visor down selection. Similarly, the visor may only be selected up when the nose has also been selected but no signal can be transmitted to the visor until the nose is up.

Fig. 11.4.1 DC-3 cockpit and the flat V-shaped windshield.
(Courtesy of Douglas Aircraft Co.)

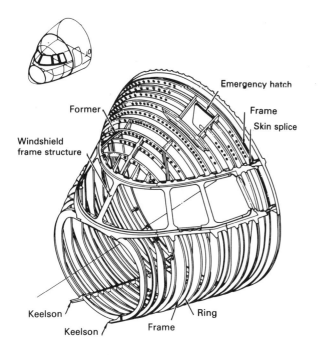

Fig. 11.4.3 Cockpit structural framework — L-1011.
(Courtesy of Lockheed Aeronautical Systems Co.)

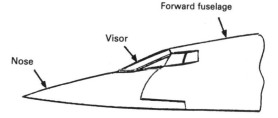

(a) Nose and visor in up position

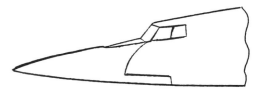

(b) Visor lowered and nose up

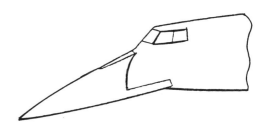

(c) Both visor and nose lowered

(Note: See structural cutaway drawing of Ref. 11.35)

Fig. 11.4.5 Visor and droop nose — Concord supersonic transport.

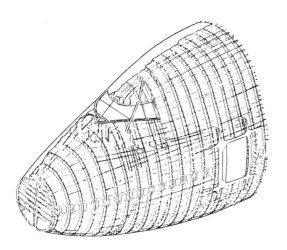

Fig. 11.4.4 Cockpit structural framework — B747.
(Courtesy of The Boeing Co.)

Design problems unique to transparent structures are mainly the result of three situations. The first, and most dramatic, is the phenomenal increase in airplane performance during recent years, accompanied by correspondingly more severe operating conditions. Prior to World War II, airplane canopies, windows, and windshields served mainly as protection from wind and weather; consequently, structural requirements were rarely critical. Later, high-altitude flight and pressurized cabins became commonplace among airliners and miltary aircraft. The heretofore lightly-loaded transparent areas now are required to withstand cabin pressure differentials. In addition, temperatures as low as −65°F (−100°F on a few occasions) are encountered at high altitude. It is not surprising that several years have been necessary for materials, attachment methods, fabrication techniques, and the general technology of transparent materials to catch up with this change in requirements. Supersonic flight is now a reality, and with it comes "aerodynamic heating" or "ram temperature rise" as it is sometimes called. Glass has superior heat-resistance, but forming limitations and weight penalties may restrict its use.

The second major problem of designing with transparent materials centers around inherent properties of the materials themselves. All of the suitable transparent plastics now available are very sensitive to service environment. Their strength is greatly affected by temperature, rate of load application, duration of loading, weather exposure, aging, and a number of other conditions. The compelling demand for minimum weight in aircraft structures has led to high refined stress analysis methods, but these procedures are futile when the structural capabilities of materials are as indeterminate as those of transparent plastics. Glass is less affected by environment, but is subject to large variations in properties between individual parts or batches of parts, so that the problem is essentially the same. This condition dictates the empirical approach; that is, new designs must be based largely upon extrapolations of successful designs of the past, making it difficult to condense the underlying principles into a few formulae and simple rules.

The third problem deals with producibility of parts. The structural integrity of a glass, plastic or glass-plastic part is determined to a large extent by the method and processes used in forming, trimming and machining, cementing on edge reinforcements, and annealing or heat-treating it.

The modern aircraft windshield affords a degree of protection for the pilot far more important than the original functions as previously mentioned. A windshield today is part of the external surface of the fuselage. In a high flying aircraft it must withstand pressurized loads; because of its location it is exposed to ice formation as well as impact by hailstones or birds. It must dependably provide the highest standard of visibility and crew protection.

The modern windshield uses the composite cross-section that combines the excellent abrasion resistance and thermal conductivity of glass with the superior toughness of certain plastics. The result is a near optimum combination using a glass surface ply and a stretched acrylic load bearing ply. In this case, the plastic load bearing ply comprises two individual acrylic plies, each able to carry independently pressurization loads in accordance with dual load path or structural redundancy fail-safe requirements. The visibility of a stretched acrylic windshield after impact of a 4 lb bird is shown in Fig. 11.4.6.

Fig. 11.4.6 Interior view of windshield after impact of a 4-lb bird at speed of 350 knots.
(Courtesy of Lockheed Aeronautical Systems Co.)

Crucial decisions affecting overall design concepts had to be made early in the program because of abbreviated schedules. A decision that contributed significantly to the eventual success of the program was one concerning material selection. A decision had to be made whether to proceed with a conservative but highly reliable design, using an acrylic canopy and a flat glass windshield, or to proceed with a contoured windshield.

A novel approach was introduced since the 1940s in the design of material used in aircraft windows and windshields. The development is an improvement of the laminated safety type of glass which makes it possible to more fully realize the advantages of this type by eliminating most of the disadvantages. Specially, the plastic inner layer has thickened and toughened and extended well beyond the edges of the glass so that the window or windshield can be installed by bolting or clamping this plastic edge directly to the window frame (see Fig. 11.4.7).

The plastic edge permits tight clamping without danger to the glass and prevents frame deformation during severe plane maneuvers from being transferred to the glass. This eliminates the necessity for elaborate metal and metal mountings and reduces the precision formerly required of this type of window frame.

Furthermore, the thick plastic inner-layer is functional in two other respects: it gives many times the impact resistance of regular laminated glass; and it acts as a diaphragm to retain cabin pressure when pressurization is used. The windows and windshields of the airplane must withstand not only pressurization but sudden depressurization as the cabin is deflated.

Regular annealed glass is polished plate glass of conventional anneal. Semi-tempered glass is polished plate glass that has been subjected to a heat treatment to increase its strength to approximately 2 to 4 times that of regular annealed glass. Fully-tempered glass is polished plate glass that has been heat-treated to increase its strength by approximately 4 to 5 times. Tempered glass looks no different than regular plate glass, but its surfaces are both in uniform high compression stress. This compression stress, which must be overcome before the glass can be stressed in true tension, accounts for the increased strength of tempered glass. All cutting, bending, and edge grinding must be completed before the glass is tempered. Laminating is, of course, done after tempering.

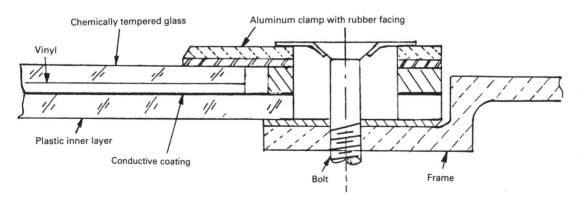

Fig. 11.4.7 Typical laminated safety type of glass.

Airframe Structural Design 401

However, new data is being compiled continuously; in order to define and develop airplane glass products and their application, the reader should pay attention to this subject frequently to use up-to-date design data.

In Fig. 11.4.8, the forward fuselage curved windshields and side panels blend smoothly into the large diameter fuselage and the windshield side posts are positioned well aft (actually, in plane of control wheels) to provide a panoramic view.

This is often referred to as a "clamped-in" design [see Fig. 11.4.9(a)]. Note that the surrounding aircraft structure is stiff by virtue of its deep section so that it can carry fuselage loads around the windshield opening, and the windshield is structurally isolated within its sub-frame by thick-walled silicone sponge bushings around the mounting bolts. In this mounting approach, it is necessary to design the surrounding structure to the airframe's fatigue life requirements without depending on the windshield for any support.

Instead of isolating the windshield from fuselage structural loads, the design shown in Fig. 11.4.9(b) purposely carries those loads through the windshield. In this manner, the windshield reinforces the surrounding structure, allowing a lighter weight structure to meet the fatigue life requirements than would otherwise be the case. In this design, the aircraft structure surrounding the windshield is specifically designed to be "soft" at the corners and in effect expands on pressurization to place the windshield in membrane tension. The windshield-to-frame and frame-to-aircraft junctures have been carefully designed and controlled to provide maximum direct structural coupling between the airframe and windshield, both to reinforce the surrounding structure and to develop maximum hoop tension in the windshield.

Canopy Design for Fighters

Recently, the material polycarbonate has been used for aircraft transparency such as for military aircraft, primarily due to its high temperature capability. Unlike stretched acrylic, polycarbonate gradually softens as temperature is increased, and tends to retain good physical properties at depressed temperatures. Another major advantage is superior toughness and ductility; this is a clue to the extraordinary impact strength of polycarbonate windshields and canopies. One other reason for selecting polycarbonate for some canopies is the ability to form it to deep compound contours.

The high maneuver aircraft such as shown in Fig. 11.4.10 are enthusiastic about the visibility and describe a sensation of flying free in space with vision unencumbered by posts and other reinforcement. However, one polycarbonate problem to deal with is its susceptibility to attack by solvents and other chemicals. This characteristic has been overcome by the use of transparent protective coatings.

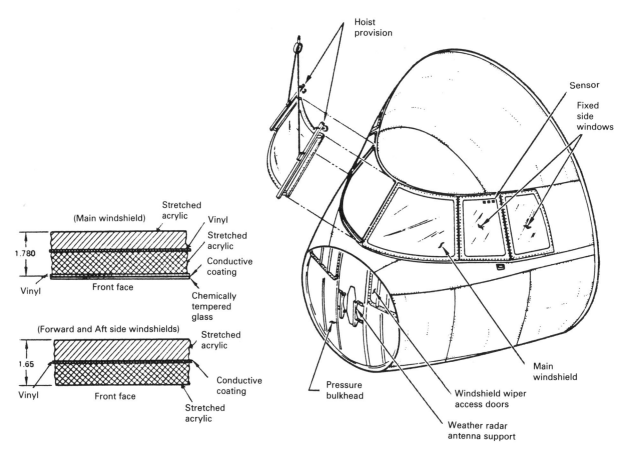

Fig. 11.4.8 L-1011 cockpit curved windshield.
(Courtesy of Lockheed Aeronautical Systems Co.)

402 Airframe Structural Design

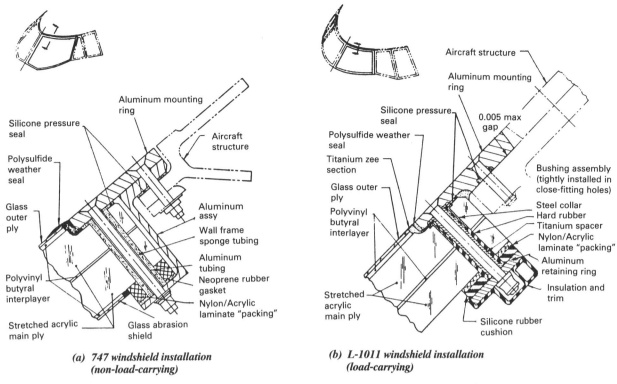

Fig. 11.4.9 Two different types of mountings of curved windshield.
(Courtesy of The Sierracin Corp.)

Fig. 11.4.10 One piece polycarbonate wraparound canopy — F-16.
(Courtesy of General Dynamics Corp.)

Airframe Structural Design 403

Most transparencies for tactical aircraft are laminates which consist of at least two and as many as six plies; current design practice includes several alternating plies with interlayers, and coated to provide some type of shielding or protection as shown in Fig. 11.4.11.

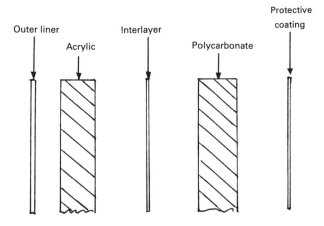

Fig. 11.4.11 *Typical configuration of transparency for fighter.*

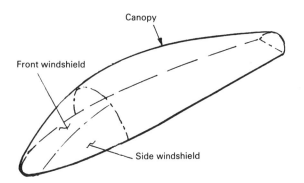

Modern ancillary cockpit systems that affect the loads and thermal environment of the transparent enclosure include the cockpit pressurization system, cockpit environmental-control system, windshield anti-ice system, windshield rain-removal system, windshield and canopy-defog system, and the canopy-jettison system. The pertinent system parameters that affect the transparency load and temperature environments are shown in Fig. 11.4.12. The design requirements for the front windshield and canopy are listed in Fig. 11.4.13.

In-service experience on aircraft indicates that extremes in the ambient environment influences the canopy lock and unlock functions. Canopy closure problems were encountered when the outside ambient temperature went below 20°F (approximately), or exceeded 95°F (approximately). Canopy thermal distortions were identified as the cause of the closure malfunction. The bimaterial (aluminum frame plus acrylics) canopy construction caused the canopy to distort differently when unlocked in low- and high-temperature environments, producing the gaps indicated in Fig. 11.4.14.

A series of tests should be conducted in which the thermal distortion characteristics of the newly designed canopy and the operating thermal envelope of the canopy lock/unlock system should be established.

Nose Radome

The radome of the aircraft consists of a general purpose nose radome housing the weather radar, glideslope antenna, etc. The modern transport radome is made of fluted-core sandwich-type construction (Fig. 11.4.15), utilizing epoxy fiber glass laminates, and includes the following design considerations:

- Optimum electrical characteristics of the antenna systems housed within the radome
- Optimum strength to weight structure
- Minimum aerodynamic drag
- Installation of permanent-type lightning strips requiring minimum maintenance

Ancillary System	Type	Designed by	Affected Transparency
Anti-ice	Hot air blast	Temperature	Outer surface (windshield)
Rain removal and repellent	Hot air blast and chemical spray	Temperature	Outer surface (windshield)
Defog	Electrical heating and hot air blast	Temperature	Inner surface (windshield)
Environmental control	Engine bleed air	Temperature	Inner surface (all)
Pressurization	—	Pressure	All
Jetison	—	Pressure	All (one piece canopy) Canopy only (if front windshield is fixed)

Fig. 11.4.12 *Fighter cockpit ancilliary system parameters.*

Design Requirement	Front Windshield	Canopy and Side Windshield
Bird impact	4-lb bird at 350 knots	
Projectile	Required by spec of bullet-resistant material	
Internal pressure	Yes	Yes
Aerodynamic load	Triangular distribution high at front, decreasing to zero at bow	
Fatigue pressure	Yes	Yes
Thermo load	Aerodynamic heating, anti-icing, rain dispersing, and defogging	Aerodynamic heating, internal heating
Under water implusion load	Yes	Yes
Canopy jettison	Yes (if one piece canopy)	Yes

Fig. 11.4.13 Design requirements for windshields and canopy-fighter.

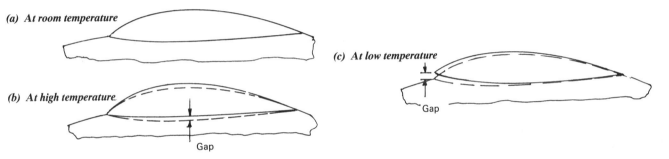

Fig. 11.4.14 Canopy thermal distortion.

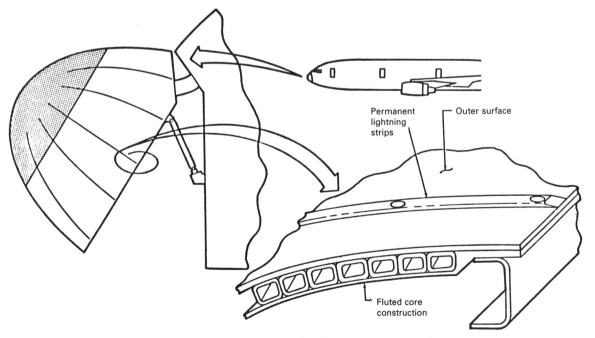

Fig. 11.4.15 Typical design of modern transport nose radome.

- Nose radome developed for use with housing for different radars, giving the customer the option of changing radars
- Easily replaceable neoprene boot or other similar materials bonded to the leading edge of the nose radome for rain erosion protection
- Fluted-core-type radome construction ensures good electrical performance and provides a high degree of hail protection

The fiber glass honeycomb and filament-wound resin-impregnated glass fiber are typical radome designs for either transports or fighters. The fiber glass wall thickness and uniformity are closely controlled for proper microwave transmission and reception.

There exists a problem, to house instruments of airborne radar in a material which requires de-ice. Obviously, electro-thermal methods cannot be used since they effect the radar signal.

11.5 Wing and Fuselage Intersection

If the airplane is of the low wing or the high wing type, the entire wing structure can continue in the way of the airplane body. However, in the mid-wing type or semi-low wing type (most for military fighter aircraft), limitations may prevent extending the entire wing through the fuselage, and some of the shear webs as well as the wing cover must be terminated at the side of the fuselage. However, for the mid-wing type, the wing box structure does not allow to carry through the fuselage structure, therefore, heavy forging structures are designed to carry through the wing loads (see Fig. 11.5.1).

Main Frames (Bulkhead)

Fuselage structures are subjected to large concentrated forces at the intersections with these airfoil surfaces and landing gears. The loads applied at these concentrated points must be distributed into the fuselage shell. Loads imposed by the wing and main landing gear form redundant or multiple load paths through the center wing and fuselage structure. Structural weight penalties to the uncut shell include not only those associated with the cutout area, but also penalties at the wing and gear supporting frames required to transfer loads from the wing to the fuselage. Typical wing and fuselage intersection views of transports are shown in Fig. 11.5.2 to Fig. 11.5.4.

Size of the fuselage cutout region, which is governed by the wing chord and landing gear volume, has a direct effect on the degree of penalty to skin panels, frames, floor struts, and lower structure within the wheel well area. Vertical location of the wing on the fuselage has a bearing on related skin panel and frame weight penalty. A low wing cuts through areas highly loaded in compression and, therefore, imposes a greater penalty to skin panels in the case of a high wing.

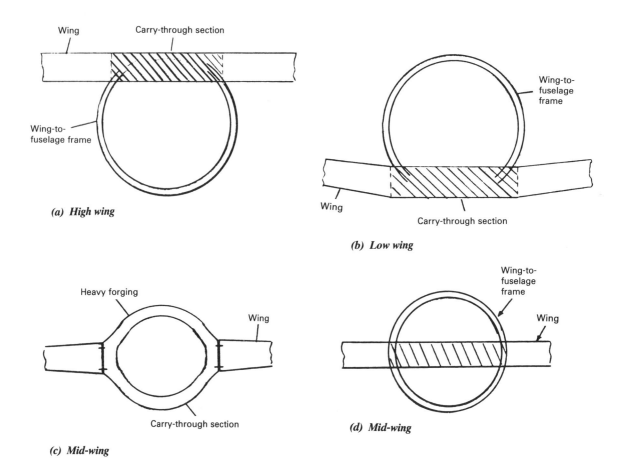

Fig. 11.5.1   The vertical location of the wing relative to the fuselage.

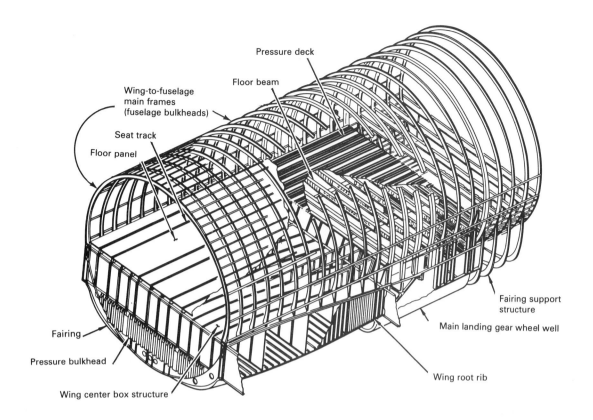

Fig. 11.5.2 Wing and fuselage intersection — L-1011.
(Courtesy of Lockheed Aeronautical System Co.)

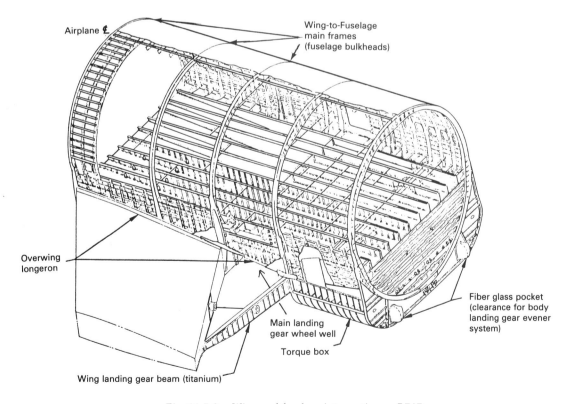

Fig. 11.5.3 Wing and fuselage intersection — B747.
(Courtesy of The Boeing Co.)

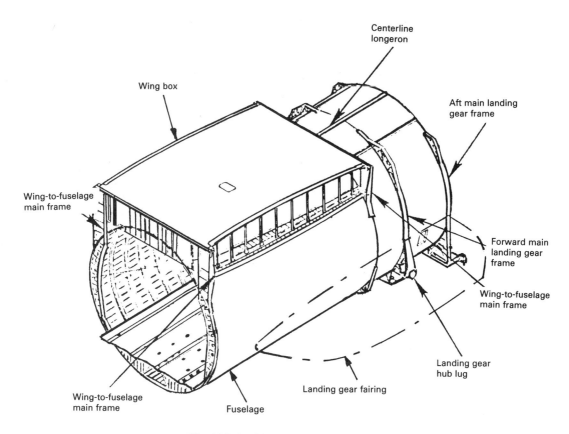

Fig. 11.5.4 Wing and fuselage intersection — C141.

This is of particular importance to transport aircraft because cargo loads added to down-tail loads can produce compression-critical lower panels. On the other hand, a high wing produces greater penalties to the frames than a low wing, because wing down loads from the wing create compression loads in wing-to-fuselage frames. In analyzing a high wing versus a low wing for minimum weight penalty, the summation of penalties to skin panels and frames may be a standoff, depending on the geometry of the airplane being studied. A wing located in the center of the fuselage probably imposes the lowest weight penalty since only shear material is affected; however, this wing location is highly impractical for transports.

Wing and Fuselage Connection
The wing connection to the fuselage presents some very interesting design problems. A design of adopting a four-pin design concept for the B707 transport during the 1950s is simple and straightforward (see Fig. 11.5.5). The lift and moment loads can be carried between the wing and fuselage by simple shear on the four pins, but there is another load vector to be considered: drag or thrust (fore and aft load) which is taken by breather web. This design allows the wing spar and fuselage bulkheads to deflect independently of each other such that no spar moment is directly transferred to the bulkheads (wing-to-fuselage frames).

As shown in Fig. 11.5.6, the small executive transport wing is dished at the center section to receive the fuselage which is attached by high tensile steel links at the intersection of the root ribs and main spars, with a

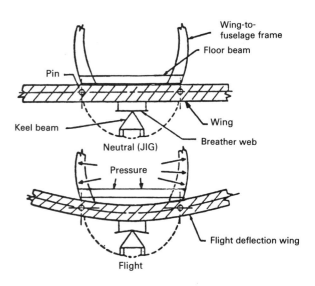

Fig. 11.5.5 B707 wing and fuselage connection.

drag and side stray spigot at the center-line rib and rear spar intersection.

Fig. 11.5.7 illustrates a wing-to-fuselage joint with truss links and lugs, and the pressure seal between fuselage and wing.

Typical design of modern transport wing-to-fuselage connection is to bolt the main frames to both front and rear spars of the wing box. This type of

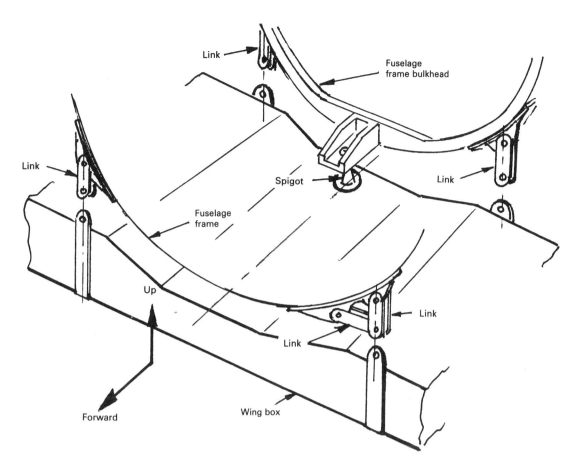

Fig. 11.5.6 Wing-to-fuselage joined by links and spigot.

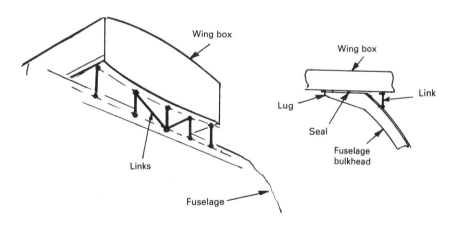

Fig. 11.5.7 Wing-to-fuselage joined by truss links.

connection has been widely adopted by aircraft designers for many years and very detailed design should be given such that it must be capable of sustaining the fatigue loads due to the deflection imposed by the wing bending. Fig. 11.5.8 shows that both spar moment and shear connections are spliced into the fuselage forward and aft bulkheads. The bulkhead and wing spar are rigidly connected together as one integral unit which has been chosen primarily to save structural weight. This type of construction utilizes the elastic characteristics of related parts and requires sophisticated finite element analysis technology.

The upper frame is a primary structure, which reacts shear loads into the fuselage shell. It can accommodate the induced rotation from the rear spar with proper detailed design. The lower portion of the bulkhead is made up of the rear spar plus extensions from it. These extensions are mostly secondary structures, but they pose a design problem because they want to deflect as an integral extension of the spar. A computer program has been used to select a design which will accommodate the deflections without failing.

Fig. 11.5.9 illustrates the fuselage frame and bulkheads connected to stub floor beams over the

Airframe Structural Design 409

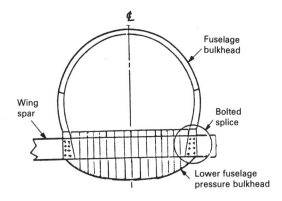

Fig. 11.5.8 Integral unit of fuselage bulkhead and wing spar.

wing center section box and wing spars. It is quite obvious that the deep floor beam imposes a much greater load on the frame, a fact proved by frame failure experienced in aircraft with relatively low service time. The point is not to overlook the loads induced by structural deflections or instability of supporting or attaching structures.

The keel beam is the most highly loaded structure in the fuselage that the wing box (center section) goes through the middle of the fuselage as shown in Fig. 11.5.10. For practical reason, the keel beam is so designed for the necessity of transporting low wing airplane loads that the main landing gear wheel well area is provided in the fuselage where the gears can be retracted into the fuselage.

In the wheel well area where the maximum fuselage bending moment occurs, the lower half of the skin/stringer system is missing. To obtain the bending integrity through this area, the load carried by the lower half of the fuselage skin/stringer panel in the forward and aft fuselage is concentrated in large longitudinal members. These are referred to as the keel beam longerons, and the complete structure required for their support and stabilization is called a keel beam (or Keelson beam).

Fig. 11.5.11 shows a triangular-shaped keel beam structure that is an independently stable box column. Another unique structural requirement in this area is the flexible shear web used to complete the shear path between the keel beam and the upper lobe. It is shown schematically in Fig. 11.5.11 with a deflected web

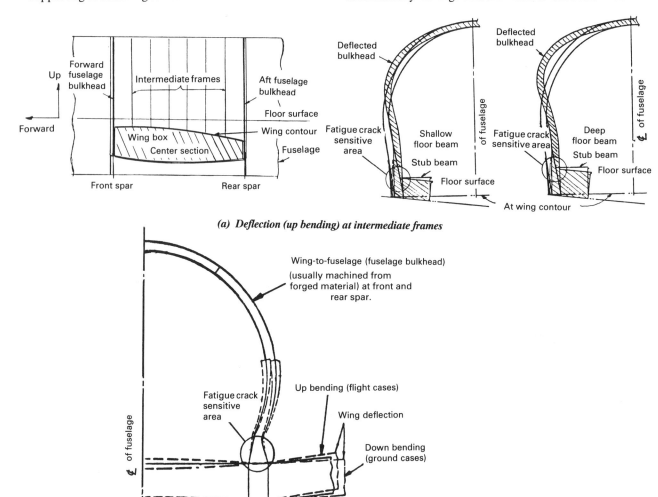

Fig. 11.5.9 Induced frame bending from wing deflection which could cause wing-to-fuselage main frame premature fatigue cracks.

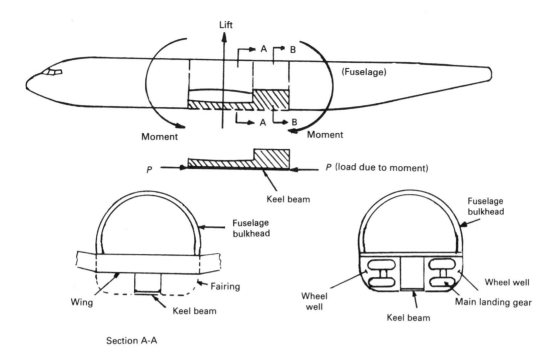

Fig. 11.5.10 Typical transport keel beam location and its function.

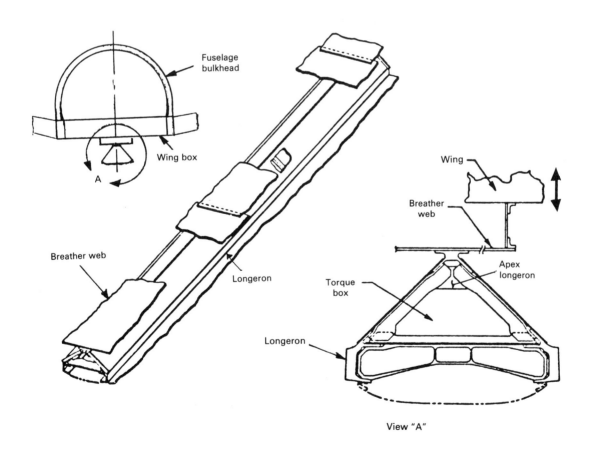

Fig. 11.5.11 Triangular torque box keel beam design with breather web.

Airframe Structural Design 411

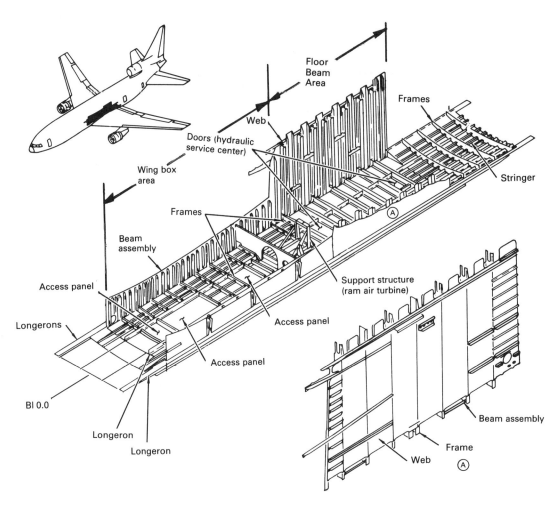

Fig. 11.5.12 Keelson beam construction — L-1011.
(Courtesy of Lockheed Aeronautical Systems Co.)

(breather web) across the apex of the keel beam torque box. The transverse stiffeners must be capable of stiffening the web as a shear member and still be flexible enough to bend as the wing deflects. This breather web also provides the structure to transfer fore and aft loads between the wing and fuselage.

Fig. 11.5.12 illustrates a rectangular box type keel beam. The inside of the box area is used to install hydraulic system equipment. Basically, it is a skin-stringer panel construction with access doors at the bottom of the box.

11.6 Stabilizer and Aft Fuselage Intersection

Aerodynamically, it is very important that the horizontal and vertical tail surfaces be so located that they are not blanketed by the fuselage. The vertical tail surfaces are most likely to be blanketed not only by the fuselage but also by the horizontal tail surfaces, especially when the airplane is at a high angle of attack. In order to minimize this effect, a position of the horizontal tail surfaces behind the vertical tail surfaces would clear both, so that neither would be blanketed by the other.

Horizontal Stabilizer

Fig. 11.6.1 illustrates several configurations of horizontal stabilizer and fuselage intersection and the modern transport stabilizer may be adjusted through a small angular displacement either on the ground or in the air from the cockpit. The adjustable stabilizer is used to change the trim angle of the airplane without displacing the elevator (the purpose to do this is to minimize the airplane trim drag with using the adjustable stabilizer rather than the elevator based on aerodynamic point of view). Modern high speed airplanes have adopted this adjustable stabilizer as shown in Fig. 11.6.2. All are hinges mounted to the aft fuselage with jackscrew installed on the forward side of the stabilizer. There is one structural problem: to design these hinges with fail-safe capability in case one of these hinges failed in flight. A jammed condition is also under consideration. In high performance fighter design, the all flying tail or taileron as shown in Fig. 11.6.1(d) is adopted typically to act as function of a powerful elevator or wing aileron. So far this configuration has not been used on any transport aircraft. However, it is not recommended on transports due to its complicated design as well as structural weight and cost.

412 Airframe Structural Design

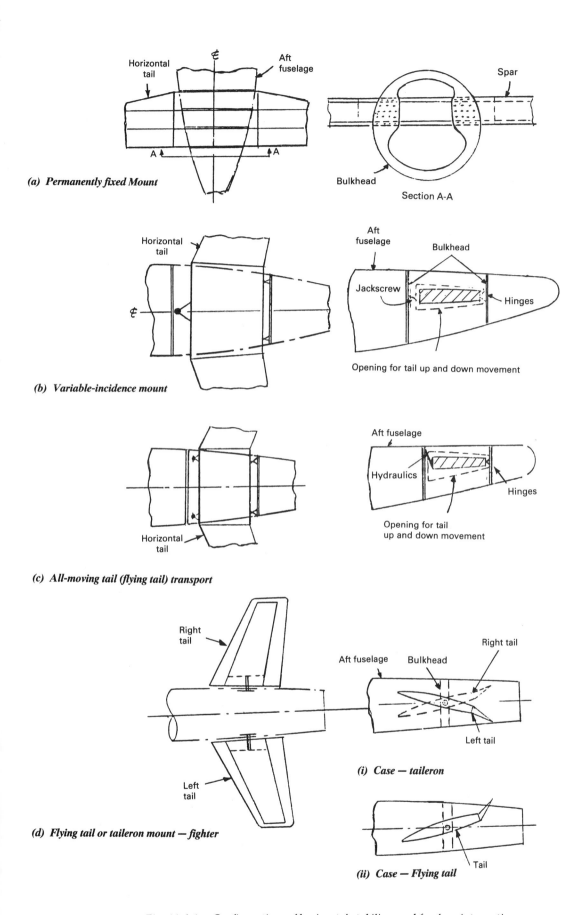

Fig. 11.6.1 Configurations of horizontal stabilizer and fuselage intersection.

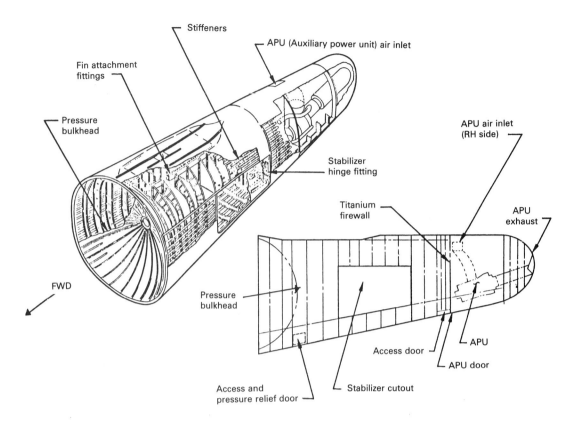

(a) Aft Fuselage Arrangement (also refer to Fig. 10.3.4)

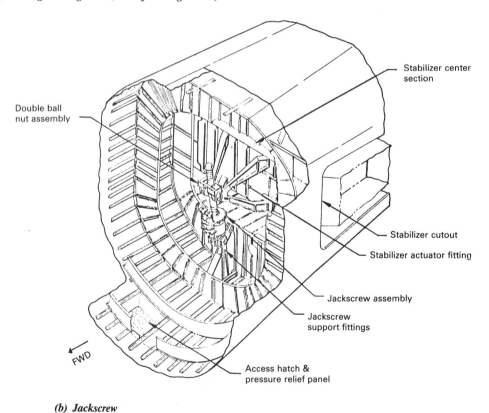

(b) Jackscrew

Fig. 11.6.2 *Aft fuselage structure — B747.*
(Courtesy of The Boeing Co.)

414 Airframe Structural Design

Vertical Stabilizer and Aft Fuselage

The vertical stabilizer is generally mounted on the aft fuselage with joints as shown in Fig. 11.6.3; the front and rear spars are attached to aft fuselage bulkheads by either a permanent joint or fittings.

As illustrated in Fig. 11.6.4, the vertical tail structure is completely integral with the aft fuselage (non-pressurized area). The spars enter the fuselage and become part of the fuselage frames and the skins tie directly to the fuselage skins. A three-spar design is employed on some of the tails to provide adequate fail-safe characteristics. The loss of any one member will not impair the structural integrity of the airplane.

Fig. 11.6.5 illustrates an aft fuselage structure that carries the center engine and both horizontal and vertical stabilizer loads to the fuselage. This structure can be divided into four major zones: above the S-duct, below the forward portion of the S-duct, around the pressure dome, and beneath the horizontal stabilizer box. This is one of the most complicated fuselage designs with a large number of compound curved surfaces and interfaces.

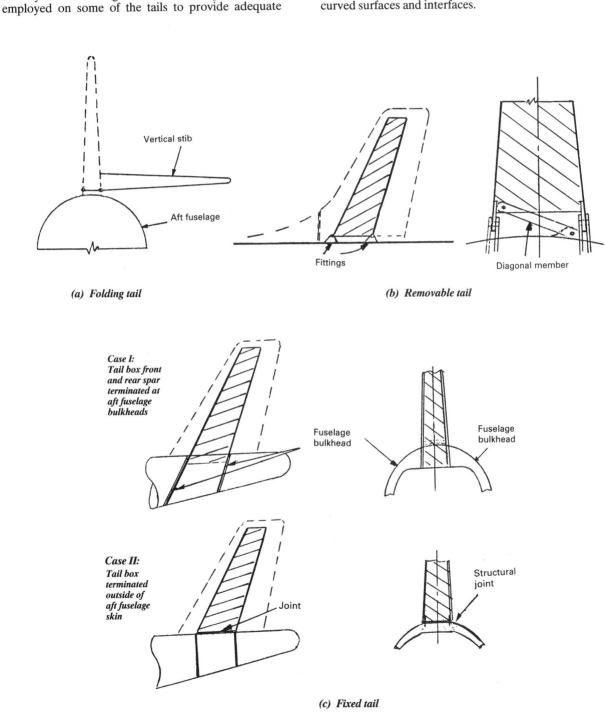

Fig. 11.6.3 Configurations of aft fuselage and vertical stabilizer intersection.

Airframe Structural Design 415

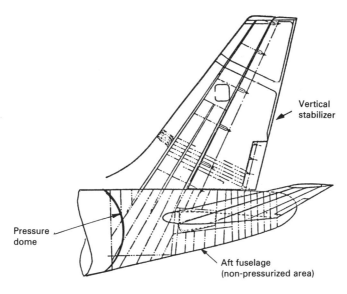

Fig. 11.6.4 Aft fuselage and vertical stabilizer intersection.

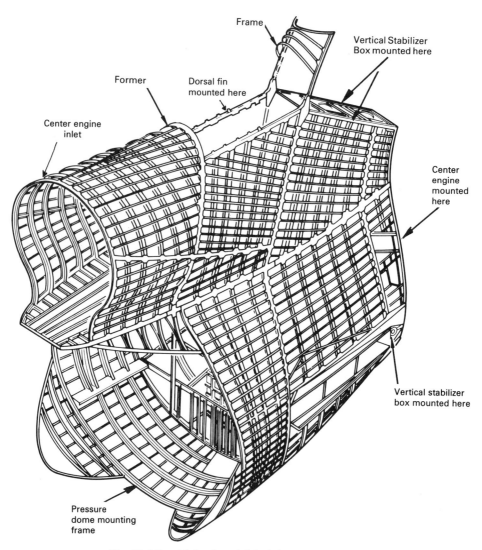

Fig. 11.6.5 Aft fuselage (aft body) structure — L-1011.
(Courtesy of Lockheed Aeronautical Systems Co.)

11.7 Fuselage Opening

Transport fuselages contain numerous cutout areas of different sizes and shapes located in various regions of the fuselage body. However, cutouts for doors, such as those big passenger doors, cargo doors, service doors, emergency exits, windows etc., often occur in regions where high loads must be resisted and, therefore, additional structures are needed to carry the loads around the openings. Such openings require the installation of jambs as well as strengthening of the abutting internal structure.

In general, doors are classified by type, which includes the manner of opening and load-carrying ability. "Plugged" or "non-plugged" are the usual terms defining type, while "stressed" or "non-stressed" designate load-carrying ability. A plugged door closes from the inside to provide safety advantages for pressurized aircraft. The passenger entrance and service doors are examples of plugged doors. Plugged-type entrance and cargo doors in the pressurized area are somewhat a combination of the two. In this case, hoop tension loads from pressure are carried through the door but, due to the quick-opening requirements, the major portion of tensile and shear loads are carried around the cutout structures. A typical side-loading main deck cargo door (above floor cargo door) used on cargo transports is shown in Fig. 11.7.1. This design is an example of a non-plugged door. However, hoop tension loads are carried through the door and transmitted into the fuselage shell by means of latches along the lower surface and hinges at the top.

Big-sized cargo doors must be designed to carry loads through the structure (see Section 6.6 of Chapter 6.0) if proper provision is made for transmitting the loads. However, such factors as loose fits and requirements for emergency operation prevent this from being an easily-established condition.

Windows are not considered to carry loads through the structure. It is true that pressure on them is transmitted, but the installation and the material of the windows are such that they are not thought of as providing continuity to the structure for stress analysis purposes.

Windows

Three factors influence window configuration: size, quantity, and shape. From a structure standpoint, they should be small, few, and round. From a passenger appeal standpoint, they should be large, many, and square. In the final configuration, it will be noted that structures came out second best.

With the size and spacing established, the design of the window cutouts presents some very interesting problems for the structure designers. The window cutouts fall in the area of the highest skin shear from fuselage bending. Since vertical bending is a flight condition, the cutout reinforcement must be designed for the combined loads of bending and pressure. Since pressure loads are involved, it is also a fatigue critical area. Therefore, the simple way to solve this problem is to use integrally-machined window belt panels or riveted on reinforced doublers (it may be added on the outside of the fuselage such as on the DC-9 and DC-10 transports) as shown in Fig. 11.7.2.

Fig. 11.7.3 shows a typical design of the complete passenger window installation. The outer pane is the primary pressure carrying pane. The mid-pane will carry the design pressure load in the event of an outer pane failure. Except for a small ventilation hole, the mid-pane is sealed against the inner flange of the forging. The remainder of the installation, dust cover and the trim are nonstructural decorative interior.

Fig. 11.7.1 Fuselage upper deck freight door (120 in × 73 in).

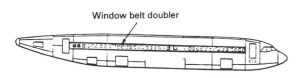

Fig. 11.7.2 Longitudinal window belt doublers.

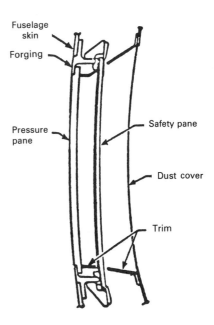

Fig. 11.7.3 Typical passenger window installation.

The safety achieved by the use of multiple structure can be seen in Fig. 11.7.4 of the window arrangement. The outer pane carries the entire pressure load with a safety factor in the order of 8. If, however, a failure should occur in the outer pane, the inner is capable of assuming the full load. An additional safety feature exists in most of the windows where a third pane has been added inboard of the other two. This pane is primarily included for acoustic reasons; however, it is of such a size and is mounted in such a way that, in the unlikely event of a failure of both the inner and outer main window panes due to a severe accident, it is available to carry full cabin pressure.

This acoustic pane also serves the purpose of protecting the inner pane from damage from inside the cabin.

The cabin windows, located midway between the fuselage frames as shown in Fig. 11.7.5 are mounted in window frame forgings that are riveted to the skin and a bonded doubler. They are of vented two-pane construction; either pane is capable of carrying the full differential pressure load, and they are made of stretched acrylic material with an air breather hole near the upper edge of the inner pane. Replacement is achieved from inside the cabin by the simple detachment of clips.

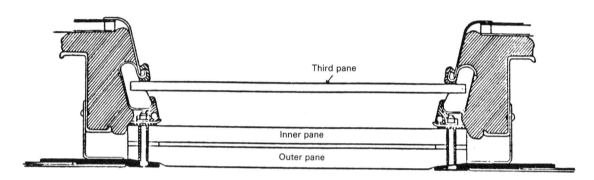

Fig. 11.7.4 *Passenger window installation — DC-8.*

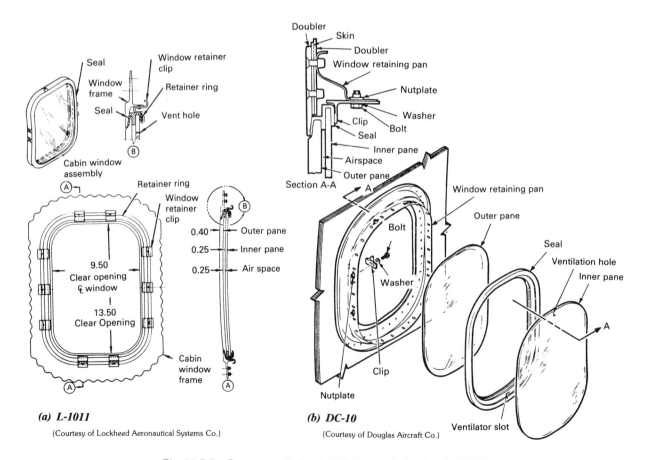

(a) *L-1011*
(Courtesy of Lockheed Aeronautical Systems Co.)

(b) *DC-10*
(Courtesy of Douglas Aircraft Co.)

Fig. 11.7.5 *Passenger window installation — L-1011 and DC-10.*

Fuselage Doors

Transport fuselages contain numerous cutout areas of different sizes and shapes located in various regions of the body. Some of the openings are small, such as those required for service interphone jacks and antennae. These necessitate only local reinforcement around the holes, usually in the form of skin doublers. In these instances, it is usually safe to assume that the material of the skin removed is equal to the material of the doubler added. However, cutouts for doors, such as those shown in Fig. 11.7.6, often occur in regions where high loads must be resisted and, therefore, additional structure is needed to carry the loads around the openings. Such door openings require the installation of jambs as well as strengthening of the abutting internal structure. A study of panel modifications for large doors on the transport revealed the average ratio of material added to that removed was approximately three to one (excluding the door and operating mechanism).

The doors and special exits for passenger aircraft must conform to certain regulations designed to provide for the safety and well-being of passengers. These regulations are established by authorized agency, i.e. FAA, and they must be followed in the design and manufacture of all certificated aircraft for passengers.

The external surface of the doors on transport aircraft must be equipped with devices for locking the doors and for safeguarding against opening in flight either inadvertently by persons or as a result of mechanical failure. It must be possible to open doors from either the inside or the outside even though persons may be crowding against the door from the inside. The means of opening must be simple and obvious and must be so arranged and marked internally and externally that it can be readily located and operated even in darkness.

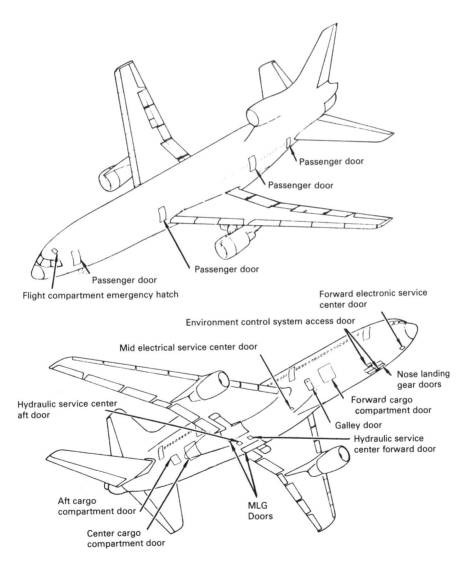

Fig. 11.7.6 Fuselage door locations — L-1011.
(Courtesy of Lockheed Aeronautical Systems Co.)

Reasonable provisions must be made to prevent the jamming of any door as a result of fuselage deformation in a major crash.

(1) Design Criteria

The following criteria shall be used for the design and analysis of the fuselage *plug-type door* (resisting internal or cabin pressure only) and *nonplug-type door* (carrying fuselage shell loads in addition to resisting internal pressure).

(a) Design ultimate factor for pressure:
- Door structures shall be designed for 3.0 factors on pressure, for tension members and splices.
- All stop fittings, door latches and hinges shall be designed for 3.0 factors on pressure. Lateral loads on stops caused by friction shall be taken into consideration.
- Door structures shall be designed for 2.5 factors on pressure for shear and compression members.
- All door structures shall be designed for a negative 1.5 psi ultimate pressure acting singularly.

(b) Design ultimate factors for flight loads plus pressure:
All door structures shall be designed for 2.0 factors on the maximum applicable operating pressure plus ultimate flight loads.

(c) Flight loads acting alone:
All door structures shall be designed for ultimate flight loads alone when this is a critical condition.

(d) Flight loads shear distributions (for shear-type doors):
The door shall be designed to withstand the following conditions.
- 100% of the ultimate shear load shall be carried across the door (assume zero shear carried by fuselage cutout).
- $\frac{2}{3}$ of the ultimate shear shall be carried across the door with $\frac{1}{3}$ being redistributed in the surrounding fuselage structure to account for wear, tolerances and misalignment.
- The effects of shear and pressure deformations of the door relative to the cutout surrounding structure shall be taken into consideration for distribution.

(e) Gust and random door loads:
- Door opening or closing — doors shall be designed to withstand an equivalent horizontal steady state wind of 40 knots while opening or closing. The ultimate design load shall be the 40 knot limit wind load times 1.5
- Door open position — doors shall be designed to withstand an equivalent horizontal steady state wind of 65 knot in the full open position. The ultimate design load shall be the 65 knot limit wind load times 1.5.
- Hand or foot loads — the door structure and actuating mechanism shall be capable of withstanding an ultimate random hand or foot load of 300 lb acting downward or a load of 150 lb acting in any other direction. These loads shall be considered acting singularly and applied to any point on the door and at any position of the door, open or closed.

(f) Actuator torque requirements:
The torque required shall be based upon the maximum envelope of door resisting hinge moments (door weight only or door weight plus 40 knot wind) plus the torque required to overcome the friction load in the linkage.

(g) Door jammed condition:
The door structure shall not fail if the door becomes jammed and full actuator power is applied.

(h) Design ditching pressures:
Design pressures for ditching shall be established during emergency landing on water surface.

(i) Fail-safe design:
The door structure shall be designed for the failure of any single member. The fail-safe design pressure shall be a differential pressure times 1.5.

(j) Materials:
2024 aluminum or other equivalent materials are recommended to be used for skins and critical tension members.

(k) Latching mechanism:
- Provisions shall be made for the visual inspection of the latches when the door is closed. The airplane shall not be capable of being pressurized unless all latches are fully engaged.
- It shall not be possible under any conditions of wear, maladjustment, or deflection of build-up tolerances sufficient to cause loss of door.

A summation of structural design criteria for pressurized fuselage doors are shown in Fig. 11.7.7.

(2) Passenger Doors

The doors for a pressurized airliner must be much stronger and much more complex than the door for a light airplane. The door usually consists of a strong framework to which is riveted a heavy outer skin formed to the contour of the fuselage. Two examples are shown in Fig. 11.7.8 and Fig. 11.7.9.

Because the door size is a major parameter, the peripheral and longitudinal dimensions, as well as total area, are realized. Vertical location on the transport fuselage body is also important from the standpoint of whether the door occurs in a region designed by compression, tension, or shear considerations. Horizontal location can be equally significant.

Door designs vary substantially. In addition to the effects of size and location, the function design of the door has considerable impact on structural weight. In general, doors are classified by type — plug or non-plug, which includes the manner of opening and load-carry ability. Should failure of the latching or hinging mechanism

Reference	Design Condition		Component	Limit Load	Safety Factor	Ultimate Load	Remarks
(a)	Maximum operating pressure		Tension members	9.0 psi	2.5	22.5 psi	
(a)			Door stop fittings and door latches	9.0 psi	3.0	27 psi	Lateral loads on door stops caused by friction shall be considered
(a)	Maximum relief valve pressure		Door structure critical in compression or shear	9.4 psi	2.0	18.8 psi	
(a)	Negative pressure differential		Door structure	−1.0 psi	1.5	−1.5 psi	
(e)	Ground gust operational loads	Door opening or closing	Door structure and actuation system	40 knot wind load	1.5	pressure	
(e)		Door open position	Door structure	65 knot wind load	1.5	pressure	
(e)	Random door loads		Door structure with door in any position			300 lb	Downward acting load
						150 lb	Load in any direction
(f)	Emergency handle loads		Door structure, door handle & actuation system			450 lb 225 lb	
(f)	Emergency loads, opening only					2500 lb 400 lb	Roller load per latch Internal load applied by passenger
(i)	Fail-safe		Door structure and stops				
(i)	Internal plus aerodynamic pressure		Internal pressure	9.4 psi	1.25	11.8 psi	Design door structure for failure of a single member
			Aerodynamic pressure	pressure	1.25	pressure	

Note: Assume cabin pressure = 9.0 psi; cabin relief valve pressure = 9.4 psi; negative pressure = −1.0 psi.

Fig. 11.7.7 Design criteria for a passenger door.

occur during flight, internal pressure would hold the door in place. All transport passenger entrance and service pressurized doors are plug-type doors.

Quick-opening access doors in the unpressurized areas of the fuselage are examples of hook and latch-type non-stressed doors, which are also used on fighter airplanes and small aviation airplanes with unpressurized fuselage.

As shown in Fig. 11.7.10, at the top and bottom edges of the door are hinged gates that make it possible, in effect, to decrease the height of the door so it can be swung outward through the door opening. The hinging and controlling mechanism of the door is rather complex in order to provide for the necessary maneuvering to

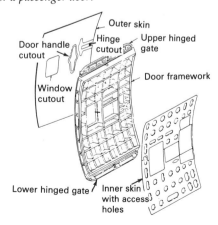

Fig. 11.7.8 Plugged-type passenger door construction — B737.

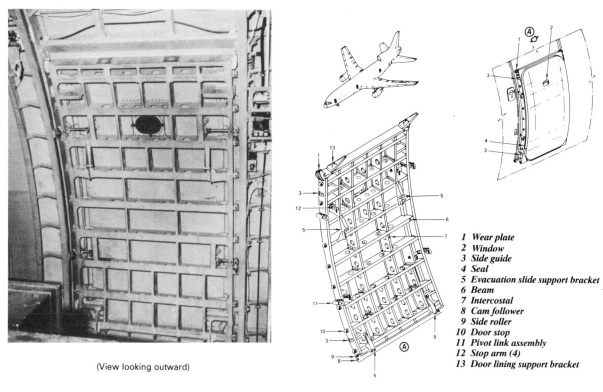

Fig. 11.7.9 Plugged-type passenger door construction — L-1011.
(Courtesy of Lockheed Aeronautical Systems Co.)

1. Wear plate
2. Window
3. Side guide
4. Seal
5. Evacuation slide support bracket
6. Beam
7. Intercostal
8. Cam follower
9. Side roller
10. Door stop
11. Pivot link assembly
12. Stop arm (4)
13. Door lining support bracket

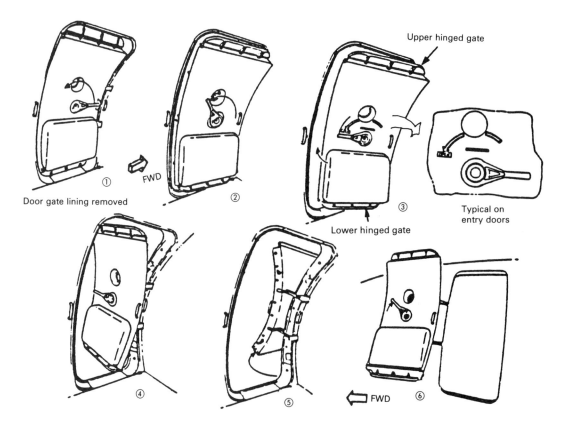

Fig. 11.7.10 Swing opening passenger door — B747.
(Courtesy of The Boeing Co.)

422 Airframe Structural Design

move the door outside the airplane when loading and unloading passengers. For safety in a pressurized fuselage, the door is designed to act as a "plug" for the door opening and the pressure in the cabin seats the door firmly in place. To accomplish this, the door must be wider than its opening and must be inside the airplane with pressure pushing outward.

Fig. 11.7.11 shows a plug-type passenger door that opens by first moving inward and then by sliding upward on rollers. Door is stowed inside, away from the weather and the possibility of damage from passenger loading equipment. This door is normally operated electrically, either from a panel inside the cabin or from an exterior control panel. The door is counterbalanced, and a mechanical disconnect is provided in the drive mechanism so that it can be manually opened with ease in case of loss of electrical power or in an emergency evacuation situation.

The passenger and service doors as shown in Fig. 11.7.12 are also a plug-type door with built-in lugs on the vertical members which, in the closed position, are located behind similar lugs on the vertical doorway frame members. The doors open and close on a parallel hinge mechanism, a door lifting mechanism is fitted to engage and disengage the lugs on the door from the lugs on the frame. A mechanical lock is fitted to prevent the door opening under negative g conditions during unpressurized flight. Under pressurized conditions the outward load on the door prevent inadvertent opening.

The main entrance doors as shown in Fig.

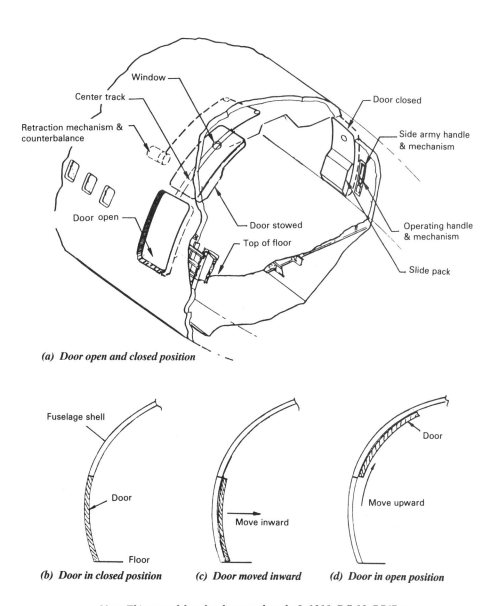

(a) Door open and closed position

(b) Door in closed position *(c) Door moved inward* *(d) Door in open position*

Note: This type of door has been used on the L-1011, DC-10, B767 etc

Fig. 11.7.11 *Vertical slide opening door.*

Airframe Structural Design 423

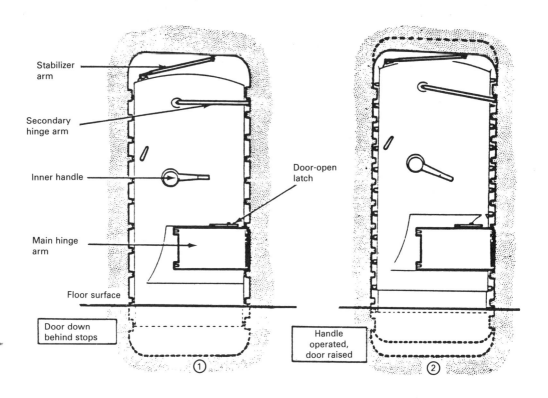
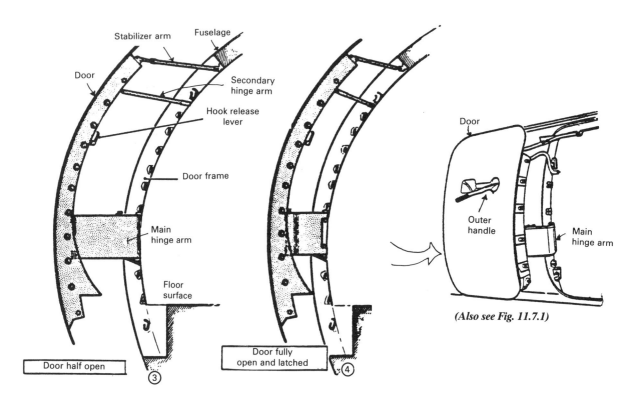

Fig. 11.7.12 Horizontal slide opening door.
(Courtesy of British Aerospace PLC)

11.7.13 are tapered plug-type doors. They are so named because of their vertical sliding action into a tapered fuselage opening. The forward edges of these doors are nearly vertical while the aft edges are canted, thus giving a wedging action when the doors are closed. Continuous vertical tracks along the forward and aft edges of the door openings restrain the main entrance and service doors in the closed position. The door operating mechanism clinches the doors down by means of a cable system, and actuates an interlock mechanism pin on the forward edge of the door to keep the door down and closed in the event of cable failure.

(3) Fuselage Lower Deck Cargo Doors

The construction of fuselage lower deck cargo door as shown in Fig. 11.7.14 is basically identical to that of upper deck passenger door with plug-type door design (except the big upper deck cargo door shown in Fig. 11.7.1). It is obvious that the function of this cargo door is for loading and unloading cargo or containers and, therefore, the door hinge design as well as the open position is different from the passenger door as discussed previously. Fig. 11.7.15 illustrates several cargo door arrangements.

Nose and Tail Cargo Loading Doors

Although many passenger transport aircraft which have been converted into freighters have big upper deck side doors, a pure freighter should have better accessibility via wide open doors in the front (see Fig. 11.7.16) or rear of the fuselage to allow loading in a longitudinal direction. Some military cargo transports such as the U.S. C-5 and Russian An-124 have used both front and rear loading doors for quick loading and unloading military equipment in war zone.

The various possibilities for the door arrangement are discussed below.

(1) Door in the fuselage nose as used on the C-5 [Fig. 11.7.17(a)], B747F [Fig. 11.7.17(b)], An-124, and Argosy 100 and Bristol freight. This design is not applicable to smaller transports because the high cockpit causes smaller transports considerable drag.

(2) A swinging fuselage nose (including the flight deck) creates considerable difficulties in carrying through cables, wires, plumbing, etc. The weight penalty is of the order of 12% of the fuselage structure weight. Its use may be considered in very special cases, e.g. the Guppy aircraft (Fig. 11.7.18).

(3) Door in the aft fuselage as used on the Andover,

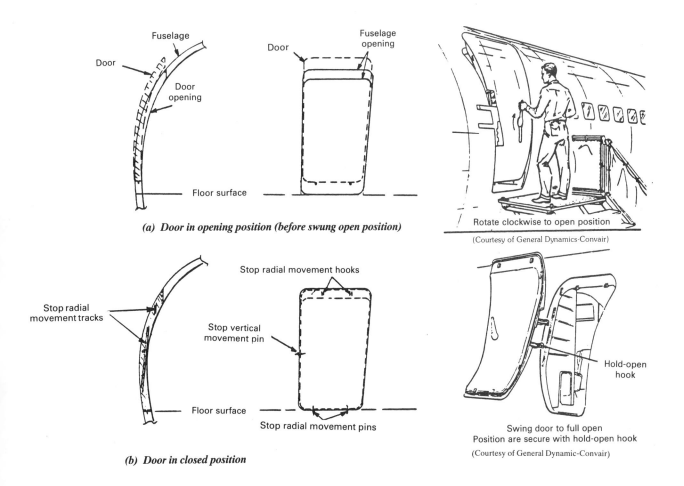

Fig. 11.7.13 Tapered door CV880.

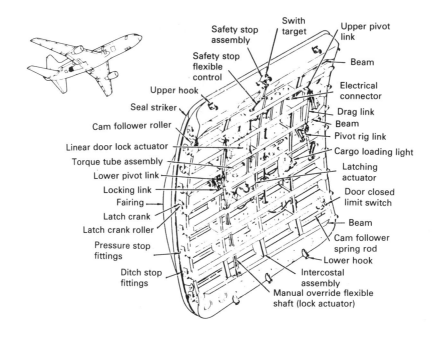

Fig. 11.7.14 *Fuselage lower deck cargo door — L-1011.*
(Courtesy of Lockheed Aeronautical Systems Co.)

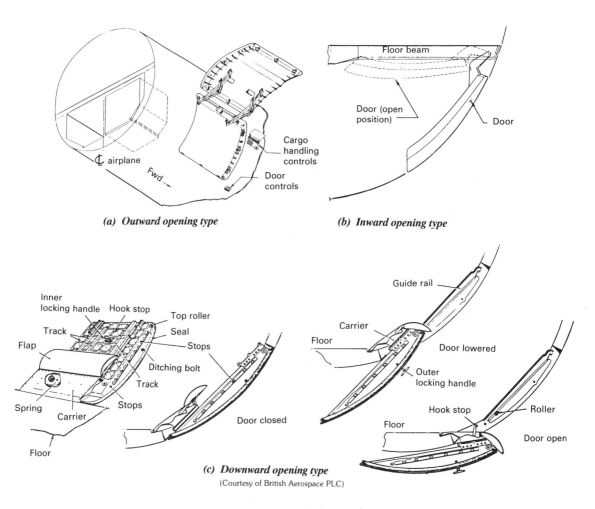

(a) **Outward opening type**

(b) **Inward opening type**

(c) **Downward opening type**
(Courtesy of British Aerospace PLC)

Fig. 11.7.15 *Lower deck cargo doors.*

426 Airframe Structural Design

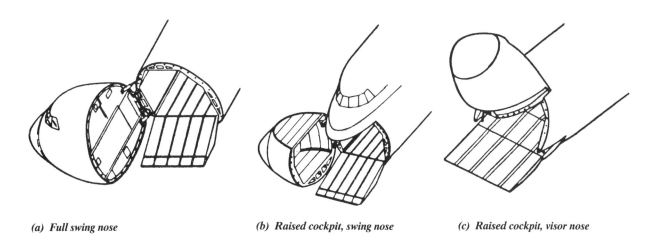

(a) Full swing nose   *(b) Raised cockpit, swing nose*   *(c) Raised cockpit, visor nose*

Fig. 11.7.16 Examples of nose loading design.

(a) C-5
(Courtesy of Lockheed Aeronautical Systems Co.)

(b) B74F
(Courtesy of The Boeing Co.)

Fig. 11.7.17 Door in fuselage nose.

Airframe Structural Design 427

Fig. 11.7.18   Swinging fuselage nose — Super Guppy.

Fig. 11.7.19   Aft fuselage loading door — C-130.
(Courtesy of Lockheed Aeronautical Systems Co.)

Short Belfast, C-5, C-141 and C-130 (Fig. 11.7.19). This design is for easy access especially in small freighters. It is essential to camber the aft fuselage upwards, thus creating an aerodynamic drag and also structural stiffness problem because of large cutouts.

(4) Tail boom layout in combination with a door in the rear part of a stubby fuselage. This configuration, occasionally seen on freighter aircraft (the H.S. Argosy, Noratlas, IAI Arava, Fairchild C-82 and C-119), offers a readily accessible freight hold and permits the use of a beaver tail for air dropping purposes, if required. The high aerodynamic drag is a disadvantage.

(5) Swing-tail, a layout proposed by Folland in a freighter project as far back as 1922. To date, the swing-tail has been implemented only on the Canadair CL-44 (Fig. 11.7.20), at the expense of a penalty of some structure weight relative to a side door. From an aerodynamic standpoint, however, the swing-tail is ideal and the structural complexity may be outweighed by a considerable reduction in fuel consumption.

Fig. 11.7.20   Swing-tail loading — CL-44.

References

11.1 Weiser George L.: *Sierracin Glass/Plastic Composite Windshields*, The Sierracin Corporation, Sylmar, Calif., 1969.

11.2 Olson, Jan B.: *Design Considerations Affecting Performance of Glass/Plastic Windshields in Airline Service.* The Sierracin Corporation, Sylmar, Calif., 1973.

11.3 Miller, William A.: *Polycarbonate Aircraft Transparencies.* The Sierracin Corporation, Sylmar, Calif., 1974.

11.4 McKinley, J.L. and Bent, R.D.: *Basic Science for Aerospace Vehicle.* McGraw-Hill Book Company, New York, N.Y., 4th Ed. 1972.

11.5 Williams, D.: *An Introduction to the Theory of Aircraft Structures.* Edward Arnold Ltd., London, 1960.

11.6 McFarland, R.A.: *Human Factors in Air Transport Design.* McGraw-Hill Book Company, Inc. New York, 1946.

11.7 Williams, D.: "Pressure-Cabin Design", *Aero. Res. Council Tech. Rep. C.P. No. 26*, (1955).

11.8 Spaulding, E.H.: "Trends in Modern Aircraft Structural Design", *SAE No. 92*, (1957).

11.9 Bates, R.E.: "Structural Development of the DC-10", Douglas paper No. 6046, (1972).

11.10 Bruhn, E.F.: *Analysis and Design of Flight Vehicle Structures.* Tri-State Offset Company, Ohio, 1965.

11.11 Sperring, Richard F.: "Glass Materials: 1974–1980," *SAE No. 750344*, (Feb., 1975).

11.12 Anon.: "Take A Seat", *Flight International*, (Jul. 8, 1965), 62–65.

11.13 Zeffert, H.: "Airliner Flight Deck Design for Crew Co-ordination," *Aeronautical Journal*, (Dec., 1973).

11.14 Anon.: "Furnishing and Finishing." *Flight International*, (Jan. 21, 1965), 91–99.

11.15 Fixler, S.Z.: "Thermostructural and Material Considerations in the Design of the F-14 Aircraft Transparencies." *Journal of Aircraft*, (Mar. 1977), 257–264.

11.16 Bateman, L.F.: "An Evolutionary Approach to the Design of Flight Decks for Future Civil Transport Aircraft." *Aircraft Engineering*, (Jul., 1978).

11.17 Mercer, Charles R.: "The Windshield Development of a Modern Wide-Body Transport." *Conference on Aerospace Transparent Materials and Enclosures.* (Apr., 1978).

11.18 Goff, W.E.: "Concorde Vision." *Flight International*, (Mar. 2, 1972), 321–322.

11.19 Neal, M.: "Tough Problems Seen Through." *Flight International*, (Nov. 12, 1964), 845-847.

11.20 Roberts, W.G.: "Glass Windshields for Wide Bodied Aircraft." *Aeronautical Journal*, (May, 1974).

11.21 Stansbury, J.G.: "Stretched Acrylic (A Transparent Glazing Material for High-speed Aircraft)." *Aeronautical Engineering Review*, (Jan., 1958).

11.22 King, B.G.: "Functional Cockpit Design." *Aeronautical Engineering Review*, (Jun., 1952), 32–40.

11.23 Anon.: "Conference on Aerospace Transparent Materials and Enclosures." *AFFDL-TR-78-168.*, (Apr., 1978).

11.24 Burce, R.S.: "Design and Development of Glass Windscreens for Wide-bodied Aircraft." *Aircraft Engineering*, (Jul., 1976), 4–8.
11.25 Vogel, W.F.: "Pressurized-cabin Elementary Frame Analysis." *Aero Digest*, (Jan. 1949).
11.26 Anon.: "Windshield Technology Demonstrator Program-Canopy Detail Design Options Study." *AFFDL-TR-78-114*, (Sept., 1978).
11.27 Saelman, B.: "Optimum Spacing of Shell Frames." *Journal of Aircraft*, (Jul–Aug. 1964), 219.
11.28 Goodey, W.J.: "The Stress Analysis of the Circular Conical Fuselage with Flexible Frames." *Journal of the Royal Aeronautical Society*, (Aug. 1955), 527–550.
11.29 Isakower, R.I.: "Simplified Calculations for Ring Redundants." *Machine Design*, (Mar. 1965).
11.30 Lee, T.H.: "New Design Formulas for Varying Section Rings." *Product Engineering*, (Jul. 1964).
11.31 Blake, A.: "How to Find Deflection and Moment of Rings and Arcuate Beams." *Product Engineering*, (Jan. 1963).
11.32 Brahney, J.H.: "Windshields: More Than Glass and Plastics." *Aerospace Engineering*, (Dec. 1986).
11.33 Yaffee, M.L.: "New Cockpit Enclosure Materials Sought." *Aviation Week & Technology*, (Jan. 20, 1975).
11.34 Anon.: "Cockpit Enclosures Performance Key." *Aviation Week &Space Technology*, (Jan. 26, 1976).
11.35 Goff, W.E.: "Droop Nose." *Flight International*, (Aug. 12, 1971), 257-260.

CHAPTER 12.0

LANDING GEAR

12.1 Introduction

Some time ago, the improvements in the aerodynamic characteristics of the airframe led designers to make the conventional landing gear with tail wheel or skid, which was then fitted to all aircraft and made retractable. Although this resulted in a weight increase owing to the installation of a retraction gear, jacks, retracting and locking mechanism, distributors, valves, etc., experience proved that the gain in speed achieved by reduction of drag largely offset the weight increase. The decrease in drag varies from 6% at a speed of 150 mph to 8-10% at a speed of 200 mph. One should bear in mind that the aircraft fitted with a conventional tail wheel has a considerable angle of incidence when on the ground. The great difference between this angle and the minimum drag angle hampers the take-off and also presents an element of discomfort in commercial aircraft (see Fig. 12.1.1).

Fig. 12.1.1 Landing gear with tail wheel of DC-3 vs. tricycle gear of DC-9.

The tricycle landing gear with nose wheel disposes of these disadvantages. It reduces the ground roll at take-off with a saving of corresponding energy. Above all, it offers greater stability, particularly in crosswind landing conditions and in making turns on the ground at high speeds. In certain aircraft of high gross weight and with multiple main wheels, the directional movement of the nose wheels is insufficient to provide an acceptable turn radius in relation to the width of the taxiway and the wheel-base of the aircraft. In consequence, some degree of lateral freedom must be provided for the rear wheels to give this steering ability.

Since the functions fulfilled by aircraft are becoming extremely diverse, it becomes apparent that each landing gear represents an individual case and the relative importance of its various functions — damping of the impact on landing, absorption of braking energy, and ground maneuvers — depends upon the duties of the aircraft. In the case of a carrier-based aircraft, where the energy to be absorbed at impact is high owing to the high rate of descent, it is conceivable that powerful brakes could be eliminated, as the braking is provided by the arrester cables in a deck landing. Similarly, the taxiing conditions for cargo aircraft operating from unprepared airfields are entirely different to those for an aircraft using international airports. The landing gear of the former will be built around low pressure tires of large sizes, while the latter will use high pressure tires requiring relatively small space.

Today, the manufacture of a landing gear necessitates close collaboration with the aircraft designer right from the design study stage. This collaboration provides joint agreement on, in addition to the positioning of the gear, such matters as the various attitudes of the aircraft, shock absorber travel in all conditions encountered, landing gear mounting points, installation of steering controls, retraction circuits — in short, everything concerned with the operation.

The search for a simple landing gear system — a question of lightness and economy — is made difficult by the problem of stowing the gear. Available space is limited by the size of the fuselage nose section and by use of relatively thin wing sections (the case of the F-104 supersonic fighter, where the main landing gear has to be stowed in the fuselage at the cost of a very complicated mechanical system).

However, the purpose of an aircraft landing gear arrangement is two fold: to dissipate the kinetic energy of vertical velocity on landing, and to provide ease and stability for ground maneuvering. The design of landing gears to perform these functions efficiently has become quite complex with the increasing loads and diminishing storage space of today's aircraft.

The designing of a landing gear system is not merely a structural problem. The landing gear of a modern airplane is a complex machine, capable of reacting the largest local loads on the airplane. Its function is to convert a relatively uncompromised airborne vehicle into a rather awkward and lumbering ground machine. In one brief moment the landing gear must make the best of returning the airplane from its natural environment to a hostile environment — the earth. In most instances all this machinery, comprising about 10% of the airplane's structural weight, must be retracted and stowed away while the airplane is airborne.

Landing gear design is an art and some parts of this chapter have been extracted from Ref. 12.32, which is one of the most valuable references in landing gear design.

Examples of Current Landing Gear Design
(1) Lockheed L-1011

The L-1011 landing gear is one of the typical commercial transport landing gear examples. As shown in Fig. 12.1.2(a), each main landing gear assembly includes the four wheel/brake/tire assemblies mounted on a truck beam (bogie); an air-oil shock strut with supporting trunnion; and folding side braces connected to the hydraulic operated retraction system. Dual springs are used for positive downlocks, with or without hydraulic pressure and there are towing lugs on each gear for airplane ground handling conditions.

Fig. 12.1.2(b) shows the nose landing gear assembly, including the two wheel/tire assemblies, the air-oil shock strut and integral side struts, articulated drag struts and jury brace with dual springs for positive downlock, hydraulic steering and retraction mechanisms. The steering mechanism acts as a shimmy damper when nose steering is not engaged. Integral towing lugs are located near the axle to accommodate standard tow bars.

(2) Transall C-160 and Breguet 941

One of the important developments in landing gear for the STOL transport is the Messier 'Jockey' twin-wheel main units as fitted to the Breguet 941, Transall, An-22, Aeritalia G222,

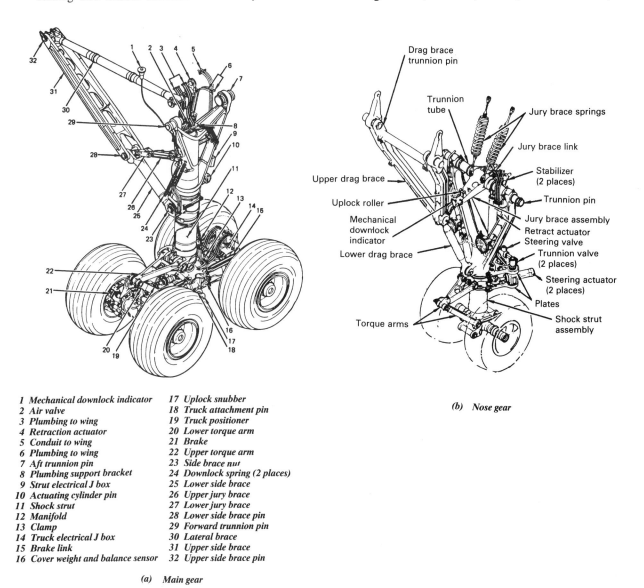

1 Mechanical downlock indicator
2 Air valve
3 Plumbing to wing
4 Retraction actuator
5 Conduit to wing
6 Plumbing to wing
7 Aft trunnion pin
8 Plumbing support bracket
9 Strut electrical J box
10 Actuating cylinder pin
11 Shock strut
12 Manifold
13 Clamp
14 Truck electrical J box
15 Brake link
16 Cover weight and balance sensor

17 Uplock snubber
18 Truck attachment pin
19 Truck positioner
20 Lower torque arm
21 Brake
22 Upper torque arm
23 Side brace nut
24 Downlock spring (2 places)
25 Lower side brace
26 Upper jury brace
27 Lower jury brace
28 Lower side brace pin
29 Forward trunnion pin
30 Lateral brace
31 Upper side brace
32 Upper side brace pin

(a) Main gear

(b) Nose gear

Fig. 12.1.2 Main gear and nose gear construction of the L-1011.
(Courtesy of Lockheed Aeronautical Systems Co.)

etc. The general principle of this gear is the coupling of two wheels in tandem and independently pivoted on trailing arms at each end of the double-acting shock absorber fitted horizontally and parallel to the axis of the aircraft. This is particularly suitable for high-wing aircraft in which the landing gear retracts into the fuselage main landing gear fairing, as shown in Fig. 12.1.3. This is an extremely simple construction which offers a number of possibilities for variations, for the trailing arm enables the aircraft to ride smoothly over any irregularities in the terrain. In addition, the tandem wheel layout gives the aircraft good taxiing qualities, and the coupling of the two wheels to a common shock absorber enables the shock forces encountered while taxiing over bumps and hollows to be balanced out in the gears rather than in the fuselage (see Fig. 12.1.4).

Among the particular advantages of the Jockey landing gear is the fact that loading of the aircraft is considerably simplified. By semi-retraction of the wheels, the fuselage can be either lowered parallel with the ground or tilted

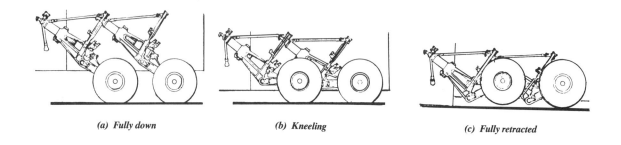

(a) Fully down   *(b) Kneeling*   *(c) Fully retracted*

Fig. 12.1.3 *Geometry of the Transall main gear.*
(Courtesy of Aerospatiale.)

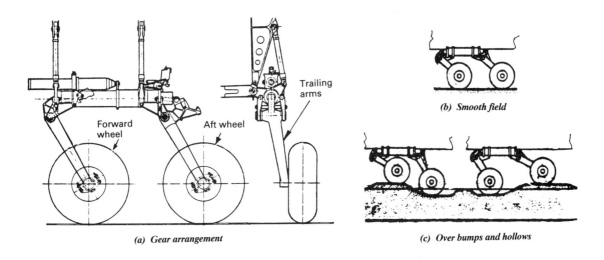

Fig. 12.1.4 Breguet 941 main gear (double-acting shock absorber).

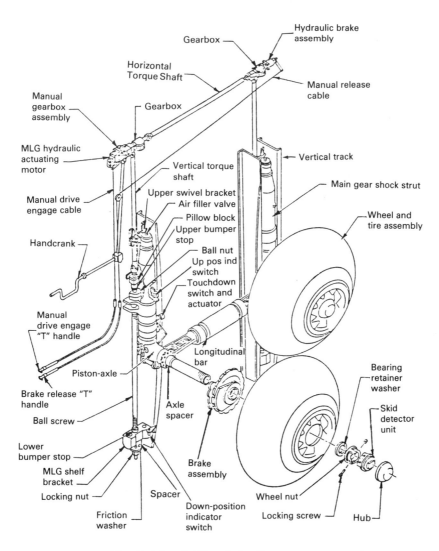

Fig. 12.1.5 C-130 main landing gear arrangement.
(Courtesy of Lockheed Aeronautical Systems Co.)

Airframe Structural Design 433

forward or backward according to requirements.

(3) Lockheed C-130

The Lockheed C-130 transport landing gear consists of four single wheel units; two in tandem on each side of the fuselage. Each wheel is on an axle offset outboard from separate oleo struts (see Fig. 12.1.5).

The struts are supported directly on the sides of the fuselage by vertical tracks; each track serving as a guide to retract the complete gear vertically upward into pods built on each side of the fuselage. Rotation of the axle and piston about a vertical axis is prevented by a longitudinal bar rigidly attached to each axle. The tandem struts are free to deflect independently since the longitudinal bars are pinned about their transverse axis, and incorporate a telescoping joint at the center. The gears are retracted along track guides by a hydraulic motor which rotates a vertical screw at each strut. The strut is raised or lowered by means of a ball bearing nut through which the screw turns. The oleo of the main landing gear is a conventional hydraulic-air system with a fixed orifice metering device supported by a plunger tube. The piston surrounds the orifice plate and, as it slides up or down, hydraulic fluid is forced through the orifice.

The landing loads are carried directly into the fuselage. A lower fitting at the fuselage floor level supports all of the vertical load and can resist side and drag loads. An upper strut support is free to slide on the vertical track, but can resist side and drag loads. Any moment about the gear vertical axis is resisted by the longitudinal torque bar in bending. Reactions to the bending of the torque bar causes side loads at the axle to bar attachment points. The forward and aft gears are identical.

(4) De Havilland Trident

The de Havilland Trident main gear units have four tires abreast (see Fig. 12.1.6); this permits a small turning circle without serious tire scrubbing and also pays large dividends with regard to stowage. The main gear legs are twisted through almost 90° during retraction; this effects the neatest stowage of the wheels within the fuselage. There are no external bulges underneath the fuselage to enclose the landing gear.

The nose gear is hinged about a position offset from the fuselage centerline in order to stow it transversely in the fuselage. Thus, the nose gear bay takes up the minimum length of fuselage, leaving ample room for a large electrical bay below the floor of the flight compartment.

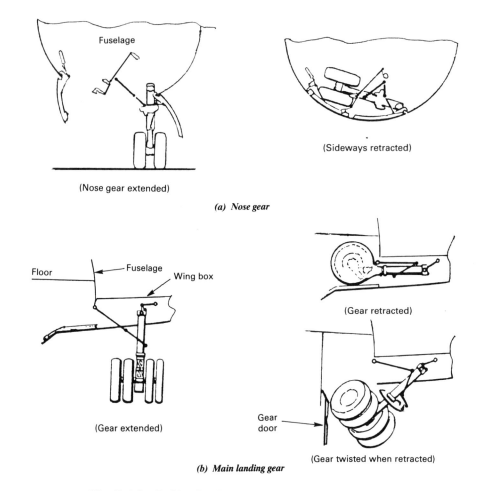

Fig. 12.1.6 De Havilland Trident main and nose landing gear.

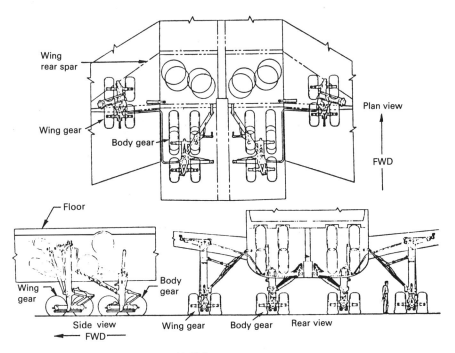

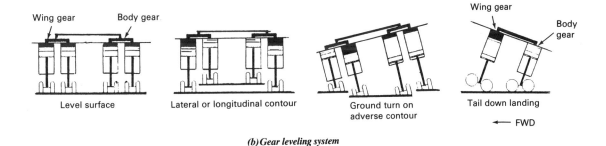

(a) Main gear arrangement

(b) Gear leveling system

Fig. 12.1.7 B747 main gear oleo-pneumatic leveling system.
(Courtesy of The Boeing Co.)

(5) Boeing B747
The gross weight of the B747 is well over twice that of the heaviest B707; the pavement bearing loads are not expected to be very much higher. This has been achieved by having four main bogies, and, although the tires are of the same size as those on the B707, the wheels are wider spaced and tire pressures are lower. The wing main gear on each side is attached to the wing rear spar and retracts inward and the bogie is twisted to lie almost transverse in the fuselage bay. The body main gear simply retracts forward and upward.

The aircraft is capable of landing on any two of the four bogies in the event of a malfunction with the others. Each leg can be lowered individually, but all four legs are retracted simultaneously. To spread the load as evenly as possible between the bogies, the oleo-pneumatic suspensions of the pair on each side are interconnected so that any uneven loading is balanced out. The action is explained in various situation diagrams as shown in Fig. 12.1.7.

(6) Lockheed C-5
The Air Force/Lockheed C-5 Galaxy is equipped with four main gears and one nose gear. The kneeling-type main gear for this giant airplane positions the cargo floor 73 inches above the ground for convenient loading, while the nose gear can be moved to incline the floor up 1.5 degrees or down 0.9 degree. The main landing gears consist of two six-wheel bogies as shown in Fig. 12.1.8(b) on each side of the fuselage with the wheels in a triangular pattern. During retraction the bogies rotate 90 degrees around the strut centerline, and the struts rotate inward and upward. Main gear pod doors open and close mechanically as the landing gear lowers or retracts. The nose landing gear as shown in Fig. 12.1.8(a), which retracts straight aft and up, kneels by retracting against a bumper.

Because of its exceptionally severe design

Airframe Structural Design 435

(a) Nose

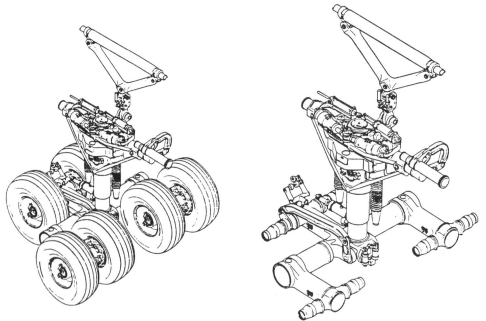

(b) Main gear

Fig. 12.1.8 *Main and nose landing gear — C-5.*
(Courtesy of Lockheed Aeronautical Systems Co.)

requirements, the landing gear of the huge C-5 military transport is probably the most advanced and comprehensive ever designed. Lockheed claims to have studied 700 different designs, deduced from a computerized parametric study of 2,600 landing variants, to determine the optimum combination which gives the highest floatation for the least overall weight to enable the aircraft to operate on dispersed bare soil or mat-covered strips, achieving the greatest versatility of military deployment. The resulting arrangement consists of 28 wheels, all the same size — four nose wheels and 24 main wheels on four six-wheel bogies.

Each of the four main gear units has three pairs of wheels arranged in a triangular pattern with the apex pointing forward. For crosswind landings, all these wheels can be swivelled up to 20 degrees either side of center. During taxiing the two rear main gear units are castored to prevent tire scrubbing and to enable the aircraft to make a complete turn-about on a 150 ft wide runway.

(7) C-141

The C-141 main landing gear is a simple design which meets the floatation requirement by using the 4 wheel-bogie type. The oleo strut has been so designed that it can provide a truck-bed-height cargo floor and can be extended several feet to provide adequate tail clearance during landing and take-off. In addition, the gear is mounted specially on a cantilevered main trunnion (or hub) attached to a main fuselage frame and simply rotated forward to retract into a relative small pod (or fairing). [See Fig. 12.1.9, also see Fig. 11.5.4 for fuselage gear hub and Fig.

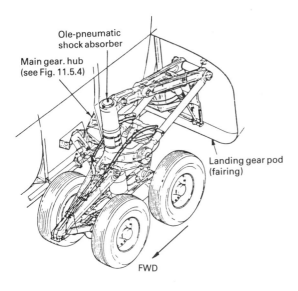

Fig. 12.1.9   C-141 main landing gear.
(Courtesy of Lockheed Aeronautical Systems Co.)

12.3.4(a) for gear retraction sequence.]

(8) Fighter Airplane Landing Gear

The landing gear design for a fighter airplane is a very big challenge because its stowage retraction has to be in a fuselage which has little room to spare. Therefore, the gear design, especially main landing gear, has to be unique for each fighter — very small or compact, rotated during retraction, etc. The retraction sequence of this unique main gear is specially designed to fit neatly into a limited space of the fuselage while at the same time leaving room for the carriage of pylon-mounted stores beneath the fuselage to meet mission requirements. Fig. 12.1.10 illustrates a few unique cases.

Design Requirements

A general outline of the many design considerations which face the landing gear engineer is presented here. A great many engineering disciplines can become rather intimately involved with landing gear development and design; therefore, much of the material presented below will be of a general nature. This information will serve as a useful guide to those engaged in landing gear design or provide improved understanding to engineers in related activities. Fig. 12.1.11 shows a list of design requirements for designing a landing gear system.

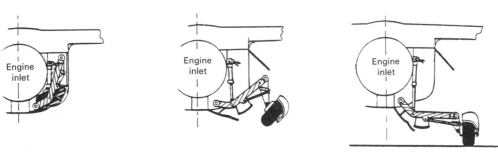

(a) MiG-23

Fig. 12.1.10   Fighter main landing gears.

Airframe Structural Design 437

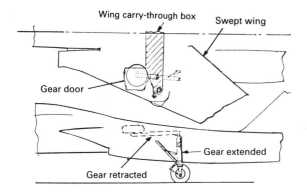

(b) F-14
(Courtesy of Grumman Corp.)

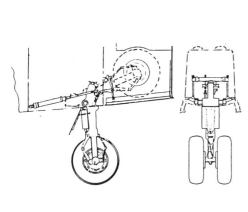

(AV-8B: an improved Hawker Siddeley Harrier)

(c) Hawker Siddeley Harrier — Harrier has its engine and hot jet outlets in the center fuselage which makes the stowage of a conventional landing gear impractical in this region. Consequently a single twin wheel unit is located in the aft fuselage and clear of the hot jet effluxes.

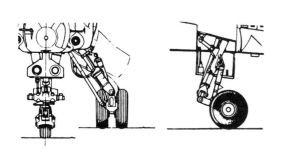

(d) Jaguar — Main landing gears are specially designed for the fighter to carry stores under the fuselage and to have the ability to operate from roughly prepared landing fields. The levered suspension main gears comprise two wheels mounted in "diabolo" fashion.
(Courtesy of British Aerospace Plc.)

Fig. 12.1.10   (Continued)

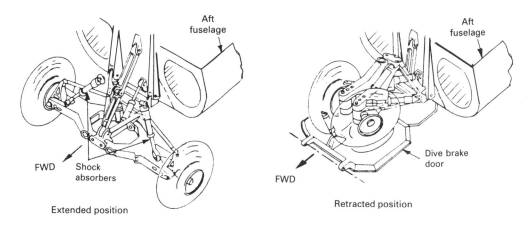

(e) F-111
(Courtesy of General Dynamics Corp.)

Fig. 12.1.10 (Continued)

(1) Preliminary Design Phase

The following functional requirements have a bearing on landing gear layout in the preliminary design phase.
- During the phases of take-off rotation and lift-off and landing flare-out and touch-down, only the wheels should be in contact with the ground. There should be adequate clearance between the runway and all other parts of the aircraft, such as the rear fuselage, the wing tips and the tips of propellers or engine pods.
- The inflation pressure of the tires and the configuration of the landing gear should be chosen in accordance with the bearing capacity of the airfields from which the aircraft is designed to operate.
- The landing gear should be able to absorb the normal landing impact loads and possess good damping characteristics. When taxiing over rough ground no excessive shocks should be transmitted by the landing gear.
- Braking should be efficient, the maximum braking force allowed by the condition of the runway being the limiting factor. During cross-wind landings and high-speed taxiing there should be no tendency to instabilities such as canting of the aircraft or ground looping.
- Suitable structural elements should be provided in the aircraft to serve as attachment points for the landing gear, and there should be sufficient internal space for retraction.

(2) General Requirements

It is understandable that in many respects the various licensing agencies have similar, or identical requirements. The following itemizes the requirements that would provide a gear acceptable airworthiness.
- Design the mechanism, doors, and support structure to permit lowering the gear at $1.6V_{s1}$, with flaps retracted and at maximum landing weight (V_{s1} = the calibrated stalling speed).
- Unless there are other in-flight deceleration devices, design the gear and doors to withstand loads with gear down at $0.67V_c$ (V_c = design cruise speed).
- The turnover angle should not exceed 63° and 54° for land-based and carrier-based aircraft respectively.
- A tail bumper or skid should be provided.
- The tail bumper should not touch the ground when the main wheel is at the static position.
- The value of θ shall not be less than α, and θ

Airframe Structural Design 439

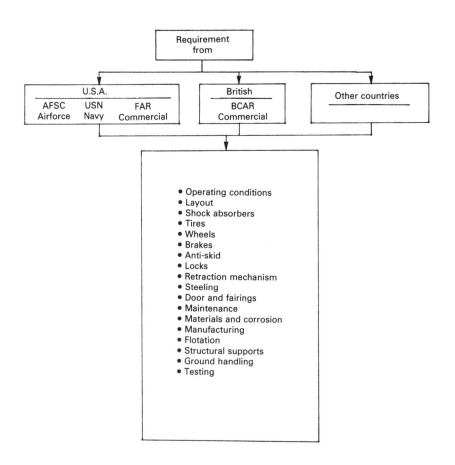

Fig. 12.1.11 *Landing gear design requirement.*

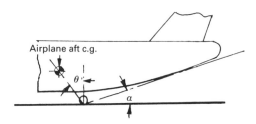

Fig. 12.1.12 *Main landing location vs. airplane aft c.g.*

shall not be less than 15° (see Fig. 12.1.12).
- Shock strut normal oil level above the orifice should be at least 125% of piston diameter, or 5 inches, whichever is less; otherwise test to demonstrate satisfactory shock absorption with performance impaired by foaming and/or leaking oil.
- The distance between the outer ends of the shock-strut bearings should be at least 2.75 times the piston diameter.
- Shock-absorber units should be interchangeable left and right.

- Drop tests should be conducted to show that the shock absorber can absorb energy due to landing at 1.2 times the specified sink speed.
- Nosewheel tire pressure should be based on allowable dynamic loads. These loads are 1.40 times static allowable for Type III tires, and 1.35 times static allowable for Type VII. (See Ref 12.2 for tire definitions.)
- Main-gear tire size should allow for 25% growth in airplane gross weight.
- Main-gear tire load rating shall not be exceeded under equal loading at maximum gross weight and critical airplane c.g. position.
- On a multiple-wheel gear, ensure that when any one tire or wheel fails, the remaining tires and wheels can withstand the overloads imposed at maximum taxi gross weight.
- Wheel bead seat temperatures from braking should not exceed 350° during normal and overload energy stops.
- Install fuse plugs to release tire pressure at, or less than, 400°F tire bead seat temperature.
- Use forged aluminum-alloy wheels.
- Normal brake energy is based on the greater of 1.15 times the recommended brake application speed, 1.0 times normal touch-down speed, or 1.1 times stalling speed in landing configuration.
- Install a parking brake capable of preventing roll on a 1:10 gradient, or on a level runway with maximum take-off power applied on one engine.
- Anti-skid systems shall be as reliable as the rest of the braking system, and cockpit warning lights shall indicate system failure.
- Uplocks shall be independent of door locks.
- Uplocks shall be releasable in an emergency by positive mechanical means.
- Downlocks should not be stressed by ground loads.
- Electrically operated locks must not be unlocked by electrical failures.
- Ground locks shall be provided, and their installation shall be foolproof.
- Retraction systems shall not use cables or pulleys, except in an emergency.
- An emergency extension system shall be provided, independent of the primary system. The latter is defined as all parts stressed by ground loads.
- Do not use an emergency system requiring hand-pumping or cranking by the pilot.
- Minimize the use of sequencing mechanisms.
- In retracting mechanisms, do not use telescoping rods, slotted links, or cables.
- The maximum retraction time shall be 10 seconds (Navy).
- The maximum extension time shall be 15 seconds (Navy).
- Eliminate the possibility of mud or other material being trapped in cavities.
- Route all service lines, and locate all mechanisms and equipment, such that they will not be damaged by dirt, mud, water, or other material thrown by rotating wheels.
- MIL-L-87139 suggests a new requirement that the loss of any landing gear fairing door shall not result in the loss of the actuation power system, i.e. wires and hydraulic lines should not be routed on the doors.
- Close the doors after gear extension (if required), and/or provide covers or guards on the gear.
- Wheel well equipment that is essential to safe operation of the aircraft must be protected from the damaging effects of burst tires or loose tread.
- Ensure that fuel tanks, lines carrying flammable fluids, and other hazard-creating items cannot be critically damaged by failure of landing gear parts.
- Stop the wheels from spinning in the retracted position after take-off.
- Provide enough power to steer the aircraft without the necessity of forward motion.
- Provide an emergency system capable of steering the aircraft without interruption if the normal steering system fails.
- Interchangeable main landing gears for ease of stocking spares.

Primary Military Specifications (U.S.A.)

MIL-A-18717	Arresting Hooks
MIL-A-83136	Arresting Hook Installations (USAF)
MIL-A-8629	Drop Tests (see also MIL-T-6053)
MIL-A-8860	Airplane Strength — General Specifications
MIL-A-8862	Ground Handling Loads
MIL-A-8865	Airplane Strength & Miscellaneous Loads
MIL-A-8866	Strength & Rigidity Reliability Requirements — Fatigue
MIL-A-8867	Ground Tests
MIL-A-8868	Airplane Strength Data & Reports
MIL-B-8075	Anti-Skid
MIL-B-8584	Brakes — Control Systems
MIL-C-5041	Tire Casings
MIL-D-9056	Drag Chute
MIL-H-5440	Hydraulic Components
MIL-H-5606	Hydraulic Fluid
MIL-H-8775	Hydraulic System Components
MIL-L-8552	Shock Absorbers — AFSC and USN
MIL-L-87139	Landing Gear Systems
MIL-P-5514	Packing — Shock Struts; also O-Rings and Glands
MIL-P-5516	Packings — Shock Struts
MIL-P-5518	Pneumatic Components
MIL-P-8585	Primer — Wheel Wells
MIL-S-8812	Steering Systems
MIL-T-5041	Tires
MIL-T-6053	Drop Tests (also MIL-A-8629)
MIL-T-83136	Tie Down Requirements
MIL-W-5013	Brakes and Wheels
MIL-STD-203	Controls and Displays in Flight Station
MIL-STD-805	Tow Fittings
MIL-STD-809	Jacking Fittings
MIL-STD-878	Tires and Rims Dimensioning and Clearances
MIL-STD-568	Corrosion Prevention and Control

Brief Summary of Landing Gear Loads
(1) General
- Maximum sink speed landing at take-off gross weight
- Maximum sink speed landing at landing gross weight
- Level landing
- Spin up
- Spring back
- Lateral drift landing
- Rebound landing
- Towing
- Jacking
- Turning
- Unsymmetrical loads on multi-wheel gears
- Deflated tires
- Braked roll
- Taxi
- Fatigue
- Depart cleanly (breakaway condition) if striking obstacle
- Brake chatter
- Shimmy
- Brake application during retraction
- Extension and retraction actuator loads

(2) Main gear only conditions
- Pivoting
- Manifolded oleo considerations for four post gears

(3) Nose gear only conditions
- Unsymmetrical braking
- Nose gear yaw
- Nose gear slapdown condition
- Nose gear steering

12.2 Development and Arrangements

Design

In order to understand the varied design considerations that face the landing gear engineer, a brief discussion pertaining to gear design is provided below:

(1) Ground Handling
- Towing provisions must be given, on the nose gear in most cases, that permit towing and pushing the airplane at full gross weight. Allowances must be made either for disconnecting the steering system as shown in Fig. 12.2.1, depressurizing the steering system, or designing the steering system to withstand being overpowered repeatedly by the tow bar. Some airplanes have tow fittings attached to the nose gear by fuse bolts designed to fail before damaging the gear or steering system. In many cases the customer prefers a specific tow fitting design so that one tow bar can be used for several different airplanes. An often neglected designed feature on landing gears is provision for jacking to permit wheel and brake changes. The jacking balls must be so located as to permit rapid tire changes. These should be high enough to provide space for a jack with all tires flat and laterally deflected, and should be integrated into the axle structure in such a way that the entire jacking pad

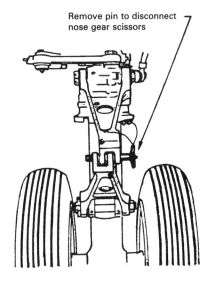

(a) Connect

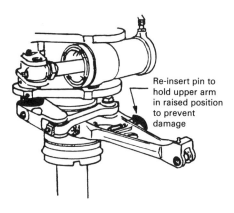

(b) Disconnect

Fig. 12.2.1 Steering disconnect for towing.

area can be severely abused without premature fatigue failures.
- It is general practice to install the ground safety locks on all gears whenever the airplane is being towed or when it is left unattended for any period of time. The ground safety locks prevent the gear from being inadvertently retracted on the ground, and are commonly used during functional retraction tests to permit the retraction of only one landing gear at a time. The locks must therefore be so designed to safely withstand full unlocking forces.
- If the airplane is backed out by means of tow tug and tow bar (an increasingly popular method), it is obvious that the disconnecting of the tow bar and restoration of the steering

system must be as simple and foolproof as possible since the ground crews are under considerable pressure; the conditions are far from ideal, and the more skillful mechanics are not always assigned this task.

(2) Take-off
- The landing gear and support structure must be dynamically stable at all ground speeds and loading conditions; therefore, take-off conditions can well be critical from a shimmy standpoint. Rubber pedal steering is best limited to small angles, and must be deactivated when the nose gear is fully extended to prevent landing gear turned and to avoid having the gear rotate in the wheel well.
- In most operational refused take-off (RTO), the abort speed is relatively low, but is initiated at or near the maximum ground speed and at high gross weight. The RTO is thus one of the prime factors in the design of the braking system.
- The lift-off presents no real problems except that as the typical shock strut extends rapidly a very annoying thump is discernible in the cabin. It has become quite common to design positive snubbing in the shock struts to raise the gear. The gear retraction should be accomplished as briskly as possible in order to "clean-up" the airplane for optimum climb speed. Six to ten seconds are practical limits for retraction times (see Fig. 12.2.2).
- A requirement is to provide a means to hold the gear up reliably throughout the flight. Since it is common practice to depressurize the gear retraction system in flight, the up-lock must hold the gear without hydraulic pressure (Fig. 12.2.3). On passenger airplanes, considerable effort must be expended to reduce or eliminate all alarming noises associated with gear retraction. It is generally desirable to stop the rotation of the wheels upon retraction, since an unbraked wheel can last for some time and it may set up an alarming thumping under the floor. Wheels equipped with brakes are usually to stop rotation automatically by light application of brake pressure or by some manner of friction paddles rubbing on the tires when the gear retracts.

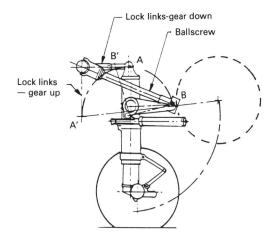

Fig. 12.2.3   C-5 nose gear lock-links.

Two general methods of door actuation are used, the gear driven method and the independently powered and sequenced method. Needless to say, the doors must be held flush and sealed tight at high speed, must permit access to the wheel well on the ground, and must not interfere with the safe extension of the gear under any circumstances.

The available power to retract a gear must take into account air speed. There does not appear to be any operational value in providing retraction capability at greater speeds. However, there may be a need for higher limits on some types of unconventional aircraft (see Fig. 12.2.4 for typical airspeed limits).

(3) Landing
- Extension: Although the gear extension is simply a reversal of the retraction cycle, it differs in one major aspect — safety. There must be no single malfunction that can prevent the capability for extending the gears; therefore, a backup system (usually manual) must be provided in the event the normal power system becomes inoperative (see Fig. 12.2.5). The manual system must ensure that the gears get down and locked. Effects of friction, aero-

Aircraft	Retract (Sec.)	Extend (Sec.)
A-10	6–9	6–9
B-52	8–10	10–12
B66	10	8
C-5	20	20
C123	9	6
C130	19	19
C135	10	10
F-5	6	6
F-100	6–8	6–8
F-105	4–8	5–9
F-111	18	26
T-37	10	8
T-38	6	6

(Note: The above are approximate values)

Fig. 12.2.2   Typical landing gear operating times (ref. 12.32).

Airframe Structural Design 443

Airplane	Airspeed Limits (knots)			
	Gear Down	Retract	Extend	Emergency Extend
A-7	224	220	220	180
A-10	220	220	220	220
B-52	305	220	305	305
B-57	200	200	200	200
B-66	250	250	250	250
F-4	250	250	250	250
F-5A/B	240	240	240	240
F-5E	260	260	260	260
F-15	300	300	300	250
F-16	300	300	300	300
F-100	230	230	230	230
F-105	275	240	275	275
F-106	280	280	280	250
F-111	295	295	295	295
T-37	150	150	150	150
T-38	240	240	240	240
T-39	180	180	180	180

Fig. 12.2.4 Typical airspeed limits (ref. 12.32).

dynamic forces, and hydraulic snubbing must be accounted for. The designers must try to balance the requirement for positive freefall, gear-down locking against the usual need for smooth and quiet lowering of the gear under powered operation.

On most high performance airplanes, the gears have a very useful function as speed brakes, should the need arise for high rate of descent in an emergency. Thus a real need exists to lower the gears at as high speed as possible.

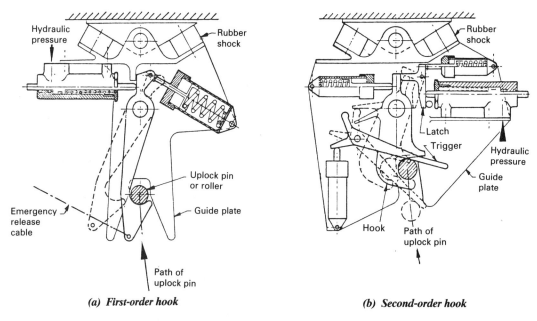

(a) First-order hook

(b) Second-order hook

Note — One of the faults of a first-order hook is that considerable force is required to push the hook from under the roller, and a way to overcome this defect is to use a second-order lock. The hook is held both open and closed by secondary latches, and if the guide plates are used, it needs little or no rigging. This design is simple, reliable, and efficient.

Fig. 12.2.5 First order uplock vs. second order uplock (ref. 12.32).

- Touch-down: Since this maneuver is subject to pilot skill, visibility, and turbulence, a good design is required to allow for a wide range of landing conditions. The major consideration is sink speed. Normal landings are considered those that fall in the category of 1-3 ft/sec (fps) sink speed. Landings above 4 fps sink speed are considered hard landings, and occur with diminishing frequency up to 10 fps. The design case for the landing gear shock absorbers is 10 fps for commercial airplanes, but might be greatly increased for carrier-based aircraft or other special types (refer to Chapter 3).

  In addition to absorbing the sink rate energy of the airplane, the strut should be designed so that the recoil or extension of the strut does not push the airplane back into the air, thus causing abnormal bouncing.

- Braking: The landing stops are performed with predictable regularity; therefore, smoothness, heat dissipation, brake life, and reliability must be accounted for.

  The landing stop is also performed on a variety of runway conditions from dry concrete, very wet, to very icy. To permit maximum braking torque without skidding, most modern aircraft are fitted with an automatic braking, or anti-skid system. These systems can automatically account for variations in runway friction coefficients and wheel loads, and instantly adjust the brake pressure to obtain optimum retardation without allowing appreciable skidding.

(4) Parking

The brakes are usually used to keep the airplane parked at through stops, and hopefully dissipate heat to sufficiently permit taking off safely and retracting the gear for the next flight. On short haul jets with about 10 minutes turnaround time, the accumulation of brake heat from one flight to the next can pose a serious design problem.

(5) Flight Cycles or Missions

It should be noted that although the number of flight cycles per day for the so-called short range jet is much greater than for the long range, the daily utilization is greater for the long range airplane. Such factors as brake life, brake heat, fatigue life, overhaul cycles, wheel life, and tire economics are all effected by flight frequencies.

Flotation

Before getting into a new gear design, considerable thought must be given to flotation, or pavement loadings. The stresses induced into runways is a function of several variables over which the designer has considerable control: strut load, tire spacing, number of tires per strut, tire size (i.e. pressure), etc. The optimum arrangements for orthodox aircraft at existing gross weights has been fairly well established. However, should the aircraft have unusual functions such as operation from unimproved surfaces, or if the aircraft is greatly larger than existing types, special solutions must be sought.

A convenient approach for establishing tire sizes for conventional airplanes is to compare the new airplane with a number of existing types, with particular emphasis on those types being currently operated on the same type of runways and route systems as those planned for the new design. Fig. 12.2.6 shows the runway thickness required for some typical aircraft.

Airplane Type	Gross Weight (pound)	Concrete Thickness at Static Load for Unlimited Operation (inch)
DC-3	31 000	5.7
DC-4	107 000	9.4
DC-6	97 000	8.7
Constellation	110 000	10.8
C-124	216 000	11.6
C-130	155 000	9.8
707	297 000	11.3
C-141	316 000	12.3
C-5	769 000	10.0
DC-8	335 000	12.4
DC-9	115 000	10.2
DC-10	410 000	11.9
747	775 000	12.8
L-1011	410 000	11.9
Concrete Flex. Stress = 400 psi, K-300 psi, Poiss., Ratio = 0.15		

The concrete thickness requirements illustrate the effects of wheel arrangements, and a classic example is the C-5 versus the Boeing 747. Both airplanes are about the same weight, and yet the unusual C-5 wheel arrangement results in a 10-inch thickness requirement versus 12.8 for the B747.

Fig. 12.2.6 Runway thinkness vs. aircraft type (ref. 12.32).

Since the gross weight is not under the control of the gear designer, it can be seen that the number of struts, tire spacing, and tire pressure can be selected to prevent the trend of increased gross weight causing a linear increase in pavement strength. Of course, the designer is under pressure to keep the number of landing gears, wheels, brakes, etc. to a minimum in the interest of cost, reliability, minimum weight, and service costs. Ref 12.2 is one of the standard references for selecting the existing tire sizes. For new or unusual tire sizes, the major tiremakers will provide recommendations and preliminary data to aid tire selection.

When a new airplane extends the gross weight beyond the scope of existing aircraft weights or if unusual flotation problems must be dealt with, a more thorough study of runway strengths must be made. Since there is little promise that major airports can be counted on to improve their taxiways and runways in the near future, the designer of increasingly larger airplanes must hold their flotation characteristics to as good as or better than current large airplanes. Fig. 12.2.7 shows some typical wheel and gear arrangements. The gear designer should have several suitable arrangements in mind during the early stages of design so that he can adapt the gear arrangement to structural supports and wheel well size and shape.

Soft and rough airfields impose requirements which conventional landing gears do not meet. Soft fields make it necessary to provide increased ground flotation, and rough fields require much more shock absorption between the airframe and the field surface to keep ground-induced airframe loads at reasonable levels. Ground flotation may be improved by employing more tires, larger tires, increased spacing between tires, reduced tire inflation pressure, or by special tires having wide cross-sections in combination with reduced rim diameters. Rough field performance is improved by using larger tires with smaller rim diameters to provide increased working section heights for step bumps or short-wavelength bumps of large amplitude. Long-stroke, double-acting shock absorbers (see Fig. 12.1.3 and Fig. 12.1.4) are essential to alleviate the high tire loads induced by rough field bump to the level induced by normal operations on paved, smooth airfields.

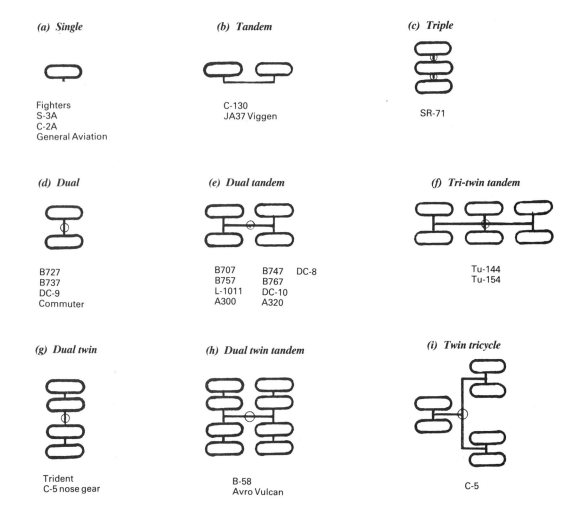

Fig. 12.2.7 Wheel arrangements.

Airplane Ground Attitude

The gear designer may not always select the airplane attitude on the ground, which is often critical for take-off performance. The Aerodynamics Group often recommends a desirable attitude other than level to enhance performance. Having selected a ground attitude, the gear designer should now step in and assert the establishment of the ground level with respect to the airplane. It is only too obvious that the airplane should sit as low as possible. This problem can only be worked in conjunction with the determination of the airplane configuration and gear locations in elevation and plan.

Propeller tips and jet engine nacelles must not be too close to the ground, because of the danger of sucking up debris or fouling obstacles (important considerations in bush and unpaved field operations).

Gears in Elevation and Plan

Since the gear locations, ground clearances, stability, and retraction schemes are all related, the gear designer is usually frustrated in determining these items until the airplane airframe design has stabilized. The following general rules have been well proved over the years and should not be deviated from without thorough study.

(1) Elevation

The main gears should be located between 50-55% of the wing mean aerodynamic chord. (a) A main gear location too far aft will result in an airplane slow to rotate on take-off, thus penalizing field length. Also, if excessive weight is borne by the nose gear, the braked wheels of the main gear are less loaded, thus detracting from the breaking force available. (b) If the main gears are too far forward, the danger obviously exists of falling back on the tail during the loading process or during reverse braking when the aircraft is being pushed backward. The nose gear location is not critical, but for minimum weight, optimum braking, and stability, it should be as far forward as possible.

The ground level is usually established next by the intersection of the tail clearance line with the main gear location. A minimum angle for the tail clearance line is about 12° with the gear in the taxi position (see Fig. 12.2.8). Dynamic studies may suggest the need for a tail bumper, but often the final determination cannot be made until the airplane has been flight tested.

(2) Plan

The arrangement of the main gears in the plan should be as close to the centerline as possible, subject to the limitations of the overturning angle shown in Fig. 12.2.9.

Support Structure

As previously stated, the landing gear loads and reactions are the largest local loads on the airplane. For this reason, transmitting such large local loads into a semi-monocoque structure such as a wing box or fuselage shell requires extensive local reinforcement. The gear designer can improve the airplane by working closely with the fuselage and wing designers from the start in an attempt to keep load paths simple and determinate-structure. Since the landing gear

Airplane	α	β
B727-200	7	10
B720B	10.4	12
B707-320C	10.2	12.2
B737-200	12	15.3
C-5	10	11.8
L-100-10	—	12.4
Electra	10	13
L-1011	13	13.6
F-104	13	15
Jetstar	14.3	17.5
DHC — Twin Otter	—	12
Aero Commander 685	11	13
Piper Super Cub	12	13
Mercure	12	14
Concorde	—	14.9
Piper Turbo Navajo	—	16
Beech B99	15	16

(Note: The above are approximate values)

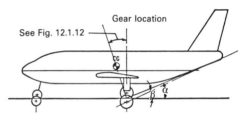

α — Angle with landing gear at static position
β — Angle with landing gear in extended position

Fig. 12.2.8 Airplane tail-down angles (ref. 12.32).

loads are large, there can be severe weight penalties in the use of indeterminate structural load paths. An indeterminate structure is one in which a given load may be reacted by more than one load path; the distribution being subject to the relative total stiffness of these load paths. The term is somewhat misleading since the manner in which the members share the load can be determined, but only when the design is finalized. Even then, there usually remains sufficient doubt regarding the load sharing that the designer and stress analyst make "overlapping" assumptions to guard against overloading any of the members. This can result in each load path being designed to carry uncertainty of the load design.

Very often the gear loads can be spread out so as to keep the local reinforcement to a minimum. Fig. 12.2.10 shows the basic gear support scheme for the types A and B. Notice that the type A gear loads are reacted by the wing in two places and by the fuselage frames. The type B gear is cantilevered from the wing box in one place, requiring massive wing reinforcement locally.

Support structure in the wing is designed to higher loads than the gear itself to ensure that in the event of impact with some obstacle during landing or taxiing the landing gear will break cleanly with the wing and not precipitate a fuel tank rupture.

Low Wing Transport	Turnover Angle (Degree)	Fighters	Turnover Angle (Degree)
Lockheed Electra	34	F-4	39
B747	39	F-104	36
A300	41	**Others**	
L-1011	43		
B737	46	Aero Commander	38
Concorde	47	Piper Turbo Navajo	43
DC-9	48	Beech B99	44
B707	49	Piper Comanche	45
B727	49	Beech U-21A	47
High Wing Transports		Bonanza	51
		Piper Super Cub	59
C-141	53		
Breguet 941	61		
C-130	61		

(Note: The above are approximate values)

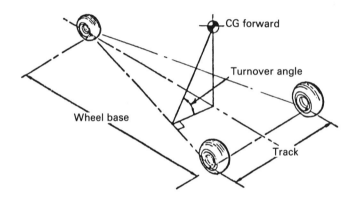

Fig. 12.2.9   Airplane turnover angles (ref. 12.32).

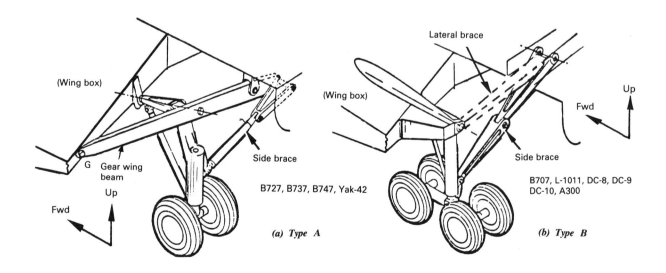

Fig. 12.2.10   Main gear support configuration (transport low wing design).

448   Airframe Structural Design

12.3 Stowage and Retraction

Almost all successful gears are simply hinged to retract. It is preferable that the hinge axis be parallel to the basic airplane axis in the interest of keeping the kinematics simple. However, the use of canted hinges can often solve the problem by getting the gear stowed in an optimum-sized wheel well that would otherwise have to be made wastefully oversized. Usually, canted axis hinges result in the left and right hand gears being opposite (rather than identical), thus doubling much tooling, spares inventories, etc.

Gears that retract fore and aft should, if at all possible, retract forward. An aft retracting gear will not free-fall down because of air force stream and requires extensive manual effort to extend in an emergency. While the emergency case occurs in flight, thus the time and energy spent in hand-cranking a gear forward into the air stream is seldom appreciated.

There are practically no limits to the ingenuity the designer can use in the design of the folding struts, down-locks, up-locks, and actuator systems. Although the retraction and lock system may appear to be a small part of the landing gear design package, often this phase of the design will consume over half the design hours for the entire gear. It is, therefore, wise to establish a workable folding and locking system early in the design to avoid being forced to extremely ingenious and complex systems.

After the designer has a general idea of how the gear must retract, it is usually fruitful to study other airplanes in an effort to learn from other designers' solutions to similar problems. (Information in Ref. 12.4 and Ref. 12.32 is strongly recommended.)

The following precautions might be helpful.
- Avoid tracks and rollers (heavy and poor life).
- Avoid hook latches if possible (noisy and cannot be free of play).
- Over-center toggles are preferable but over-center distance should be kept to a minimum to avoid noise and reduce power required to unlock.
- Avoid having one actuator perform more than one function. The apparent economy is often offset by the difficulty in snubbing all the motions to keep loads and noise minimum.
- Provide generous space for oversized bearings.
- Keep the mechanism simple.
- Allow adequate gear clearance in the wheel well to account for gear motion in flight.

Fig. 12.3.1 illustrates an assortment of common kinematic concepts of gear retraction and extension. Case (a) is used in many airplanes due to its simplicity, and there are many variants of this type.

The most usual variation involves the use of an extra bracing link extending from the top of the shock strut to the drag or side brace elbow. This provides extra support for the brace, and thereby minimizes structure weight. Fig. 12.3.2 shows how this extra brace works.

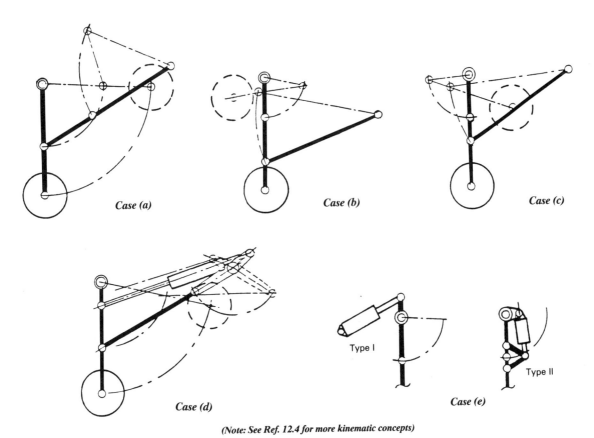

(Note: See Ref. 12.4 for more kinematic concepts)

Fig. 12.3.1 *Common types of kinematic concepts.*

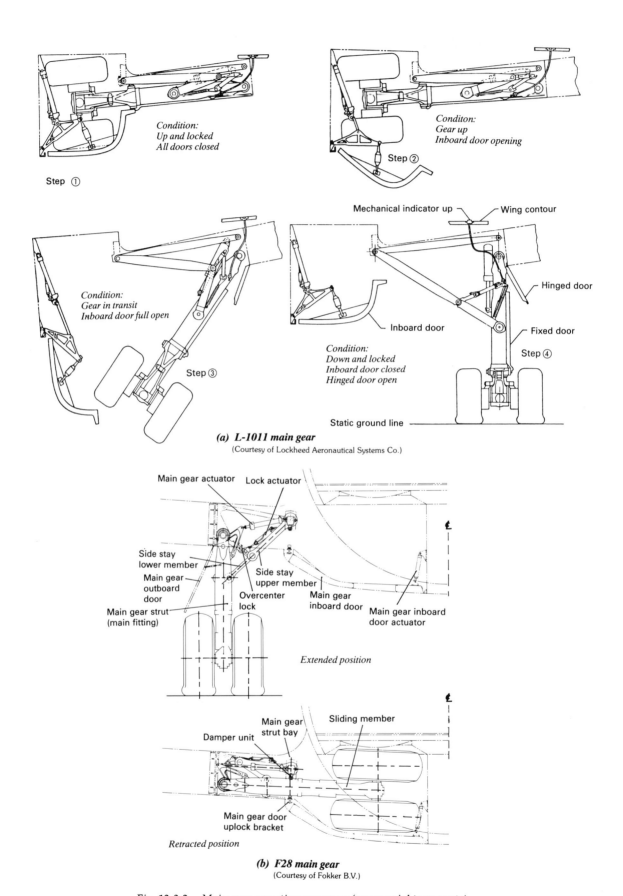

Fig. 12.3.2 Main gear operation sequences (commercial transports).

450 Airframe Structural Design

Cases (b) and (c) are similar, and can be used whenever it is required to retract the wheel into a cavity almost vertically above the down position (see Fig. 12.3.3).

Case (d) shows how the retraction actuator can be incorporated into the kinematics such that the loads are balanced out within the gear structure, as opposed to case (a) where the actuator must be mounted on the airframe.

Case (e) shows two methods of rotating the top of a case (b), and there are many variants of this. An example of such variations is shown in Fig. 12.3.4.

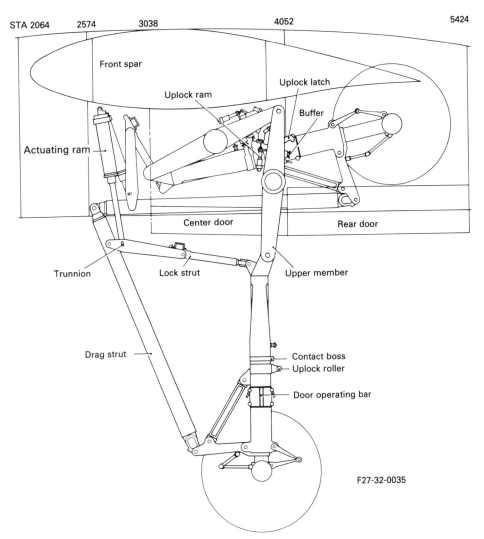

Fig. 12.3.3 F.27 main landing gear retraction.
(Courtesy of Fokker B.V.)

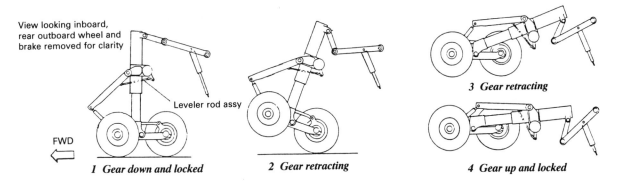

Fig. 12.3.4 Rotating the top concept cases — C-141.

Airframe Structural Design 451

In many cases, the wheels or bogies must be rotated to fit inside the available space, and as with the linkages, there are many ways to do this. Some degree (as high as 90°) of wheel rotation can be accomplished by appropriate choice of a skewed axis and examples of this are found on the F-14, F-16, A-7, S-3A, Trident etc.

A folding bogie design may be needed so that it occupies minimum space when retracted in the fuselage. Fig. 12.3.5 shows a complex but efficient arrangement to accomplish the design requirement.

Ramps are sometimes used to rotate the bogies as shown in Fig. 12.3.6. In case (a), as the gear retracts, the forward tire encounters the ramp and pushes the bogie over into the required position. But tire sizes vary considerably and this would create a variation in the gear-up position. Also, with a large gear being retracted quickly, tire-bounce would be severe, and for these reasons a roller is used as shown in case (b). Case (b) is to eliminate bounce and to accurately position the retracted gear; this has been used on the C-5 main gear systems.

The simplest method of retraction of a gear is the best of all. An example is shown in Fig. 12.3.7, which has only two basic parts — the shock absorber and an actuator (as a side brace) with an internal downlock.

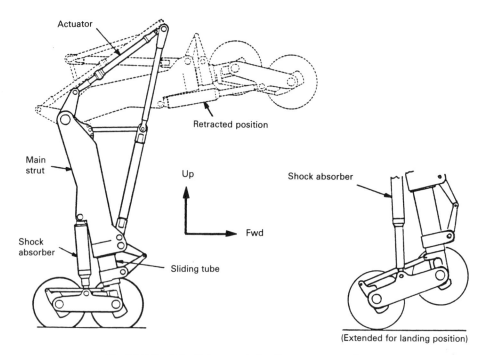

Fig. 12.3.5 *BAe Vulcan main landing gear retraction.*
(Courtesy of British Aerospace Plc.)

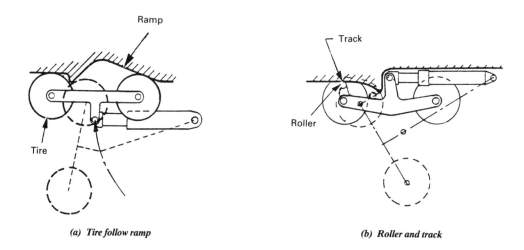

Fig. 12.3.6 *Ramps used for bogie rotation.*

452 Airframe Structural Design

Kinematic Guidelines

- Use computer graphics such as CADAM to layout the kinematics as early as possible in the design stage.
- Ensure that satisfactory moment arms are provided throughout the travel.
- Use the simplest possible kinematics.
- Actuator "dead length" must be approximated in preliminary kinematic layout [see Fig. 12.3.8(a)]:
  — No internal lock, dead length = 6-7 inches
  — One internal lock, dead length = 8-11 inches
  — Two internal locks, dead length = 12-15 inches

  The lower and higher values generally apply to smaller and larger diameter actuators, respectively. Also, the above values include an estimated one inch for the actuator end fitting. This can be deducted if a trunnion mount as shown in Fig. 12.3.8(b) is used, but on a hydraulic or pneumatic actuator, this type of mount is relatively expensive. Offset mounts, as shown in Fig. 12.3.8(c), should be avoided, since they cause undesirable stresses and deflection.
- Whenever possible, the landing gear doors should be moved by the gear actuator such that the doors and gear move together.
- Torque links (as shown in Fig. 12.3.9) should be designed such that their included angle is not more than 135° when the gear is extended.

Gear Lock Design Guidelines

- Keep it simple; a complex lock increases manufacturing tolerances and assembly and installation errors, resulting in poor reliability.
- Minimize rigging because it can be misrigged.
- Avoid having the lock mechanism, other than primary hook or plunger, subjected to ground loads.
- Make a careful check of clearances and tolerance build-ups to ensure that no more than two faces abutt against each other simultaneously.
- Structural and functional deformation must be recognized and appropriate allowances made.
- Up-locks must include a straightforward emergency release device to ensure that the lock can be released if the primary release system fails.
- Down-lock may be categorized as:

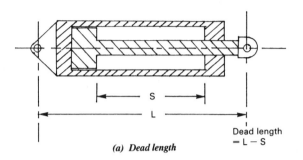

(a) Dead length

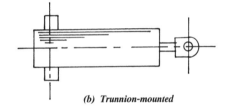

(b) Trunnion-mounted

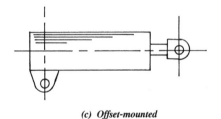

(c) Offset-mounted

Fig. 12.3.8 Actuator.

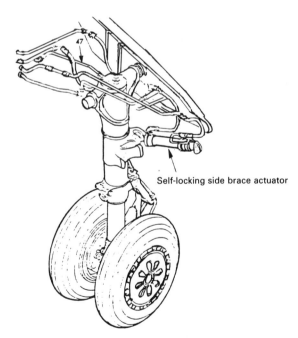

Fig. 12.3.7 *Main landing gear construction — Jetstar.*
(Courtesy of Lockheed Aeronautical Systems Co.)

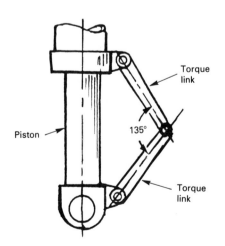

Fig. 12.3.9 Torque link included angle.

Airframe Structural Design 453

— Internal lock in telescopic brace or actuator
— Spring-loaded plunger engaging detent in top of shock strut
— Spring-loaded catch engaging fixed gear structure

Always remember that, of all the landing gear parts, it is most important that the locks work properly. If the up-lock jams, for instance, preventing the gear from lowering, the aircraft may be destroyed.

12.4 Selection of Shock Asorbers

The airplane during landing comprises the static and dynamic loads and dividing the dynamic loads by the static loads to obtain the landing gear load factor. The load factor's value ranges from 0.75 to 1.5 for large aircraft, to 3.0 for small utility aircraft, and to 5.0 for some fighters and military trainers (see Fig. 12.4.1). Its magnitude is usually determined by the airframe structure design requirements. Therefore, the shock absorber must be designed such that, upon landing, the load factor is not exceeded; otherwise, the structural integrity of the wing and/or fuselage will be jeopardized.

There are essentially two types of shock absorbers: those using a solid spring such as steel or rubber, and those using a fluid such as air, oil, or air/oil. Fig. 12.4.2 compares the various types, and illustrates the superiority of the oleo-pneumatic unit from the efficiency and weight standpoints. Other types are used, however, where cost, reliability, and maintainability are predominant factors.

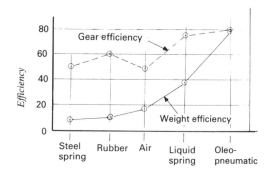

Fig. 12.4.2 Shcok absorber efficiency (ref. 12.32).

Light planes often use simple spring or rubber type shock absorbers because of the economics. For more efficient landing gears, this is relatively insignificant. However, as aircraft size and weight increase, steel and rubber type shock absorbers become impractical due to weight penalty and gear size.

Steel Coil Spring
These were used during World War II, but are rarely considered in present-day aircraft due to their extremely high weight and low efficiency. They weigh about 7 times as much as a comparative oleo-pneumatic unit and have an efficiency of about 50 percent.

Steel Leaf Spring
These are used in some light aircraft today, and as noted previously, they may be ideally suited to such aircraft from the weight, simplicity, reliability, and cost standpoints.

Rubber
Shock absorber efficiency is dependent upon the degree to which the shock absorbing medium is uniformly stressed. To obtain an efficiency of about 60 percent, therefore, rubber is usually used in the form of discs. These discs are vulcanized to plates and are stacked one above the other as shown in Fig. 12.4.3. These discs are in general no more than 1.5 inches thick to permit satisfactory vulcanizing. They have been widely used, notably by de Havilland on the Mosquito, Hornet, and Dove, and on the DHC Beaver and Twin Otter (Fig. 12.4.4). The Lockheed-Georgia Hummingbird is an example of more recent usage. As noted previously, they make excellent shock absorbers on lighter aircraft, where cost and reliability are overriding factors.

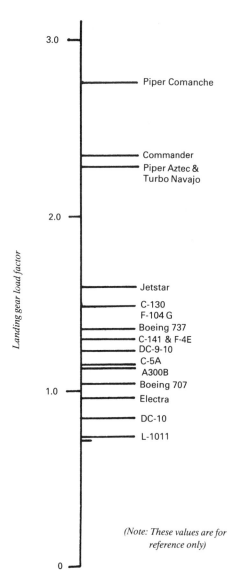

Fig. 12.4.1 Landing gear load factor comparison (ref. 12.32).

Air

Completely pneumatic shock absorbers have been used on landing gears, but not in recent years. They are just as complex as oleo-pneumatic units, are heavier, less efficient, and considerably less reliable. The so-called liquid springs and oleo-pneumatic units have an inherent means of lubricating the bearings. The pneumatic unit does not, and consequently the design of a leak-proof bearing is extremely difficult.

It is also worth noting that AIRIDE spring, as shown in Fig. 12.4.5, could be used advantageously in some design. The AIRIDE spring is made of nylon-tire-chord-reinforced neoprene rubber. Compressed air is trapped inside one or more bags (stacked in series if necessary), so that the end product is essentially a pneumatic spring which is devoid of the normal pneumatic strut problems (leaking seals and high friction).

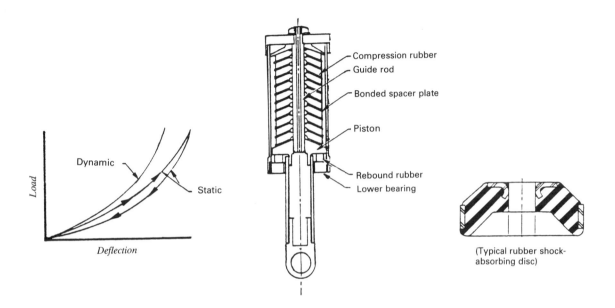

Fig. 12.4.3   Typical rubber shock absorber.

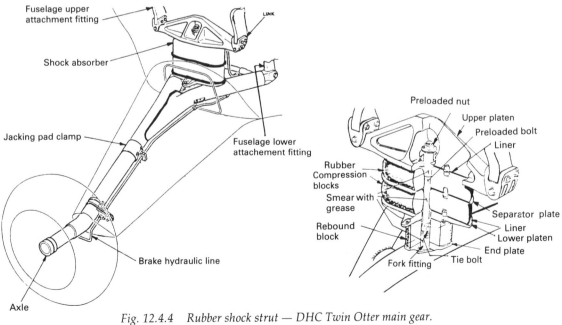

Fig. 12.4.4   Rubber shock strut — DHC Twin Otter main gear.
(Courtesy of De Havilland Aircraft of Canada.)

Airframe Structural Design 455

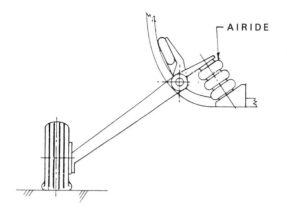

Fig. 12.4.5 Possible AIRIDE spring application.

Oil (Liquid Spring)

Fig. 12.4.6 shows a typical liquid spring. These have about 75 to 90 percent efficiency; they are as reliable as an oleo-pneumatic unit, but slightly heavier due to the robust design necessitated by high fluid pressures. The advantages of a liquid spring are: few fatigue problems due to robust construction, elimination of inflation/deflation, and relatively small size. Disadvantages are: fluid volume changes at low temperature affect shock absorber performance; the shock absorbers can only be pressurized while the aircraft is on jacks (i.e., with gear extended) due to the pressure levels required; high pressures must be sealed; and the unit has high mechanical friction and stick-slip action.

The liquid spring, as the name implies, uses the compressive properties of liquids as a springing medium. The same fluid volume is used in a dash-pot effect to control the recoil stroke. The liquid spring,

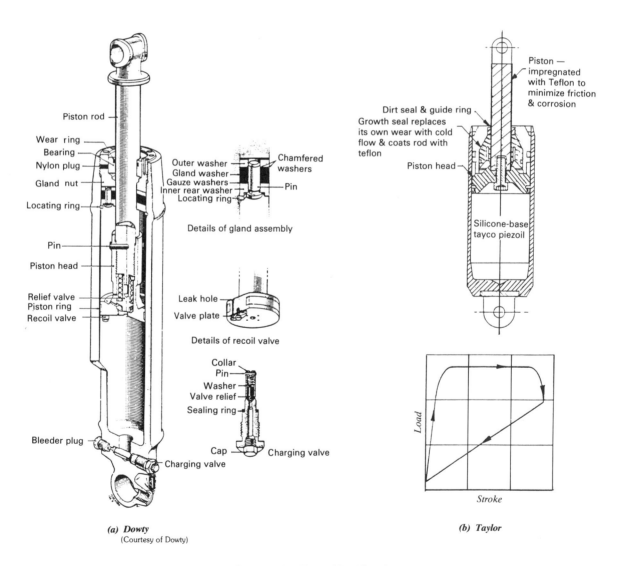

Fig. 12.4.6 Typical liquid spring.

456 Airframe Structural Design

illustrated in Fig. 12.4.6, is simple in construction, comprising a cylinder, piston rod, piston, and gland. Spring motion is accomplished by forcing the piston rod into the cylinder, displacing fluid volume thereby compressing the liquid. Energy is dissipated during the compression stroke by transferring fluid to the opposite side of the piston passing the central spring-loaded valve and a smaller open orifice in the piston. On the recoil stroke, the spring-loaded valve closes, restricting flow to the small open orifice, thus damping the outward movement of the piston rod (see Fig. 12.4.7).

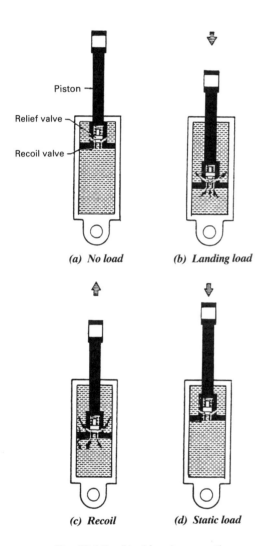

Fig. 12.4.7 Liquid spring operation.

The spring-loaded valve gives the shock absorber different spring characteristics for high and low dynamic loading. This is especially important on nosewheel landing gears, which require a hard shock absorber when taxiing to prevent a slow pitching movement while requiring a softer shock absorber to avoid exceeding the maximum permissible airframe reaction on landing. The velocity of shock absorber closure when taxiing is low, of the order of 1-2 feet per second. Consequently, the pressure drop across the small open orifice is not large, since the pressure drop is proportional to the square of the velocity. This pressure differential is not enough to overcome the spring behind the valve, and thus the orifice used for taxiing is the small one; the shock absorber is hence hard and can successfully resist pitching. On landing, the velocity of closure is much higher, and the pressure differential across the piston builds up until it reaches a predetermined value, overcoming the spring. The valve then opens and the large orifice comes into use, giving a soft shock absorber.

Air/Oil (Oleo-Pneumatic)

Most aircraft use oleo-pneumatic shock absorbers (Fig. 12.4.8). The purpose of the shock strut is to alleviate load on the airframe and to cushion impact. Fig. 12.4.8(a) is a typical load/stroke curve for an oleo-pneumatic unit, and the high efficiency under dynamic conditions means that stroke is minimized for a given sink speed and load factor. With efficiencies as high as 90 percent, it is an almost perfect device for absorbing the kinetic energy due to sink speed. The oleo-pneumatic unit not only has the highest efficiency of all types of shock absorbers, but it is also the best in terms of energy dissipation. Unlike a coil spring, it does not store the energy and then release it, causing the aircraft to bounce down the runway. Instead, the oil returns to its normal static condition at a controlled rate such that rebound does not occur [see Fig. 12.4.8(b)]. The ideal situation is one in which an aircraft can make a hard landing, after which the rebound characteristics of the shock strut will ensure that the wheels stay on the ground.

Thus, the oleo-pneumatic unit has the highest efficiency, and it is also an excellent energy dissipator with good rebound control. They are obviously more complex than other types of units, but constant refinement during the last 60 years has resulted in high reliability. As illustrated in Fig. 12.4.8, oil (Such as MIL-H-5606) is poured in with the strut compressed. The space above the oil is then pressurized with dry air or nitrogen. When the aircraft lands, fluid is forced from the lower chamber to the upper chamber through an orifice. Although this orifice could be merely a hole in the orifice plate, most American designs have a pin extending through it, and by varying the pin diameter the orifice area is varied. This variation is adjusted so that the strut load is fairly constant under dynamic loading [Fig. 12.4.8(b)]. If this could be made constant, the dynamic load curve would be a rectangle, and efficiency would be 100 percent. In practice, this is never obtained and efficiencies of 80 to 90 percent are more usual. The final value is not known until the completed strut has been drop tested, and possible adjustments have been made to the metering pin size.

Fig. 12.4.9 illustrates various types of oleo-pneumatic shock absorbers. They absorb energy by pushing oil in the lower chamber and compressing air in the upper chamber. Energy is dissipated during this process by oil being forced through one or more orifices. After initial impact, rebound must be controlled. During rebound, the expanding air pressure causes oil to flow back into the lower chamber through one or more recoil orifices. If oil flows back too quickly, the aircraft will be bounced back into the air again. If oil does not flow back quickly enough, the

Airframe Structural Design 457

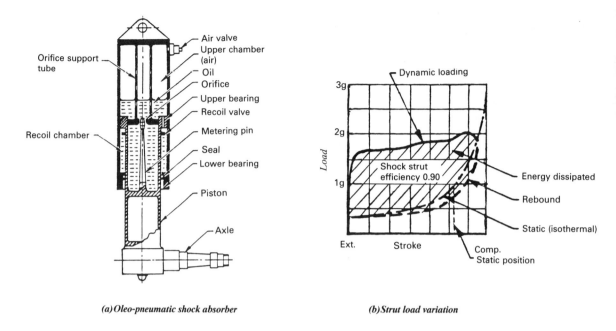

(a) Oleo-pneumatic shock absorber *(b) Strut load variation*

Fig. 12.4.8 *Oleo-pneumatic shock absorber.*

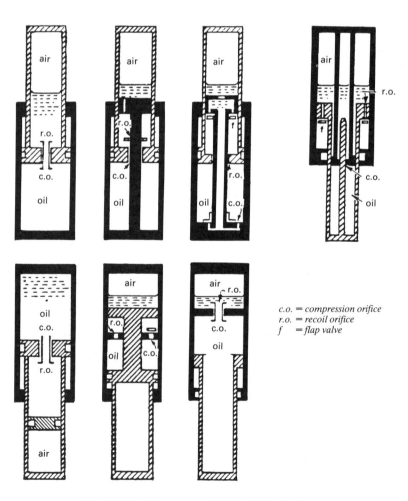

c.o. = *compression orifice*
r.o. = *recoil orifice*
f   = *flap valve*

Fig. 12.4.9 *Oleo-pneumatic shock absorber types.*

458 Airframe Structural Design

frequent short wavelength bumps encountered during taxiing will not be adequately damped, since the shock absorber will not respond quickly enough in restoring the wheels to their static position. The objective then is to design recoil damping such that the tires stay in contact with the runway upon landing impact, and also respond quickly enough to taxiing conditions.

The distance from the static to the fully compressed position is largely a matter of choice. It is good practice to call for an inflation pressure giving not less than one-third extension at maximum weight and not more than one-half extension at light load. If this cannot be arranged, then strut pressures must be adjusted to suit the prevailing weight conditions. However, an examination of the values shown in Fig. 12.4.10 shows that aircraft such as the Piper Comanche, Aztec and Navajo, Beech 99, and Aero Commander all have extensions of 35 to 45 percent, while transport aircraft have extensions of about 16 percent. The latter gives a harder ride while taxiing, but it tends to prevent wallowing — an important factor in large transport aircraft. In addition, with the static extension point being so far "up" the load-deflection curve, weight changes during loading do not result in substantial deflections of the shock strut. To sum up, the designer usually selects an extension which has been successfully used on similar aircraft operating under similar conditions. The shock strut characteristics are calculated and the original assumptions then modified as required.

Aircraft weight may change appreciably between take-off and landing, and to allow for this, calculations should be made for both conditions to verify that performance is satisfactory. In addition, calculations should be made to determine initial inflation pressures for varying airplane weights. This information is then quoted on a plate affixed to the shock strut so that ground crew can ensure that strut pressures are appropriate to the airplane weight.

It is almost impossible to stipulate a precise all-encompassing method for calculating the sizes and characteristics of an oleo-pneumatic shock absorber. Initial assumptions have to be made concerning some or all of the following: static position, compression ratio, air volume in the compressed position, maximum g-force applied to the strut, and maximum and minimum pressures inside the strut.

Piston diameters are generally chosen on the basis of having the maximum static strut pressure around 1,500 psig. Higher pressure will result in high dynamic pressures on the seals and also smaller diameters for the entire strut leg, thus providing a strut that is inefficient in bending and torsion. Lower pressures provide a strut of large diameter with thin walls resulting in efficient bending and torsion sections but lower bending allowables due to the high $\frac{D}{t}$ (where D = piston diameter and t = piston wall thickness). The thin wall strut also tends to ovalize and permit seal leakage.

Having selected the strut stroke and piston diameter, the designer can now determine what type of air spring, or air curve, is most suitable. Since one function of the air spring is to reliably push the piston out after take-off, it is apparent that the extended pressure must be high enough to overcome seal and bearing friction. The extended pressure of 65 psig is probably as low as practical. On the other hand, if the extended pressure is too high, the load required to begin the stroking of the strut on landing becomes excessive. Thus when a very low sink rate landing is made, the strut will compress only slightly and the airplane will bounce. Experience has indicated that the extended pressure should not exceed 250 psig, particularly for a single axle gear. A truck type (or bogie) gear, and also multi-main gears, will have less tendency to bounce since all the tires do not contact the ground simultaneously.

It is obvious that all the best features can not be combined in one strut having a simple air spring. If struts were built with double air chambers, double acting shock struts, or mechanical devices that provide all the desirable characteristics — low extended pressure, soft taxi ride, etc., they had seldom proved popular, especially for commercial transports, due to complicated servicing procedures, noise, leakage, and higher cost and weight.

However, landing gear design for the military cargo transports fulfill airfield roughness specifications for operation on both standard and substandard runways. Substandard conditions include three-inch high step bumps and 1-cosine wave bumps at specified wave length and amplitude. To minimize the transmission of these loads to the structure, double-acting shock

Airplane	Total Stroke (in.)	Distance (in.) Static to Compressed
DC-9	16	.88
DC-10	26	2.5
F-4	15.9	1.5
C141	28	3
Electra	20	2.2
L-1011	26	3.5
B707	22	3
B720	20	3
B737-200	14	2.1
B727-200	14	2.5
Jetstar	15.5	3.5
C-130	10.5	3
Beech U-21A	10.8	3.3
Piper Turbo Navajo	8	2.8
Piper Aztec	8	3.1
Beech 99	12	4.8
Aero Commander	8.8	3.5
F-104	13.8	5.6
Piper Comanche	6.1	2.8

(Note: The above values are approximate)

Fig. 12.4.10 *Shock strut static extension comparison (ref. 12.32).*

struts (as shown in Fig. 12.4.11) are used in both main and nose landing gears. As the name imples, the double-acting shock strut [Fig. 12.4.11(c)] has two air chambers instead of the conventional single-chamber [Fig. 12.4.11(b)]. One chamber is the normal air chamber of a conventional air-oil shock absorber; the second chamber is inside the primary piston beneath a floating secondary piston. The chamber is preloaded by compressed air in excess of maximum static requirements.

During taxi operations over moderate or long wavelength undulations, the double-acting shock strut acts as a conventional shock absorber. The secondary piston serves as the equivalent of a hydraulic surge chamber, absorbing unsprung mass momentum and attenuating peak loads.

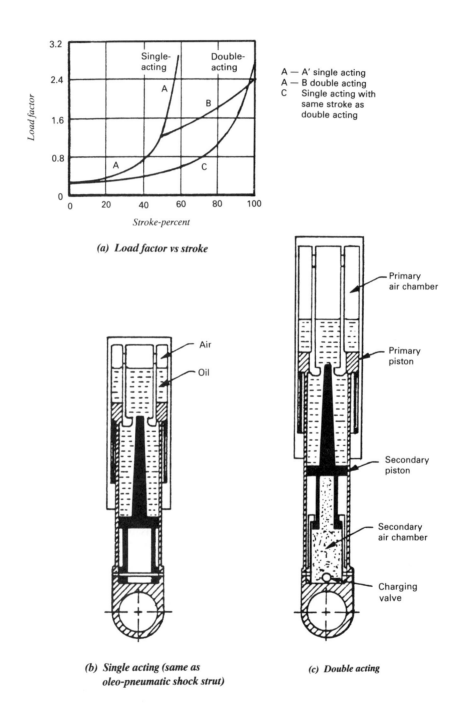

Fig. 12.4.11 *Double-acting shock strut vs single-acting shock strut.*

460 Airframe Structural Design

12.5 Wheels and Brakes

Any ground vehicle generally has to have wheels to roll and brakes to stop and go. On the ground, an aircraft is no exception: to use brakes to steer by different ways, to hold the aircraft stationary when parked, to hold the aircraft while running up engines prior to take-off, and to control speed properly while taxiing.

In airframe design, it always has space problems for stowing landing gears; therefore, the wheels and brakes have to be designed compactly. In addition, it requires that kinetic energy absorption capacity of the brakes be adequate to accomplish the number of stops without replacement of brake linings or other parts except as noted within the design requirements. Fig. 12.5.1 shows a cut-away of typical assembly of wheel, brake and tire. Fig. 12.5.2 lists a summary of typical wheel and brake parts.

Wheel Design

Wheels as shown in Fig. 12.5.3 are usually made from forged aluminum alloy, such as 2014-T6. It is important to design the forging such that optimum grain flow is obtained, with particular attention to the tire bead seat area. Photostress and stress lacquer techniques are used to show the general stress distribution and to ensure that the item is free from harmful stress concentrations.

Inboard Wheel Half	Torque Plate Assembly
Outboard Wheel Half	Fluid Inlet Connection
O-Ring Seal	Bleed Screw
Excluder	Piston Assembly
Wheel Bolts	Pressure Plate Assembly
Outer Grease Seal	Stator Plate Assembly
Outboard Bearing	Segmented Rotor
Inboard Bearing	Torque Tube
Inner Grease Seal	Backing Plate Assembly
Rotor Drive Key	Automatic Adjuster
Fusible Plug	Piston Return Spring
Inflation Valve	Pad Wear Indicator

Fig. 12.5.2 Summary of wheel and brake parts.

(a) Cutaway view

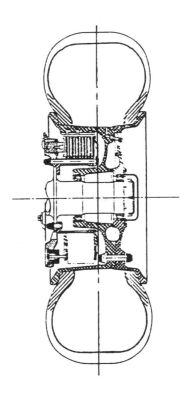

(b) Cross-section view of a wheel, tire and brake

Fig. 12.5.1 Typical assembly of wheel, brake and tire.
(Courtesy of Dunlop Ltd.)

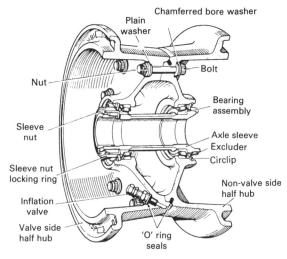

Fig. 12.5.3 Typical wheel construction.
(Courtesy of Dunlop Ltd.)

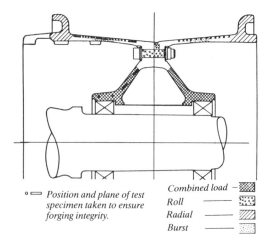

Fig. 12.5.4 Critical stress areas in wheels.
(Courtesy of Dunlop Ltd.)

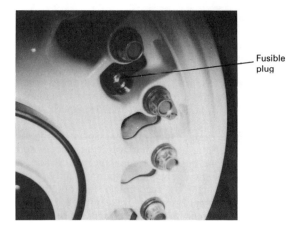

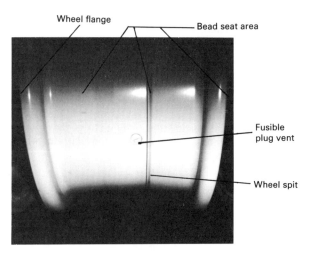

Used on the tubeless wheels of large aircraft to relieve excessive pressure and prevent blowouts due to excessive brake heat; fusible plugs are generally not removed during a routine tire change unless they are defective. They are, however, removed and inspected during wheel assembly overhaul.

Fig. 12.5.5 Wheel and fusible plugs.
(Courtesy of Loral Systems Group)

Fig. 12.5.4 illustrates the critical areas of stress concentration. The rim contour is in accordance with international standards. Static and fatigue loads design the flange bead ledge and bead seat area, with the flange acting as a torsion ring to hold the tire bead in position. The flange must also distribute the shear loads, from the ground reaction, to the rest of the wheel.

The two wheel halves are joined together by a number of tiebolts. This area of the wheel is designed for high stiffness. They are lubricated prior to assembly to minimize torque/tension variation, and are then torqued to very precise values in order to optimize fatigue life. The hub is designed to house the wheel bearings, and in many cases sufficient material is left so that oversized bearings can be installed if required. The bearings are of the taper-roller type, and are sealed to ensure their grease is not ejected at high speed, as well as to protect the bearing from contamination. A standard tire inflation valve is installed in the outboard wheel, usually near the tiebolt flange. Fusible thermosensitive pressure release plugs are also installed in the wheel in this area. This plug releases the tire pressure if local temperature reaches a predetermined level. Each plug is sealed by an O-ring and consists of a hollow casing housing a eutectic insert, a solid piston, and a rubber seal (see Fig. 12.5.5).

Other items that have to be considered include the rotor drive keys or blocks, a heat shield if required, and possibly a tire change counter. The drive blocks are high-strength steel and are dovetailed into the wheel half surrounding the brake. Heat shields are sometimes provided to minimize heat transfer from the brake, and the tire change counter is sometimes specified to record tire changes.

Brake

Until about 1963, all brake heat-sink materials were made from steel. Beryllium was introduced and was selected for certain aircraft to save weight. Beryllium is a good heat-sink material (see Fig. 12.5.6) because of its heat-absorbing characteristics, such as a high specific heat and excellent thermal conductivity which

Property	GY4000 Carbon	Beryllium	Steel	Desired Characteristics
Density (lb/cu. in.)	0.061	0.066	0.283	High
Specific Heat @ 500°F (BTU/lb/°F)	0.31	0.56	0.13	High
Thermal Conductivity @ 500°F	100	75	24	High
Thermal Expansion @ 500°F ($10^{-6} \times$ in./in./°F)	1.5	6.4	8.4	Low
Thermal Shock Resistance Index ($\times 10^5$)	141	2.7	3.5	High
Temperature Limit, °F	4000	1700	2100	High

Fig. 12.5.6 Comparison of heat-sink materials (ref. 12.32).

can provide a more uniform and rapid heat transfer throughout the disc stack. However, the use of beryllium has been discouraged because of its crack formation by thermal stresses and toxicity.

The low density of beryllium means that a brake disc would require more space than a steel brake. Its low impact strength requires special design consideration for keyways and thermal relief slots. The high thermal expansion coefficient and low tensile strength values require extensive thermal relief slots to avoid crack formation by thermal stresses. Despite all these problems, and its high cost, beryllium has been used as the brake heat-sink material on many contemporary aircraft.

In recent years, carbon brake material has been used, and although it is the lightest known heat-sink material, it is also the most expensive. However, the cost difference between carbon and other materials is decreasing. The development of carbon brakes started in 1966 with the objective of developing heat-sinks that had most of the advantages of beryllium but not its disadvantages. Carbon has properties that make it highly desirable as a heat absorber and the high specific heat is also important in the interest of light weight. Carbon's high temperature capability is nearly twice that of steel.

Fig. 12.5.7 illustrates an assembly of a brake and its main parts — stators and rotators.

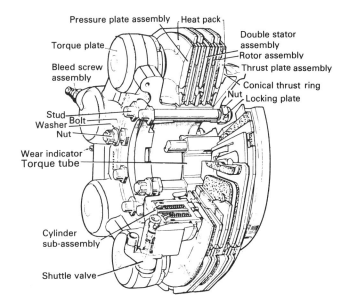

Fig. 12.5.7 Typical brake assembly.
(Courtesy of Dunlop Ltd.)

Tire Selection

The step-by-step selection process is as follows:
- Determine the maximum static load on the main gear tire. If more than one tire is used, determine the individual tire load by simple static analysis.
- Determine the static and dynamic loads on the nose gear tire. If the c.g. position is unknown, assume that 90 percent of the aircraft weight is reacted by the main gear (in a conventional arrangement).
- Using manufacturers' tables, list all candidate tires meeting the load and speed conditions. Tread thickness decreases with speed increase, and hence tire life is reduced.
- From experience or from calculation, decide which wheel size will be used. This eliminates some of the possible tires.
- Based upon flotation, customer preferences, or instinct, determine what maximum pressure is acceptable. This will eliminate more tires.
- From a study of space considerations, determine which of the remaining tires is the optimum choice.
- If a degree of roughness is specified, ensure that the tire has sufficient section height to absorb the bump. In operation, the tire encounters the bump and immediately deflects; the shock absorber, being somewhat slower in reacting, compresses after the tire has compressed.

Subsequent analysis may indicate that brake heat or flotation, for example, demand a change to the selected tire. Such changes are quite normal in the development of a gear, and the designer must be flexible enough to accept such changes as a matter of course.

As noted in the discussion on requirements, the condition of one tire or wheel failing on a multiwheel gear must be considered, and after selecting the optimum tire this condition should be checked. Make sure that the tire load resulting therefrom does not exceed the tire bottoming load.

There is one other factor that occasionally has to be considered — that of an instantaneous peak load, such as running over a deck cable or hitting a sharp bump. For short-period applications the tire is capable of withstanding loads beyond the bottoming load. At this point, the tire is not a pneumatic shock absorber; it is virtually a case of rubber in compression.

12.6 Detail Design

While the general rule in aircraft structural design is that multiple load paths be provided to give fail-safe capability, most landing gear structures do not lend themselves to this concept. Accordingly, the gear must be designed and proved of satisfactory safe-life capability. The essence of the safe-life concept is that the fatigue life of the gear parts can be safely predicted or that the growth of cracks is slow enough to permit detection at normal inspection intervals. Naturally, the ideal goal for landing gear parts should be that the safe-life should equal the expected life of the airplane.

It is inevitable, of course, that landing gears must present a formidable challenge in achieving high reliability since they are a complex mixture of various systems, structures, and mechanisms. Unlike basic airframe structure as mentioned previously, it is usually impractical to apply fail-safe principles to landing gear parts. Instead, they must be designed for safe-life capability and selection of optimum material is the first step in this regard.

Material Selection

Material selection guidelines are as follows:
- Where steel forgings are used, use only vacuum arc remelt parts.
- The preferred method of cold straightening of steel parts, hardened to tensile strengths of 200,000 psi and above, would be to temper the parts while in a straightening fixture.
- Magnetic particle inspection should be performed on all finished steel parts which are heat treated in excess of 200,000 psi tensile strength.
- Many parts are received with forging laps, inclusions, etc., that were in the part at the time of manufacture. These defects may not be detrimental to the service life of the part; however, when the part is magnetically inspected at the depot after service, inspectors cannot determine that these indications are forging laps and not fatigue cracks and, therefore, the part may be rejected.
- Bushings should be limited to non-ferrous materials for the principal static and dynamic joints.
- All joints should be bushed to facilitate depot rework.
- All surfaces, except holes under $\frac{3}{4}$-inch diameter, of structural forgings forged from stress-corrosion-susceptible alloys, which, after final machining, exhibit exposed transverse grain, should be shot-peened or placed in compression by other means.
- Areas of components considered to be critical in fatigue should have a surface roughness in the finished product not to exceed 63 rhr, as defined by ASTM B 46.1, or should be shot-peened, with a surface roughness prior to peening of not over 125 rhr. Unmachined aluminum die forgings should be approximately 250 rhr, except surfaces where flash has been removed.
- Efforts should be made to reduce stress concentration, such as using relief heat treatments (except aluminum alloys), to optimize grain flow orientation, using "wet installed" inserts and pins, and extensive use of surface cold working.
- Avoid cross-drilling of joint pins. Drilling operations result in material surface damage and stress risers that are difficult to control.

(1) Steel

The most common landing gear steels are 4130, 4340, 4330V, and 300-M. Where stiffness for minimum cost is important, use 4130. For maximum strength/weight ratios, 4340 and 300-M are used, with the former being used primarily in the 260–280 ksi range, and the latter in the 280–300 ksi range. In the last few years, 300-M has been used with great success, for such items as bogies, pistons, braces, and links.

It has about the same fatigue properties as 4340, excellent ductility at very high strength, and since material can be interrupted quenched, distortion due to heat-treat is greatly reduced. The maximum section size appropriate to heat-treated 300-M (280 ksi) is approximately twice that at which 4340 can attain 260 ksi. Although air-melt materials have been widely used, vacuum-melt materials should be used in all high-heat-treat applications. Fig. 12.6.1 shows steel applications.

Material	Ultimate Stress F_{tu} (ksi)	Application
4340	260	B707, DC-8, C-141, Electra, L-1011
300M	275	B720, B727, B737, B747, C-5A, DC-10
D6AC	260	F-111
H11	260	B-70
35NCD16	270	Mirage, F-101, Concorde, A300
Marging	260	Harrier, Super VC 10
Hy-Tuf	220	DC-9

Fig. 12.6.1 Steel materials for gear applications.

(2) **Aluminum Alloys**
Due to the unpredictable nature of stress corrosion, most critical aluminum parts are now made from 7075 forgings and the reduction in properties is accepted. Aluminum alloys are most economical where the loads are lower and the parts must be quite long, where if steel were used the sections would become too thin and warpage would be excessive. Also, if the parts can be used substantially as forged, the aluminum parts are much cheaper to fabricate than high strength steels.

(3) **Titanium**
Titanium alloys have not found great acceptance in landing gear design. The strength/weight ratio is not better than steel, machining is difficult (expensive), the material is very costly, and threads and bearing surfaces tend to gall very readily. The material has excellent corrosion resistance and good fatigue life.

Gear Joint Design

In designing landing gear joints, the following guidelines should be used:
- Fit bushings in all joints to prevent contact of mating structural parts, and to greatly simplify correction of deficiencies at the joint.
- Use bushing material different from pin or structure material to prevent galling. One particularly good combination is an aluminum bronze bushing and a chrome-plated steel pin.
- Surveys indicate that aluminum-nickel bronze and stainless steel (17-4PH) are proving to be the best bushings in current airline usage.
- Bushings should be installed by shrinking rather than as a press fit, since the latter may remove some corrosion protection.
- All joints should be lubricated, using either grease or self-lubricated bushings. This improves pin-removal capabilities, and fights corrosion.
- Ensure that corrosion-causing cavities are eliminated. For instance, do not install shouldered bushings in each side of a hole unless lubricant is injected into the space between them.
- Avoid shims and spacers as much as possible. They get lost, and they are a potential cause of trouble due to being inadvertently forgotten by the ground crew.
- Allow sufficient material (0.06 on the radius), if possible, around the joint to allow for rework of the pin hole, and accept a larger bushing.
- The lug hole and faces must be properly protected against corrosion and wear. Cadmium plate and dry-film lubrication are inadequate for this.
- Chromium-plate all pins a minimum of 0.002 inch thick, and corrosion-resistant pin material should be considered.
- Ensure that the grease passage is located such that a fatigue stress riser is not introduced.
- Do not lubricate more than one point from one grease fitting.
- Always use protruding type grease fittings. The flush types are hard to find on landing gears covered in dirt.
- Provide generous fillet radii, and ensure that all transitions are smooth. Avoid any sharp corners.

It is generally agreed that probably no phase of detailed design is more critical from a fatigue standpoint than the landing gear joints. It is good practice to size the pins and lugs early in the gear layout since the size of pins and lugs will influence their locations and subsequently the geometry of major structural members. In design, the bushing allowables as shown in Fig. 12.6.2 should be considered and the lug thickness is no more than half the pin diameter. This means the pins will be large enough to be very stiff in bending, substantially hollow, and the lugs properly proportioned to equalize stress distribution (see Fig. 12.6.3).

Bushing Material	F_{br}, Maximum Static Capacity (ksi)
4130 Steel (180 ksi)	115
17-4 Steel (AMS 5643)	90
Beryllium Copper (Fed Spec QQ-C-530)	90
Al-Ni-Bronze (AMS 4640 & 4880)	60
Al-Bronze (Fed Spec QQ-C-465)	60

Fig. 12.6.2 Bushing material bearing allowables (ref. 12.32).

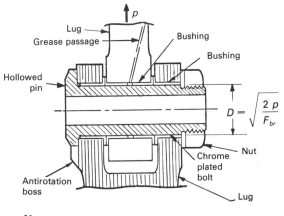

Notes:
(1) *P — Static loads*
   F_{br} — Maximum static capacity bushing bearing allowable, see Fig. 12.6.2
(2) *Lug sizing, see Section 7.5 of Chapter 7*

$$D = \sqrt{\frac{2p}{F_{br}}}$$

Fig. 12.6.3 Typical landing gear lug configuration.

Concerning load-life values, steel bushings are satisfactory for a limited number of cycles, but aluminum-nickel (Al-Ni) bronze or aluminum bronze bushings are far better if appreciable motion is present.

As noted earlier, bushings should be installed by shrinking since this does not remove any of the corrosion protection. This type of fit is accomplished by cooling or heating parts so that the resulting contraction or expansion permits assembly without metal-to-metal interference. A dry ice and methanol bath is capable of chilling parts to $-120°F$, but liquid nitrogen is the preferred coolant and can provide $-320°F$.

The following guidelines can be used in bushing design:
- Chrome plate (hard) all pins (or use corrosion-resistant material).
- Do not install shouldered bushings from each side of a hole unless grease is injected into the cavity where the two bushings meet.
- Do not use non-corrosion-resistant steel bushings.
- If possible, allow the bolts to rotate somewhat inside the bushing. This helps prevent crankshafting and corrosion.

Other Considerations
(1) Lubrication
All dynamic joints must have greasing provisions. Al-Ni bronze is commonly employed for bushings, with careful attention paid to grease distribution passages in the bearing surface. The grease can be injected from the O.D. (lug) or the I.D. (pin). If the grease passage is through the lug, it must be placed in an area of minimum stress to avoid fatigue stress raisers. If grease passages are drilled in the pin, the holes must be accounted for in checking pin strength. It is unwise to expect to lubricate more than one bushing with only one grease fitting since the grease will seek out the shortest path and starve the other bushings. It is best practice that all grease fittings on the gear be the protruding Zerk type. The so called flush grease fittings are difficult to find and will often go unattended if used in combination with the Zerk type fittings.

(2) Transitions
The transitions between the various elements of a part must be smoothly blended to avoid stress concentrations. For complex shapes it is often desirable to model the parts in three dimensions so that a better feel for the load paths can be visualized. Large corner radii, fillets, and blends must be employed to achieve even stress distribution. Use photostress techniques for determining stress levels in complex shapes.

(3) Finishes
Adequate finishing of landing gear parts is essential since the joints and materials must be protected against moisture, salt, Skydrol, cleaning solvents, and power plant deposits. The finishes staff should be consulted for the latest and best techniques and processes. Particular attention must be paid to:
- Hollow members. Plated if possible, drainage and breathing capability preferred, with two coats of primer plus corrosion prevent compound.
- Joints. Chrome plate or flanged bushings to protect primary members.
- Paint. All possible areas of steel and aluminum parts should be primed and painted.
- Holes. Controlled plating specified for all holes where size permits.

The following summarizes most of the finishes of concern to the landing gear design:
(a) Non-corrosion-resistance alloy steel: The surface should be cadmium-titanium plated, or chromium plated on wearing surfaces heat treated to 220 ksi and above, or nickel and chromium plated on wearing surfaces heat treated below 220 ksi. The organic finish is one coat of MIL-C-8514 wash primer, one coat of MIL-P-23377 epoxy primer, two coats of STM37-307 polyurethane white, with no paint on the functioning or wearing surfaces.
(b) Non-clad 2000 and 7000 series aluminum alloys (including all aluminum alloy castings): The surface should be sulfuric acid anodized, and the organic finish is the same as in item (a).
(c) Clad aluminum alloys and non-clad aluminum alloys [other than those in (b)]: The surface should be color conversion treated, and the organic finish is same as in (a).
(d) Titanium and titanium alloys: The surface should be cleaned and no organic finish is required, but if paint is required for appearance, use the same finish as in (a).
(e) Fiberglass (covers, shields, etc): No surface finish is required. If paint is required for appearance, finish with one coat of STM37-307 white polyurethane.

12.7 Testing

Landing gear testing is one of most extensive testing programs in aircraft design and many testing tools are used both during and after landing gear design to substantiate strength, life, and performance.

Structural Tests
(1) Photostress
The use of photostress has become standard practice to determine stress levels and direction in the transition areas of all complex gear parts. Usually the complete gear is built out of plastic as soon as the drawings are available and checked by photostress, followed by strain gaging. This permits local changes to be made before the tooling is completed for the parts. The photostress process is repeated when the first gear is loaded prior to fatigue testing.

(2) Fatigue Test
Since the landing gear is usually a safe-life structure, both commercial and military aircraft require substantiation of gear life by fatigue testing (see Fig. 12.7.1). All the various gear loading conditions are applied in blocks to simulate those loads expected to be typical for normal service. The gears are completely dismantled and inspected periodically to check for cracks, wear, galling, etc.

(3) Static Test
Since most parts of the gear are designed by fatigue conditions, it is common to find generous margins of safety under ultimate load; therefore,

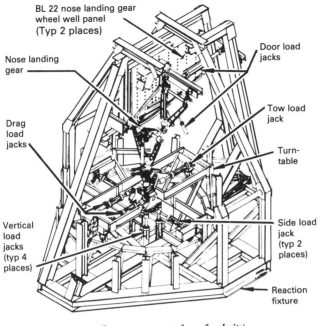

(Some structure not shown for clarity)

(a) Nose landing gear

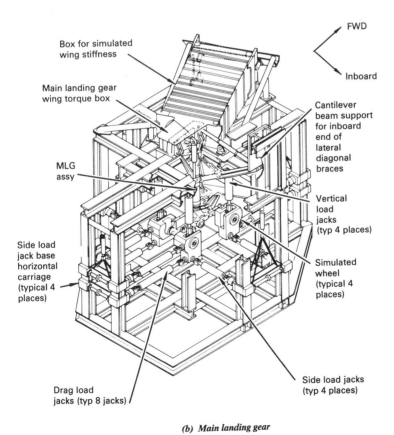

(b) Main landing gear

Fig. 12.7.1 Landing gear fatigue tests — L-1011.
(Courtesy of Lockheed Aeronautical Systems Co.)

Airframe Structural Design 467

static testing probably does not justify the expense involved.

(4) Detailed Tests

Detail parts are often fatigue tested or static tested separately from the above tests. Gear breakaway or fuse detailed are often tested in detail if the analysis is in doubt and where precise failure modes are required.

System Tests

(1) Drop Test

The drop testing of the gear is required to develop and substantiate the energy absorption characterisitcs of the shock strut (see Fig. 12.7.2). The number of drop test conditions can vary from as few as 20 to perhaps 60. The variables include:

- Landing weight
- Sink speed
- Level attitude
- Tail down attitude
- Wheel speeds.

Drop testing is accomplished by loading the gear in a jig mounted in a vertical tower. The jig can be loaded with varying weights. It is hoisted to appropriate heights, (depending on the desired sink speed) and allowed to drop free, the wheels landing on a calibrated platform to determine vertical and drag loads. The effect of spinning up the wheels upon landing is simulated by spinning the wheels to the desired landing speed before the drop.

The drop test is also used to substantiate the capability of the gear to sustain overload landings without structural failure and to obtain data points for fatigue analysis.

(2) Retraction Test

As rapidly as production parts can be acquired, both the main and nose gear retraction systems are laboratory tested. All production parts involved in the retraction are tested, including lock and door systems. The parts are mounted in a jig fixture so that the entire airplane system can be operated. Parts that are not designed or affected by retraction loads are usually dummies. The gear weights must be simulated, and compressed air cylinders are commonly used to simulate aerodynamic forces on the gear and doors. These tests serve three functions:

(a) The rig will be used for as long as 2 or 3 months to tune and adjust the hydraulic and mechanical systems so the gear operates smoothly without large impact forces (thus noise). The manual extension system is proved out, and often changes are made to locking springs, snubbing orifices, etc. to ensure positive gear operation under all conditions. Needless to say, this phase of the testing should be completed as early as possible so that all important design changes can be implemented prior to first flight.

(b) The second use made of the retraction test is to prove the service life of moving parts. It also demonstrates the wearing parts such as

(a) Nose gear   *Fig. 12.7.2 Nose and main landing gear drop tests — L-1011.*   *(b) Main gear*

(Courtesy of Lockheed Aeronautical Systems Co.)

seals, bearings, bushings, locks, etc. will survive one overhaul in airline service. Frequent teardown inspections are conducted in the early cycling for evidence of premature wear.

(c) Experience has shown that the failure to test and prove fatigue life of many parts in the retraction mechanisms can have serious safety implications. Strain gaging is employed in many locations in the tests to provide the stress group with improved data for fatigue analysis. However, due to the complex shapes of many mechanism parts, the only final proof of fatigue life is actual cycling.

The extended fatigue test cycling will also prove useful to continue comparison tests of various different bearings and optimize the final production selection before too many airplanes have been produced.

Shimmy Test

Landing gear shimmy is an unstable condition caused by the coupling of the torsional mode with side bending mode of the gear. The shimmy characteristics of a gear are analyzed by structural dynamics by use of the computer as soon as reliable spring rate information is available. The exposure to shimmy is great on all nose gears and main gears and an additional tool for checking gear shimmy is the flywheel dynamometer as shown in Fig. 12.7.3.

The entire landing gear must be used and mounted by actual or simulated support structure. Combinations of load, yaw, tire unbalance, gear slop, and speed must be run, with an abrupt torsional excitation artificially applied by one brake or an abrupt punch to the end of the axle.

Wheel and Brake Test

The flywheel dynamometer is the basic tool for testing wheels and brakes. The dynamometer consists of a large wheel that is spun up to the desired speed by an electric motor. The mass of the wheel can be varied to obtain the desired kinetic energy. One wheel, brake, and tire is mounted on an axle, which is attached to a hinge arm. The dynamometer is brought up to speed with the tire off the wheel. When the desired wheel speed is established, the wheel is allowed to coast. The mandrel arm is then "landed" on the wheel with sufficient force to obtain a predetermined rolling radius and the brake is applied. By this means all necessary data is obtained to assess brake performance. This information including continuous plots of speed, torque, and distance, and a large number of brake temperature readings.

Wheels are static tested to ultimate and yield load with tires installed, and are usually roll tested several thousand miles to establish fatigue life. Fig. 12.7.4 shows a wheel-roll test machine.

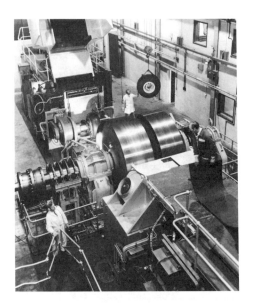

Fig. 12.7.3 Dunlop dynamometer rig.
(Courtesy of Dunlop Ltd.)

Fig. 12.7.4 Dunlop's wheel-roll test machine.
(Courtesy of Dunlop Ltd.)

References

12.1. Anon: *Airdrome Manual, Part 2, 7920 AN/865/2,* ICAO, 2nd ed., Ottawa, 1965.
12.2. Anon: *The Tire and Rim Association Year Book,* Akron, Ohio.
12.3. Nichols, D.E.: 'Overall Braking for Jet Transports.' *A.S.M.E. Papter No. 60-AV-2.*
12.4. Conway, H.G.: *Landing Gear Design.* Chapman & Hall Ltd., London, 1958.
12.5. Milwitzky, Ben.: "Analysis of Landing-gear Behavior." *NACA Rep. 1154,* (1953).
12.6. Flugge, W.: "Landing-gear Impact." *NACA TN2743,* (Oct. 1952).
12.7. Flugge, W.: "The Influence of Wheel Spin-up on Landing-gear Impact." *NACA TN3217,* (Oct. 1954).
12.8. Anon: *Aircraft Engineering,* (Jan. 1968).
12.9. Nightingale, J.: "The Determination of Orifice

Parameters for Shock Absorbers." *Aircraft Engineering,* (Sept. 1951), 261−262.
12.10 Burger, F.E.: "Practice of Shock-absorber Design." *Aircraft Engineering,* (Dec. 1949), 384−385.
12.11 Orloff, G.: "A problem of Nose-wheel Undercarriage Reactions." *Aircraft Engineering,* (May 1950), 129−131.
12.12 Conway, H.G.: "The German View of the Tricycle Undercarriage." *Aircraft Engineering,* (May 1947), 147−149.
12.13 Dirac, G.A.: "The Explicit Determination of Orifice Parameters in shock Absorbers." *Aircraft Engineering,* (Aug. 1947), 258−262.
12.14 Blinkhorn, J.W.: "Undercarriages for Deck Landing." *Aircraft Engineering,* (Oct. 1948), 304−305.
12.15 Conway, H.G.: "Undercarriage Retraction Mechanisms." *Aircraft Engineering,* (Jun. 1945), 167−176.
12.16 Drury, G.W.: "Appreciation of Landing Problems Part I & II." *Aircraft Engineering,* (Jul. 1945), 188−191, and (Aug. 1945), 218−223.
12.17 Taylor, J.L.: "The Problem of Landing Shock." *Aircraft Engineering,* (Aug. 1946), 270.
12.18 Wernitz, W.: "Tricycle Undercarriage Development." *Aircraft Engineering,* (Jan. 1941), 6−11.
12.19 Andrews, J.: "Tricycle-Undercarriage Take-off." *Aircraft Engineering,* (Jul. 1941), 180−183.
12.20 Burger, F.E.: "Theory of Shock-absorber design." *Aircraft Engineering,* (Feb. 1943), 51−54.
12.21 Walker, P.B.: "Tricycle Undercarriage Design." *Aircraft Engineering,* (Jun. 1940), 171−173.
12.22 Holden, D.: "Tyre Selection for Modern Aircraft." *Aeroplane,* (Mar. 1965).
12.23 Anon.: "Modern Aircraft Tire Development." *Interavia,* (1961), 1257.
12.24. Norman S. Currey: "C-5 High Flotation Landing Gear." *Lockheed-Georgia Quarterly,* (Jun. 1968).
12.25 Bingham, A.E.: "Liquid Springs: Progress in Design and Application." *The Institution of Mechanical Engineers,* (May 1955).
12.26 Saelman, B.: "Designing Cylinders and Struts for Maximum Strength." *Machine Design,* (Aug. 1953).
12.27 Kraus, H. and Saelman B.: "Determination of Shock Absorbing Distance." *Design News,* (Aug. 1958).
12.28 Saelman, B.: "Hydraulic Actuating Systems." *Machine Design,* (May 1957).
12.29 Leclercq, J.: "Evolution or Revolution in Undercarriages." *Interavia,* (Jul. 1962), 868.
12.30 Willitt, A.A.J.: "Today's Undercarriage Design Problems." *Interavia,* (Sept. 1961), 1254.
12.31 Kemp, T.M.: "Liquid Spring Principle Applied to Landing Gear Design." *Aero Digest,* (Aug. 1956).
12.32 Currey, N.S.: *Landing Gear Design Handbook.* Lockheed − Georgia Co., 1982.
12.33 Clifton, R.G. and Leonard J.L.: "Aircraft Tyres − an analysis of performance and development criteria for the 70s." *Aeronautical Journal,* (Apr. 1972), 195−216.

12.34 McBearty, J.F.: "A Critical Study of Aircraft Landing Gears." *Journal of The Aeronautical Sciences,* (May, 1948).
12.35 Liming, R.A.: "Analytical Definiton of a Retractable Landing Gear Axis of Rotation," *Journal of The Aeronautical Sciences,* (Jan. 1947).
12.36 Hadekel, R.: "Shock Absorber Calculations. (A Method of Estimating the Performance of Oleopneumatic Struts." *Supplement to Flight International,* (Jul. 25, 1940), 71−73.
12.37 Burger: "Practice of Shock-Absorber Design (Steps in the Design of an Oleo-pneumatic Undercarriage Leg)." *Aircraft Engineering,* (Dec. 1949), 384−385.
12.38 Cameron-Johnston, A.: "The Undercarriage in Aeroplane Project Design." *Aircraft Engineering,* (Feb. 1969), 6−11.
12.39 Imrie, W.M.: "Ultra High Tensile Steel Landing Gear Components − Material Selection and Manufacture." *The Aeronautical Journal of the Royal Aeronautical Society,* (Feb. 1971), 139−152.
12.40 Anon: "Landing Gear." *Aircraft Engineering,* (Jul, 1986).
12.41 Smith S.M.: "Aircraft Wheel Design and Proving." *Aircraft Engineering,* (Jul, 1986).
12.41 Best K.F.: "High Strength Materials for Aircraft Landing Gear." *Aircraft Engineering,* (Jul. 1986).
12.42 Ahivin, R.G.: "Developments in Pavement Design in the U.S.A.-Flexible Pavements. *Paper presented at the Third International Conferences on the Structural Design of Asphalt pavements,* 11−13 Sept. 1972, London.
12.43 Anon.: "Aircraft Dynamic Loads from Substandard Landing Sites." *U.S. Air Force Flight Dynamics Laboratory Report DG-16190,* (Nov. 1965).
12.44 Kraft, D.C., Hoppenjans, J.R., and Edelen, W.F.: "Design Procedure for Establishing Aircraft Capability to Operate on Soil Surface." *AFFDL-TR-72-129,* (Dec. 1972).
12.45 Ladd, D. and Ulery, H.: "Aircraft Ground Flotation Investigation." *AFFDL-TR-66-43,* (Aug. 1967).
12.46 Williams, W.W., Williams, G.K., and Garrard, W.C.J.: "Soft and Rough Field Landing Gears." *SAE No. 650844,* (Oct. 1965).
12.47 Kraft, D.C. and Phillips, N.S.: "Landing Gear/Soil Interaction Development of Criteria for Aircraft Operation on Soil During Turning and Multipass Operation. *AFFDL-TR-75-78,* (Oct. 1975).
12.48 Anon.: *Wheel and Brake Design Guide for Airframe Engineers.* B.F. Goodrich Company, Troy, Ohio.
12.49 Pazmang, L.: *Landing Gear Design for Light Aircraft.* Pazmang Aircraft Corp., San Diego, CA92138 (1986).
12.50 Anon.: "Landing Gear Achieves Advanced Design Goals." *Aviation Week & Space Technology,* (Dec. 14, 1987), 75−79.
12.51 Anon.: "Technical Advances in Tyres, Wheels and Brakes." *Aircraft Engineering,* (Nov. 1987).
12.52 Currey, N.S.: Aircraft Landing Gear Design: Principles and Practices. AIAA Education Series. 1988.

CHAPTER 13.0

ENGINE MOUNTS

13.1 Introduction

To least affect the aerodynamic characteristics of the wing, it would be desirable to locate the nacelle below the wing [see Fig. 13.1.1(a)]. To reduce torsional loads imposed on the wing structure by the eccentric thrust-line position, it would be desirable to locate the nacelle more or less with its axis in line with the wing chord line. Usually, however, the governing condition for the low-wing monoplane is the required propeller clearance with the ground as shown in Fig. 13.1.1(b).

(a) Wing-pod (pylon) mount of a jet engine — L-1011

(b) Wing mount of propeller engine — Lockheed Electra (L-188)

Fig. 13.1.1   Wing mount of engines — low wing configuration.
(Courtesy of Lockheed Aeronautical Systems Co.)

Airframe Structural Design 471

For jet engines, the wing-pod mount is preferred; since fuel is carried in the wing, the location of the jet pod below the wing is a primary consideration. The torsional moment imposed on the wing is desirable to offset the wash-out of the wing ocurring at high angles of attack and under accelerating conditions. Fig. 13.1.2 illustrates common engine mounts for modern jet airplanes.

Rear Fuselage Mount vs Wing-pod Mount
The question is whether wing-pod mounts [Fig. 13.1.1(a)] or rear fuselage mounts (Fig. 13.1.3) is the better solution.
(1) The favorable claims for fuselage mount are:
- The clean wing allows a high C_L maximum and therefore a shorter take-off field.
- The small spanwise offset alleviates the control problems in the "engine-out" case.
- No engine ground clearance limitation, therefore can use shorter gear struts.
- There is less noise in the cabin.
- The airplane gross weight penalty is minimized.
- The gross drag penalty is minimized.

(2) The counter arguments are:
- The smaller yawing moment of an engine out is largely cancelled by the smaller tail arm resulting from the airplane C.G. problem.
- Less noise in the cabin is granted; however all jet aircraft can be made sufficiently quiet in the cabin.
- The rear-engined layout carries a weight penalty; the fuselage is heavier, all the engine systems are heavier, the fin is much heavier, because of the need for a high tailplane, and no engine bending relief is obtained on the wing.
- The concentrated weight of the engines placed well aft of the airplane C.G. not only has a destabilizing inertia effect, but makes it difficult to achieve indiscriminate positioning of the payload.

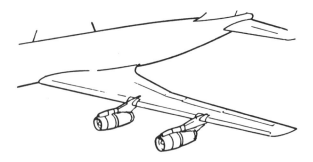

(a) Wing-pod mount-subsonic

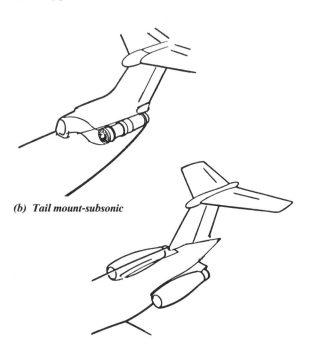

(b) Tail mount-subsonic

(c) Rear fuselage mount-subsonic

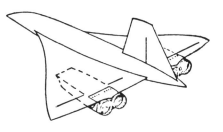

(d) Wing mount-supersonic

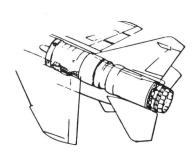

(e) Fuselage mount (single engine)- supersonic

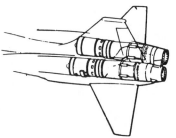

(f) Fuselage mount (twin engines) — supersonic

Fig. 13.1.2 Common engine mounts for modern jet airplanes.

So far as wing mounting is concerned, the top requirement is clearly to try and achieve favorable aerodynamic interference with the wing or, at worst, to minimize the unfavorable interference.

Wing-pod Mount Study

For positions of the engine nacelle centerline (exhaust plane) relative to the wing leading edge (see Fig. 13.1.4), $\frac{X}{C} >$ approximately 0.20, it is found that favorable interference is obtained with increasing vertical displacement $\frac{Z}{D}$ up to 1.1. Only at forward nacelle locations, $\frac{X}{C} < 0.20$, when it is beneficial to move the nacelle up toward the wing to obtain highly favorable results.

Therefore, the interactive effects between the flow fields of the wing and nacelle combination is a highly important consideration in the attainment of a low drag installation. In addition, small differences in nacelle aftbody contour, aftbody length, and jet efflux characteristics can introduce interactions leading to large effects on wing-nacelle drag. Every small change in aftbody contour or length must receive extensive wind-tunnel evaluation before commitment to final design. Finally, careful flight test evaluation must be conducted on an operating prototype configuration and may require post-flight modification.

The effect of each turbofan engine nacelle configuration on the total installed weight, airplane drag, and engine performance plus their combined effects on aircraft range are discussed as follows, (see Fig.13.1.5).

(a) The integral nacelle configurations (1) and (2) are heavier than the corresponding pylon mounted nacelles [Conf. (3) through (6)]. This is particularly true for the aft wing mounted version due primarily to the structural weight penalty.

(b) The integral nacelle [Conf. (1)] located in the mid-chord position has a high interference drag. The gearbox located in the gas generator section creates a loss in net thrust and/or specific fuel consumption resulting from the high duct velocities and seal leakage through the fan duct access door to the accessory compartment.

(c) The long fan duct [Conf. (3)] is heavy and offers a potential engine performance advantage inherent in efficient exhaust mixing; but achieving this efficiency involves an intensive development effort. The increased engine performance estimated for the long duct is slightly more than that required to compensate for the increased weight of the installation.

Fig. 13.1.3   Rear fuselage mount.

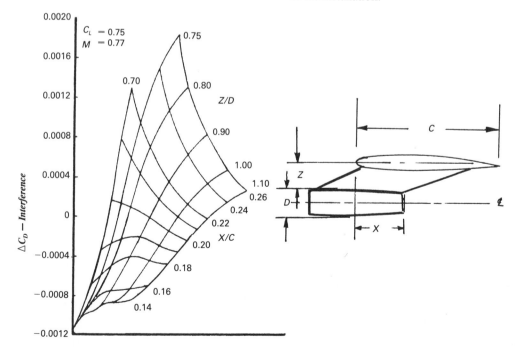

(For preliminary design)

Fig. 13.1.4   Wing-pod mount nacelle position effect on drag ($C_L = 0.45$).

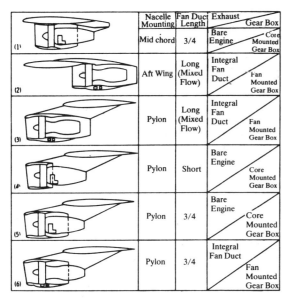

Fig. 13.1.5 Wing-pod mount nacelle/engine configurations.

(d) The aft mounted nacelle [Conf. (2)] offers the lowest drag of any configuration. However, the saving is not sufficient to overcome the high weight penalty.
(e) The short fan duct [Conf. (4)] is the lightest configuration. However, it has disadvantages of reduced engine performance due to reduced fan nozzle efficiency which is caused by close-coupled flow turning and internal pylon interference and increased scrubbing drag. The weight saving gain in Conf. (4) is not sufficient to make up for the reduced performance levels.
(f) On examining the $\frac{3}{4}$ length fan cowl with the core mounted gear box [Conf. (5)], it is seen that even though it has a small weight and drag advantage, the thrust loss due to access door seal leakage and high duct velocity results in an appreciable and unacceptable airplane range loss.
(g) Conf. (6) is the common choice; major advantages summarized below:
- Power plant installation design — The configurations with the external fan mounted gear box offer design simplicity in such areas as accessories and auxiliary systems. The long fan duct cowl configuration offers the advantage of a single target reverser for both the fan and primary streams.
- Maintenance — The configurations, with externally mounted accessories offer advantages in maintenance. The long fan duct nacelle provides reverser simplicity with its inherent maintenance advantages.

Loads

Once the aircraft engine and location has been determined, the next step is installation on the aircraft. The thrust and inertia loading on the engine and the air loading on its attached structure are carried back to the aircraft body via the engine mounts. The engine mounts and support structure will react loads in any direction as P_x (thrust), P_y (side loads), P_z (vertical loads) and the corresponding three moments M_x, M_y, M_z (see Fig. 13.1.6).

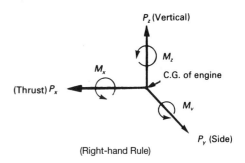

Fig. 13.1.6 Engine loads.

The nacelle, nacelle strut, and engine mounts shall be designed for the following inertia load conditions for preliminary sizing of commercial transports (for additional requirements or higher load factors for military airplanes, refer to Chapter 3):

Condition	Ultimate load factors (n)
Vertical	6.5
	$6.5 + 1.5\ T(C)$
	-3.5
	$-3.5 + T(C)$
Thrust	$3.0\ T(\text{max}) + 3.0$ vertical
	$3.0\ T(\text{max}) + 1.5$ vertical
	$3.0\ T(R)$
	$3.0\ T(R) + 3.0$ vertical
Side	± 3.0
Gyroscope	± 2.25 rad/sec yaw $+ 1.5\ T(C)$ $+ 1.5$ vertical
	± 2.25 rad/sec pitch $+ 1.5\ T(C)$ $+ 3.75$ vertical
Engine seizure	Torque equivalent to stopping mass in approximately 0.60 sec
Where: $T(\text{max})$	= maximum take-off thrust at sea level
$T(C)$	= cruise thrust (maximum or minimum, whichever is critical)
$T(R)$	= reverse thrust

Besides these engine mounts to react all the engine loadings, the other design requirement is to allow the thermal expansion force in such a way that it does not jeopardize the engines support structure.

Engine Break-away Design

It shall be established by structural analysis that in the event of a wheels-up emergency landing (as shown in

Fig. 13.1.7) or in the event of a sudden centrifugal imbalance caused by the loss of fan blade material, fuel tank rupture shall not occur at the area of separation for the wing-mounted engine installation. Usually, the break-away points are designed between the engine pod and pylon location either using specially designed shear-fuse bolts or shear-off structure. To establish the strength of these break-away loads, usually the test is conducted which gives the exact values used for designing pylon structure and local wing box back-up structures. A minimum 15% margin is maintained in the analysis for this structure to ensure that the wing structure has strength adequate to prevent fuel tank rupture for emergency break-away loads.

In addition, the same attention also should be given to the engine break-away during flight conditions in case of engine vibration due to broken fan or blade.

Other Considerations

The vibratory frequency imposed on the jet engine mounts by the turbo-jet engine is usually well above the range of vibration of the aircraft mounting system and structure. In addition, the vibratory loads in all directions are so small that vibration isolators or absorbers are frequently not required. The problem of mounting the jet engine consists of a straightforward design to take thrust, torque, gyroscopic, flight, landing, and any other loads peculiar to the intended operation of the aircraft.

All engines, auxiliary power units, fuel-burning heaters, and other combustion equipment which are intended for operation in flight as well as the combustion, turbine, and tail-pipe sections of turbine engines must be isolated from the remainder of the airplane by means of firewalls, shrouds, or other equivalent means.

Firewalls and shrouds must be constructed in such a manner that no hazardous quantity of air, fluids, or flame can pass from the compartment to other portions of the airplane. All openings in the firewall or shroud must be sealed with close-fitting fireproof grommets, bushings, or firewall fittings. Firewalls and shrouds must be constructed of fireproof material and be protected agsinst corrosion.

13.2 Propeller-driven Engine Mounts

The engine-mount supports virtually the entire power plant in the usual case, although for engines of very large horsepower the engine mounts do not extend more than a few inches beyond the rearmost accessory. It is a better engineering design to use tubular steel supports as shown in Fig. 13.2.1, which may be enclosed with a suitable cowling because of the necessity of gaining access to various parts of the engine and its accessories.

An engine mount is a frame that supports the engine and holds it to the fuselage or nacelle. It may be made of built-up sheet metal, welded steel tubing, or some other suitable material. Engine mounts vary widely in appearance and construction, although the basic features of construction are similar and well standardized. They should be designed so that the engine and its accessories are accessible for inspection and maintenance. Engine mounts may be built as individual units which can be detached easily and quickly from the supporting structure. In many of the large transport aircraft, the engine mounts, the engine, and its accessories are removed and replaced as a single complete power-unit assembly. this makes maintenance and overhaul simpler as well as the time required for engine change much shorter.

The vibrations that originate in reciprocating engines are transmitted through the engine mount to the airplane structure; hence mounts for such engines must be arranged with some sort of rubber of syn-

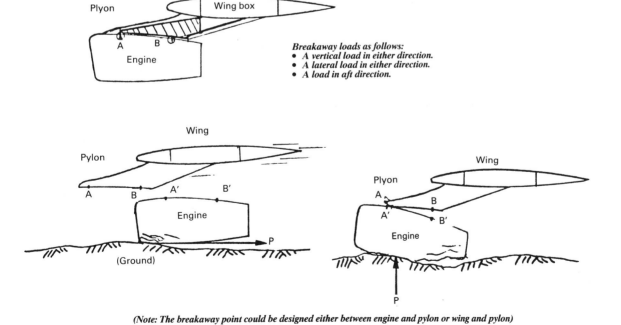

(Note: The breakaway point could be designed either between engine and pylon or wing and pylon)

Fig. 13.1.7 Engine breakaway case in emergency landing.

Airframe Structural Design 475

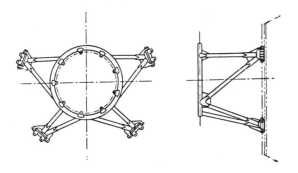

Fig. 13.2.1 Engine mount of a welded tubing truss structure.

thetic rubber bushings, as shown in Fig. 13.2.2, between the engine and mounts attaching structure for damping these vibrations. These bushings are often a part of the engine-mounting bracket and may be installed on the engine at the factory. The maximum vibration absorption is obtained when the mounting bolts are tightened so that the engine can move within reasonable limits from any fore-and-aft movement.

The torsional motion is then damped by the restraining action of the pads or cushions and the friction of the metal surface held by the bolts. If these bolts are too tight, the mount tends to vibrate with the engine, which is obviously undesirable.

The typical turbopropeller engine installation is shown in Fig. 13.2.3.

Fig. 13.2.4 shows a detailed turboprop engine mount known as the QEC (QEC is an abbreviated notation for "Quick Engine Change"). It is a semi-rigid structure made up of frames, shear panels, and truss members. The QEC is divided into three general structural subdivisions as follows:

(1) The nose cowl: It surrounds the engine air intake and gear box, and supports a complex of engine mounts, which includes two main mounts symmetrically located on either side of the QEC vertical centerline, two auxiliary mounts near the top centerline, and one auxiliary mount near the bottom centerline. The two main engine mounts will react loads in any direction while the top and bottom auxiliary mounts each react only fore-and-aft (X direction) loads. (The top mount is pin-ended and the lower mount is very soft in all but the X-direction). The nose cowl distributes

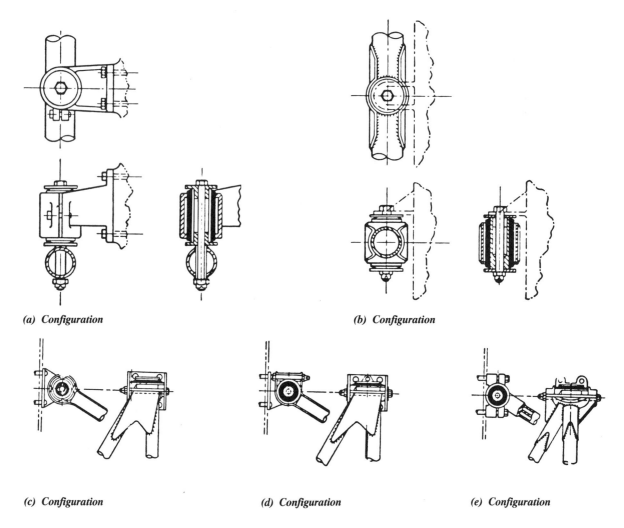

Fig. 13.2.2 Engine mounting lugs and isolators.

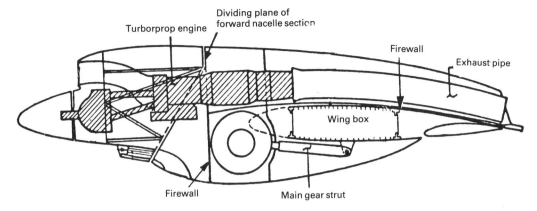

(a) Over-wing mount (Lockheed Electra)

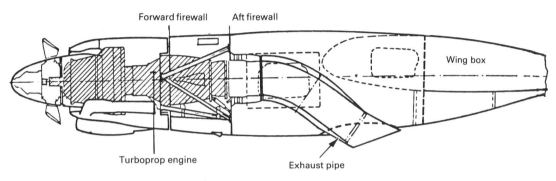

(b) Mid-wing mount (Aerospatiale Fregate)

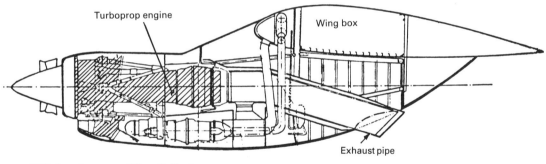

(c) Under-wing mount (Short Belfast)

Fig. 13.2.3 *Turbopropeller engine installation.*

the engine mount loads to the truss and shear panel members of the QEC structure.

(2) The truss members: These are top and bottom longeron members of fabricated metal construction plus tubular vee-frame members which form two substantially vertical truss planes symmetrical about the QEC longitudinal centerline. As separate trusses, each is capable of carrying both fore-and-aft (X-direction) and vertical (Z-direction) loads between the nose cowl and supporting structure (nacelles).

(3) The shear panels: There are two of these and they form the top and bottom cowls. They are substantially horizontal, of typical tension field shell construction, and are capable of transferring lateral (Y-direction) load from the nose cowl to the supporting structure (nacelles). The upper and lower longerons, all of which have simultaneous use as elements of the trusses mentioned above, form the side frames of these shear cowls. The two side cowls are non-structural doors which provide major access to engine components on each side. The QECs are interchangeable between all four nacelles.

Airframe Structural Design 477

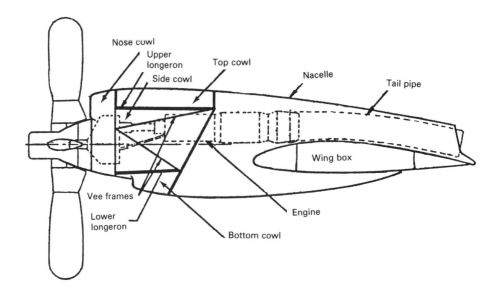

Fig. 13.2.4 Detail configuration of QEC engine mount — Lockheed Electra.

Another turboprop engine installation is the Short Belfast transport [see Fig. 13.2.3(c)]. Each engine is mounted in a nacelle of semi-monocoque construction underslung from the wing. The forward end of the nacelle terminates in a stainless steel fireproof bulkhead; forward of this the engine is carried by a steel tubular mounting which attaches to the engine through four anti-vibration bushes with synthetic rubber snubbing pads. Although the lines of the aft portion of the nacelles differ at the inner and outer positions due to the changing wing section, all equipment and installations, including the jet pipe, are identical for all engines.

13.3 Inlet of Jet Engine (Fighter)

Matching the air inlet system of the airframe to the turbojet and turbofan engine requirements presents a more difficult problem as speeds move into the transonic and supersonic region. Some of the design problems encountered are:
- Need for high inlet pressure recovery.
- Undesirable flow instability at Mach numbers over 1.5.
- Flow pattern distortions caused by the forward part of the induction system.
- Duct rumble due to flow separation ahead of side inlets at low air flows, separation at inlet lips at high angles of attack.
- Excessive spillage.
- Reasonable efficiency of airflow and energy recovery.

The advent of the various types of jet engines requiring such enormous amounts of air has increased the importance of the design of the induction systems, especially for modern turbofan engines. Not only have important gains been made in obtaining good ram recovery, but there has also been a marked impact on the amount of drag (in relation to the entire airplane design) that may be accepted. Airplanes powered with jet engines obviously require well-coordinated and integrated aerodynamic, thermodynamic, and structural designs.

For supersonic induction systems, additional classification may be made, such as:
- The external compression type, using either a ramp or a conical inlet.
- The internal compression type, employing either the so-called two-dimensional or three-dimensional design.

Nose Inlet

This is located in the nose of a fuselage as shown in Fig. 13.3.1. The apparent favorable features of this type of inlet are:
- It is free of boundary layer effects.
- Except for large angles of attack, there are few flow complications.
- Excellent energy recoveries are possible.
- High critical Mach numbers may be obtained.
- It is considered an optimum design insofar as inlet aerodynamic characteristics are concerned.
- It is ideal for nacelle installation, especially when nacelles are below the wing.

The disadvantages are:
- The duct system may be excessively long.
- The duct system may be of an undesirable shape.
- The design may lead to excessive weight.
- The design may lead to structural complications.
- If it is used in the nose of a fuselage, vision is

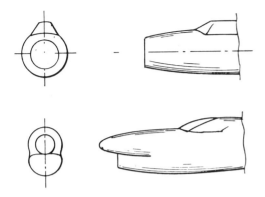

Fig. 13.3.1 *Nose inlet in fuselage.*

restricted since a more aft location of the cockpit and wider nose of fuselage are usually required.
- Radar equipment cannot be located in the fuselage.

Advanced Fighter Inlet Arrangement

One of the many problems facing the aircraft designer is that of integrating the airframe and inlet to obtain the best overall advanced fighter performance. Should the inlet be placed ahead of the wing, above it, or below the wing? Should the inlet be circular or rectangular? Each of these approaches works and has been successfully used. Fig 13.3.2 shows eight configurations which are commonly used on current fighter airplanes.

The synthesis and performance evaluation of several designs have been presented below:
- No one particular combination of inlet type, inlet location, and airframe design gives the best performance under all conditions.
- Configurations of conventional design with a side-by-side engine arrangement have favorable characteristics for transonic design conditions because their weight and friction drag are less than designs with separated engines.
- Configurations with separated engines have desirable features for super-sonic design conditions; the wave drag is less than for configurations with a side-by-side engine arrangement.
- Axisymmetric inlets have favorable characteristics for transonic design conditions because their spill drag is low.
- Two-dimensional (2-D) inlets performed well under supersonic design conditions owing to good pressure recovery at a reasonable level of inlet drag.

The engine of SR-71 is so installed to get the inlet away from the wing (see Fig. 13.3.3) and fuselage effects within the limits of the shock patterns developed by the fuselage nose, and to get the nozzle to work in a field to minimize drag.

13.4 Wing-pod (Pylon) Mounts

The pylon (wing-pod mount structure) is illustrated in Fig. 13.4.1. This is basically applied on subsonic jet transports. Engines are supported by box-beams of aluminum, titanium or steel construction. Doors are provided for systems access and inspection. The forward engine mount bulkhead and lower spar act as firewalls and the aft engine mount bulkhead is a secondary fire seal, all of titanium or steel alloy. The pylon (pod) leading edge is stiffened with transverse ribs and is quickly removable for systems access. Pylon structure may be identical for left and right pylons, thereby minimizing spare parts required. The pylon is attached to the wing front spar and lower skin panel. Pylon loads are distributed to the wing structure in such a manner that wing box secondary deformations are minimized. Vents and drain holes are provided to preclude the accumulation of flammable vapors and liquids.

The wing pylon structure as illustrated in Fig. 13.4.1 (a) is a cantilever box beam consisting of two upper and two lower longerons. Two side skins transmit the vertical shears and a lower skin primarily carries the lateral shear and also acts as a firewall. Forward and aft mount bulkheads are included to transfer the engine loads to the pylon structure and the bulkheads are included to transfer the engine loads to the pylon structure and the bulkheads take the pylon loads onto the wing box structure via lug attachments to the wing front spar of the pylon upper longerons and utilizes a rear drag strut to transfer the pylon lower longeron loads to a point between the wing front and rear spar.

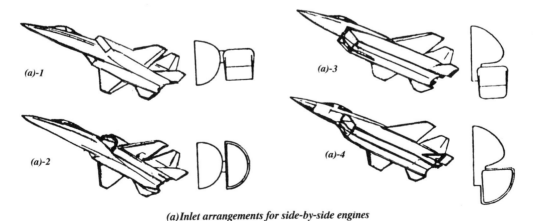

(a) *Inlet arrangements for side-by-side engines*

Fig. 13.3.2 *Advanced supersonic fighter engine inlet arrangements.*

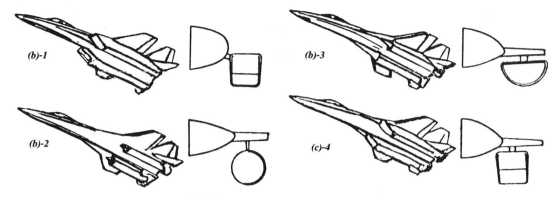

(b) Inlet arrangements for separated engines

Fig. 13.3.2 (continued).

Fig. 13.3.3 Engine mount of Mach 3+ SR-71 airplane.

Fig. 13.4.1(b) shows the pylon box beam design, which is to extend the box structure beyond the wing front spar fitting and ends at the aft pylon fitting, which is attached between the wing front and rear spars. This design is to put more weight on the pylon, but save weight on the wing box and minimize some potential fatigue problems at the wing lower surface.

The pylon is attached to the wing, through a fitting on the wing front spar for vertical and side loads, to a fitting beneath the front spar on the wing lower surface for thrust loads, and to a fitting attached to wing box structure on the wing lower surface at the

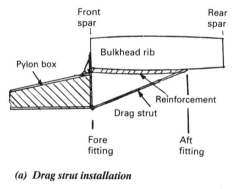

(a) Drag strut installation

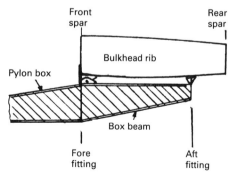

(b) Box beam installation

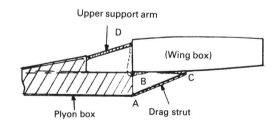

(c) With upper support arm (redundant support)

Fig. 13.4.1 Wing-pod (wing-pylon) mount configurations.

480 Airframe Structural Design

end of the pylon for vertical and side bending loads. Spherical bearings are used at the pylon-to-wing attachments to avoid over constraint to the wing lower front spar.

Side fairing panels, with attached bulb seals, cover the gap between the pylon structure and wing lower skin. The pylon structure is identical left and right and is interchangeable to minimize spare parts. However, this design will complicate the design at the fitting on the wing front spar. The pros and cons depend on trade-off results.

Fig. 13.4.1(c) illustrates an installation which is a redundant support structure.

The advantages of such a structure are:
- It is the most efficient structure to react the moment loads due to the overhanged engine; the moment arm A-D is obviously greater than A-B and, therefore, a lighter structure is achieved.
- The most efficient configuration transfers the engine moment loads into the wing box structure and therefore further weight saving is obtained.
- This benefits the design of engine position closer to the wing lower surface for the purpose of engine-to-ground clearance [see Fig. 13.1.1(a)] or for externally blown flap (EBF) propulsive-lift concept as used in the McDonnell Douglas YC-15 transport (Fig. 13.4.2 and Fig. 9.3.9).
- It inherently has the structural fail-safe feature due to the redundant design.
- The engine position can be located further forward without severe structural weight penalty.

The disadvantages are:
- Complicated structural analysis due to its redundant design.
- More rigging problems to ensure the proper structural load distributions.
- Interference with wing leading edge control systems such as control cables, rods, hydraulic tubes, heating ducts for de-icing, etc.
- Complexity of mounting and dismounting.

Design Example of Fig. 13.4.1(a)
Fig. 13.4.3 illustrates the L-1011 wing-pylon engine mount. Fig.13.4.3(a) shows detailed engine mounts to pylon; the forward mount transmits fore and aft loading (x-direction), lateral loading (y-direction) and vertical loading (z-direction). The aft mount transmits lateral, vertical, and torque (M_x).

Fig. 13.4.3(b) illustrates the pylon structure and longerons made of titanium and the wing pylon main attachments are electronic beam-welded on upper longerons. The drag strut is made of steel to meet stiffness requirements for wing flutter.

Design Example of Fig. 13.4.1(b)
Fig. 13.4.4(a) illustrates the A300 wing-pod mount and Fig. 13.4.4(b) is the DC-10 wing-pod mount.

The Lockheed C-141 engine is underslung from the pylon by a three point statically determinate attachment as shown in Fig. 13.4.5.

The C141 pylons are of conventional box beam construction, with four longerons connected by upper and lower cover panels and side panels. There are five bulkheads within the box beam. The bulkheads serve to distribute loads into the box beam due to engine mounts and forward wing-to-pylon attachments in addition to maintaining the shape of the box beam. The upper and lower cover panels and the side panels are of conventional skin-stiffener construction. The engine is attached to the pylon at three points (two forward points and one aft point) by a statically determinate system of links and fittings. All fore and aft loads are carried by the left hand forward mount through a thrust link and introduced into the lower left longeron and bulkhead. Vertical and side loads are beamed to the forward and aft mounts. Moments about the fore and aft axis (M_x) are carried as a couple on the forward two links. Moments about the vertical (M_z) and horizontal (M_y) axes are carried as couples between the forward and aft mounts. The forward mount introduces loads into the box beam through the power plant bulkhead. The aft mount introduces loads into the box beam through another bulkhead.

The pylon is attached to the wing at five points, (see Fig. 13.4.5). The four forward points (A, B, C, D) attach to the wing front beam while the aft point attaches to the wing at a point (E) between the front and rear wing beams. The two extreme forward pylon points (A and B) are attached to the upper cap (H and J) of the wing front beam by the N truss. The outboard point (D) underneath the front beam carries loads in all three directions while the inboard point C carries loads in the drag-vertical plane only. The aft attach point (E) carries loads in all three directions.

The S-3A engine is mounted to the pylon at three locations (see Fig.13.4.6). The aft mount consists of sway braces that take the loads produced by roll moment as well as vertical applied loads. The thrust pin is located immediately in front of the sway braces, which takes fore and aft and side loads. The forward

Fig. 13.4.2   Over-wing mount — YC-15.

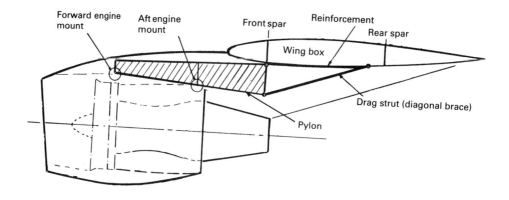

(a) Wing-pylon mount

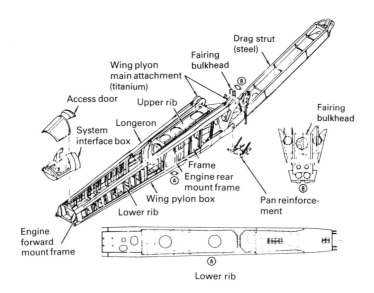

(b) Pylon structure

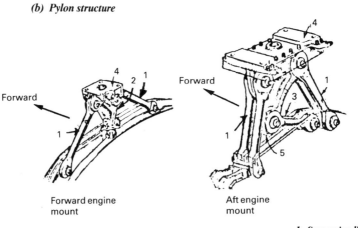

(c) Engine mounts

1. *Suspension links*
2. *Thrust link*
3. *Sway link*
4. *Mount fitting*
5. *Fail-safe pad*

Fig. 13.4.3 Wing-pylon mount configuration — L-1011.

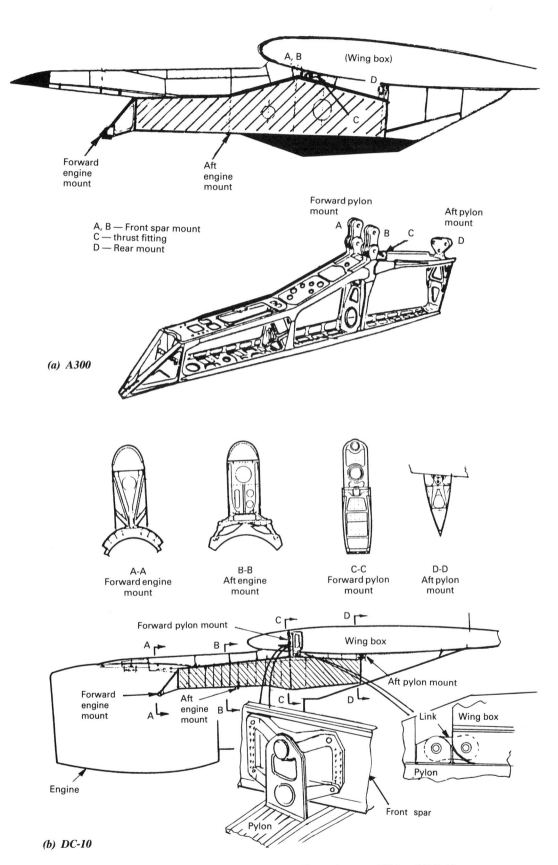

Fig. 13.4.4 *Wing-pylong mount configurations — A300 and DC-10.*

Airframe Structural Design 483

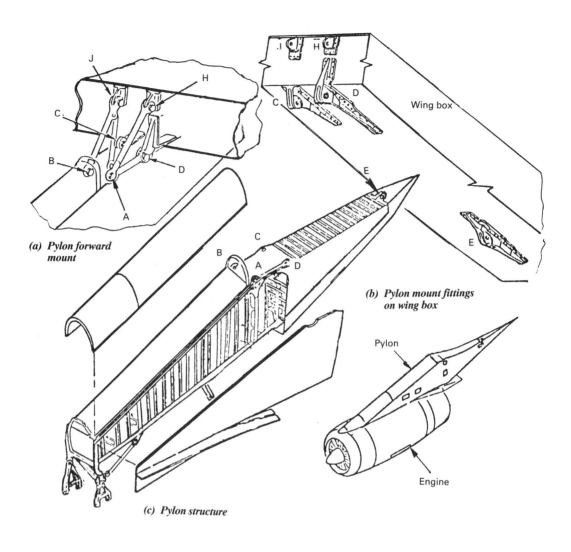
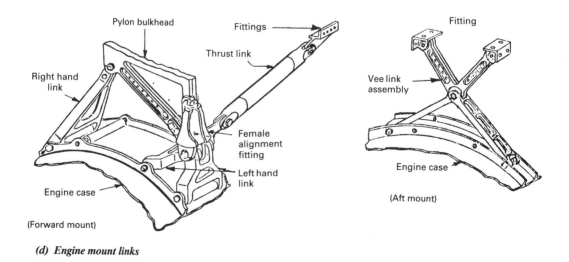

Fig. 13.4.5 Wing-pylon mount configuration — C-141.

mount is a swing link that takes only vertical and side loads. Despite the complexity of the engine mount system, it is a statically determinate structure.

The S-3A pylon is basically of truss design. Some of the outside skins carry shear by virtue of being loaded through differential strains in the truss members. Steel skin and members are used in the lower forward portion of the pylon where they are utilized as fire barriers in case of engine fire.

The pylon is rigidly mounted to the wing at the front and aft attachment points, thereby being capable of taking loads in all three directions at both points. The aft attachment has an eccentric bushing which is locked in place once the pylon is installed on the wing. This mounting system was chosen over one with a swing link at the aft attachment point after studies indicated a substantial stiffness gain.

Design Example of Fig. 13.4.1(c)
Fig. 13.4.7 illustrates the Boeing 747 pylon and its mountings. The pylon has three spars (longerons) — upper, middle and lower- and three major bulkheads, and is attached to the wing at four primary point. These are two mid-spar fittings, an upper link and a diagonal brace (drag strut). The attachment pins are secured with "fuse" bolts which are hollow carbon steel devices that have been heat-treated to shear-fail at a defined load.

In the landing breakaway condition (wheels-up landing), the sequence is designed to fail the upper and lower links so that the pylon rotates around the mid-spar and upward.

The wing-pylon design provides considerable load path redundancy such that an upper link can fail, partially or completely, and there is an alternate path — lower diagonal brace.

Fig. 13.4.7(c) shows the engine mounts. The forward mount takes vertical and side loads, and the aft mount takes engine thrust load, torque moment (M_x), vertical load and side load.

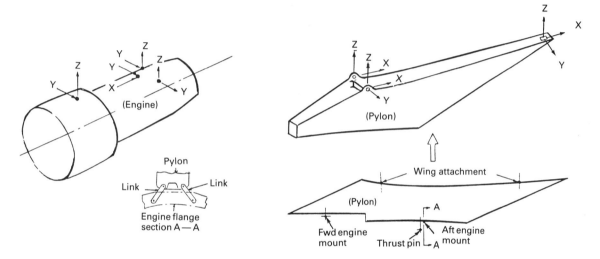

Fig. 13.4.6 Wing-pylon mount configuration — S-3A.

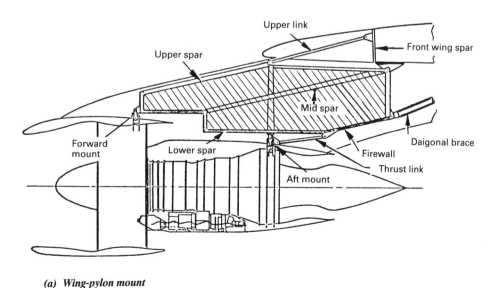

(a) Wing-pylon mount

Fig. 13.4.7 Wing-pylon mount configuration — Boeing 747.
(Courtesy of The Boeing Co.)

Airframe Structural Design 485

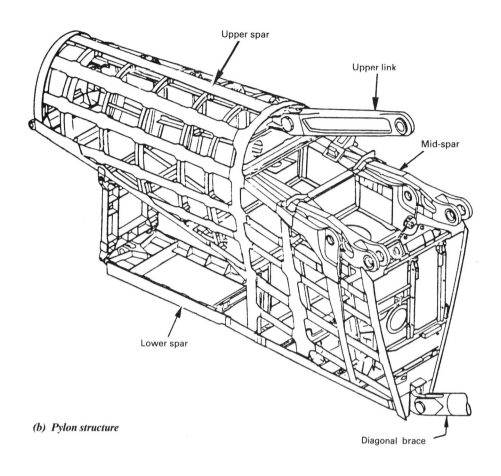

(b) Pylon structure

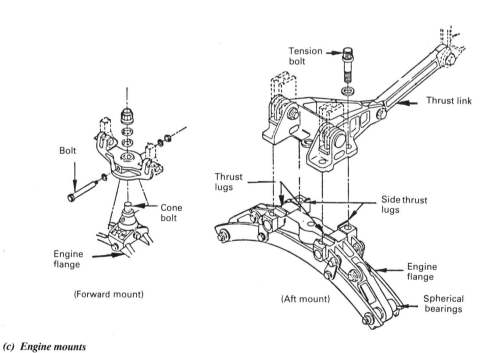

(c) Engine mounts

Fig. 13.4.7 (continued).

486 Airframe Structural Design

13.5 Rear Fuselage Mount and Tail Mount

Rear Fuselage Mount

Fig. 13.5.1 shows rear fuselage mount where provisions may be made for mount features on either side of the engine.

(a) Fig. 13.5.1 (a) shows the lightest engine mount but results in a heavy cowling (part of the cowling is used as engine support structure) but this configuration is one of the best designs to suit rear fuselage dual engine mounts (two engines per side of the fuselage) as shown in Fig. 13.5.2 for instance.

(b) Fig. 13.5.1(b) produces a heavier engine mount with the lightest cowling. Such an example is shown in Fig. 13.5.3.

The two pairs of arched cantilever beams, which are bolted on the extremities of the forward and rear engine mounting beams in the fuselage, are machined from steel forgings and are safe-life components.

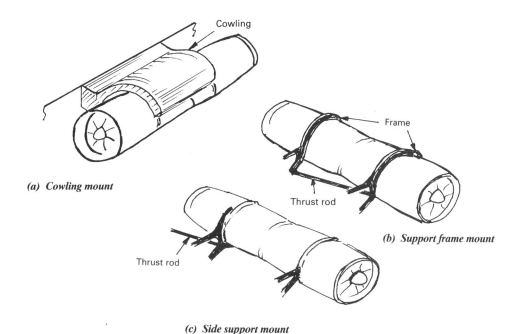

Fig. 13.5.1   Rear fuselage mount cases.

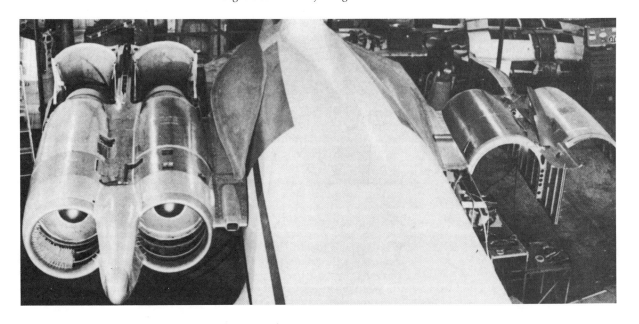

Fig. 13.5.2   Rear fuselage dual-engine mount — Lockheed Jetstar.
(Courtesy of Lockheed Aeronautical Systems Co.)

The engines are each attached to the airframe at three mounting points. To achieve interchangeability of the engine in the three installed positions, five pick-up fittings are provided on each engine. These fittings comprise a trunnion on each side capable of transmitting load in any direction, two lugs between the trunnions on the upper half of the casing and a rear suspension lug on top of by-pass casing above the rear bearings of the engine.

On the side engines the attachment points to the front cantilever beam are the inboard trunnion and a single link to one of the lugs on the upper part of the compressor casing. The rear cantilever is bolted at its outer end to the rear suspension lug on the engine. Thrust is transmitted from the engine into the mount through the front inboard trunnion only. A tubular steel strut carries the forward and reverse thrust loads from the trunnion, diagonally back into a fitting at the root end of the rear cantilever mounting beam, from which they are diffused forward into the fuselage skin by a heavy stringer.

(c) Fig. 13.5.1(c) shows the engine attached at the side and supported by a beam extending part way over the top of the engine. An added advantage of this mount and the one shown in Fig. 13.5.1(b) is that the beam can be used to carry a gantry for ground handling the engine into the nacelle.

The rear fuselage mount is basically similar to that of the wing-pod mount, and the engine attached pod mounting at three points (two forward and one aft or vice versa) by a statically determinate system of links and fittings. Some designs use diagonal struts to diffuse the engine thrust load into the fuselage which gives the lightest pod design. Fig. 13.5.4 shows the engine thrust load transmitted by a pod (pylon) rather than diagonal strut.

Tail Mount

The tail mount system is also similar to that of a wing-pod mount by a statically determinate structure of links and fittings. Mountings for the tail engine (for transports) is shown in Fig. 13.5.5. A torque box is cantilevered off the fin and aft fuselage bulkhead which picks up both engine forward and aft mounts. These are tri-engine transports and, therefore, it is convenient to use the same engine mounts as the wing-pod mount to avoid additional engine mount provisions.

All tail mount engines [inside fuselage as shown in Fig. 13.5.5(a)] are designed to withstand forward decelerations of at least 12.0g in those directions which would allow a less securely mounted engine to enter the passenger compartment.

The tail mount as shown in Fig. 13.5.6 for example, consists of the four corners, and the aft light weight fairing. The tail engine is mounted to the main torque box at two pads located forward and aft on the mating plane. Attachment is achieved with high strength tension bolts in clearance holes and captive, replaceable nuts. The forward pad (mount) transmits vertical

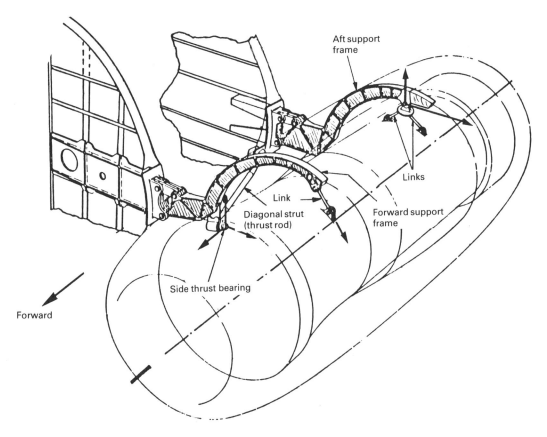

Fig. 13.5.3   Rear fuselage mount — Fokker F-28.
(Courtesy of Fokker B.V.)

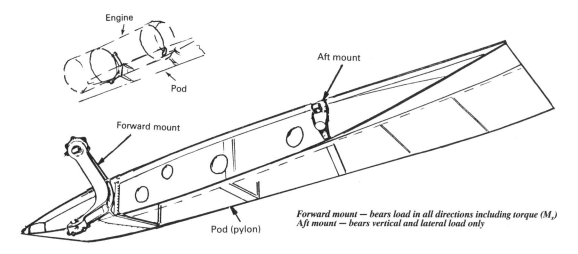

Fig. 13.5.4 Rear fuselage mount — Boeing B727.

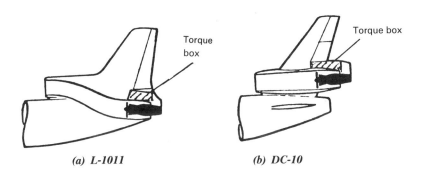

Fig. 13.5.5 Two typical tail mount configurations.

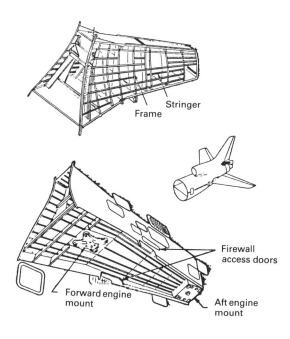

Fig. 13.5.6 Tail mount torque box — L-1011.
(Courtesy of Lockheed Aeronautical Systems Co.)

side, and thrust loads while vertical side, and torsional loads are transmitted through the aft pad (mount) in the same way as the wing-pod engine mount (see Fig. 13.4.3).

13.6 Fuselage Mount (for Fighters)

Mounting jet engines on the airplane structrue is somewhat simpler than in the case of reciprocating engines. The gas turbine can be installed within the airplane fuselage (basically for military fighters) rather than ahead of it, and attachment to the basic airplane structure can be accomplished with the addition of a minimum of interconnecting structures. The engine mount must be designed to prevent airplane deflections from introducing loads into the powerplant, and must permit thermal expansion of the engine both axially and radially. A typical mounting system is shown in Fig. 13.6.1.

The major portion of the vertical loads is carried on two trunnions located near the engine C.G. Side loads are taken out on the trunnion on one side only, the other being free to move laterally to allow for thermal expansion. The forward mount (some design are located in the aft) is a universal joint capable of carrying vertical loads only. Since the trunnions are located near the C.G., therefore, the major forces imposed on the front support arising from the gyro-

Airframe Structural Design 489

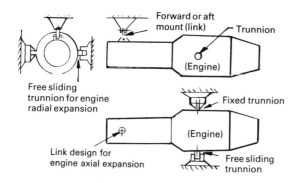

Fig. 13.6.1 *Typical fuselage mount arrangement.*

scopic couple caused by angular velocity in yaw and the inertia moment caused by angular acceleration in pitch are small. At present, most jet engines are attached rigidly to the airplane structure. Since the moving parts in a gas turbine have a simple rotary motion, and since combustion is continuous rather than intermittent, the unbalanced forces which might excite vibration are few in number and small in magnitude.

Some fuselage engine mounts for modern fighter jet engines are shown in Fig. 13.6.2.

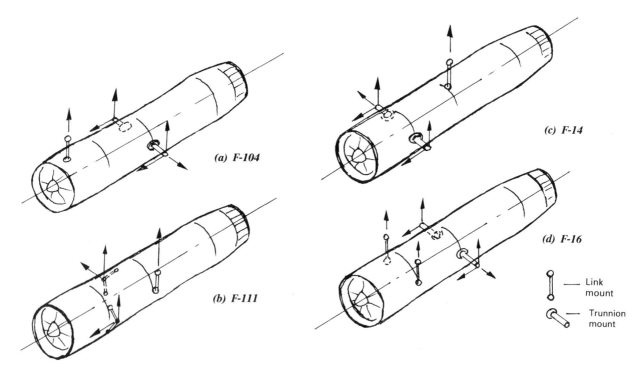

Fig. 13.6.2 *Examples of fuselage engine mounts — fighter airplanes.*

References

13.1. Hill, P.W.: "Airframe-inlet Integration". *AIAA paper No. 70–933*, (July. 1970).
13.2. Jordan, D.J.: "Design of Turbojet Installations". *Aero Digest*, (Oct. 1948).
13.3. Garbett, G.H.: "The Soviet YAK-40". *Aircraft Eng.*, (Dec. 1975).
13.4. Ramsd, J.M: "Lockheed, Douglas, Boeing and P&W in a Week". *Flight International*, (Nov. 13, 1969).
13.5. Technical Editor: "Boeing 747", *Flight International*, (May 9, 1963).
13.6. Harrison, N.: "Boeing 747", *Flight International*, (Dec. 12, 1986).
13.7. Technical Editor: "DC-9, The Douglas Family of Short- and Medium-haulers". *Flight International*, (Mar. 3, 1966).
13.8. Technical Editor: "Lockheed C-5A". *Flight International*, (Feb. 10, 1966).
13.9. Technical Editor; "Trident Structural Design". *Flight International*, (Jun. 1964).
13.10. Neal, M.: "VC-10, Vickers-Armstrongs' Long-range Jet Airliner", *Flight International*, (May 10, 1962).
13.11. Wilson, M.: "Transall C-160". *Flight International*, (Apr. 25, 1968).
13.12. Torenbeek, E.: *Synthesis of Subsonic Airplane Design*. Delft University Press, 1976
13.13. Ellis, V. and Keech, W.A.: "Flexible Mountings for Turbines." *Aero Digest*, (Jul. 1947).
13.14. Laser: "Design probe. (How many engines, and where?)" *Flight International*, (Aug. 8, 1968).
13.15. Anon.: "Two or Three Engines?" *Flight International*, (Sept. 18, 1969), 446.
13.16. Higgins, R.W.: "The Choice Between One Engine or Two." *Aircraft Engineering*, (Nov.1968).
13.17. Benedict, M. C. and Gabriel, A. W.: "Installation of Jet Turbines." *Aero Digest*, (Jan. 1947).
13.18. Graham, C.D. and Tembe, N. R.: "Torque on Engine Mountings." *Aircraft Engineering*, (Jun. 1942), 162–163.
13.19. Anon. "Aerodynamic of Power-plant Installation." *AGARDOgraph 103*. Part I and II, (1965).

13.20. Anon.: "Prevention of Engine Mount Failure". *Aero Digest*, (Dec. 1942).
13.21. Lawson, K.S.: "The Influence of the Engine on Aircraft Design." *Aircraft Engineering*, (Aug. 1969), 12–16
13.22. Russel, A.E.: "Some Factor Affecting Large Transport Aeroplanes with Turbo-porp Engines." *Aircraft Engineering*, (Mar. 1950), 76–84 & (Apr. 1950), 114–118.
13.23. Anon: "Integration of Aft-Fuselage-Mounted Flow through engine Nacelles on an Advance Trasport Configuration at Mach No. 0.6 to 1.0." *NASA TM X-3178*, (1975).
13.24. Moss, G.M.: "Some Aerodynamic Aspects of Rear-mounted Engines." *Journal of The Royal Aeronautical Society*, (1964), 837–842.
13.25. Anon.: "Airframe/Engine Integration." *AGARD-LS-53*, (1972).
13.26. Dugan Jr., J.F.: "Engine Selection for Transport and Combat Aircraft." *Von Karman Institute of Fluid Dynamics Lecture series 49*, (Apr. 1972).
13.27. McIntire, W.L.: "Engine and Airplane — Will it be a Happy Marriage?" *SAWE Paper No. 910*, (May 1972).
13.28 Cleveland, F.A. and Gilson, R.D.: "Development Highlights of the C-141 Starlifter." *Journal of Aircraft*, (Jul.–Aug. 1965).
13.29 Magruder, W.M.: "Development of Requirement, Configuration and Design for the Lockheed 1011 Jet Transport." *SAE Paper No. 680688*, (Oct. 1968).
13.30 Hesse, W. J. and Mumford Jr. N.V.S.: *Jet Propulsion for Aerospace Applications.* Pitman Publishing Corp., New York, N.Y. 1964.
13.31. Ward, P.A., Mucklow, P.A. and Herbstritt, K.: "Design for Installation of New transport Engines." *SAE 680334*, (1968).
13.32. Morrison, J.A.: "747 Engine Installation Features." *SAE 680335*, (1968).
13.33. Herring, H.W., Lee, G.G. and Reynolds, B.I.: "C-5A Propulsion System Installation." *SAE 680333*, (1968).
13.34 Gunston, W.T.: "Pod Pros and Cons. (Some reflections on American and British Engine-installation Method)." *Flight International*, (Sept. 11, 1953).

CHAPTER 14.0

ADVANCED COMPOSITE STRUCTURES

14.1 Introduction

Modern composites owe much to glass fiber-polyester composites developed over the 1940s, to wood over the past centuries, and to nature over millions of years. Numerous examples of composites exist in nature. For filamentary composites, an example can be found in bamboo.

Another example is seaweed which has been in existence for at least 150 million years. It is interesting that the concept of lamination to support planar configurations was utilized for at least that length of time.

Through the years, wood has been a common natural composite and the properties with and against the grains vary significantly. Such directional or anisotropic properties have provided design approaches to take advantage of the superior properties while suppressing the undesirable ones through the use of laminates.

Plywoods, for example, are made with an odd number of laminas. Such stacking arrangement is necessary in order to prevent warping. In the language of modern composites, this is referred to as the symmetric lay-up or zero extension-flexure coupling.

The emergence of boron filaments gave birth to a new generation of composites in the early 1960s. The composites that employ high modulus continuous filaments like boron and graphite are referred to as advanced composites. This remarkable class of material is cited as one of the most promising developments that has profoundly impacted today's and future technologies on airframe design. The term composite or advanced composite material will be defined as a material consisting of small-diameter, high-strength, high-modulus (stiffness) fibers embedded in an essentially homogeneous matrix. This results in a material that is anisotropic; and it has mechanical and physical properties that vary with direction. Terms commonly used in describing advanced composites are provided in the glossary of this section.

Serious development work with advanced composite materials started in the mid '60s with boron/epoxy as mentioned earlier. Since that time, a host of new materials have been added, including three types of graphite, organic materials such as Kevlar, and new matrix materials including polyimide, thermoplastics, and even metals such as aluminum, titanium, and nickel. Due to the remarkable specific properties of composite materials, component weight savings up to 30% were achieved. However, the resulting structures were generally much more expensive than the metal counterparts, due in part to the very high raw material costs and the fact that the major emphasis is on maximum weight savings. To accomplish this objective, the design approaches should emphasize structural simplification, reduced part count, and elimination of costly design features as illustrated by the example in Fig. 14.1.1.

Composite materials are ideal for structural applications where high strength-to-weight and stiffness-to-weight ratios are required. Aircraft and spacecraft are typically weight-sensitive structures in which composite materials are cost-effective. When the full advantages of composite materials are utilized, both aircraft and spacecraft will be designed in a manner much different from the present.

The study of composite materials actually involves many topics, such as manufacturing processes, anisotropic elasticity, strength of anisotropic materials, and micromechanics. Truly, no one individual can claim a complete understanding of all these areas. In this chapter, the emphasis is hence on practical design purpose rather than on theoretical analysis which can be found in either composite publications or references of this chapter. It is worthwhile to mention that both Ref. 14.23 and Ref. 14.60 provide valuable data for those who are interested in home-built composite airframes.

Over the past decades, a variety of composite materials have been developed which, when combined with the appropriate matrix, offer mechanical properties that are competitive with common aluminum and steel but fractions of their weight. Fig. 14.1.2 gives a comparison of the properties of several different composites with conventional metallic materials.

Matrix materials used in advanced composites to interconnect the fibrous reinforcements are as varied as the reinforcements. Resins or plastic materials, metals, and even ceramic materials are used as matrices. Today, epoxy resin is the primary thermoset composite matrix for aerospace applications.

In all thermoset materials, the resin matrix is cured by means of time, temperature, and pressure into a dense, low void content structure in which the reinforcement is aligned in the direction of anticipated loads. It is possible for the designer to locate and orient the reinforcement in sufficient quantity and in the proper direction, even in very localized areas to withstand the anticipated loads. This concept has been enhanced by the development of computer analyses, which can totally optimize a structure for

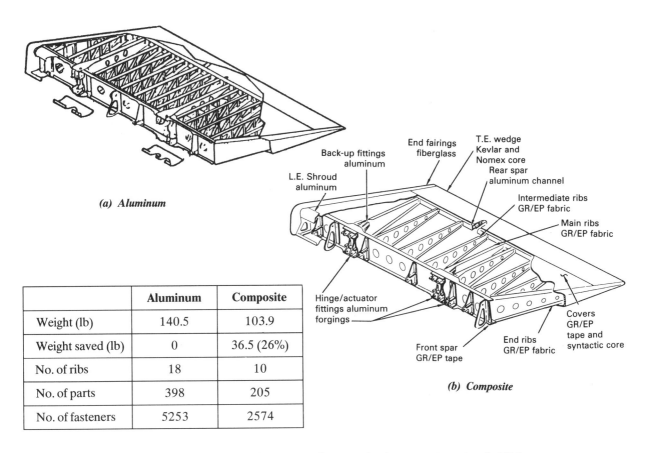

Fig. 14.1.1 Comparison of composite aileron to aluminum counterpart — L-1011.
(Courtesy of Lockheed Aeronautical Systems Co.)

	Graphite/Epoxy (Unidirectional)		Kevlar/Epoxy (Woven cloth)	Glass/Epoxy (Woven cloth)	Boron/Epoxy	Aluminum	Beryllium	Titanium
	High Strength	High Modulus						
Specific strength, 10^6 in	5.4	2.1	1	0.7	3.3	0.7	1.1	0.8
Specific stiffness, 10^6 in	400	700	80	45	457	100	700	100
Density, lb/in^3	0.056	0.063	0.05	0.065	0.07	0.10	0.07	0.16

Fig. 14.1.2 Composite vs. conventional materials.

weight.

Historically, aluminum materials have been the primary material for aircraft and spacecraft construction. Today, structural weight and stiffness requirements have exceeded the capability of conventional aluminum, and high-performance payloads have demanded extreme thermo-elastic stability in the aircraft design environment. During the past decades, advanced composite materials have been increasingly accepted for aircraft and aerospace structural materials by numerous developments and flight applications. To achieve the best composite structure design, the composite engineers should be trained to obtain the basic knowledge as well as experience about metal structures. As matter of fact, composite engineers shall not consider composite materials as a panacea, however, some areas in aircraft structures the metal material is still the most cost effective.

As mentioned previously, composite material costs are high compared with common aircraft metals. Design costs are higher due to these new materials and components testing; certification and documenta-

Airframe Structural Design 493

tion testing are even more costly. Furthermore, production and prototype tooling costs are higher than with conventional metals. Quality control, especially non-destructive inspection (NDI) is another high cost operation.

The maturation of composite technology is still in progress, but the basis of understanding has broadened significantly. Fig 14.1.3 illustrates the diversity of developmental experience now contributing to the recent commercial transport structures. However, application of advanced composite materials in civil aircraft has generally lagged behind military usage because:
- Cost is a more important consideration.
- Safety is a more critical concern, both to the aircraft manufacturer and government certifying agencies.
- A general conservatism due to past experiences with financial penalties from equipment downtime.
- Absorb military service and flight experience

The use of materials in military fighter aircraft construction has changed in the past and is expected to change further in the future [see Fig. 14.1.4(a)]. The development of advanced composites in the 1960s resulted in a quantum jump in weight saving potential. This trend will continue with the introduction of new high strain and high toughness composite materials such as thermoplastics which have been selected for use on the future U.S. advanced tactical fighter (ATF) [Fig. 14.1.4(b)].

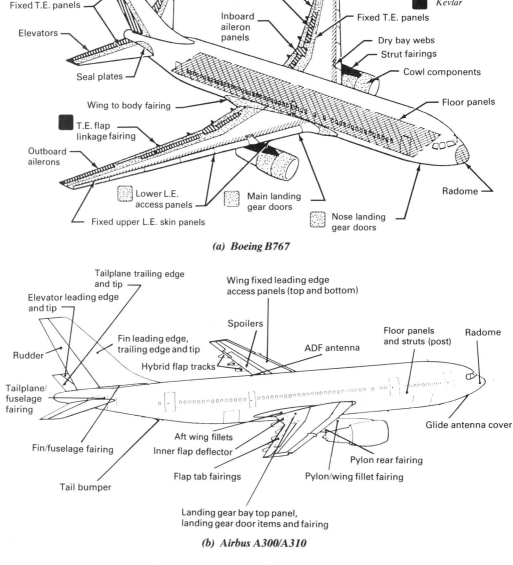

Fig. 14.1.3   Composite materials applications on commercial airplanes.

494   Airframe Structural Design

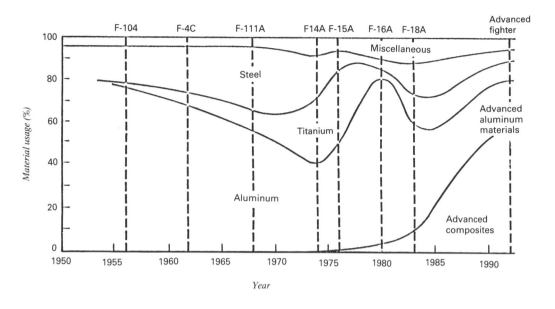

(a) Progress in using composite materials

(b) Lockheed artist's concept for an advanced tactical fighter (ATF) features using 60% of thermoplastic composite material on airframe structure

Fig. 14.1.4 Use of composite materials in military fighters.

Airframe Structural Design 495

Characteristics of Composites

The commonly and most often used advanced composite materials are graphite fiber, Kevlar fiber, and boron fiber. Graphite (or carbon) fibers are manufactured by pyrolysis of an organic precursor such as rayon or PAN (polyacrylonitrile), or petroleum pitch. Generally, as the fiber modulus increases, the tensile strength decreases. Among these fibers, graphite fiber is the most versatile of the advanced reinforcements and most widely used by the aircraft and aerospace industries. Products are available as collimated, preimpregnated (prepreg) unidirectional tapes or woven cloth. The wide range of products makes it possible to selectively tailor materials and configurations to suit almost any application.

An important element in determining the material behavior is the composition of the matrix resin that binds the fibers together. The resin formulation selected determines the cure cycle and affects such properties as creep, compressive and shear strengths, thermal resistance, moisture sensitivity, and ultraviolet sensitivity, all of which affect the composite's long-term stability. Fig. 14.1.5 presents a composite resin selection summary.

Resin composites can also be affected structurally by exposure to a moist environment. Since it is the resin and not the fiber that exhibits these hygroscopic characteristics, the matrix sensitive properties are seriously reduced, especially at high temperatures. For aircraft structures, which experience rapid changes of environment, this loss of mechanical performance due to moisture absorption can be a serious problem.

Kevlar (Aramid) is the trade name for a synthetic organic fiber and a density of 0.052 lbs/in^3 which gives Kevlar a specific tensile strength higher than either boron or graphite. Compared to other composite materials such as graphite and boron, Kevlar has poor compressive strength and this is an inherent characteristic of Kevlar which represents an internal buckling of the filaments. However, Kevlar demonstrates a significant increase in resistance to damage compared to other composite materials.

The boron family of reinforcement materials was among the earliest used in composite applications. Boron filaments are produced by the chemical vapor deposition of boron on a substrate filament, tungsten or carbon. Borsic is a modification wherein a surface coating of silicon carbide is deposited over the boron to provide a somewhat higher temperature capability. The boron/epoxy materials have very high tensile and compressive strength and stiffness comparable to most graphite/epoxy materials. A major drawback of boron/epoxy is the difficulty in handling the material. The fibers are extremely stiff and brittle, difficult to work with, and limit the minimum radius around which they can be wrapped. Another drawback involves ply thickness, which is determined by the filament diameter and desired fiber volume fraction. Most aircraft structures require thin skin and need thinner ply to tailor the ply orientation to optimize the light weight. But, the boron filaments are an order of magnitude larger than graphite fibers. The thickness of individual plies will always exceed fiber diameter in unidirectional laminates unless the fiber volume fraction is sufficiently low to permit nesting.

(1) Metal Matrix Composites

Metal matrix composites (MMC) are constructed in much the same way as resin composites. MMCs are produced by combining a reinforcing material in a ductile matrix of metal such as aluminum, magnesium, titanium, etc. The resulting MMC is lighter, stronger and stiffer than

---

- Epoxy:

    Most widely used; best structural characteristics; maximum use temperature of 200°F; easy to process; toughened versions now available

- Bismaleimide:

    Maximum use temperature of 350°F; easy to process; toughened versions becoming available

- Polyimide:

    Variety of resin types; can be used up to 500−600°F; difficult to process; expensive

- Polyester:

    Relatively poor structural characteristics limit usage to non-structural parts; easy to process

- Phenolic:

    Same limitations as polyesters; more difficult to process, but provide higher use temperature and low smoke generation

- Thermoplastics:

    Greatly improved toughness; unique processing capabilities, but also have processing difficulties

Fig. 14.1.5 Resin selection summary.

aluminum. The advantages of MMC are:
- The transverse strength is very high compared to resin composites. This is due to the fact that transverse strength is essentially the same as the strength of the matrix material, and, obviously, metals are much stronger than resins.
- A high temperature performance which is determined by the matrix alloy and they have been demonstrated at temperatures above 2000°F.

A variety of reinforcement-matrix combinations are used for MMCs; some of these are presented in Fig. 14.1.6. Each group of materials can have a range of properties depending upon the fiber and matrix.

Much of the MMC development work has been government funded; the major requirements of MMCs are good strength at high temperature, good structural rigidity and dimensional stability, light weight, and processing flexibility.

(2) Thermoset vs. Thermoplastic Matrix Resins

A brief comparison between thermoset and thermoplastic composites has been given in Section 4.6 of Chapter 4. Thermoplastic matrix resins are tough, resistant to high temperatures and solvents, and have low moisture sensitivity. Thermoplastic resins' major advantage over thermoset resins is their shorter fabrication cycle and do not need a chemical cure.

Even though a lot of fabrication problems have to be overcome, most structural designers still believe that thermoplastic composites will be the next generation of composites.

Composites vs. Aluminum Alloys

For some time it seemed as if composite materials would replace aluminum as the material of choice in new aircraft designs. This has put pressure on the aluminum developers to improve their products, and the result was aluminum-lithium. The first aluminum-lithium alloy, called 2020, was actually developed in the 1950s for the U.S. Navy RA-5c Vigilante.

Aluminum-lithium has several goals to meet:
- Damage tolerance similar to 2024 aluminum alloys but with 10% lower density
- High strength similar to 7075-T6 aluminum alloy but with 8% lower density
- Same properties as 2024-T3 but with a 10% lower density
- Price not more than two to three times the conventional 2000 and 7000 series alloys

The 25–35% weight savings composites offer over aluminum constructions plus a substantial reduction in the number of parts required for each application represents a major attraction of these composites. The obstacles to a wider use of these composite materials are their high acquisition cost compared with aluminum, the labor-intensive construction techniques and substantial capital costs of buying a new generation of production equipment. However, the labor-intensive construction can be solved by automation of the manufacturing process which is the key technology in developing composite materials. The use of tape laying machines, for example, can cut the time and cost of constructing composite components by a factor of ten or more.

The use of composites in the U.S. began in the early 1970s under USAF funding and in the late 1970s NASA helped kick some life into composite technology and produced the desired results — aircraft manufacturers became more comfortable with the materials and more efficient construction techniques were developed; the increased demand led to lower costs of composite materials.

Composite Standards

In the aircraft and aerospace (not limited to) industries, composite materials will undoubtedly play a major role. As the variety of composite materials and the number of producers continue to grow, the aircraft industry has responded by developing some industry standards.

The selection of composite materials for specific applications is generally determined by the physical and mechanical properties of the material, evaluated for both function and fabrication. The functional considerations include such items as the strength, weight, hardness, or abrasion resistance of the finished part. Fabrication considerations include cure cycle (time, temperature, pressure), quantity of parts, tooling costs, equipment, and availability of facilities.

There are a number of standards and specifications

Fiber	Matrix	Applications
Boron	Aluminum	Simple members for high tension or compression load. Beef-up aluminum member for additional strength
	Titanium	Higher strength structures
Borsic	Aluminum	Higher strength structures
	Titanium	Higher strength and high temperature
Graphite	Aluminum	Aerospace
	Magnesium	Aerospace

Fig. 14.1.6 Common group of MMC materials.

which are purposed to ensure repeatable results by carefully defining either the technical requirements of a material or the specific steps used in the manufacturing process.

(a) Military specifications: These are issued by the Department of Defense (DoD) to define materials, products, or services used only or predominantly by military entities.

(b) Military standards: These provide procedures for design, manufacturing, and testing, rather than giving a particular material description.

(c) Federal specifications and standards: These are similar, except that they have come out of the general services administration, and are primarily for federal agencies. However, in the absence of military specifications and standards for a given product, federal specifications and standards are acceptable.

(d) Federal Aviation Regulations (FAR): In addition to military specifications, the Federal Aviation Administration (FAA) has specifications for materials and part fabrication mothods used in commercial aircraft. The FAA specification includes several areas;
- Test plan — A unified program and schedule for tests that verify design allowables. These tests for composites might include a coupon test, static full-scale test for durability, environmental tests, stress analysis, and tests for subcomponents of a major structure.
- Process specifications — This includes both a material specification to be used to help select a commercial product and a process specification.
- Quality assurance plan — This details process inspection of fabricated parts, acceptance and repeatability tests for material.
- Report submission — Final report submission, audit tests, how tests are acccomplished, and who witnesses them.

(e) Company specifications — In many cases, companies feel that military standards and specifications do not reflect the most up-to-date materials and processing techniques. So the company have developed specifications that will ensure all the requirements for fulfilling the military contracts.

(f) International standards — The International Standards Organization Technical Committee 61 and its subcommittee 13 covers the reinforced composites.

Among the latest efforts from the DoD itself are two materials specifications:
- For thermoplastic composite (MIL-P-46179A), covers polyamide-imide, PES, PEEk, Polysulfone, PEI, PPS, etc.
- For high-temperature thermosetting composites (MIL-P-46187)

To date, the greatest problem is how to get users to agree on test measures and shrink the number of tests, now that every company has its own proprietary products and established its own specifications and handbooks. In composite technology, a composite standard is not only a must but essential for cutting cost.

Certification of Composite Structures

In August 1982, the FAA put out an Advisory Circular AC 20-107 on the certification of composite aircraft structures. It is a brief document stating that the evaluation of a composite should be based on achieving a level of safety at least as high as that currently required for metal structures. It also emphasizes the need of test for the effect of moisture absorption on static strength, fatigue and stiffness properties for the possible material property degradation of static strength after the application of repeated loads. In addition, it requires the development of more damage-tolerance with particular reference to effects of moisture and temperature and better strength on crashworthiness. Other data include flammability, lightning protection, weathering, ultraviolet radiation and possible degradation by chemicals and fuel; and also specifications covering quality control, fabrication techniques, continuing surveillance and repair.

In August 1982, the FAA published special "rules" which were applicable solely to the Lear Fan 2100 aircraft as shown in Fig. 14.1.7. This aircraft is made

Fig. 14.1.7   Lear Fan 2100 business aircraft airframe is made of 77% graphite/epoxy composite material.

of advanced composite material with extensive use of bonding in assembly. This material and assembly is completely different from the typical semi-monocoque aluminum airframe.

Glossary

Advanced Composites Advanced composites are defined as composite materials applicable to aerospace construction and made by imbedding high-strength, high-modulus fibers within an essentially homogeneous matrix.

Advanced Filaments Continuous filaments made from high-strength, high-modulus materials for use as a constituent of advanced composites.

Angleply Same as *crossply*.

Anisotropic Not isotropic; having mechanical and/or physical properties which vary with direction relative to natural reference axes inherent in the material.

Autoclave A closed vessel for producing an environment of fluid pressure, with or without heat, to an enclosed object while under-going a chemical reaction or other operation.

B-Stage An intermediate stage in the reaction of a thermosetting resin in which the material softens when heated and swells in contact with certain solvents but does not entirely fuse or dissolve. Materials are usually precured to this stage to facilitate handling and processing prior to final cure.

Balanced Laminate A composite laminate in which all laminae at angles other than 0° and 90° occur only in ± pairs (not necessarily adjacent).

Composite Material Composites are considered to be combinations of materials differing in composition or form on a macroscale. The constituents retain their identities in the composite; that is, they do not dissolve or otherwise merge completely into each other although they act in concert. Normally, the components can be physically identified and exhibit an interface between one another.

Constituent In general, an element of a larger grouping. In advanced composites, the principal constituents are the fibers and the matrix. Glass scrim cloth, where used, is also considered to be a constituent, although of seccondary importance.

Crossply Any filamentary laminate which is not uniaxial. Same as *angleply*. (In some references, the term crossply is used to designate only those laminates in which the laminae are at right angles to one another.)

Cure To change the properties of a thermosetting resin irreversibly by chemical reaction, i.e., condensation, ring closure, or addition. Cure may be accomplished by addition of curing (cross-linking) agents, with or without catalyst, and with or without heat.

Fabric A material constructed of interlaced yarns, fibers, or filaments, usually a planar structure. Nonwovens are sometimes included in this classification.

Fiber A single homogeneous strand of material, essentially one-dimensional in the macro-behavior sense, used as a principal constituent in advanced composites because of its high axial strength and modulus.

Fiber Content The amount of fiber present in a composite. This is usually expressed as a percentage volume fraction or weight fraction of the composite.

Fiber Direction The orientation or alignment of the longitudinal axis of the fiber with respect to a stated reference axis.

Filament A variety of fibers characterized by extreme length, such that there are normally no filament ends within a part except at geometric discontinuities. Filaments are used in filamentary composites and are also used in filament winding processes, which require long continuous strands.

Filamentary Composites A major form of advanced composites in which the fiber constituent consists of continuous filaments. Filamentary composites are defined here as composite materials composed of laminae in which the continuous filaments are in nonwoven, parallel, uniaxial arrays. Individual uniaxial laminae are combined into specifically oriented multiaxial laminates for application to specific envelopes of strength and stiffness requirements.

Filament Winding An automated process in which continuous filament (or tape) is treated with resin and wound on a removable mandrel in a pattern.

Filament Wound Pertaining to an object created by the filament winding method of fabrication.

Fill Yarn oriented at right angles to the warp in a woven fabric.

Glass In composite materials, all reference to glass will refer to the fibrous form of glass as used in filaments, woven fabric, yarns, mats, and chopped fibers.

Hybrid A composite laminate comprised of laminae of two or more composite material systems.

Interlaminar Shear Shearing force tending to produce a relative displacement between two laminae in a laminate along the plane of their interface.

Isotropic Having uniform properties in all directions. The measured properties of an isotropic material are independent of the axis of testing.

Lamina A single ply or layer in a laminate made of a series of layers.

Laminate A product made by bonding together two or more layers or laminae of material or materials.

Laminate Orientation The configuration of a crossplied composite laminate with regard to the angles of crossplying, the number of laminae at each angle, the exact sequence of the lamina layup.

Matrix The essentially homogeneous material in which the fibers or filaments of a composite are imbedded.

Orthotropic Having three mutually perpendicular planes of elastic symmetry.

Prepreg, Preimpregnated A combination of mat, fabric, nonwoven material, or roving with resin, usually advanced to the B-stage, ready for curing.

Pultrusion A process for producing continuous lengths of shapes with given cross-section by pulling resin-impregnated fibers through a die where curing occurs.

Quasi-Isotropic Laminate (0/±45/90) layup filament has equal properties in any direction and is optimum for combined biaxial and shear loading.

Resin An organic material with indefinite and usually high molecular weight and no sharp melting point.

Resin Content The amount of matrix present in a composite either by percent weight or percent volume.

Symmetrical Laminate A composite laminate in

which the ply orientation is symmetrical about the laminate midplane.
Thermoplastic A plastic that can repeatedly be softened by heating and hardened by cooling through a temperature range characteristic of the plastic, and that in the softened state can be shaped by molding or extrusion.
Thermoset A plastic that, after having been cured by heat or other means, is substantially infusible and insoluble.
Tow A loose, untwisted bundle of filaments.
Warp The longitudinally oriented yarn in a woven fabric; a group of yarns in long lengths and approximately parallel.
Yarn Generic term for strands of fibers or filaments in a form suitable for weaving or othewise intertwining to form a fabric.

14.2 Composite Materials

Composite materials have gained their reputation among structural engineers during the last decade. The performance of a composite depends upon:
- The compostion, orientation, length and shape of the fibers;
- The properties of the material used for the matrix (or resin);
- The quality of the bond between the fibers and the matrix material.

The most important contribution from fibers is probably that of orientation. Fibers can be unidirectional, crossed ply, or random in their arrangement and, in any one direction, the mechanical properties will be proportional to the amount of fiber orientation in that direction as shown in Fig. 14.2.1. It is evident that with increasingly random directionality of fibers, mechanical properties in any one direction are lowered.

The fibers and the resin material interact and redistribute the loads. The effectiveness of such transference, however, depends largely upon the quality of the bond between them. Encapsulation by the resin should be achieved without imposing excessive internal strain on the fibers by shrinkage.

Note: There is no composite material allowable data given in this chapter because numerous varieties plus many improved products are available every year. Much precaution should be given to the selection of suitable materials.

Fibers

While composite materials owe their unique balance of properties to the combination of resin (matrix) and fibers (reinforcement), it is the fiber system that is

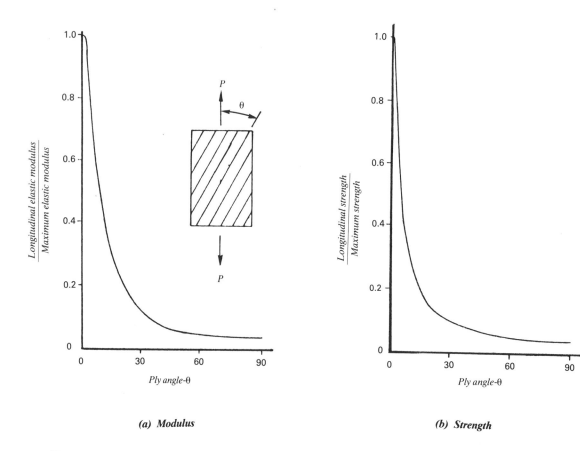

Fig. 14.2.1 Modulus and strength of composites drop steeply as the angle between the fibers and the direction of load increases.

primarily responsible for such structural properties as strength and stiffness. However, the fiber dominates the field in terms of volume, properties, and design versatility. Fig. 14.2.2 shows the comparison for several common composite fibers.

(1) Fiber glass

The most widely used fiber is unquestionably fiber glass, which has gained acceptance because of its light weight, high strength, and non-metallic characteristics. Fiber glass composites have been widely used on aircraft parts that do not have to carry heavy loads or operate under great stress. They are used principally for interior parts such as window surrounds and storage compartments, as well as for wing fairing and wing fixed trailing edge panels. In aircraft structural applications, there is a trend to replace fiber glass with higher strength and lighter weight Kevlar materials (see Fig. 14.2.3).

The two most common grades of fiber glass are "E" (for electrical) and "S" (for high strength). E-glass provides a high strength-to-weight ratio,

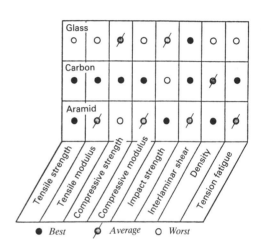

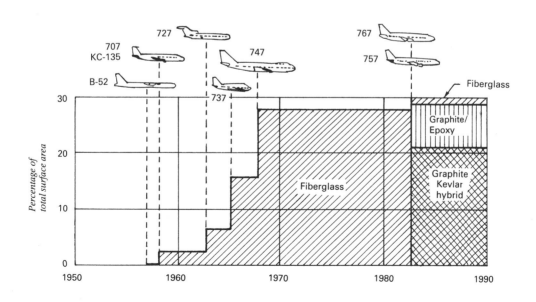

Fig. 14.2.2   Composite fiber comparison.

Fig. 14.2.3   Composite usage of commercial aircraft.

Airframe Structural Design 501

good fatigue resistance, outstanding dielectric properties, retention of 50% tensile strength to 600°F, and excellent chemical, corrosion, and environmental resistance. While E-glass has been proved highly successful in aircraft secondary structures, some applications required higher properties. To fill these demands, S-glass was developed, which offers up to 25% higher compressive strength, 40% higher tensile strength, 20% higher modulus, and 4% lower density. This glass also has higher resistance to strong acids than E-glass.

Designing with fiber glass is much simpler than designing with some other composite systems because of the large volume of empirical data collected over the years and the availability of standard systems from many manufacturers with well-documented properties.

(2) Kevlar (Aramid fiber)

Kevlar fiber has been used for structural applications since the early 1970s. Combining extremely high toughness, tensile strength, and stiffness with low density (the lowest in recently developed advanced composite materials) Kevlar fiber offers the highest specific tensile properties among all fibers available (see Fig. 14.2.4).

As shown in Fig. 14.2.4, Kevlar 29 and 49 have very impressive specific tensile strengths. This provides the basis for the claim that Kevlar on a pound-for-pound basis is five times as strong as steel. Fig. 14.2.5 illustrates tensile stress/strain curves for tensile loading. Like most other composite materials, Kevlar has a classically brittle response with a tensile strength a little greater than 200×10^3 psi for a typical unidirectional composite. When Kevlar is under compression the behavior is quite different from the tensile response. At a compressive load about 20% of the ultimate tensile load, a deviation from linearity occurs. This is an inherent characteristic of Kevlar 49 fiber representing an internal buckling of the filaments. This unusual characteristic of Kevlar 49 fiber has made fail-safe designs possible because fiber continuity is not lost in a compressive failure.

When tensile and compressive loadings are combined in a flexural bending, instead of the brittle failure encountered with glass and graphite fibers, the bending failure of Kevlar 49 is similar to what is observed with metals (see Fig. 14.2.6). This helps to explain the outstanding toughness and impact resistance of composites reinforced with Kevlar.

Another area in vibration damping, Fig. 14.2.7 shows the decay of free vibrations for various materials; Kevlar is less prone to flutter and sonic fatigue problems.

Kevlar fibers also offer good fatigue, cut, and chemical resistance, and retain their excellent tensile properties to relatively high temperature. Because of their high specific properties and less handling problems, these fibers have replaced glass fibers in many applications. However, relatively low transverse and compressive strengths have kept them out of many aircraft primary structures. New methods for machining Kevlar are also needed because the fibers are too tough to cut with conventional tools.

(3) Graphite (carbon)

Graphite fibers are among the strongest and stiffest of the composite materials being used in matrix systems for high-performance structures. The outstanding design properties of graphite/resin composites are their high strength-to-weight and stiffness-to-weight ratios. With proper selection and placement of fibers, the composites can

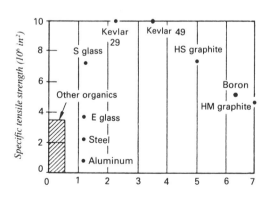

Fig. 14.2.4 Specific tensile strength vs. specific modulus of composite fibers.

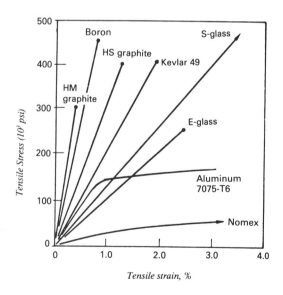

Fig. 14.2.5 Stress and strain curves of various fibers/epoxy.

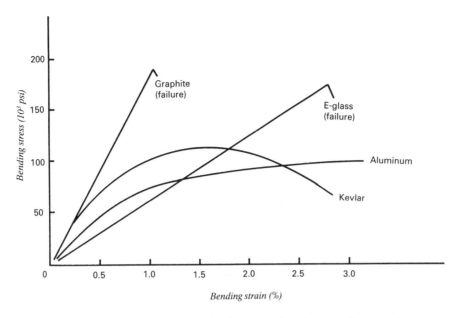

Fig. 14.2.6 *Unidirectional composite bending stress/strain curves (epoxy resin).*

Material	Loss factor × 10^{-4}
Cured polyester resin	400
Kevlar 49/Epoxy	180
Fiberglass/Epoxy	30
Graphite/Epoxy	17
Stainless steel	6

Fig. 14.2.7 *Loss factor from vibration decay.*

be stronger and stiffer than equivalent steel parts at less than half the weight. Fig. 14.2.8 shows comparison of graphite with other materials.

Graphite fibers are produced either from organic precursor fibers (PAN-polyacrylonitrile) or pitch.

- PAN-derived fibers have been available for many years; for several of the lower modulus varieties, large data bases have been developed through their use in aircraft and aerospace programs. These fibers are generally selected for their high strength and efficient property translation into the composite.
- The pitch-based fibers are newer and, while

Type of Material	Unidirectional Strength, ksi		Unidirectional Tensile Modulus, 10^6 psi	Density lb/in³
	Tension	Compression		
Graphite AS4/Epoxy	288	242	21.0	0.056
Graphite HMS/Epoxy	199	124	30.0	0.059
S-901 glass/Epoxy	400	—	8.6	0.072
E-glass/Epoxy	180	—	7.6	0.072
Aramid Kevlar-49/Epoxy	350	70	12.0	0.050
Aluminum (7075-T6)		83	10.0	0.100
Titanium (6Al-4V)		160	16.5	0.160
Steel (4130)		80–200	30.0	0.289

Fig. 14.2.8 *Strength and tensile modulus comparison of graphite composite vs. other materials.*

Fibers Property	PAN Fibers			Pitch Fibers	
	Thornel 30	Thornel 50	Celanese GY-70	Thornel VSB-32	Thornel VS-0053
Fiber strength (10^3 psi)	400	350	300	300	300
Tensile modulus (10^6 psi)	33	55	75	55	75
Composite strength (10^3 psi)	220	175	110	150	150
Composite modulus (10^6 psi)	20	32	44	32	44

Note: All are unidirectional fibers.

Fig. 14.2.9 Graphite material property comparison (PAN vs. pitch fibers).

they are not as strong as the PAN fibers, the ease with which they can be processed to high modulus makes them attractive for stiffness-critical applications. Cost projections for volume production also are favorable for the pitch fibers because of lower raw-material cost.

Fig. 14.2.9 shows a property comparison of these two fibers.

Graphite composite laminates offer fatigue limits far in excess of aluminum or steel, along with superior vibration damping. Further, the thermal expansion coefficients of graphite composite fibers (see Fig. 14.1.2) become increasingly negative with increasing modulus. This allows the design of structures with virtually no thermal expansion or contraction across widely ranging thermal cycles. As with fiber glass, graphite fiber products are available as prepreg, molding-compound, and other standardized product forms.

A major concern in joining of metals to graphite composites is galvanic corrosion. Graphite is cathodic in nature, so the best joiners are cathodic metals such as titanium or passive stainless steel. Aluminum, however, is highly anodic, so care must be exercised when joining it to a graphite composite. Graphite is generally available in three forms;

- HTS — High-tensile-strength fiber
  $F_{tu} = 350$ ksi and $E = 30 \times 10^6$ psi
- HM — High-modulus fiber
  $F_{tu} = 200$ ksi and $E = 50 \times 10^6$ psi
- UHM — Ultra-high-modulus fiber
  $F_{tu} = 150$ ksi and $E = 70 \times 10^6$ psi

(4) Boron

Boron fibers have found limited application and their high stiffness had made possible the early use of composites in primary aircraft structures. So far, the relatively high cost and large fiber diameter of boron have kept them from high volume application.

Resins (Matrix)

Many types of resins have been used for composite matrices, both thermoset and thermoplastic. The resin is responsible for the integrity of the composite structures. It binds the fibers together to allow effective distribution of loads, as well as to protect the notch-sensitive fibers from self-abrasion and externally-induced scratches. The resin also protects the fibers from environmental moisture and chemical corrosion or oxidation, which may lead to embrittlement and premature failure. And although the fibers provide much of the tensile and flexural strength and stiffness, a composite shear, compression, and transverse tensile properties are usually resin(matrix) dominated.

With any fiber, the material used for the resin must be chemically compatible with the fibers and should have complementary mechanical properties. Also, for practical reasons, the resin material should be reasonably easy to handle.

Because it is the resin which holds the fibers together, the general thermomechanical behavior of the composite is dominated by the resin's heat resistance. The development of high strength and high thermal resistance is frequently accompanied by intractability in thermo-plastics, or complex cure procedures or brittleness in thermosets. Overcoming these obstacles has proven the key to developing viable composite resins, with processing/fabrication constraints of fiber wet-out, prepreg shelf life, tack and drape, cure shrinkage, etc, adding to the complexity.

Compared to thermoplastics, thermoset resins offer lower melt viscosities, lower processing temperatures and pressures, and are more easily prepregged. On the other hand, thermoplastic resins offer indefinite shelf life, faster cycles, simple fabrication, and generally do not require controlled-environment storage or post curing.

(1) Thermoset Resins

Thermoset resin systems have been dominating the composite industry because of their reactive nature. These resins allow ready impregnation of fibers, malleability into complex forms, and a means of achieving high-strength, high-stiffness crosslinked networks in a cured part. (Refer to Fig. 4.6.1 of chapter 4 for comparison of several thermoset resins.)

(a) Epoxy (modified)

Epoxy systems generally provide outstanding chemical resistance, superior adhesion to fibers, superior dimensional stability, good hot/wet performance, and high dielectric properties. Epoxy can be formulated to a wide range of viscosities for different fabrication processes and cure schedules. They are without void-forming volatile evolution, have long shelf lives, provide relatively low cure shrinkage, and are available in many thoroughly-characterized standard prepreg forms. Applications such as aircraft primary structures, filament-wound pressure vessels and fuselage bodies, all requiring high reliability, are frequently manufactured with epoxy resins. The epoxy family is the most widely used resin system in the advanced composite field. Its service temperature is usually 200°F–250°F in aircraft applications.

(b) Polyimides (higher service temperature resins)

Polyimides are thermo-oxidatively stable and retain a high degree of their mechanical properties at temperatures far beyond the degradation temperature of many polymers, often above 600°F.

Bismaleinides (BMI), a special polyimides system, operate around a 350°F to 400°F upper limit. BMI offer good mechanical strength and stiffness, but are generally brittle and may have cure-shrinkage.

(c) Polyesters

Polyesters resins can be cured at room temperature and atmospheric pressure, or at a temperature up to 350°F and under higher pressure. These resins offer a balance of low cost and ease of handling, along with good mechanical, electrical, and chemical properties, and dimensional stability. To date, epoxy dominates very high-performance composite projects, and the polyesters dominate industrial areas.

(d) Phenolics

Phenolics are the oldest of the thermoset plastics, and have excellent insulating properties and resistance to moisture. Chemical resistance is good, except to strong acids and alkalis.

(2) Thermoplastic Resins

Thermoplastic resins are not new to the aircraft industry. They have been used for many years for various components mainly in the interior and other non-structural components. A comparison between the thermoset and thermoplastic resins has been given in Section 14.1. Fig. 14.2.10 illus-

Material Properties	Relative Advantage		
	Thermoplastics	Thermosets	Metal
Corrosion resistance	xxx	xxx	x
Creep	xxx	xxx	x
Damage resistance	xx	x	xxx
Design flexibility	xxx	xxx	x
Fabrication	xxx	xx	x
Fabrication time	xxx	xx	x
Final part cost	xxx	xx	x
Finished part cost	xxx	xx	x
Moisture resistance	xx	x	xxx
Physical properties	xxx	xxx	xxx
Processing cost	xxx	xx	x
Raw material cost	x	xx	xxx
Reusable scrap	xx	—	xxx
Shelf life	xxx	x	xxx
Solvent resistance	xxx	xx	x
Specific strength	xxx	xxx	x
Strength	xxx	xxx	x
Weight saving	xxx	xx	o

Note: xxx-best; xx-good; x-fair; 0-not applicable

Fig. 14.2.10 Relative advantages of thermoset, thermoplastics and metals.

trates some of the relative property advantages of thermosets, thermoplastics and metals.

The thermoplastic resins have higher continuous service temperatures, from 250°F to 400°F depending on the system, higher resin melting temperatures, higher viscosity which leads to higher processing temperature (about 700°F), and higher mold pressure in autoclave operation.

Advanced high-performance thermoplastics are on the way for aircraft primary structure. The increased temperature resistance is making these resins more suitable for the aerospace industry (high temperature applications). Their resistance to chemicals, especially the solvents, fuels, and oils used in the aerospace industry, makes them attractive for use in areas affected by these chemicals.

Polyarylene sulfide (PAS), polyphenylene sulfide (PPS), polyetheretherketone (PEEK), polyamideimide (PAI), polyetherimide (PEI), polysulfone (PS) and thermoplasticpolyimide (TPI) are available for research or commercial use.

(3) Metal Matrix Composite (MMC)

During the last decade, MMCs have been concentrated on boron/aluminum (B/Al), graphite/aluminum (GR/Al), and silicon carbide/aluminum composites but other types of matrix materials are also being studied, including titanium and magnesium. There remains some difficulties to be solved:

- High cost probably the most serious problem
- Can not be extruded or forged
- Lack of machining and joining techniques
- Lack of nondestructive testing techniques
- Need to improve the adhesion of the fibres to the matrix

Material Forms

Choice of material form for continuous fiber often depends upon the fabrication process selected. Filament-wound processes usually dictate the use of lower filament count yarns to minimize the catenary effects. Autoclave, vacuum bag, and compression molding of relatively flat or simple curvature parts can use the oriented unidirectional tapes.

The following definitions and descriptions apply to fibers (refer to Fig. 14.2.11).

- Filament — The basic structural fibrous element, which is continuous or at least very long compared to its average diameter (usually 6 to 10 microns).
- Yarn — A small, continuous bundle of filaments, generally fewer then 12,000. The filaments are lightly stranded together so they can be handled as a single unit and may be twisted to enhance bundle integrity.
- Tow — A large bundle of continuous filaments, generally 10,000 or more, not twisted.

(1) Unidirectional Tape

Properties of a composite transverse to the fiber direction depend largely upon the matrix material but, in any case, are very weak compared to the longitudinal properties. Consequently, in the design of most structures that are subject to both longitudinal and transverse loadings, the fibers must be oriented in specific directions to withstand these loads. Massive design data are available from many sources that dictate layup patterns required for combining layers of unidirectional tape to achieve desired direction properties which presented in carpet plots as shown in Fig. 14.2.12.

The carpet plots represent various combinations of symmetric and balanced laminates that contain 0°, 90° and ±45° plies. The plots may be used to find:

- Properties for a given laminate
- Various laminates that satisfy a particular property requirement

Example:
Given a laminate panel (0°/±45°/90°)
   70% of 0° plies
   20% of ±45° plies

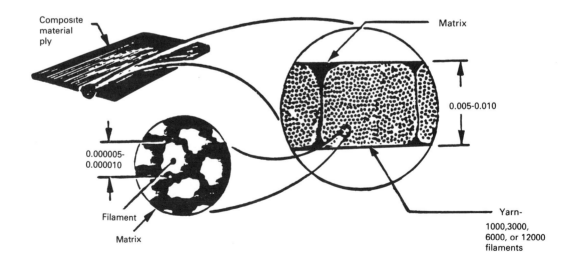

Fig. 14.2.11 Contents of composite materials.

506 Airframe Structural Design

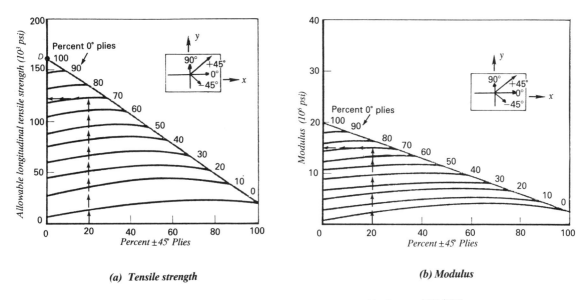

(a) Tensile strength *(b) Modulus*

Fig. 14.2.12 *Families of example curves of graphite/epoxy (GR/EP).*

10% of 90° plies
From Fig. 14.2.12, read vertical scale of chart (a),
$F_{11} = 125000$ psi
and read vertical scale of chart (b),
$E_{11} = 15 \times 10^6$ psi

(2) Woven Fabrics
Woven fabrics are more expensive than unidirectional tapes. However, significant cost savings are often realized in the molding operation because layup labor requirements are reduced. Complex part shapes for processes requiring careful positioning of the reinforcement can benefit from the use of the more handleable woven forms of fiber. Some fabrics are essentially unidirectional and these fabrics are oriented in one direction and held in position by tie yarns of a non-structural nature as shown in Fig. 14.2.13(a).

Other fabrics are of a plain [Fig. 14.2.13(b)] or

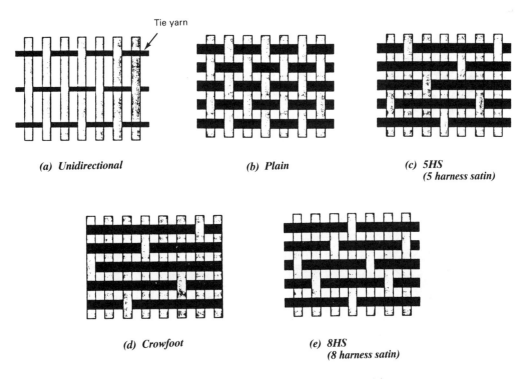

(a) Unidirectional *(b) Plain* *(c) 5HS (5 harness satin)*

(d) Crowfoot *(e) 8HS (8 harness satin)*

Fig.14.2.13 *The most common weave styles of material forms.*

Airframe Structural Design 507

satin construction [Fig. 14.2.13(c) and (e)]. In particular the commonly used 8-harness satin retains most of the fiber characteristics in the composite and can be easily draped over complex mold shapes. Plain-weave fabrics are less flexible and are suitable for flat or simply-contoured parts, but a slight sacrifice in fiber property translation.

Fabrics are generally described according to the type of weave and the number of yarns per inch, first in the warp direction (parallel to the length of the fabric), then in the fill direction (perpendicular to the warp)

- Plain weave — In this construction, one warp end is repetitively woven over one fill yarn end under the next. It is the firmest, most stable construction, which provides minimum slippage. Strength is uniform in both directions.
- Satin weave — In this construction, one warp end is woven over several successive fill yarns, then under one fill yarn. A configuration having one warp end passing over four and under one fill yarn is called a five-harness satin weave [Fig. 14.2.13(c)]. Satin weaves are less open than other weaves, strength is high in both directions.
- Knitted fabrics — Knitted fabrics retain more of the strength and stiffness to fibers through the elimination of crimps in woven fabrics, while reducing both material use and production times associated with woven fabrics. These better looking, stronger laminates at lower part weights are finding their way into aerospace applications. These fabrics are knitted together with a non-reinforcing binder fiber, eliminating the crimp caused by weaving the fibers over and under one another. This means that the fibers can bear full loads, providing efficient translation of stress from resin to fiber. Fig 14.2.14 shows a knitted fabrics.

A comparison between unidirectional tape and woven fabric material forms is presented in Fig. 14.2.15.

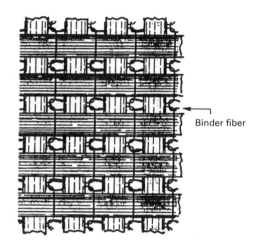

Fig. 14.2.14   Knitted fabrics.

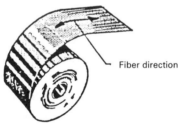

(a) Tape

- Advantages of tape
  - Can be tailored more easily to match loads
  - Aerodynamically acceptable surface (requires less surface treatment than fabric)
  - No splice overlaps parallel to fibers
  - Less porous than fabric
  - Higher allowable strength and stiffness
  - Lower raw material cost
- Recommended for use where advantages justify increased fabrication costs, and where contours permit

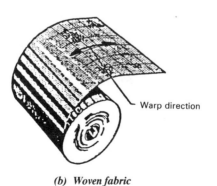

(b) Woven fabric

- Advantages of fabric
  - Lower fabrication costs
  - Less material handling damage
  - Easier forming on contours and corners
  - More resistant to surface breakout and delamination
- An acceptable aerodynamic surface is harder to obtain on fabric than on tape

Fig. 14.2.15   Material form comparison between unidirectional tape and woven fabric.

(3) Prepreg

Thin plies of unidirectional or bidirectional (woven fabric) fibers, impregnated with B-staged resin, can be laid up in plies and cured into typical composite components. In prepreg, the partly cured resin holds the fibers in alignment and, in sheet form, allows the pre-impregnated material to be handled easily, cut to shape for layup and assembled for curing with minimum production difficulty. Also, the proportion of fiber to resin is closely controlled, giving, in turn, close control of strength and weight in the finished component.

14.3 Design

The advanced composite materials are uniaxial in their single-ply state, having very high mechanical properties along their longitudinal axis, and low-to-moderate properties along their transverse axis. This is the primary difference, from a structural design and analysis standpoint, between advanced composites and metals.

Metals are normally homogeneous and isotropic in nature, and their reaction to an applied load can be defined by knowing two of the three basic elastic constants (the modulus of elasticity E, the modulus of rigidity G and the Poisson's ratio ν). A basic unidirectional lamina, or any balanced symmetric laminate, on the other hand, is orthotropic in nature, having three mutually perpendicular planes of elastic symmetry. For planar applications, this type of materials can be defined by four of five basic elastic constants for orthotropic materials. It should be noted that there are twice as many independant planar elastic constants for orthotropic materials as for isotropic materials because of the different properties in the planes of symmetry. For many composite applications, the laminate is not even orthotropic, but is only anisotropic. This occurs when an orthotropic laminate is loaded in a direction which does not coincide with one of the principal axes, or when the laminate layup is symmetric but not balanced about a principal reference axis. A $[0°/\pm 45°/90°]_s$ laminate is balanced and quasi-orthotropic, while a $[0°/\pm 45°]_s$ laminate is orthotropic. This type of laminate requires six elastic coefficients for definition. As a general rule, all laminates should be symmetrically laid up about their midplane; coupled laminates should be avoided. Fig. 14.3.1 shows the difference between the different types of laminates, while Fig. 14.3.2 shows the differences between a symmetric laminate and one which is not symmetric.

The directional nature of composite laminae provides the ability to construct a material which can meet specific loads and/or stiffness requirements without wasting materials by providing strength and stiffness only where they are needed. If the design requirement is simply to provide axial strength or stiffness, the majority of the material should be unidirectionally oriented. If this material is adhered on restrain members, such as shown in Fig. 14.3.3, all of the fibers may be so oriented. If the composite material is unconfined, it is wise to provide a nominal amount of transverse reinforcement to account for any off-axis loading that may occur, either during fabrication or by induced loading and to reduce the Poisson's effects.

An example of such an unrestrained application would be any plate or skin, whether or not it is on an elastic base, or a stiffener application. A rule-of-thumb is to make the number of laminae in the transverse direction equal to 10% of the total number of laminae in the part.

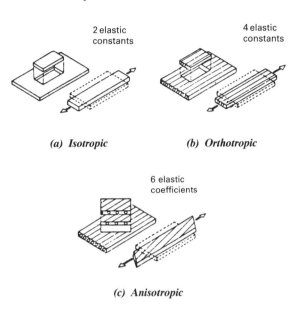

Fig. 14.3.1   Composite laminate types.

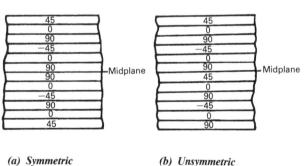

(a) Symmetric   (b) Unsymmetric
(45, 0, 90 means 45°, 0°, 90° respectively)
Fig. 14.3.2   Laminate symmetry.

If shear loading or shear stiffness is the primary design consideration, then most of the material should be oriented at ±45° to the longitudinal axis, as this provides the highest shear properties. However, care must be taken to evaluate any loading in the longitudinal or transverse directions, since the strength in these directions is quite low. Conditions which would not normally be critical in metal design may approach or exceed the strengths available in these directions when using a pure [±45°] layup. This consideration may necessitate the inclusion of a sufficient number of 0° and/or 90° laminae.

In addition to the pure axial or pure shear case just mentioned, there are many applications which require the ability to withstand a combination of loadings. Although it is possible to determine an optimum orientation sequence for any given loading condition, it is felt more practical to limit the number of orientation to a few specific families which can then be characterized by tests.

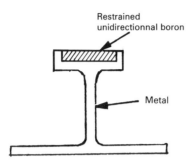

Fig.14.3.3 Restrained unidirectional boron applications.

In the foregoing discussion, we have seen that the anisotropic nature of composite materials, while allowing the designer to tailor material more closely to the design requirements, imposes the problem of selecting the proper orientation for application. This is a consideration which does not arise in metal design, and the designer must be aware that the traditional methods of design and analysis have to be developed to higher orders of refinement for anisotropic materials, not only to provide a basis for selecting the proper orientation, but even for defining stresses and margins of safety.

In the design of large structures, one of the basic ground rules is to establish the ultimate gross area cutoff stress to be used in designing in tension- and compression-critical areas. This cutoff value automatically covers many design considerations, such as the practical economic life in terms of cracks, high local stress areas, joints of various kinds, and structural integrity in terms of crack growth and fracture. In the application of composite material to structures, the allowable levels (expressed in strain instead of stress) are low because of the following limits:
- Tolerance for impact damage (this is the dominant failure mode in compression)
- Flaw growth resistance
- Stress concentration associated with cutouts, joints, etc.
- Strength at temperature with moisture content

These factors restrict design ultimate gross area strains to about 50% of the composite material failure strain depending on loading and laminate orientation. Fig. 14.3.4 illustrates the primary factors which govern the design strains of composite materials.

For composite laminates with a given flaw or hole size, strength retention in general increases as the percentage of ±45° plies present in an orientation is increased (see Fig. 14.3.5). For example, the stress concentration factor is more than three times in unidirectional graphite/epoxy than in a laminate consisting of only ±45° plies for a drilled hole.

Fig. 14.3.4 shows the relative effect of the environment on laminate structures.
- Under tension loading the governing consideration is relative to notches (holes).
- Compression loading takes over when laminate damage occurs as the result of impact.

Since damage due to impact is likely to occur during the life time of a structure, the design ultimate strain has to be restricted. The types of damage are;
- Type 1 damage — Non-visible damage will not grow to degrade the strength to less than design ultimate load.
- Type 2 damage — Visually detectable damage must be found and repaired before residual strength degrades to less than design limit load.

Fig. 14.3.6 shows the impact damage to laminates, and non-destructive inspection (NDI) methods is used to detect the scope of delamination under dent impact.

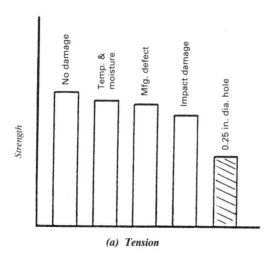
(a) Tension

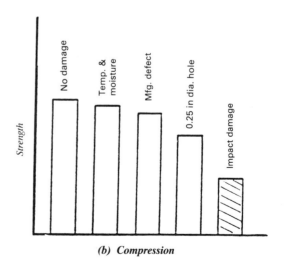
(b) Compression

Fig. 14.3.4 Factors affecting design strength.

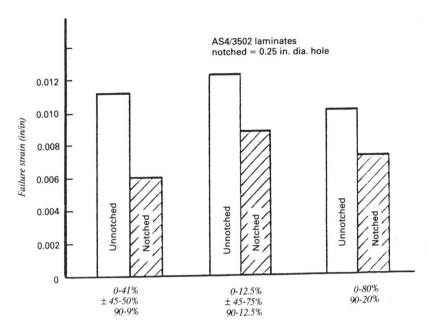

Fig. 14.3.5 Notch sensitivity vs. percentage of ±45 plies.

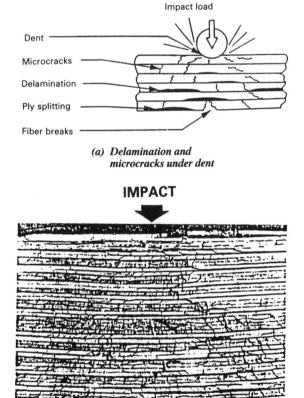

Fig. 14.3.6 Delamination after impact damage.

Laminated Strength Design Consideration

With the advent of advanced fiber reinforced composite materials, there is an opportunity for materials design to be integrated into structural design as an added dimension. A basic understanding of the interaction of fiber, resin, and fiber-resin interface in the composite will be a valuable aid to designers and materials and structural analysis. The modern science of micro-mechanics, the study of structural material interactions, is of particular importance for advanced composite analysis.

The design of a laminate for strength presents a complex laminate design problem. The design for simple loading conditions such as pure uniaxial or shear loads can be satisfied efficiently through the use of unidirectional or ±45° orientations. The relationship between angle-ply and structural strength is shown in Fig. 14.3.7.

> ±45° plies give buckling stability
>   0° plies give column stability
> ±45° plies carry shear
>   0° plies carry tension or compression
> 90° plies carry transverse loads and reduce Poissons effects
> 90° plies help stability better than 0° in long narrow panels

Fig. 14.3.7 Structural strength as function of ply orientation.

However, the laminate which is required to withstand numerous different load conditions and environments cannot be efficiently designed with such case. Techniques of using combinations of 0°, ±45°,

Airframe Structural Design 511

and 90° laminae can produce a simple design of relatively good efficiency. The best efficiencies, however, are achieved by computerized techniques. These computerized techniques are based on both empirical test data and constituent material properties along with micro-mechanics theories.

The following preliminary study indicates some of the expressions relating the longitudinal and transverse mechanical properties of an unidirectional fiber reinforced composite to the properties of the constituents.

The basic equations used in this study to predict the longitudinal modulus and strength of fiber reinforced composites are the parallel element mixture equations. These are based on reasonable assumptions and they do not violate theories of elasticity (see Fig. 14.3.8).

$$E_{11} = E_f V_f + E_m (1 - V_f) \quad (14.3.1)$$

$$F_{11} = F_f V_f + F_m (1 - V_f) \quad (14.3.2)$$

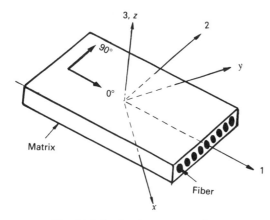

Fig. 14.3.8 *Lamina axes rotation.*

where E_{11} = modulus of elasticity parallel to fiber length
F_{11} = material strength parallel to fiber length
E_f = modulus of elasticity of fiber
F_f = strength of fiber
V_f = volume of fiber
E_m = modulus of elasticity of resin (matrix)
F_m = strength of resin (matrix)

The assumptions on which these two equations are based are as follows:
- The fibers are completely surrounded and wetted by resin material and accordingly are not allowed to contact one another.
- The transfer of load from the resin to the fiber occurs across the interfacial surfaces comprising wetting areas.
- The strength, size, shape, orientation, and bonding of the fibers are as uniform as possible.

Major Poisson's ratio of the composite can also be predicted using an equation of similar form.

$$\nu_{12} = \nu_f V_f + \nu_m (1 - V_f) \quad (14.3.3)$$

where ν_{12} = Poisson's ratio of the composite

ν_f = Poisson's ratio of the fiber
ν_m = Poisson's ratio of the resin

These three equations are fairly straightforward and yield good correlation with test results.

Predictions of the tensile modulus normal to the fiber direction and the shear modulus in a unidirectional composite are difficult to make because of their sensitivity to voids, and their dependence on accurate knowledge of both the resin modulus and the details of the fiber-resin packing.

The two equations given below are used to approximate the transverse tensile modulus and shear modulus of unidirectional fiber reinforced composite materials.

$$E_{22} = \cfrac{1}{\cfrac{V_f}{E_f} + \cfrac{1 - V_f}{E_m}} \quad (14.3.4)$$

$$G_{12} = \cfrac{1}{\cfrac{V_f}{G_f} + \cfrac{1 - V_f}{G_m}} \quad (14.3.5)$$

Note that Equations 14.3.1–14.3.5 represent but a few of the many micro-mechanical expressions developed to predict the behavior of fiber reinforced composite materials (ref. 14.66 for further information).

Composite Laminates

It is recognized that one of the outstanding features of filamentary composite structure is its ability to be tailored, through crossplying, to match individual loading or stiffness requirements. This being so, it follows that large numbers of individually different crossplied laminates are likely to be encountered from one application to another. Each of these laminates is unique in its properties and characteristics and, hence, must be distinctly identified whenever it is to be associated with specific quantitative or numerical data.

A laminate orientation code has been devised which provides both concise reference and positive identification to the laminate. In this code, the crossply angles are listed in ascending numerical order, separated by a slash or comma, with the entire listing enclosed within brackets. Where there is more than one lamina at any given angle, the number of laminae at that angle is denoted by a numerical subscript (eliminate degree notation for simplicity):

$[0_5/90_2]$ or $[0_5, 90_2]$

(indicating 5 plies in 0° direction and 2 plies in 90° direction)

A distinction is made between (+) and (−) angles; however, when there are the same number of (+) and (−) angles of the same magnitude, they may be combined into a (±) type of notation in the listing:

$[0_5/\pm 45_3/90_2]$ or $[0_5, \pm 45_3, 90_2]$

(indicating 5 plies in 0° direction, 3 plies in +45° direction, 3 plies in −45° direction and 2 plies in 90° direction)

It should be noted that these codes only signify the number of plies in each direction. They do not indicate in any way the stacking sequence of the plies

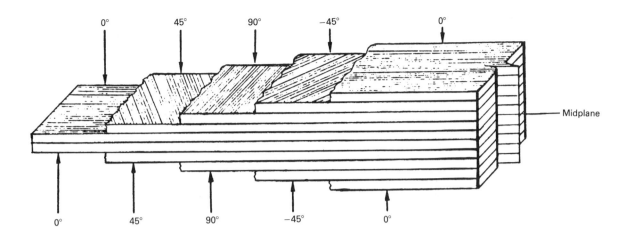

Fig.14.3.9 Symmetrical balanced layup — two sets of $[0/-45/90/45/0]_s$.

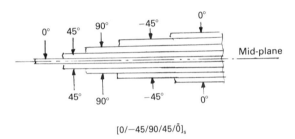

Fig. 14.3.10 A common single ply in midplane of laminates.

within the part.

If it is a symmetrical laminate (frequently used to prevent warpage) as shown in Fig. 14.3.9, the orientation code and stacking sequence are:

$[0/-45/90/45/0]_s$ or $[0, -45, 90, 45, 0]_s$

(indicating 4 plies in 0° direction, 2 plies in +45° direction, 2 plies in −45° direction and 2 plies in 90° direction)

If the symmetric laminate with an odd number of plies, the center ply is overlined to indicate this condition. Stacking with this ply, the rest of the code indicates a mirror image of the laminate layup as shown in Fig. 14.3.10.

After the selection of the fiber/resin material, the design process concentrates on the lamination rationale: for a particular section of the structure, how many plies are required and what are their angular orientations and their stacking sequence? Some have proposed that a standard quasi-isotropic $[0/\pm45/90]$ laminate be used over the majority of the structure, but this approach mitigates the primary attribute of advanced composites, that of high specific directional properties. There exists no universal lamination geometry which effectively satisfies the various loading requirements. The lamination geometry must be based on the stress state, i.e., magnitude, direction and combined biaxial and shear, and the strength and/or stiffness requirements to realize the potential structural efficiency of the composite. The stiffness requirements may be based on laminate buckling (flexural), static structural deformations (in-plane), or aeroelastic restraints (inplane). The degree-of-freedom of both the material properties and the design restraints should be sufficient to require a systematic approach to determine the lamination geometry.

When selecting the stacking sequence for a laminate the following should be considered:

(1) Plies of a laminate must be stacked symmetrically, and overall balance must be maintained to avoid bending-stretching-torsion coupling. Balance implies that, for every +45° ply, there exists a −45° ply in the laminate. Symmetry requires a mirror image of ply stacking about the midplane. Fig. 14.3.11 shows how simple loads result in unusual deformations because of coupling actions.

(2) Stacking order severely affects the flexural stiffness and, consequently, the buckling behavior of the laminate. A high aspect ratio panel with all edges supported has its highest buckling strength when its 90° and/or ±45° plies are at/or near the outer surfaces. A wide column has its highest buckling strength when its 0° plies are at or near the outer surface. (See Fig. 14.3.12.)

(3) The ply adjacent to a bonded joint should be oriented with the fibers parallel to the direction of loading. Joints with plies oriented 90° to the loading direction have minimum strength.

(4) Adjacent plies should be oriented (when possible) with no more than 60° (except woven fabrics) between them. Studies have shown microcracking can occur from curing stresses if adjacent plies are oriented at greater than 60°. The same rule applies to the transfer of interlaminar shear stresses. While not normally affecting static strength, this can affect fatigue strength.

(5) The stacking sequence can cause interlaminar

Airframe Structural Design 513

(a) 0/90 stacking — Because of different thermal expansion characteristics in each layer, this stacking deforms into a "saddle" when heated.

(b) 0/90 stacking — This arrangement bends under pure tension because the modulus-weighted centroid is not coincident with the geometric centroid, resulting in an offset load path.

(c) ±θ at any angle — Opposing shear deformations in the plus and minus plies result in stretching-torsion interaction.

Fig. 14.3.11 *Symmetry effects on deflection of composite laminates.*

normal stresses to occur at the free edge of the laminate (see Fig. 14.3.13). Interlaminar tension stresses can cause delamination under both static and cyclic loading. The sign (tension or compression) of the normal stress depends both on the sign of the laminate in-plane loading and the stacking sequence. A given ply set can be stacked in such a way that maximum or minimum tension or compression σ_z can be obtained. In classical lamination theory, no account is taken of interlaminar stress such as σ_z, τ_{zx}, and τ_{zy}. Accordingly, classical lamination theory is incapable of providing predictions of some of the stresses that actually cause failure of a composite material. Interlaminar stresses are one of the failure mechanisms uniquely characteristic of composite materials. Moreover, classical lamination theory implies values of τ_{xy} where it cannot possibly exist, namely at the edge of a laminate. Physical grounds will be used to establish that:

- at the free edge of a laminate, sides of a laminate or holes, the interlaminar shearing stress is very high (perhaps even singular) and would therefore cause the debonding that has been observed in such regions.
- layer stacking sequence changes produce differences in tensile strength of a laminate even though the orientations of each layer do not change (in classical lamination theory, such changes have no effect on the stiffnesses). Interlaminar normal stress (σ_z) changes near the laminate boundaries are believed to provide the answer to such strength differences.

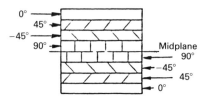

(a) A given laminate stacking sequence

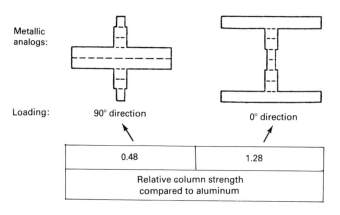

(b) Equivalent flexural stiffness

Fig. 14.3.12 *Effects of stacking sequence on stability.*

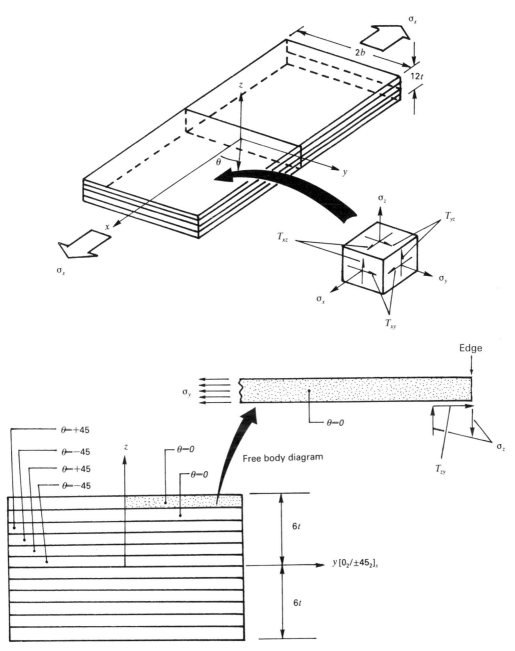

Note: See Fig. 14.3.14 for interlaminar normal σ_z stress distribution

Fig.14.3.13 *Interlaminar geometry and stresses.*

The maximum negative σ_z is obtained by placing the plies with the maximum lateral (σ_y for applied load in the x direction) tension stress at the center and then, placing the remaining plies in descending order until all the plies with lateral tension are finished. Then continue the layup in ascending order of lateral compression stress finishing with the plies with maximum lateral compression on the outside. The complete reverse will give maximum positive σ_z. To minimize σ_z place the plies with the highest lateral stress at the center, alternate tension and compression plies, and continue in descending order until the outer plies have the minimum lateral stress. The effect of stacking sequence on the interlaminar stress σ_z is shown in Fig. 14.3.14.

(6) If possible, avoid grouping 90° plies; separate them by 0° or ±45° plies (0° is direction of critical load) to minimize interlaminar shear and normal stresses.

(7) Wherever possible maintain a homogeneous stacking sequence and avoid grouping of similar plies. If plies must be grouped, avoid grouping more than 6 to 8 plies of the same orientation together to minimize edge splitting.

Airframe Structural Design 515

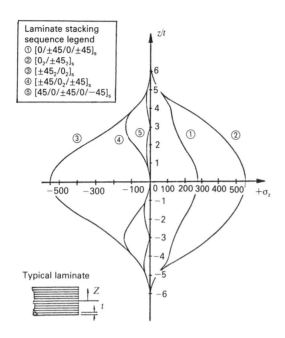

Fig. 14.3.14 *The effect of laminate stacking sequence on free-edge interlaminar normal stresses (under axial compressive $\sigma_x = 48$ ksi).*

(8) Exterior surface plies should be continuous and 45° (not 0° or 90°).
(9) If possible, ply drop-offs should be symmetric about the laminate midplane.
(10) Damage tolerance may be augmented by:
 • using the least strength critical plies (i.e. soft skin — large percentage of ±45° plies) at or near the cover outer mold line surface, or the entire cover skin of a strength critical component, such as wing and empennage covers;
 • minimizing grouping any particular ply fiber orientation;
 • if possible, using bi-directional woven outer surface plies.
(11) All laminates should contain a minimum of 10% reinforcement (fibers) in the 0°, ±45°, 90° directions.

Other Design Considerations
(1) Crashworthiness
The crashworthiness of an airframe structure is measured by three major capabilities:
 • The reduction of mechanical forces upon impact with the ground debris or other objects
 • The capability of the structure to remain intact to provide the occupants with protection in the event of a post-crash fire
 • The maintenance of fuel tank integrity in a crash

However, airframe structure fabricated of composite material must provide at least the same level of safety as conventional metal structures. Advanced composite material in general is considered inferior, because its high degree of brittleness makes it less crashworthy. If the aluminum structure (mainly fuselage shell) is replaced with an advanced composite material, energy absorption would be reduced and more structural break-up would be expected to occur unless some innovative designs are incorporated.

Fig. 14.3.15 shows that aluminum sustains more than 24 times the deformation and possesses more than 65 times the energy absorption capability of composite material.

There is encouraging evidence that in a fire, composite material systems would provide greater burn-through protection; but pyrolysis of the resin may impose an additional threat due to the toxic gas release.

Experience has proven that considerable protection from critical damage can be provided for fuselage frames and center-wing fuel tank of transport aircraft during survivable accidents by appropriate support structure and the composite

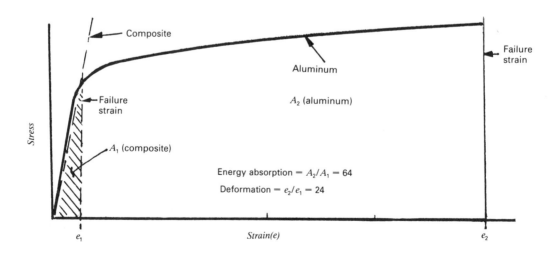

Fig.14.3.15 *Stress and strain relationship of graphite/epoxy and aluminum material.*

structure would provide the same level of protection as conventional structures.

(2) Lightning Strike Protection

Boron and graphite filament organic resin composites are susceptible to lightning damage, and do not provide electromagnetic shielding. Exposure of an unprotected graphite or boron laminate to direct lightning strike can result in severe laminate damage. The basic lightning strike protection systems that have been tested and found suitable are:
- Aluminum flame spray approximately 6 mils thick
- Aluminum foil 5 mils thick
- Aluminum wire mesh (120 × 120 mesh, 0.003 wire diameter; and 200 × 200 mesh, 0.0021 wire diameter)

These protection systems are described in Fig. 14.3.16.

Aircraft protruding tips, leading edges, and trailing edge are the exterior mold line surfaces most likely to be primary lightning strike zones; other airfoil surfaces are secondary strike zones (refer to Section 8.2 of Chapter 8.0). Both must be conductive to facilitate lightning "streamering" and the dissipation of static electricity to ground or to the static dischargers. Conductive paint and spaced metal "diverter" strips have been investigated to provide the desired surface conductivity. The lightning protection system selected for use in advanced composite applications should satisfy the following requirements:
- The advanced composite structure requiring protection shall not be electrically exposed to

Protection* System	Weight (lb/ft^2)	Installation Method	Advantages	Disadvantages
Aluminum flame spray (6 mils)	0.070–0.080	Cocured	1. Independent of surface shape and size 2. Repairable 3. Low maintenance 4. Partial environmental seal of composite surface	1. Coating weight and quality is operator-dependent 2. Aluminum flame spray quality cannot be determined prior to part cure 3. Limited long-term service fatigue experience record
Aluminum foil (5 mils)	0.070	Cocured	1. Environmental seal of composite surface 2. Uniform surface conductivity 3. Surface material completely replaceable	1. Foil stock width limitations 2. Difficult to install on compound contours 3. Poor repairability characteristics 4. Poor part handle ability characteristics 5. Heaviest system
Aluminum wire mesh (3 mils)	0.030–0.035	Cocured	1. Minimum shape constraint 2. Lightest-weight system 3. Repairable 4. Low maintenance 5. Lowest cost system (mesh cocured with laminate)	1. Mesh stock with limitations 2. Inadequate environmental seal for composite
Copper wire mesh (4 mils)	0.040	Cocured		

* Minimum or lighter weight coating designs are possible for secondary lightning strike zones.

Fig. 14.3.16 *Some of the basic lightning strike protection systems.*

the effects of lightning strike.
- The selected protective system and its application process should be considered neither to enhance nor detract from the advanced composite material properties.
- The system design shall provide for the prevention of electrical arcing when dissipating high-impulse, short duration (microsecond) and high-current electrical energy.
- A conductive-surface-to-metallic-substructure joint is required to provide for electrical grounding.
- The surface protective system must withstand the mechanical forces involved in dissipating high electrical (lightning) energy loads.
- The surface protective system shall permit the dissipation and flow of static electricity toward static dischargers (pigtails).
- The lightning protection conductive surface design should provide adequate shielding from electromagnetic interference.
- The surface protective material should be repairable and require a minimum of maintenance.
- The conductivity characteristics of the surface and the electrical grounding joint must not significantly degrade with time or operational environment exposure.
- Aircraft wings for carrying fuel will either have to use bladders or other means to prevent metal fastener arcing inside the wing.

(3) Hail and Foreign Objects
The leading edge of wing and empennage structures and some lower surface components of aircraft are subject to hail and foreign object damage. The use of metal matrix composites and metal thin skin on organic resin (matrix) composites will afford some level of impact protection. Complete impact protection, of course, is impossible. The designer, consequently, should take the following precautions:
- Design for repairability
- Design for replaceablity
- Avoid minimum gage designs

(4) Humidity
Epoxy composites undergo a decrease in strength when exposed to high humidity. For example, flexural strength may be reduced by 5—10% at room temperature, and as much as 30—40% at 350°F. These humidity effects on epoxy composites can be reversed by exposure in service to temperature above 200°F. Edge sealing and application of face protection will minimize humidity effects. Aluminum matrix composites do not exhibit this humidity sensitivity.

Calculation Examples
Example 1
To demonstrate some of the differences between designing with metals and advanced composites, consider the case of a cylindrical tube loaded by internal pressure (Fig. 14.3.17).
Strength material theory gives:

circumferential stress: longitudinal stress:

$$f_y = \frac{(pr)}{t} \qquad f_x = \frac{f_y}{2} = \frac{(pr)}{2t}$$

where r is the average tube radius and t is the tube wall thickness (assume the thickness is reasonably small compared with tube radius). Therefore, for any given pressure and radius, the required wall thickness can be directly calculated for any given material.

For composites, however, the designer can determine an optimum orientation so that the tube is stronger in the direction of maximum load (f_y) and in the direction of the lesser load ($f_x = \frac{1}{2} f_y$). It is apparent that, since shear strength is not a requirement, that some combination of 0° plies (along the tube) and 90° plies (around the tube) will provide the optimum structure.

The solution to this design is shown in Fig. 14.3.18. This verifies what would be intuitive for this type of loading case, in that 67%, or two-thirds, of the laminae should be oriented in the 90° direction, since

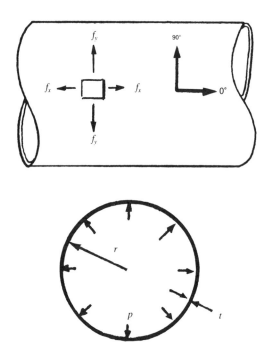

Fig. 14.3.17 Cylindrical tube with internal pressure.

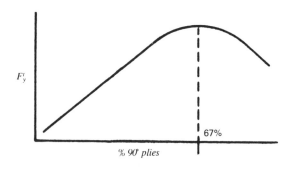

Fig. 14.3.18 F_y^t versus percent 90° plies for $f_y/f_x = 2$.

the stress in that direction is twice the stress in the 0° direction. The required wall thickness can then be calculated.

Thus, the composite tube is designed in a more efficient manner than the metal tube, with the biaxial strength of the tube matched to the type of loading applied.

Example 2
(a) Given load: $N_x = 7000$ lb/in; $N_y = 0$;
$N_{xy} = 4500$ lb/in

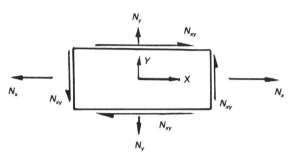

Requirements: $GT = 600000$ lb/in
$ET = 1.5 \times 10^6$ lb/in
where $T =$ total thickness of laminate

(b) Given Graphite/Epoxy material allowables:

Unidirectional Ply	±45° Ply
$F_{11}^t = 98$ ksi	$F_{11}^{45} = F_{22}^{45} = 15$ ksi
$F_{11}^c = 74$ ksi	$E_{11}^{45} = E_{22}^{45}$
$E_{11}^t = 20.5 \times 10^3$ ksi	$= 3 \times 10^3$ ksi
$E_{11}^c = 18 \times 10^3$ ksi	$F_{12} = 38$ ksi
$F_{22} = 0$	$G_{12} = 5 \times 10^6$ ksi
$E_{22} = 0$	
Ignore the strength from epoxy contribution	
Ply thickness = 0.005 in	

(c) Define number of plies in each direction and stack.

(i) Find number of ±45° plies:

Based on shear requirement,

$N_{xy} = 4500$ lb/in

$$\frac{N_{xy}}{t F_{12}} = \frac{4500}{0.005 (38000)} = 23.68 \text{ plies}$$

Based on shear stiffness requirement, $GT = 600000$ lb/in

$$\frac{GT}{t G_{12}} = \frac{600000}{0.005 (5)(10^6)} = 24 \text{ plies}$$

A minimum total of 24 of ±45° (+45° and −45°) plies are required to satisfy above requirement and its total thickness $t_{45} = 0.005 (24) = 0.12$ in.

(ii) Find number of 0° plies:
Based on axial load requirement, $N_x = 7000$ lb/in

$$\frac{N_x - F_{11}^{45} t_{45}}{t F_{11}^t} = \frac{7000 - (15000)(0.12)}{(0.005)(98000)}$$

$$= 10.6 \text{ plies}$$

Based on flexural stiffness requirement, $ET = 1.5 (10^6)$ lb/in

$$\frac{ET - E_{11}^{45} t_{45}}{t E_{11}^t} = \frac{1.5 (10^6) - 3 (10^6)(0.12)}{0.005 (20.5)(10^6)}$$

$$= 11.17 \text{ plies}$$

A minimum total of 12 of 0° plies are required to satisfy above requirement and its total thickness $t_0 = 0.005 (12) = 0.06$ in.

(iii) Find number of 90° plies:

Given $N_y = 0$

Four 90° plies are arbitrarily added (about 10%) to enhance lateral stability crack propogation inhibition as well as to relieve the Poisson's ratio effect.

(iv) Total number of plies and stack of the laminate:

Total number of plies = $0_{12}/\pm 45_{12} 90_4$ (40 plies)

Laminate thickness = 0.005 (40) = 0.200 in

Recommended stacking: $[\pm 45_2/0_2/\pm 45_2/0_2/\pm 45_2/0_2/90_2]_s$

(d) Find allowable strength in y-direction
From 90° plies:
$$(t_{90})(F_{11}^t) = [4(0.005)](98000) = 1960 \text{ lb/in}$$

From ±45° plies:
$$(t_{45})(F_{22}^{45}) = (0.12)(15000) = 1950 \text{ lb/in}$$

From 0° plies:
$$(t_0)(F_{22}) = (0.06)(0) = 0$$

The allowable strength:
$$N_y = 1960 + 1950 + 0 = 3910 \text{ lb/in}$$

or $F_y = \dfrac{3910}{0.200} = 19550 \text{ psi}$

14.4 Structural Joint Design

Joints, which must be present when any two components, are assemblied, are major source of stress concentrations. In the case of bonded joints, stress concentrations occur to maintain strain compatibility between components. In the case of mechanical joints, they are a result of the decreased area at the hole and the loaded hole itself. The primary purpose of this section is to acquaint the designer with some of the problem areas encountered, introduce some of the joint design allowables generated on the subject, and show a few example of how typical problems have been solved.

To realize the full potential of advanced composites in lightweight aircraft structure, it is particularly important to ensure that the joints, either bonded or bolted, do not impose a reduced efficiency on the structure. This problem is far more severe with composite materials than with conventional metals, such as aluminum, titanium and steel, because the high-specific-strength filaments are relatively brittle. They have very little capacity to redistribute loads and practically none of the forgiveness of a yielding metal to mask a multitude of design approximations. This is the reason why greater efforts are devoted to understanding joints in composite materials and to providing reliable design techniques, particularly for the thicker section and for multiple fastener pattern design cases.

There are six basic items to be considered in the design of a joint:
- The loads which must be transferred.
- The region within which load transfer must be accomplished.
- The geometry of the members to be joined.
- The environment within which the joint must operate.
- The weight/cost efficiency of the joint.
- The reliability of the joint.

The first four items are generally prescribed. It remains then to satisfy the last two items in some optimal manner. The first decision should concern the class of joining techniques which should be studied. In general, adhesive joints are proved to be more efficient for lightly loaded joints, while mechanically fastened joints are more efficient for highly loaded joints.

Some general rules for composite joints are that the most efficient joints are scarf and stepped lap joints in which there is relatively little change in the load path. The double-lap and single-lap joints are quite a bit less efficient, the former is worse than the latter. To realize maximum efficiency from adhesives, joints should be specifically designed for adhesive bonding.

Mechanically Fastened Joints

Failure modes for advanced composite mechanical joints are similar to those for conventional metallic mechanically fastened joints. Fig.14.4.1 presents typical simplified representations of these failure modes; i.e., shearout, net tension, bearing, and combined tension and shearout. The shearout failure mode can also be sometimes characterized by a single-plane "cleavage" failure, where the apparent laminate transverse tensile strength is less than the corresponding in-plane shear strength. In addition, bearing or shear failure of the fastener, and bolt pulling through the laminate are other possible failure modes.

The following equations should be used to determine allowable joint strengths:

Bearing: $P^{br} = DtF^{br}$ (14.4.1)

Shear-out: $P^s = 2(\dfrac{e}{D} - 0.5) DtF^s$ (14.4.2)

Net tension: $P^t = 2(\dfrac{s}{D} - 0.5) DtF^t$ (14.4.3)

where F^{bu} = design bearing strength
F^{su} = design shearout strength
F^{tu} = design net tension strength
D = fastener diameter
t = Laminate thickness
e = edge distance
$s = \dfrac{\text{edge distance or fastener spacing}}{2}$

The equation (14.4.2) for predicting shearout strength utilizes an equation applicable for both shearout and cleavage failure, since F^s has been empirically obtained to cover both cases. Fig. 14.4.2 represents an element of a typical mechanical joint and defines the key dimensions by illustration. Although many assembly problems have been solved with adhesive-bonding techniques, there are many cases where only mechanical joints are capable of meeting design requirements. Examples include parts requiring replacement or removal for ease of fabrication or repair, assemblies joining materials of dissimilar elastic or thermal properties, access covers, and joints subjected to complex loadings.

(1) Some of the obvious advantages of mechanical joints are:
- Utilization of conventional metal-working tools and techniques, as opposed to adhesive-bonding procedures.
- Ease of inspection.
- Utility of repeated assembly and dis-assembly for fabrication replacement, or repair.
- Assurance of structural reliability.

(2) Offsetting the advantages are some disadvan-

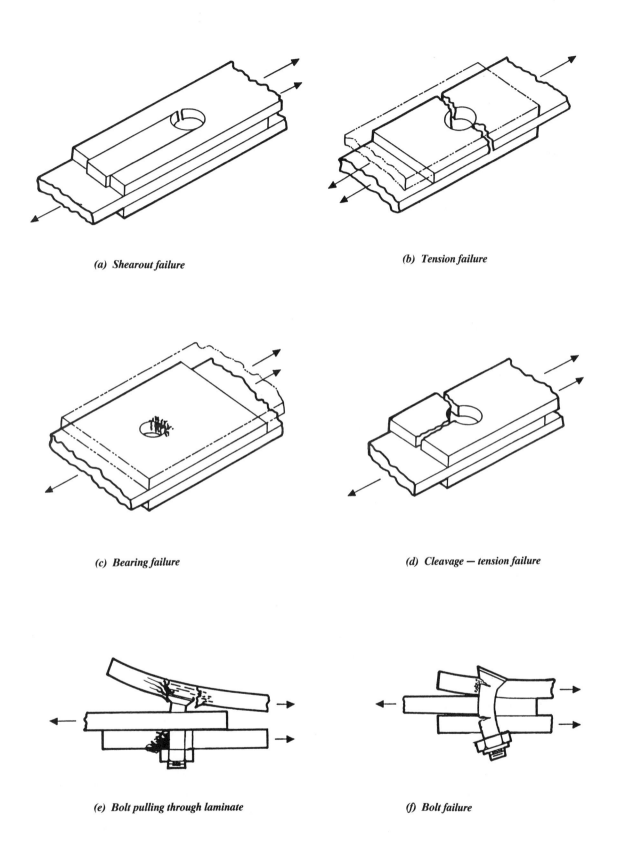

Fig. 14.4.1 Failure modes of advanced composite mechanical joints.

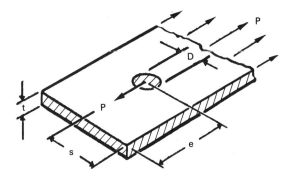

Fig.14.4.2 Typical mechanical joint element.

tages, which are:
- Strength degradation of the basic laminate and a resultant weight penalty.
- Necessity of additional loose parts (fasteners) for assembly.
- Need for more careful design than used with conventional metals because of the lack of ductility to relieve local stress concentrations and because of the unequal directional properties of the laminate.

(3) The following design practices of mechanically fastened joints are recommended:
- Stress concentrations exert a dominant influence on the magnitude of the allowable design tensile stresses. Generally, only 20—50% of the basic laminate ultimate tensile strength is developed in a mechanical joint.
- Mechanically fastened joints should be designed so that the critical failure mode is in bearing, rather than shearout or net tension, so that catastrophic failure is prevented. This will require edge distance to fastener diameter ratio (e/D) and side distance to fastener diameter ratio (s/D) relatively greater than those for conventional metallic materials. At relatively low e/D and s/D ratios, failure of the joint occurs in shearout at the ends or in tension at the net section. Considerable concentration of stress develops at the hole, and the average stresses at the net section at failure are but a fraction of the basic tensile strength of the laminate.
- Multiple rows are recommended for unsymmetrical joints, such as single shear lap joints, to minimize bending induced by eccentric loading.
- Local reinforcing of unsymmetrical joints by arbitrarily increasing laminate thickness should generally be avoided because the increased resulting eccentricity gives rise to greater bending stresses which counteract or negate the increase in material area.
- Since stress concentrations and eccentricity effects cannot be calculated with a consistent degree of accuracy, it is advisable to verify all critical joint designs by testing representative sample joints.

(4) Mechanical joint design guides
(a) Shearout [see Fig. 14.4.1(a)] and cleavage [see Fig. 14.4.1(d)] failure:
- Use large fastener edge margin than in aluminum design, such as $e/D \approx 3$.
- Use a minimum of 40% of ±45 plies. See Fig. 14.4.3 for the effect of layup on the bearing stress at failure.
- Use a minimum of 10% of 90° plies.

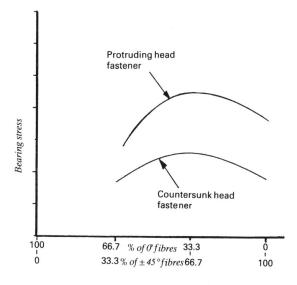

Fig. 14.4.3 The effect of layup on the max. bearing stress.

(b) Net tension failure [see Fig. 14.4.1(b)]:
- Use larger fastener spacing than in aluminum design, such as $s/D \approx 6$.
- Padup to reduce net stress.
(c) Fastener pull-through from progressive crushing/bearing failure [see Fig. 14.4.1(c)]:
- Design joint as critical in bearing.
- Use padup.
- Use a minimum of 40% of ±45 plies.
- Use washer under collar or wide bearing head fasteners.
- Use tension protruding heads when possible.
(d) Fastener shear failure:
- Use larger diameter fastener.
- Use higher shear strength fastener.
- Should not (or never) be shear failure in joint design.
(e) Conventional push through interference fit installation of fasteners will damage laminate:
- Use hole size clearance of 0.000 to +0.002 for normal structural joints.
- Use standard close-ream holes for critical joints.
- Use special designed fastener to obtain

interference fit such as expanded sleeve type fastener (see Fig. 7.2.10).
- Be careful to use interference or transition fit holes.
- Do not use taper-loks or cold-work holes.

(f) Use two row joints when possible. The low ductility of advanced composite material confines most of the load transfer to the outer rows of fasteners.

(g) The use of graphite composites in conjunction with aircraft metals is a critical design factor. Improper coupling can cause serious corrosion problems to metals. Materials such as titanium, corrosion-resistant steels, nickel and cobalt alloys can be coupled to graphite composites without such corrosive effects. Aluminum, magnesium, cadmium plate and steel will be most adversely affected because the difference of electrical potential between these materials and graphite. Fig. 14.4.4 shows the galvanic compatibility of fastener materials with graphite composites.

(h) The choice of optimum layup pattern for maximized fastener strength is simplified by the experimentally-established fact that the quasi-isotropic patterns $(0/\pm 45/90)_s$ or $(0/45/90/-45)_s$ are close to optimum. This reduces the experimental costs and simplifies

Fastener Material	Compatibility with Graphite/Epoxy and Application Guidelines
Titanium, Ti Alloys, Ti-CP	Fasteners of these materials are compatible with graphite/epoxy composites. Permanent fasteners should be sealed to prevent water intrusion but removable fastener may be used with no supplement protection.
MP-35N (AMS 5758) Inco 600 (AMS 5687)	These materials are compatible with graphite/epoxy components. This probably also applies to other high nickel and cobalt alloys.
A286 (AMS 5731, AMS 5737) PH13-8Mo (AMS 5629)	These CRES alloys and some other austenetic and semi-austenetic alloys are marginally acceptable in contact with graphite composites. In a severe marine/industrial corrosion environment, superficial rusting and stains develop on the fastener. Although loss of fastener integrity has not been established, this staining is usually objectionable. Permanent fasteners that can be installed with sealant and overcoated with sealant are usually satisfactory. Removable fasteners are not acceptable to some design activities.
Monel	Marginally acceptable in contact with graphite/epoxy composites. Significant current flow and material loss.
Low Alloy Steel, Martensitic Stainless Steels	Not compatible with graphite/epoxy materials. Severe rusting.
Silver Plate, Chromium Plate, Nickel Plate	These plating materials are compatible with graphite but are not adequate to protect steel in contact with graphite/epoxy composites. Silver plated A286 or PH13-8Mo would be compatible with graphite and suitable if there is no aluminum or titanium in the joint.
Cadmium Plate, Zinc Plate, Aluminum Coatings	Not compatible with graphite/epoxy composite materials. Rapid deterioration of plating or coating.
Aluminum, Aluminum Alloys, Magnesium Alloys	Not compatible with graphite/epoxy composite materials. Not feasible to adequately protect fasteners of these materials from severe corrosion if in contact or close proximity to graphite.

Fig.14.4.4 Galvanic compatibility of fastener materials with graphite composite.

the analysis and design of most fastened joints.

(i) One of the key factors governing fastened joint behavior in advanced composite structure is concerned with the vast difference between double-lap and single-lap efficiencies. The eccentricity in the load path for single-lap joints leads to non-uniform bearing stresses across the thickness of the laminate. This, in turn, leads to the development of the critical bearing stress and bypass stress around hole at the laminate interface, at an even lower average bearing stress because of the brittle composite materials. Fig. 14.4.5 shows the bearing stress distribution at fastener hole, and the use of the bearing reduction factor to account for this effect. It is difficult to define the reduction factor because it is a function of fastener material characteristics, composite material and its layup sequence, fastener fittnes, etc. Currently, an arbitrary value of 1.5 to 2.0 is used for the reduction factor until the result from test for each particular design case is established.

(j) Develop a bearing/bypass stress interaction envelope curve (function of laminate material, laminate thickness and plies layup sequence or stacking, fastener diameter, etc.) to size mechanical joint as shown in Fig. 14.4.6.

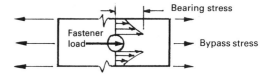

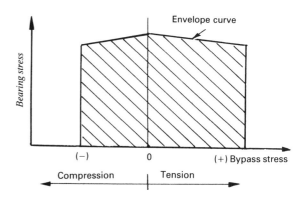

Fig. 14.4.6 *Bearing/bypass stress interaction envelope curve (notched and wetted test data).*

Adhesive Bonded Joints

In order to ascertain the efficiency of a joint, its failure modes must be known. For an adhesive-bonded joint, the micro-mechanical failure modes are:
- In-plane axial failure of the laminate.
- Interlaminar shear failure of the laminate between two laminae.
- Adhesive failure of the adherend/adhesive interface.
- Cohesive failure of the adhesive.

The prime function of adhesive-bonded joints is to transfer load by shear and most analysis of bonded joints are so oriented. However, some joints develop associated peel stresses because of eccentricities in the shear load path through the joint while others are actually subjected to externally applied loads inducing peel loads as shown in Fig. 14.4.7. Because bonded joints are inherently weak in peel and composite laminates are even weaker in interlaminar tension, it is extremely important to minimize these adverse influences in design and to make sure that the most critical condition is accounted for in analysis.

For maximum effectiveness and confidence, adhesive bonds should be designed in accordance with following general principles:
- The bonded area should be as large as possible.
- A maximum percentage of the bonded area should

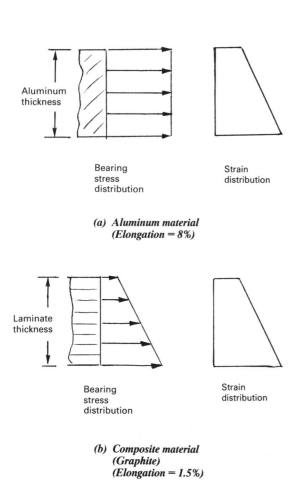

Fig. 14.4.5 *Ultimate bearing stress distribution of aluminum material vs. composite material (single-lap joint).*

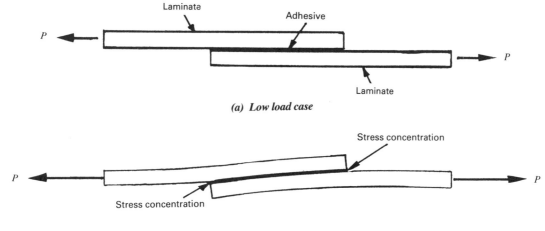

(a) Low load case

(b) Deformation under high load case

(c) Interlaminar failure case

Fig. 14.4.7 *Single-lap bonded joint vs. load cases.*

contribute to the strength of the joint.
- The adhesive should be stressed in the direction of its maximum strength.
- Stress should be minimized in the direction in which the adhesive is weakest.

The strength of bonded single-lap and double-lap joints depends primarily on the lap length and the extensional stiffnesses of the laminates for a specific adhesive system. Thoery and test have generally shown that the highest strengths are attained when the Et of the two laminates are equal to one another.

Fig. 14.4.8 illustrates the designs of different bonded joint types.

In advanced composite structures, there are commonly two bonding methods used in adhesive bonded joints:
- Cocuring method
- Secondary bonding method by adhesive

The cocuring method gives the strongest joint strength and is always recommended in composite construction to eliminate mechanical fasteners. The use of structural adhesives is becoming more prominent in the aerospace industry with new airframe designs, but so far only in the secondary structures. With continued improvements in materials and processing, designers foresee wider use of cocuring construction in future primary airframe application to replace the conventionally fastened composite structure (called *black aluminum design*) by the design of

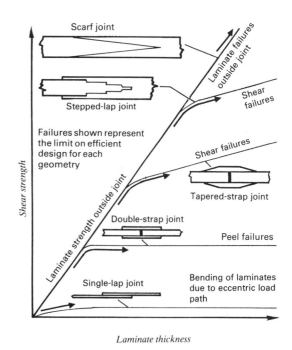

Fig. 14.4.8 *Bonded joint strength vs. laminate thickness.*

Airframe Structural Design 525

eliminating all or most fasteners (called *true composite design* stage or *second generation composite design*).

One of the main shortcomings in using either adhesive or cocuring method for aircraft manufacturing is that there is no means of testing nondestructively to verify the strength of a bond to date. But a new NDI device is being developed and may be available soon to solve this problem.

Bonded joint design guidelines are as follows:

(a) Short joint lengths are more efficient in avoiding shear failure due to peak shear stress (see Fig. 14.4.9).

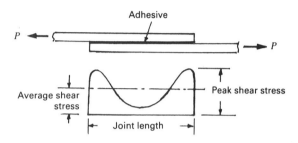

Fig. 14.4.9 Shear stress distribution along bonded joint.

(b) Reduce joint eccentricity.
   - Thick laminates are affected more than thin laminates.
   - Double shear joints reduce peeling effect.
   - Chamfer or taper laminate thickness to reduce peeling.
(c) Do not use 90° plies on outer surfaces of the laminate (use ±45° plies).
(d) Thermal effect — This occurs when thermal stresses are induced between the laminate materials. Such stresses are characteristically negligible at the cure temperature (350°F) but become progressively worse in proportion to the square of the temperature differential between curing and operating temperatures. However, the bonded joint laminates of composite materials and metal such as titanium should give attention to their thermal expansion difference, especially in high temperature applications.
(e) Because adhesive bonding is a surface phenomenon, surface preparation is one of the keys to successful bonding. The durability of an adhesive bond is very much dependent on the surface treatment of the laminates.

14.5 Manufacturing

Manufacturing processes and procedures are the controlling elements of the cost of a composite component and, therefore, it is mandatory that they should be an integral part of the design process. Usually, composite components contain fewer parts than their metallic counterparts. This feature, plus the reduced number of mechanical fasteners required in most composite designs, is a basis for cost reduction. Of course, the part size, geometry, complexity, and required quantity are all considerations in the selection of a fabrication process.

In the past, the most labor-intensive step in composite fabrication has been the uncured ply layup of tape segments, which is normally done by hand. In areas amenable to the tape form, advantage can be taken of the automated methods of dispensing and laying tape with pre-determined orientation and ply sequencing. Recently, certain companies have directed a large portion of their research efforts into the manufacturing process such that the implementation of automated systems could be cost-effective. Some of the advances are beginning to be incorporated into the industry's building blocks for the so-called "Factories of the future". One of their processes which has received a great deal of attention in manufacturing research is automated system in composite. As a matter of fact, the automated system is a must in composite manufacturing process to reduce cost.

Curing Methods

The most commonly used method of applying heat and pressure simultaneously to the layup assembly on the mold is the pressurized autoclave as shown in Fig 14.5.1. The autoclave is usually employed for parts of complex or double curvature, which it can more easily accommodate and process. The press is essentially for curing flat components, although curved and more complex assemblies can be processed in them with special tooling. Fig. 14.5.2 illustrates a typical layup and bagging for a composite laminate.

Fig. 14.5.1 Typical autoclave.

Why vaccum and pressure are used during cure cycles requires some explanation. The primary purpose of applying vacuum is to hold the laminate parts in position and to position the vacuum bag during cure. At the beginning of the cure cycle, a vacuum is drawn from the inside of the vacuum bag. Autoclave pressure is applied to the outside of the bag at the same time vacuum is drawn from the inside. The pressure presses and holds the mating surfaces of the assembly in close contact while the adhesives, melt, flow, and cure.

The bonding tool serves as a bed and causes a back pressure to be imposed against the bottom surfaces of the assembly.

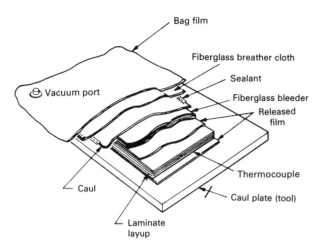

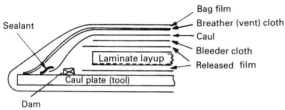

Fig. 14.5.2 *Typical composite layup and bagging.*

If the assembly were not bagged and sealed, vaccum could not be drawn. In this case, the autoclave pressure would be the same at all points in the assembly. The pressure pushing outward on the surfaces would equal that pushing inward. No difference in pressure would exist and the pressure would serve no purpose at all. Fig. 14.5.3 shows a typical autoclave cure cycle for thermoset composite laminates.

Precise cure control can directly reduce part scrappage and energy consumpton. More importantly, such control can enable the production of larger structures complete with integral stiffeners, thereby minimizing post-cure assembly and machining, and reducing the handling costs associated with bagging/debulking and curing of numerous small parts.

As mentioned previously, advanced composite systems based on thermoplastic resins (matrices) are being considered as replacements for sheet metals and thermosets in future aerospace structures. One of the major benefits offered by thermoplastics is the rapid transformation of raw materials into finished parts. The time required for this rapid conversion is limited only by two factors:
- The rate at which heat can be added to the thermoplastic resin to bring it to processing temperature (usually at or above its melting point).
- The rate at which the heat can be removed from the material once the forming process has been completed.

The actual forming step represents a small portion of the total time for heat-up and cool-down. Processing time is not governed by the time required to complete chemical reactions, as with thermosets.

Other curing methods beside the autoclave are as follows:
- Ovens
- Presses
- Heating blankets
- Heat lamps
- Sunlight cure resin
- Induction heating coils
- Microwave cure (for pultrusion)
- Exothermic reaction
- Room-temperature curing (for homebuilt airplane, windmill blades, boats, etc.)

Filament Winding

Most filament-winding technology has been developed for the rocket-motor industry, the cylindrical shape and large size of rocket-motor cases make filament winding particularly attractive for this application.

Providing the highest strength-to-weight ratio, filament winding consists of feeding reinforcement filament, or roving through a resin bath or using preimpregnated roving and winding it on a mandrel as shown in Fig. 14.5.4. Special winding machinery lays down the impregnated roving in predetermined patterns, giving maximum strength where required. After the appropriate layers are applied, the wound mandrel is cured, and the molded part removed from the mandrel. Filament winding provides the greatest control over orientation and uniformity, but is frequently restricted to surfaces of revolution.

Two methods of applying resin to the continuous fiber are used in filament winding:
- In the wet system, the fiber picks up resin as it travels through a trough. A variation of this system is the "controlled wet" procedure where in the resin is metered onto the fiber. Although the wet system is the lower-cost method, it is messy, slow, wastes resin, and requires protection of personnel against odor and fumes.

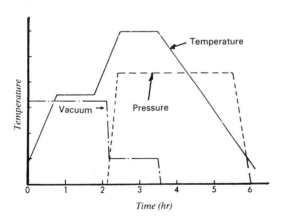

Fig. 14.5.3 *Typical autoclave cure cycle — thermoset composite.*

Airframe Structural Design 527

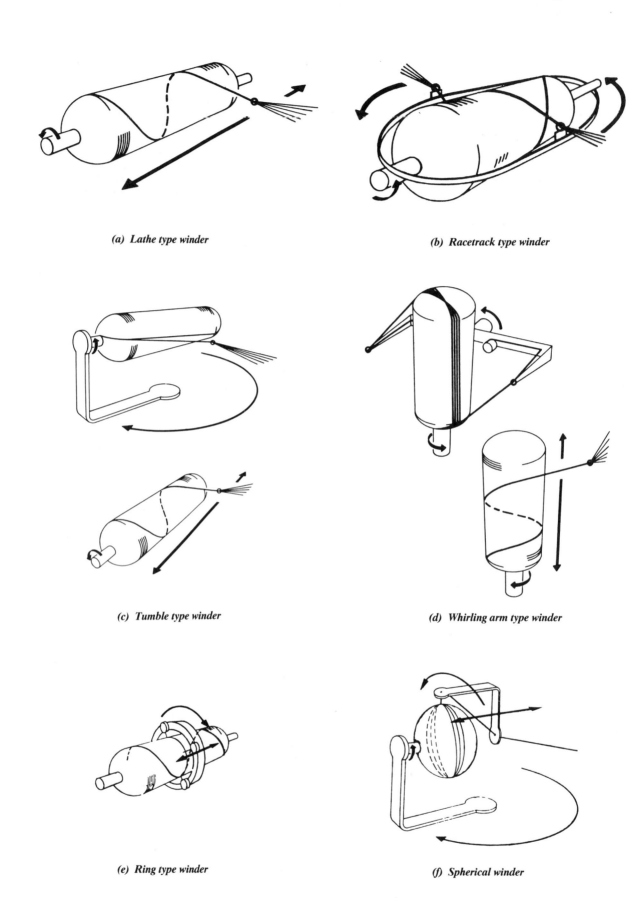

Fig. 14.5.4 Types of filament winding process.

- The prepreg system overcomes the limitations of wet system, but prepreg material costs more. Also, cold storage (for thermosets) is required for prepreg material to prevent resin advancement.

Although the fibers used in filament winding and in layup/autoclave processing are the same, characteristics of the composites are significantly different as shown in Fig. 14.5.5.

Fig. 14.5.6 shows two typical cases for filament-wound applications, one of them is the Beechcraft Starship fuselage which is the first filament-wound application in large fuselage to date. This method has been seriously considered in future transport fuselages to take the filament winding advantages to

Filament-Wound Structures	Layup/autoclave Pressure-cured
Basically for rocketmotor industry requirements	Aircraft industry requirements
Generally wet wound	Generally preimpregnated materials
Moderate to high void content	Low void content (0–1%)
Minimal cure pressure	High cure pressure
Moderate environmental requirements	Severe environmental requirements
Good pressure-vessel performance, i.e. fuselage	Good structural properties, i.e. wing, empennage, etc.
Low viscosity resin	High viscosity resins
Moderate to high elongation resins	Low elongation resins
High fiber volume composite	Moderate fiber volume composite
Low temperature cure resin systems	High temperature cure resin systems

Fig. 14.5.5 Filament-wound vs. autoclave-cured composites.

(a) Solid rocket booster

(b) Beechcraft Starship fuselage
(Courtesy of Beech Aircraft Corp.)

Fig. 14.5.6 Filament-wound applications — rocket shell and fuselage.

Airframe Structural Design 529

reduce fuselage structural weight and cost.

Pultrusion

In pultrusion, resin-impregnated filaments are fed into a heated die (i.e. microwave heating). The cured section emerging from the die is grasped and the remaining filaments are pulled through at a constant rate as shown in Fig. 14.5.7. The process is widely used for making complex shapes, but has been limited to items with constant cross-section. This restriction may soon be eliminated by further development of variable section pultrusion.

For thermoplastic resin pultrusion, the die should be heated above the resin melting point; in some cases, it might be cooled in the downstream zone. Because thermoplastics shrink much less than thermoset resins, expect higher drag forces in the die. One of the key advantages of thermoplastic pultrusion (also thermoplastic filament-wound) is the ability to post-form the part after consolidation.

Thermoforming and Stamping

(1) Thermoforming

Thermoforming (for thermoplastic resins) consists of laminates in flat-sheet form, which are assembled from individual plies of woven prepreg fabrics or unidirectional prepreg tapes to the desired thickness and reinforcement (fibers) orientation, and then consolidated with heat and pressure. This allows considerable flexibility in arranging the plies to achieve isotropic or anisotropic mechanical properties, and the sheet can also have non-uniform thickness if desired. However, the thermoforming process has three key elements (see Fig. 14.5.8):

- A laminate support frame, which carries the laminate into the heat source, supports the laminate during and after softening of the matrix, rapidly transfers the softened part from the heat source to the forming tool, and then releases the laminate onto the lower tool.
- A heat source capable of evenly heating the laminate to its processing temperature in a short period of time.
- A thermoformer capable of rapid closing speeds with sufficient clamp pressure to form the laminate.

Since the laminate must slip into the tool, there can be difficulties in maintaining an adequate gas seal to hold vaccum below the laminate and positive air pressure above it. Therefore, an elastomeric or thin aluminum bladder or diaphragm clamped over the laminate. Pressure could then be applied directly to the diaphragm (hydrostatic forming or hydroforming process).

(2) Stamping

Stampable thermoplastic composite sheets are

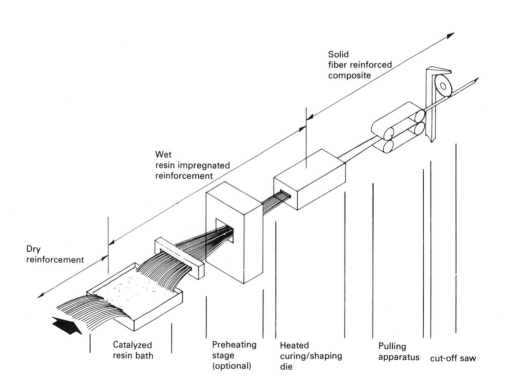

Fig. 14.5.7 Pultrusion processing.
(Courtesy of Shell Chemical Co.)

expensive matched-steel molds, similar in design to those used for compression molding thermosets. Unlike the thermoforming, stampable composites are capable of a good deal of flow and are recommended for parts requiring good surface finish.

Compression Molding

Bulk-molding compound (BMC) or sheet-molding compound (SMC) pre-form, or mat molding may all be performed by compression molding with heat and pressure. Part configurations and cross-sections are extremely flexible, inserts and attachments may be accommodated, and very high tolerances may be met.

Automatic Tape-laying Machine

The tape-laying machine as shown in Fig. 14.5.9, using computer-controlled automation techniques to replace manual layup methods, greatly reduces the cost of using tape materials in large, very thick and contoured aircraft structures. It also gives more precise, automatic control of layup parameters that alter quality and performance.

In practice, the machine dispenses tape from a supply roll, places it directly on the underlying mold or tape-covered surface, applies shoe pressure to seat and debulk the tape, takes up and stores the release paper, and at the end of each movement across or along the mold, cuts the tape to the required length and angle. It also changes the angle from layer to layer so as to give the advantages of cross-ply construction. In addition, it adds extra tape lengths at selected locations/angles to give local reinforcement or attachment provisions.

Tooling

Advanced composite materials are well established in the aerospace industry for the manufacture of light weight, dimensionally stable components. Layers of prepreg material are laid onto a mold tool to form the component, which is then cured in an autoclave at temperature under a consolidating pressure.

The demands placed upon a mold tool can be very severe and the ideal tool should:
- reproduce the pattern or master model with a high degree of dimensional accuracy;
- have thermal expansion characteristics identical to those of the components to be produced;
- withstand autoclave cures without deterioration of surface finish, loss of gas tightness, or distortion;
- retain a large proportion of its stiffness and strength at the high working temperature.

Composite laminate surfaces vary with the material, assembly method, and tooling mold surface. The bag surface, as opposed to the mold surface, will be rough. The fabrication process does not ensure an acceptable mating surface flatness on the bag side. Surface smoothness can be improved by the use of a caul sheet (or plate) in some cases. However, a prime consideration for selection of materials for fabrication of large tools used for curing composite structures is compatibility of thermal expansion between the tool and part. Fig. 14.5.10 shows a list of thermal expansion coefficients for several selected materials for comparison. Aluminum tooling can be used for small and slightly contoured laminates (i.e. graphite). For a large tool, titanium or steel is a better choice than

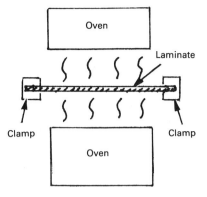

(a) Heating unit

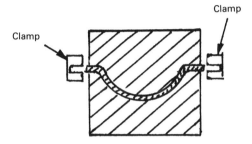

(b) Matched tools–laminate is unclamped during forming, so it can slip into the tool

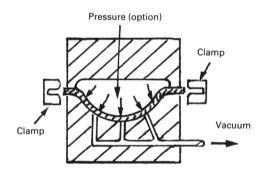

(c) Vacuum forming–the laminate is free to slip into the tool, maintaining a gas seal may be difficult, unless a bladder or diaphgram is laid over the laminate

Fig. 14.5.8 Thermoforming process.

reinforced with continuous swirl mat or chopped-fiber mat, rather than woven or unidirectional fabrics. The stamping process is basically the same as for thermoforming except it requires higher pressure forming equipment and more

Airframe Structural Design 531

Fig. 14.5.9 Cincinnati milacron automatic tape-laying machine.
(Courtesy of Cincinnati Milacron Inc.)

aluminum but still not totally compatible with thermal expansion. High-temperature graphite/epoxy tooling is the ideal tool to cure graphite/epoxy laminates because it has excellent strength-to-weight characteristics as well as thermal and dimensional stability. Also because the tools eliminate heat-sink considerations during part curing and subsequent molding, autoclave time is reduced significantly and manufacturing cost is cut.

Tooling cost should be kept as low as possible, particularly in a prototype program. The use of sculptured metal shapes should be limited. The use of plastic-reinforced tooling can also be expensive if the tooling cast or molds are used to layup and cure; the final tool has to be high-temperature-resistant and compatible in thermal expansion.

Steel has been used as the predominant tooling material because of availability, low cost, and compatible coefficient of thermal expansion compared with composites. Stainless steel is used extensively when severe radius forming is required.

The elastomeric tooling concept as show in Fig. 14.5.11 has been used successfully to manufacture cocured integrally stiffened panels.

A comparison of tooling material is provided in Fig. 14.5.12.

Layup-over-foam Method

This is generally for home-built aircraft. If one or a few of parts are to be fabricated, it is advantageous to build the parts by making a foam core and layup the composite materials over the core. It is because the method of wrapping reinforced materials around a given foam core is the easiest way without additional tool (actually foam core is a tool). The foam core is cut by a hot-wire saw as shown in Fig. 14.5.13.

The hot-wire saw is a piece of stainless steel safety wire, stretched tight between two pieces of tubing. The wire gets hot when an electric current passes through it and this thin, hot wire burns (cuts) through the foam. To get a smooth accurate cut, templates are required and the hot-wire should be guided around the templates with light pressures. Pushing too hard against the template may move them or flex the foam

Tooling Materials	Coefficient in/in/F x 10^{-6}	Composite Materials	Coefficient in/in/F x 10^{-6}
Graphite/Epoxy	−0.5 ~ 1.5	Graphite/Epoxy	−0.5 ~ 1
Monolithic graphite	−0.5 ~ 1.5	Boron/Epoxy	2.2 ~ 5
Cast ceramic	0.45	Glass/Epoxy	4 ~ 6
Ti-6al-4v alloy	5		
Steel	6.3		
Nickel (electroformed)	7		
aluminum	13		
Silicone rubber	45 ~ 200		
Beryllium	7.5		

Fig. 14.5.10 Coefficient of material thermal expansion.

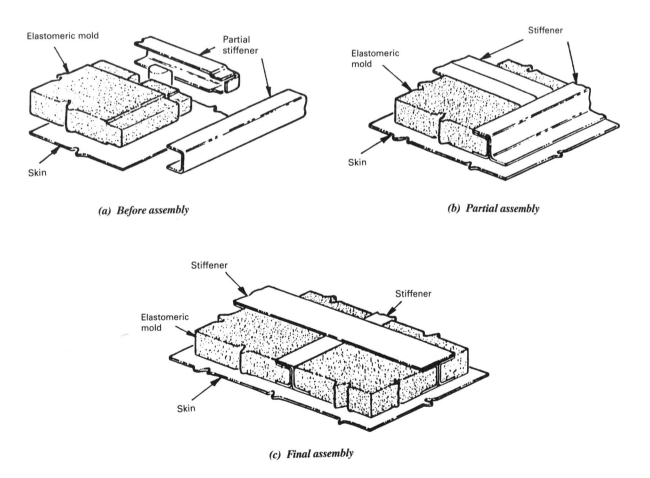

(a) *Before assembly*

(b) *Partial assembly*

(c) *Final assembly*

Fig. 14.5.11 The elastomeric tooling concept.

Airframe Structural Design 533

Tooling Material	Advantages	Disadvantages
Aluminum	• Good heatup rate • Easily machined • Low weight and mass • Stable with temperature • Readily adaptable to combination type fixtures	• Incompatible coefficient of expansion • Hard jig pick-up points cannot be provided • Must provide slotted rather than hard jig pick-up points
Steel	• Compatible coefficient of expansion • Dimensionally stable with temperature • Provides hard surface for jig pick-up points • Readily adaptable to combination type fixtures	• Machining is slow • Slow heatup rate • High tool weight and mass
Silicone rubber	• No shape restrictions • Transmits pressure readily • Cheaper for duplicate tooling	• Pressure hard to control, predict, or measure • Loses dimensional stability with repeated use
Graphite/Epoxy	• Excellent dimensional stability • Good heatup rate • Very good compatible coefficient of expansion • Light weight	• Must build master model • Durability • Not feasible for molding cocured stiffened panels
Monolithic graphite	• High temperature cure for thermoplastic composite • Easily machined	• High cost • Needs nickel electro-deposited tooling surfaces • Needs back-up structure

Fig.14.5.12 Tooling material comparison.

block which results in an undercut foam core. Proper wire tension and temperature should be maintained for good cutting.

After the core is cut, remove surface irregularities with sandpaper and seal the surface with micro, wet layup laminates and then cured at room temperature (Refer to Ref. 14.60).

Manufacturing Guidelines
(a) Size limitation — The available facilities impose limits on the size of advanced composite assemblies, such as autoclave size (diameter and length).
(b) Components shape — The shape limitations of construction result partially from the drape quality of the tape and fabric. Avoid tight radii and abrupt changes in surface features, which usually cause bridging between plies. Always provide access for layup and tooling, as well as inspection.
(c) Surface smoothness and flatness — Required surface smoothness and/or flatness can be obtained

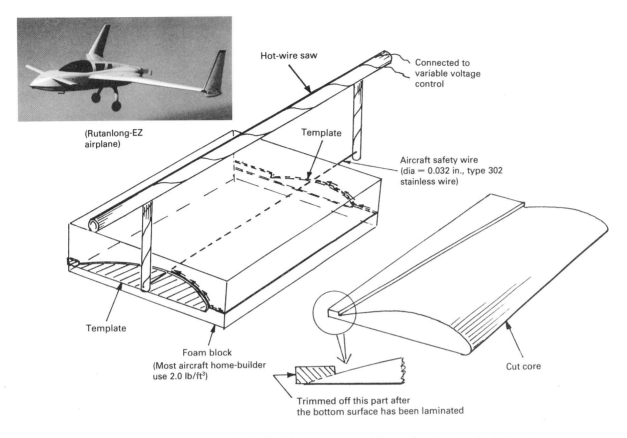

Fig. 14.5.13 Hot-wire cut method (which has been successfully used on home-built airplanes).

by means such as:
- Specifying the tool surface side of the composite laminates;
- Specifying requirements for a defined area.

(d) Laminate thickness — For most parts, drawing control of thickness to tolerances closer than that provided by specification is not necessary.

(e) Tool selection — There are three basic configurations used for tooling: male, female and matched male/female (see Fig. 14.5.14).

(f) Drilling/countersink — Drilling and countersinking of graphite/epoxy laminates require use of special carbide tools (Kevlar/epoxy and thermoplastic composites may need another special drilling procedure and tool).

(g) Drawing practices — Basic requirements apply to standard drawing practices:
- Ply orientation symbol
- Ply orientation callout
- Tooling surface designation
- Ply stacking section view
- Do not callout laminate thickness
- Flat patterns (if any)
- Ply stacking sequence usually start from tooling surface for convenience.

Quality Assurance (QA)

Quality assurance is one element in the manufacturing cost that exists in both composite and metallic constructions and represents another high-cost driver.

Typical defects that occur in composite laminates fall into one of the following types:
- Broken fibers — Broken or mislocated fibers in the internal or external layers of laminates.
- Burned — A condition caused by excessive heat during cure.

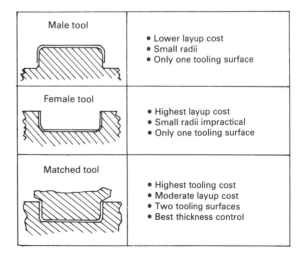

Fig. 14.5.14 Tooling selection.

Airframe Structural Design 535

- Delamination — The separation of adjacent layers within a multi-layer structure.
- Disbond — The lack of a bond in a joint area between two separate components.
- Distortion — Distortion/wrinkles may occur during the cure cycle from improper part layup.
- Inclusions — These include foreign materials, such as separator backing paper, tape, and other solid contaminations, inadvertently in the layup.
- Incorrect ply count — The incorrect number of plies.
- Incorrect ply orientation — Ply orientation not within angularity tolerance specified on drawing.
- Incorrect ply stacking — Incorrect location of a ply, or plies, called for in a layup.
- Resin variations — This means resin-rich or resin-starved which may occur under condition of improper compaction or improperly controlled bleeding during cure.
- Void — This is an empty, unoccupied space in an laminate.

Ultrasonic and radiographic inspections are the NDI methods commonly specified in the composite process:
- Ultrasonic inspection — Ultrasonic inspection has proven to be very useful for detecting internal delaminations, voids and composite structure inconsistencies.
- Radiography — Radiography is a very useful NDI method in that it essentially allows a view into the interior of the part. It is relatively expensive and special precautions must be taken because of the potential radiation hazard (x-ray).

References

14.1 Anon.: *Advanced Composites Design Guide. Vol. I–IV.* U.S. Air Force Materials Laboratory, 1983.

14.2 *MIL-HDBK-17A*: 'Plastics for Aerospace Vehicles, Part I. — Reinforced Plastics.' (1987).

14.3 Tsai, S.W., Halpin, J.C. and Pagano, N.J.: *Composite Materials Workshop.* Technomic Publishing Company, Stamford, Conn., 1968.

14.4 Ashton, J.E. and Halpin, J.C.: *Primer on Composite Materials: Analysis.* Technomic Publishing Company, Stamford, Conn., 1969.

14.5 Calcote, L.R.: *The Analysis of Laminated Composite Structures.* Van Nostrand Reinhold Company, New York, N.Y., 1969.

14.6 Sendeckyj, G.P.: *Composite Materials — Volume 2, Mechanics of Composite Materials.* Academic Press, New York, N.Y., 1974.

14.7 Noton, B.R.: *Composite Materials — Volume 3, Engineering Applications of Composites.* Academic Press, New York, N.Y., 1974.

14.8 Chamis, C.C.: *Composite Materials — Volume 7, Structural Design and Analysis, Part I and Part II.* Academic Press, New York, N.Y., 1975.

14.9 Jones, R.M.: *Mechanics of Composite Materials.* Scripta Book Company/McGraw Hill Book Company, New York, N.Y., 1975.

14.10 Bert, C.W. and Bergey, K.H.: 'Structural Cost Effectiveness of Composites.' *SAE Paper No. 730338*, (Apr. 1973).

14.11 Murrin, L. and Erbacher, H.: 'Design Problems Associated With The B-1 Composite Horizontal Stabilizer.' *SAWE Paper No. 1178*, (May 1977).

14.12 Shimizu, H., Dolowy, Jr, J.F., Taylor, R.J. and Webb, B.A.: 'Metal-Matrix Composites Behavior and Aerospace Applications.' *SAE Paper No. 670861*, (Oct. 1967).

14.13 Logan, T.R.: 'Costs and Benefits of Composite Material Applications to a Civil STOL Aircraft.' *AIAA Paper No. 74–964*, (Aug. 1974).

14.14 Dittmer, W.D.: 'Boron Composite — Development and Application Status.' *SAWE Paper No. 992*, (Jun. 1973).

14.15 Taylor, R.J.: 'Weight Prediction Techniques and Trends for Composite Materials Structure.' *SAWE Paper No. 887*, (May 1971).

14.16 Tetlow, R.: 'Design in Composites Application to Aircraft Structures.' *SAWE Paper No. 994*, (Jun. 1973).

14.17 Lovelace, A.M.: 'Advanced Composites.' *Journal of Aircraft*, (Sept. 1974), 501–508.

14.18 Anon.: 'Advances in Graphite and Boron Improve Savings and Performance.' *Product Engineering*, (Mar. 1971).

14.19 Stambler, I.: 'Bright Future Forecast for Composites in Aerospace.' *Interavia*, (Dec. 1972), 1363–1366.

14.20 Roselines, D.A. and Wood, H.A.: 'Military Application and Experience with Composites.' *AIAA Paper No. 710408*, (1971).

14.21 Fanti, R.: 'Composite Technology in Perspective.' *Shell Aviation News*, (1969).

14.22 Phillips, L.N.: 'Quiet Revolution — An Update on Carbon Fibres for Aircraft Structures.' *Shell Aviation News, No. 443*, (1977).

14.23 Hollmann, M.: *Composite Aircraft Design.* 11082 Bel Aire Court, Cupertino, Calif. 95014, 1982.

14.24 Anon.: 'B-70 Spurs Wider Use of Sandwich Design.' *Space/Aeronautics*, (Mar. 1961) 46–49.

14.25 Troxell, W.W. and Engel, H.C.: 'Column Characteristics of Sandwich Panels Having Honeycomb Cores.' *Journal of The Aeronautical Science*, (Jul. 1947).

14.26 Niu, Michael C.Y.: *Composite Drafting Manual.* Lockheed Aeronautical Systems Company, 1988.

14.27 Hooker, D.M.: 'PRD-49, A New Composite Material — Its Characteristics and Its Application to the BO-105 Helicopter.' *SAWE Paper No. 915*, (May 1972).

14.28 McKinney, J.M.: 'A Crack Stopper Concept for Filamentary Composite Laminates.' *Journal of Composite Materials*, (1972), 420–424.

14.29 Eisenmann, J.R. and Kaminski, B.E.: 'Fracture Control for Composite Structures.' *Engineering Fracture Mechanics*, (1972), 907–913.

14.30 Wolffe, R.A. and Etc.: 'New PRD-49 Products for Aerospace And Commercial Applications.' *Paper for Presentation at National SAMPE, Los Angeles, California*, (Apr. 1973).

14.31 Belbin, G.R.: 'Thermoplastic Structural Composites — a Challenging Opportunity.' *Proc Instn Mech Engrs, Vol. 198c*, (1984).

14.32 Measuria, U. and Cogswell, F.N.: 'Thermoplastic Composites in Woven Fabric Form.' *I Mech E (C21/86)*, (1986).

14.33 Newby, G.B. and Theerge, J.E.: 'Long-term Behavior of Reinforced Thermoplastics.' *Machine Design*, (Mar. 1984), 171–177.

14.34 Anon.: 'Impact Damage to Composite Structures.' *AGARD-R-729*, (1986).

14.35 Anon.: 'Design of Bolted Joints in Composites.' *AGARD-R-727*, (1986).
14.36 Meyer, R.W.: *Handbook of Pultrusion Technology.* Chapman and Hall, 29 West 35th St., New York, N.Y. 10001, 1985.
14.37 Ratwani, M.M: 'Impact of Composite Materials on Advanced Fighters.' *SAMPE Quarterly*, (Jan. 1986).
14.38 Cole, R.T., Bateh, E.J. and Potter, J.: 'Fasteners for Composite Structures.' *Composites*, (Jul. 1982).
14.39 Rouse, N.E.: 'Making Reliable Joints in Composites.' *Machine Design*, (Mar. 1985).
14.40 Bunin, B.L.: 'Critical Joints in Large Composite Primary Aircraft Structures.' *NASA CR-3914*, (1985).
14.41 Anon.: 'New Composites Expand Action for Processors.' *Plastic World*, (Dec. 1985).
14.42 Wood, A.S.: 'Patience: Key to Big Volume in Advanced Composites?' *Modern Plastics*, (Mar. 1986).
14.43 Dastin, S.: 'Repairing Advanced Composite Materials.' *Machine Design*, (Feb. 1986).
14.44 Myhre, S.H. and Laber, J.D.: 'Repair of Advanced Composite Structures.' *AIAA Paper No. 80-0776*, (1980).
14.45 Anon.: 'Tooling in Graphite, for Graphite.' *Advanced Composites*, (Nov./Dec. 1986).
14.46 Bulloch, C.: 'Certificating Composite Structures.' *Interavia*, (Dec. 1982).
14.47 Anon.: 'Filament Winding: Beyond the Symmetrical.' *Advanced Composites*, (Jan./Feb. 1987).
14.48 Smillie, D.G.: 'The Impact of Commposite Technology on Commercial Transport Aircraft.' *Aircraft Engineering*, (May. 1983).
14.49 Irving, R.R.: 'Metal Matrix Composites Pose A Big Challenge to Conventional Alloys.' *Iron Age*, (Jan. 1983).
14.50 Hartness, J.T.: 'Polyetheretherketone Matrix Composites.' *SAMPE Quarterly*, (Jan. 1983).
14.51 Donoghue, J.A.: 'Composites vs. Aluminum Alloys: "It's a Horserace".' *Air Transport World*, (Mar. 1984).
14.52 Anon.: 'Composites: Looking for A Breakthrough.' *Aviation Week &Space Technology*, (Oct. 21, 1985).
14.53 Anon.: 'Materials: Technology Keeps Pace with Needs.' *Aviation Week & Space Technology*, (Oct. 13, 1986).
14.54 Christian, T.F., Stowers, M.K., Schweinberg, W.H. and Oyler, G.W.: 'Operational Experience of U.S. Air Force with Structural Composites.' *AIAA Paper No. 86-0946*, (1986).
14.55 Anon.: 'Protecting Electronics Against Lightning Gets Harder.' *Aerospace America*, (May 1986).
14.56 DeMels, R.: 'Lightning Protection for Aircraft Composites.' *Aerospace America*, (Oct. 1984).
14.57 English, L.K.: 'Fabricating the Future with Composite Materials, Part I; The Basics.' *Mechanical Engineering*, (Nov. 1986).
14.58 English, L.K.: 'Fabricating the Future wiht Composite Materials, Part II: Reinforcements.' *Mechanical Engineering*, (Jan. 1987).
14.59 English, L.K.: 'Fabricating the Future with Composite Materials, Part III: Matrix Resins.' *Mechanical Engineering*, (Feb. 1987).
14.60 Anon.: *Long-EZ Plans.* Rutan Aircraft Factory Inc. Bldg 13, Mojave Airport, CA 93501. 1980.
14.61 Lubin G.: *Handbook of Composites.* Van Nostrand Reinhold Co., 1982.
14.62 Reinhart T.J.: *Engineering Materials Handbook, Vol. 1: Composites.* Available from AIAA Marketing Dept. ASM. 1633 Broadway, New York, NY10019, 1987.
14.63 Sheppard L.M.: 'The Revolution of Filament Winding.' *Advanced Materials & Processing*, (July, 1987), 31–41.
14.64 MIL-A-87221 — Durability Design and Qualification Requirements for Composite Structure (July, 1985)
14.65 MIL-STD-1530A — The USAF Aircraft Structural Integrity Program (ASIP) — Program for Full Scale Development of Metal and Composite Structure.
14.66 Tsai, S.W. Composite Design, *Think Composites*, 3033 Locust Camp Road Dayton, OH 45419 (1987).
14.67 Federal Aviation Administration (FAA) Advisory Circular (AC 20-107A) — *Composite Aircraft Structure* (1984)
14.68 Anon.: 'Designing with Composites' (Nov. 26, 1987) 'Building Composites with Computers' (Nov. 12, 1987) 'Optimizing Composite Design' (Feb. 25, 1988) 'The Challenge of Manufacturing Composites' (Oct. 22, 1987) 'Spotting flaws in Advanced Composites' (Dec. 10, 1987). The preceding articles were printed in above issues of *Machine Design*. Denton Publishing Inc. 1100 superior Ave. Cleveland, OH 44114.
14.69. Niu, Michael C.Y.; Composite Airframe Structures. CONMILIT Press Ltd., 1992.

CHAPTER 15.0

FATIGUE, DAMAGE TOLERANCE AND FAIL-SAFE DESIGN

15.1 Introduction

There is a pattern of approach and an accumulation of knowledge and additional safeguards that are developing to provide from catastrophic effects of fatigue failure. Thus while an accurate detail solution of the fatigue life problem is now recognized as fundamentally impossible, nonetheless there are ample steps which can be taken to minimize the hazard. This recognition alone is a most important forward step. The design principle of "fail-safe", that is, adequate safety after some degree of damage, has reduced the fatigue problem from the safety level to the economic level. While fail-safe concepts have reduced the status of fatigue from the frighteningly hazardous to a more calm economic problem, fatigue is nevertheless of first rank importance. Aircraft structures must be demonstrated to have a satisfactory fatigue strength either by comparative experience, or by analysis and test, even though the structure may be designed to be fail-safe. The mode of failure of structures associated with design criteria are shown in Fig. 15.1.1.

Mode of Failure	Design Criteria	Allowables Data
Static strength of undamaged structure	Structure must support ultimate loads without failure for 3 seconds	Static properties
Deformation of undamaged structure	Deformation of the structure at limit loads may not interfere with safe operation	Static properties and creep properties for elevated temperature conditions
Fatigue crack initiation of undamaged structure	1. Fail-safe structure must meet customer service life requirements for operational loading conditions 2. Safe life components must remain crack free in service. Replacement times must be specified for limited life components	Fatigue properties
Residual static strength of damaged structure	1. Fail-safe structure must support 80-100% limit loads without catastrophic failure. 2. A single member failed in redundant structure or partial failure in monolithic structure	1. Static properties 2. Fracture toughness properties
Crack growth life of damaged structure	1. For fail-safe structure inspection techniques and frequency must be specified to minimize risk of catastrophic failures 2. For safe-life structure must define inspection techniques and frequencies and, replacement times so that probability of failure due to fatigue cracking is extremely remote	1. Crack growth properties 2. Fracture toughness properties

Fig. 15.1.1 Design criteria for sizing aircraft structures.

The aircraft design process involves a number of interrelated phases of work which are illustrated in the information flow diagram shown in Fig. 15.1.2. The customer in conjunction with the manufacturer prepare design specifications and contract requirements in the conceptual stage of development. In U.S. military aircraft, preliminary design definition contracts are usually awarded to several bidders to finalize contract requirements and design specifications; for commercial aircraft this phase of the work is initially funded by the manufacturer, but later paid for by the customer from aircraft sales.

During the design development stage, design criteria is established, basic external applied loads are computed, internal element or part loads are calculated, and allowable element or part strengths and margins of safety determined. Laboratory tests of materials and components help to establish design allowables to ensure that the aircraft will pass the qualification and certification test program.

The proof and qualification stage consists of static, fatigue, fail-safe, and flight test programs which must meet or exceed the customer requirements or the requirements of the certifying agency such as the FAA of the United States.

After completion of all test requirements the aircraft is put into service. During the operation stage the airplane must be maintained by using specified inspection, maintenance, and repair procedures. To ensure that an aircraft will perform satisfactorily in service, the structure is designed for four main failure modes which are:

- Static ultimate strength (including yield strength)
- Fatigue life of the airframe (crack initiation)
- Fatigue life of damaged structure (inspection interval)
- Static residual strength of damaged structure

How each of these failure modes are accounted for in the aircraft design and qualification process is illustrated in Fig. 15.1.3. Designing for these failure modes will provide a structure which will meet the static strength and life requirements of any aircraft. Of these failure modes it will be concerned primarily with the last three modes, i.e., fatigue, crack propagation, and residual strength of damaged structure.

Fatigue performance has been the focal point of the preceding remarks but from time to time mention has been made of fail-safe design. While use of the terms safe-life and fail-safe structures may indicate there are two paths for building certifiable structure, there is really no such clearcut opportunity. Both safe-life and fail-safe structural design concepts are equally necessary to create a structurally safe and operationally satisfactory airplane. Just as the ultimate strength design must consider and combine all sources of uniaxial, biaxial, or shear stresses for strength assessment, so must adequate fatigue evaluation include both the resistance of structure to fatigue damage initiation and the resistance of fatigue damage growth to the point of catastrophic structural failure. Thus, the designer has really but one overall objective relative to fatigue design philosophy: the design of a structure which has a high degree of structural reliability and safety during the intended service life of the structure.

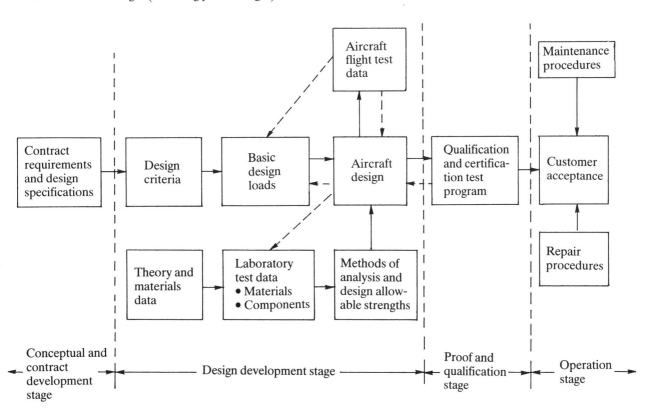

Fig. 15.1.2 Aircraft design, development and certification.

Airframe Structural Design 539

Mode of Failure	Specification	Design Criteria	Design Loads	Airframe Design	Proof Tests - Material and Components	Proof Tests - Airframe	Structural Data Delivery
Static strength-stiffness-undamaged	• Requirements • Environments • Performance	• Performance envelope • Factors of safety	• Static • Dynamic • Stiffness requirements	Selection of • Material • Configuration • Sizing	• Materials data • Envir. & process effects • Design allowable • Comp. develop.	• Proof tests • Flight loads tests	Performance strength & operating limits
Deformation undamaged	• Life • Performance	Environment • Thermal • Chemical	• Steady • Cyclic • Temperature	Selection of • Material • Configuration • Sizing	• Materials data • Envir. & process effects • Design allowable • Comp. develop.	• Operating life tests	• Inspection techniques
Residual static strength-stiffness-damaged	• Damage tolerance goals	Fail-safe • Performance envelope • Damage size	• Fail-safe loads • Stiffness requirements	Selection of • Material • Configuration • Sizing	• Materials data • Envir. & process effects • Design allowable • Comp. develop.	• Fail-safe damage tolerance tests	• Inspection technique • Damage limits • Repair instructions
Fatigue crack initiation-undamaged	• Life • Routes • Operations	• Route analysis • Fatigue quality standards	• Life spectrum of operating loads	Selection of • Material • Configuration • Sizing	• Materials data • Envir. & process effects • Design allowable • Comp. develop.	• Airframe full-scale fatigue tests	Inspection • Locations • Techniques Repair instructions
Crack propagation-life-damaged	• Operations • Inspection • Maintenance • Repair	• Inspection techniques • Damage size limits	• Limited spectrum-operating loads	Selection of • Material • Configuration • Sizing	• Materials data • Envir. & process effects • Design allowable • Comp. develop.	• Extended fatigue tests • Arbitrary damage tests	Inspection • Locations • Techniques • Intervals Repair instructions

Fig. 15.1.3 Failure modes accounted for the aircraft design and qualification process.

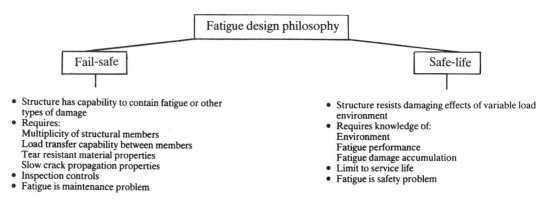

Fig. 15.1.4 Fatigue design philosophy.

Therefore, as illustrated in Fig. 15.1.4, the total fatigue design philosophy includes both fail-safe and fatigue resistant safe-life concepts.

Fatigue is a progressive failure mechanism and material degradation initiates with the load cycle. The degradation/damage accumulation progresses until a finite crack is nucleated, then the crack propagates until the failure process culminates in a complete failure of the structures (see Fig. 15.1.5). The total life, from the first cycle to the complete failure, can be divided into three stages:

(a) Initial life interval during which a complete failure can occur only when the applied load exceeds the design ultimate strength, i.e., time to initiate a crack which will tend to reduce the design ultimate strength capability. This time interval is usually defined as the fatigue life or the safe-life interval.

(b) Life interval, after safe-life interval, during which a complete failure will occur even when the applied load is below the ultimate design load and the strength reduction, due to a small crack, is a function of the material fracture toughness properties (refer to Chapter 4).

(c) Final life interval, during which a complete failure will occur even when the applied load is below

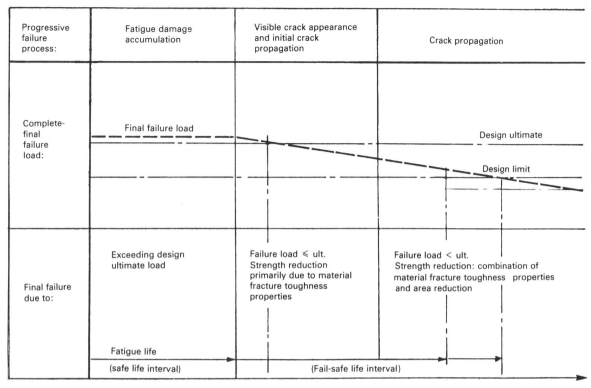

The current requirement is 100% limit load

Fig. 15.1.5 *Progressive failure of a structural element.*

the ultimate design load and the strength reduction is a function of the material fracture toughness properties and area reduction due to a growing crack.

(b) and (c) combine to form a time interval which may be called the "fail-safe" life. The length of this life is a function of the residual strength reduction rate, crack propagation rate and the fail-safe design criteria which limits the residual strength to the limit load established by certifying agency. The fail-safe life corresponds to the time interval between inspections. This means that a crack which may initiate after an inspection should not propagate to a critical length; that is, the residual strength should not decrease below the fail-safe design load before the next inspection, during which the crack should be detectable.

Structures which exhibit a very short fail-safe life interval and where structural redundancy cannot be practically provided for such as nose landing gear and main landing gear are designated as safe-life structures. On the other hand, structures which have a finite fail-safe life, and usually contain structural redundancy such as wing skin-stringer and fuselage skin-stringer panels are designated as fail-safe structures.

The primary goal of a good fatigue design is to attain a safe-life interval equal to the projected lifetime of the aircraft. Based on the statistical aspects of fatigue, the safe-life may be defined as the initial life interval during which the probability of crack initiation is an acceptable low value. Fatigue failure during this interval implies a fatigue crack initiation and not a complete failure. An optimum fatigue design should exhibit a high reliability safe-life for the purpose of aircraft availability and economical operation and a reasonably long fail-safe life for safety, and to a certain extent economical operation by minimizing the inspection frequency.

In summary, fatigue performance is a multi-variate phenomenon. Hence, design criteria must be pointed towards controlling the many features of design and manufacturing affecting the realization of fatigue performance. Design planning and execution, manufacturing quality control, analysis, test demonstration, inspection, and service monitoring of the aircraft experience and usage provide means to produce or maintain a high level of fatigue performance. The very detailed nature of fatigue response requires considerable dedication to task as well as cultivation of detail analysis beyond the practice necessary for ultimate strength design.

Glossary

Cycle — Time function which is repeated periodically.

Design Life — The life specified by the contractor or the customer that the aircraft or components of the aircraft must remain free of fatigue problems in service.

Endurance Limit (fatigue limit) — The maximum stress below which a material can withstand an infinite number of cycles of stress reversal, that is, the stress at which the S-N curve becomes horizontal and appears to remain so (see Fig. 15.1.6).

Fail-Safe — This means that the structure will support designated loads with any single member failed or partial damage to extensive structure. Sufficient stiffness shall remain in the damaged structure to prevent flutter, divergence, severe vibrations or other uncontrolled conditions at any flight within the normal design flight envelope.

Fatigue — The phenomenon of the failure of materials due to cyclic variations of applied stress.

Fatigue Life — The number of repetitions of load or stress or strain of a specified character that a specimen sustains before a specific failure occurs. The repetitions are expressed in terms of cycles or flights.

Fatigue Quality Index K — A measure of the fatigue quality of a local design detail of a structure or component.

Flight-by-Flight Fatigue Test — A fatigue test procedure wherein the representative flight (stress, strain, load, temperature) history is applied in sequence for each flight. In this type of test the transition loading from ground to air to ground occurs every flight.

Ground-Air-Ground Cycle — A cycle which is defined by the transition from the minimum or mean (stress, strain, load, temperature) to the maximum or mean (stress, strain, load, temperature) during a flight. The transition from minimum flight or ground loading condition to maximum flight or ground loading condition during a flight is referred to as the once per flight ground-air-ground cycle.

Life Reduction Factor — A factor, negotiated with the certifying agency, used to reduce the test life to arrive at a certified life for safe-life components. The magnitude of the life reduction factor is a function of the number of specimens tested, the type of test, the type of material, and the magnitude of the applied loads.

Safe-Life — Safe-life components are those components whose failure would result in catastrophic collapse and loss of the airplane. Safe life components must remain crack free during service life. The number of flights in service of a part or component is obtained by fatigue analysis and/or fatigue test results divided by an appropriate life reduction factor which indicates that premature failure is extremely remote.

Service Life — The number of flights or flight hours a component, part, assembly, or structure will remain useable in service.

S-N Curve — When the alternating or maximum stress is plotted versus the number of cycles to failure (fatigue life), the curve will have the typical shape shown in Fig. 15.1.6. These curves are referred to as S-N curves.

Spectra Fatigue Test — Variable amplitude loading fatigue test as distinct from a constant amplitude loading fatigue test.

Stress Cycle — The smallest part of the stress-time function which is repeated periodically and identically. Fig. 15.1.7 illustrates in a simple cycle with following terms:

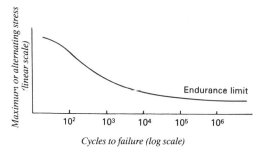

Fig. 15.1.6 Typical S-N curve.

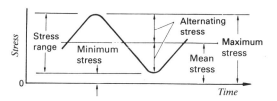

Fig. 15.1.7 Stress cycle diagram.

- Alternating Stress (or Variable Stress), S_a — One-half the range of stress; that is $S_a = \dfrac{(S_{max} - S_{min})}{2} = \dfrac{S_r}{2}$.

- Maximum Stress, S_{max} — The highest algebraic value of stress in the stress cycle, tensile stress being considered positive and compressive stress negative.

- Mean Stress, S_m — The algebraic mean of the maximum and minimum stress in one cycle, this is

$$S_m = \frac{(S_{max} + S_{min})}{2}.$$

- Minimum Stress, S_{min} — The lowest algebraic value of stress in the stress cycle, tensile stress being considered positive and compressive stress negative.

- Range of Stress, S_r — The algebraic difference between the maximum and minimum stress on one cycle, that is $S_r = S_{max} - S_{min}$.

- Stress Ratio — The algebraic ratio of the minimum stress to the maximum stress in one cycle, that is

$$R = \frac{S_{min}}{S_{max}}$$ or the algebraic ratio of the alternating stress to the mean stress in one cycle, that is $A = \dfrac{S_a}{S_m}$.

Test Life — The number of simulated service hours or flights which a specimen sustained in a fatigue test at the time the initial failure of a specific nature was detected.

Unlimited Life — Unlimited life means that the life of an adequately inspected and maintained structure will not be limited by fatigue problems.

15.2 Performance and Functions

Design, manufacturing, and usage of a product all have their share in the ultimate service behavior of aircraft structures. Engineers are directly associated with those phases connected with the detail design and the subsequent manufacture of the structure. Physical environment and load environment encompass vital segments of the fatigue process that can only be estimated during the design and fabrication of the structure. The operator of the product has direct control of these functions influencing the actual realization of the fatigue performance.

Design and Manufacturing

These two areas are grouped together because of the interrelationships of design and subsequent manufacturing. In the area of fatigue, the designer cannot ignore the potential influence of not only his work in configuration, but also the very details of a fabrication process. For instance, chem-milling may be a perfectly satisfactory material removal process. However, the detail process of this fabrication process may alter the estimated fatigue performance by several factors or even an order of magnitude (i.e., a factor of 10) or greater. Both the chemistry and the resultant process output have significant effects on potential fatigue performance. In many cases the engineer must rely upon laboratory testing or existing service data to establish the likelihood of the delivered structure meeting its service goals. Major areas for focusing fatigue design effort or control in these fields of design and manufacturing are outlined below.

(1) Material Selection

During the design of the airframe, a number of steps are taken to ensure that the structure will have a practically unlimited life. One of the first and very important steps is the selection of the proper material for each part of the structure. A very extensive evaluation program is conducted to compare all the properties of the materials that affect the strength and life of the structure including the effects of the physical and chemical environments to which the materials will be exposed during service. These environments include various combinations of stress, temperature and time in the presence of contaminants, such as humidity, fuels, hydraulic fluids and industrial and sea atmospheres. Some of the properties considered include:
- Static strength and stiffness — the ability to endure static loads without failure or excessive structural deformation.
- Corrosion and stress corrosion — the ability to endure static steady and residual loads and chemical environments, considerig time and temperature, without failure or deterioration of the material by chemical or electrochemical reaction within its environment.
- Fatigue strength — the ability of the material to withstand cyclic variations of applied stress without producing fatigue cracks.
- Crack growth — the material's ability to retard the growth of cracks under cyclic variations of stress.
- Residual strength (fracture toughness) — the ability of the material to support static loads in the presence of structural damage.

In fatigue critical fail-safe structure the fatigue strength, crack growth, and residual strength properties receive primary consideration because these properties actually determine the size of the primary structure. Corrosion characterictics are also important, but cladding, sealants or coatings can usually be provided to assure adequate protection from the corrosive environment (for more detail see Chapter 4). Fuselage and wing skin materials are clad and anodized to provide adequate corrosion pretection from the atmospheric environment. The 2024-T3 aluminum alloy has good fatigue properties and has exhibited satisfactory service experience. The age-stabilized heat treatments such as T73 and T76 are used to improve the stress corrosion properties at the expense of some loss in static strength properties. The age-stabilized aluminum alloys generally have fatigue properties which are equivalent or slightly less than the same aluminum alloys with the T6 temper. Titanium alloys, particularly forgings, are receiving increased attention for fatigue critical structure.

(2) Joint Configuration

The designer and stress analyst have within their domain the capability to increase fatigue performance significantly. Of course, the absolute limit for design improvement is the material itself in the fabricated condition or the design limited detail. For instance, there isn't much room for improvement in fatigue performance of a basic open hole. Perhaps the use of residual stresses through coldworking the hole and immediate area can extend service life, but there are limitations imposed by the total load environment which may blank out any advantages of coldworking a hole. Reduction in operating stress level is probably the only simple positive approach. In considering the operating stress level in a joint, both the local stress amplification of load transfer and the fastener hole add to whatever stresses which may originate from load path eccentricities that occur through the joint region. Several points should be observed. First, it is apparent that strength is not a measure of fatigue performance. The lap and the double scarf joints have the same approximate strength but are, respectively, the worst and the best performers in fatigue. Secondly, the joints having the fewer fasteners in a row have better fatigue performance.

Elimination of the number of fasteners in a row and the limited number of joint parts, plus the local thickening-up or padding in the skin and stiffener, all aid in the achievement of improved fatigue life. Reduction of local stress in a joint by adding material is usually a positive means of improving fatigue life.

Fastener selection, fastener arrangement, and most of all, the transition configuration and symmetry of the joint, place an upper limit to the practical use of such a design technique. Each joint being designed for optimum fatigue performance should be checked for the local stresses

Airframe Structural Design 543

due to load transfer, notches or stress concentrations and load path eccentricities to ensure the best utilization of the added material.

Fastener flexibility is an important part of the load transfer and local stress levels within a joint. Fastener shear deflection and bending deflection make up part of the fastener flexibility characteristics. The joint material local bearing deflection and overall stress field deflection between fasteners also contribute to the fastener flexibility and joint deflection characteristics or load transfer behavior. (For more detail, see Chapter 7.)

(3) Fasteners

As indicated above, fastener flexibility characteristics and hole filling characteristics along with the joint configuration have a primary influence upon local stress levels. Tapered shank fasteners and interference fit cylindrical shank fasteners provide a beneficial effect on fatigue performance, not entirely through increased fastener rigidity or reduced flexibility but also through local residual or prestress effects.

Considering the interference fastener, an improvement in fatigue performance of several times can be obtained. Fastener locations are generally a fatigue critical item, since not only do problems of material section change (i.e., the fastener hole) and load transfer exist but also the fact that the basic overall stress field is changing such as that in the region of an access door, stiffener run-out, load input fitting, etc.

(4) Stress Levels

The design stress level is usually selected for each project based on past experience, coupon and component testing, and fatigue analysis of representative points in the structure. The stress level selected will depend on the severity of the loading spectrum. The maximum design ultimate stresses for aluminum structures are in the neighborhood of 45,000 to 55,000 psi in areas which are fatigue critical (for transport wing covers).

Of course, the design stress level selected for a particular aircraft is based on the premise that a certain fatigue quality level is achieved in the final design. The equivalence of design stress level and fatigue quality index is illustrated by the curve in Fig. 15.2.1 for aluminum alloys plotted for constant life. This curve shows that the ultimate design tension stress level for $K = 4.0$ and 2024-T3 material is 43,000 psi for a given life. Higher K values and the same life requirements would require a lower design stress level. The minimum design level of $K = 4.0$ (see Chapter 7 for detailed analysis of K values) is recommended to take care of such uncontrollable factors as:
- Manfacturing damage
- Holes drilled and fastener installation
- Repairs of mild to major damage
- Normal nicks and scratches

(5) Fretting

The phenomenon known as fretting corrosion has a serious influence on fatigue life. Fretting itself is the wear or tearing down of faying surfaces by the relative sliding motion of the contact surfaces. Galling of bearing surfaces might be thought of as an extreme case of fretting. When corrosion of

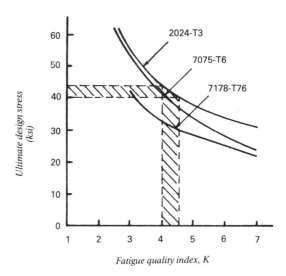

Fig. 15.2.1 Design ultimate stress vs. K for transport wing structures.

the tearing or shearing particles and their surfaces takes place, fretting corrosion is the result. Fretting at faying surfaces can be minimized by lubrication, differential hardness, protective finishes, or best by separation of the parts. Because of the minute relative motions precipitating fretting, faying surface regions of load transfer, including fastener hole surfaces, even with bushings, can be susceptible to fretting without proper precautions.

(6) Manufacturing

The step between design and delivery of a product is a very vital stage in the development of potential fatigue performance. Fit-up can induce increased stress levels. Omitted or reduced fillet radii on machined parts, heat treatment, protective finish or cleaning operations, material substitution, grain direction, type of material (i.e., forging, extrusion, sheet or plate), all can have a significant effect on whether the potential fatigue performance is ever realized.

Mechanical Environment

The actual load exposure of a structure has a primary influence on realization of predicted fatigue life. While satisfactory estimates can be made on the probable load environment, the actual loading environment may really be different. In some operations a counting accelerometer is used to monitor experience; however, the general practice is to collect data for only a statistical assessment of the exposure rather than for the actual exposure of each aircraft. Gusts, maneuvers, landing impact conditions, ground roll or taxi, climb cruise, descent, etc. all have a part in defining the load history input. The major areas to consider in the environmental input to fatigue evaluation are itemized below.

(a) Flight Profile

Many parts of the aircraft receive a load cycle by virtue of just becoming airborne and returning to the ground regardless of any other maneuver, turbulence exposure, or source of variable loading. Fig. 15.2.2 illustrates the flight profile comparison for a transport airplane and fighter airplane.

Altitude is significant because of turbulence and temperature conditions. Cabin differential pressure, of course, is effected by the ambient air pressure at altitude. Flight length is significant because of the ground-air-ground (GAG or G-A-G) cycle as shown in Fig. 15.2.3. The stress range of the GAG cycle for wing structure or the pressure structure of a pressurized fuselage is probably the largest and most frequently applied single load cycle for most transport aircraft. Every flight regardless of extent of maneuvering and turbulence exposure adds load variability and fatigue damage to the structure. Maneuvers and turbulence are merely superimposed upon the 1.0g levels on the ground or in the air so that the maximum stress range of the combined loads is taken as a measure of the GAG cycle. Hence, short flights have a preponderance of GAG load cycles while long range flight have a lesser

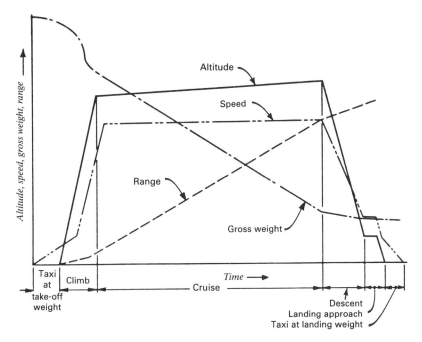

(a) Transport or cargo airplane

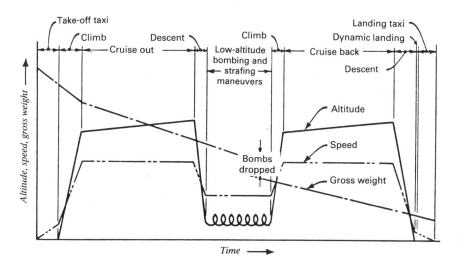

(b) Fighter airplane

Fig. 15.2.2 *Typical mission and flight profiles.*

Airframe Structural Design 545

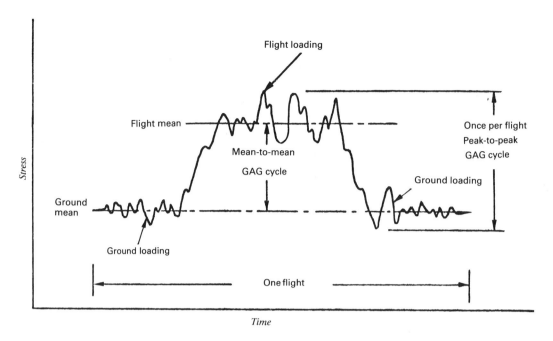

Fig. 15.2.3 Ground-air-ground (GAG) cycle definitions for fatigue analysis.

exposure in terms of GAG cycles per hour. The flight length data presented in Fig. 15.2.4 indicates a usage that has a recognizable difference between the mean or the median flight length (i.e., median has 50% of the flights either greater or less than itself). It should be apparent that the GAG cyle for a 50-mile flight occurs about four times as frequently as for a 200-mile flight. Hence, fatigue damage for the shorter flight in terms of cycles per mile is about four times that of the longer flight usage.

(b) Load Environment

A power-spectral gust load analysis has been used by airplane designers and this is a rationalization of the turbulence exposure in which the atmosphere is defined in terms of its own characteristics as discussed in Chapter 3 (see Ref. 3.43 for further information). This analysis has

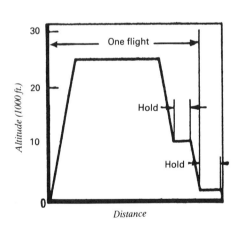
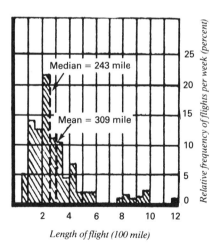

- *Altitude affects turbulence level and cabin differential pressure*
- *Holding distance affects load environment exposure*
- *Flight length affects GAG cycle*

Fig. 15.2.4 Airplane flight profile as related to load environment.

two approaches: one is the generalized description of the atmosphere in terms of frequencies and the coincident power in each frequency level; and other requires definition of the elastic response characteristics of the airplane in terms of a transfer function which will translate the defined gust spectrum into a specific airplane response.

Similar acceleration measurements describe the ground loads during the landing impact and roll, the take-off roll, and taxi conditions. Other ground handling conditions, such as turning and braking are also well defined. Relative to runway and taxi strip loadings, considerable work is done to characterize runway surfaces in a similar manner to that of the atmosphere through use of power spectral analysis techniques.

(c) Other Environments

An important segment of fatigue performance and its realization lies with the physical exposure of parts. Temperature and corrosion are the major considerations in this area. Corrosion is taken to include the exposure to the ambient environment such as fluids or air in its various conditions of humidity. The detrimental effects of corrosion can reduce fatigue performance by several factors.

15.3 Design Criteria and Ground Rules

The nature and requirements of fatigue design criteria are related to and can be resolved by analysis and consideration of fundamental material behavior in the fields of fatigue damage initiation, fatigue crack growth and the residual strength or brittle fracture characteristics of fatigue damaged material. Stress analysis and good engineering design in these three areas of material behavior provide a key to the solution or elimination or at least minimization of fatigue problems. The major areas of fatigue design criteria and ground rules are briefly discussed below:

(a) Service Goals and Usage

Success in transport aircraft design has become more than mere flight, payload, or speed capability. Day in and day out performance of transportation tasks is expected that subject structure to an existence of true drudgery. The rehabilitation effort should be practically nil over the total and variable or extendable usage period.

Likewise, customers who used to think 20 years of service was a good goal for the aircraft are not necessarily looking for more years but are thinking in terms of more total hours before retiring the aircraft. The utilization of the jet transport aircraft has exceeded the probable expectation of the designers during the developmental stage.

(b) Identification of Critical Areas

The many disappointments in fatigue performance generally have characteristics of oversight or deliberate but unintentional design neglect.

The basic structure, loading or attaching points, joints, and geometry should be carefully identified in detail before the final design stage.

(c) Fatigue Stress Analysis

One very clear fact concerning fatigue performance is the importance of the very local stress, its stress field or gradients, and its variation with time. A stress analysis, recognizing the principles of elasticity of materials, equilibrium of forces, compatibility of deformations, and fastener load transfer and deflection characteristics, must be accomplished. Either exact analytical analyses or finite element matrix methods of analysis should be used to calculate typical local stresses under the fatigue loading environment. In the use of finite element matrix methods of analysis, the element configuration should be fine enough locally to recognize real stress gradients and fields such as the elastic stress field around a hole.

Naturally all load levels within the operational load envelope and/or the limit load strength envelope should be reviewed for calculating local detail stresses. Special attention should be given to stresses at 1.0g loads as well as those involving accelerated flight or ground conditions which cause the most calculated damage. The maximum loads should include those levels which can lead to microscopic fatigue damage.

In other words, a very few high load stress cycles can precipitate fatigue damage which can grow at the very large number of lower stress level loads found in the typical flight environment.

(d) Developmental Testing

As a partner to fatigue or microstress analysis, developmental testing is a part of fatigue design criteria. Such testing confirms or may even substitute for a fatigue stress analysis regardless of the designers' intent. This testing is generally pointed towards locating the critical areas, measuring the fatigue performance and obtaining crack growth data. Fatigue performance of basic structure is a very important standard to develop since it represents the major part of the structure. Any premature fatigue damage initiation throughout this portion of the structure practically forecasts early retirement of the structure or a severe economic maintenance burden. Stress levels for checking or developing the fatigue performance of structural details should include GAG cycles as well as representative flight and ground stresses as necessary to define and verify the allowable fatigue performance. The GAG cycle is defined within the practice as the maximum range of stress encountered over the complete flight profile, including ground and flight conditions.

(e) Airplane and Component Testing

Full scale testing of the airplane or its components has a very important place in the development and demonstration of potential fatigue performance. Actual stress distributions and representative manufacturing processes are automatically gained in this testing. Flight-by-flight testing is essential to gain the fatigue damage input of the ground-air-ground cyclic load. Generally, full-scale component testing is a brute force operation because of the load control problem. More information regarding structural testing is discussed in Section 15.8.

(f) Fatigue Performance Analysis

There is a definite need for a complete fatigue

analysis of the structure. Integration of probable usage and the probable level of allowable fatigue performance can focus attention on the potential problem areas in service. The Palmgren-Miner cumulative damage hypothesis (refer to Section 15.4) is the analysis tool, but consideration of the complete prediction problem is necessary as indicated previously. Gust, maneuver, ground, and other loadings peculiar to the specific model or product are necessary inputs to the analysis. Particular attention must be given to the allowable fatigue performance data for the structure.

(g) Fatigue Performance Variability (Scatter)
Being a statistical function or phenomenon, fatigue values can only be established by sampling through either a specific or available experiment. Of course, besides material variability, one encounters technological variability as reflected in the extent and accuracy of stress analysis and loading history. To define the variability factor or fatigue scatter factor, these following considerations are involved:
- A confidence level factor due to the size of the test sample establishing the fatigue performance.
- Number of test samples.
- An environmental factor that gives some allowance for environmental load history.
- A risk factor that depends on whether the structure has a safe-life or fail-safe capability.

The product of all above factors is the total fatigue scatter (or variability) factor.

(h) Design and Manufacturing Processes
Some design and processing controls which need criteria identification and/or drawing identification are listed below:
- Fastener selection and installation; both initial planning and design execution need full attention in this area.
- Structural penetration of countersunk head fasteners.
- Faying surface fretting protection.
- Corrosion protection (i.e., protective finish).
- Fastener edge margins; 2.0 rivet diameter is specified as a minimum and retention of interference stresses between hole and fastener.
- Fastener row spacing and stagger; loads parallel or normal to fastener involve considerations of net area, interaction between fasteners and basic stress flow through the fastener rows as in a wing spanwise splice.
- Material selection and treatment against stress corrosion from processing, fit-up or installation; both planning and design functions have considerations in this area.
- Surface mechanical finish from surface roughness to residual stress effects from plating, chemical action (chem-milling), rolling or shot-peening.
- Forging, bar or plate usage, and interchange.

(i) Aircraft in Service Experience
The actual service experience on aircraft structure provides an excellent yardstick for forecasting or extrapolating the available fatigue performance in the structure. Loading and physical environmental exposure and cumulative fatigue damage are specifically and automatically integrated. The most likely critical details become identified along with a typical service time.

However, service experience on one model can be used to alert the designer to likely critical areas and details on another model. The development of fatigue performance beyond that gained in service requires a laboratory comparison of the fatigue performance of the critical detail and the proposed improvement.

Improvement of a critical service detail in terms of ratios of laboratory fatigue performance is the simplified form of a comparative fatigue analysis. Such an analysis can be made independently of the Palmgren-Miner type analysis.

(j) Inspection Intervals
The inspection of aircraft structure is vital to the control of its integrity. Where fatigue damage initiation means termination of the operational life of the structure, either no fatigue damage is tolerable or pre-critical detection of the fatigue damage is necessary. In fail-safe structures, the initial inspection time can be estimated on the basis of the calculated time to the first detectable crack at the specific location. To cover environmental variability, successive inspections must be geared to the time increment to develop the fail-safe crack length from the first detectable crack length. This reinspection interval is dependent upon the level of inspection set up for the detection of the first crack, since the fail-safe crack length is generally well defined. Naturally, a factor of safety on crack growth is necessary just as in the associated fatigue performance of structure. At least a factor of safety or variability factor of two should be used to cover the probability that an inspection may miss a marginally detectable crack. Concerning the first inspection time, two considerations must be met. First, there is the possibility that first crack initiation may be less than that estimated for many reasons. Hence, inspections should begin before the defined initiation time. Second, the time interval between these "pre-initiation time" inspections should be similar to the crack growth control inspection times.

15.4 Structural Life Estimation

Prediction of fatigue performance is a simple process but one beset by several complexities. Load magnitude and sequence are very important elements of the process. Although many techniques have been devised to satisfy specific conditions, the simplest and most practical technique is the Palmgren-Miner hypothesis. In 1924 Palmgren devised the techniques relative to calculating the service life of ball bearings. In 1945 Miner presented a paper reporting on its application to structural elements. Generally speaking, the process can work quite adequately. The Palmgren-Miner method merely proposes that the fraction of fatigue life used up in service is the ratio of the applied number of load cycles at a given level divided by the allowable number of load cycles to failure at the same variable stress level. If several levels of variable stresses are applied to a detail, then the sum of the respective cycle ratios is the fraction of fatigue life used up. When the cycle ratio sum equals unity, all of the potential service life has been used. It is important to note that there is just so much potential fatigue life available for operational utilization. The extent of realized service life depends on the rate at which it is used. The well known S-N curve emphasizes the importance of stress level and the exponential fatigue life behavior of structure in regard to stress level. Fig. 15.4.1 illustrates the fatigue life calculation process for structure.

Certain fatigue computer programs have been prepared to do these calculations in great detail and extent for aircraft structure. One point to note in Fig. 15.4.1 is that the cyclic life of a spectrum of loads may be reduced to a single equivalent load cycle to facilitate interpretation of test results. However, it is also extremely important to note the fatigue performance depends upon actual local stresses rather than the calculated average stress. The actual fatigue test of a detail does respond to the critical local stresses although the fatigue performance is defined in terms of the gross or average stress. Local stress response is not necessarily consistent throughout the load range of the structure. Therefore, testing at one level only may not completely solve the fatigue performance determination problem.

Palmgren-Miner Method of Analysis

Of the several cumulative damage fatigue theories known, the one most widely used and best known is the one suggested by Palmgren and later independently by Miner. The Palmgren-Miner hypothesis is that the fatigue damage incurred at a given stress level is proportional to the number of cycles applied at that stress level divided by the total number of cycles required to cause failure at the same level. This damage is usually referred to as the cycle ratio or cumulative damage ratio. If the repeated loads are continued at the same level until failure occurs, the cycle ratio will be equal to one. When fatigue loading involves many levels of stress amplitude, the total damage is a sum of the different cycle ratios and failure should still occur when the cycle ratio sum equals one.

$$D = \sum_{i=1}^{k} \left(\frac{n_i}{N_i} \right) = 1.0 \tag{15.4.1}$$

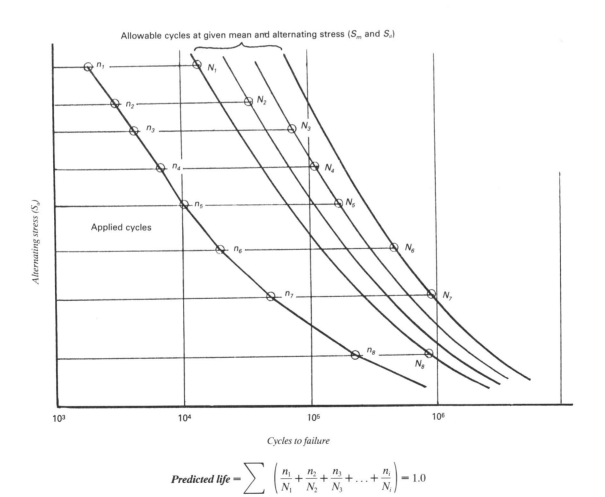

Predicted life $= \sum \left(\frac{n_1}{N_1} + \frac{n_2}{N_2} + \frac{n_3}{N_3} + \ldots + \frac{n_i}{N_i} \right) = 1.0$

Fig. 15.4.1 Structural fatigue life calculation.

where n_i = number of loading cycles at the ith stress level

N_i = number of loading cycles to failure for the ith stress level based on constant amplitude S-N data (from Ref. 4.1 or other sources) for the applicable material and stress concentration factor (from Ref. 7.10)

k = number of stress levels considered in the analysis

Fatigue crack initiation is assumed to occur when the D value is equal to 1.0. There are three parameters which affect the magnitude of the summation of the cycle ratios.

- First, there is the effect caused by the order of load applications. Consider for example, two different stress levels, f_1, and f_2, and their cyclic lives, N_1 and N_2, respectively. If f_1 is greater than f_2 and if it is applied first, the life will be shorter than if f_2 is applied first.
- The second effect on the summation of cycle ratios is due to the amount of damage caused by continuous loading at the same level. The summation of cycle ratios for different stress levels is accurate only if the number of continuous cycles at each stress level is small. For most aircraft applications, the loading is random and the stress level is constantly changing. Hence, the number of continuous cycles at a particular level should be small and the summation of cycle ratios should be fairly accurate.
- The third parameter which affects the summation of the cycle ratios is whether or not the fatigued part is notched (such as fastener holes, etc.) or unnotched. The unnotched part generally gives a summation less than one, while the notched part gives a summation greater than one. Since practically all of the structural fatigue failures originate in some form of a notch, it indicates that 1.5 would be a good average value to use for the cyclic ratio to predict failure of typical aircraft major structural members, such as wings. For simpler structural members, a cycle ratio of 1.0 should be used.

Scatter Factors

There are as many philosophies about scatter factors as there are government bureaus; civil regulations for commercial aircraft and Air Force and Navy specifications for military aircraft.

The fatigue strength requirements are of a very general nature without specifying life requirements or scatter factors. Specifically, the requirement states: those parts of the structure whose failure could result in catastrophic failure of the airplane must be evaluated either for fatigue strength (safe-life) or fail-safe strength. The fatigue strength evaluation must show by analysis and/or testing that the structure can withstand the typical loading spectrum expected in service. For design certification, commercial transport aircraft structures, with the exception of landing gears, is usually designed as fail-safe. However, aircraft manufacturers design all structures for high structural reliability by designing for both: crack free life for economy and fail-safe for safety.

With respect to fatigue life verification testing of safe-life structures, it is understood that the U.S. Federal Aviation Regulation (FAR) adheres to the use of the following scatter factors:

Number of test specimens	Scatter factor
1	3.0
2	2.58
3	2.43
4	2.36

This means that the test should verify the safe-life times the factor without failure, or if failure occurs earlier, the safe-life is designated as test life divided by the scatter factor.

Furthermore, in the fatigue life analysis of safe-life structures the FAR adheres to the use of a factor 3.0 on stress strength. This factor is intended to account for fretting, clamped assembly stresses, size and surface effects, cumulative damage inaccuracies and other factors which affect fatigue life but are not accounted for by the material S-N data (see Ref. 4.1) used in the analysis. If any of these factors are reflected by the S-N data or can be proven that they do not exist for a given part, the factor of 3.0 may be reduced appropriately.

U.S. Air Force structural reliability requirements are set forth in military specification MIL-A-8866, and further detailed in contract specifications. They specify the required design safe-life and the expected utilization and mission profiles. For full-scale safe-life verification tests, under typical average loads spectrum, a scatter factor of 4.0 is required. U.S. Navy fatigue strength requirements are also specified in MIL-A-8866 and contract specifications. Full-scale safe-life verification test, under most critical airplane configuration loads spectrum, requires testing to twice the design life. The above short review of scatter factor requirements do not necessarily reflect the latest thinking.

Example

Analysis of a part that has a fatigue quality index, K equal to 4.0 at a fatigue critical point in aluminum 2024-T3 material.

The stress spectrum to be applied locally to the part consists of 5 loading sequences arranged in a flight-by-flight sequence as shown in Fig. 15.4.2. The constant amplitude fatigue properties of the material for $K = 4.0$ are given in Fig. 15.4.6. Use the Palmgren-Miner linear cumulative damage equation (15.4.1)

$$D = \sum_{i=1}^{k} \left(\frac{n_i}{N_i}\right) = 1.0$$

to predict the fatigue life of this given structure. Assume a scatter factor equal to 3.0.

Calculation of the GAG stresses at the 1000 flights (as defined in Fig. 15.4.2):

$$S_{max} = 14.7 \text{ ksi and } S_{min} = 2.5 \text{ ksi}$$

(read from Fig. 15.4.4)

$$S_m = \frac{S_{max} + S_{min}}{2} = 8.6 \text{ ksi};$$

Mean Stress S_m (ksi) ①	Varying Stress S_a (ksi) ②	Number of Cycles (n) ③	Maximum Stress S_{max} (ksi) ④ = ① + ②	Minimum Stress S_{min} (ksi) ⑤ = ① − ②
8.0	3.7	70	11.7	4.3
	7.0	10	15	1.0
	8.3	150	16.3	−0.3
	9.1	30	17.1	−1.1
9.5	2.1	50	11.6	7.4
	7.9	140	17.4	1.6
	13.0	150	22.5	−3.5
	13.8	40	23.3	−4.3
2.5	12.5	70	15.0	−10.0
	13.5	60	16.0	−11.0
	15.1	40	17.6	−12.6
	15.8	40	18.3	−13.3
14.0	7.0	150	21.0	7.0
	7.8	70	21.8	6.2
5.0	9.6	180	14.6	−4.6
	10.3	120	15.3	−5.3
	10.9	50	15.9	−5.9
	11.7	30	16.7	−6.7
		Total 1450 cycles		

Fig. 15.4.2 Local spectrum stresses (ksi) occurring every 1000 flights.

Maximum Stress (ksi)	Number of Cycles	Σ Number of Cycles	Minimum Stress (ksi)	Number of Cycles	Σ Number of Cycles
23.3	40	40	−13.3	40	40
22.5	150	190	−12.6	40	80
21.8	70	260	−11.0	60	140
21.0	150	410	−10.0	70	210
18.3	40	450	−6.7	30	240
17.6	40	490	−5.9	50	290
17.4	140	630	−5.3	120	410
17.1	30	660	−4.6	180	590
16.7	30	690	−4.3	40	630
16.3	150	840	−3.5	150	780
16.0	60	900	−1.1	30	810
15.9	50	950	−0.3	150	960
15.3	120	1070	1.0	10	970
15.0	70	1140	1.6	140	1110
15.0	10	1150	4.3	70	1180
14.6	180	1330	6.2	70	1250
11.7	70	1400	7.0	150	1400
11.6	50	1450	7.4	50	1450

Fig. 15.4.3 Tabulation of the max. and min. stresses vs. Σ cycles.

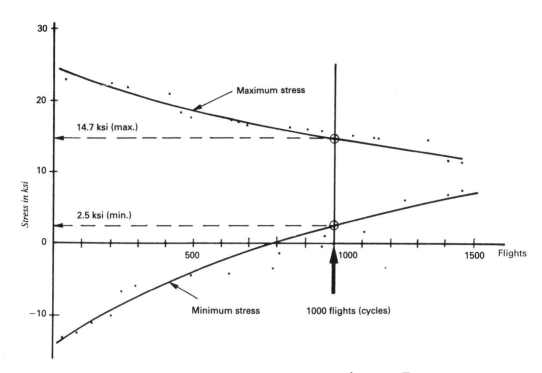

Fig. 15.4.4 Plots from Fig. 15.4.3 of the max. and min. stresses vs. Σ flights.

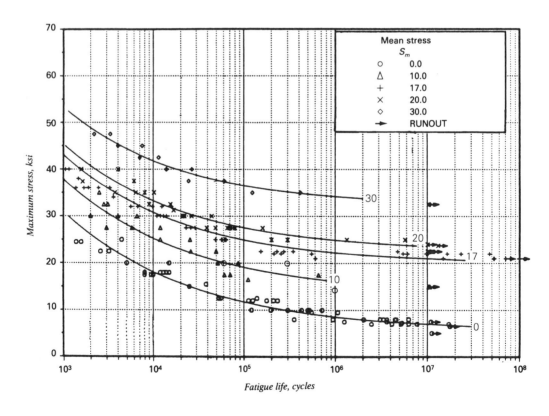

Fig. 15.4.5 Best-fit S-N curves for notched, K = 4.0 of 2024-T3 aluminum alloy sheet, longitudinal direction (Ref. 4.1).

552 Airframe Structural Design

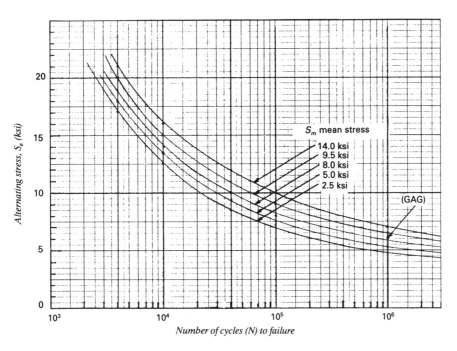

Fig. 15.4.6 S-N curves of 2024-T3 aluminum alloy for K = 4.0 (These curves are generated from Fig. 15.4.5).

Mean Stress S_m (ksi)	Varying Stress S_a (ksi)	Number of Cycles(n)	Number of Cycles to Failure (N)	n/N
	(From Fig. 15.4.2)		(Fig. 15.4.6)	
8.0	3.7	70		
	7.0	10	270000	0.00
	8.3	150	100000	0.0015
	9.1	30	64000	0.00047
9.5	2.1	50		0.00
	7.9	140	240000	0.00058
	13.0	150	18000	0.00833
	13.8	40	14000	0.00286
2.5	12.5	70	10500	0.00667
	13.5	60	8200	0.00732
	15.1	40	6000	0.00667
	15.8	40	5000	0.008
14.0	7.0	150	1100000	0.00014
	7.8	70	430000	0.000163
5.0	9.6	180	36000	0.005
	10.3	120	26000	0.00462
	10.9	50	21000	0.00238
	11.7	30	16000	0.00188
	(GAG)			
8.6	6.1	1000	1000000	0.001
				D = Σ 0.05761

$S_a = \dfrac{S_{max} - S_{min}}{2} = \pm\ 6.1$ ksi

The calculated life $= \dfrac{1000}{0.05761} = 17358$ flights

The best estimate of predicted fatigue life $= \dfrac{17358}{3.0}$

$= 5786$ flights

Airframe Structural Design 553

15.5 Fail-safe Design

It has been long recognized that statically indeterminate and/or multiple element structures are inherently safer when asked to carry load after being damaged. Any airplane should be capable of reaching the ground safely even after certain parts of it have failed in the air. It is impossible to guard completely against the effect of breakage of any part, for that would mean either complete duplication of all parts or such an increase in structure weight to give the necessary load factor that the airplane should be too heavy for any useful purpose. The designer, therefore, can only duplicate or strengthen such parts liable to accidental breakage or, should any particular part of the structure fail, the remainder will be able to carry the extra loads from adjacent failed part until the airplane has landed.

A large percentage of airplane structures exhibit fail-safe capability without having been specifically designed for such a characteristic. Aircraft have continued safe operation or have been safely landed after various degrees of known or unknown damage had been inflicted on the structure. In the past, this fail-safe characteristic of the primary structure of various airplanes has been evidenced by the safe return after damage under circumstances.

Fail-Safe Concept
The application of the fail-safe concept to structural design is based on the fundamental idea that a failure or obvious partial failure of a single principle structural element shall not cause the loss of the aircraft while in flight through:
- Complete structural element collapse.
- Large deflections resulting in loss of control such as
  — Large wing deflections which may make flight impossible
  — Flap assymetry
- Flutter, whether it is of a fixed or movable surface.
- A component failure such as a turbine engine bleed duct which in turn could blow up a wing, fuselage or empennage.
- A change in the aerodynamic characteristics such that continued flight is impossible. An example could be the complete loss of a wing leading edge where a partial loss would be survivable.

Structural failures can be the result of a wide variety of misfortunes that have already been mentioned. The degree or of a single principle structural element failure has not been defined, however some interpretations have been made thereto. If a structural assembly such as a wing box were made up of a relatively large number of elements such as integrally stiffened planks comprising the wing surfaces, a single failure would be the complete severence of one "plank". Similarly, a single failure could involve the severence of a beam cap or a shear web of the wing beam. Each of the examples cited above are single elements in themselves, and when joined with other elements form a composite structure. The definition of a single structural failure is obvious in these cases.

There are instances where the above failure concept cannot be applied so vigorously. In the case of a pressurized fuselage where the cover material is made up of large sheets covering an extensive peripheral area, a complete element failure involving the entire panel requires reevaluation.

A similar situation exists in the wing where an entire lower surface may be made up of several large sheets suitably reinforced with stiffeners and ribs; there again, some qualification must be made with the idea of failure of a complete single structural element.

Since the failure of a complete element as defined above cannot be sustained without involving an unreasonable weight penalty or design complication, it is apparent that some other definition of sustainable damage is necessary.

Many instances have occurred wherein only partial failure of these large elements has been sustained. These partial failure have been observed and repaired before they progressed to catastrophic failure. It is evident that these partial failures must be easily detected by an inspection that is much more casual than that normally given airplane structure. Thus, in addition to the design requirement for the complete failure of a single element, the alternate design requirement of an obvious partial failure of a large element is permissible for fail-safe.

In establishing the degree of obvious partial failure for design purposes, the following factors are considered in establishing their degree:
- Past service experience on similarly constructed airplanes operated in similar patterns and environment. For example, if a 12 inch crack has been discovered on a fuselage skin panel on a pressurized transport in airline service, it would seem logical that a partial failure equal to or greater than this should be selected as a basis for fail-safe design on a new airplane of a similar category.
- Some areas may be difficult to inspect such as the sandwiched member in a double shear joint. The only means of detecting a crack in such a member would be by means of very specialized inspection techniques such as the use of X-ray. It is obvious that the hidden member must be designed so that a fatigue failure will always occur first on the exterior members and never on the hidden element.
- Rate of crack propagation. Crack growth is slow in terms of loading cycles or airplane hours for a particular crack length depending on the stress level in the area. As the crack approaches its critical length, growth becomes more rapid. The critical length for design purposes must be selected such that the growth rate is still slow when the failure is very easy to detect by a "walk around" type of inspection.

The application of the fail-safe idea has been discussed as well as its definition. Some particular portions of the airplanes' primary structure are specifically exempt insofar as their application is concerned. These include:
- Engine supporting structure. Engine installations have fallen from multi-engined airplanes in many instances in the past and safe landings have been made. On the basis of past experience it is evident that the fail-safe requirement need not be applied here. However, it must be demonstrated by tests and/or analysis that the engines can break away without tearing out some critical primary structure.
- Landing gear. Modern airplanes require crash-

worthiness to accommodate emergency landings under circumstances where the gear may not be extended. The intent of the requirement in protecting the airplane occupants then is satisfied by the emergency landing or crash load requirements. Landing gears are generally designed to the safe life requirements because it would add excessive weight to the structure to make them fail-safe. However, some components of the landing gear are designed fail-safe.

- Seats and equipment. Here again the crash load requirements achieve the objective of occupant protection.
- Functional items such as the hydraulic system, electrical system, controls system, flap actuators, etc., are exempted unless their failure would directly cause the loss of the aircraft in flight. Any critical systems are duplicated so that one system will operate with another system out. Dual systems should be separated so that a turbine disk failure or other type of damage induced in a local area would not fail both systems.

Requirements

Fail-safe design requirements are illustrated in Fig. 15.5.1. The amount of damage that a structure can sustain for static ultimate design loads is very small unless the section under consideration has a large margin of safety. However, because loading conditions in service seldom exceed limit design loads it is not necessary to design the structure to be fail-safe for ultimate design loads. The policy for aircraft manufacturers is to design the structure to provide ultimate strength for permissible extent of fail-safe damage for 100% limit load conditions to meet the fail-safe strength may not be less than FAR 25.571 and 25.573 requirements for commercial transport aircraft. Prescribed fail-safe ultimate design load level represents limit airplane strength. No specific requirement is in effect by the FAA for such items as the vertical tail, flaps and flap supporting structure, control surface hinges, tab actuation devices, etc. whose failure would cause loss of the airplane in flight. It has been practice, however, to design these items for limit load. For example, if the vertical tail were critical for an engine failure (structural) condition, as in the case of a propeller airplane, some other conditions would be used such as a lateral gust condition or the dynamic overswing condition for the definition of the fail-safe load.

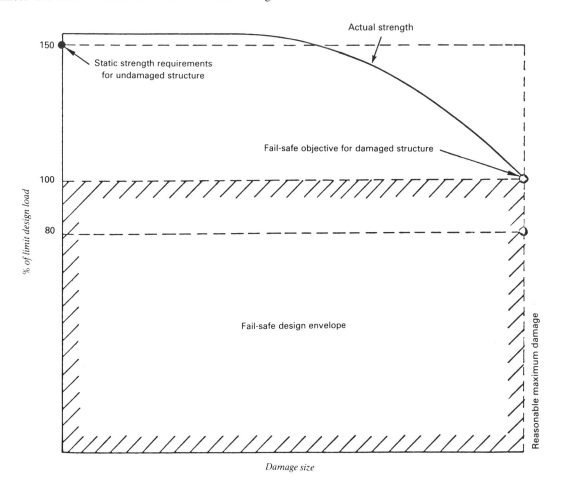

Fig. 15.5.1 Fail-safe design requirements.

For fail-safe structure the occurrence of corrosion and fatigue cracks are more of concern with regard to maintenance costs than flight safety. To insure that fatigue or corrosion cracks are detected before they reach the fail-safe damage limits requires the establishment of inspection intervals based on previous service experience or crack growth histories obtained from analysis and/or tests as illustrated in Fig 15.5.2. The time to initiate a crack requiring repair is indicated by point ①. The period of time between point ① and point ② is the time available for inspection and repair. The critical areas of the structure susceptible to fatigue and corrosion damage must be inspected at intervals which are a fraction of the time interval between ① and ② to ensure that the crack will be discovered and repaired before the damage size reaches the fail-safe limit. For civil aircraft, the inspection frequency and procedures are worked out jointly between the manufacturer, customer and certifying agency.

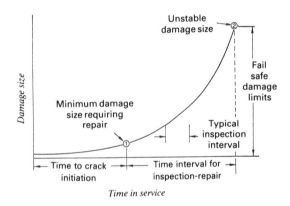

Fig. 15.5.2 Typical structural crack growth vs. time.

Military airplane requirements refer to the airplane damage tolerance requirements per Ref. 15.59.

The design procedures followed to meet the fail-safe design requirements are listed below:
- Establish fail-safe design criteria and required strength levels
- Specify fail-safe damage limits
- Select materials with high fracture toughness and slow crack growth characteristics
- Specify fail-safe design stress allowables
- Specify methods of fail-safe analysis
- Establish fail-safe design load levels
- Design structure to meet load/stress-damage limit requirements
- Proof test damaged components and structure

Damage Tolerance Requirements

The fail-safe requirements of the FAA (FAR 25.571) state that "it must be shown by analysis, or tests, or both that catastrophic failure or excessive structural deformation that could adversely affect the flight characteristics of the airplane are not probable after fatigue failure or obvious partial failure of a single principal structural element". It is the aircraft manufacturer's policy to meet their requirements for obvious partial failure by designing fail-safe structure for the following types of assumed damage:

- Any single member completely severed. For fail-safe purposes, a single member is any redundant structural member or that part of any member of several elements where the remaining part can be shown to have a high probability of remaining intact in the event of the assumed failure. It must be demonstrated that the damage to the assumed severed part must be discoverable by normal inspection methods.
- Extensive structure severed between the boundaries of the effective crack barriers. A mechanical splice (not welded or brazed) or major structural members (frames or stringers) are considered to be effective crack barriers.
- In extensive structure, a major structural member attached directly and continuously to the skin fractured together with the skin between adjacent crack barriers.
- In cutout regions, a single element attached directly and continuously to the skin failed and a skin crack extending to an effective barrier. Additionally, for a skin crack alone extending from the corner of the cutout to an effective crack barrier. The size of the cutout and the size of the crack are considered simultaneously in the analysis.
- In all of the above cases it will be demonstrated by analysis and/or test that damage will propagate slowly under normal operational loads so that detection and repair are ensured.
- All fail-safe joints and skin splices will be designed to have sufficient shear lag to distribute loads from the failed section. This can be achieved by
  — designing the joint to be bearing critical,
  — providing sufficient margin in fastener shear strength so the progressive failure of the fasteners will not occur prior to skin and reinforcement failure.
- All adhesive bonded lap or butt joints will contain sufficient rivets to carry ultimate design loads without benefit of the adhesive.
- Bonded structure for the purpose of fail-safe must be considered monolithic. Cracks will be assumed completely through all bonded layers.
- For members with no positive crack stoppers, reduction in fail-safe damage limits shall be reduced to partial damage of less extent where crack lengths can be shown to be of a size ensuring detection and repair. For these cracks it must be demonstrated that a detectable damage size (1-inch crack length minimum) will not grow to 67% of the critical damage size in less than two normal structural inspection intervals.

The occurrence of any single damage case listed above will not result in flutter, divergence, uncontrollable vibration, or control reversal at speeds up to the V_D boundary.

In addition to the above fail-safe requirements, the general policy is to demonstrate the capability of primary structure for extensive damage under certain emergency conditions. This damage could occur in some areas from turbine disk penetration. In others, the cause could be accidental impact or explosion. For these demonstrations, which may be analytical and/or test, reduced loads will be used based upon maneuver and flight conditions consistent with cautious emergency operation to safe landing of the airplane.

Calculations for Damaged Structures

(1) The basic calculation for the ultimate strength of damaged structure is given in

$$S_g = \frac{\beta K}{\left(\sqrt{\frac{\pi}{2}}\right)(\sqrt{\ell})\,\alpha} \quad (15.5.1)$$

The above equation is identical to that of Eq. 4.2.2 except β which is the reinforcement factor and can be obtained from Fig 15.5.3.

Example

Determine the damaged or fail-safe allowable stress for a 2024-T3 aluminum skin panel stiffened by 7075-T6 aluminum stringers, 10 inch. apart and assume the crack length is 10.0 inch.

$$A_e = \frac{A}{1+\left(\frac{y}{\rho}\right)^2} = \frac{1.19}{1+\left(\frac{1.5}{1.16}\right)^2} = 0.45 \text{ in}^2$$

Then

$$\left(\frac{\Sigma A_e}{t}\right)(b_e) = \left(\frac{2 \times 0.45}{0.1}\right)(10.0) = 90$$

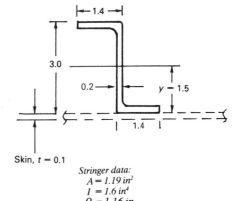

Skin, $t = 0.1$

Stringer data:
$A = 1.19 \text{ in}^2$
$I = 1.6 \text{ in}^4$
$\rho = 1.16 \text{ in}$

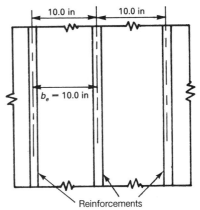

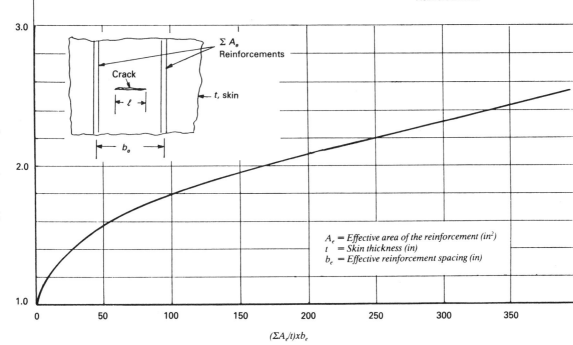

Fig. 15.5.3 Reinforcement factor (β).

and from Fig. 15.5.3 obtain reinforcement factor $\beta = 1.75$

$K = 60.0$ ksi (from Fig. 4.2.4)

The damaged allowable stress from Eq. 15.5.1 is

$$S_g = \frac{1.75 \times 60}{\left(\sqrt{\frac{\pi}{2}}\right)(\sqrt{10})(1.0)} = 26.5 \text{ ksi}$$

If the crack increases to 20.0 in and the stringer over crack skin is broken,

$$\left(\frac{\Sigma A_e}{t}\right)(b_e) = \left(\frac{2 \times 0.45}{0.1}\right)(20.0) = 180$$

and $\beta = 2.05$

$$S_g = \frac{2.05 \times 60}{\left(\sqrt{\frac{\pi}{2}}\right)(\sqrt{20})(1.0)} = 21.95 \text{ ksi}$$

(2) Crack Growth Calculations

To determine the structural crack growth behavior start with the given initial crack size, the stress spectra at the location being analyzed, and the crack propagation and fracture toughness properties of the material. (Usually the crack growth properties are determined by tests conducted on cracked panels with cyclic loads and crack growth measurements taken.)

Crack growth rate depends on:
- Material property
- Environment — grows faster with increase in humidity and corrosive content of environment
- Loading frequency
- Crack geometry — surface flaw, through crack, crack at edge of hole, etc.
- Structural arrangement — crack growth slows down when tip of crack approaches stiffeners or integrally stiffened risers, etc.
- High loads — periodic high loads tends to slow down or retard the crack growth due to loads following a high load.

(a) Constant Amplitude Loading

Crack growth calculations are performed for increments of crack growth over which it is assumed that the crack length remains constant, i.e. Δk and S_{max} remain constant. The calculations are made incrementally as shown in Fig. 15.5.4.

$$\text{Crack growth rate} = \frac{d\ell}{dN} = \frac{\Delta \ell}{\Delta N}$$

or $\quad \Delta N = \frac{\Delta \ell (dN)}{d\ell}$

where $\frac{d\ell}{dN}$ is the crack propagation rate for the material, then

$$N_2 = N_1 + \Delta N$$

Example

Analyze the crack growth behavior for a through crack in a flat skin 7075-T6 aluminum with crack growth properties shown in Fig. 4.2.6 of Chapter 4. Assume a maximum design cyclic stress of 15 ksi at a stress ratio of $R = 0$ (zero-to-tension). Predict the number of cycles required to grow the crack from 1.0in to 4.0in. (See tabulation in Fig. 15.5.5).

Use Eq. 4.2.3 as

$$\Delta K = S_{max}(1-R)^m \sqrt{2a}$$

or $\quad \Delta K = 15(1-0)^{0.5}\sqrt{2a} = 15\sqrt{\ell}$

(A geometric factor such as curvature or reinforcement should be included in above equation. However, it can be assumed to be 1.0 for a short crack in a flat skin panel.)

(b) Spectrum Loading

For spectrum loading the crack growth is calculated in crack length increments per flight basis as shown in Fig. 15.5.6.

The calculations are then performed in the following steps:
- Assume crack length, l_1 (in)
- Calculate ΔK for each cycle in the spectrum as shown in Fig. 15.5.7.

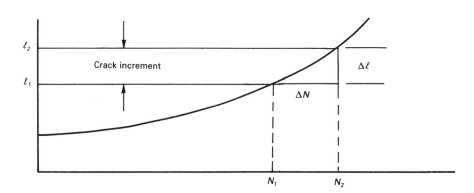

Fig. 15.5.4 Crack growth rate and crack growth remains constant.

ℓ (in)	$\sqrt{\ell}$ (in)	ΔK (ksi $\sqrt{in}$)	$d(2a)/dN$ (in/cycle) Fig. 4.2.6	avg. $d\ell/dN$ (in/cycle)	$\Delta \ell$ (in)	ΔN (cycles)
①	②	③	④	⑤	⑥	⑦ = ⑥/⑤
1.0	1.0	15	2×10^{-5}			
1.5	1.22	18.4	6×10^{-5}	4×10^{-5}	.5	12500
2.0	1.41	21.2	8×10^{-5}	7×10^{-5}	.5	7143
2.5	1.58	23.8	2×10^{-4}	1.4×10^{-5}	.5	3517
3.0	1.73	26	3×10^{-4}	2.5×10^{-4}	.5	2000
3.5	1.87	28.1	4×10^{-4}	3.5×10^{-4}	.5	1430
4.0	2.0	30	6×10^{-4}	5×10^{-4}	.5	1000
						Total cycles = 27640

Fig. 15.5.5 Example calculation of cycles required to grow the crack from 1.0 in. to 4.0 in.

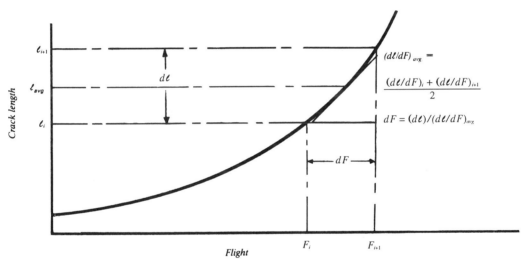

Fig. 15.5.6 Flight and crack length relationship curve.

- Calculate number of cycles for each ΔK_i

$$\text{Each flight} = \frac{\text{Total No. of cycles}}{\text{No. of flights}} = \text{cycles/flight}$$

- Determine $\dfrac{d\ell}{dN}$ for each ΔK

- Determine $\Delta \ell_i = [\dfrac{d\ell}{dN} \times (\text{cycles/flight})]$ for each stress cycle in the spectrum

- Sum $\Delta \ell_i$'s for all cycles in the flight $\left(\dfrac{\Sigma \Delta \ell_i}{\text{flight}}\right)$

- Repeat calculation for crack length ℓ_2

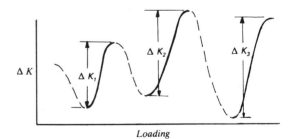

Fig. 15.5.7 ΔK vs spectrum loading.

- Calculate the average crack growth rate per flight

$$\frac{\left(\dfrac{d\ell_1}{dF}\right) + \left(\dfrac{d\ell_2}{dF}\right)}{2} = \left(\dfrac{d\ell}{dF}\right)_{avg}$$

- Calculate number of flight to grow increment of crack growth

$$\text{No. of flight} = \frac{(\Delta \ell)}{\left(\frac{d\ell}{dF}\right)_{avg}}$$

- Repeat calculations for all increments of crack growth

Example

Calculate the crack growth for one increment of growth in 7075-T6 material as shown in Fig. 4.2.6. The stress spectra are shown in Column ①, ② and ③ in Fig. 15.5.8. (Calculate flight oor 1.0 inch to 1.5 inch increment of growth and neglect the effect of compression stress in crack growth).

Flight	Smax (ksi) ①	Smin (ksi) ②	No. of Cycles in 8000 Flights ③	Cycles per Flight ④	R ⑤	ΔK ⑥ = ① $(\sqrt{1-R})$
Maneuver	12.5	7.2	102 × 10³	12.750	0.576	8.14
	16.1	7.2	57 × 10³	7.125	0.447	11.97
	19.6	7.2	38 × 10³	4.75	0.367	15.59
	23.0	7.2	28 × 10³	3.5	0.313	19.06
	26.6	7.2	14 × 10³	1.75	0.27	22.71
Gust	13.5	−0.6	16.2 × 10³	2.025	0	13.5
	14.8	−2.1	1.73 × 10³	0.216	0	14.8
	16.0	−3.5	0.5 × 10³	0.063	0	16.0
	17.8	−4.8	0.13 × 10³	0.016	0	17.8
GAG	25.0	−1.0	8.00 × 10³	1.0	0	25

xx.xxx — For crack length ℓ = 1.0 in
(xx.xxx) — For crack length ℓ = 1.5 in

Flight	ΔK (ksi $\sqrt{in}$) ⑦ = ⑥ × $\sqrt{\ell}$	$(d\ell/dN) \times 10^{-5}$ (in/cycle) Fig. 4.2.6 ⑧	$\Delta \ell \times 10^{-5}$ (in/flight) ⑨ = ④ × ⑧
Maneuver	8.14	0.6	7.65
	(9.97)	(1.0)	(12.75)
	11.97	1.5	10.69
	(14.66)	(2.2)	(15.68)
	15.59	3.0	14.25
	(19.09)	(6.0)	(28.5)
	19.06	6.0	21.0
	(23.34)	(11.0)	(38.5)
	22.71	11.0	19.25
	(27.81)	(13.0)	(22.75)
Gust	13.5	1.0	2.03
	(16.53)	(2.2)	(4.46)
	14.8	2.5	0.54
	(18.13)	(4.5)	(0.97)
	16.0	3.0	0.19
	(19.6)	(6.0)	(0.38)
	17.8	4.0	0.06
	(21.8)	(10.0)	(0.16)
GAG	25	20.0	20.0
	(30.62)	(40.0)	(40.0)
		$\Sigma \Delta \ell =$	95.7 × 10⁻⁵ (164.2 × 10⁻⁵)

Average crack growth rate = (95.7 + 164.2) × 10⁻⁵/2 = 129.95 × 10⁻⁵ (in/flight)
No. of flight to grow crack from 1.0 in to 1.5 in = 0.5/129.95 × 10⁻⁵ = 385 flights

Fig. 15.5.8 Example calculation of crack growth (7075-T6).

15.6 Detail Design

The structural design of a new aircraft requires the selection of structure and materials leading to the highest strength/weight ratio consistent with safety, long life, and low maintenance cost.

The role fatigue plays in design requires that the airframe be as trouble-free as possible, needing only minor maintenance, which can be accomplished with a minimum amount of airplane down-time. This can be attained only by diligent effort toward developing the best detail design possible and following the detail design throughout manufacturing to insure proper fabrication and installation.

The operator's maintenance requirements provide the following objectives for the aircraft designer:
- A practical crack-free service life.
- Aircraft structures designed to retard or restrict progression of a crack or tear.
- Aircraft structures fabricated so that the integrity is maintained.
- Aircraft structures designed for maximum intervals between inspections.

Detail design is more important than either the loading spectrum or operating stress level. This is because it is possible to miss the desired life by a factor of 100 or more. The practical design will inevitably have holes, splices, local attachments, clips and many reasons to interrupt the continuity of the material. The ingenious designer and structural engineer will search for the design simplicity that avoids interruptions yet accomplishes the functional task. It is necessary to give careful consideration to the remaining few unavoidable interruptions, splices, joints, etc. to reduce the influence of all effects detrimental to fatigue life.

Ground Rules for Fatigue Design

Examples of design botches start off with notches in various forms, including designs with no radii, short fillet radii, and sharp bend radii. Special cases of notches include square holes and feathered edges. Feathered edges may more logically be blamed on the shop, since most standard shop procedures include rounding off all corners. Nevertheless, where sharp edges are anticipated, special pains should be taken to see that the drawing clearly spells out rounding off all corners. This is especially critical where careless handling might result in nicking sharp exposed edges. Poor load distribution is another design botch, and closely related to this is the case of superimposed stress concentration in its various forms.

(1) Notches
  (a) Sharp notch — The manufacturer aim at saving a penny, but the user will always pay heavily, and often the the aircraft manufacturer will pay later in the form of warranties. In any event, machine time saved is peanuts when compared with overall structural integrity costs — and, in many cases, using machine cutters with proper edge radii will avoid sharp notches at no extra cost. See Fig. 15.6.1.
  (b) Dings — Dings can be intentional or accidental. Precautions should be taken in handling parts to avoid denting and nicking. Should dings be discovered, they should be called to the attention of the foreman and fatigue specialist.
  (c) Fillet radius — A good policy is to use as large a radius as space and practicality permit, especially at changes of section thicknesses.
  (d) Sharp bends or nuisance failures — Use a bend radius that is as generous as possible without destroying functional performance. While failure of clips may never constitute more than a nuisance, the replacement cost is appalling. See Fig. 15.6.2.
  (e) Square holes — There is usually a purpose in making square holes. If it is possible, make an elliptical hole encompassing the rect-

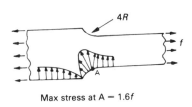

Max stress at A = 1.6f

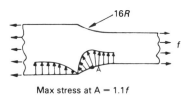

Max stress at A = 1.1f

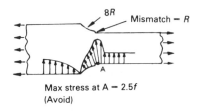

Max stress at A = 2.5f
(Avoid)

Fig. 15.6.1 Stress concentration comparison at fillets.

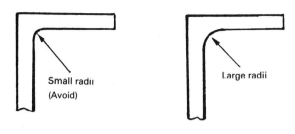

Fig. 15.6.2 Use generous corner radii.

angular hole (or round hole instead of square hole). Fatigue life is five times that with the rectangular hole. Passenger, baggage, and cargo door cutouts with corner radii in the range of $\frac{3}{8}$ to 1.25 inch have in the past produced considerable fatigue cracks at the corners. Patches and steel doublers at the corners did not stop cracking. Sharp corner radii must therefore be avoided. The corner radii should be proportioned to the size of the cutout, (see Fig. 15.6.3) and the material should be distributed so as to avoid eccentricities and discontinuities.

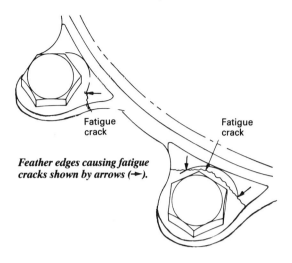

Fig. 15.6.4 Spotface feathered edges.

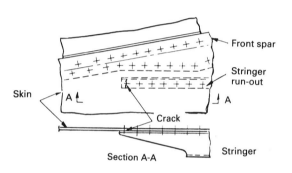

Fig. 15.6.5 Fatigue crack at wing stringer run-out.

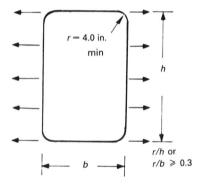

Fig. 15.6.3 Cutouts for passenger, baggage, and cargo doors.

(2) Feathered Edge and Sharp Corners
No one in his right mind plans to have sharp corners or feathered edges. They usually occur because the designer didn't pay attention to what would happen when two or three drawing views were integrated. It would be surprising to find raw edges of the kind that caused service failure in these examples.

Fig. 15.6.4 shows a thin flange from the spotface resulting in feathered edges which might ultimately cause a fatigue failure.

(3) Load Distribution
- Stringer run-out — These stringers failed (Fig. 15.6.5) because there is too much stress at the end fasteners. Since the failure here is caused by a number of items, including axial loading, bending, and fretting, a reduction in any or all would improve life. An attempt is to reduce stress at the end fastener by;
  — using softer fastener (if possible)
  — thinning or tapering the stringer flange to relieve the end fastener stress
  — replacing the end fastener with interference fit fastener
- Backup structure — One of the fundamentals of science is that for every action, there has to be a reaction. As shown in Fig. 15.6.6(a), a husky control bracket is attached to a flimsy spar web structure which results in cracks. This is typical problem where one person designs a component while another designs the supporting structure. In this particular case, it is doubtful that much thought is given to supporting structure. The correct design is shown in Fig. 15.6.6(b).

(4) Superimposed Stress Concentration
- Intersecting holes — Fig. 15.6.7 is a case where a hole is tapped into the highly stressed region of another hole in tension flange of a fitting. This is similar to the case of grease fitting hole for a bearing, i.e. landing gear trunnion fittings. If it must have a hole, the idea is to move it to the least damaging position. A little local reinforcing (beef up) is also helpful.

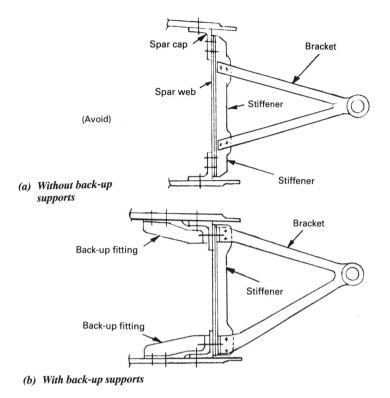

(a) Without back-up supports

(b) With back-up supports

Fig. 15.6.6 *Bracket support on wing box rear spar.*

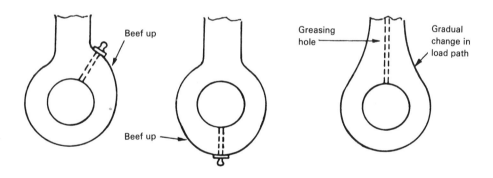

Fig. 15.6.7 *Locate greasing hole to minimize effect on fatigue life.*

- Radius at change of section — There always has to be a radius at any change of section. However, don't make one radius right on top of the other [Fig. 15.6.8(a)]. The radius for change in section should have been made at another location [Fig. 15.6.8(b)].

  If this is impossible, both radii should have been enlarged to permit a more gentle transition.
- Rough surface finish — It is not acceptable to have rough surface finish of a machined part, especially with tool marks normal to the direction of loading.
- Cross grain — The parts shown in Fig. 15.6.9 failed for two reasons: rough surfaces, and the material's grain structure is normal to the direction of loading.

(5) Hard Plating surfaces

This is a typical example of where a part is chrome-plated to make it more wear-resistant. The cracks in the plating act as stress raisers that eventually fail the part while the plating is supposed to protect. Shot peening prior to plating is a common inhibitor of fatigue cracking in chrome-plated parts. It is unwise to chrome-plate parts for dimensional buildup or wear resistance without the help of the fatigue specialist.

(6) Mismatch

Even excellent machinists often machine a curved surface that doesn't meet its straight counterpart, leaving what amounts to a superimposed stress raiser. While it is not so bad where the two surfaces are convex, the concave ones usually result in failure as shown in Fig. 15.6.10. While it

Airframe Structural Design 563

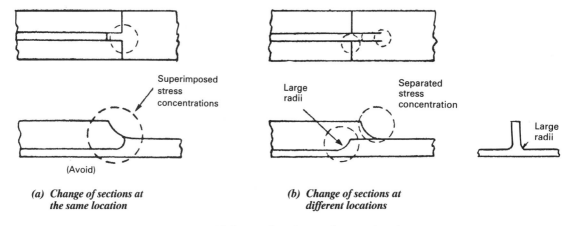

(a) Change of sections at the same location

(b) Change of sections at different locations

Fig. 15.6.8 Avoid change of sections at the same location.

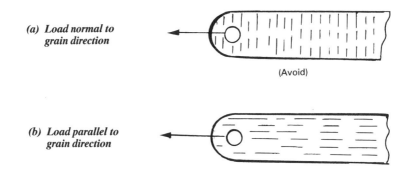

Fig. 15.6.9 Direction of material grain.

would be virtually impossible to define the amount of mismatch that can be permitted in every case, the rule of thumb is to use extreme care with concave surfaces.

(7) Excessive Clamping
The bolt on this part shown in Fig. 15.6.11 is tightened without having the proper spacer bushing. Fatigue failure finally set in, as might be expected. Make sure to use the right length bushing and the right length bolt. (Refer to Section 7.8, Chapter 7.0)

(8) Wing Box Design
 • Wherever load direction must change or cross-section areas must shift at wing chordwise joint, the eccentricities should be reduced to the minimum (see Fig. 15.6.12).
   Straps or doublers should taper out to less than $\frac{1}{3}$ of the thickness of adjacent concentric structural material or less than $\frac{1}{4}$ of the thickness of adjacent material if strap in single shear. (See Fig. 15.6.13.) Rivet sizes should be proportional to doubler thicknesses.
 • Access holes with a door cover are best designed without structural benefit of the cover plate. At low loadings (typical for fatigue), joint deformations are greater than adjacent material so that the door just does not load up as expected. Strain gages on one typical door, designed for full effectiveness, measure only 50% of its design load at limit loadings and approximately 10% at lower loadings.

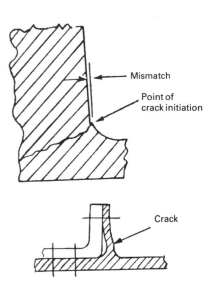

Fig. 15.6.10 Radius mismatch.

564 Airframe Structural Design

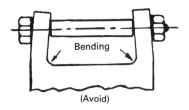

(a) Bushing too short induce flange bending

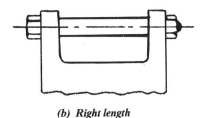

(b) Right length bushing

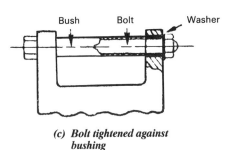

(c) Bolt tightened against bushing

Fig. 15.6.11 *Use right length of bushing.*

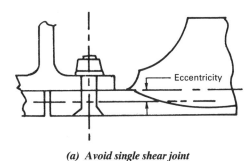

(a) Avoid single shear joint

(b) Double shear

Fig. 15.6.12 *Use double shear joint to avoid eccentricity.*

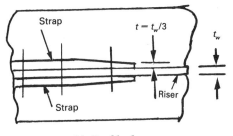

(a) Double shear

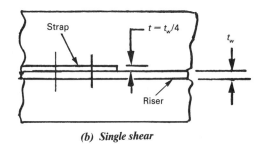

(b) Single shear

Fig. 15.6.13 *Reinforced strap thickness.*

- Padding spanwise joints does not produce the corresponding stress reduction as padding transverse joints, since the additional material is strained just about as much as adjacent material. Reduction in bearing stress of attachments, and elimination of feather edges in countersinks are useful results of padding longitudinal splices.
- Feather edges in countersinks produces K ranging as high as 10. By maintaining material thickness $= 1.5 \times$ countersink head height, the effective stress concentration factor may be no worse than normal riveting.
- Blind areas are to be avoided if at all possible. When blind areas as shown in Fig. 15.6.14 are proven necessary all effort must be expended so that no crack may develop within the covered area. It is legislated policy that fatigue test failure in blind area is unacceptable. Failure must occur in base material outside the blind area. Arbitrary reduction in allowable stress values along with mildest possible K values may accomplish this goal.
- Shear webs may fail in fatigue because the low initial buckling allows the panel to work continually with the result that damage from bending stresses in the wrinkling processes are

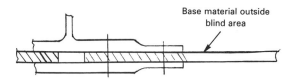

(Blind area as shown in shaded area)

Fig. 15.6.14 *Blind areas of a double shear joint.*

Airframe Structural Design 565

accumulated over many small loads during normal usage, and when too severe, cracking will eventually develop. To delay this type of failure and to offer better aerodynamic smoothness, initial buckling criteria are usually established for each airplane. Current popular rules are:
- No buckling at 1.0g steady flight, normal operating loads. (Military specifications sometimes call for 1.50g.)
- Ratio of ultimate shear load to initial buckling load not to exceed 5.0.
- Arbitrary panel size limitations may be specified for lightly loaded or secondary structure.

• Secondary structures (aft trailing edge structure, flaps, ailerons, tips, ect.) are not dangerous when minor cracks develop, but are a nuisance and an economic liability for the operator to repair. Attention to continuity of material, panel cutouts, panel sizes (oil-canning), type of attachments, etc. will reap benefits if a noticeable reduction in cracking propensity can be achieved.

• All transverse joints and any other local areas deemed critical are required to be fatigue tested to prove the quality level achieved in detail design. A minimum acceptable quality level is established and, if necessary, redesign is accomplished until it is guaranteed that this quality level is met.

Fuselage

The pressure cabin has been the source of the most spectacular disasters directly attributed to fatigue in aircraft. The contribution to the failure of the fuselage shells are:
— Cutouts in shell structures create high local stresses.
— Cut countersink rivets adjacent to the edge of the cutout compounds the stress concentration effects.
— Aluminum materials with high yield to ultimate strength ratios are prone to rapid tearing at low stress levels.

The third item is especially important in view of the large stored energy in compressed air inside the fuselage shell. With improper structure, a small fatigue crack may trigger off disastrous rupture of the entire fuselage.

Pressurization creates two major stresses in a cylindrical shell as shown in Fig. 15.6.15, and general rules in fuselage design are:

• Simplifying assumptions usually are made that the skin supports the pressure loads without benefit of the stringers or rings. Poisson's effect operates to reduce the degree of approximation involved; however, panel pillowing or quilting may create additional bending stresses across rivet lines due to stringer ring stiffness.

• To preclude primary material or joint fatigue failure, current fuselages are limited to gross area hoop tension stresses of 12,000 to 14,000 psi at the design operating pressure level (see Fig. 11.3.7 of Chapter 11.0). This corresponds to approximately 20% of the material strength, based upon hoop tension loads.

• 2024-T3 sheet material is usually specified for fuselage skin material because its fatigue allowable is as good as the strongest and its tear resistance is superior.

• For sizeable cutouts, corner strains and corresponding stresses are controlled primarily by the corner radius and the gross fuselage shell structure.

• Riveting adjacent to a cutout edge should have ample edge distance to reduce the compounding of the stress concentration factors.

• Careful assessment of the placement of reinforcing structure must be made. Too much overhang of the skin from frames and stringers may allow buckling of the compression corners with consequent increase in load transfer to the tension corners, and increased surface stresses from the buckles.

• All loading conditions must be carefully consid-

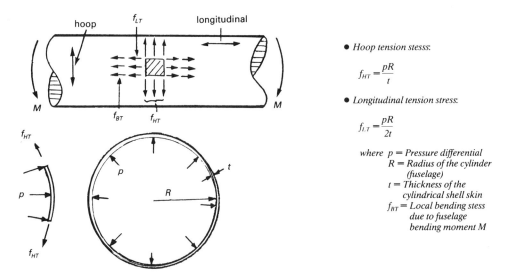

• Hoop tension stesss:

$$f_{HT} = \frac{pR}{t}$$

• Longitudinal tension stress:

$$f_{LT} = \frac{pR}{2t}$$

where p = Pressure differential
 R = Radius of the cylinder (fuselage)
 t = Thickness of the cylindrical shell skin
 f_{BT} = Local bending stess due to fuselage bending moment M

Fig. 15.6.15 Fuselage skin stresses due to cabin pressure.

ered in choosing the critical design conditions. Pressurization stresses combine with flight maneuvers and gusts, and stresses in corners of cutouts combine algebraically (see Fig. 11.7.7)
- Door opening on the outside require major attention to the positive locking mechanisms. Plug-type door (refer to Section 11.7) design is preferred and plug-type lock is illustrated in Fig. 15.6.16.

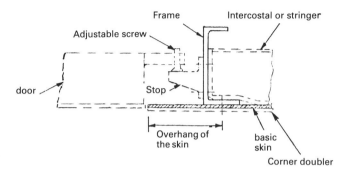

Fig. 15.6.16 Plug-type door lock.

Mechanisms and Structural Fittings

Mechanisms for this discussion may include most mechanical parts of the airplane. Structural fittings are included in this category because the design principles are similar. Some of the more important ones are:
(a) Landing gear*
(b) Control systems
 - Primary flight controls including boost
 - Secondary controls
   — flap*
   — tab
   — dive brake*
   — retracting and extension systems*
   — engine and systems operating mechanisms*
(c) Pressure door latches
 - outward opening doors (vital)*
 - inward opening (plug) doors (non-vital)
(d) Structural fittings
 - longeron or spar cap joint fittings
 - stringer fittings
 - major attach fittings
   — tip tank to wing
   — nacelle to engine
   — nacelle to wing
   — wing to gear*
   — wing to fuselage
   — any other major structural joints
(e) Parts with high ratio of operating to design loads. The above items marked with a (*) form a special class of parts in which operating loads may be a large ratio of limit design loads. Members of this class are especially prone to fatigue problems; at repeated limit loads, the life may be as low as 1000 cycles.
(f) Pin bending in the thicker plates is definitely a factor in increasing K values and reducing fatigue life.
(g) Tight fit pins, and tightly clamped parts show considerable improvement in fatigue allowable over clearance fit and loose clamped parts.

15.7 Sonic Fatigue Design and Prevention

Fatigue of aircraft structures has long been a problem and has in instances resulted in considerable structural damage and often in costly redesign and retrofit programs. With the advancement in airframe and engine technology, and the resulting high performance aircraft, an additional fatigue phenomenon, sonic fatigue, began to occur in service. As implied, this is fatigue of structural assemblies resulting from the random noise and/or vibration associated with engine and equipment operation, localized inflight boundary layer turbulence and cavity response.

The consideration of acoustically-induced fatigue failure in aircraft has been a design consideration for over 30 years. Since the early investigations of the development of sonic fatigue design criteria, it has been realized that such failures can substantially increase the maintenance burden and life cycle cost of the aircraft. Sonic fatigue failures have resulted, however, in unacceptable maintenance and inspection burdens associated with the operation of the aircraft. In some instances, sonic fatigue failures have resulted in major redesign efforts of aircraft structural components. As with any topic of concern to aircraft design, much progress towards establishing acceptable prediction techniques and design methods was realized early in the investigations although some of the techniques tended to introduce conservatism into the designs. The conservatism was expressed, as is usual with aircraft design, in terms of increased weight.

The parallel development of improved testing techniques and data analysis capabilities has resulted in both prediction techniques and design methods that yield acceptable structural configurations in terms of weight, ease of manufacture, and cost. Hence, the designer is only faced, today, with applying these results to his particular aircraft requirements. The main problem facing the designer is simply accumulating and assessing the vast amount of data available that relates to sonic fatigue design. Hence, the designer is required to continually utilize bits and pieces of data resulting from all sources in the fields of acoustic excitation, structural response, and fatigue life estimation.

Structural items for which critical acoustic loading occurs only during take-off and/or landing will be designed to withstand period exposure to maximum acoustic levels experienced during take-off and period exposure to maximum acoustic levels experienced during landing. A typical exposure time for a air superiority fighter aircraft is shown in Fig. 15.7.1 with an estimated service life of 4000 hours (8,000 to 10,000 hours has been used on later fighter designs).

Reseach has shown that certain types of construction are more sensitive to acoustically-induced fatigue than others and a susceptibility index has been devised to broadly indicate comparative rates at which sonic fatigue may occur in some of the more common methods of fabrication and construction. These are listed in Fig. 15.7.2.

In general, structural response to acoustic excitation decreases with increase in structural damping. Hence, other factors being equal, more highly damped structures are less susceptible to sonic fatigue than

Airframe Structural Design 567

lightly damped structures and the damping ratio (δ) are listed in Fig. 15.7.3.

Flight Mode	Time	
	Percent	Hours
Group run-up	5	200
Take-off	5	195
Climb	8	300
Cruise	51	2065
Acceleration	4	170
Combat	6	275
Descent	8	300
Loiter	8	300
Landing	5	195

Fig. 15.7.1 Estimated cumulative exposure times for an air superiority fighter aircraft.

Description	Susceptibility Index
Wet lay-up edge fiberglass honeycomb	1 (best)
Braxed steel honeycomb construction	2
Heavy plate rib (conventional) construction	2
Integrally stiffened panels	2
Flat bonded beaded panels	3
Riveted corrugated construction	4
Clips, angles, brackets and gussets	5
Riveted trusses	8
Lightweight plate rib construction	11 (worst)

Fig. 15.7.2 Sonic fatigue susceptibility index.

Structural Type	Damping Ratio (δ)
High aspect ratio riveted panels	0.010 — 0.020
Longerons and ribs	0.020 — 0.025
Laminated panels	0.025 — 0.035
Double skin bonded-beaded	0.020 — 0.030
Honeycomb sandwich	0.020 — 0.040

Fig. 15.7.3 Damping ratio vs. structural types.

Stress Concentrations

Sonic fatigue failures will normally originate in regions of high stress concentration. The following design principles should be followed to minimize the stresses in critical areas:
- Avoid abrupt changes in cross-section.
- Avoid joggles in members such as stringers.
- Avoid attachment of secondary elements at points of high stress in primary members.
- Make bend radii as large as possible.
- Provide generous fillets at the intersection of all structural elements.
- Bevel all sharp corners.
- Where cutouts are necessary in skin panels or webs, provide the largest radius possible in each corner.

Materials

Composite materials, either in sandwich or laminated form, is very resistant to acoustical fatigue because of high vibration-damping characteristics inherent in the material.

Experience has shown that fabricated construction of certain metals, e.g. magnesium, are highly susceptible to sonic fatigue due to their high sensitivity to "stress raisers" such as rivet holes, scratches, notches, etc. Therefore, the use of such materials for conventional types of construction is not recommended in areas where the overall sound pressure levels exceed 140 decibels unless adequate sonic fatigue testing is performed.

Fabrication

In the fabrication of structural assemblies, the observance of certain priniciples, listed below, will have a marked effect on the sonic fatigue-resistant properties of the components:
- In general, fatigue life is shortened by fabrication or assembly operations which involve spot-welding, press-forming of flanges, beads and stiffeners, skin dimpling, and work-hardening during installation of high heat-treat fasteners. Machining and heat treatment processes which minimize residual stresses should be used, as should stress relieving treatments, where appropriate. Surface finish should be as smooth as practicable; rough or high-speed grinding should be avoided to eliminate possible local surface overheating.
- Drilled holes are preferable to punched holes in reducing fatigue potential.
- Extruded parts usually are superior to roll-formed parts of the same shape.
- Other factors to consider include the location and method of marking part numbers, to ensure that they are either applied by a non-damaging process or located where the lowest stress is expected; the application of appropriate protective treatments to minimize corrosion; and, wherever practicable, the requirement for grain direction to be aligned with the direction of strain (load).

Joints and Splices

In designing joints, adherence to the practices listed below, while not providing an absolute guarantee against sonic failure, will aid in achieving a longer fatigue life:
- Use simple fastener patterns; minimize fretting

between the faying surfaces by selecting proper tolerances and fits for the holes and fasteners; avoid mixing different types of fasteners (bolts, rivets, etc.) in any one joint.
- Use close tolerance holes for bolts wherever possible; where considerations such as interchangeability dictate other than close tolerance holes, specify washers under bolt heads and nuts.
- In general, correct design should be considered as more important in a joint than appropriate material selection. Wherever possible, joints should be symmetrical. Long splice joints should be avoided; scarf or fishtail splices used instead of stepped splices.

Panel Construction

Implicit in the construction of the sonic fatigue design is the assumption that through proper attention to detail design, sonic-induced stresses in the panel at points of attachment to substructure will not result in structural fatigue. The relationship of panel configuration to the problem of sonic fatigue for each of several typical designs is discussed below. In general, it should be noted that vibratory edge stresses usually are greater for a square panel than for a rectangular one of the same area and that, regardless of the type of panel, its boundaries must be investigated carefully since any discontinuity tends to accentuate the fatigue problem.

(1) Stringer-Stiffened Panels.
    The use of symmetrical stringers and symmetrical attachments will prolong sonic fatigue life of stringer-stiffened panels. Where manufacturing considerations require the use of unsymmetrical stiffeners, adequate tie-in support of the stiffeners must be provided to offset the effect of eccentric loading.

(2) Integrally-Stiffened Panels.
    Integral stiffening of a panel offers an advantage over stringer-stiffened panels in allowing the use of additional material in areas of high stress concentration without abrupt changes in cross-sectional properties.

(3) Bonded-Beaded Inner Skin Panels.
    Beads frequently employed over large areas of unsupported paneling are characterized generally by poor termination. If large areas remain unsupported, the panel will vibrate as on large diaphragm, causing cracks to form rapidly because of the heavy working of the ends of the beads. Use of a double-ended bead, or the addition of doubler at the end of the bead, can result in a 50 to 100 percent increase in service life.

(4) Corrugation-Stiffened Panels.
    The primary sonic fatigue problem associated with corrugation stiffened panels is that of having only a limited amount of area available for transmitting the panel loads into the substructure. In the design of corrugated panels for sonic fatigue, particular emphasis must be placed on distributing the loads as uniformly as possible over the available attachment area. Attachment to substructure should be made through symmetrical members.

(5) Skin-rib structure
    This structural configuration is the most widely used in aircraft design. Conventional skin-rib structure is critical from a sonic fatigue design standpoint at the fastener row in the skin, at the bend radius of the rib, or the fasteners may fail in tension. Therefore, particular attention must be given to the edge conditions of each panel or bay and also to the rib thickness. It is most desirable to have sonic fatigue failures in the skin rather than in the ribs, if any occur, since routine aircraft inspections will most probably uncover any skin cracks. However, support structure damage is not easily detectable and may progress with accumulation of flight time and result in major structural failures.

(6) Honeycomb sandwich design
    Honeycomb sandwich offers one of the best structural approaches in designing for resistance to sonic fatigue due to its high stiffness-to-weight ratio. The edges of a honeycomb sandwich panel are the most vulnerable areas for sonic fatigue damage, and such, they must be critically examined to insure a satisfactory service life. This involves the proper selection of panel face sheets, edge doubler thickness, and edge closure member.
    Tests, including the placing of panels directly behind an actual jet engine, have substantiated that honeycomb sandwich is the superior type of construction in resisting sonic fatigue as well as providing the best weightwise approach where high intensity sound pressure level must be combated. However, for the lower levels of sound intensity, fabrication and maintenance requirements usually render honeycomb sandwich panels less advantageous than other types of construction.

Monographs for the various structural configurations are presented in Ref. 15.28 which will serve as a guide to the designer who is required to develop a structure that is resistant to sonic fatigue.

Sonic Fatigue Prevention

A new set of criteria has been developed for sound pressure environments through testing and successful designs on airplanes in service. These criteria are based on bonding the detail assemblies such as honeycomb panels and laminated sheet for rib webs and spars. Conventional riveting and bolting detail assemblies are satisfactory provided attention is paid to detail design and eliminating sharp reductions in cross-section notches and sharp edges loaded by sonic pressures acting normal to web or skin.

The lightest and cheapest skin structure in sonic environment is honeycomb with fiberglass face sheets and a derivation core (i.e. Nomex). This is particularly true if local aerodynamic loads are only involved and deflections due to the low modulus of elasticity of fiberglass are no problem. If deflections are a factor, square edge honeycomb with stronger face sheets and core should be used.

Ground rules for sonic fatigue design:
- All sound pressure levels where shown in db are referenced to 0.0002 dynes per square centimeter RMS.
- All edges of aluminum materials of webs and cutouts must be machined except: (a) edges may be

sheared in "0" condition prior to heat-treat, and (b) edges may be sheared in heat-treat condition provided the area is overdesigned by 7 db.
- Countersinks with knife edges (feathered edges) are not allowed unless the area is overdesigned by 7 db, except that a knife edge on the internal faying surface of laminated sheets is allowed.
- Dimpling is not allowed unless the area is overdesigned by 9 db.
- Spotwelds are not allowed unless the area is overdesigned by 15 db.
- Holes in webs with diameters under 0.25 inch are allowed, provided the hole is reinforced with a doubler or the area is overdesigned by 4 db. Small holes filled with tight rivets do not require any reinforcement.
- Holes or cutouts in webs with diameters over 0.25 inch require the addition of stiffeners on each side of hole or overdesign of the area by 7 db. Overdesign by only 4 db is required if the hole is flanged.
- Beaded stiffeners on webs may be used, but they do not provide the full support of a separate stiffener. When using beads as stiffeners overdesign the web by 5 db.
- Lightning holes, tool holes, and the periphery of parts fabricated from aluminum material 0.063 inch or thicker in the "T" condition shall not be pierced, punched, blanked, or sheared unless followed by a minimum of 0.03 inch cleanup all around. Blanking, etc. will be permitted provided it is done prior to heat treatment.
- If blind fasteners are unavoidable in a sonic environment, a bond should be designed for the faying surface as shown in Fig. 15.7.4.

Attention should be paid to details, especially notches. Fig. 15.7.5 is a typical example of an unsupported corner detail that will experience an early failure due to web diaphragming from sound pressures normal to the web.

If a clamp or hard point is installed in the middle of a web (Fig 15.7.6) the natural frequencies of the web will be changed and diaphragming amplitudes will increase. An early fatigue failure will result. Instead, install the clip on a stiffener or chord (spar cap) member.

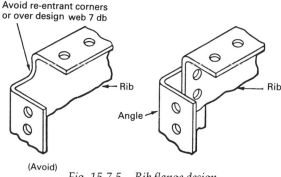

Fig. 15.7.5 Rib flange design.

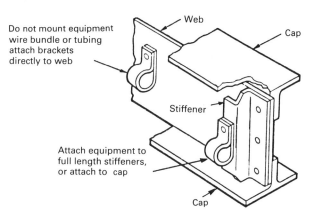

Fig. 15.7.6 Bracket mount.

15.8 Verification tests

In terms of total impact, the airplane fatigue test is probably the most important in the test program. As previously mentioned, service life policies on the structure are becoming increasingly demanding, and fatigue analysis is the most difficult of the total structural analysis problem.

The objectives of the test are shown in Fig.15.8.1. Because of the shortcomings of present fatigue analysis, test knowledge of the critical areas is very important. Quite often, relatively minor changes can be incorporated early in the production line to vastly improve the fatigue performance. Also, the test is extremely valuable in developing inspection and maintenance procedures. In some cases, preventive maintenance may be used, such as oversizing fastener holes, to obtain a sizable increment in fatigue life.

In the airplane fatigue test, it is, of course, necessary to apply a spectrum of loads representative of the intended usage. These loads must be applied in a rapid manner such that the test conducted over a period of a few months represents as near as possible the complete lifetime load spectrum. This is derived by evaluating the intended usage, including gross weight, speed, altitudes, payloads, etc. In the case of military airplanes, this usage is either specified or negotiated with the contractor early in the design. By the use of an analysis procedure, the total lifetime spectrum is condensed into a fatigue test spectrum. The flight profile is usually divided into different

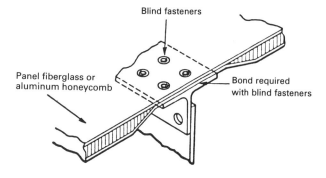

Fig. 15.7.4 Bond doubler in blind fastener area.

- To provide basic data on the methods of analysis used

- To determine the effects of various factors affecting the fatigue life of the structure such as materials and processes, fabrication procedures, environmental conditions etc

- To establish design allowable stresses for the materials to be used in the structure including processing effects

- To evaluate and compare the fatigue quality of structural design details (development testing)

- To substantiate that the structure will meet or exceed the design life requirement of the vehicle (qualification and certification testing)

- Locate fatigue critical areas of airplane at earliest possible time for production incorporation

- Develop inspection and maintenance procedures for customer

- Provide test data for analytical studies to assist in determining times for inspection and maintenance

Fig. 15.8.1 Fatigue test objectives.

segments for deriving the stress spectrum, as shown in Fig. 15.8.2. For each critical location on the airplane, the fatigue damage is calculated using statistical gust data, estimated maneuvers, ground loads, etc., together with the expected fatigue quality to arrive at the calculated fatigue life. The compressed fatigue test spectrum is then derived and the structural life in a specified number of flights can be obtained.

It is important that the airplane experiences during the test a reasonable representation of the various levels of stress, such as ground stresses and maneuver stresses, and the corresponding ground-air-ground cycle shall be included. In the case of certain military aircraft, for example, the maneuver spectrum is specified for design based upon previous similar aircraft experience. When applicable, cabin pressure cycles are also applied in the test spectrum.

Cyclic test equipment used in the complete airplane test has progressed significantly over the last 30 years. Automatic systems including load programmers and monitoring devices are in wide use to reduce the time of testing. One of the structural test problems is to design the test equipment to withstand the repeated loading application. During the testing, test equipment

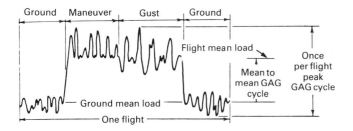

(a) *Flight-by-flight random loading sequence*

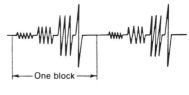

(c) *Block loading sequence without GAG cycles*

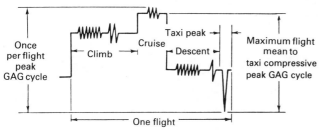

(b) *Flight-by-flight low-high ordered loading sequence*

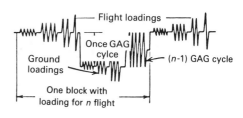

(d) *Block loading sequence with GAG cycles*

Fig. 15.8.2 Typical spectrum test loading sequences.

(a) Iron bird functional test facility

(b) Functional test of flap system

Fig. 15.8.3 Use iron bird to do functional test for L-1011.
(Courtesy of Lockheed Aeronautical Systems Co.)

failures were not uncommon; and several sets of loading cables or hydraulic loading jacks were used. Fatigue tests on regular transports usually are planned on the basis of approximately one to two years in length.

Warranties on structural performance are usually negotiated in customer contracts and therefore, even with fail-safe design, it is practical to conduct fatigue tests to verify the analysis and to aid in checking life estimates.

Fatigue tests may be divided into four categories:
- Flight and ground loads;
  These loadings simulate the stress history expected from normal flight and ground loadings.
- Sonic;
  Tests are conducted simulating the environment from engine or boundary layer noise.
- Fail-Safe;
  Although this is not really a type of fatigue test, it is usually conducted with the fatigue test to establish the static strength of a partially failed structure.
- Functional;
  The title indicates this test is a functional test, but quite often turns out to be a fatigue test of various parts of the mechanism. Examples are flaps and landing gear actuation systems. Fig. 15.8.3 shows an "Iron Bird" for functional test.

Fatigue testing may be classified either by the type of specimen being tested, i.e., coupon, panel, component, complete airframe or by the type of loading applied (S-N, block spectra, flight-by-flight spectra). Each of these types of tests are discussed below.

Coupon Testing

Coupon fatigue testing (see Fig. 15.8.4) is performed when the fatigue strength of a local feature or critical point is required under representative loads or stresses for initial design information, or when comparisons are needed between the fatigue performance of materials for a particular component. The problems of load selection are minimized when attention can be focused on a local area.

Some of the types of unnotched and notched specimens used for coupon fatigue testing are shown in Fig. 15.8.5. These include notched and unnotched specimens as well as some simple joint type specimens. Simple joint specimens are included here since these types of specimens are usually tested in the same test equipment as the unnotched and notched coupons.

Unnotched fatigue specimens are often used to evaluate the effects of surface treatment to the material. However, it must be kept in mind that fatigue failures almost never occur in an unnotched section of the structure. Comparisons between surface treatments based on unnotched coupon tests can often be misleading when applied to notched areas of the structure. The results of some tests conducted on anodized aluminum test coupons (Fig. 15.8.6) illustrate this point. This shows that the fatigue strength is reduced 50% for an unnotched coupon, 10—25% for $K_t = 3$ and 4, and no reduction when notch is as high as $K_t = 7$. Therefore, it is important to evaluate the effect of surface treatment in combination with the severity of the notch that will be present in the actual structure.

Engineers should keep in mind the distinction

Fig. 15.8.4 Coupon test machine.

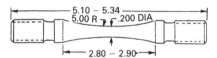

(a) Unnotched round specimen

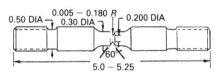

(b) Notched round specimen

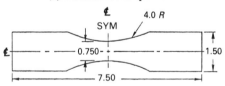

(c) Unnotched sheet specimen

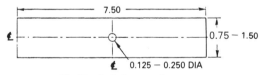

(d) Notched sheet specimen

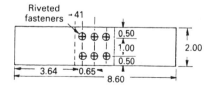

(e) Lap joint fatigue specimen

Fig. 15.8.5 Coupon fatigue testing specimens.

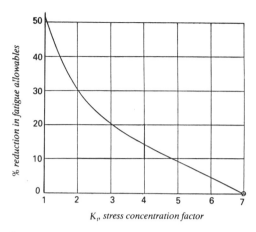

Fig. 15.8.6 Reduction in fatigue allowable due to anodizing aluminum.

between flat and round specimens. Test results for flat and round specimens for same K_t value are not the same. This is due to the difference in the state of stress at the notch. Essentially a plane stress state exists in the vicinity of notches in thin flat sheets. However, in notched round specimens a triaxial state of stress exists due to the restraint of the material on either side of the notch. Therefore, again the type of specimen should be selected to simulate the type of notch in the structure. To simulate notches in a shell structure, flat sheet types are usually used and to simulate notches in heavy forgings and landing gear type structures, notched round specimens are generally used.

Panel Tests

Panel tests are conducted to check and/or develop design details for shell structures. The panel specimens usually include some or all of the following: stringers and splices, beam cap and rib attachments, cutouts, holes, and sometimes dihedral if applicable. Any detail that will affect the fatigue life of the panel area represented should be included in the test specimen.

The middle third of the panel is usually considered the test section. The end third section of the panel provides the transition for distributing the loads from the loading attachments to the test section. The design of the transition section of the panel is very important so that the fatigue failure will occur in the test section. This means that the design detail of the end attachments must be better than the design detail of the test panel. Therefore bonded doublers are often used on the ends as reinforcements. Also friction clamping is often used which is another way to transfer the load into the specimen without introducing severe notches of high stress concentrations. However, when friction clamping is used, fretting fatigue between the mating surfaces must be considered.

Since panels usually cover a considerable portion of the structure, it is not possible to reproduce the desired stress at all points. Loads to be applied to the panel are usually calculated based on the net area stresses that will occur on the minimum test section. If this does not produce the desired stress at other critical sections, the loads can be adjusted accordingly. Even though the stresses are not reproduced exactly as desired, fatigue analysis of the test results can be made to determine if an adequate life has been achieved.

Test panels should always be instrumented so that the actual stresses applied to critical areas of the panel can be determined. A number of instrumentation techniques are available for this purpose. Load cells are used to monitor the overall loads applied to the panels. Strain gages are installed at locations away from notches (unless the stresses in the notch is of interest) so that net area and gross area stresses at critical cross-sections can be determined. Stress coat or photo stress can be applied to the panels so that an assessment of the local stress concentration can be made. The use of stress coat or photo stress usually pays off since failures in areas outside the test section can often be avoided.

Component Tests

Component tests are usually conducted for the development or substantiation of fatigue critical parts of the structure. Some of these tests are shown in Fig. 15.8.7.

Structural analysis should be made to ensure that the correct stress distributions are achieved in the component being tested.

Instrumentation of the component as discussed previously will also verify that the proper stresses for each loading condition are achieved.

The loads applied to components should produce the desired stresses as closely as possible at all critical points in the structure. For a specimen as complicated as a component, it is not usually possible to get the desired stress at all critical locations. The magnitudes and occurrences of the various loading conditions are usually modified to obtain the best compromise distribution. The component should then be instrumented so that the stress at all critical points in the structure can be determined. When a fatigue failure occurs, the applied stress spectrum at the point of failure can then be determined. This will permit evaluation of the failure by the use of fatigue analysis to determine if the life requirements have been met in case the failure location was over-loaded.

Full Scale Aircraft Tests

A fatigue test program on a full scale specimen is the only means that can be employed to account for the true loading environment of an airframe structure. Such a test allows for the proper interaction of all the structural components. In addition, full scale testing provides a realistic basis for the determination of in-service inspection and repair procedures.

The full-scale fatigue test program is a vital step in the development of the airframe structure. How well such a program is planned and conducted will determine to a great extent the adequacy of the structure in service. Therefore, the challenge to the test engineer is to ensure that as high a degree of simulation is obtained as is technically feasible.

Fatigue testing of full-scale (complete airframe) or major segments of airframe is usually required by both commercial and military customers for any major aircraft program even though not required by U.S. FAA certification.

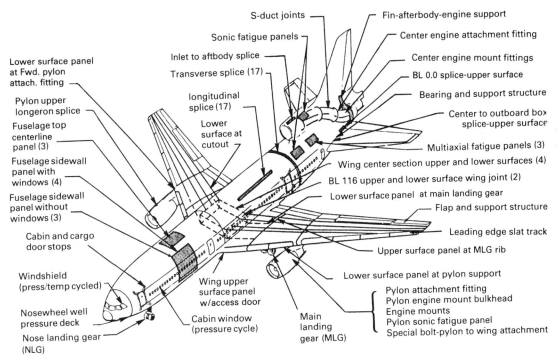

Fig. 15.8.7 Components structural fatigue tests — L-1011.
(Courtesy of Lockheed Aeronautical Systems Co.)

A full-scale airframe fatigue testing of a commercial transport airplane is shown in Fig. 15.8.8. It included pertinent operating systems such as flight controls, hydraulic, fuel, and pneumatic. Dummy engines and landing gears were used since these components were fatigue tested separately. A total of 200 servo-controlled hydraulic jacks were used to apply the loads. All significant ground and flight loadings, including cabin and fuel pressurization, were applied on a flight-by-flight basis for approximately 20 years of service utilization. The tests, conducted on a 24 hour basis, took approximately two years to complete.

During fatigue testing the aircraft is being inspected periodically for fatigue cracks or other sources of damage. The inspection techniques developed during the test program will help establish inspection procedures for service aircraft. Structural modifications and repairs will be made as required if any fatigue cracks develop before the life goals have been reached. The

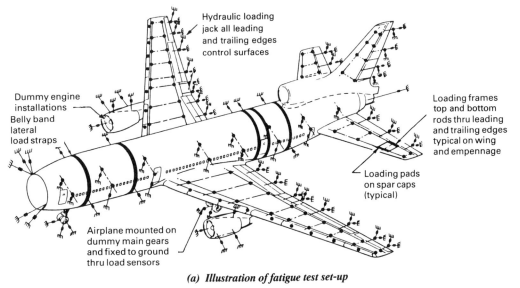

(a) Illustration of fatigue test set-up

Fig. 15.8.8 Complete airframe fatigue tests (indoor) — L-1011.
(Courtesy of Lockheed Aeronautical Systems Co.)

Airframe Structural Design 575

(b) Fatigue test airframe surrounded by steel framework

Fig. 15.8.8 (continued)

completion of the full scale fatigue test will provide substantiation that fatigue cracks requiring major structural rework or repair are unlikely to occur during the economic life of the aircraft.

Fig. 15.8.9 shows the setup in three separate fatigue test sections so that all three portions of the structure could be tested simultaneously but independently. In addition to these test sections, separate fatigue tests were conducted on nose and main landing gear and other components.

The multi-section full-scale fatigue testing was based on

- Fewer compromises in load spectrum
- Only critical loads used
- Other sections can continue when one section is down for repair
- Early test completion and therefore more rapid incorporation of failure into production
- Less complex and costly

Obviously multi-section full-scale testing has its advantages in cost and elapsed time for testing a full-scale structure. Although dividing the airframe into three sections plus constructing and assembling additional end supports create additional expense, it is

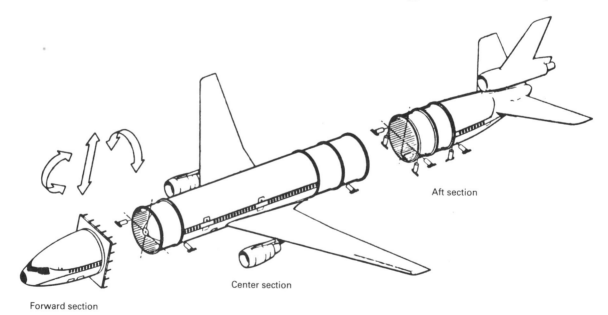

Fig. 15.8.9 *Multi-section full-scale fatigue test arrangements.*

offset by the reduction in personnel required to conduct the tests.

Fig. 15.8.10 shows an outdoor setup of a transport airplane fatigue test airframe to simulate the actual weather environment, which is one of the most important factors affecting the airframe fatigue life.

Fail-Safe Tests

Periodically during the fatigue test program, fail-safe loads were applied to the airplane. These loads were approximately 80% of design limit load times 1.15 dynamic factor, and are specified by the U.S. FAA. Their purpose is to prove that in the event of failure

(a) Outdoor fatigue test rig

(b) Test rig surround wing

Fig. 15.8.10 Airframe fatigue test (outdoor test) — B757.
(Courtesy of the Boeing Co.)

Airframe Structural Design 577

of any major structural element the remaining structure will be capable of carrying the maximum expected operating load for a partially failed structure condition. After completion of the fatigue test, artificial cut simulating cracks were started at different locations considered to be crtiical from a fail-safe standpoint. After the cuts were made, cyclic loading was applied to extend the cracks in length.

A limited number of fail-safe tests are conducted on structural assemblies to substantiate that the structure will meet the fail-safe requirements. At selected locations in each assembly, damage will be artificially produced and fail-safe loading and representative environmental conditions applied. In general, the damage shall consist of cutting members to simulate the damage levels for which the structure was designed.

For instance, in the wing, the member may be a spar cap, a shear web or a surface panel. In the fuselage, the damage may be a cut through a window or door frame element, a circumferential cut in the skin passing through a stiffener, a longitudinal skin cut through a ring or a skin cut extending between rings (see Fig. 15.8.11). Successful application of the test conditions without significant change in specimen deformation will, in general, be followed by repair and retest at another location in the specimen. The intent of these tests is to show compliance with the damage tolerance requirements.

Fig. 15.8.11 *Fuselage simulated damage — 40 in. sawcut — through frame (indicated by arrows) — L-1011.*

(Courtesy of Lockheed Aeronautical Systems Co.)

For cracks or damage limits not confined by crack barriers, it will be demonstrated by simulated flight-by-flight fatigue testing that a detectable damage size (1 inch crack length minimum) will not grow to 67% of the critical damage size in less than two normal structural inspection intervals. The inspection intervals must be consistent with the maintenance and inspection procedures established for normal servicing of the aircraft. Normal commercial transport structural inspection intervals are approximately 10,000–12,000 flight hours.

Sonic Fatigue Test

The depth and scope of the sonic fatigue test program required to establish proof of design will depend upon the severity of the anticipated service environment and on the amount of experimental data available on the sonic fatigue resistance of similarly constructed structure. The sonic fatigue test program comprises three phases: (1) development tests, (2) component tests, and (3) prototype airplane tests. The objectives and general procedures of each of these phases are discussed below.

(1) Development Tests

There are two main objectives of sonic fatigue development tests: (a) to determine the feasibility of a given structural configuration to withstand its anticipated acoustic environment; and (b) to achieve the optimum quality of detail design for the structural configurations chosen. In order to be most effective, these tests must be performed as early as possible in the design phase of the program.

To ensure meaningful results of the test program, the following rules should be followed in the selection and design of test specimens:
- Test specimens should include structure containing representative joints and splices.
- Test specimens should be mounted to the test jig by means of structure representative of aircraft substructure.
- Test specimens should contain a minimum of two bays.

Development tests should be conducted in the progressive wave test facility. Initial tests should be sinusoidal sweep tests to determine the frequencies, mode shapes, damping and linearity of response for each of the significant panel resonances. The strobe light should be used in the modal surveys to supplement instrumentation supplying quantitative data since it often reveals details of panel response not discernible by other means. For orthotropic panels, these preliminary tests should be conducted for both possible panel orientations relative to the direction of wave propagation in the test facility.

The main objective of the sinusoidal sweep tests is to isolate and identify the structural mode or modes believed to be most critical from the standpoint of sonic fatigue. This information is then used to:
- Check the adequacy of test specimen instrumentation.
- Determine the most realistic test procedures (e.g. intensity and spectrum of acoustic excitation and panel orientation in the test cell).
- Interpret test results in terms of service failures.
- Suggest means of improving the design to reduce susceptibility to sonic fatigue.
- Evaluate the validity of the analytical procedures.

Following the sinusoidal tests the panel should be subjected to a suitable acoustic excitation to determine its capablilty to resist sonic fatigue. Test levels should be chosen so that a test time of 10 hours will be equivalent to service life from the standpoint of fatigue damage. Whenever possible, test input levels should be within the range of linear structural response. If no failure occurs within 10 hours, the input level should be increased by 3 db and the test repeated. Structural response should be monitored throughout the test in order to detect the onset of structural failure. For some tests it may be desirable to continue testing after a failure is detected to observe the manner in which the failure progresses.

(2) Component Tests

Component tests are those tests performed on sections of the full scale airplane such as sections of fuselage, wings, and control surfaces. The objective of the component sonic fatigue tests is to provide additional assurance that the structure will have a satisfactory sonic fatigue life.

The sonic fatigue component test program should be closely integrated with the overall structural test program since the same test specimens can often be used for sonic fatigue and other proof tests. Strain gage locations on the test specimens should be based on information gained from the development tests. These should be selected on the basis of where fatigue cracking is most likely to occur and where modal response characteristics can be best determined. Particular consideration should be given to the location of strain gages on any light weight substructure which was not realistically represented in the development tests.

The test specimens should be subjected to simulated service acoustic excitation in the reverberation test chamber. Test input levels should be chosen so that 100 hours of test time is equivalent to airplane service life from the standpoint of sonic fatigue cracking. During the test, adequate data should be taken to define:

- The acoustic environment over the entire test specimen
- The predominant modes of structural vibration
- RMS values of stress in potentially critical areas

Provisions should be made to simulate any service loading conditions during the test which will have any appreciable effect on the sonic fatigue life of the structure.

(3) Prototype Airplane Tests

The objective of the prototype airplane sonic fatigue tests is to provide final proof of design for sonic fatigue. Even in those cases where the actual service acoustic environment is well known it is not possible to reproduce in the test facility all of the salient features of this environment from the standpoint of structural response. Provisions must therefore be made to measure the acoustic environment and resulting structural response on the prototype airplane under actual service conditions. The acoustic environments should be measured as a basis for accounting for differences in structural response characteristics between the test and service acoustic environments. Structural response should be measured throughout portions of the structure where sonic fatigue is a potential problem. Strain gages should be placed in the same locations as in the component tests, as well as other locations which may be critical for sonic fatigue, so that resulting response data can be correlated with that from the component tests.

References

15.1 Grover, H.J.: *Fatigue of Aircraft Structures*. U.S. Naval Air Systems Command publication (NAVAIR 01-1A-13), 1966.

15.2 Sines G. and Waisman, J.L : *Metal Fatigue*. McGraw-Hill Book Company, Inc., N.Y., 1959.

15.3 Osgood, C.C.: *Fatigue Design*. John Wiley & Sons, Inc., New York, N.Y. 1970.

15.4 Smith, C.R.: *Tips on Fatigue*. U.S. Bureau of Naval Weapons publication (NAVWEPS 00-25-559), 1963.

15.5 Anon.: *Shot-peening Applications*. Metal Improvement Co., Los Angeles, Calif., 4th ed.

15.6 Anon.: *Fatigue and Stress Corrosion — Manual for Designers*. Lockheed-California Co., Burbank, Calif., 1968.

15.7 Anon.: *Fatigue Prediction Study*. U.S. Air Force Systems Command publication (WADD TR 61-153), 1962.

15.8 Walker, P.B.: 'Estimation of The Fatigue Life of A Transport Aircraft.' *Journal of The Royal Aeronautical Society*, (Oct. 1953.) 613—617.

15.9 Anon.: 'Design Against Fatigue — Design Principles.' *Engineering Sciences Data Item No. 71009*, Engineering Sciences Data Unit, London.

15.10 Walker, P.B.: 'Fatigue of Aircraft Structures.' *Journal of The Royal Aeronautical Society*, (1949).

15.11 Walker, P.B.: 'Design Criterion for Fatigue of Wing.' *Journal of The Royal Aeronautical Society*, (Jan. 1953).

15.12 Landecker, F.K.: 'Shotpeening.' *SAE Quarterly Transactions*, (Apr. 1948).

15.13 Ekvall, J.C., Brussat, T.R., Liu, A.F. and Creager, M.: 'Preliminary Design of Aircraft Sructures to Meet Structural Integrity Requirements.' *Journal of Aircraft*, (Mar. 1974.), 136—143.

15.14 Hardrath, H.F.: 'Fracture Mechanics.' *Journal of Aircraft*, (Jun. 1974.)

15.15 O'Lone, R.G.: 'Boeing Bolsters Fail-safe Data.' *Aviation Week &Space Technology*, (Jan. 1978), 24—26.

15.16 Vann, F.W.: 'A 300B Static and Fatigue Tests.' *Aircraft Engineering*, (Dec. 1973).

15.17 Ramsden, J.M.: 'The Geriatric Jet Problem.' *Flight International*, (Oct. 22, 1977).

15.18 Hibert, C.L.: 'Factors and Prevention of Corrosion.' *Aero Digest*, (Apr. 1955).

15.19 O'Lone, R.G.: 'Airlines Begin 707 Wing Program.' *Aviation Week &Technology*, (Feb. 5, 1968).

15.20 Lundberg, B.: 'Bear-up Requirements for Aircraft.' *Aero Digest*, (Dec. 1947).

15.21 Kaplan, M.P. and Reiman, J.A.: 'Use of Fracture Mechanics in Estimating Structural Life and Inspection Intervals.' *Journal of Aircraft*, (Feb. 1976).

15.22 Anon.: 'U.S. Sets Pattern for Fatigue Standards.' *Aviation Week &Space Technology*, (Mar. 28, 1977).

15.23 Fonk, G.J.: 'Investigation of The Fail Safe Properties of Civil Aircraft.' *Paper presented at the 7th ICAF Symposium in London*, (Jul. 1973).

15.24 Stone, M.: 'Airworthiness Philosophy Developed From Full-Scale Testing.' *Douglas Paper No. 6099. Biannual Meeting of The International Committee on*

Aeronautical Fatigue, London, England, (Jul. 1973).
15.25 Sanga, R.V.: 'The 747 Fail-Safe Structural Verification Program.' *Paper presented at 7th ICAF Symposium in London,* (Jul. 1973)
15.26 Bates, R.E.: 'Structural Development of The DC-10.' *Douglas Paper No. 6046,* (May, 1972).
15.27 Mackey, D.J. and Simons, H.: 'Structural Development of The L-1011 Tri-Star.' *AIAA Paper No. 72-776,* (1972).
15.28 Anon.: 'Refinement of Sonic Fatigue Structural Design Criteria.' *AFFDL-TR-67-156,* (Jan. 1968).
15.29 Hardrath, H.F.: 'Fatigue and Fracture Mechanics.' *Jorunal of Aircraft,* (Mar. 1971).
15.30 Fisher, A.P.: 'Fatigue Aspects of Structural Design.' *Journal of The Royal Aeronautical Society,* Vol. 60.
15.31 Little, R.E.: 'How to Prevent Fatigue Failure.' *Machine Design,* (Jun.—Jul. 1967).
15.32 Combes, Jr., R.C.: 'Design for Damage Tolerance.' *Journal of Aircraft,* (Jan.—Feb. 1970).
15.33 Craig, L.E. and Hummel, K.H.: 'The Value of Fleet Experience in Exploiting the Full Economic Potential of Aircraft Structures.' *SAE Paper No. 760913,* (1976).
15.34 Anon.: 'B-52 Lifetime Extension Effort Pushed.' *Aviation Week &Space Technology,* (May 10, 1976).
15.35 Branger, J. and Berger, F.: 'Problems with Fatigue in Aircraft.' *ICAF Doc. No. 801, Proceedings of the ICAF Symposium.* (Jun. 1975).
15.36 Speakman, E.R.: 'Fatigue Life Improvement Through Stress Coining Methods.' *Douglas Paper No. 5506,* (Jun. 1969).
15.37 Anon.: *Prevention of the Failure of Metals under Repeated Stress.* John Wiley, New York, N.Y. 1941.
15.38 Forrest, P.G.: *Fatigue of Metals.* Pergamon Press, 1962.
15.39 Heywood, R.B.: *Designing Against Fatigue.* Chapman and Hall, London, 1962.
15.40 Almen, J.O. and Black, P.H.: *Residual Stress and Fatigue in Metals.* McGraw-Hill Book Co., Inc. New York, N.Y. 1963.
15.41 Madayag, A.F.: *Metal Fatigue-Theory and Design.* John Wiley Book Co., Inc. New York, N.Y. 1969.
15.42 Barrois, W., and Ripley, E.L.: *Fatigue of Aircraft Structures.* The Macmillan Company, New York, N.Y. 1963.
15.43 Plantema, F.J. and Schijve, J.: *Full-scale Fatigue Testing of Aircraft Structures.* Pergamon Press, New York, N.Y. 1961.
15.44 Anon.: 'Proceedings of The Symposium on Fatigue of Aircraft Structures.' *WADC Technical Report No. 59—507,* (Aug. 1959).
15.45 Walker, P.B.: 'The Fatigue Situation for Civil Aircraft.' *The Aeroplane,* (Apr. 1952).
15.46 Anon.: 'Sonic Fatigue design guide for Military Aircraft.' *AFFDL-TR-74-112,* (May 1975).
15.47 Anon.: 'Specialists Meeting on Design Against Fatigue.' *AGARD-CP-141,* (Oct. 1973).
15.48 Osgood, C.C.: 'A Basic Course in Fracture Mechanics.' *Machine Design,* (Jul./Aug./Sept. 1971).
15.49 Anon.: 'A Guide for Fatigue Testing and The Statistical Analysis of Fatigue Data.' *ASTM Special Technical Publication No. 91-A (1963),* American Society for Testing and Material, 1916 Race St., Philadelphia Pa., U.S.A.
15.50 Stone, M.: *DC-10 Full-Scale Fatigue Test Program.* Douglas Aircraft Company, McDonnell Douglas Corporation.
15.51 Anon.: 'Structural Design for Acoustic Fatigue.' *ASD-TDR-63-820.* (Oct. 1963).
15.52 Wilhem, D.P.: 'Fracture Mechanics Guidelines for Aircraft Structural Applications.' *AFFDL-TR-69-111,* 1970.
15.53 Frost, N.E., Marsh, K.J. and Pook, L.P.: *Metal Fatigue.* Clarendon Press, Oxford, 1974.
15.54 Sih, G.C. and Chow, C.L.: *Proceedings of an International Conference on Fracture Mechanics and Technology.* Sijthoff and Noordhoff International Publishers, Netherlands. (1977)
15.55 Swanson, S.R.: *Random Load Fatigue Testing: A State of the Art Survey.* Materials Research & Standards. 1968.
15.56 James, D.: 'Fatigue — Life Establishment and Detection.' *SAE Paper No. 680342,* (1968).
15.57 Anon.: 'Nondestructive Testing for Aircraft.' *AC 20-61. Department of Transportation, Federal Aviation Administration,* (1969).
15.58 Anon.: 'Crashworthiness of Airframes.' *AGARD-R-737,* (1985).
15.59 *MIL-A-83444 (USAF)*: 'Airplane Damage Tolerance Requirement.' (Jul. 1974).
15.60 Anon.: 'Fatigue Design of Fighters.' *AGARD-AG-231,* (1968).
15.61 Anon.: 'Problems with Fatigue in Aircraft.' *Proceedings of the Eighth ICAF Symposium (ICAF Doc. 801) Held at Lausanne,* 2—5 June, 1975.
15.62 Tada, H., Paris, P.C., and Irwin, G.R.: *The stress Analysis of Crack Handbook.* Del Research Corp., Hellertown, Pa., 1973.
15.63 Hudson, C.M.: 'Effect of Stress Ratio on Fatigue Crack Growth in 7075-T6 and 2024-T3 Aluminum Alloy Specimens.' *NASA TN D-5390,* (Aug. 1969).

CHAPTER 16.0

WEIGHT CONTROL AND BALANCE

16.1 Introduction

It is the firm conviction of many aircraft designers that one of the most important functions of aircraft design is weight control. The engineering department generally depends solely on the weight group for basic weight estimates on new designs and assumes contractual obligations on these estimates. Both under-estimation and over-estimation of the gross weight have their pitfalls. If the weight estimate is too high, it is impossible to compete with other companies; and if too low, serious financial penalties are assessed and no production orders are received, which usually is the prime issue at stake. These items affect the preliminary design section. Other sections of the engineering department are vitally affected by the weight estimate. The aerodynamics section must have certain specified center of gravity limits for stability and controllability calculations. If the estimator misses these c.g. limits too far, necessary design changes result. If the gross weight goes over the estimate, performance guarantees can not be met, and, conversely, if the gross weight goes under, a contract may be lost because the performance guarantees are too low.

The effect of weight on performance is not merely a question of converting weight into performance, but is indirectly obtained from the variations in wing span, power-loadings, etc., caused by the changing weight. The problem that confronts designers of high performance aircraft is essentially one of compromise among those factors that, as a whole, are the prerequisites of a good aircraft. To favor any one of these factors will generally be at the expense of one or more of the others. Thus, what is gained in one direction will have its compensating loss in another, resulting usually in an airplane of one good characteristic, but definitely limited as to others. The relative importance of these factors is highly controversial; therefore, no attempt at their evaluation is made. Nevertheless, a few of the more prominent performance factors are arbitrarily listed: speed, rate of climb, striking power, range, maneuverability, take-off distance, and landing speed. Viewing the problem as a whole, it is necessary to consider two distinct categories separately: namely, military and commercial. Although certain fundamental requirements apply equally to both fields, the specific mission for which an airplane is intended governs how each design factor should be evaluated in order to arrive at the optimum design.

If the center of gravity (c.g.) is allowed to be too far ahead of the center of pressure the aircraft will dive out of control. If the c.g. is allowed to be too far behind the center of pressure the aircraft will stall. There is a range, generally ahead of the center of pressure, in which the center of gravity must be located to preserve safety of flight. In this range there is an optimum point at which the aircraft is the most stable, most controllable, and most effective. Any location of the c.g. either fore or aft of this optimum point endangers safety by reducing the stability, controllability, and effectiveness of the aircraft to an extent comparable to its distance from the optimum point.

Weight affects flying speed only slightly but has a marked influence upon take-off and landing speeds. More thrust is required for a heavier weight and if the greater thrust cannot be developed within the limits of the take-off runway and its obstacles, safety is definitely at stake. As a result of the development of higher stalling speed, additional weight will increase the landing speed. Safety is thus diminished, for a given aircraft, with faster landing speeds. Extra weight may make it impossible to land an aircraft safely within the limits of the landing field and its surrounding obstacles. Overweight diminishes safety margins of the aircraft structure by imposing loads heavier than those for which the aircraft is designed. Such reduction in margins may limit safety quite critically when the aircraft encounters rough air or poor landing fields. Rate of climb and climbing angle are both decreased by additional weight. Such impairing of climbing characteristics may mean the difference between success and failure in military operations, or may spell disaster for either commercial or military operations over mountainous terrain.

Overweight also lowers the ceiling of the aircraft and may jeopardize safety critically in military operations where the higher ceiling is a distinct tactical advantage, or in any operations where high mountains must be cleared.

If balance is not properly controlled, the c.g. may be in such a position that it will impose loads upon the structure of the airplane that are substantially higher than those computed for the normal c.g. location. For example, an extreme nose heavy condition of imbalance may cause collapse of the nose landing gear structure.

Keeping the weight of the unloaded aircraft (MEW — Manufacturer's Empty Weight) to a minimum is the prime and most fruitful means of developing load-carrying capacity. A pound can be added to the load for every pound that is whittled out of the unloaded aircraft. For military operations this means more ammunition, more bombs, etc., and greater striking power which may provide the decisive margin. For commercial transport operations weight control develops larger revenues and profits.

The project designer must be fully aware of potential developments in the fields of aerodynamics, system, structures, materials and propulsion as well as in future trends in operators and airworthiness requirements. It must minimize the adverse effects of weight without jeopardizing the future prospects of the aircraft. Any new design must incorporate a certain amount of weight growth potential, even though this may involve some initial weight penalty. Stretched versions of civil aircraft tend to sell better than the initial design — even when the original has been designed to a specification which the airlines were convinced would exactly suit their requirements. It is therefore prudent to avoid built-in constraints at the initial project stage.

For is a short haul turboprop aircraft such as the one shown in Fig. 16.1.1. Weight changes have a more significant effect on aircraft performance than changes in cruising, drag, or fuel consumption. This is because flight time is short and fuel load is relatively low, while empty weight is a high proportion of take-off weight. In addition, airworthiness requirements make a big impact on fuel weight, and the fuel breakdown is shown in Fig.16.1.1. and it is shown that for a 230 nautical mile range, as much fuel is required to meet the requirements for stand-off, diversion, and contingency as is actually used en-route. An unforeseen increase of 600 lb in the basic weight would completely destroy this airplane's payload-range curve. As range increases, drag and fuel consumption have an increasing influence on aircraft weight. Consequently, in assessing the relative merits of alternative power plants it is necessary to compare the weight of power plant plus fuel for a given operation. On the Concorde supersonic aircraft, for example, weight of fuel reserves alone actually exceeds the weight of the payload.

The purpose of this chapter is to present a perspective of aircraft weight prediction and control; covering both preliminary design and project phases of airplane design. The relationships between airplane weight and performance will be outlined to give a better understanding of the important consequences of actual weight of an airplane and tight weight control to the success of an airplane program.

Weight Engineering

Weight prediction is, simply, the engineering task of accurately predicting the weight of an airplane, well in advance (usually several years) of the time the actual weight can be determined by placing the real airplanes on scales.

It is a job that begins on the day that a management decision is made to proceed with exploratory and preliminary design studies of a new airplane with many facets, among which are the following considerations:

• Of the several possible wing designs that fulfill

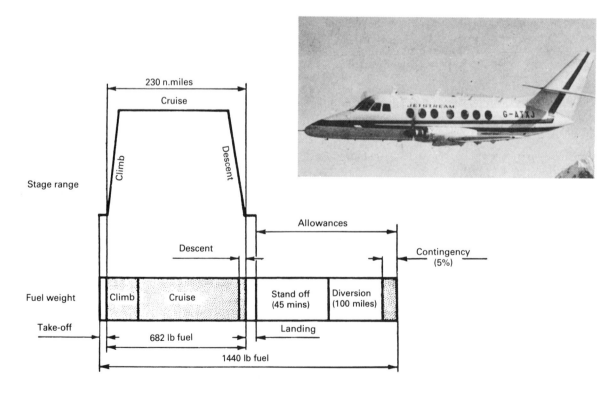

Fig. 16.1.1 Fuel breakdown vs. flight profile — Jetstream airplane.

582 Airframe Structural Design

equal aerodynamic performance requirements, which one is the lightest?
- Which wing results in the lightest airplane to do the mission?
- Which fuselage? Single or double deck?
- Where should the wing be located to make the airplane balance for all the fuel and payload loading conditions?
- How much does the c.g. shift as fuel is being used?
- What is the airplane weight effect if the wing airfoil thickness is reduced from 10% to 8%?
- How much lighter will the airframe be if Material B is used instead of Material A?
- What if the fuel tank and rib location are changed in the wing?
- Should the landing gear be cantilevered from the wing box or should a separate beam be used to support the gear trunnion (Fig. 12.2.10)?
- If the fuselage cross-section is not quite circular for aerodynamic drag reasons, how much more will it weigh than a circular one?
- How much do the systems and equipment weigh, which comprise about half of the empty weight of the airplane?

At every stage in the design, it is the responsibility of the weight engineer to predict accurately what that airplane would weigh if it became real hardware on the scales.

Importance of Weight Prediction
Most performance characteristics are adversely affected by increasing weight, and improved by decreasing weight. In a like manner, the cost of making an airplane increase as it gets heavier.

Weight data is a tool used by engineering management as well as potential customers in arriving at good design decisions and product choices; therefore, timely and accurate evaluation of weight is a necessary part of every design decision as shown in Fig. 16.1.2.

In most cases, it is the total mass properties characteristics of a system, not just component weights, that are of interest in configuration development. For example, a heavy tail coupled into an airplane with an aft center of gravity problem is especially undesirable.

In the case of an all moving tail surface a heavier tail may be more desirable than a lighter versions, if the inertia of the heavier tail is less. Lower inertia may allow a significantly reduced weight of hydraulic and flight control system due to reduced flutter stiffness requirements. Thus, in the total picture, the heavier, low-inertia tail configuration is lighter. Therefore, the search for factors that affect the total airplane system weight has to include not only the first order effects of a design decision, but also the second and third order effects.

"Weight control" can be described as the process by which the lightest possible airplane is derived within the constraints of the governing design criteria. Entire on reasonable and sound weight estimates can go "sour" in the design phase if the various staff and design specialists are allowed to forget the overall objective of compromise design.

Airplane design is a series of compromises. Every design alternative has a weight effect. Good knowledge of these weight effects is a necessary ingredient in making good design decisions. Good design decisions do not necessarily result in minimum weight.

- $R = k \dfrac{L}{D} \cdot \dfrac{V}{C} \log \dfrac{W_1}{W_2}$ (Breguet equation)
- Take-off field length $= f\left(\dfrac{T}{W} C_L \dfrac{W}{S}\right)$
- Landing field length $= f\left(\dfrac{W}{S} C_L\right)$
- Product costs $= f(W + \text{complexity})$
- Operating costs $= f(W + \text{performance})$
- Loadability = C.G. range = user flexibility

where:
k — Constant
R — Range in nautical miles
L — Lift; D — Drag
V — Velocity (knot)
C — Specific fuel consumption; lb fuel/lb-thrust-hour
W_1 — Initial weight; W_2 — Final weight
W — Weight of the airplane
T — Engine thrust
C_L — Lift coefficient
S — Total planform area of wing

Fig. 16.1.2 The role of weight prediction and control in airplane design.

Airplane Weight and Center of Gravity
Weight and center of gravity prediction of an aircaft serves a multitude of purposes, the more important of which are listed below:
- Design weight and center of gravity data must be provided to other engineering groups. Data is required for analysis of flutter and dynamic problems, by the Basic Loads and Stress Groups for calculation of structural loads and stresses, and by the Aerodynamics Group for calculation of performance and control characteristics.
- The Manufacturing Organization must be furnished weight and center of gravity data for use in designing assembly fixtures and for safe and efficient handling of large components.
- Engineering management must be informed at all times of the weight and center of gravity represented by the current airplane design configuration. It is only with this information constantly at hand that the design effort can be guided to meet guaranteed payload and performance capabilities.
- Technical data must be forwarded to the customer in accordance with contractual requirements. Data submitted during the design stage is needed by the customer for evaluation of contractor progress in meeting weight and center of gravity commitments. Final data submitted is compiled with similar data from other manufacturers to form a basis for the analysis of future airplane proposals.
- Weight data must be compiled for each airplane produced in order to improve the accuracy of weight estimation methods and to provide the only data base for the estimation of future airplanes.

- Component weights are needed by the Price Estimating Department. A weight/cost relationship, derived from previous airplane designs, is often utilized in conjunction with other data in establishing the airplane price.

Other Aspects

The more important aspects of weight engineering are listed below:
- Accurately predict weight and balance characteristics of all airplanes from the time of original conception to time of delivery and recommend corrective action for unfavorable trends.
- Determine weight for proposed products consistent with mission requirements and assist in establishing guaranteed weights.
- Promote lightweight design practices throughout the life of the product in order to maintain the most saleable product possible.
- Cooperate with Project Engineering, Staff Engineering, Equipment and Standards Engineering and other organizations to achieve weight control.
- Weigh aircraft for verification of predicted data and to ensure meeting all flight safety criteria.
- Ensure that all weight and balance contractual commitments to customers and certifying agencies are satisfactorily met.

Weight Terminology

The purpose of this section is to acquaint personnel with some of the more common descriptive terminology (Fig. 16.1.3) and symbols they will encounter in the process of working in the weight group. It should be noted that there are limitations and pit-falls concerning terminology. Some of these areas will be emphasized later in the discussion. The main point is that the exact meaning should be known and understood if it is to be correctly applied. However, there are particular terms that are used by just about everyone.

Manufacturer's Empty Weight (MEW) — is the weight of the airframe structure, powerplant installation and fixed equipment. It is essentially a "dry" weight, excluding unusable fuel and oil, anti-icing fluid, portable water and chemicals in toilets.

Basic Empty Weight (BEW) — is composed of airframe structure weight, propulsion weight and the weight of airframe equipment and standard items.

Operational Empty Weight (OEW) — is the weight of the airplane without payload and fuel.

Maximum Zero Fuel Weight (MZFW) — is the OEW plus payload. (MZFW is used to denote the airplane weight less fuel in the wing). This is important in the case of airplanes with fuel tanks in the wing at positive load factors, as there is no relieving load due to the fuel mass which produces the the maximum wing load. The MZFW must be sufficiently high to ensure that this structural limit will not create a payload restriction for the usual payload and loading conditions. An increment in MZFW increases the loading flexibility at the cost of a structure weight penalty, mainly in the wing and wing-support structure. Usually this also entails an increase in the MLW.

Maximum Landing Weight (MLW) — is the maximum weight authorized at landing. It generally depends on the landing gear strength or the landing impact loads on certain parts of the wing structure. The MLW should always be higher than the MZFW

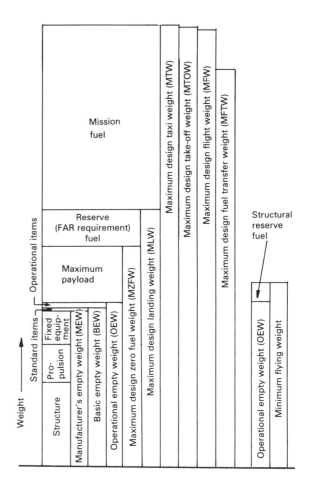

Fig. 16.1.3 Weight groups and characteristic weight terminology.

plus the regular fuel reserve, otherwise the payload will frequently be limited by the MLW. For certain categories of light airplanes the MLW and the MTOW are equal; for others the MLW may be slightly lower, namely down to 95% MTOW. This is of particular concern to operators flying short route sectors without refueling. For short-haul aircraft this type of flight execution is common practice and a fairly high MLW is desirable, namely of the order of 90—95% of the MTOW. A fuel jettisoning system is required on most multi-engine aircraft unless the airplane meets certain landing and approach climb requirements at a weight equal to the MTOW minus the fuel consumed during a specified 15-minute flight. This weight designs the gears and support structure in the landing descent mode based usually on the sink rate of the airplane, the flaps and support structure, portions of the wing, the horizontal tail, and the aft body for certain critical payload combinations. The difference between maximum landing weight and maximum zero fuel weight must be at least equal to the undumpable fuel weight, otherwise an emergency landing might have to be made above the maximum landing weight.

Maximum Taxi Weight (MTW) — This MTW is usually critical for wing down bending cases of dynamic taxiing, especially for those of wing mount powerplants or tip fuel tanks. This weight designs the gears and their support structure both in the static mode and ground turn mode.

Maximum Take-off Weight (MTOW) — is the maximum weight authorized at take-off brake release. It is frequently fixed by structural requirements in the take-off and occasionally by the maximum en route weight. (The MTOW is generally determined by structural design, and occasionally by ground handling criteria). In the pre-design stage the MTOW is calculated by adding together the OEW, the estimated normal payload and the total fuel load required at the design range.

Maximum Transfer Weight — This is the highest weight at which the outboard reserve tank fuel may be transferred to the outboard main tank thus removing the structural relief needed at higher weights (see Fig. 8.8.4). This weight limit, of course, only applies to those airplanes having a reserve tank. The selected weight is the maximum attainable when the wing is designed to meet the other critical design parameters; it is not usually a design parameter in itself. However, if the reserve tank is required to be empty at maximum landing weight, because of fatigue or dutch roll considerations, then it may well become a design factor.

16.2 Weight Prediction

Weight engineering consists of four distinct phases: preliminary design, project, flight and research.

(1) Preliminary Design Stage

The preliminary design phase of aeronautical engineering is defined as the phase in which a new aircraft is conceived. It is during this phase that size, arrangement, and propulsion characteristics as well as mission requirements and flight performance are established.

Preliminary estimation of weight is one of the first steps in the evolution of a new design. The preliminary weight, available power, and required performance are the basic parameters which determine the size and general configuration of the aircraft.

This initial estimate depends primarily on empirical equations derived from analysis of existing aircraft. These equations determine the first approximate gross weight for the aerodynamicist. As preliminary design progresses, it should proceed to refine the weight estimate to the extent that time, skill, and the definition of the aircraft permit.

Since the final size and cost of the aircraft are strongly influenced by the estimated weight at this stage, accuracy is of importance. In a typical aircraft one pound of excess empty airplane weight may result in an addition of ten pounds to the gross weight of the aircraft requires to maintain the same performance. In missiles and space vehicles the addition is often even more. If the weight estimate is too low, it may be impossible to achieve the weight guaranteed to the customer — with interesting economic consequences.

(2) Project Stage

When the preliminary design effort results in a decision to proceed with production, design is referred to as the project stage. To produce a competitive airplane, performance and payload capabilities must be maintained at the highest possible level consistent with operational requirements. Strictly speaking, it is these "airplane weight" which are sold to the customer. Since for a fixed gross weight each additional pound in the basic empty airplane weight results in either a decrease in payload or a penalty in performance, therefore structure and system weights must be held to a minimum. "Weight Control" engineering is devoted to this end.

It is apparent that creation of an efficient design is a compromise between the aerodynamicist, designer, structure engineer, and other specialists who must all have as a goal the minimum weight design that will meet their particular requirements to the task of maintaining the highest possible performance and payload capabilities of the airplane. This is a continuous task from airplane conception through final delivery.

(3) Flight Stage

Flight weight engineering deals with "operational" weight and balance control. This specialized group is responsible for the weight and balance control of the airplane from the time it leaves the assembly manufacturing area until it is delivered to the customer. Functions of the group may include weighing the aircraft, providing weight and balance control during the engineering and production flight test periods, and preparing an aircraft weight and balance report for delivery with each airplane.

Each airplane is weighed to establish an accurate base for weight and balance control in the flight test stage. In establishing the basic weight and center of gravity of each airplane a complete equipment inventory is conducted. Subsequent changes are closely monitored and the airplanes are reweighed as necessary to ensure accuracy.

(4) Research (Methods and Data)

Weight research and development may be defined as "effort expended in providing the weight engineer the tools, data and methods to accurately predict the weight effects of design decisions".

The research and development function is always performed to a certain extent by the various preliminary design, project, and flight weight groups. Typical duties of such a group are presented below:

- Obtain weight data on existing airplanes and understand it and why it occurred.
- Maintain a library of the latest materials, procedures, techniques of design, etc., for use by weight personnel.
- Generate and study new ideas for weight reduction in cooperation with other technologies.
- Establish new methods of weight estimation and constantly improve existing methods.
- Develop and coordinate training programs for weight personnel.
- Develop methods that effectively utilize computer to get better and/or quicker answers.

Airplane Major Weight Systems

The major weight systems are airframe, airplane

systems, mission systems and useful load:
- Airframe
  - Wing
  - Fuselage
  - Empennage
  - Landing gear
- Airplane systems/equipment
  - Propulsion
  - Fuel
  - Flight controls
  - Electrical
  - Hydraulic
  - Pneumatic
  - Flight deck
- Mission systems/equipment
  - Passenger accommodations
  - Cargo handling
  - Air conditioning
  - Avionics
  - Ordnance
  - Operational items
- Useful load
  - Fuel
  - Payload

(1) The airframe is the basic strength and pressure hull. It is the part of the airplane that is particularly responsive to structural considerations, such as strength, stiffness, and fatigue. It is the area where weight engineers have tended to focus their attention.

(2) The airplane systems and equipment grouping is intended to complete, when added to the airframe, an airplane as a flying machine. It includes:
- Propulsion and fuel support and controls
- Flight controls and control surfaces (elevator, rudder, aileron, flaps, etc.)
- Landing gear and controls
- Mechanical, electrical, hydraulic and pneumatic power supplies
- Flight deck

(3) The mission systems and equipment grouping includes those in and on the airplane that are there because of the use or mission of the airplane.

(4) The useful load is composed of fuel and payload. The first three groups, taken together, represent the Operating Empty Weight (OEW).

Weight Estimating Methodology

Preliminary estimation of weight is one of the first steps in the evolution of a new airplane design. The preliminary weight, available power, and required performance are the basic parameters which determine the size and general configuration of the airplane.

Most airplane manufacturers develop their own methods of empty weight prediction, usually basing them on extensive experience with a limited and well-defined category of airplanes. The more generalized and simpler methods presented in current reports have a predictive accuracy of the order of 2–5% error for the major structural groups.

Although a wealth of detailed published information on weight engineering can be found in various technical papers, consistent and up-to-date methods for calculating the empty weight of the various categories of modern airplanes are very limited indeed. This is due to the high proprietary value placed in such methods in the highly competitive aerospace industry. The information contained in Fig. 16.2.1. may

Airplane Group	MTOW (kips)	Wing (%)	Fuselage (%)	Tail (%)	Edges (%)	Landing Gear (%)	Nacelle Pylon (%)
Light single propeller engine	1.5–3	9.0–11.0	7.0–11.0	2.0–2.8	1.1–2.0	4.5–7.0	1.1–2.1
Light twin propeller engines	5.0–10	9.0–11.0	7.0–8.5	1.9–2.5	1.2–1.8	4.5–6.5	3.5–4.5
Executive jet	15–30	9.0–10.0	7.7–12.0	1.8–3.0	1.0–2.5	3.0–4.4	1.8–2.6
Turbopropeller 2 to 4 engines	13–45	7.5–10.0	7.0–14.0	1.7–3.0	0.7–1.8	3.5–5.0	1.0–4.1
Jet transports 2 engines	41–300	10.0–14.0	10.0–13.0	1.8–2.8	1.4–2.3	3.3–4.5	1.2–2.3
Jet transports 3 engines	110–450	9.0–11.0	10.0–12.0	2.1–2.8	1.3–1.9	3.8–4.7	1.5–2.4
Jet transports 4 engines	300–850	8.5–12.0	6.5–10.0	1.5–2.5	0.7–1.0	3.4–4.2	1.3–2.2

Fig. 16.2.1 Aircraft weight breakdown in percentage of maximum take-off weight (MTOW).

be helpful for checking the accuracy of the methods presented in this chapter by comparing the results with data on several existing airplanes.

Preliminary aircraft design weight prediction is always a mixture of rational analysis and statistical methods. Statistical weight equations for many components are usually written in the exponential form which will be seen in weight prediction.

The selected preliminary weight prediction equations presents in this chapter include the wing, tail, fuselage, landing gear, and engine nacelle and mounting only. Refer to the related data, such as the references of this chapter, for other weight prediction methods for other components.

(1) Wing Weight Estimation

A reasonably accurate wing weight estimate can be made in preliminary design as the loads on the wing are fairly well known at the design stage. Usually the bending moment in flight is assumed to be decisive for most of the primary structure. For a certain category of high-speed aircraft, however, torsional stiffness requirements may become dominant, and the extra structural weight required to safeguard against flutter may amount to as much as 20% of the wing weight. The location of the inertia axis of the wing plus wing-mounted engines is important. A fairly large portion of the wing weight is also made up of secondary structures and non-optimum penalties, such as joints, non-tapered skin, landing gears support structure, etc.

(a) Transport Aircraft

Eq. (16.2.1.) represents a statistical wing weight equation that represents essentially transport aircraft category comprising turbo-prop, turbo-jet, and turbo-fan powered configurations in the medium to high speed ranges and medium to high wing loadings. Using this as a basis, increments are applied for the various high-life concepts.

$$W_w \text{(lb)} = 0.81 \, CWI \left(\frac{NW_G S_W}{TCE} \right)^{0.6} \times$$

$$\left(1 - \frac{W_R}{W_G} \right)^{0.4} \times$$

$$(1 + TR)^{0.4} \frac{(AR)^{0.5}}{[(TS)^{0.2} (\cos Q)^{1.2}]} +$$

$$3.3 \, S_p + 3.28 \, (S_{LED})^{1.13} \quad (16.2.1)$$

where

CWI = 1.0 for MLG (Main Landing Gear) in wing;
= 0.95 for MLG in fuselage
N = Ultimate load factor
W_G = Design gross weight (lb)
S_W = Wing area (ft²)
W_R = Wing fuel + wing engines + external tanks (lb)
TR = Taper ratio = $\left(\frac{C_T}{C_R} \right)$

AR = Aspect ratio
TS = Design tension stress (psi)
Q = 50% Chord sweep angle
S_p = Spoiler area (ft²)
S_{LED} = Leading edge device area (ft²)

TCE = Effective wing thickness ratio $\frac{T}{C}$ (%)
$= \frac{4}{5} [TCR\,(\%) + TCT\,(\%)]$

where

$TCR\,(\%)$ = Wing thickness ratio at root in %
$TCT\,(\%)$ = Wing thickness ratio at tip in %

(b) Light Utility Aircraft

Eq. (16.2.1) will predict unrealistic weight for light utility aircraft. The following equations are recommended for the low to moderate performance light utility aircraft.

$$W_W \text{(lb)} = 96.948 \left[\left(\frac{W_{TO} N}{10^5} \right)^{0.65} \times \right.$$

$$\left(\frac{AR}{\cos \Lambda_{1/4}} \right)^{0.57} \left(\frac{S_W}{100} \right)^{0.61} \times$$

$$\left. \left(\frac{1 + \lambda}{2 t/c} \right)^{0.36} \left(1 + \frac{V_e}{500} \right)^{0.5} \right]^{0.993}$$

(16.2.2)

where

W_{TO} = Take-off weight (lb)
N = Ultimate load factor
AR = Aspect ratio
$\Lambda_{1/4}$ = Wing quarter chord sweep angle
S_W = Wing area (ft²)
λ = Taper ratio
t/c = Wing thickness ratio
V_e = Equivalent maximum airspeed at sea level (knot)

(c) Fighters (Ref. 1.4):
- Air force:

$$W_W \text{(lb)} = 3.08 \left(\frac{K_{PIV} N W_{TO}}{t/c} \times \right.$$

$$\left\{ \left[\text{Tan } \Lambda_{LE} - \frac{2(1-\lambda)}{AR(1+\lambda)} \right]^2 + \right.$$

$$\left. \left. 1.0 \right\} \times 10^{-6} \right)^{0.593} \times$$

$$[(1 + \lambda) AR]^{0.89} (S_W)^{0.741}$$

(16.2.3)

- Navy:

$$W_W \text{(lb)} = 19.29 \left(\frac{K_{PIV} N W_{TO}}{t/c} \times \right.$$

$$\left\{ \left[\text{Tan } \Lambda_{LE} - \frac{2(1-\lambda)}{AR(1+\lambda)} \right]^2 + \right.$$

$$\left.1.0\right\} \times 10^{-6}\right)^{0.464} \times$$

$$[(1 + \lambda) AR]^{0.70} (S_W)^{0.58} \quad (16.2.4)$$

where
- K_{PIV} Wing variable sweep structural factor
  = 1.00 fixed wings
  = 1.175 variable swept wings
- t/c = Wing thickness ratio
- W_{TO} = Take-off weight (lb)
- Λ_{LE} = Leading edge sweep angle
- λ = Taper ratio
- AR = Aspect ratio
- S_W = Wing area (ft^2)
- N = Ultimate load factor;
  11 for fighter aircraft (based on a design limit load factor of 7.33);
  4.5 for bomber and transport aircraft (based on a design limit load factor of 3.0)

(2) **Tail Weight Estimation**

The tail weight is only a small part of the MTOW (about 2%) but on account of its remote location it has an appreciable effect on the position of the airplane c.g. Accurate weight prediction is difficult due to the wide variety of tailplane configurations and the limited knowledge. If the tailplane (both horizontal and vertical stabilizers) area is not yet known, the total tailplane weight may be assumed about 4% of the OEW.

(a) Transport and Executive Jets (see Fig. 16.2.2) (Ref. 1.3)

$$W_h = k_h S_h \left[\text{function of} \left(\frac{S_h^{0.2} V_D}{\sqrt{\cos \Lambda_h}} \right) \right]$$
for horizontal tail (16.2.5)

$$W_v = k_v S_v \left[\text{function of} \left(\frac{S_v^{0.2} V_D}{\sqrt{\cos \Lambda_v}} \right) \right]$$
for vertical tail (16.2.6)

where
- V_D = expressed in terms of equivalent air speed (KEAS).
- k_h = 1.0 for fixed stabilizer; 1.1 for variable incidence tails (flying-horizontal stabilizer; differential tail-plane or taileron).
- k_v = 1.0 for fuselage-mounted horizontal tailplane; $1.0 + 0.15 \left(\frac{S_h b_h}{S_v b_v} \right)$ for fin-stabilizers such as T-tail or cruciform tailplane.
- h and v = sweepback angle of horizontal and vertical tail respectively.
- S_h and S_v = gross planform area of horizontal and vertical tail respectively
- b_h and b_v = span of horizontal and vertical tail respecitvely (see Fig. 16.2.3 for definition of b_v)

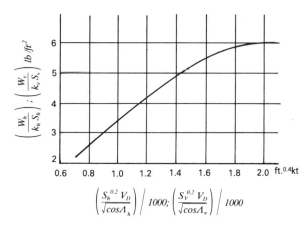

Fig. 16.2.2 *Curve for estimating horizontal and vertical tail weight (Ref. 1.3).*

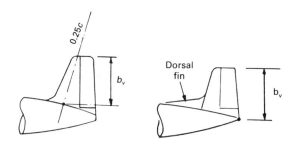

Fig. 16.2.3 *Definition of vertical tailplanes span (b_v).*

(b) Fighters (Ref. 1.4):

The following equations are the same ones as for the wing weight relationship; the constants and exponents are, however, different.

- Air force:

$$W_t \text{(lb)} = 5.4 \left(\frac{K_{PIV} N W_{TO}}{t/c} \left\{ \left[\text{Tan } \Lambda_{LE} - \frac{2(1 - \lambda)}{AR(1 + \lambda)} \right]^2 + 1.0 \right\} \times 10^{-6} \right)^{0.455} \times [(1 + \lambda) AR]^{0.683} \times (S_w K_{TAIL})^{0.569} \quad (16.2.7)$$

- Navy:

$$W_t \text{(lb)} = 9.56 \left(\frac{K_{PIV} N W_{TO}}{t/c} \left\{ \left[\text{Tan } \Lambda_{LE} - \frac{2(1 - \lambda)}{AR(1 + \lambda)} \right]^2 + 1.0 \right\} \times 10^{-6} \right)^{0.427} \times [(1 + \lambda) AR]^{0.611} \times (S_w K_{TAIL})^{0.534} \quad (16.2.8)$$

where
- W_t = Total weight (lb) of both horizontal and vertical tail.
- K_{TAIL} = Between 0.3 and 0.6; use 0.4 for fixed wing; 0.53 for variable swept wing
  For other definitions, see Eq. (16.2.3) and Eq. (16.2.4).

(3) **Fuselage Weight Estimation**

The fuselage includes the complete shell, excluding the nose radome, but including windshield, windows, doors (including main landing gear doors), flooring and supports, cargo compartment lining and provisions for powered container handling system. The fuselage makes a large contribution to the structural weight, but it is much more difficult to predict by a generalized method than the wing weight. The reason is that the large number of local weight penalties in the form of floors, cutouts, attachment and support structures, bulkheads, doors, windows and other special structural features.

(a) Jet transports:
$$W_f (\text{lb}) = k_1 k_2 \{2446.4 \, [0.5 \, (W_g + W_L)]^{0.3} \times$$
$$\left(1 + \frac{1.5 \, \Delta P}{4}\right)^{0.5} S_g^{0.325} \times$$
$$[0.5 \, (H_f + B_f)]^{0.5} \, L^{0.6} \times 10^{-4} - 678\}$$
(16.2.9)

where
- L = Total fuselage length (ft)
- W_g = Design flight gross weight (lb)
- W_L = Design landing weight (lb)
- H_f = Maximum fuselage depth (ft)
- B_f = maximum fuselage width (ft)
- ΔP = Design cabin pressure differential (psi)
- S_g = Total fuselage wetted area, excluding wing and tail intersections (ft²)
- k_1 = 1.05 for fuselage mounted main landing gear; 1.0 for wing mounted main landing gear;
- k_2 = 1.1 for fuselage mounted engines; 1.0 for wing mounted engines

Note: W_f (lb) includes weight of baggage compartment lining and provisions

(b) Fighters (Ref. 1.4):
- Air force:
$$W_f (\text{lb}) = 10.43 \, (K_{INL})^{1.42} \, (q \times 10^{-2})^{0.283}$$
$$\times (W_{TO} \times 10^{-3})^{0.95} \left(\frac{L}{H_f}\right)^{0.71}$$
(16.2.10)

- Navy:
$$W_f (\text{lb}) = 11.03 \, (K_{INL})^{1.23} \, (q \times 10^{-2})^{0.245}$$
$$\times (W_{TO} \times 10^{-3})^{0.98} \left(\frac{L}{H_f}\right)^{0.61}$$
(16.2.11)

where

- q = Maximum aerodynamic pressure (lb/ft²)
- L = Fuselage length (ft)
- H_f = Maximum fuselage height (ft)
- K_{INL} = 1.25 for inlets on fuselage
  = 1.0 for inlets in wing root or elsewhere
- W_{TO} = Design take-off weight (lb)
- $\frac{L}{H_f}$ = 8.0 to 12.0 for the air force
  = 6.0 to 11.0 for the navy

(4) **Landing Gear Weight Estimation**

The weight of conventional landing gears may be found by summation of the main gear and the nose gear, each predicted separately with the following equation:

(a) for Transport and Executive Jets (Ref. 1.3)
$$W_{1g} (\text{lb}) = k_{1g} [k + k_1 (W_{TO})^{\frac{3}{4}} + k_2 W_{TO} + k_3 (W_{TO})^{\frac{3}{2}}]$$
(16.2.12)

where
- W_{1g} = Weight (lb)
- k_{1g} = 1.0 for low-wing airplanes
  1.08 for high-wing airplanes
- W_{TO} = Design take-off weight (lb)

For values of k, k_1, k_2 and k_3, refer to Fig. 16.2.4.

Gear Configuration	k	k_1	k_2	k_3
Fixed nose gear	25	0	0.0024	0
Fixed main gear	40	0.1	0.019	0
Retractable nose gear	20	0.1	0	2.0×10^{-6}
Retractable main gear	40	0.16	0.019	15×10^{-6}

Fig. 16.2.4 *Values for the calculation of the transport landing gear weight.*

In many transport aircraft the critical load is formed by the landing impact load and the MLW should therefore be used to predict the landing gear weight. A reasonable approximation for the weight of retractable landing gears is 5% of the MLW.

(b) Fighters (Ref. 1.4):
- Air force:
$$W_{1g} (\text{lb}) = 62.21 \, (W_{TO} \times 10^{-3})^{0.84}$$
(16.2.13)

- Navy:
$$W_{1g} (\text{lb}) = 129.1 \, (W_{TO} \times 10^{-3})^{0.66}$$
(16.2.14)

Note: W_{1g} includes landing gear controls; W_{TO} is design take-off weight (lb)

Airframe Structural Design

(5) Engine Nacelle and Mounting Weight Estimation
The following statistical data may be used:
  (a) For aircraft with turboprop engines:
    $W_n = 0.14$ lb per take-off ESHP (Equivalent Shaft Horsepower). Add 0.04 lb per ESHP if the main landing gear is retractable into the nacelle and 0.11 lb per ESHP for overwing exhausts (i.e. Lockheed Electra).
  (b) For aircraft with pod-mounted turbojet or turbofan engines:
$$W_n = 0.055\ T_{TO} \qquad (16.2.15)$$
    where T_{TO} = maximum thrust at airplane take-off
  (c) For aircraft with high by-pass turbofans with short fan duct:
$$W_n = 0.065\ T_{TO} \qquad (16.2.16)$$
    The above value includes the pylon weight and extended nacelle structure for a thrust reverser installation. In the absence of thrust reverser a reduction of 10% is assumed. For a typical "quiet" turbofan pod, acoustic lining may be required over the nacelle area. A typical weight penalty is 20% of the nacelle weight, apart from the extra weight of the engine itself.

Engineering depends upon both data and methods being available. Weight prediction is no exception to this rule, and improvements in both are essential if improved weight predictions are desired.

Because weight engineering embraces every structural component and every system of the airplane, several engineering curriculums provide appropriate foundations.

To effectively carry out assigned functions, the weight engineer must:
- Be able to utilize empirical and analytical equations, structural analysis techniques, and computer equipment.
- Possess extensive knowledge of materials and how they can be employed to the best weight advantage.
- Know the effects of environment (heat, sonic, etc.) on a part or the whole.
- Be well versed in cost and ease of maintenance.
- Have a working knowledge of systems design and production techniques.
- Have a working knowledge of electronics, mechanical and hydraulic systems, and propulsion systems.
- Possess a thorough knowledge of design pratices and airplane design.
- Possess a thorough knowledge of methods and equipment for experimental determination of weight, center of gravity and moment of inertia.
- Remain abreast of the latest developments in the above areas.

16.3 Performance and Configuration Influences

Aerodynamic

In evaluating parameter changes the weight effect of changing a single design variable in one component on other airplane components is often desired. In other words the total effect on OEW must be considered at all times for any single parameter.

Take wing sweep for example. In addition to the obvious parameters that vary with sweep, such as aeroelastic effects, and changes in longitudinal stability (pitch-up) etc., a secondary effect is that for low wing configuration as sweep decreases it becomes more difficult to support the landing gear in the wing. This results in an increase in the weight of the landing gear support structure.

Also if we assume a constant amount of c.g. travel required for payload, as sweep increases the amount of c.g. travel required for fuel increases. If the same loadability is maintained, a greater amount of c.g. travel is required which in turn requires a larger tail resulting more weight and drag and in turn requires increased gross weight to perform the same mission, etc.

Another factor that is related to wing sweep and wing area is weight of high-lift systems. For some jet transports high-lift systems run as much as six percent of manufacturers' empty weight and 15 to 20 percent of total wing weight. As sweep and speed increase, wing loading has tended to increase, requiring higher C_{Lmax} to maintain reasonable approach speed. In addition to being a large part of wing weight, high-lift systems may add to primary wing loads and may affect tail size required. It may also affect fuselage design loads.

Dead Weight Distribution

The objective is to distribute the variable dead weight load, i.e., fuel and payload in such a manner as to establish the most critical shears, moments and torsion the airplane will experience when combined with appropriate load factors, air loads, tail loads and factors of safety.

This is done by dividing the airplane into panels in the following manner:
- A panel load OEW is developed which takes cognizance of such things as interior arrangement variables, galley weights and location, customer options and weight growth. This is necessary since for obvious practical reasons, the OEW panels are continuously varied or it would be impossible to develop a single set of loads for the airplane.
- In distributing both payload and fuel, the weight engineer must keep in mind operational restrictions that may result from the manner in which these items are or can be distributed. For example, it may be possible to distribute fuel and payload in a manner that produces very favorable dead weight moments. This may in turn result in very complicated fuel management, passenger seat assignment and other operating restrictions. Therefore, dead weight distribution must be a compromise between favorable dead weight loads and ease of operation for the user.

Airplane Center of Gravity

It is concerned that in the end an airplane will have maximum utility for minimum weight and it begins with proper location of airplane center of gravity.

Increasing wing sweep increases the total amount of center of gravity travel required for fuel if the fuel is in the wing. Increasing the taper ratio on the other hand increases the amount of center of gravity travel required for fuel. Aspect ratio has little effect on the amount of center of gravity travel required for fuel, but does have a significant effect on fuel volume.

Lack of stability may penalize loadability. The main landing gear must be located far enough aft of the aft center of gravity limit to prevent upsetting the airplane during normal loading and unloading operations and to prevent excessive pitch-up during ground roll and the resulting loss in nose gear steering. Generally, an airplane with a tricycle gear will have sufficient stability if the centerline of the main landing gear is approximately 10 to 15 percent of total c.g. travel aft of the aft center of gravity limit.

Material Selection

One important consideration often used in structural material selection is the ratio of strength to density. However, many factors other than strength enter into material selection or a particular airframe application (Fig. 16.3.1). Advanced materials such as composite materials (see Chapter 14.0) offer potential for structure weight reduction of future airplanes. However, their uses require joint considerations of structure engineer, manufacturing, finance and project personnel before a reliable weight estimate of their use can be evaluated.

Property	Titanium	Stainless Steel	Aluminum
Density (lb/in^3)	0.16	0.29	0.10
Young's modulus (psi)	16×10^6	30×10^6	11×10^6
Thermal expansion (in/in/°F)	5.3×10^{-6}	7×10^{-6}	12×10^{-6}
Thermal conductivity (BTU/hr-ft-°F)	5	10	73
Specific heat (BTU/lb-°F)	0.13	0.11	0.23

Fig. 16.3.1 Properties of materials for airframe design.

Structural Fatigue Performance

In addition to ultimate static strength requirements, most airplanes have some service life requirements. These service life requirements can have considerable weight effect. It will in many instances limit the choice of materials which can be used to perform a given function and can also determine the allowables for a given material. In most cases this allowable is expressed in terms of a maximum 1.0g stress level. Some of the considerations which effect weight are 1.0g tension stresses for wing lower surfaces. Sometimes ultimate design tensile stresses and skin hoop tension stresses are used for pressurized fuselage. Typical of another environment is the sound-pressure level considerations. Structure and equipment are subjected to high sonic and vibrational fatigue loads and are evaluated using past and present test data. Particular emphasis must be paid to secondary structure that in these areas the major portion of the structural maintenance occurs.

Product Growth

We are all familiar with inflation, as time goes by things become more expensive, and if we have fixed incomes we gradually experience a decrease in our capability to do the things we want to do.

A similiar phenomenon occurs with airplanes. After they are initially designed, they experience a weight growth which is the result of weight additions made by the manufacturer and the user.

The weight growth attributed to the manufacturer is usually the result of product improvement changes, cost reduction, fixes to design deficiencies, etc. The changes the customer adds are usually confined to interior additions, maintenance equipment, etc.

What does all this mean to the weight engineer? These weight increases can stimulate airplane growth programs that are intended to maintain or improve range payload capability. This continued weight growth also means constant monitoring of the effect of these weight increases on loadability, airplane design dead weight loads, and future delivery guarantee status.

16.4 Balance and Loadability

Payload, range, and operating cost receive the most attention, as they have a marked effect on airline profit. However, while the airplane is in service, payload and range are generally matched to the routes flown, while balance is a factor which the operator must continually assess.

If an airplane has restrictive loading procedures, or is constantly in need of last minute load adjustment, then the airplane is considered to have poor loading flexibility. Defining and predicting these situations in the early stage of an airplane design is quite difficult. Furthermore, an adverse situation is seldom catastrophic. It is not surprising, therefore, that balance is given only a cursory consideration by the design team and it is the weight engineer's responsibility to present management with clear and timely data whenever a balance problem exists.

Airplane loading flexibility is the degree of freedom afforded the airline with the useful load combinations that they actually encountered during daily operations. These combinations can be graphically shown on the balance grid and their compatability with the operational limits assessed (Fig. 16.4.1).

Operational limits are never greater and are generally less than the maximum structural limits. They are established by current airplane performance, weather, and airport conditions, and arbitrarily imposed by the airlines to simplify their loading schedule and their turn-around time. As the latter two items are often necessary to keep the airline competitive, it is most important that the maximum possible loading flexibility is obtained in all new and growth design configurations. Weight penalties, of course, must be kept to an acceptable level.

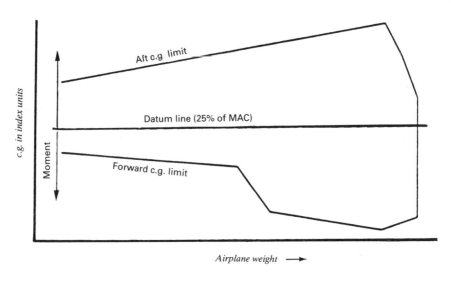

(a) Horizontal type

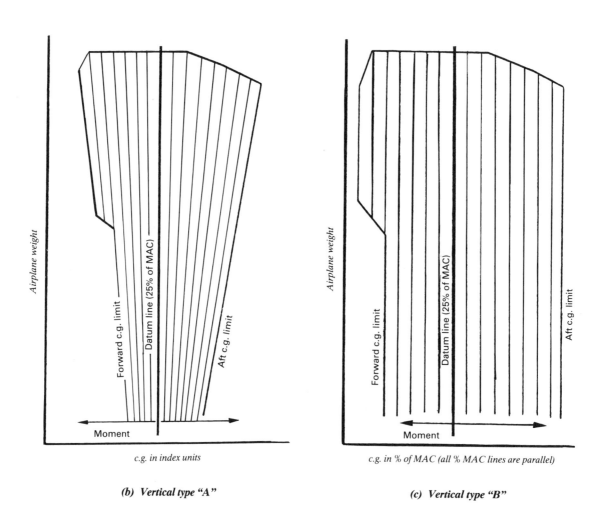

(b) Vertical type "A"

(c) Vertical type "B"

Fig. 16.4.1 Typical grid forms used for airplane balance analysis.

592 Airframe Structural Design

It is interesting to note that the balance grid is used during the whole life span of an airplane from concept to retirement. Therefore it is very important to understand the grid and its many uses. These uses may include, but are not limited to, the followng areas:
- Initial preliminary design — positioning of wings, engines, fuselage, tail, etc.
- Project stage — monitoring the design to ensure that the original balance predictions remain valid to ensure that new concepts do not curtail loading flexibility beyond acceptable limits.
- Fuel management — establishing the optimum fuel tank ends, and the optimum usage procedures which are often a trade-off between wing bending relief, flutter requirements and loading flexibility.
- Fuel slosh — determining the slosh effect on airplane center-of-gravity and recommending baffle rib placement.
- Load analysis — the balance grid is not only used as a work tool to investigate and select critical load points but it is also used for pictorial presentation of selected condition in the substantiation documents after the loading points have been mathematically checked for correctness.
- Passenger interiors — determining the effect of interior seat layouts, galley locations, and position of toilet complexes on loading flexibility.
- Balance presentations — presentations to both management and customer are usually given on balance grids. A whole host of useful load combinations can be shown clearly on just a few show cards. With the use of clear plastic overlays, it is often possible to run through new conditions during the meeting in which they are initially discussed rather than hold the subject over until a later time period.
- Finally, the balance grid appears in one form or another in most of the loading schedules which are used to establish the correct horizontal stabilizer trim for take-off, to ensure correct load distribution during all phases of the flight, and to assist in initial load planning of the airplane.

If the weight and moment coordinates (or c.g. in % of MAC) of a balance grid start at zero then the working area of the grid becomes far too small for practical use. This is clearly illustrated by the shaded area. Fig. 16.4.2. shows the effect of moving the datum to an arbitrary point midway between the fore and aft limits at approximately the wing quarter chord (25% of MAC). The moment scale is increased and the weight scale is also increased and cut off below OEW. The sole purpose of these changes is to gain the maximum working area for a given size of paper. Although zero weight is seldom shown it converges at the intersection of the moment datum line and zero weight. It is a wise precaution to check scales and datum for consistency when transferring plotted data from one grid to another or interchanging overlays. Plotted data or overlays are seldom transferable. Most balance grids and useful load overlays are clearly marked by a scale block.

However, grids used in certifying documents such as the flight manual and the weight and balance manual, detailed specifications, and loading schedule which are used by the customers may take a different form as shown in Fig. 16.4.1. The reason for the

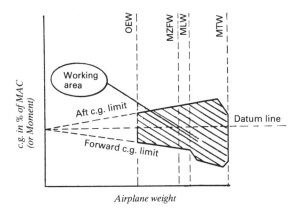

Fig. 16.4.2 Working area of the grid.

change from the Fig. 16.4.1(a) type grid to the Fig. 16.4.1(b) type grid is presumably the more natural feeling that is given by the latter when the weight scale goes up as weight increases and the airplane faces towards the left edge of the paper.

Mean Aerodynamic Chord

The coordinates of a balance grid are weight and moment. Weight can be readily assessed but moment is far more difficult to evaluate. First, the location of the zero datum can drastically affect the numerical value of a moment, and moments can be taken about any convenient datum provided the same one is used throughout the calculation. Second, the effect of a moment is dependent not only upon its own magnitude, but also upon the weight of the airplane to which it is applied and this is not reflected in the numerical value of a moment. For these reasons moment is considered a poor balance indicator and only acts as a necessary mathematical link in the balance calculation.

A more meaningful indicator is one that identifies the location of an imaginary fulcrum on which the airplane rests in static equilibrium. Now, if we apply a moment to the airplane, its effect can be expressed as the distance the imaginary fulcrum must move fore or aft to maintain equilibrium. Distance, like weight, is easier than moment to evaluate.

The scale used for fore and aft measurement is balance arm. The zero point is usually ahead or close to the airplane nose to keep most values positive. Unfortunately, balance arms have no direct relationship with the aerodynamic characteristics of an airplane, and therefore, require further modification. A more satisfactory scale is one showing percent of mean aerodynamic chord (MAC). The chord length (MAC) and its leading edge balance arm (LEMAC) are established by wing geometry and are closely related to aerodynamic characteristics. The percent MAC scale parallels the balance arm scale, is located in the general area of the wing, and can be plotted on the balance grid as a series of lines [see Fig. 16.4.1(c)]. Percent MAC and index unit have most of the qualities needed for a balance indicator and is the most common balance unit found on a balance grid. The relationship between percent MAC and balance arm is shown in Fig. 16.4.3.

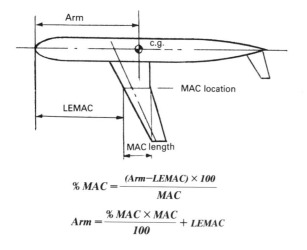

$$\% MAC = \frac{(Arm - LEMAC) \times 100}{MAC}$$

$$Arm = \frac{\% MAC \times MAC}{100} + LEMAC$$

Fig. 16.4.3 Airplane percent MAC vs. balance arm.

Center of Gravity

An airplane must be designed in such a way that good stability, control properties and adequate flexibility in loading conditions are obtained. By suitable arrangement of the design layout and acceptable tailplane size, acceptable fore-and-aft limits of the center of gravity must be established, taking into account the following aspects:
- Fore-and-aft position of the wing relative to the fuselage
- Provision of suitable locations for payload and fuel
- Design of the horizontal tailplane, the elevator, and the longitudinal flight control system
- Location of the landing gear legs.

The c.g. must be established in both the longitudinal and the vertical direction. The airplane designer must demonstrate that in all likely loading conditions the actual c.g. will remain inside the fore-and-aft limitations, without undue penalties in the form of loading restrictions.

Balance Grid Diagram Construction

Fig. 16.4.4. shows many of the limiting parameters that establish the envelope boundaries. A particular airplane model or configuration will have a weight and center-of-gravity envelope based on several of these parameters but certainly not all of them. Furthermore, both the reason given for a particular cutoff and its stated location is in general terms and is only given to assist the weight engineer in anticipating possible areas of critical balance.

Airplanes with fairly extensive growth histories usually possess a more complicated envelope. Improvements in payload and range with minimum change in weight, cost and complexity is often accepted by the airline at the expense of loading flexibility. This is manifested by the chopped corners of the envelope. Once again, the importance of achieving maximum loading flexibility during the initial design stage is clearly emphasized. Airplane sales, after the break-even point has been reached, often require the airplane to have a state-of-the-art improvement in payload and range, and continued good balance characteristics will make future sales much more attractive.

Fig. 16.4.4 illustrates some of the important areas which influence the modern commercial airplane design as listed below;

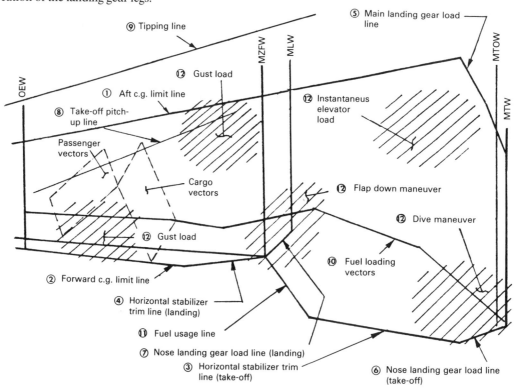

Fig. 16.4.4 Balance grid diagram vs. airplane compatability — transport.

594 Airframe Structural Design

① Aft c.g. limit line
This line may be a take-off and landing limit with gear and flaps down or a flight limit with gear and flaps up. Effect of gear and flap retraction, if appreciable, must be considered and the limit clearly defined for either take-off and landing flight, or possibly both. This however, is a secondary effect of airplane pitch stability and control requirements which generally establishes the aft c.g. limit line.

② Forward c.g. limit line
This line may be a take-off and landing limit with gear and flaps down or a flight limit with gear and flaps up. Structural areas of the airplane, which may be affected by this c.g. limit, is the aft fuselage section during a light weight gust condition. Gusts are particularly critical in this area of the grid as the airplane inertia is at a minimum and the point of rotation is well forward.

③ Horizontal stabilizer trim line (take-off)
Extreme forward c.g. at high airplane weight requires a high down tail load. This increases the wing and flap loads, increases drag, and requires a maximum airplane nose up horizontal stabilizer trim setting. These factors degrade takeoff performance considerably, inhibit rotation and usually establish the forward limit at high take-off weight. However, cutting back this forward limit is seldom possible as long range fuel is usually located in the center section tank. This is located well forward on a swept wing airplane and demands plenty of forward limit at high airplane weights. The use of cells or bays in the center tank is often due to the desire to keep the center tank fuel as far aft in the airplane as possible.

④ Horizontal stabilzer trim line (landing)
With flaps set for landing, the center of pressure on the wing moves inboard; and due to the wing sweep, it also moves forward. This tends to relieve the tail down load but this effect is more than offset by the very large nose down pitching moment caused by the flap surfaces. This nose down moment has to be counterbalanced by an increase in tail down load. The distribution of load is thus similar to that produced by the forward take-off c.g. limit at higher weights. However, the higher flap angle results in higher loads at landing restricting the most forward c.g. limit to a value well aft of the limit set for high take-off and flight weights.

⑤ Main landing gear load line
This will give several plot points on the static constant load line. However, in the ground turn condition the dynamic loads may be limiting. This will depend on the location of the airplane vertical c.g. Sometimes, if payload is added to gain additional weight, the vertical c.g. will rise sufficiently high to give a critical over turning load.

⑥ Nose landing gear load line (take-off)
As in the case of the main landing gear, this is a constant moment line, based on static loads. It may be limited by gear structure, support structure, or tires.

⑦ Nose landing gear load line (landing)
This is a line similar to the nose landing gear constant load line at take-off except that the higher loads obtained during touch down are offset by restricting the most forward c.g. limit to a value well aft of the c.g. limit set at the high take-off weights.

⑧ Take-off pitch-up line
Directional control during the initial take-off ground roll is dependent almost entirely on nose landing gear steering. At low speeds the vertical tail and rudder offer little help and the use of differential braking or engine settings is not desirable. Sufficient load must therefore be maintained on the nose landing gear to prevent scrubbing during nose gear steering by establishing the most aft ground c.g. limit that is acceptable for take-off. This limit is the take-off pitch up line and is seldom a balance problem as the airplane usually has sufficient fuel on board to keep it well forward of this line, and on landing normal braking assures an adequate nose gear load.

⑨ Tipping line
On tricycle landing geared airplanes, this line represents the balance arm of the main landing gear wheel centerline and is the most aft point the airplane c.g. can move to before the airplane, from a level and static condition, will tip on its tail. Numerous factors must be considered when setting a safety margin forward of this point. This safety margin may include, but is not limited to, the following items:
- Nose up ramp slope — A 3% slope (1.7 degrees) is considered to be a reasonable maximum for airport aprons and taxi ways.
- Wind — A 40 knot gust is usually considered a satisfactory limit.
- Snow load — Depths in excess of 6 inches are seldom permitted to accumulate; wet snow weighs approximately 8 lb per ft^3.
- Fuel slosh due to wing droop may be encountered on large airplanes which have swept flexible wings.
- Tolerance for unknown deviations from the loading and unloading assumptions used in OEW and payload balance calculations.
- Dynamic effect of towing, taxiing, and braking.

⑩ Fuel loading line
An arbitrary line which traces the fuel loading vector. It starts at the aft c.g. limit and maximum zero fuel weight and continues to a more forward c.g. at a higher weight, thus, cutting off part of the aft area of the envelope. If this deleted area is unaccessible, its removal will not affect loading flexibility and it does avoid the necessity to flight test the airplane in that part of the grid that is not going to be used. However, it should be noted that a subsequent increase in maximum zero fuel weight, which might otherwise require a simple revision to the structural analysis, would require additional flight test. It might have been far less costly to conduct the additional flight test during the initial program. This is why growth potential should always be considered even though, for economic reasons, it may not be provided for.

⑪ Fuel usage line
As in the case of the fuel loading line, it is an arbitrary line which follows an established fuel usage vector so that it does not limit loading

Airframe Structural Design 595

flexibility. It forms a transition line back to the most forward c.g. limit at maximum landing weight or maximum zero fuel weight.

⑫ Shaded areas
Specific load cases tend to fall into these general shaded areas. However, while one case may fall on the c.g. limit line at one location during the start of flight and not reach the c.g. limit line at the low fuel end, quite the reverse may be true for other critical cases. Fuel tank ends, lower cargo locations, passenger interior arrangements, all will affect the selection of the critical load conditions. These items affect the selection to such a degree that it is not possible to establish a set of simple ground rules. However, a general overall understanding of the major parameters are essential to establish the weight and center-of-gravity envelope which is to become a useful engineering tool and meaningful balance studies.

References

16.1 Anon.: *Publications and Technical Papers Index — Society of Aeronautical Weight Engineers. Inc.* 7771 South Race Street, Littleton, Colorado 80120.

16.2 Blythe, A.A.: 'The Hub of The Wheel — A Project Designer's view of Weight.' *SAWE Paper No. 996*, Society of Aeronautical Weight Engineers, Inc. P.O. Box 60024, Terminal Annex, Los Angeles, Calif. 90060. (1973).

16.3 Patterson, R.W.: 'Weight Estimates for Quiet/STOL Aircraft.' *SAWE Paper No. 1001*, Society of Aeronautical Weight Engineers, Inc. P.O. Box 60024, Terminal Annex, L.A., Calif. 90060. (1974).

16.4 Howe, D.: 'Empty Weight and Cruise Performance of Very Large Subsonic Jet Transports.' *SAWE Paper No. 919*, Society of Aeronautical Weight Engineers, Inc. P.O. Box 60024, Terminal Annex, L.A. Calif. 90060. (1972).

16.5 Taylor, R.J.: 'Weight Prediction Techniques and Trends for Composite Material Structure.' *SAWE Paper No. 887*, Society of Aeronautical Weight Engineers, Inc. P.O. Box 60024, Terminal Annex, L.A. Calif. 90060. (1971).

16.6 Torenbeek, E.: 'Prediction of Wing Group Weight for Preliminary Design.' *Aircraft Engineering*, (July, 1971).

16.7 Kasten, H.G.: *Proceedings of the Fifth Weight Prediction Workshop for Advanced Aerospace Design Projects.* Air Force Systems Command, Wright-Patterson Air Force Base, Ohio. Oct., 1969.

16.8 Burt, M.E.: 'Weight Prediction for Wings of Box Construction.' *R.A.E. Report No. 186*, (Aug., 1955).

16.9 Saelman, B.: 'The Growth Factor Concept.' *SAWE Paper No. 952*, Society of Aeronautical Weight Engineers, Inc. P.O. Box 60024, Terminal Annex, L.A., Calif. 90060. (1973).

16.10 Banks, J.: 'Preliminary Weight Estimation of Canard Configured Aircraft.' *SAWE Paper No. 1015*, (May, 1974).

16.11 Saelman, B.: 'Methods for Better Prediction of Gross Weight.' *SAWE Paper No. 1041*, (May, 1975).

16.12 Station, R.N.: 'Weight Estimation Methods.' *SAWE Journal*, (Apr.–May, 1972).

16.13 Caddell, W.E.: 'On the Use of Aircraft Density in Preliminary Design.' *SAWE Paper No. 813*, (May, 1969).

16.14 Ahl, W.H.: 'Rational Weight Estimation Based on Statistical Data.' *SAWE Paper No. 791*, (May, 1969).

16.15 Hopton-Jones, F.C.: 'A Practical Approach to the Problem of Structural Weight Estimation for Preliminary Design.' *SAWE Paper No. 127*, (May, 1955).

16.16 Marsh, D.P.: 'Post-Design Analysis for Structural Weight Estimation.' *SAWE Paper No. 936*, (May 1972).

16.17 Sanders, K.L.: 'A review and Summary of Wing Torsional Stiffness Criteria for Predesign and Weight Estimations.' *SAWE Paper No. 632*, (May 1967)

16.18 Ritter, C.R.: 'Rib Weight Estimation by Structural Analysis.' *SAWE Paper No. 259*, (1960).

16.19 Sanders, K.L.: 'High-lift Devices; A Weight and Performance Trade-off Technology.' *SAWE Paper No. 761*, (May, 1969).

16.20 Hammitt, R.L.: 'Structural Weight Estimation by the Weight Penalty Concept for Preliminary Design.' *SAWE Paper No. 141*, (1956).

16.21 Robinson, A.C.: 'Problems Associated with Weight Estimation and Optimization of Supersonic Aircraft.' *SAWE Paper No. 234*, (1959).

16.22 Green, L.D.: 'Fuselage Weight Prediction.' *SAWE Paper No. 126*, (May, 1955).

16.23 Tobin, E.W.: 'A Method for Estimating Optimum Fuselage Structural Weight.' *SAWE Paper No. 152*, (May, 1957).

16.24 Simpson, D.M.: 'Fuselage Structure Weight Prediction.' *SAWE Paper No. 981*, (Jun., 1973).

16.25 Kraus, P.R.: 'An analytical Approach to landing Gear Weight Estimation.' *SAWE Paper No. 829*, (May, 1970).

16.26 McBaine, C.K.: 'Weight Estimation of Aircraft Hydraulic Systems.' *SAWE Paper No. 128*, (1955).

16.27 Cate, D.M.: 'A Parametric Approach to Estimate Weights of Surface Control Systems of Combat and Transport Aircraft.' *SAWE Paper No. 812*, (May, 1969).

16.28 Burt, M.E. and Ripley, E.L.: 'Prediction of Undercarriage Weight.' *RAE Report Structures No. 80*, (Jun, 1950).

16.29 Philips, J.: 'A Method of Undercarriage Weight Estimation.' *RAE Report Structures No. 198*, (Mar. 1956).

16.30 Burt, M.E. and Philips, J.: 'Prediction of Fuselage and Hull Structure Weight.' *RAE Report Structures No. 122*, (Apr. 1952).

16.31 Ripley, E.L.: 'A Method of Fuselage Structure Weight Prediction.' *RAE Report Structures No. 93*, (1950)

16.32 Burt, M.E.: 'Weight Prediction of Ailerons and Landing Flaps.' *RAE Report Structures No. 116*, (Sept., 1951).

16.33 Ripley, E.L.: 'A Method of Wing Weight Prediction.' *RAE Report Structures No. 109*, (May, 1951).

16.34 Ripley, E.L.: 'A simple Method of Tail Unit Structure Weight Estimation.' *RAE Report Structures No. 94*, (Nov., 1950).

16.35 Grinsted, F.: 'Prediction of Wing Structure Weight.' *RAE Report Structures No. 15*, (1948).

16.36 Grinsted, F.: 'Simple Formulae for Prediction the Weight of Wing, Fuselage and Tail Unit Structures.' *SAE Report Structures No. 24*, (1948).

16.37 Currey, N.S.: 'Structure Weight.' *Interavia*, (Feb., 1949).

16.38 Grinsted, F.: 'Aircraft Structural Weight and Design Efficiency.' *Aircraft Engineering*, (Jul. 1949).

16.39 Carreyette, J.F.: 'Aircraft Wing Weight Estimation.' *Aircraft Engineering*, (Jan., 1950) 8–11; (Apr., 1950) 119.

16.40 Solvey, J.: 'The Estimating of Wing Weight.' *Aircraft Engineering*, (May, 1951).
16.41 Rosenthal, L.W.: 'The Weight Aspect in Aircraft Design.' *Journal of the Royal Aero. Soc.*, (Mar., 1950).
16.42 Taylor, J.: 'Structure Weight.' *Journal of the Royal Aero. Soc.*, (Oct., 1953).
16.43 Hyatt, A.: 'A Method for Estimating Wing Weight.' *Journal of the Aero. Sciences*, (Jun, 1954).
16.44 Green, L.D. and Mudar, J.: 'Estimating Structural Box Weight.' *Aero. Engineering Review*, (Feb., 1958).
16.45 Tye, W. and Montangnon, P.E.: 'The Estimation of Wing Structure Weight.' *ARC R.&M. (Aero. Research Council Reports and Memoranda) No. 2080*, (1941).
16.46 Maswell, R.: 'A Loadability Comparison: L-1011/DC-10-10' *SAWE Paper No. 1094*, (May, 1975).

Appendix A:
Conversion Factors (U.S. unit vs. SI unit)

The units used in this book are given in customary units of U.S. measure. The conversion factors selected in the following figures may be used to assist those readers who are using SI unit in their industries.

Physical Quantity	U.S. Customary Unit	Conversion Factor	SI Unit
Area	inch^2 (in^2)	645.16	Millimetre^2 (mm^2)
Length	inch (in) feet (ft)	25.4 0.3048	Millimetre (mm) Meter (m)
Mass	pound (lb)	0.4536	Kilogram (kg)
Modulus of Elasticity Stress Pressure	psi (lb/in^2)	6894.757	Newtons/meter^2 (N/m^2 or Pa) (MPa = Pa x 10^6) (GPa = MPa x 10^3)
Speed	mile/hour (mph) knots	0.44704 0.5144	Meter/sec (m/s) Meter/sec (m/s)
Moment	inch-lb (in-lb)	0.1130	Meter-newtons (N-m)
Temperature	Fahrenheit (°F)	5/9 (°F + 459.67) 5/9 (°F − 32)	Kelvin (°K) Celsius (°C)
Thermal Expansion	in/in/°F	9/5	m/m/°C
Stress Intensity Factory	ksi $\sqrt{in}$	1.0989	MPa $\sqrt{m}$
Density	lb/in^3	27.68	Megagram/meter^3 (Mg/m^3)

Fig. A

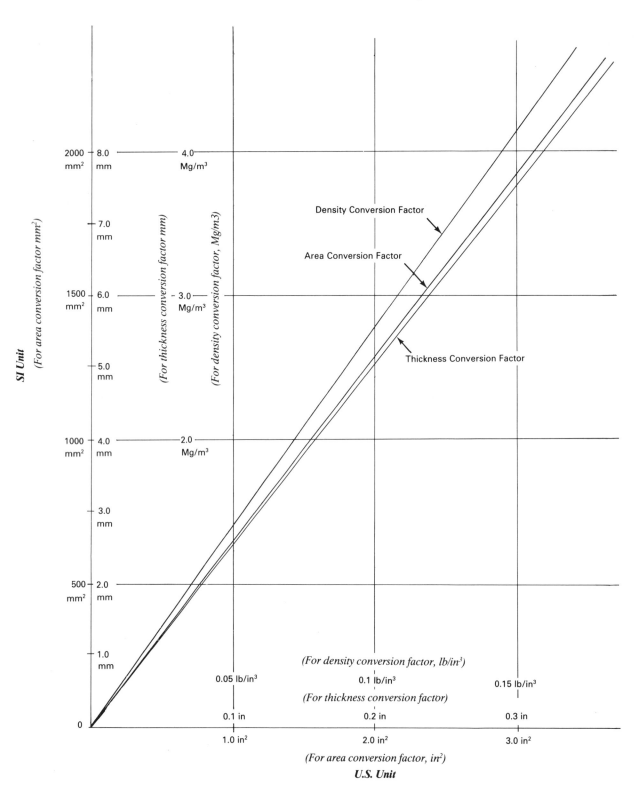

Fig. B

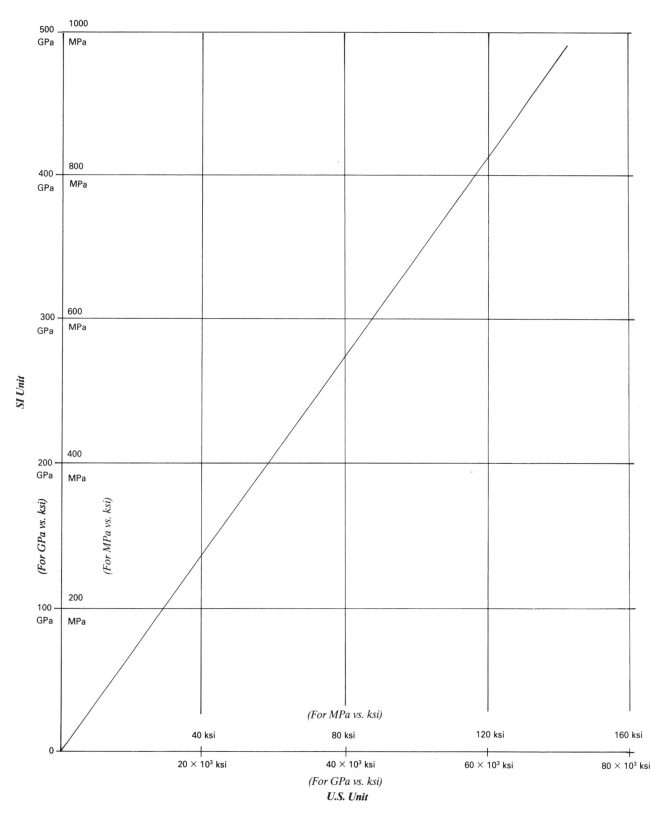

Fig. C

Appendix B:
List of Cutaway Drawings

Structural cutaway drawings are very valuable references in designing airframe structures. Owing to its copy right as well as large quantity, it is impossible to illustrate them in this book. However, a list of the cutaway drawings (approximate 500) is included in this appendix for convenience. They are divided into seven groups:

- COMMERCIAL TRANSPORTS
- MILITARY TRANSPORTS: transports, bombers, patrol and AEW airplanes
- COMMUTER, EXECUTIVE AND BUSINESS AIRPLANES
- SMALL AIRPLANES (four seats or less)
- FIGHTER, ATTACK, TRAINERS, INTERCEPTOR AND RECONAISSANCE AIRPLANES
- WORLD WAR II PROPELLER FIGHTERS
- HELICOPTERS
- MISCELLANEOUS

Document Sources:

A — Air International   F — Flight International   I — Interavia   • — Amphibious Airplane

Aircraft Name	Reference Document Dated	Page
Commerical Transports		
A300	F (Dec. 11, 1976)	1724
A300B-2	A (Sept., 1974)	130
A300-600R	A (May, 1987)	220
A310	F (Feb. 27, 1982)	480
	A (Jun., 1979)	284
	A (Sept., 1982)	114
A320	A (Jun., 1985)	288
A320-200	A (May, 1987)	232
Accountant	F (Jul. 6, 1956)	36
	F (Aug. 12, 1955)	228
Avro Ashton	F (Dec. 7, 1950)	526
A. W. Apollo	F (Aug. 26, 1948)	236
B 707	F (Jul. 25, 1958)	146
	F (Jul. 6, 1956)	13
	F (Nov. 20, 1959)	584
B 727-200	A (May, 1978)	218
B 737-200	A (Aug., 1979)	64
B 737-300	A (May, 1985)	230
B 747	F (Dec. 12, 1968)	986
B 747-200	A (Oct., 1976)	172
B 747-236	F (May 21, 1977)	1416
B 747-300	A (Sept., 1985)	118
B 747SP	F (May 22, 1975)	820
B 757-200	F (Jan. 2, 1982)	16
B 767-200	A (Feb., 1980)	64
BAC 111	F (Jan., 1979)	8
Britannia	F (Aug. 12, 1955)	228
	F (Jul. 6, 1956)	25
Caravelle	F (Jul. 25, 1958)	102

Aircraft Name	Reference Document Dated	Page
	F (Jul. 6, 1956)	33
	F (May 20, 1955)	690
Comet 4	F (Jul. 6, 1956)	27
	F (Jul. 25, 1958)	123
	F (Nov. 20, 1959)	598
Concorde	F (Mar. 4, 1965)	320
	F (May 18, 1967)	816
	F (Dec. 7, 1967)	957
	F (Jan. 17, 1976)	9
	A (Feb., 1976)	77
Convair 600	F (Nov. 20, 1959)	590
	F (Nov. 18, 1960)	796
CV 880	F (Jul. 25, 1958)	138
CV 990	I (Sept., 1962)	1152
Dakota IV (Douglas)	Australian Aviation & Defense Review (Jun., 1980)	12
Dart Herald	F (Aug. 19, 1955)	262
	F (Jul. 6, 1956)	60
	F (Jul. 25, 1958)	132
DC7C	F (Jul. 30, 1954)	145
	F (Jul. 6, 1956)	53
DC8	F (Jul. 6, 1956)	15
	F (Nov. 20, 1959)	594
DC9	F (Mar. 3, 1966)	344
	F (Mar. 18, 1967)	814
DC9-80	F (Jun. 23, 1979)	2268
	A (Jun., 1980)	268
DC10	F (Mar. 13, 1969)	409
	A (Dec., 1973)	274
	F (Sept. 2, 1978)	806

602 Airframe Structural Design

Aircraft Name	Reference Document Dated	Page
Electra	F (Jul. 6, 1956)	41
	F (Jul. 25, 1958)	136
F. 22	A (Nov., 1971)	294
F. 27	F (Jun. 3, 1955)	764
	F (Jul. 6, 1956)	48
	F (Jul. 25, 1958)	104
F. 28	A (Feb., 1973)	66
	F (May 1, 1975)	695
	A (May, 1977)	234
Handely page Hastings	F (Jun. 6, 1950)	55
Il-18	F (Jul. 25, 1958)	110
	F (Jul. 1, 1960)	14
Il-28	A (Dec., 1971)	354
Il-62	The Aeroplane and Aeronautics (Sept. 15, 1966)	5
L-1011	F (Oct. 1, 1970)	530
L-1011-500	F (Jun. 4, 1977)	1658
	A (Apr., 1980)	166
L-1649A	F (Jul. 6, 1956)	55
L-2000 SST	F (Jan. 20, 1966)	96
Marathons	F (May 19, 1949)	594
Mercure	A (Mar., 1971)	122
	F (May 20, 1971)	722
	A (Mar., 1977)	122
Princess*	F (Sept. 26, 1952)	413
Safari	F (Jul. 6, 1956)	64
Space Shuttle	F (Dec. 25, 1975)	
Trident (HS121)	F (Feb. 11, 1965)	208
	A (Jul., 1971)	88
TU-104	F (Nov. 16, 1956)	784
TU-114	F (Jul. 25, 1958)	108
TU-154	The Aeroplane and Aeronautics (Oct. 6, 1966)	16
Universal Freighter	F (Mar. 23, 1950)	376
VC10	F (Nov. 20, 1959)	618
	F (May 10, 1962)	735
Vicker-Armstrongs Varsity, T. Mk. I	F (Mar. 30, 1950)	408
Vickers Vanguard	F (Jul. 6, 1956)	43
	F (Nov. 20, 1959)	616
Viscount 810	F (Jul. 6, 1956)	20
	F (Jul. 25, 1958)	124
Yak-42	F (Mar. 19, 1977)	713

Military Transports: Bombers, Patrols and AEW

Aircraft Name	Reference Document Dated	Page
A3D Skywarrior	F (Feb. 15, 1957)	212
Aeritalia G. 222 L(ST)	A (Apr., 1979)	172
	A (Jul., 1972)	12
AEW Defender	F (Mar. 21, 1987)	
AN-2	A (Mar., 1983)	130
Avro Lancaster Mk III	A (Dec., 1981)	284
Avro Lincoln	F (Jun. 6, 1950)	53
Avro Shackleton G.R.1	F (May 18, 1950)	614
B-1	F (Dec. 26, 1974)	913
	A (Feb., 1975)	60
	A (Mar., 1975)	122
B-1B	A (Aug., 1986)	68
B-17C	A (Dec., 1974)	282
B-29	Aero Digest (Aug. 15, 1945)	62-66
B-52	F (Nov. 15, 1957)	772
B-52G	F (May 5, 1979)	1498
B-70	F (Jul. 2, 1964)	22
BAe Coastguarder	F (Apr. 21, 1979)	1255
Boeing E-3A	A (Jan., 1975)	20
	F (Jun. 26, 1975)	998
Blohm und voss Bv138c-1*	A (Nov., 1979)	234
Blohm und voss Bv222 Wiking*	A (Apr., 1981)	182
C-5A	F (Feb. 10, 1966)	224
C-5B	A (Feb., 1984)	64
C-130A	F (Sept. 14, 1956)	492
C-130H	A (Nov., 1974)	228
C-133	F (Sept. 26, 1958)	520
C-141B	A (Mar., 1983)	120
C-160	F (Apr. 25, 1968)	618
	A (Jun., 1981)	284
CASA C. 212	F (Mar. 31, 1979)	1003
CP-140	F (Jul. 9, 1977)	154
CX-84 (Canadair)	F (Oct. 1, 1970)	536
Dassault-Breguet Atlantic ANG	A (Nov., 1981)	214
DHC-5D	A (Aug., 1976)	62
Dornier Do 18D*	A (Apr., 1980)	182
E-2C (Grumman)	A (Jan., 1977)	12
Electric Canbera, B.I.	F (Jan. 26, 1950)	114
F-111A	F (Feb. 24, 1966)	302
	A (Dec., 1986)	282
F. 27 Maritime Enforcer	F (Dec. 27, 1980)	2328
F. 27 MPA	A (Dec., 1980)	262
	F (Jun. 8, 1985)	44
Fiat BR. 20 Cicogna	A (Jun., 1982)	292
Fokker TV	A (Nov., 1986)	246
Heinkel He 111B-2	A (Jun., 1979)	300
	A (Aug., 1987)	78
Heinkel He 111H-3	A (Oct., 1987)	182
Heinkel He 111P-2	A (Sept., 1987)	130
Heinkel He 115B	A (Feb., 1987)	98
Hercules C Mk 3	A (Feb., 1983)	72
Hurel-Dubois H.D. 32	F (Jun. 18, 1954)	794
Ilyushin Il-4	A (Mar., 1986)	136
Ilyushin Il-28	A (Dec., 1971)	354
Kawanishi (H6K4) Type 97*	A (Dec., 1985)	296
KC-135	A (Nov., 1980)	224
KC-10A	A (Dec., 1982)	266
	A (Mar., 1987)	152
L-400	F (Feb. 2, 1980)	324
Lockheed (MCE) Tristar Mk 1	A (Dec., 1985)	276
Lockheed Hudson I	A (Nov., 1985)	244
Martin B-26C Marauder	A (Jan., 1988)	26
Martin-Baker M.B. 5	A (Feb., 1979)	78
Messerschmitt Bf 110G-2	A (May, 1986)	244
Messerschmitt Bf 110 G-4b/R3	A (Jul., 1986)	30
Messerschmitt Me323E Gignat	A (May, 1983)	234
Messerschmitt Me410A-1	A (Oct., 1981)	184
Mitsubishi Ki-46-Hei	A (Nov., 1980)	228
Mitsubishi Ki-67-Hiryu	A (Jul., 1983)	26
Mitsubishi G4M2A	A (Dec., 1984)	300
Mitsubishi Ki-21-II-ko, Army type 97	A (Aug., 1986)	78
Nimrod AEW Mk 3	A (Mar., 1978)	114
	A (Aug., 1980)	60
	F (Feb. 7, 1981)	362
	A (Jul., 1981)	12
P-3C	A (Sept., 1975)	122
Petlyakow Pe-8	A (Aug., 1980)	78
Piaggio P. 108kB I	A (Dec., 1986)	300
Pilatus Britten-Norman, AEW Defender	F (Mar. 21, 1987)	26
Potez 63. 11A3	A (Aug., 1983)	76
S-3A (viking)	A (Jul., 1974)	8
	F (Jun. 22, 1977)	907

Airframe Structural Design 603

Aircraft Name	Reference Document Dated	Page
	F (Nov. 7, 1979)	642
	A (Jul., 1986)	44
Shin Meiwa US-1	A (Feb., 1982)	76
Short S.A. 4 Sperrin	F (Jan. 21, 1955)	80
Short Sunderland III*	A (Sept., 1981)	128
	A (Jul., 1984)	30
Sud-Est Le0451	A (Oct., 1985)	186
TU-26 (Backfire)	F (Sept. 6, 1980)	1024
Tu-95	A (May, 1987)	224
Tu-142	A (Sept., 1984)	138
	A (Apr., 1987)	176
	A (May, 1987)	
VC10K Mk 2	A (Oct., 1980)	162
YB-49 (Northrop)	F (Nov. 25, 1948)	629
YC-14	A (Sept., 1973)	116
	F (Jan. 30, 1975)	148
	A (Nov., 1976)	230

Commuter, Executive & Business Airplanes

Aircraft Name	Reference Document Dated	Page
Aeritalia AP68TP	A (Nov., 1982)	216
Aerostar 601P	F (Dec. 31, 1977)	1910
AN-2	F (Jul. 13, 1956)	97
AN-28, PZL Mielec (Antonov)	F (Feb. 11, 1984)	387
Andover C. 1	F (Aug. 31, 1967)	334
Arado Ar196A-3	A (Jan., 1979)	26
AVRO 748	F (Jun. 3, 1960)	759
	F (Nov. 18, 1960)	784
ATP	A (Jan., 1986)	12
	F (Aug. 9, 1986)	22
ATR 42	F (Feb. 15, 1986)	26
	A (Mar., 1987)	124
Beechcraft 1900	F (Nov. 20, 1982)	1502
Beech Duchess	F (Jun. 30, 1979)	2338
Beechcraft Duke B60	F (Mar. 27, 1976)	788
Beech Starship 1	F (May 3, 1986)	20
Beech Sper King Air 200	F (Jun. 20, 1974)	798
	A (Jan., 1980)	8
Canadair CL-13 Sabre Mk 4	A (Apr., 1971)	200
Canadair CL-84	A (Jan., 1971)	16
Canadair CL-215	A (Oct., 1975)	164
Canadair CL-246	A (Jan., 1971)	14
	A (Jan., 1972)	14
	A (Jan., 1980)	8
CASA 201 Aviocar	A (Apr., 1973)	166
CASA-Nurtanio CN-235	A (Dec., 1983)	272
Cessna 421C	F (Feb. 21, 1976)	420
Cessna Citation II	F (Apr. 7, 1979)	1062
Cessna Citation III	F (Nov. 24, 1979)	1762
CASA-Nurtanio CN-235	F (Sept. 3, 1983)	648
CL-41	F (Jan. 29, 1960)	142
CL-600 (Challenger)	A (Sept., 1978)	112
Commander 690	F (Nov. 2, 1972)	614
Commander 700	F (Mar. 11, 1978)	708
Comet 1	A (Apr., 1977)	176
Dassault-Breguet HU-25A	A (Apr., 1981)	176
DeHavilland Dash 7	F (Sept. 7, 1977)	833
DeHavilland Dash 8	A (Sept., 1980)	118
	F (Apr., 1983)	1008
	A (Jul., 1983)	8
DeHivilland Dash 8-300	A (Feb., 1988)	62
DeHavilland D.H. 84 Dragon	A (Mar., 1984)	132
DeHavilland D.H. 91 Albatross	A (May, 1973)	236
DeHavilland Heron	F (Jan. 22, 1954)	98
DHC-6	A (Feb., 1975)	96
Dornier 228-200	A (Oct., 1987)	164
Donier Do335B-2	A (Jan., 1973)	18

Aircraft Name	Reference Document Dated	Page
EMB-111	A (Apr., 1978)	166
EMB-120	A (Nov., 1983)	218
EMB-121	F (May 13, 1978)	1470
	A (Sept., 1983)	112
F. 27 Mk500	A (Apr., 1984)	176
Fairchild 300	F (Sept. 8, 1984)	574
Falcon 50B	F (Jun. 12, 1975)	937
	F (Jan. 29, 1977)	236
Gates Learjet 28/29	F (Feb. 10, 1979)	402
Gates Learjet 54/55	F (Aug. 18, 1979)	516
Grumman GA-7 Cougar	F (Aug. 5, 1978)	406
Gulfstream 3	A (Dec., 1981)	270
Gulfstream 4	F (Sept. 14, 1986)	
	F (Nov. 1, 1986)	
Gulfstream SRA-1	A (Jul., 1985)	12
Hurel-Dubois H.D. 32	F (Jun. 18, 1954)	794
HS125-700	A (Sept., 1977)	116
HS125-800	F (Jun. 4, 1983)	1684
	A (Mar., 1985)	114
HS146	F (Oct. 18, 1973)	627
	A (Sept., 1974)	114
HS146-300	A (Jun., 1987)	270
HS748	A (Jun., 1976)	270
	A (Feb., 1978)	60
HS1182	A (May, 1973)	226
IAI1125 Westwind Astra	F (Jan. 5, 1985)	24
In the Wind (Vicker 1000)	F (Mar. 26, 1954)	357
Islander	F (Aug. 24, 1967)	296
Jet Provost 5	F (Aug. 24, 1967)	302
Jetstar (CL-329)	F (Sept. 20, 1957)	490
Jetstar II	F (Jul. 2, 1977)	26
Jetstream	F (May 11, 1967)	740
Jetstream 31	F (Oct. 10, 1981)	1088
	A (Jul., 1982)	8
Kawanishi N1K1-J Shiden-Kai	A (Apr., 1973)	184
Learavia Lear Fan	F (Feb. 26, 1981)	1891
Learjet 35/36	F (Sept. 12, 1974)	318
Learjet 50	A (May, 1980)	
Learstar 600	F (Jul. 10, 1975)	37
Martin Midget	F (Nov. 19, 1954)	735
Mitsubishi MU-300 Diamond 1	F (Jul. 18, 1981)	168
	A (Jun., 1982)	268
Mitsubishi Mu-2J	A (Nov., 1973)	220
Mitsubishi J2M3	A (Jul., 1971)	70
Nomad 22	A (Jun., 1973)	284
Nomad N24A	A (May, 1982)	226
Partenavia P68C	A (Nov., 1982)	220
Piper Seneca	F (Dec. 16, 1971)	968
Piper T-1040	A (Sept., 1981)	122
SC. 1	F (Jun. 10, 1960)	794
Scandia	F (Jun. 3, 1948)	608
SF-340	F (Oct. 30, 1982)	1276
	A (Jun. 1983)	270
SD3-30	A (Mar., 1976)	120
Short 360	A (Aug., 1984)	64
Short C-23A Sherpa	A (Nov., 1984)	226
Short Sealand	F (Dec. 4, 1947)	634
Slingsby T. 53	F (May 18, 1967)	786
SR/45	F (Jan. 15, 1948)	
Swearingen Merlin IIIB	F (May 19, 1979)	1656
Swearingen Metro II	A (Jan., 1978)	12
Trislander	A (Sept., 1974)	112
	A (Jul., 1975)	44
Twin Pioneer	F (Mar. 18, 1955)	358
Vicker-supermarine Seagull	F (Dec. 11, 1947)	664
WFW614	A (Dec., 1971)	370
Yak-23	A (May, 1973)	232

Aircraft Name	Reference Document Dated	Page
Small Airplanes — 4 Seats or Less		
Ayres Turbo Thrush	F (Sept. 16, 1978)	1086
Caproni Vizzola C-22J	A (Feb., 1981)	62
Cranfield A1	F (Sept. 18, 1976)	908
Firecracker	F (Jul. 23, 1977)	276
Ipanema EMB-201A	F (Apr. 22, 1978)	1105
Man-powered Aircraft	F (Oct. 29, 1977)	1254
Microjet 200	F (Aug. 14, 1982)	361
NDN6 Fieldmaster	F (Aug. 29, 1981)	650
Piper PA-36 Pawnee Brave	F (Jan. 3, 1976)	24
Robin ATL	F (Jun. 22, 1985)	28
Robin HR100/285 TIARA	F (Dec. 12, 1974)	844
Sheriff	F (Apr. 5, 1980)	1054
Spartair F-5	F (Jun. 11, 1970)	976
Fighter, Attack, Trainer, Interceptor, & Reconnaissance Airplanes		
A3D Sky Warrior	F (Feb. 15, 1957)	213
A-4M Skyhawk	A (Nov., 1981)	222
A-6 Intruder	A (Jan., 1976)	
	F (May 25, 1985)	
	A (Nov., 1986)	232
A-7 Corsair	A (Aug., 1987)	84
	A (Apr., 1982)	174
A-10A Thunderbolt	F (Mar. 20, 1976)	712
	A (Jun., 1979)	268
	F (Dec. 1, 1979)	1846
AD-1 Sky Raider	F (Jan. 26, 1950)	126
Aeritalia-Aermacchi-Embraer (AMX)	A (Sept., 1986)	138
Aermacchi MB-326K	A (Jun., 1978)	268
Aerospatiale Epsilon	A (Jan., 1987)	12
Ajeet	A (Jan., 1977)	284
Alpha Jet	F (Feb. 28, 1974)	267
	A (Jun., 1984)	272
AMX	A (Aug., 1981)	70
	A (Oct., 1983)	
	A (Aug., 1986)	
	A (Sept., 1986)	138
AV-8A	A (Apr., 1976)	
	A (Dec., 1978)	
	A (Apr., 1983)	
	A (Aug., 1985)	
AV-8B	A (Jun., 1977)	270
	F (Dec. 29, 1979)	2130
	A (Feb., 1982)	
	A (Mar., 1984)	
	A (Oct., 1986)	178
A.W. 650 (Argosy)	F (Nov. 20, 1959)	580
Bell P-59A Airacomet	A (Mar., 1980)	134
Bell X-22A	F (Mar. 23, 1967)	446
BAC167 Strikemaster	F (Oct. 12, 1972)	502
	A (Nov., 1977)	
	A (Sept., 1978)	
	A (Dec., 1984)	
	A (Sept., 1986)	
BAC221	F (Jul. 23, 1964)	136
Buccaneer S Mk 2B	A (Aug., 1982)	62
C22 (Caproni Vizzola)	A (Feb., 1981)	
C-101 CASA Aviojet	A (Aug., 1978)	68
CL-44D-4	F (Nov. 20, 1959)	588
CL-1200 Lancer (Lockheed)	A (Sept., 1971)	176
EAP	F (Apr. 19, 1986)	28
	A (Jun., 1986)	304
EMB-312 (T-27)	A (Jun., 1980)	
	A (Jan., 1983)	10
Embraer T-27	F (Aug. 30, 1986)	
	A (Jun., 1980)	290
Enaer T-35 Pillan	A (Apr., 1985)	174
F-1 (Mitsubishi)	A (Mar., 1980)	130
F-4D-1 (F-6A) Skyray	A (Oct., 1982)	180
F-4E Phantom	A (Nov., 1978)	214
	F (Jun. 30, 1966)	1094
	A (Dec., 1985)	290
F-4K	F (Jun. 30, 1966)	1094
F-5	F (Jan. 8, 1960)	44
	F (Oct. 17, 1981)	1146
F-5G (Tigershark)	A (Mar., 1982)	112
F-9F-5 Panther	A (Mar., 1982)	130
F-9F-8 (F-9J) Cougar	A (Jul., 1982)	26
F-14	F (Sept. 27, 1973)	510
	A (Jan., 1982)	26
F-15	F (May 1, 1975)	708
	A (Aug., 1981)	66
	A (Jul., 1987)	8
F-16	F (Feb. 7, 1974)	174
	F (Aug. 21, 1975)	263
	A (Nov., 1977)	220
	A (Oct., 1981)	166
	F (Sept. 4, 1982)	696
F-18	F (Aug. 21, 1975)	263
	A (Dec., 1978)	260
F-80	F (Jan. 26, 1950)	130
F-86E	F (Jan. 30, 1953)	142
	A (Jun., 1985)	316
F-100 Super Sabre	F (May 20, 1955)	682
	A (May, 1985)	242
F-101 Voodoo	A (Aug., 1985)	78
F-102A Delta Dagger	A (Jan., 1986)	32
F-104	F (May 30, 1958)	740
	F (Oct. 19, 1961)	624
	F (Sept. 2, 1978)	803
	A (Dec., 1986)	310
F-105D Thunderchief	A (Apr., 1986)	190
F-106A Delta Dart	A (Oct., 1986)	202
Fanliner	A (Feb., 1977)	
	F (Sept. 15, 1979)	
FH-1 Phantom	A (Nov., 1987)	234
FMA IA. 58 Puscara	A (Oct., 1977)	168
FMA IA. 63 Pampa	A (Feb., 1987)	62
Folland Gnat Trainer	F (Nov. 27, 1959)	642
G4 (Super Galeb)	A (Jan., 1985)	
Gloster Meteor F. 8	F (Jun. 6, 1950)	49
Granfield A1	F (Sept. 18, 1976)	908
Harrier	F (Nov. 2, 1967)	728
	F (Sept. 28, 1972)	424
	F (Sept. 2, 1978)	846
	A (Aug., 1985)	60
Hawk	F (Sept. 7, 1972)	319
	F (May 23, 1974)	671
	F (Oct. 29, 1977)	1267
	A (Nov., 1977)	228
	A (Dec., 1984)	290
	F (May 31, 1986)	50
	A (Sept., 1986)	142
Hawker Hunter F(GA) Mk9	A (Jul., 1981)	26
Hawker P. 1052	F (Feb. 9, 1950)	192
Hawker Sea Hawk	F (Jan. 26, 1950)	103
HF-24 (Marut)	A (Sept., 1976)	
HiMAT	F (Jul. 5, 1980)	
IA.58 (Pucara)	A (Oct., 1977)	
IA. 63 (Pampa)	A (Feb., 1987)	
JA37 Viggen (SAAB)	F (Apr. 20, 1967)	638
	A (Jul., 1980)	
	F (Sept. 27, 1980)	1262
JAS39 Gripen (SAAB)	A (Jun., 1986)	268

Airframe Structural Design 605

Aircraft Name	Reference Document Dated	Page
	A (Nov., 1987)	228
Jaquar	A (Dec., 1979)	272
Kfir-C2	I (Sept., 1976)	782
	A (Nov., 1976)	782
Jet Provost	F (May 6, 1955)	584
L-80TP	A (Mar., 1986)	
Lightning F. 53 (BAC)	F (Sept. 5, 1968)	374
Lavi (IAI)	A (Jun., 1986)	274
Marage IIIE	A (Jul., 1985)	26
Marage IVP	A (Apr., 1987)	166
Marage 2000	A (Sept., 1980)	112
MiG15	F (Jan. 28, 1955)	113
	A (Jun., 1985)	318
MiG19	A (Jan., 1983)	18
MiG21	F (Sept. 25, 1975)	445
	Australian Aviation & Defense Review (Jun., 1979)	42
MiG23	A (Sept., 1984)	134
	A (Apr., 1986)	164
MiG25	F (Apr. 23, 1977)	1123
	A (Nov., 1979)	248
MiG27	A (Jan., 1987)	22
MiG29	A (Jun., 1987)	276
MB-339	A (Jun., 1978)	272
	F (Dec. 9, 1978)	2095
Meteor 8	F (Oct. 6, 1949)	466
N-156F (Northrop)	The Aeroplane and Aeronautics (Dec. 4, 1959)	574
	F (Jan. 8, 1960)	44
P. 180 (Avanti)	F (Jun. 21, 1986)	
PC-7 (Pilatus)	A (Sept., 1979)	114
PC-9 (Pilatus)	A (Nov., 1985)	224
PZL-130 Orlik	A (Oct., 1985)	168
PZL-Mielec TS-11, Iskra-Bis B	A (Mar., 1979)	130
RA-5C (Vigilante)	F (Aug. 13, 1964)	254
Rafale	A (Jun., 1986)	272
RFB Fantrainer	F (Sept. 15, 1979)	902
	A (Feb., 1986)	74
RF-8G Crusader	A (Sept., 1987)	124
SAAB-29	F (May 4, 1950)	557
Saunders-Roe, S.R./A.I.	F (Dec. 14, 1950)	554
Sea Hawk	F (Nov. 10, 1949)	616
Sea Hawk FGA Mk4	A (Dec., 1982)	284
Sea Hornet F Mk20	A (Oct., 1982)	196
Short Seamen	F (Feb. 20, 1956)	82
Spey 25R	F (Nov. 16, 1967)	809
SR-71 (Blackbird)	A (Feb., 1985)	60
SU-7	A (Sept., 1982)	134
SU-15	A (Jan., 1988)	20
SU-17	A (Sept., 1986)	126
SU-24	A (Sept., 1987)	112
SU.25	A (May, 1986)	224
Super Etendard	A (Feb., 1986)	64
Supermarine 510	F (Mar. 2, 1950)	290
Supermarine Attack FB Mk2	A (May, 1982)	234
T-34 (Turbo Mentor)	A (May, 1976)	
T-46A (Fairchild)	F (Apr. 13, 1985)	
	A (Jun., 1985)	276
Tornado GR. 1	F (Aug. 7, 1976)	332
	A (Nov., 1979)	218
	A (Sept., 1984)	118
	A (May, 1986)	218
U-2 (Lockheed)	F (Jun. 16, 1984)	1564
	A (Oct., 1984)	188
Valmet L-80	A (Mar., 1986)	112
Vampire	F (Nov. 23, 1951)	414
	F (May 8, 1953)	578

Aircraft Name	Reference Document Dated	Page
Vickers-Supermarine Attacker	F (Jan. 26, 1950)	103
Victor B. Mk2	F (Dec. 30, 1959)	467
Westland Wyvern	F (Jun. 15, 1956)	778
Wastee Talent	F (Oct. 5, 1944)	364
X-15	F (Dec. 23, 1978)	2259
X-29A	F (Jun. 16, 1984)	1564
XFV-12A	A (Feb., 1974)	
XF-103	A (Jul., 1985)	28
YAK-38	A (Aug., 1986)	84

World War II Propeller Fighters

Aircraft Name	Reference Document Dated	Page
Aichi D3A1 Navy Type 99 Mode 111	A (Dec., 1987)	286
Arado Ar 196A-3*	A (Jan., 1979)	26
Bell P-39D Airacobra	A (Jan., 1982)	36
Blackburn Firebrand TF Mk10	A (Jul., 1978)	26
Blackburn Skua	A (Nov., 1977)	236
Bloch 152C1	A (Apr., 1978)	182
Bristol Beaufort I	A (Nov., 1978)	232
Chance Vougth F7U-3 Cutlass	F (Oct. 12, 1956)	607
Chance Vougth F4U-1 Corsair II	A (May, 1979)	234
Curtiss SB2C-4 Helldriver	A (Dec., 1979)	284
DeHavilland Mosquito B Mk XVI	A (Jan., 1983)	26
DeHavilland Mosquito NF Mk II	A (Feb., 1983)	78
DeHavilland Sea Mosquito TR Mk33	A (Jun., 1984)	288
Douglas SBD-3 Dauntless	A (Oct., 1979)	182
Fairey Albacore I	A (Mar., 1978)	130
Fairey Battle I	A (Mar., 1981)	128
Fairey Firefly FR Mk4	A (Jul., 1979)	26
Fairey Flycatcher	A (May, 1978)	234
Fairey Spearfish 7D Mk1	A (Jan., 1978)	22
Grumman F7F-3 Tigercat	A (Apr., 1984)	184
Grumman F8F-1 Bearcat	A (May, 1980)	234
Handley Page Hampden Mk I	A (Nov., 1984)	248
Hawker "Hardy"	F (Sept. 2, 1978)	802
Hawker Hurricane	A (Jul., 1987)	30
Hawker Sea Fury FB Mk11	A (Feb., 1980)	96
Hawker Sea Hurricane F Mk IV	A (Sept., 1979)	130
Hawker Sea Hurricane II	A (Sept., 1979)	130
Hawker Spanish Fury	A (Jan., 1980)	286
Hawker Tempest II	A (Feb., 1981)	94
Henschel Hs129B-2	A (Dec., 1980)	280
Grumman TBM-1C Avenger	A (Jun., 1978)	290
Junkers Ju878-2	A (Oct., 1984)	192
Junkers Ju188E	A (Apr., 1982)	182
LaGG-3	A (Jan., 1981)	26
Lockheed P-38J Lightning	A (May, 1981)	234
	F (Jun. 22, 1944)	666
Martin-Baker M.B. 5	A (Feb., 1979)	78
Macchi C. 200 (Serie XIX) Saetta	A (Dec., 1977)	288
North American F-82G Twin Mustang	A (Nov., 1983)	236
North American P-51 Mustang	A (Sept., 1983)	134
Messerschmitt Bf110G-2	A (May, 1986)	244
Nakajima Ki. 44-II-Hei; Shoki	A (Jul., 1972)	22
Petlyakov Pe-2	A (Aug., 1979)	78
Potez 63. 11A3	A (Aug., 1983)	76

Aircraft Name	Reference Document Dated	Page
Republic P-47D-10 Thunderbolt	A (Feb., 1978)	84
Supermarine Seafire XVII	A (Oct., 1978)	178
Supermarine Spitfire 1	A (Feb., 1985)	78
Supermarine Spitfire 21	A (Apr., 1985)	184
Supermarine Spitfire IX	A (Mar., 1985)	136
Vickers Wellesley Mk I	A (Jul., 1980)	30
Vultee V-11-GB	A (Jul., 1972)	40
Westland Lysander Mk III	A (Jan., 1984)	26

Helicopters

Aircraft Name	Reference Document Dated	Page
Aerospatiale AS332 Super pume	A (Jan., 1984)	8
Aerospatiale Westland SA341 Gazelle	A (Dec., 1977)	280
Agusta A. 105 Hirundo	F (Jan. 25, 1973)	128
AH-64 Apache	A (May, 1984)	228
	F (Oct. 6, 1984)	876
AS. 355	F (Jan. 10, 1981)	82
Bell 212	F (Mar. 2, 1972)	328
Bell 214 ST	F (Jun. 30, 1979)	2347
	A (Oct., 1982)	164
Bell 222	A (Oct., 1979)	172
	F (Feb. 17, 1979)	482
Bo105C	F (May 4, 1972)	638
Boeing Vertol CH-47D Chinook	A (Jul., 1979)	8
Boeing Vertol Model 234	F (Mar. 22, 1980)	930
Boeing Vertol Model 360	F (Apr. 18, 1987)	26
CH-47C	F (Apr. 22, 1971)	550
CH/HH-53	F (Apr. 22, 1971)	551
Chinook HC Mk 1	F (Sept. 23, 1978)	1174
EH Industries EH101	A (Dec., 1987)	280
Foche Achgelis Fa-223E	A (May, 1984)	260
Kaman SH-2F Seasprite	A (Feb., 1988)	72
Lynx AH Mk I	F (Aug. 29, 1974)	20
Lynx HAS Mk 2	A (Mar., 1979)	116
MBB-Kawasaki BK117	F (Jan. 22, 1983)	188
Mil Mi-24 Hind-E	A (May, 1984)	230
Mil Mi-24 Hind-D/E	A (Jul., 1986)	18
Mil Mi-26	A (Jun., 1983)	314
SA. 365F Dauphin 2	F (May 28, 1983)	1520
SH-60B Seahawk	A (Jul., 1984)	12
Sikorsky CH-53E Super Stallion	A (Mar., 1981)	110
Sikorsky S-76	F (May 6, 1978)	1380
Sikorsky S-76 Spirit	A (Mar., 1980)	112
Sikorsky SH-60B	F (Feb. 25, 1978)	510
Sikorsky UH-60A	F (Feb. 25, 1978)	506
Westland 30 Series 200	A (Aug., 1983)	60
Westland Sea King HAS Mk 5	A (May, 1981)	218
Westland Sea King HC Mk 4	A (Mar., 1979)	120
YAH-63A	F (May 22, 1976)	1366
YUH-60A	F (Aug. 21, 1976)	451
YUH-61A	F (Aug. 21, 1976)	450

Miscellaneous

Aircraft Name	Reference Document Dated	Page
A-10 Cockpit Arrangement	F (Dec. 1, 1979)	1846
A300 Fuselage Floor Venting	F (Dec. 4, 1976)	1647
Airscrew Seat	F (Sept. 21, 1972)	397
Apollo	F (Sept. 2, 1978)	804
Atlas(ICBM)	F (Sept. 2, 1978)	805
B707 — Horizontal Tail Box Root Joint	F (Jan. 1, 1985)	31
B747 — Engine Mount	F (Dec. 12, 1968)	991
B747 — Fuselage Floor Venting	F (Dec. 4, 1976)	1649
B747 — High Lift Devices	F (Dec. 12, 1968)	981
B747 — RB. 211-524B engine mounts and pylon	F (May 7, 1977)	1264
B747 — Main Landing Gears and Center Fuselage	F (Dec. 12, 1968)	990
	F (Sept. 2, 1978)	808
B747 — CF6 engine mounts and pylon	F (Aug. 9, 1973)	248
B2707 — SST Wing Pivot	Aviation Week and Space Technology (Oct. 17, 1966)	45
B7X7 — Wing Leading and Trailing Edge Detail	F (Jun. 19, 1976)	1636
BAe Olympus	F (Jan. 11, 1986)	26
BL755 Cluster Bomb	F (Feb. 14, 1974)	222
BN-2T Engine Installation	F (Nov. 14, 1981)	1509
C-5A — Main Landing Gear	F (Feb. 10, 1966)	226
C160 (Transall) — Main Landing Gear	F (Apr. 25, 1968)	616
Concorde — Visor Structures	F (Aug. 12, 1971)	259
DC-9 — Main Landing Gear	F (Mar. 3, 1966)	348
DC-10 — Cargo Door Lock System	F (Apr. 4, 1974)	422-433
	F (May 22, 1976)	1345
DC-10 — Fuselage Floor Venting	F (Dec. 4, 1976)	1646
Doodlegur Flying Bomb	F (Sept. 2, 1978)	804
Dowty Rotol's V-P Fan	F (Sept. 21, 1972)	398
F-14 — Wing Center Box	I (Feb., 1979)	143
Islander — Shrouded Fan Engine Mount	F (Jul. 16, 1977)	211
JA37 — Wing Main Spar	F (Apr. 24, 1976)	1087
L-1011 — Cargo Door Lock System	F (Feb. 6, 1975)	209
Metro III — PT6A-45R Engine Mount	F (Mar. 2, 1985)	25
Sketch Book — HS. 146 Engineering: Wing root joint; Rear cabin; Petal-type airbrake; Interior design; ALF502 engine mounts; Main landing gear	F (Apr. 11, 1974)	458
Sketch Book — YF-17 inlet; Ted crew seat; Speery gimballed twin nozzle; C-141 Main gear; Wing tip pod air refuelling; Norman's fire-fighter water-bomb; PZL Kruk's wing and spray; YC-15 propulsion nozzle and thrust-reverser cascades; ML portable power-pack; The manually operated under-carriage of the IS-28M2 powered sailplane; The RB189's thrust-reverser buckets.	F (Sept. 18, 1976)	891
Space Transportation System	F (Sept. 25, 1976)	978
Tankers, DC-10, KC-135 & B747 proposal	F (Dec. 3, 1977)	1674
Trident — Outbd Wing Joint Detail	F (Aug. 13, 1977)	467
Viscount — Passenger Door Detail	F (Feb. 22, 1957)	242
Viscount — Wing	F (Mar. 20, 1953)	378

Index

Abrasion 95
Absorber 454
Access hole 162, 263, 297, 355, 363, 564
Adhesive bonding joint 227, 524
Advanced composite 112
Aerodynamic 21, 36
Aeroelasticity 30
Aileron 59, 73, 349
Aircraft loads 21
Airfoil 250
Air inlet 478
All-moving tail 360
Alternating stress 542
Alclad 93
Aluminum alloy 96, 101, 497
Angle of attack 30
Anisotropic 499, 509
Anodize 356, 466, 574
Autoclave 499, 526
Automation 497, 531
Aspect-ratio 46
Assembly 17
Asymmetric engine thrust 65

Bagging 527
Balance grid 592, 594
Balanced laminate 499, 513
Balanced weight 350, 372
Beaded web 172, 569
Beam (or spar, rib) 165, 270
  built-up- 270, 275
  integrally machined- 270, 271
  diagonal-tension- 271
  keelson (or keel)- 410
  shear web- 269
  tapered- 271
  truss- 270, 279
Beam-column 120, 128
Bearing 226, 524
  -distribution factor- 240
  spherical- 226
  -stress concentration factor 239
Black aluminum 525
Blind fastener 211
Bolt 215, 286
  12 point- 216
  blind- 212
  hexagon- 216
Bolted joint 207
Bogie 431
Bonded joint 227, 524
Boron 496, 501, 504
Brazing 230
Breakaway design 278
Brakes 461, 462
Brittle 524
B-stage 499
Buckling 118, 137
  compression- 141
  curved plates- 138, 140
  diagonal tesion- 165
  flexural- 121, 148
  local instability- 146
  initial- 148
  inter-rivet- 144, 147
  shear- 139
  torsional- 121, 128
Buffet 63
Bulkhead 277, 363, 376, 396, 406
Bushing 226, 465
Bypass load 238, 524

Cable 311, 328
Cabin pressure 67, 378, 399
CAD/CAM system 7
Canopy 74, 402
Carbon 496
Caul plate 527
Center of gravity envelope 39, 594
Certification program 4, 539
Cladding (alclad) 93, 115
Clip 259, 271, 274, 277, 390
Cockpit 399
Coefficient
  beam-column- 133
  compression buckling- 138, 139
  end-fixity- 122
  shear buckling- 139, 140
  thermal expansion- 18, 533
  wing section lift- 55
Cold work 240
Collars (fastener) 212, 213, 214
Column 118, 120, 121
Compressibility 24, 45
Composite material 112, 492, 500
Compression 118, 141
Computer-aid 6
Control surfaces 58, 73, 347, 358, 365
Core
  fluted- 404
  foam- 532
  honeycomb- 346, 355, 367
Corrosion 98, 113
  -prevention and control 113, 116
  exfoliation- 114
  fretting- 115, 116, 544
  galvanic- 115, 116
  intergranular- 114, 116
  pitting- 113
  stress- 100, 114, 116
  surface- 113
Crack
  -growth 96, 556
  -propagation 97
Crashworthiness 516
Cripping stress 134
Crossply 499
Crushing load 278
Cumulative damage 549
Cure 499, 526
Curved panel 154, 186
Cutout 162
  beam- 165, 173
  circular- 165, 175
  fuselage- 186, 204
  rectangular- 173
  skin-stringer panel- 177
  wing cover- 163, 177, 263

Damage tolerance 510, 538, 556
Damping 567
Damping ratio 568
Debulk 531
Delamination 511, 536
Diagonal-tension (tension field) beam 165, 271
Discrete gust 30, 56
Distortion 350, 536
Dome 396
Door 74, 263, 419
  access- 162, 263, 297, 355, 564
  cargo- 417, 425
  -design criteria 420
  -loads 420
  -stop 197, 421
  fuselage- 419
  non-plug type- 417, 420
  non-stressed- 263, 417, 420
  nose loading- 425
  passenger- 420, 421
  plug type- 417, 420
  pressurized- 421
  stressed- 263, 417
Doubler 231
Droop nose 303, 332
Ductility 215
Dynamic load 56
Dynamic pressure 24

Eccentricity 119, 149, 257, 526
Eccentuator 333
Effective
  -depth 165
  -length 122
  -width 142
Efficiency
  shock absorber- 454
  static strength- 95
  structural- 96, 379, 513
Elastic support 123, 133
Elastromeric mold 533
Elevator 371
Elongation 91
Embrittlement 98
Empennage (tail) 59, 358
Endurance limit 542
Engine
  -cowling 475
  -inlet 478
  -loads 67, 474
  -mounts 471
  -naccelle 475
  -pylon (pod) 479
Engine mount 471
  fuselage- 489
  isolator 476
  rear fuselage- 472, 487
  tail- 487, 488
  wing- 473
Epoxy 112, 496, 505
Euler-Engesser 125
Euler equation 125
Extrusion 94

Fabric 112, 499
Factor
  Beam-column magnification- 130
  bearing distribution- 240
  bearing stress concentration- 239
  composite bearing reduction- 524
  fitting- 209
  gust alleviation- 44
  hole condition- 240
  hole filling- 240
  landing gear load- 71
  load- 21
  lug tension efficiency- 221
  lug shear-bearing efficiency- 220
  Farrar's efficiency- 148
  scatter- 232, 550
  safety- 23, 118
  severity- 238
  stress concentration- 162, 239, 510, 561
  stress intensity- 98

Fail-safe 538, 540, 542, 554
  -design 276, 293, 312, 363, 554
  -loads 555
  -straps (tear straps) 187, 382, 383
Failure modes in compression 121
Farrar's efficiency factor 148
Fastener 150, 207, 544
  blind- 211
  -allowables 208
  -feather edge 208, 562
  -fits 216, 544
  -spring constants 234
  Hi-Lok- 212
  Hi-shear- 212
  permanent- 210
  removable- 214
  Taper-Lok- 213
Fatigue 95, 293, 538, 591
  -design 230, 282
  -life 241, 245, 542, 549
  -loads 544
  -quality index 242, 542
  sonic- 344, 363, 567, 569
  -tests 570
Feather edge 208, 562
Fiber 500
Fiberglass 96, 501
Filament 499, 506
Filament winding 527
Fin 358, 369
Finite element 6, 235
Fish-mouth splice 228
Fittings 209, 271, 567
Fixed leading edge 352
Fixed trailing edge 352
Flap 58, 303
  double slotted- 337
  fowler- 58, 339
  -tracks 339, 342
  -construction 344
  single slotted- 336
  triple slotted- 337
Flight profile 544
Flight station 390
Floatation 445
Floor beam 384, 391
Flutter 37, 250, 293, 348, 350, 360, 363
Flying-tail 363
Forced bending 349
Forging 16, 94,
Former 344, 390
Forming 16, 17, 530
Fowler flaps 339
Fracture 96
Frame 384
Fuel tank 74, 296, 299
Fuselage 376, 412, 566
  -bulkheads 390, 396
  -configurations 379
  -cutouts 186, 204
  -design requirements 379
  -doors 419
  -fail-safe straps 383
  -floors 384, 391
  -frame 384
  -loads 66, 378
  -material 389
  -venting system 391
  -windows 417
  -windshields 399

Galvanic corrosion 115, 116, 523
Gears (landing) 430
General aviation 361
Glass 400
Grain direction 94, 98
Graphite 96, 112, 496, 502
Griffith-Irwin theory 97
Ground-air-ground load 546
Ground handling load 30
Gust 43
  discrete- 30, 56
  dynamic- 56
  -alleviation factor 44
  -envelope 43
  -loads 43, 63
  lateral (side)- 59, 63, 65
  power spectral- 57

Harness satin 507
High lift device 58, 303
High angle of attack 56, 258
Hinge 312, 350, 372, 421
Hole
  access- 263, 265
  crawl- 277
  fastener- 240
  -condition factor 240
  -filling factor 240
  lightening- 165, 277
Honeycomb structures 346, 355, 367, 569
Horizontal stabilizer 59, 363, 412
Hot-wire cutting 535
Humidity 518

Impact damge 510
Inertia force (loads) 72, 278, 350, 474
Inlets 478
Inserts 217
Instability 146, 377
Interaction design curves 152, 153, 154, 159
Interference fit 231, 249
Integrally-stiffened panels 155, 163
Intergranular corrosion 114, 116
Inter-rivet buckling 144, 147
Iron bird 572
Isotropic 499, 509

Jam condition 312, 420
Johnson-Euler column 125
Joint 207, 282, 465, 520, 543, 568
  adhesive- 227
  bolted- 207, 231, 282, 520
  bonded- 227
  co-cured- 525
  folding- 282
  -life prediction 241
  lug- 219
  pivot- 290
  riveted- 231
  welded- 227
  wing root- 282

Keel beam (keelson) 410
Kevlar 96, 496, 501, 502
Knitted fabrics 508
Kruger flaps 329

Laminate 499, 512
Landing gear 430
  -bogie 431
  -brakes 462

Jockey- 432
  -design requirements 437, 466
  -drop tests 468
  -flotation 445
  -joints 465
  -kinematic 453
  -leveling system 435
  -loads 69, 442
  -load factors 454
  -locks 443, 453
  -materials 463, 464
  -retraction 449
  -shock absorbers 454
  -steering 442
  -stowage 449
  -tires 463
  -torque link 453
  -wheels 461
Landing load 57, 69
Leading edge 303, 326, 352
Lift
  additional- 53
  basic- 55
Lightening hole 165
Lightning strike 254, 255, 517
Limit load 23
Link 408
Liquid spring 456
Loadability 591
Loads
  bypass- 238
  crushing- 278
  door- 74, 421
  flight- 30
  fuel slosh- 278
  fuselage- 66, 378
  ground handling- 30, 73
  gust- 43
  hinge- 74
  horizontal tail- 59, 363
  inertia- 69
  landing gear- 69, 442
  limit- 23
  -factor 21, 350
  -criteria 23
  maneuver 21, 33, 43, 293
  pitching- 33, 62
  propulsion- 67, 474
  rolling- 34, 59
  taxi- 57
  tail- 59, 363
  ultimate- 23
  vertical stabilizer- 63, 363
  wing- 53, 250
  yawing- 64
Lug 219, 282 287

Mach number 25, 30
Magnesium alloys 96
Maintainability 14, 263
Manufacturing 11, 526, 534, 543, 548
Mass balance weight 350, 372
Matched tool 535
Material 90, 267, 268, 389, 463, 464, 492, 534
  aluminum- 96, 101, 465
  composite- 112, 492, 506
  -properties 90, 96, 463
  steel- 96, 110, 464
  titanium- 96, 109, 465
Matrix 112, 499, 504
Metal matrix 112, 496, 497, 506

610 Airframe Structural Design

Mismatch   245, 563
Modulus   90, 500
  -of elasticity   122, 512
  secant-   90, 120
  tangent-   90, 120
Monolithic graphite   534

Nacelle   56, 473, 475
Nondestructive inspection (NDI)   510, 536
Notch sensitivity   511
Numerical control (NC)   16
Nuts   215
Oleo-pneumatic   457
Openings (cutouts)   162, 417
Orthotropic   499, 509

Palmgren-Miner method   549
PAN   503
Panel   141, 146, 155, 569
  integrally-stiffened-   155, 262
  -design curves   153, 154, 159
  skin-stringer-   141, 258
Phenolic   112, 496, 505
Photostress   466
Pin bending   224
Pitch maneuver   33, 62
Pivot   282, 290, 367
Plain flaps   58, 335
Plane strain   97
Plane stress   97
Poisson's ratio   91, 512
Polyester   112, 496, 505
Polyimide   112, 496, 505
Pod (pylon)   479, 473
Powerplant mounts   471
Power-spectral density   45, 57
Precision forging   16, 94
Pre-impregnated   499
Prepreg   499, 509
Pressure bulkhead   396
Proportional limit   90
Propulsion loads   67, 474
Producibility   12, 13, 100
Pull-up stress   244
Pultrusion   499, 530
Pylon (pod)   473, 479

Quality assurance   535
Quasi-isotropic   499

Radiography   536
Radome   74, 404
Ramberg-Osgood Formula   91
Range (Brequet equation)   583
Resin   112, 496, 499, 504
  Bismaleimide-   496
  epoxy-   112, 496, 505
  phenolic-   112, 496, 505
  polyester-   112, 496, 505
  polyimide-   112, 496, 505
  thermoplastic-   112, 496, 497, 505
  themoset-   112, 497, 504, 505
Rib   277, 363
  root-   251, 283, 367
  function of-   278
  -arrangement   280
  shear web-   278
  truss-   279
  -spacing   150, 279, 344, 363
Rivet   150, 210
  blind-   211
  -allowable loads   209
  -materials   210
  -requirements   210
Robots   17
Rolling maneuver   34
Rudder   63, 371

Safe-life design   464, 540, 542
Safety factor   23, 118
Scatter factor   232, 548, 550
Screws   215
Seal   297, 300, 355
Semimonocoque structures   376
Severity factor   238
Shear lag   181
Shear pin   224
Shim   243
Shimmy   469
Shock absorbers   454
  coil spring-   454
  double-acting-   432, 460
  leaf spring-   454
  liquid spring-   456
  oleo-pneumatic-   457
  rubber-   454
Shock strut   459
Shot-peening   17
Silicone rubber   534
Skin-stringer panel   141, 146, 163, 250
Slat   304, 326
Slotted flap   58, 334, 336, 337
Slots   333, 336, 337
Soldering   230
Sonic fatigue   344, 363, 567, 569
Spar   269, 358
  multi-   248, 256, 369
  single-   248, 252
  -types   270
  -arrangements   248, 256, 280
  -fail-safe design   276
  -design rules   271
  tapered-   271
  truss type-   270
S-N curves   542
Spigot   408
Splice   231, 262, 568
Split flap   58, 335
Spoilers   347
Stability   118
Stacking squence   512, 514
Stamping   530
Standard atmosphere data   26
Steel alloys   96, 110, 464
Stiffness   40, 348, 350, 360
Stop   197, 371, 421, 567
Strain   90
Stress   90
  alternating-   542
  photo-   466
  pull-up-   244
  -allowable   95
  -corrosion   100, 114, 116
  -cycle   542
  -intensity   96
  -level   544
  -concentration   162, 239, 266, 510, 561
  -ratio   542
  ultimate-   91
  yield-   91
Stress-strain curve   91

Stringer (stiffener)
  Beam-   168
  fuselage panel-   381
  -runout   263
  Wing panel-   259, 261
Structural influence coefficients   36
Superplastic forming   16
Superposition   129
Symmetrical laminate   499, 509, 513
Swept wing   31

Tail   59, 358
Tail-down angle   447
Taileron   360
Tail load   59, 363
Tail-off   59
Tape   508
Tape-laying   531
Taper ratio   46
T-tail   360
Tangent modulus   90, 120
Taxi load   57, 250, 442
Tests   5, 570, 573
  airframe-   5, 570
  component-   574, 579
  coupon-   573
  drop-   468
  fail-safe-   578
  fatigue-   466
  full-scale-   574
  functional-   572
  landing gear-   466
  shimmy-   469
  -program   5
  sonic fatigue-   578
  static-   466
  system-   468
Thermal expansion   18, 463, 533
Thermoforming   530
Thermoplastic   112, 496, 500
Thermoset   112, 500, 504
Thread   215
Tire   71, 463
Titanium alloy   96, 109, 249, 465
Tolerance   13, 293
Tooling   15, 18, 531, 534, 535
Torque link   453
Torsional instability   121
Toughness   96
Tow   500, 506
Trailing edge   303, 335, 352
Transparent materials   400
Trunnion   369
Turn-over angle   447

Ultimate loads   23
Ultrasonic inspection   536
Undercarriage (see landing gears)   430
Upright   168, 271

Variable camber-kruger flap   326
Variable swept wing   249, 288
Vertical stabilizer   369, 415
V-g diagram   41
Visor   400
V-n diagram   41

Weight   5, 38, 581
  non-optimun-   587
  -balance   591
  -distribution   590

-prediction 583, 585
Warp 500
Wedges 346
Welded joint 227
Wheels 446, 461
Window 417
Windshield 399
Wing 247
  delta- 247
  multi-spar- 256
  single spar- 248, 252
  spoilers 347
  variable swept- 249, 288
  -additional lift (loading) 53
  -airloads 53
  -basic lift (loading) 55
  -box design 251
  -carry-through box 294
  -control surfaces 347
  -cover cutouts 163, 177, 183, 263
  -deflection 33, 349
  -fuel tank 296
  -fuel tank sealing 296, 300
  -high-lift devices 58, 303
  -layout 254
  -leading edges 263, 303, 326, 352
  -design loads 259
  -engine breakaway 278, 475
  -main gear breakaway 278
  -material 267, 268
  -ribs 277
  -root joint 282
  -spars 269
  -splice design 262
  -stiffness 250, 289
  -stringer run-out 263
  -trailing edges 263, 303, 335, 352
Woven fabrics 507, 508
Wrinkling 147

Yarn 500, 506
Yield strength 91
Young's modulus 90